System Theory

A UNIFIED STATE-SPACE
APPROACH TO CONTINUOUS AND
DISCRETE SYSTEMS

89254

LOUIS PADULO *Stanford University, Palo Alto, California*

MICHAEL A. ARBIB *University of Massachusetts, Amherst, Massachusetts*

1974

W. B. SAUNDERS COMPANY   Philadelphia   London   Toronto

W. B. Saunders Company: West Washington Square
Philadelphia, Pa. 19105

12 Dyott Street
London, WCIA 1DB

833 Oxford Street
Toronto, Ontario
M8Z 5T9, Canada

System Theory   ISBN 0-7216-7035-0

Last digit is the Print No.   9   8   7   6   5   4   3   2   1

For
KATHY and PRUE

Preface

Since its birth about twenty-five years ago, system theory has developed into a scientific and engineering discipline which seems destined to have an impact upon all aspects of modern society. First studied as control theory by mathematicians and engineers, system theory is now found increasingly in the curricula of economics, business, biology, political science, sociology, and psychology departments. What is system theory and what makes it so important to such a wide variety of specializations?

System theory or system science is a body of concepts and techniques which are used to analyze and design systems of various types, regardless of their special physical natures and functions. The single most important aspect of system analysis and design is the development of a quantitative model which describes the cause and effect relationships and interactions between the system variables. This means that the language of system theory is mathematics and that any serious attempt to learn system theory must be accompanied by the acquisition of mathematical precision and understanding. Accordingly, textbooks and journals devoted to system and control theory have evolved more and more into presentations of the underlying mathematical and analytical concepts.

The first generation of textbooks on system theory dealt almost exclusively with differential equations, Fourier and Laplace transforms, transfer functions, and frequency-domain techniques. The second generation of systems books introduced the state-variable ideas of Bellman and Kalman and revitalized the time-domain analysis of linear systems. This volume is a third-generation systems textbook, integrating the first and second generation of transform and state-variable methods into a framework which unifies continuous-time and discrete-time systems and embraces both classical linear systems and automata.

Philosophy Of The Book

We have tried in writing this book to develop a pedagogically sound introduction to system theory which will be accessible to readers with little background beyond elementary calculus and matrix theory, and at the same time be stimulating to those readers who are mathematically or system-theoretically sophisticated. We tried to write a book which teachers and readers without any system theory background at all could use and enjoy.

In presenting our unified approach, we provide the first fully developed textbook exposition of a view of system theory given early expression by Zadeh and Desoer in their *Linear System Theory* (1963), and developed at the advanced monograph level by Kalman, Falb, and Arbib in *Topics in Mathematical System Theory* (1969). This approach shows that many concepts are available for nonlinear systems as well as linear ones; that discrete-time systems are not the fearsome objects that their placement at the rear of conventional texts would seem to indicate; and that there are some problems for linear systems that are more easily solved by our unified approach than by matrix manipulation. This is not to deny that there are powerful and useful techniques available for linear systems in particular, and these are given a thorough exposition throughout the book.

As do most texts, this book grew out of the authors' classroom lecture notes and, during the five years in which it was evolving, we have each taught from the text several times to a variety of audiences. We have presented this material most often in a first-term graduate course in system science to electrical and aerospace engineers at Stanford, but we have also taught it in first-year graduate courses in operations research, computer science, and mathematics departments. In the course of all this teaching, we tried to get a feeling for what students do and do not know and to incorporate improvements suggested by colleagues teaching from our notes in other institutions and other departments.

As a result, we completely reworked the book to make it more self-contained and less specialized. For example, we deliberately toned down the use of electrical engineering jargon and examples so that, say, an industrial engineer would not get scared away because he hadn't had a course in circuit theory. For the reader who has had undergraduate courses in calculus and differential equations and who at least once in his life has multiplied two matrices correctly, we develop all the necessary mathematics in the text right where we need it. We put in some exercises to provide the inexperienced reader with various definitions and experience with matrices which would clutter the main text; other exercises range from drill on the material just covered to extensions of that material for the sophisticated reader to play with.

We make extensive use of examples and pictures to help the reader develop a feeling for and a reliable intuition about the abstract ideas of system theory, and from time to time we draw upon freshman physics (mostly circuit theory or mechanics) for these examples. Such examples have been so placed that students from mathematics, operations research, economics, or computer science will be able to understand the material even if they omit the examples.

Organization Of The Book

The eight chapters of the book may be broken down as follows: Chapters 1 through 3 provide a view of general systems which places the theory of linear systems into perspective. Chapters 4 through 7 then provide the heart of the book, placing central topics of linear system theory into the perspective of our unified approach to systems in general. Chapter 8 and Appendix A–3 then provide "windows" to two advanced areas which build upon the preceding theory: algebraic system theory and optimization theory. To be more specific, here is a chapter-by-chapter breakdown of the book's contents:

Chapter 1, "Systems and the Concept of State," introduces the state as the mediator of the relationship between a system's past and future behavior, and provides a framework for discussing deterministic systems, be they continuous-time or discrete-time, time-varying or time-invariant, nonlinear or linear.

Chapter 2 shows how the effect of inputs upon the states of a system may be described "locally" (in the sense of a locality in time) by one-step transition functions for discrete-time systems, and by differential equations for continuous-time systems. Section 2–3, "Vector Spaces and the Notion of a Basis," may be used simply as a reference section for students with a background in linear algebra, but will see effective classroom use with less well-prepared students. (This background linear algebra is completed in Section 3–3, "Matrices: Tools for Linear System Characterization.") Section 2–4, "Some Useful Mathematical Spaces," introduces such spaces as Banach spaces, for much system theory can be done in this generality without any work beyond that required for the finite-dimensional case treated in most conventional textbooks. (The same reasoning motivates our introduction of Hilbert spaces in Section 4–4, "Inner-Product Spaces and Adjoints.") Once he understands how simple the concepts of Hilbert and Banach spaces really are, the reader will be able to tackle portions of the research literature that would otherwise be inaccessible. We close Chapter 2 by showing how various real physical and mechanical systems may be fitted into our formalism.

Chapter 3 introduces the notion of a linear system, and shows how the process of linearization enables us to approximate many nonlinear systems by linear systems, at least over a restricted range of behavior. The notion of derivatives in Banach spaces introduced in our discussion of linearization will also prove useful in our later study of stability and optimization in Chapter 7 and Appendix A–3.

Section 4–1, "The Story So Far," summarizes the key aspects of Chapters 1 through 3 for the reader whose main interest is in linear systems, and who already has a fair background in linear algebra. The remainder of the chapter shows how simple it is to define reachability, controllability, and observability in our general framework, and then specializes the general definition to prove the duality of reachability and observability for linear systems. The chapter closes with an introduction to the Z-transform and transfer functions for discrete-time systems.

We start Chapter 5, "Global Behavior of Continuous-Time Systems," by noting how easy it is to integrate the one-step transition functions of discrete-time systems and, using this, to motivate the integration of the differential equations which describe continuous-time (possibly time-varying) systems. Section 5–2 then provides a rigorous derivation of this integration procedure by using the fixed-point theory of contractive mappings to solve vector differential equations, be they linear or nonlinear, time-varying or time-invariant; and Section 5–3 specializes the general theory to study the solutions of linear vector differential equations and develops the important properties of the state transition matrix. We then pass on, in Sections 5–4 and 5–5, to the input-output characterization of systems, and to conditions under which systems with different state-space descriptions may be equivalent "from the outside."

Chapter 6, "Constant Linear Systems," is the bread and butter of system theory. Here we tie our unified approach back to such second-generation state variable topics as the role of the matrix exponential in finding complete solutions for system behavior, and the structure theorems involving diagonalization or the Jordan canonical form. Also, we place in perspective such classical first-generation frequency-domain topics as the role of transform methods, and the use of signal-flow graphs, with special attention to the case of single-input, single-output systems.

Chapter 7, "Controllability, Observability and Stability," builds on Chapter 6 to give a thorough state-variable approach to the role of canonical forms to obtain state estimation and feedback compensation. It also provides a brief overview of the problem of optimal control.

Chapter 8, "Algebraic Approaches to Systems Realization," provides an introduction to algebraic system theory embracing the Nerode realization theory from automata theory and the Arbib-Zeiger adaptation of that theory to linear systems, which we use as a framework for a presentation of Ho's algorithm and Risannen's recursive algorithm for linear systems. All of this is related to Kalman's module-theoretic approach to linear systems. We then discuss group machines to stress that a more general approach is needed. In the final two sections, then, we present the Arbib-Manes theory of machines in a category, which provides the sought-for generalization in a pleasingly compact form.

In Appendices A–1 and A–2 we present useful formulas and transform pairs of the Laplace transform and the Z-transform. In Appendix A–3 we outline the mathematical theory leading to the three major tools of optimal control: the linear regulator, bang-bang control, and the Pontryagin minimum principle. We make extensive use of the Banach space derivatives introduced in Section 3–4 and developed for the optimal control problem in Section 7–6.

Teaching From The Book

A balanced instructional program in system theory not only needs the present volume to provide a unified theoretical perspective, but should also be supplemented by a course of practical engineering case studies, and advanced courses in optimal control, stochastic filtering and estimation, and computation. Having thus placed our book in perspective, let us now sketch some specific courses for which it is ideally suited. Although the whole volume may be used as a text for a one-year course, we expect the following to be the four courses for which the book will prove most useful.

I. Introduction to Deterministic Systems and Control

A. A **two-quarter** to two-semester course for students who require additional exposure to linear algebra, and who are to be exposed to the methodological bases of the systems concepts they employ:

1. Systems and the Concept of State . . . . 5 lectures
   Sections 1–1, 1–2, 1–3, 1–5, 1–6.
2. System Dynamics and Local Transition Functions . . . . 6 lectures
   Sections 2–1, 2–2, 2–3, 2–4, 2–5.
3. Linear Systems and Linearization . . . . 5 lectures
   Sections 3–1, 3–2, 3–3, 3–4.
4. Reachability and Observability. 8 lectures
   Sections 4–2, 4–3, 4–4, 4–5, 4–6.
5. Global Behavior of Continuous-Time Systems . . . . 6 lectures
   Sections 5–1, 5–2, 5–3, 5–4, 5–5.
6. Constant Linear Systems . . . . 7 lectures
   Sections 6–1, 6–2, 6–3, 6–4, 6–5, 6–6.
7. Controllability, Observability, and Stability and An Introduction
   to Optimization . . . . 14 lectures
   Sections 7–1, 7–2, 7–3, 7–4, 7–5, 7–6.
8. Algebraic Approaches to System Realization . . . . 6 lectures
   Sections 8–1, 8–2, 8–3.

Such a course would be suitable for first-year graduate students in engineering or operations research or for senior-year undergraduate mathematics majors. By omitting certain sections, the first seven chapters which comprise the heart of the subject can be covered in just one term:

B. For example, in a **one-quarter**, 4-credit introductory graduate engineering course at Stanford emphasizing continuous-time systems, one of the authors used his thirty-six lectures to cover:

Chapter 1 . . . . 5 lectures
   Sections 1–1, 1–2, 1–3, 1–5, 1–6.
Chapter 2 . . . . 5 lectures
   Sections 2–1, 2–2, 2–4, 2–5.
Chapter 3 . . . . 4 lectures
   Sections 3–1, 3–2, 3–3, 3–4.

The class consisted of first-year graduate students in electrical engineering who would take a separate course in digital control, so the only discrete-time theory presented to them consisted of Section 2–1, especially Exercises 7, 8, 9, 10, 11 and 14; Section 3–2, especially Example 7 and Exercises 2, 3 and 16; and Section 5–1.

II. Linear System Theory

A **one-quarter** to one-semester course in linear systems for students who know elementary linear algebra (Sections 2–3 and 3–3 being available as reference material). Chapters 4 and 5 still give the student a taste of the unified perspective, but the teacher can stress the solid development of standard topics in linear system theory given in Chapters 6 and 7:

    4. Reachability and Observability (4–1, "The Story So Far," is designed to aid readers who wish to start serious study of the book at this point.)
    5. Global Behavior of Continuous-Time Systems
    6. Constant Linear Systems
    7. Controllability, Observability and Stability

Such a course might well be offered as a "Second Course in Linear Systems" by Electrical Engineering Departments and can easily be extended to a full two terms by adding the following material:

    3. Linear Systems and Linearization (omitting material such as Section 3–3 on matrices, which is assumed already known).
7–6 and A–3 An Introduction to Optimization.

III. Algebraic System Theory

An advanced course for students who have already had a conventional course in linear system theory, and wish to master such algebraic approaches to system theory as Kalman's module theory and the Arbib-Manes theory of machines in a category. Such a course should prove particularly attractive to Mathematics and Computer Science Departments, and to coding theorists:

Finally, the book may be used with the companion volume, *Discrete Mathematics: Applied Algebra for Computer and Information Science* by L. S. Bobrow and M. A. Arbib, as the text for a four-semester Junior-Senior course in Applied Mathematics which presents both modern algebra and advanced calculus together with their application in the computer, information and system sciences.

Acknowledgments

We are indebted to many writers and colleagues whose work has deepened our understanding of system theory and contributed to the point of view expressed in this book. In particular, we are grateful to Gene Franklin (who started out as our co-author but succumbed to other demands on his time) for making available to us numerous examples gleaned from his many years in control theory. Both authors benefitted a great deal from their association with the Information Systems Laboratory at Stanford University and the ambience of understanding produced by the interest of associates such as Kalman, Kailath, W. Linvill, Luenberger, Gopinath, Smallwood, Abramson, Widrow, and Franklin in the teaching of system theory.

Arbib would like to take this opportunity to thank the people who have aided his development as a system theorist: Michael Athans and Peter Falb introduced him to modern control theory and the work of R. E. Kalman; John Westcott asked him to speak at an optimal control conference when he knew only automata theory, thus getting him to work on the rapprochement of which this book is an expression; Lotfi Zadeh and Charles Desoer wrote the book, *Linear System Theory,* which provided key conceptual ideas for that rapprochement; Rudolf Kalman gave him a Research Associateship at Stanford and introduced him to the theory of Section 8–4; Paul Zeiger collaborated in bringing the automata-theoretic ideas to bear on algorithmic linear system theory (Section 8–2); Ernest Manes collaborated in developing the theory of Machines in a Category (Sections 8–5 and 8–6); and, of course, Louis Padulo. Arbib is also grateful for the support for this approach to mathematical system theory that was provided by the U.S. Army Research Office, Durham, N.C., under Contract DAHC 04-70-C-0043.

Padulo acknowledges a special debt to B. J. Dasher, who brought him to Georgia Tech and supported his research; to Gabor Szego, Zuhair Nashed, and Eric Immel, who taught him mathematics; to John Linvill for bringing him to Stanford and President Hugh Gloster of Morehouse College for letting him go there; to Michael Arbib and Robert Newcomb for enlarging his concept of systems; to George Dantzig for inviting him to lecture in the operations research department and for encouraging him to make system theory accessible to non-engineers; and to his good friend Tom Kailath for pointing out so many shortcomings.

Reviewers Robert Kasriel, San Wan, John Gill, and Franco Preparata aided us immensely in their detailed critiques and suggestions for improvement. Former students Patrick Bergmans, Jean Pierre Tranvouez, Dov Rosenfeld, Brad Dickinson, Adrian Segall, Martin Morf, and Martha Sloan found many errors and checked the exercises for their suitability; and, with his customary enthusiasm, Gursharan Sidhu proofread the final manuscript and was a persuasive lobbyist for further improvements right up to the final deadline. Colleagues George Dantzig, Bostwick Wyman, Charles Hutchinson, Tom Cover, Patrick Hirschler, Patrick Bergmans, Edward Kamen, and Gene Franklin have all taught from the manuscript and contributed to its usefulness as a text. Any further suggestions for improvement from readers and teachers are most cordially solicited.

Finally, both authors thank their wives and families for putting up with all the pressures, inconveniences, and spoiled vacations which dogged the writing of this book.

Louis Padulo

Michael Arbib

To the Student

How To Read This Book

This book was written so that someone with just a little math and no systems experience at all could start at the beginning and go straight through the logical development of the subject. If you are already very good at mathematics and know linear algebra and analysis, you will be able to go through this book quite a bit faster, but should still read the mathematical sections to make sure you are familiar with any special notation we may introduce. If you only know math but have no background in the natural sciences or engineering, you should not be discouraged by our occasional use of examples and vocabulary from other fields. For instance, the subject of our very first example in the book is a black widow spider. Instead of saying, "Oh no! I've never had a rigorous course in biology or the theory of small insects," read the example anyway. Almost always, you will still be able to get the point of the example and often you will learn something about the unfamiliar field, too.

If you already know a little system theory, perhaps from earlier courses in engineering or physics, chances are that you may have to work a little harder in the beginning to get used to modern mathematical ideas and notation. You should work Section 1–2 diligently and try all the exercises there, no matter how strange and prone to nit-picking they seem. A little practice there will go a long way in helping you to express yourself with precision and to understand us when we must define things very precisely in later sections. When you encounter a section containing some mathematics that is new to you, work through the examples and do lots of exercises and you will learn the new topic as well as if you had had a course in it. You will come out at the end of this book knowing not only system theory but quite a bit of useful mathematics, too.

Whatever your previous background, you will encounter in the text and exercises some subjects which you or your teacher know that you don't need to learn for your particular discipline. In most cases you won't be hurt if you skip them and go on with the text. If they are needed anywhere later, we refer back to them in the later discussion and you can react accordingly.

Since there are so many exercises, we realize that few readers will attempt them all. In any case, you should *read* them all to broaden your perspective on the textual material. We put a ✔ next to each exercise we consider a must for you to try. Even if you are not able to complete all of these successfully, by trying them you will be developing a proper mind set for the next topic and be much better prepared to understand it. Those exercises which have **COMP** next to them are readily solved on a digital computer and have been very instructive for students who know how to program; those with **EE** next to them are especially suitable for readers with an electrical engineering background.

We use the symbol $\lozenge$ to denote the end of an example, $\square$ for the end of a proof, and $\bigcirc$ for the end of a definition; the symbol $\triangleq$ means *is defined to be equal*. Figures are numbered sequentially throughout each chapter, but within each section we number all theorems, examples, exercises, and equations starting from scratch. Thus, when we refer to Example 4–3–2, we mean Example 2 of Section 4–3.

Contents

CHAPTER 1

Systems and the Concept of State

The business of this book is to develop a mathematical framework within which we can not only talk about a large class of systems, but can also derive important results which help us to analyze and synthesize systems that behave in some desired way. Central to all this will be the notion of a model of a system, and in Section 1–1 we shall see how any attempt to understand a system in the real world requires us to abstract from that real situation relatively few variables of interest to us. This discussion will be extended in Section 1–3, where we shall see that our description of a system requires not only that we specify the inputs, or control signals, that we expect to be applied to it, and the outputs which represent the observations that we can make upon the system (or the effects that we expect that system to have upon other systems), but must also require intervening variables which constitute the internal state of the described system. To state these concepts in a general yet succinct form, we shall use the modern mathematical notation of set theory. We summarize this notation in Section 1–2, and the unprepared reader may find the array of notation there somewhat overwhelming. However, it is basically simple, and the reader is encouraged to work through Section 1–2 carefully once, though without worrying too much about memorizing the details, and then keep referring back to it as he progresses through the book. By the end of Chapter 4, all the notation should have become so familiar as to become a part of the reader's intuition, even though it seemed

1

so very nonintuitive at a first reading. In Section 1–4 we make a brief aside to note that there may be systems in which the current state and incoming inputs do not determine the future state precisely, but only within certain probabilities. However, in this book we shall emphasize the class of systems, taken up in Section 1–5, which are deterministic in the sense that a knowledge of the current state completely determines the current output; while knowledge of the state at time t_0 and knowledge of the input from time t_0 to some later time t_1 completely determines the state at time t_1. We also make explicit what we mean by continuous-time and discrete-time systems. Then, in Section 1–6, we analyze when a system will be time-varying and when it will be time-invariant. For easy reference, the key definitions of this chapter are summarized in Section 1–7. It is to be emphasized that the concepts introduced here are of great generality, and that, in particular, they hold whether or not the system considered is linear. In fact, it will not be until Chapter 3—other than in the examples—that we shall study linear systems at all.

1–1   Systems, Models, and Mathematics

In the real world we encounter actual, physical objects such as highways, nations, electronic circuits, frogs, space ships, and computer programs. We shall call these real life objects **physical systems.**

Now, as soon as we start to think about one of these physical objects, we formulate a **model** of the object by focusing, often unconsciously, upon some small number of the properties and attributes by which we envision it. For example, when we hear "consider the physical system called the 'black widow spider,'" we probably envision a poisonous little object that scurries about under its own power, eats other insects, and scares people. We probably don't go on and on for another half hour listing every little fact known to mankind about black widow spiders. For everyday life the brief listing of attributes just mentioned is probably an adequate model for the spider. However, a research journal article entitled "The Effects of Marijuana Addiction on the Tensile Strength of Southern Black Widow Spider Webs" suggests a rather different model of the same little spider; suddenly we envision a placid, almost comical little insect churning out webbing like a machine and in some mysterious fashion revealing a southern background.

The point is that, for a given physical system, we select a model of it which is appropriate to those attributes of the system that we care about in a given situation.

Once we have chosen a model for a system of interest, we then have to decide upon how we are going to talk about the model and what we are going to do with it. If we wish to analyze the model, and study it mathematically, we will need some mathematical **representation** of it, that is, some set of mathematical equations that describe the model. Even if the equations describe the model perfectly, if the model is not a complete description of the physical system, the equations might not tell us very much about the system, no matter how carefully we solve them.

We illustrate this point with a device we shall see a great deal more of throughout this book, the **analog computer,** which uses the basic operations of addition, multiplication, and integration to simulate systems and solve differential equations.

Example 1

A modern electronic analog computer contains a device which takes a given time-varying signal, say u (which takes the value $u(t)$ at time t), as its input and produces an output signal, say y, whose value at t is $y(t)$, which looks like the integral of the input over a certain range of operation. We model this real electronic circuit with the **ideal integrator** shown in Figure 1–1a, where the values of the input signal and output signal at time t are $u(t)$ and $y(t)$ respectively and $y(t_0)$ denotes the initial value of y at time t_0.

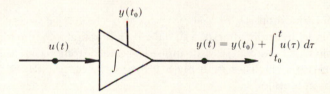

Figure 1–1(a)   The ideal integrator with input u, output y, and initial condition $y(t_0)$ at time t_0.

The initial condition term is included in the model, since we will often use an integrator to "undo" differentiation* and, by the Fundamental Theorem of Calculus,

$$\int_{t_0}^{t} \dot{y}(\tau)\, d\tau = y(t) - y(t_0)$$

so that

$$y(t) = y(t_0) + \int_{t_0}^{t} \dot{y}(\tau)\, d\tau$$

If we remember that **the initial condition term must always be present,** we may leave it out of our diagrams and use the simplified model of Figure 1–1b.

---

*Throughout this book we will use the dot over a function to denote the time derivative of the function, $\dot{x}(t) = dx(t)/dt$. Then $x^{(n)}(t)$ will denote the nth time derivative $\left(\dfrac{d}{dt}\right)^{n} x(t)$.

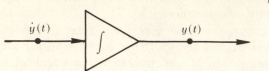

Figure 1–1(b)   The ideal integrator undoes differentiation (initial condition terms not shown).

Thus our model, the ideal integrator, has two different mathematical representations:

(i) Input and output are related by the *integral equation*

$$y(t) = y(t_0) + \int_{t_0}^{t} u(\tau)\, d\tau \qquad\qquad \textbf{(1)}$$

(ii) Input and output are related by the *differential equation*

$$\dot{y}(t) = u(t) \qquad\qquad \textbf{(2)}$$

with initial condition $y(t_0)$.

We model multiplication of a signal u by a function a by the **ideal scaler** or **multiplier** of Figure 1–1c with input $u(t)$ and output $y(t)$ at time t. The quantity $a(t)$ by which we multiply $u(t)$ to get the output $y(t)$ is called the **gain** of the multiplier.

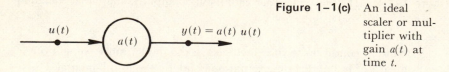

Figure 1–1(c)   An ideal scaler or multiplier with gain $a(t)$ at time t.

We model the addition and subtraction of signals by the **ideal adder** or **summer** of Figure 1–1d, which uses minus signs on the input lines to indicate when subtraction instead of addition is going on.

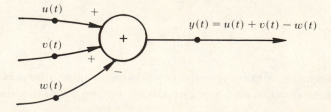

Figure 1–1(d)   An ideal adder or summer.

Now, all three of the models of Figures 1–1b, c, and d accurately describe the operation of the actual electronic devices in the computer as long as the signals involved stay within specified levels. If we start with $y(t_0) = 0$ and put in an input $u(t) = +1$ volt, at times $t \geqslant t_0 = 0$ our ideal integrator model says we'll get out the signal

$$y(t) = \int_0^t 1 \, d\tau = t,$$

for all t. According to our model, at $t = 10^6$ seconds, our output will be exactly 10^6. Do we really believe that our real electronic integrator will have one million volts as its output when we put one volt in and wait? Of course not; we expect the real output to climb like t until it gets bigger than the design range, perhaps 200 volts, at which point it stops climbing like t and levels off to some saturation value.   ◊

We shall introduce other operations of the analog computer, such as multiplication of two signals and function generation, when we need them. Let us now present a simple system from freshman physics and its model, the pure capacitor.

Example 2

An electrical device known as a Leyden jar or condenser, illustrated symbolically in Figure 1–2a, stores electric charge on the two metal foil strips shown by the dotted lines. Also shown is a battery of voltage $\mathcal{E}$, which causes a current i to flow in the direction shown by the arrow and a voltage v across the jar terminals when the switch S is closed.

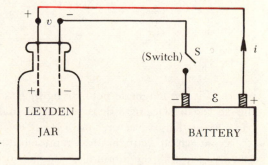

Figure 1–2(a)   A drawing of an actual physical system.

We have a pencil and paper model of the Leyden jar; it is a fictional circuit element called the **pure capacitor** and denoted by the symbol shown in Figure 1–2b, where i stands for the current flowing through the capacitor, v stands for the voltage across the capacitor, and the constant C is called the capacitance. The arrow and the $+$ and $-$ signs indicate the reference direction or polarity of i and v at any instant of time.

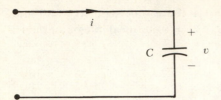

Figure 1–2(b) A pure capacitor with capacitance C.

One mathematical representation for the model of Figure 1–2b is the equation discovered by Faraday:

$$q(t) = Cv(t) \tag{3}$$

where $q(t)$ denotes the electric charge, $v(t)$ is the voltage in the model at time t, and $i(t) \overset{\Delta}{=} dq(t)/dt$.

If we differentiate both sides of Equation (3) with respect to t, we get a second mathematical representation,

$$i(t) = \frac{dq(t)}{dt} = \frac{d}{dt} Cv(t) = C \frac{dv(t)}{dt} \tag{4}$$

which relates the current and voltage of our model.

A third, perfectly respectable, mathematical representation is the set of equations due to Maxwell:*

$$\nabla \times H = j + \frac{\partial D}{\partial t}$$

$$\nabla \times E = -\frac{\partial B}{\partial t}$$

$$\nabla \cdot D = \rho \tag{5}$$

$$\nabla \cdot B = 0$$

where H is the magnetic field strength, E is the electric field strength, D is the electric displacement density, B is the magnetic flux density, j is the electric current density vector, and ρ is the electric charge density.

Now, which mathematical representation would you rather use if we asked you to find the amount of charge stored in the Leyden jar when the voltage applied across it is 5 volts? If you are smart enough to solve them, all three representations, (3), (4), and (5), will yield the charge q for our capacitor model of the Leyden jar. Obviously, the first representation is much easier to work with than the differential equations of (4) or (5). In fact, it gives the answer, $q = 5C$, right away with no solving required.

---

*Don't let these scare you; we won't use them ever again in this text.

If we now change the question to, "What is the charge stored in a Leyden jar if a voltage of five billion (5×10^9) volts is put across it," representation (3) is still a cinch to solve and says that $q = 5 \times 10^9 C$. It just might happen, however, that if we try to put five billion volts across a little Leyden jar, it will break down or perhaps even melt, so that the correct answer is really that $q = 0$ is stored in the jar, no matter what our model says.   ◊

The good system theorist must be able to pick good models and good representations for the real systems he cares about. In the rest of this book we will be studying various useful models and their mathematical representations. We will try to use very simple models to illustrate techniques and concepts with which we may analyze the complicated models needed to study real systems.

The Need for a Precise Vocabulary

In the ensuing pages we shall provide a rigorous framework within which to study modern system theory. We shall be forced to use the language and notation of mathematics in order to formulate definitions crisply and precisely.

Example 3

If we showed the proverbial man on the street the analog computer diagram of Figure 1–3a, he might well declare that we have a time-varying system* because the gains vary with time. In the same breath he might mumble something about a system being time-invariant if its impulse response $g(t, \tau)$ is a function of $(t - \tau)$ only. Imagine his consternation when we show him that our system has exactly the same impulse response as the system with constant gains shown in Figure 1–3b, which he would pronounce time-in-variant.   ◊

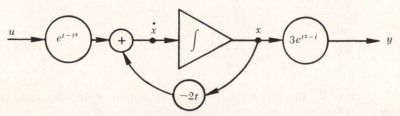

Figure 1–3(a)   A system with time-varying gains.

---

*The reader with no background in control theory need not despair; the terms used in this example will all be defined in due course of our exposition.

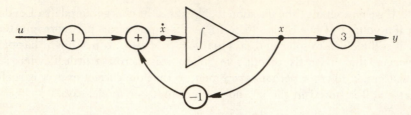

Figure 1–3(b)   A system with constant gains, having the same impulse response as Figure 1–3(a).

In Sections 1–6 and 5–2 we will give precise, mathematical definitions of time invariance and impulse response which let us decide unambiguously whether a system is time invariant or not.

Often, in studying real systems, we encounter "one-way" or unidirectional devices such as valves, ratchets, electronic diodes, and ordinary doors which open when pushed in one direction but stay closed when pushed in the opposite direction. We shall find it convenient to model such devices by the **ideal diode** shown in Figure 1–4a, with reference directions of current i through the diode and voltage v across the diode shown by the arrow and $+$ and $-$ signs. The operation of the diode is represented by the L-shaped graph of current versus voltage of Figure 1–4b, which shows that i is zero when v is negative and i can be any value when v is nonnegative. That is, the ideal diode is like a perfect switch: "off" (open) when v is negative and "on" (closed) when v is nonnegative.

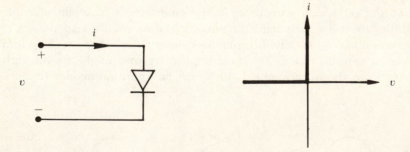

Figure 1–4(a)   The ideal diode.   **(b)**   The voltage-current relationship for the ideal diode.

In Section 3–2 we will rigorously define the notions of linearity and linear systems, but for the present informal discussion let us observe that the ideal diode is not linear since it is not true that for all voltages v_1 and v_2 we have $i(v_1 + v_2) = i(v_1) + i(v_2)$. To see this, let $v_1 = +1$ and $v_2 = -5$ so that $v_1 + v_2 = -4$. Then $i(v_1 + v_2) = i(-4) = 0 \neq i(+1) + i(-5) = i(v_1) + i(v_2)$.

Example 4

The poor man on the street who memorizes little rules like "a system is nonlinear if it contains a nonlinear element" is in hot water again with the system of Figure 1–5a, where R is a pure resistor. The net effect of the two diodes in parallel with opposite polarities is that one of them will always be "on" no matter what the sign of v is, and they are equivalent to a perfect wire always providing a conducting path between terminals A and B, as shown in Figure 1–5b.   ◊

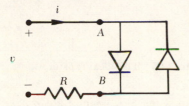   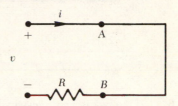

Figure 1–5(a) A system with two   **(b)** An equivalent linear system.
non-linear elements.

In order to give a rigorous definition of linearity, we have to review and develop the precise mathematical ideas of sets, vector spaces, functions or mappings, and compositions of mappings. We shall for the most part develop the particular mathematics we need right at the place where we need it. In the next section we pause for a quick, general, mathematical tune-up to introduce the language and notation with which we shall develop the central ideas of state variables and deterministic systems.

EXERCISES FOR SECTION 1–1

✔ **1.**   For the ideal integrator of Example 1, convince yourself that if you know the value of the output y at time t_0 and the input u from t_0 on, then you can tell what the output is at any time $t \geq t_0$. [In Section 1–3 we will see that this entitles the output of an integrator to be the **state** of the integrator.]

✔ **2.**   For the pure capacitor of Example 2, integrate both sides of Equation (4) to get an integral equation relating voltage $v(t_1)$ to initial voltage $v(t_0)$ and the current $i(t)$ over the interval $t_0 \leq t < t_1$. Is knowledge of voltage at a particular time and knowledge of input current from that time up to a later time enough to know the voltage at that later time?

✔ **3.**   **(a)**   For the analog computer system of Figure 1–3a, find the differential equation which describes the relationship between output y and input u. [*Hint:* The signal $\dot{x}$ at the input of the integrator is the output of the summer, so that $\dot{x}(t) = (e^{t-t^2})u(t) + (-2t)x(t)$.]

  (b)   For the system of Figure 1–3b, show that the input-output behavior is described by

$$\dot{x}(t) = -1x(t) + 1u(t)$$
$$y(t) = 3x(t)$$

4.   The pure **resistor** R shown in Figure 1–6 is a model of an actual electrical device which we say is represented by the equation (Ohm's law)

$$v(t) = R\,i(t)$$

The heat dissipated in a resistance of R ohms at time t is proportional to $[i(t)]^2$, the square of the current through it. What do you think about the usefulness of this model for large values of i?

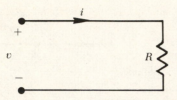

Figure 1–6   The ideal resistor.

EE 5.   For readers who enjoy circuit problems: show that the circuit of Figure 1–7a is equivalent to that of Figure 1–7b if the capacitor C is initially uncharged when the voltage v is applied.

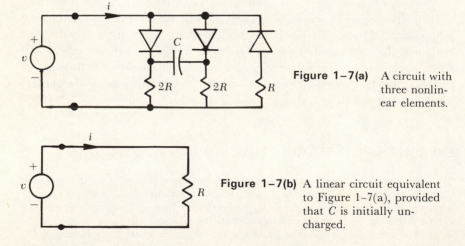

Figure 1–7(a)   A circuit with three nonlinear elements.

Figure 1–7(b)   A linear circuit equivalent to Figure 1–7(a), provided that C is initially uncharged.

✔ 6.   Draw an analog computer diagram, using integrators, multipliers, and summers as illustrated in Examples 1 and 3, whose behavior satisfies the **matrix** differential equations:

$$\begin{bmatrix} \dot{x}_1(t) \\ \dot{x}_2(t) \end{bmatrix} = \begin{bmatrix} 0 & 1 \\ 2 & a(t) \end{bmatrix} \begin{bmatrix} x_1(t) \\ x_2(t) \end{bmatrix} + \begin{bmatrix} b_1 & 0 \\ 0 & b_2 \end{bmatrix} \begin{bmatrix} u_1(t) \\ u_2(t) \end{bmatrix}$$

$$y(t) = \begin{bmatrix} c_1(t) & c_2(t) \end{bmatrix} \begin{bmatrix} x_1(t) \\ x_2(t) \end{bmatrix} + \begin{bmatrix} 5 & 6 \end{bmatrix} \begin{bmatrix} u_1(t) \\ u_2(t) \end{bmatrix}$$

[*Hint:* Perform the first matrix multiplication and get two scalar (coupled) first-order differential equations involving x_1 and x_2. Draw two integrators (leaving plenty of room between them) and label one output x_1 and the other x_2. Then their respective inputs must be $\dot{x}_1$ and $\dot{x}_2$. Your scalar equations tell how you can construct $\dot{x}_1$ and $\dot{x}_2$ out of x_1, x_2, u_1, and u_2 using multipliers and summers. Finally, multiplying in the last matrix equation shows how to construct y from appropriate amounts of the signals x_1, x_2, u_1, and u_2 which are available at various nodes in your diagram.]

1–2   Some Mathematical Notation

To build up a mathematical theory which elaborates the ideas presented in Section 1–1, we shall need a notation which allows us to express our ideas concisely and to manipulate them clearly. This section introduces such notation.

We assume that the reader is familiar with real and complex numbers. We shall use $\mathbf{Z}$ to denote the set of all integers, $\mathbf{R}$ to denote the set of all real numbers, and $\mathbf{C}$ for the set of all complex numbers. We use $\mathbf{N}$ for the set of all nonnegative integers and $\mathbf{R}_+$ for the nonnegative reals.

We also assume that the reader needs no special exposition to use the following notations:

> $\in$ for "belongs to": as in the statement $3 \in \mathbf{Z}$
> $\notin$ for "does not belong to": as in the statement $-3 \notin \mathbf{N}$
> $\subset$ for "is a subset of": as in the statement $\mathbf{R} \subset \mathbf{C}$
> (Note: Some authors use $\subseteq$ where we use $\subset$, and reserve $\subset$, as we do *not*, to indicate proper inclusion.)
> $\not\subset$ for "is not a subset of": as in the statement $\mathbf{Z} \not\subset \mathbf{N}$
> $\{x \,|\, P(x)\}$ for "the set of x for which $P(x)$ is true": as in the statement
> $\quad \mathbf{R}_+ = \{x \,|\, x \in \mathbf{R} \text{ and } x \geq 0\}$
> $A \cup B = \{x \,|\, x \in A \text{ or } x \in B\}$, the **union** of A and B
> $A \cap B = \{x \,|\, x \in A \text{ and } x \in B\}$, the **intersection** of A and B
> $A - B = \{x \,|\, x \in A \text{ and } x \notin B\}$, the **difference** of A and B

Sometimes, when we are lazy, we shall use abbreviations such as **iff** or $\Leftrightarrow$ for **if and only if** and $\Rightarrow$ for **implies** and $\Leftarrow$ for **is implied by.** On rare occasions, usually in exercises, we might use $\forall$ for **for all** and $\exists$ for **there exists.**

Example 1

The statement "For all $x \in \mathbf{R}$, there exists $y \in \mathbf{R}$ such that $y > x$" expresses the fact that there is no largest real number, and can be written quite cryptically as

$$(\forall x \in \mathbf{R})(\exists y \in \mathbf{R})(y > x) \qquad \qquad \Diamond$$

The reader has no doubt encountered many functions or operations which associate one number with another—such as the operation of squaring, which takes a number in $\mathbf{R}$ to a number in $\mathbf{R}_+$, or the operation of taking the positive square root, which takes any number in $\mathbf{R}_+$ to give another number in $\mathbf{R}_+$, but is not defined for numbers in $\mathbf{R} - \mathbf{R}_+$. We shall often need to consider functions which operate on other sets besides sets of real or complex numbers.

Let A and B be any two sets of objects. A **mapping** from A to B is an assignment to every element of A of a unique element of B. We write $F: A \to B$ to mean "F is a mapping from A into B," and sometimes we use **function, operator, transformation,** or for brevity, just **map** as synonyms for *mapping.*

If $F: A \to B$ is a mapping and if a is an element of A (written $a \in A$) then we denote by $F(a)$, or sometimes Fa, the **value** of F at a, also called the **image** of a under F. Two maps f and g from A to B are said to be **equal** if $f(a) = g(a)$ for every a in A. Often we write $a \mapsto F(a)$ to indicate that the action of the map F is to transform the specific element a to the specific value $F(a)$. We call A the **domain** of F and B the **codomain** of F.

If W is any subset of A, written $W \subset A$, the set of all elements $F(w) \in B$, where w ranges over all the elements of W, is called the **image** of W under F and is written $F(W) = \{F(w) \mid w \in W\}$. In particular, the set of all elements of the form $F(a)$ for a in A is called the **range of F** or the **image of F** and is denoted by $F(A)$ or $\mathfrak{R}(F)$.

Example 2

Let $A = \{a, b, c\}$ and $B = \{1, 2, 3, 4\}$, and let $f: A \to B$ be defined by the rule:

$$a \mapsto 2$$
$$b \mapsto 2$$
$$c \mapsto 4$$

which can be expressed by the table

x	$f(x)$
a	2
b	2
c	4

or pictorially, by the diagram of Figure 1–8. The domain of f is $\{a, b, c\}$ and the codomain of f is $\{1, 2, 3, 4\}$. The range or image of f is the subset $\{2, 4\}$ of the codomain:

$$\mathfrak{R}(f) = f(\{a, b, c\}) = \{2, 4\} \qquad \diamond$$

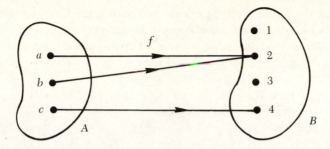

Figure 1–8   A map from $\{a, b, c\}$ into $\{1, 2, 3, 4\}$, showing the values of $f(a), f(b),$ and $f(c)$.

The **Cartesian product** of two sets A and B is written $A \times B$, read "A cross B," and is the set $A \times B = \{(a, b) \mid a \in A \text{ and } b \in B\}$ of all **ordered pairs** (a, b) whose first entry, a, belongs to A and whose second entry, b, belongs to B. Similarly, the set of all ordered **n-tuples** $(a_1, a_2, \ldots, a_n)$ with each entry a_i belonging to the set A_i, is denoted by $A_1 \times A_2 \times \cdots \times A_n$ and is called the Cartesian product of the n sets $A_1, A_2, \ldots, A_n$. When all the A_i are equal, say $A_i = A$, we write the product of n terms $A \times A \times \cdots \times A$ as A^n.

Example 3

If $A = \{x, y\}$ and $B = \{1, 2, 3\}$, then

$$A \times B = \{(x, 1), (x, 2), (x, 3), (y, 1), (y, 2), (y, 3)\}$$
$$B \times A = \{(1, x), (1, y), (2, x), (2, y), (3, x), (3, y)\}$$
$$A^2 = A \times A = \{(x, x), (x, y), (y, x), (y, y)\}$$

and

$$A^3 = A \times A \times A = \{(x, x, x), (x, x, y), (x, y, x), (y, x, x),$$
$$(y, y, x), (y, x, y), (x, y, y), (y, y, y)\}.$$

Notice that $A \times B$ and $B \times A$ have the same number of elements but are not equal; in the present example they don't have even one element in common.   ◊

The reader will appreciate the terminology "Cartesian product" by considering the diagram of Figure 1–9 and recalling that it was Descartes'

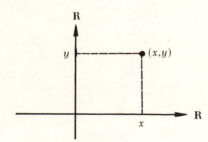

Figure 1–9   The plane is the Cartesian product $\mathbf{R} \times \mathbf{R}$.

insight that the plane could be represented as $\mathbf{R}^2$, the Cartesian product of two lines, which led to the development of analytic geometry.

In freshman calculus, when we studied "real-valued functions of several real variables" such as $z = f(x,y)$ and $\omega = g(x,y,z)$, we could have written

$$f:\mathbf{R} \times \mathbf{R} \to \mathbf{R} \quad \text{and} \quad g:\mathbf{R} \times \mathbf{R} \times \mathbf{R} \to \mathbf{R}$$

or

$$f:\mathbf{R}^2 \to \mathbf{R} \quad \text{and} \quad g:\mathbf{R}^3 \to \mathbf{R}$$

to indicate that a pair of real numbers (x,y) is plugged into f and that a triple* (x,y,z) of real numbers is plugged into g.

More generally, we shall encounter maps F which operate on n different kinds of variables, say $x_1, x_2, \ldots, x_n$ with $x_1 \in A_1, x_2 \in A_2, \ldots, x_n \in A_n$, and which associate with each ordered n-tuple $(x_1, x_2, \ldots, x_n)$ an object in some set C. Thus the domain of such a map is $A_1 \times A_2 \times \cdots \times A_n$, the codomain is C, and we write

$$F:A_1 \times A_2 \times \cdots \times A_n \to C$$

to express that $(x_1, x_2, \ldots, x_n) \mapsto F(x_1, x_2, \ldots, x_n)$.

Example 4

Consider the sets $A = \{x,y\}$, $B = \{1, 2, 3\}$, and $C = \{a, b\}$. A mapping $F:A \times B \to C$ may be defined by the table whose entries show the action of F on each point (x_1, x_2) of $A \times B$.

(x_1, x_2)	$F(x_1, x_2)$
$(x, 1)$	a
$(x, 2)$	b
$(x, 3)$	b
$(y, 1)$	b
$(y, 2)$	a
$(y, 3)$	b

$\Diamond$

When dealing with intervals of real numbers, we shall use the following notation:

---

*Sometimes called a 3-tuple.

$$(a, b) = \{x \in \mathbf{R} \,|\, a < x < b\}$$
$$[a, b) = \{x \in \mathbf{R} \,|\, a \leq x < b\}$$
$$(a, b] = \{x \in \mathbf{R} \,|\, a < x \leq b\}$$
$$[a, b] = \{x \in \mathbf{R} \,|\, a \leq x \leq b\}$$

relying on the context of a particular situation to avoid confusion of the **open interval** (a, b) with the ordered pair (a, b). We call $[a, b]$ a **closed interval** and $[a, b)$ **half-open** on the right.

When we have a mapping $F:A \to B$ but we wish only to consider the action of F on some subset W of the domain A, we write $F_W: W \to B$ or $F\,|\,W: A \to B$ and call F_W the **restriction** of F to W. Then $F_W(w) = F(w)$ for $w \in W$. For example, let $u: \mathbf{R} \to \mathbf{R}$ represent an input signal to a circuit, with the number $u(t)$ representing the current applied to the circuit at time t. If we wish only to study the effect of u on the circuit for times t greater than or equal to t_0, we are considering the restriction of u to the interval $[t_0, \infty) = \{t \,|\, t_0 \leq t < \infty\}$, and we write $u_{[t_0, \infty)}$.

We shall use $(d/dt)u$ or $\dot{u}$ to denote the derivative of a function $u: \mathbf{R} \to \mathbf{R}$, and $u^{(n)}(t_0)$ to denote the nth derivative of u evaluated at $t = t_0$; we shall *not* use primes for this purpose.

A map $f:A \to B$ is said to be **onto** (**surjective**) if and only if each point in B has an ancestor in A which f sends to b; that is, for every $b \in B$, there exists $a \in A$ such that $f(a) = b$. We say f is **one-to-one** (**injective**) if and only if $f(x) = f(y)$ implies that $x = y$; that is, f sends distinct points x, y to distinct images $f(x), f(y)$.

Example 5

Consider the following four mappings of $\mathbf{R} \to \mathbf{R}: x \mapsto e^x$ is one-to-one but not onto; $x \mapsto x \sin(x)$ is onto but not one-to-one; while $x \mapsto x^2$ is neither, and $x \mapsto 2x + 3$ is both, as is readily seen by sketching their respective graphs.

$\Diamond$

When $f:A \to B$ is both onto and one-to-one, we say f is a **one-to-one correspondence** or a **bijection.** The two sets A and B have the same number of elements (the same **cardinality**) if and only if there exists a bijection $f:A \to B$.

If $f:A \to B$ and $g:B \to C$, we define the **composition** (product) mapping $g \circ f: A \to C$ by the rule $x \mapsto g[f(x)]$; i.e., given $x \in A$, we first apply f to obtain $f(x) \in B$, and then apply g to the result to obtain $g[f(x)] = (g \circ f)(x) \in C$.

Sometimes we abbreviate $g \circ f$ by gf when there is no risk of confusion.

Example 6

Let $A = \{x, y\}$, $B = \{1, 2, 3\}$, and $C = \{a, b\}$, and consider the maps $f:A \to B$ and $g:B \to C$ shown by the diagrams of Figure 1–10a. Notice that

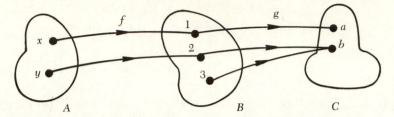

Figure 1–10(a)   Two mappings, $f: A \to B$ and $g: B \to C$.

f is one-to-one but not onto, and g is onto but not one-to-one. The composition map $g \circ f: A \to C$ is easily worked out:

$$(g \circ f)(x) = g[f(x)] = g(1) = a$$
$$(g \circ f)(y) = g[f(y)] = g(2) = b$$

as is shown in Figure 1–10b. Observe that $g \circ f$ is a bijection, even though neither of its constituents is.   ◊

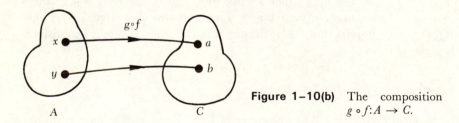

Figure 1–10(b)   The   composition $g \circ f: A \to C$.

Throughout this book we will encounter the interesting question of how various properties of individual maps are inherited by their composition. For example, in Exercise 3 we study how the one-to-one and onto properties are inherited, and in Section 3–2 we will see that the composition of two linear maps is linear.

Since the composition of two maps is clearly a cascade or serial operation (first do f, then do g), we see that composition is the framework by which we may describe systems connected in series (cascaded). For example, we might envision the operation $f: A \to B$ of Figure 1–10a as a system which we wish to study, and the operation $g: B \to C$ as our measuring or observation system. What we really see then, as a result of our measurements, is not the behavior of f but rather $g \circ f$; our "known" instrumentation or data collection process g masks the "unknown" f and sticks us with $g \circ f$. We are thus quite interested in knowing how to extract information about f from the known g and the observed $g \circ f$.

EXERCISES FOR SECTION 1–2

✔ **1.**   Let A have m members and B have n members.

    (a)   How many maps are there from A into B? How many of these are one-to-one? How many are onto (careful!)?

    (b)   How many elements are there in $A \times B$?

    (c)   How many subsets are there of A?

    (d)   How many subsets are there of $A \times B$?

Check your answers with the A and B of Examples 2 and 3.

✔ **2.**   The three sets $\mathbf{R}^2 \times \mathbf{R}$, $\mathbf{R} \times \mathbf{R}^2$, and $\mathbf{R} \times \mathbf{R} \times \mathbf{R}$ are *not* the same; e.g.,

$$\mathbf{R}^2 \times \mathbf{R} = \{((a, b), c) \,|\, a, b, c \in \mathbf{R}\}$$
$$\mathbf{R} \times \mathbf{R}^2 = \{(a, (b, c)) \,|\, a, b, c \in \mathbf{R}\}$$
$$\mathbf{R} \times \mathbf{R} \times \mathbf{R} = \{(a, b, c) \,|\, a, b, c \in \mathbf{R}\}$$

However, they certainly "look a lot alike."

    (a)   Construct the obvious map $f: \mathbf{R}^2 \times \mathbf{R} \to \mathbf{R} \times \mathbf{R} \times \mathbf{R}$ such that f is onto and one-to-one.

    (b)   Define an addition $\oplus$ on $\mathbf{R}^2 \times \mathbf{R}$ and an addition $\boxplus$ on $\mathbf{R} \times \mathbf{R} \times \mathbf{R}$ in a "component-wise way"; and similarly give "sensible" definitions of multiplication by a scalar $\alpha \in \mathbf{R}$ for the two spaces $\mathbf{R}^2 \times \mathbf{R}$ and $\mathbf{R} \times \mathbf{R} \times \mathbf{R}$.

    (c)   Test your map f in part (a) to see whether

$$f(((a, b), c) \oplus ((x, y), z)) = f((a, b), c) \boxplus f((x, y), z)$$

and whether

$$f(\alpha((a, b), c)) = \alpha f((a, b), c)$$

If your f meets all these requirements, we call it an **isomorphism** and say that $\mathbf{R}^2 \times \mathbf{R}$ and $\mathbf{R} \times \mathbf{R} \times \mathbf{R}$ are **isomorphic,** a more precise way of saying that they "look alike" not only as sets, but also under the action of such operations as addition and scalar multiplication.

✔ **3.**   The following exercises study the composition of maps:

    (a)   Show that f one-to-one and g one-to-one implies that $g \circ f$ is one-to-one.

    (b)   Show that f onto and g onto implies that $g \circ f$ is onto.

    (c)   Show that $g \circ f$ bijective implies g surjective and f injective.

✔ **4.**   Let $C^k[a, b]$ stand for the set of all functions $[a, b] \to \mathbf{R}$ which have a continuous kth derivative on the closed interval $[a, b]$. We define four mappings by the following rules:

E is defined by $[E(u)](t) = [u(t)]^2$

F is defined by $[F(u)](t) = \dfrac{d}{dt} u(t)$

G is defined by $G(u) = \displaystyle\int_0^1 u(\tau)\, d\tau$

H is defined by $[H(u)](t) = \displaystyle\int_0^t u(\tau)\, d\tau$

(a)   Let u be the function in $C^k[a, b]$ defined by

$$u(t) = \frac{1}{2} t^2 \text{ for } a \leq t \leq b$$

Compute $E(u)$, $F(u)$, $G(u)$, and $H(u)$.

(b)   Specify a codomain for each mapping. Find a set of values of a, b, and k so that each of these maps is well defined.

(c)   What maps—if any—are one-to-one? Onto?

(d)   Define a mapping from $C^k[a, b]$ which is both one-to-one and onto.

(e)   Give the rules which define the maps $E \circ H$, $H \circ E$, $H \circ F$, $F \circ H$, $F \circ G$, and $G \circ F$. Does $G \circ G$ make sense?

5.   Write out in everyday English the statements:

(a) $(\forall x)(\exists y)(x + y = 0)$ and $(\exists y)(\forall x)(x + y = x)$. Which of these two are true for $x, y \in \mathbf{Z}$? For $x, y \in \mathbf{N}$? For $x, y \in \mathbf{R}$?

(b) $x \in \mathbf{N} \Leftrightarrow x \in \mathbf{Z}$ and $x \in \mathbf{R}_+$. Is this a true statement?

(c) $x \in \mathbf{R}_+ \Rightarrow x \in \mathbf{N}$. Is this true?

(d) $(\forall \varepsilon > 0)(\exists \delta > 0)(|x - a| < \delta \Rightarrow |f(x) - f(a)| < \varepsilon)$. What do you call an $f: \mathbf{R} \to \mathbf{R}$ for which this statement is true?

✔ **6.**   In freshman calculus you studied sequences of real numbers $(x_0, x_1, x_2, \ldots, x_n, \ldots)$. Letting $\mathbf{N} = \{0, 1, 2, \ldots\}$, the nonnegative integers:

(a)   Convince yourself that a real sequence $S = (x_0, x_1, x_2, \ldots)$ is a mapping, $S: \mathbf{N} \to \mathbf{R}$, where x_n is just the value $S(n)$ of the mapping.

(b)   Generalize (a) to define a sequence of any objects, say from a set U, as a mapping with domain $\mathbf{N}$.

(c)   When we tag each object in the set U with a subscript, say $u_0, u_1, u_2, \ldots, u_n, \ldots$, by defining a mapping $S: \mathbf{N} \to U$, so that $u_n \overset{\Delta}{=} S(n)$. we say we have **indexed** the set U by $\mathbf{N}$.

Interpret the set of functions $V = \{f_a \mid a \in \mathbf{R}\}$, where for each $a \in \mathbf{R}, f_a(t) = e^{at}$, as a set indexed by the uncountably infinite set $\mathbf{R}$.

✔ **7.**   Let $f: A \to B$ and let $\mathcal{R}(f) = f(A)$ be the range of f. Prove that f is onto if and only if $f(A) = B$. [*Hint:* We know that $f(A) \subset B$, so we need only show that $B \subset f(A)$ iff f is onto to prove $B = f(A)$. A point t is in $f(A)$ iff $t = f(s)$ for some s in A.]

✔ **8.**   This exercise provides practice with **matrices** and **vectors,** which we shall study extensively in Section 2–3 and Section 3–3.

Let $f: \mathbf{R}^n \to \mathbf{R}^n$ by the rule

$$f(x) = Ax \text{ for all } x \in \mathbf{R}^n$$

where A is an $n \times n$ real matrix and Ax denotes matrix multiplication of the $n \times 1$ vector x by A.

(a)   Prove that f is one-to-one iff A is an invertible matrix.

(b)   Prove that f is onto iff A is invertible.

(c)   Prove that $\mathcal{R}(f) = \{x_1 A_1 + x_2 A_2 + \cdots + x_n A_n \mid \forall x_1, \ldots, x_n \in \mathbf{R}\}$ where $A_1, A_2, \ldots, A_n$ are the **columns** of A.

[Note: This shows that the set of all possible linear combinations of the columns of A, called the **column space of** A, is the range of F.]

(d)   Prove that f is onto iff the columns of A span $\mathbf{R}^n$.

(e)   Prove that f is one-to-one iff the columns of A are linearly independent vectors in $\mathbf{R}^n$.

(f)   Give the rule defining $f \circ f$.

1–3   The Concept of State

We may start by considering a system as a "black box," into which we feed inputs, and out of which we receive outputs, as illustrated by Figure 1–11. For a factory assembly line, the inputs might be raw materials and the outputs assembled products; for a computer, the inputs and outputs might be information encoded on punched cards or magnetic tape.

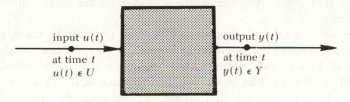

input $u(t)$
at time t
$u(t) \epsilon U$

output $y(t)$
at time t
$y(t) \epsilon Y$

Figure 1–11   A system as a black box.

A real choice is required in specifying the set U of **inputs,** and the set Y of **outputs,** which are to enter into our mathematical analysis of a system. Consider a computer. In an analysis of normal computation by the machine, we might consider the input to be a string of numbers which can be encoded upon a punched card. If, however, we are trying to design a computer which can compensate for mispunchings, then the input set must be extended to include all possible punchings on the card, and not simply those which encode numbers. An even further analysis might consider the physical condition of the card, and include variables corresponding to spindling, mutilating, and folding! Thus, our choice of the sets U and Y will depend upon the range of problems we expect to be covered by our model of the system—our model only encompasses that which we believe to be relevant. In many control systems, we assume that the power supply is sufficiently reliable that it need not enter into the analysis as an explicit input variable, but in more detailed studies of dependability, power supply may well be a crucial input variable.

Having decided (and at some stages in analysis, we may have to revise this decision) what the appropriate input and output sets U and Y are for our model, we must now determine on what **time scale** we shall analyze the system. If we consider a simple electrical circuit, in which the input is a particular voltage at any time, and the output is a current reading at any time, then we may think of the input and output variables as changing continuously, and so our model becomes a **continuous-time system** in which the time set T is indexed by the set **R** of all real numbers. On the other hand, a continuous-time description of a digital computer is inappropriate, and we prefer to model it as a **discrete-time system** in which the time set T is the set **Z** of all integers, where successive integers encode the times at which successive instructions are executed or "words" are read in or out of the computer.

Sometimes it may not be appropriate to consider the whole set **R** or the whole set **Z**—we may restrict our attention to all times later than or equal to 0, if, for instance, 0 is the time at which we start studying our system. Again, we need not think of the successive times of a discrete-time system as occurring regularly with respect to some real time—they may correspond to periodic sampling intervals, or they may simply correspond to successive actuations of a machine, with one particular pair of actuations being separated by a millisecond, while perhaps days separate some other pair of actuations.

The differential and integral calculus will provide the main tools for continuous-time systems, whereas algebra will provide the main tools for the analysis of discrete-time systems. Many readers will find themselves better prepared in calculus than in algebra, but we shall strive to intertwine our development of continuous- and discrete-time systems, for often the results for discrete-time systems, though couched in a less familiar mathematics, are much easier to understand, and help provide intuition for the analytical details required to prove corresponding properties for continuous-time systems.

In fact, in computing control schemes for real-world continuous-time systems, we are forced to treat them, in some sense, as discrete-time systems, since we can make only finitely many measurements of the output values in any time interval. Similarly, in computing what inputs we should apply to a system to control its behavior in some desired way, we shall generally use a digital computer, which is a discrete-time system, and so can only generate distinct control signals at discrete times. Thus, an important question will be that of how to treat **sampled data systems**—i.e., systems which we believe should be treated as continuous-time, but with which we can only interact at discrete "sampling" times.

Having specified the input and output sets, U and Y, and the time set T, let us now specify a particular input function which is to be applied to the system:

$$u : T \to U$$

That is, u is the name of a function $t \mapsto u(t)$, operating upon the set T of times to provide values in the set U of inputs, providing the specific input $u(t)$ at time t.

If we set a system going at some time t_0, and apply the selected input u from time t_0 through t_1, monitoring the corresponding outputs of the system during that time, can we expect the outputs to be uniquely determined by the inputs? In general, the answer is no. For example, the outputs we get from a computer during some period from t_0 to t_1 will depend not only upon what data we feed in, but also upon what program, and what other data, were already in the machine at time t_0. In other words, to specify how the inputs determine the outputs, we must add a description of some "internal state" of the system, which would correspond to the program and data already in a computer at the time we start using it.

We must thus add to our specification of the sets T, U, and Y of time, inputs, and outputs, a set X of internal states or "memory configurations," as illustrated in Figure 1–12. (The reader is warned that there are almost as many notational conventions as there are authors in system theory, and one writer may use X for the set of inputs just as facilely as we use it for the set of states.)

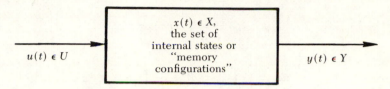

$u(t) \in U$

$x(t) \in X$,
the set of
internal states or
"memory
configurations"

$y(t) \in Y$

Figure 1–12   An internal description is needed.

Example 1

Let us consider a resistor and a capacitor, each as a system with current as input and voltage as output. These two systems are described by the equations in Figures 1–13a and b. The resistor of Figure 1–13a is "memoryless," since the output at any instant is determined solely by the input at that instant. However, for the capacitor, the input $i(t)$ from t_0 to t is not enough; we must also know the initial voltage $v(t_0)$ in order to say what the voltage $v(t)$ will be at time $t > t_0$. Thus, the capacitor requires an explicit state variable, and we see that $v(t_0)$ provides the necessary internal description or state at time t_0. If we assume that the initial voltage could have any real number as its value, we can specify the set of states by $X = \mathbf{R}$.   ◊

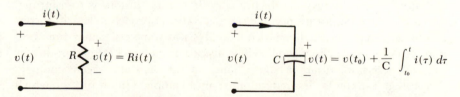

$i(t)$

$+$

$v(t)$

R $+$ $v(t) = Ri(t)$ $-$

$-$

$i(t)$

$+$

$v(t)$

C $+$ $v(t) = v(t_0) + \dfrac{1}{C} \displaystyle\int_{t_0}^{t} i(\tau)\, d\tau$

$-$

Figure 1–13(a)   A memoryless system.   **(b)** A system with memory.

The computer and circuit examples allow us to be somewhat more specific about the notion of state:

*The **state** is some compact representation of the past activity of the system complete enough to allow us to predict, on the basis of the inputs, exactly what the outputs will be, and also to update the state itself.*

For example, in the case of the capacitor of Example 1, from

$$i = C \frac{dv}{dt}$$

we get

$$v(t) = \frac{1}{C} \int_{-\infty}^{t} i(\tau)\, d\tau$$

if we assume that the capacitor was initially uncharged or relaxed at $t = -\infty$.
    Thus

$$v(t_0) = \frac{1}{C} \int_{-\infty}^{t_0} i(\tau)\, d\tau$$

so that

$$v(t) = \frac{1}{C} \int_{-\infty}^{t_0} i(\tau)\, d\tau + \frac{1}{C} \int_{t_0}^{t} i(\tau)\, d\tau$$

$$= v(t_0) + \frac{1}{C} \int_{t_0}^{t} i(\tau)\, d\tau$$

and we see that specifying $v(t_0)$ is more economical than recording all fluctuations in input current from $-\infty$ up to t_0.
    Similarly, if we receive a computer from the factory, "in its 0 state,"
and then start using it, we expect that after three months its state will have
changed. One way of describing the way in which this state has changed
is simply to provide a record of all input history up to the time at which
we come to analyze the machine. But this incredible record of data is far
too redundant, for we know that to describe the way a computer will act,
it is sufficient to specify for its state the present contents of its registers and
memory locations.
    Whereas classical analyses of linear systems, especially by transform
methods,* stress the way in which a complete input history is transformed
into a complete output history, the state variable approach to system theory
emphasizes that we usually start our analysis at some particular time and
that we do not care about the previous history of the machine except insofar
as it will determine future behavior. Thus, we use the state as some compact
description, usually far less redundant than a complete input history, which
allows us to predict the future behavior of the machine.

EXERCISES FOR SECTION 1–3

1.   Suppose you were the president of your country about to deliver your "state
of the union" address. What quantities would you select as the state variables?

✔ **2.**   Review Example 1 and Exercise 1 of Section 1–1 and show that the *output
of an integrator* may be taken as the state variable. If the output is y, define the variable
$q = 6y$ and show that q is a perfectly respectable, different choice for the state of

---

*We survey these methods in Section 6–3.

the integrator. Is there anything special about the 6? Do you see that there are often many ways to select the internal description of a system?

✔ **3.**   Could the input u of the ideal integrator of Example 1–1–1* be taken as the state? [*Hint:* Does knowledge of $u(t_0)$ and the input from t_0 to t exactly determine the output $y(t)$ at times $t > t_0$?]

✔ **4.**   For the electrical circuit of Figure 1–7a in Exercise 1–1–5, let input be v and output be i, and let x be the voltage across the capacitor. Show that x is an acceptable state variable for this system.

5.   Suppose you want to automatically administer an anti-rejection drug to heart transplant patients in an intensive care ward of a modern hospital. What quantities would you wish to monitor (use as state variables) to have an adequate "internal description" of the patient (the system)?

6.   For the ideal capacitor discussed in Example 1 and in Exercise 1–1–2, let $\omega(t)$ be the **energy** stored in the capacitor at time t and let $p(t)$ be the **power.** From freshman physics we know that

$$p(t) = \frac{d\omega(t)}{dt}$$

$$p(t) = i(t)v(t)$$

and

$$i(t) = C\frac{dv(t)}{dt}$$

(a)   Derive an expression for the energy $\omega(t)$ in terms of the current i as input and voltage v as output.

(b)   Express the output $v(t)$ as a function of $\omega(t)$.

(c)   What do you think about the suitability of energy as a state variable for the capacitor?

1–4   Stochastic Systems

We shall detour briefly to make a distinction between deterministic and stochastic systems, and thereafter deal almost exclusively with deterministic systems. We have just given examples of *deterministic* or *state determined* systems in which, having specified the state and the subsequent inputs, we can determine exactly the subsequent states and outputs. But one might also imagine systems in which, no matter how fully we specify the state, and no matter how carefully we specify subsequent inputs, we cannot determine exactly what the subsequent states and outputs will be—at best we might determine tight probabilities on the possible range of the states and outputs, and in that case, we shall speak of a **stochastic** system.

Before we return to our exclusive concentration on deterministic systems it might be worth considering two reasons why stochastic systems are worthy of study: one is that some systems are studied at the atomic and molecular levels, where many physicists believe that Heisenberg's uncertainty principle

---

*This is the first time we use our code for cross references. Example 7–5–3 refers to Example 3 of Section 7–5.

decrees, as an inescapable reality of the physical universe, that no matter how carefully we specify the system we can at best make probabilistic predictions about its future behavior; the second is that there are systems, which can appropriately be described in terms of classical deterministic Newtonian mechanics, but which nonetheless have so many "background" variables that they are better analyzed as stochastic systems.

To understand this second point, consider a collection of particles moving under the laws of Newtonian mechanics. The system has "no" input, and the mathematician finds it convenient to model this by letting the set U of inputs have only one member [just as it will be convenient to model a resistor as having exactly one state]—since there is only one member, we need never specify the "input" at any time, and we may regard this "input" as simply being a guarantee of the system's continued existence! For the output of a system of n particles, we might consider the positions of the n particles. However, we know from Newtonian mechanics that a measurement of position alone is not enough to allow us to predict the subsequent behavior of the particles, and must be augmented by a knowledge of the velocities, or momenta, of the particles. Thus, we may regard a Newtonian system of n particles as having a one-element input set, as having for a state set the set of all possible positions and momenta of the particles, and as having for an output set the set of all possible positions of the particles.

Now, suppose we have a very large number of particles, and consider some subcollection of the particles as the system of interest. Interactions with other particles provide inputs for this new subsystem. However, in describing such a subsystem, we do not usually deem it worthwhile (nor do we often find it feasible) to keep track of the effects of all particles in the external universe in decreeing the present input. Even though the environment and the system are completely deterministic, by looking only at certain gross properties of the detailed pattern of inputs, we remove some of the determinism from our analysis. Those variables that we ignore become "noise," and can only be included in our analysis by introducing statistical perturbations of the dynamics.

Thus, even deterministic systems are often best analyzed as stochastic systems—the gain is that we reduce immensely the number of variables that need be considered. Of course, it is crucial to such an analysis that the variables we do consider are central to the problems we are studying, and that the other variables are sufficiently uninfluential that they can be lumped together into relatively simple "noise" terms.

To conclude our digression on stochastic systems, then, we may see that a stochastic treatment is worthwhile either because we are analyzing systems at the quantum level, or because we are analyzing macroscopic systems which lend themselves to a stochastic description when we ignore fluctuations in microscopic variables.

1–5   Deterministic Systems

We say that a system is **deterministic** or **state determined** if, having specified sets T, U, Y, and X, we can also specify two functions η and ϕ whereby, given the state at some time t_0, we can specify the output corresponding to it at time t_0; and further, whereby, given the state at some time t_0, and the input to the system from time t_0 to some subsequent time t_1, we may also specify what the new state will be at time t_1.

Readers familiar with classical linear system theory should note that this discussion is completely general, and has nothing to do with linearity. It is for this reason that we have stressed the example of the computer, for it is not usual to impose a linear structure on the set of symbols that can be punched upon computer cards. Further, the computer is discrete-time, as distinct from the continuous-time systems of classical system theory. Our analysis might apply to even so homely a system as a blender, where the input set might simply consist of an apple, a cherry, and a pear, and the output set is some rating of the flavor of the liquid in the container at the time of sampling! It is only later that we shall force our input, state, and output sets to be in some sense numerical, and start to emphasize the computational advantages of demanding that the functions that specify the new states, and the outputs, are linear.

In looking at the possible input functions which may be applied to the system, we may expect that in the discrete-time case, all possible sequences of inputs will be admissible—the proper functioning of a computer does not depend on any special relationships holding between consecutive cards fed into the machine, as long as they provide data germane to the current program. However, when we use calculus to study a continuous-time system, our analysis may well depend upon continuity or even differentiability properties of the input function. We may wish to exclude wild variations in the input. For this reason, it is usual to specify, for a continuous-time system, not only the set U of instantaneous inputs, but also a set Ω of **admissible input functions** from T to U. Thus, if u is in Ω, then u is a *function* and each *value* of u, say $u(t)$, lies in U.

Given two input functions u and u', we may well wish to use u up to some time t_0 and u' thereafter; it is usual in this case to demand that the class Ω be **closed under splicing:** i.e., if u and u' are admissible, so too should be the function

$$u \, \textcircled{t_0} \, u' : t \mapsto \begin{cases} u(t) \text{ if } t \leq t_0 \\ u'(t) \text{ if } t > t_0 \end{cases}$$

which equals u up to time t_0 and u' thereafter, as indicated in Figure 1–14, in which $u \, \textcircled{t_0} \, u'$ is represented by the heavy portions of the curves.

Since we have defined $u \, \textcircled{t_0} \, u'(t)$ to be $u(t)$ up to and including $t = t_0$, we sometimes refer to this operation as *left-splicing* instead of just *splicing*.

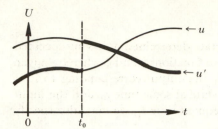

Figure 1-14 Splicing two functions at time t_0.

Typical choices for the collection of admissible input functions include the piecewise continuous functions and the piecewise differentiable functions—functions which can be decomposed into continuous, or differentiable, functions on each of a finite number of subintervals of the real line. The function u t_0 u' in the above diagram is clearly piecewise differentiable, but is not differentiable because of the discontinuity at the point t_0. In the case of the capacitor of Example 1-3-1, any integrable input function i would be admissible, so that

$$\int_{t_0}^{t} i(\tau)\, d\tau$$

makes sense—and clearly the integrable functions meet our "splicing" condition.

A subtle mathematical point: If we take a computer which is in its "0 state" with all registers set to 0, and do not apply any input, then, in fact, we *have* applied an input—the 0 input! Just as it was a great mathematical invention to introduce the number 0 so that one could just write "$A + B$," it being understood that one value of B could be 0, rather than writing "$A + B$, and in the special case that you do not add anything to A, then . . . ," so it is often valuable in analyzing a system to include in the set U of inputs a special element, which corresponds to the "null" or "zero" input. Thus, in our discussion in Section 1-4 of a classical "autonomous" Newtonian system, we saw that we could fit such a system into our new framework of mathematical system theory by considering it to have only one element, a null input, in its set of inputs.

Having fixed upon a set Ω of admissible input functions, let us now describe how an admissible input function affects the state of the system and the output of the system.

Since $x(t_1)$, the new, "updated" state at time t_1, is determined by t_0, $x(t_0)$, and the input function u over the time interval $[t_0, t_1)$, we express this functional dependence by $x(t_1) = \phi(t_1, t_0, x(t_0), u)$. This function ϕ is called the **state transition map,** and it tells us that if we specify two times, t_1 and t_0, a state $\tilde{x}$, and an admissible input function u, then, if we start the system in state $\tilde{x}$ at time t_0 and apply the input function u, the system will end up in state $\phi(t_1, t_0, \tilde{x}, u)$ at time t_1.

Using the set-theoretic notation of Cartesian products, we write $\phi: T \times T \times X \times \Omega \to X$ where $T \times T \times X \times \Omega$ is the set of ordered quadruples whose first two entries come from T, whose third entry comes from X, and whose fourth entry is an admissible input function from Ω.

Example 1

For the capacitor of Example 1–3–1, the function ϕ is given by the formula

$$v(t_1) = \phi(t_1, t_0, V, i) = V + \frac{1}{C} \int_{t_0}^{t_1} i(\tau)\, d\tau$$

while for the resistor of the same example

$$v(t_1) = \phi(t_1, t_0, V, i) = R\, i(t_1)$$

where we use voltage v for the state, current i for the input, and V for the initial value of voltage, $v(t_0) = V$. ◊

We have said that the function ϕ tells us how the input applied over a period from t_0 to t_1 serves to update the state of the system. This implies that the function ϕ cannot be completely arbitrary but must satisfy several **consistency conditions.** First, if we start in state $\tilde{x}$ at time t_0, it is clear that, no matter what input we subsequently apply, the state at time t_0 must, of course, be $\tilde{x}$:

$$\phi(t_0, t_0, \tilde{x}, u) = \tilde{x} \tag{1}$$

for all times t_0, all states $\tilde{x}$, and all admissible input functions u.

It is obvious that our state-transition equation for the capacitor given in Example 1 satisfies this condition, since with $t_1 = t_0$, the integral term drops out.

For our second condition, suppose that at time t_0 we start in state $\tilde{x}$ and then apply the input function u. Then, at time t_1 the state of the system will be $\phi(t_1, t_0, \tilde{x}, u)$. Now let us continue to apply the input function u from time t_1 until time t_2. Then, we may compute the new state in two ways. We may either look at the updating of the state $\phi(t_1, t_0, \tilde{x}, u)$ at time t_1 by the input function u from t_1 until t_2 to yield the new state $\phi(t_2, t_1, \phi(t_1, t_0, \tilde{x}, u), u)$ or we may simply start with the state $\tilde{x}$ at time t_0 and use input u to update the state over the whole time period from t_0 to t_2 to obtain the state $\phi(t_2, t_0, \tilde{x}, u)$. Since the state at time t_2 cannot depend upon the way in which we compute it, but only depends upon the system, we end up with the equality

$$\phi(t_2, t_1, \phi(t_1, t_0, \tilde{x}, u), u) = \phi(t_2, t_0, \tilde{x}, u) \tag{2}$$

Example 2

For the capacitor's ϕ in Example 1 we see that

$$v(t_2) = \phi(t_2, t_1, \phi(t_1, t_0, V, i), i) = \phi(t_2, t_0, V, i)$$

since

$$\left[V + \frac{1}{C} \int_{t_0}^{t_1} i(t) \, dt \right] + \frac{1}{C} \int_{t_1}^{t_2} i(t) \, dt = V + \frac{1}{C} \int_{t_0}^{t_2} i(t) \, dt$$

so the second consistency requirement is satisfied.   ◊

The final restriction we place on ϕ is that it be **causal;** in other words, we must demand that the transition from the state $\tilde{x}$ at time t_0 to the state $\phi(t_1, t_0, \tilde{x}, u)$ at time t_1 can only depend on u through the values that u takes from time t_0 to time t_1. If two input functions agree during the time from t_0 to t_1, then they must lead to the same state transition in that time:

$$\phi(t_1, t_0, \tilde{x}, u) = \phi(t_1, t_0, \tilde{x}, u')$$
$$\text{if } u(t) = u'(t) \text{ for } t_0 \leq t < t_1 \tag{3}$$

Reminder: Throughout this book we shall use $\dot{u}$ to denote the derivative of u with respect to t; thus, u' denotes "any old function," *not* the derivative of a given function u.

Example 3

Again using the capacitor of Example 1, if $i(t) = i'(t)$ for $t_0 \leq t < t_1$, then

$$V + \frac{1}{C} \int_{t_0}^{t_1} i(t) \, dt = V + \frac{1}{C} \int_{t_0}^{t_1} i'(t) \, dt$$

so this ϕ is causal.   ◊

To emphasize that it is only the behavior of u during the interval $[t_0, t_1)$ which can influence the transition from $x(t_0)$ to $x(t_1)$, we sometimes write

$$x(t_1) = \phi(t_1, t_0, x(t_0), u_{[t_0, t_1)})$$

where $u_{[t_0, t_1)}$ is the **restriction** of the function u to the subset $[t_0, t_1) \subset T$ of its domain.

Example 4

If we apply this third condition to the resistor of Example 1, where $v(t) = R\,i(t)$, we get more insight into the need for restricting our analysis of a system to include the response only to certain admissible input functions. Our causality condition demands that $\phi(t_1, t_0, v(t_0), i)$ depends only on input values $i(t)$ for $t_0 \leq t < t_1$, yet our equation

$$v(t_1) = R\,i(t_1)$$

makes it depend on the value $i(t)$ for the time $t = t_1$ (which is outlawed in the causality condition)! The resolution is, however, a simple one—we shall say that a function $i : \mathbf{R} \to \mathbf{R}$ is an admissible input for our resistor only when i is continuous-from-the-left; i.e., for each time t, $i(t-) = i(t)$ where $i(t-)$ is the **limit-from-the-left** *

$$\lim_{\tau \uparrow t} i(\tau)$$

In other words, we expect our model $v(t) = R\,i(t)$ to properly describe the response of a resistor only to a smoothly-varying current, and do not guarantee that our equation will describe the system "instantaneously" after a jump in the input.   ◊

Note that the collection of all left-continuous functions does indeed satisfy our left-splicing condition.

Example 5

Perhaps the most familiar example of a left-continuous function which is not continuous is the **Heaviside function,** or **unit-step function, 1,** which is piecewise constant with a unit step at $t = 0$, as shown in Figure 1–15. (Note that the different function which takes the value 1 for $t \geq 0$ and is zero elsewhere is *not* left-continuous.)

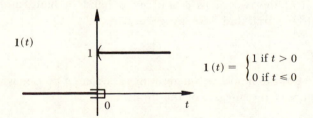

$$\mathbf{1}(t) = \begin{cases} 1 \text{ if } t > 0 \\ 0 \text{ if } t \leq 0 \end{cases}$$

Figure 1–15   The unit step function $\mathbf{1}(t)$.   ◊

---

*The symbol $\tau \uparrow t_1$ means $\tau \to t_1$ through values $\tau < t_1$ as is implicit in $u_{[t_0, t_1)}$.

Our causality condition says two things. First, the input prior to t_0 has no effect on $\phi(t_1, t_0, \tilde{x}, u)$—this says that the state variable contains enough information, since further information about the input prior to obtaining that state is irrelevant. Second, the input after t_1 has no effect on $\phi(t_1, t_0, \tilde{x}, u)$ —this says that the system cannot foresee the future, but can only respond to inputs it has already received. This does *not* deny that a system may be predictive—it may well use past input to *estimate* future input and respond accordingly, but it is nonetheless reacting only on the basis of past input, and may well behave very inappropriately should there be a later unexpected fluctuation in the input.

Now that we have carefully studied the state-transition function, we should look more carefully at the output map. It is usual to think of the output of a system as an *observation* of certain aspects of the state of the system. In this formulation, the state is considered to manifest all the dynamics and essential character of the system, and the output is just an instantaneous "read-out" of those characteristics deemed relevant to the particular use of the system. Thus, the input affects the output only indirectly through its effect in updating the state.

The **output mapping** $\eta: T \times X \to Y$ specifies that if the system is in state $\tilde{x}$ at time t_1, the output at that very time will be $\eta(t_1, \tilde{x})$.

Example 6

Again using Example 1, the output map for the capacitor is given by $\eta(t_1, v(t_1)) = v(t_1)$, and that for the resistor is given by $\eta(t_1, v(t_1)) = v(t_1)$, both being pure read-outs of the state v at time t_1.   ◊

In formulating a deterministic system model for an actual dynamic system, once we have specified the sets T, U, Ω, and Y, we then try to choose the state set X so that the state variable gets updated by a function ϕ consistent with the conditions (1), (2), and (3) and can be read out instantaneously by some appropriate output function η. If we think of the state as an internal (perhaps fictional) description of a system whose input-output behavior concerns us, we can see that the choice of a state variable and the specification of the two maps ϕ and η go hand in hand and are often determined by a trial and error procedure.

Example 7

Suppose we want to model as a deterministic system a process whose output y is the square of the integral of the input u:

$$y(t) = \left[\int_{t_0}^{t} u(\tau)\, d\tau \right]^2 \text{ for } t \geq t_0$$

As an obvious first stab at an internal state model, we might try to define the value of the output $y(t)$ as the state at time t, started from zero at time

t_0 and updated by the function ϕ defined by

$$\phi(t_1, t_0, \tilde{x}, u) = \tilde{x} + \left[\int_{t_0}^{t_1} u(\tau)\, d\tau \right]^2$$

But this ϕ does *not* satisfy our second consistency condition; in fact,

$$\phi(t_2, t_0, \tilde{x}, u) = \phi(t_2, t_1, \phi(t_1, t_0, \tilde{x}, u), u) + 2 \int_{t_0}^{t_1} u(\tau)\, d\tau \cdot \int_{t_1}^{t_2} u(\tau)\, d\tau$$

$$\neq \phi(t_2, t_1, \phi(t_1, t_0, \tilde{x}, u), u)$$

Yet it seems plausible that a system could compute

$$\left[\int_{t_0}^{t_1} u(\tau)\, d\tau \right]^2$$

and update its state by adding this quantity to the initial value $\tilde{x}$. What goes wrong? We see that to obtain $\phi(t_2, t_0, \tilde{x}, u)$ given the state $\phi(t_1, t_0, \tilde{x}, u)$, we need not only this state information but also

$$\int_{t_0}^{t_1} u(t)\, dt$$

but this is unavailable since it was already squared and added onto the initial state $\tilde{x}$. The solution is then to have

$$\int_{t_0}^{t_1} u(\tau)\, d\tau$$

kept separately available. In fact, the easiest way to arrive at a deterministic system which computes

$$\left[\int_{t_0}^{t_1} u(\tau)\, d\tau \right]^2$$

is to choose*

$$x(t) = \int_{t_0}^{t} u(\tau)\, d\tau$$

and then to start in state 0 at time t_0 the system described by the functions

$$\phi(t_1, t_0, \tilde{x}, u) = \tilde{x} + \int_{t_0}^{t_1} u(\tau)\, d\tau$$

$$\eta(t, x(t)) = (x(t))^2 \qquad\qquad \Diamond$$

---

*This choice of state variable would have been obvious if we had tried to draw an analog computer diagram to generate

$$y(t) = \left[\int_{t_0}^{t} u(\tau)\, d\tau \right]^2.$$

In the last example we started with an output specification

$$\left[\int_{t_0}^{t} u(\tau)\, d\tau \right]^2$$

and found that it did not contain enough information to be used as a state variable. We chose a new state variable to allow us to update the state consistently. In this case, we found that the original output variable was no longer required as a state variable, but could be obtained as a simple function of the current state variable. In Section 2–4 we shall see that, for many mechanical systems, the output variables (namely positions) are, by contrast, an actual subset of the state variables (which comprise both position and velocity [or momentum] variables).

It should be pointed out that some authors prefer to have the output depend explicitly upon the input as well as on the initial state, and use a mapping $\bar{\eta} : T \times X \times U \to Y$ instead of $\eta : T \times X \to Y$. Such authors would consider our resistor of Example 1 to have a single state 0, called the zero state, and to have its output depend on both the input i and (trivially) the state 0 by the formula

$$\bar{\eta}(t_1, 0, i(t_1)) = R\, i(t_1)$$

Given any theory by such authors using $\bar{\eta} : T \times \bar{X} \times U \to Y$, we can always embrace it in our theory by enlarging their state space $\bar{X}$ to include the input $u(t)$ as a component of the state vector in the new state space X, so long as admissible functions u in Ω are left-continuous. We illustrate with an example from freshman physics.

Example 8

The RC circuit shown in Figure 1–16 is driven by input voltage u and has output voltage y. By Kirchoff's voltage law, at every instant t, the applied voltage u must equal the sum of all the voltage drops across the capacitor

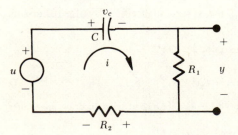

Figure 1–16   A circuit from freshman physics.

C and the two resistors R_1, R_2:

$$u(t) = v_c(t) + v_{R_1}(t) + v_{R_2}(t)$$

Since

$$i(t) = C\frac{dv_c(t)}{dt}$$

$$v_{R_1}(t) = R_1 i(t)$$

and

$$v_{R_2}(t) = R_2 i(t)$$

we get the first order linear differential equation

$$(R_1 + R_2)C\frac{dv_c(t)}{dt} + v_c(t) = u(t)$$

The reader may readily check that the solution of this differential equation is given by

$$v_c(t) = e^{-(t-t_0)/RC}v_c(t_0) + \frac{1}{RC}\int_{t_0}^{t} e^{-(t-\zeta)/RC}u(\zeta)\,d\zeta$$

where $R \overset{\Delta}{=} R_1 + R_2$, and $v_c(t)$ is uniquely determined for all t if $v_c(t_0)$ and an admissible u are known. Since knowledge of $v_c(t)$ for all t determines i, and hence $y = R_1 i$, it would appear that v_c is a natural choice for the state of this system.

Using v_c as the state, the above solution of the differential equation provides us with the state transition map $\bar{\phi}$:

$$\bar{\phi}(t_1, t_0, \bar{x}_0, u) = e^{-(t-t_0)/RC}\bar{x}_0 + \frac{1}{RC}\int_{t_0}^{t} e^{-(t-\zeta)/RC}u(\zeta)\,d\zeta$$

With this choice of state $\bar{x} = v_c$, however, the output is

$$y = R_1 i = \frac{R_1}{R_1 + R_2}(u - v_c) = \frac{R_1}{R_1 + R_2}(u - \bar{x})$$

which, while simple enough, is in the form $y(t) = \bar{\eta}(t, \bar{x}(t), u(t))$ rather than our preferred $y(t) = \eta(t, x(t))$ for an appropriate selection of state variable x.

We have seen that the answer is to expand $\bar{X}$ to the larger state-space $X = \bar{X} \times U$. However (cf. Example 4), a subtlety emerges here. In updating the state from time t_0 to time t_1 we may make use of only $u_{[t_0, t_1)}$, the input *before* t_1.

But how, then, are we to obtain $u(t_1)$ to insert as the second component of our augmented state vector $x(t_1) = (\overline{x}(t_1), u(t_1))$? The answer is, as in Example 4, to demand that our inputs be continuous from the left so that $u(t_1) = \lim_{t \uparrow t_1} u(t)$. The symbol $t \uparrow t_1$ means $t \to t_1$ but $t < t_1$, as implied by $u_{[t_0, t_1)}$. With this choice of state our new state transition map ϕ is constructed from

$$x(t_1) = (\overline{x}(t_1), u(t_1)) = (\overline{\phi}(t_1, t_0, V, u), u(t_1))$$

and so

$$\phi(t_1, t_0, (V, u(t_0)), u) = \left(e^{-(t_1 - t_0)/RC} V + \frac{1}{RC} \int_{t_0}^{t_1} e^{-(t_1 - \zeta)/RC} u(\zeta) \, d\zeta, \lim_{t \uparrow t_1} u(t) \right)$$

and now

$$y(t) = \eta(t, x(t)) = \eta(t, (v_c(t), u(t))) = \frac{R_1}{R_1 + R_2} [u(t) - v_c(t)]$$

so that $\eta: T \times X \to Y$ as required.   ◊

To summarize our discussion in this section, we present the graphs in Figure 1–17, which may help the reader to visualize the input, state, and output trajectories of a deterministic system.

In our later study of linear, continuous-time systems,* we shall see that the input-output behavior of such systems is often expressible as an integral,

$$y(t) = \int_{-\infty}^{\infty} g(t, \zeta) u(\zeta) \, d\zeta \qquad \qquad \textbf{(4)}$$

called the *superposition integral,* where $g(t, \zeta)$ is called the *impulse response* of the system.

We close this section by showing that our general definition of causality, Eq. (3), reduces to a simple condition on the impulse response for systems characterized by Eq. (4). For simplicity, we will restrict ourselves here to systems for which $g(t, \zeta)$ is a continuous real function.

---

*We do this in Section 5–4. For now, don't worry about what these adjectives mean; just suppose a system is described by an integral like Eq. (4) and see what causality implies for such systems.

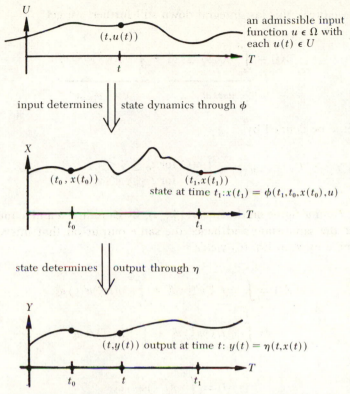

Figure 1–17   Input, state, and output trajectories of a deterministic system.

Example 9

For a system whose input-output behavior is described by Eq. (4), we may see from

$$y(t) = \int_{-\infty}^{\infty} g(t, \zeta)u(\zeta)\, d\zeta$$

$$= \underbrace{\int_{-\infty}^{t_0} g(t, \zeta)u(\zeta)\, d\zeta}_{\text{uses } u_{[-\infty, t_0)}} + \underbrace{\int_{t_0}^{\infty} g(t, \zeta)u(\zeta)\, d\zeta}_{\text{uses } u_{[t_0, \infty)}} \qquad \textbf{(5)}$$

that if we restrict the system to a set Ω of admissible inputs which are zero before the time t_0, $u(\zeta) = 0$ for $\zeta < t_0$, we have

$$y(t) = \int_{t_0}^{\infty} g(t, \zeta)u(\zeta)\, d\zeta \quad \text{for } t \ge t_0$$

If we break this last integral down still further, we get

$$y(t) = \int_{t_0}^{t} g(t, \zeta)u(\zeta)\, d\zeta + \int_{t}^{\infty} g(t, \zeta)u(\zeta)\, d\zeta \qquad (6)$$

$$\underbrace{\qquad\qquad}_{\text{uses } u_{[t_0, t)}} \quad \underbrace{\qquad\qquad}_{\text{uses } u_{[t, \infty)}}$$

Now let u' be defined by

$$u'(\zeta) = \begin{cases} u(\zeta) & \text{for } t_0 \le \zeta < t \\ 0 & \text{for } t \le \zeta < \infty \end{cases}$$

so that u and u' agree on the interval $[t_0, t)$. By causality, u and u' must then produce the same state and hence the same output on that interval. But replacing u by u' in Eq. (6) yields

$$y(t) = \int_{t_0}^{t} g(t, \zeta)u'(\zeta)\, d\zeta + \int_{t}^{\infty} g(t, \zeta)u'(\zeta)\, d\zeta$$

$$= \int_{t_0}^{t} g(t, \zeta)u(\zeta)\, d\zeta + \int_{t}^{\infty} g(t, \zeta)\, 0 \, d\zeta$$

whence

$$y(t) = \int_{t_0}^{t} g(t, \zeta)u(\zeta)\, d\zeta \qquad (7)$$

By equating (6) and (7), which must hold for all input functions u, we get

$$\int_{t_0}^{t} g(t, \zeta)u(\zeta)\, d\zeta + \int_{t}^{\infty} g(t, \zeta)u(\zeta)\, d\zeta = \int_{t_0}^{t} g(t, \zeta)u(\zeta)\, d\zeta$$

from which

$$\int_{t}^{\infty} g(t, \zeta)u(\zeta)\, d\zeta = 0$$

for all $u \in \Omega$.

Since this last expression holds for all functions u on the interval $[t, \infty)$, it must hold for the particular function

$$\widehat{u}(\zeta) \triangleq g(t, \zeta), t \le \zeta < \infty$$

whence

$$\int_{t}^{\infty} |g(t, \zeta)|^2\, d\zeta = 0$$

But the fact that the integral of the nonnegative function $|g(t, \zeta)|^2$ vanishes implies that the integrand must vanish (almost everywhere) on $[t, \infty)$ so

$$g(t, \zeta) = 0 \qquad \text{for} \qquad t < \zeta \qquad \qquad \textbf{(7)} \ \Diamond$$

Thus, *our definition of causality implies that the impulse response* $g(t, \tau)$ *of a linear system must be zero for* $t < \tau$, a criterion perhaps familiar to those readers who have had previous exposure to elementary linear system theory.

EXERCISES FOR SECTION 1–5

✔ **1.**   An "exact-change" lane in a 10¢ toll bridge accepts nickels, dimes, and pennies. This system has one state variable and two input variables u_1 and u_2 for the coin basket and the exit-pressure plate in the road. The function η maps the state into either a red light, green light, or warning bell for short change. What are U, Ω, X, Y, T, ϕ, and η for this system? (You can ignore reset to zero after a short-change condition.)

2.   What are U, Ω, X, Y, T, ϕ, and η for the system of Figure 1–18? The letter L stands for **inductance** and

$$v_L(t) = L \frac{di_L}{dt}(t)$$

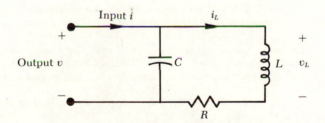

Figure 1–18   An *RLC* circuit.

✔ **3.**   Ask an electrical engineer why the function $g(t, \zeta)$ in the integral equation (4) is called the **impulse response** of the system and why our causality condition (7) can be stated colloquially as "the impulse response is zero before the impulse is applied."

✔ **4.**   This exercise should be attempted by all readers familiar with the Laplace Transformation. If you are not familiar with it, just relax; we will discuss the Laplace Transform in Section 6–3 and you won't need it until then.

Consider the series RLC circuit of Figure 1–19, which is driven by a voltage input $u(t)$ with output voltage $y(t)$.

(a)   Use Kirchoff's voltage law around the loop to get an integro-differential equation satisfied by the current $i(t)$. Assume the capacitor is "relaxed" or uncharged at $t = -\infty$.

(b)   Using the **ordinary Laplace Transform:**

$$F(s) \overset{\Delta}{=} \mathcal{L}\{f(t)\} = \int_0^\infty f(\tau) e^{-s\tau} \, d\tau \qquad \text{[from Appendix A–1]}$$

and letting $U(s)$, $I(s)$, and $Y(s)$ be the respective transforms of $u(t)$, $i(t)$, and $y(t)$, take the transform of both sides of the equation in (a). Solve the transform equation for $I(s)$ in terms of $U(s)$ and the values $i(0)$ and $v_c(0)$.

$$\left[Answer:\ I(s) = \frac{U(s) - \dfrac{1}{s}v_c(0) + Li(0)}{R + Ls + \dfrac{1}{sC}} \right]$$

Also give $V_c(s)$ in terms of $U(s)$, $i(0)$, and $v_c(0)$.

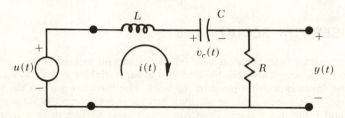

Figure 1–19   A series *RLC* circuit.

(c)   Notice that knowledge of $i(t)$ and $v_c(t)$ at time $t = 0$ and knowledge of $u(t)$ for $t \geq 0$ permit us to calculate $i(t_1)$ and $v_c(t_1)$ for any time $t_1 \geq 0$. Thus, **inductor current $i(t)$ and capacitor voltage $v_c(t)$ are good candidates for state variables for this circuit.**

Using the values $L = C = 1$ and $R = 2$, inverse-Laplace transform to get the complete time domain expressions for $i(t)$ and $v_c(t)$ for $t \geq 0$ in terms of $i(0)$, $v_c(0)$, and $u(t)$.

(d)   What is so special about time $t = 0$ that our answers depend upon the values of i and v_c at that time? [*Answer:* Nothing! We got stuck with needing to specify $i(0)$ and $v_c(0)$ only because the lower limit on the integral in the Laplace Transform is zero.]

From the equations for $i(t)$ and $v_c(t)$ in part (c), set $t = t_0$ and solve for $i(0)$ and $v_c(0)$ in terms of $i(t_0)$ and $v_c(t_0)$.

(e)   Use the results from (d) to express $i(t)$ and $v_c(t)$ in terms of $u(t)$, $i(t_0)$ and $v_c(t_0)$. This shows explicitly how knowledge of i and v_c at *any time* t_0, along with the input $u(t)$ for $t \geq t_0$, determines $i(t)$ and $v_c(t)$ for all $t \geq t_0$. [*Answers:*

$$i(t) = i(t_0)[1 - (t - t_0)]e^{-(t-t_0)} - v_c(t_0)(t - t_0)e^{-(t-t_0)}$$

$$+ \int_{t_0}^{t} (1 - \tau)e^{-\tau}u(t - \tau)\, d\tau$$

$$v_c(t) = i(t_0)(t - t_0)e^{-(t-t_0)} + v_c(t_0)[1 + (t - t_0)]e^{-(t-t_0)}$$

$$+ \int_{t_0}^{t} \tau e^{-\tau}u(t - \tau)\, d\tau$$

for $t \geq t_0$.]

(f)   Suppose that the inductor current and capacitor voltage are observed at $t = 3$ sec. to be $(i(3), v_c(3)) = (a, b)$. Write the formulas for $i(t)$ and $v_c(t)$ in **matrix** form

$$\begin{bmatrix} i(t) \\ v_c(t) \end{bmatrix} = \begin{bmatrix} \phi_{11}(t) & \phi_{12}(t) \\ \phi_{21}(t) & \phi_{22}(t) \end{bmatrix} \begin{bmatrix} a \\ b \end{bmatrix} + \begin{bmatrix} g_1(t, u(t)) \\ g_2(t, u(t)) \end{bmatrix}$$

by specifying the functions ϕ_{11}, ϕ_{12}, ϕ_{21}, ϕ_{22}, g_1, g_2. [*Answers:* $\phi_{11}(t) = [1-(t-3)]e^{-(t-3)}$, $\phi_{21}(t) = -\phi_{12}(t) = (t-3)e^{-(t-3)}$, $\phi_{22}(t) = [1+(t-3)]e^{-(t-3)}$,

$$g_1(t, u) = \int_3^t (1 - \tau)e^{-\tau}u(t - \tau)\, d\tau, \quad g_2(t, u) = \int_3^t \tau e^{-\tau}u(t - \tau)\, d\tau]$$

(g) Reflect on how messy it was even for this simple one-loop circuit to solve for the equations which show how the state variables get updated by the input u from their initial values at t_0.

(h) Now that we know that inductor current and capacitor voltage are good state variables, write two coupled, scalar differential equations for $i(t)$ and $v_c(t)$ in the matrix form:

$$\begin{bmatrix} \dfrac{d}{dt} i(t) \\[2ex] \dfrac{d}{dt} v_c(t) \end{bmatrix} = \begin{bmatrix} -\dfrac{R}{L} & -\dfrac{1}{L} \\[2ex] \dfrac{1}{C} & 0 \end{bmatrix} \begin{bmatrix} i(t) \\[2ex] v_c(t) \end{bmatrix} + \begin{bmatrix} \dfrac{1}{L} \\[2ex] 0 \end{bmatrix} u(t)$$

$$y(t) = [R \quad 0] \begin{bmatrix} i(t) \\ v_c(t) \end{bmatrix}$$

1-6   Shifting and Time-Invariance

The final concept we wish to present in this introduction is that of a stationary or time-invariant system. You will note that in our approach to updating the state we let the new state $\phi(t_1, t_0, \tilde{x}, u)$ depend not only upon the initial state $\tilde{x}$ and the input u, applied over some interval of time, but also upon the time t_0 at which the interval started. However, in many systems of real interest, such as a computer, we expect the new state not to depend upon the initial time—we expect the output of a computer to depend only upon the program and the string of data we supply it, not on the time at which we start using the computer.

To make mathematically precise the notions involved here, it will be convenient to introduce the notion of a **translation operator** $z^{-\tau}$ which associates with each function f a translated version of f, written $z^{-\tau}f$, defined by the property

$$(z^{-\tau}f)(t) \triangleq f(t - \tau) \text{ for all } t, \tau \text{ in the domain of } f$$

Notice that we have not spelled out precisely what kind of function f must be in order to have $z^{-\tau}f$ make sense. If $f: V \to W$, then the domain V of f must have enough mathematical structure on it so that for $t \in V$ and $\tau \in V$ subtraction is defined and $t - \tau \in V$. If V is any *vector space*, there is enough structure. Usually we will apply the operator $z^{-\tau}$ to members of Ω, the space of admissible inputs, to characterize the operation of **delaying** an input function u by τ time units. Elements of Ω are functions of the form $u: T \to U$, where the time set T is $\mathbf{Z}$ for a discrete-time system and $\mathbf{R}$ for a continuous-time system. Often we desire the shifted input itself to be an admissible input function to the system, so that Ω must be closed under the translation operation $z^{-\tau}:\Omega \to \Omega$.

Let us now define what we mean by the instruction "Apply the same sequence of inputs starting at $t_0 + \tau$ as if we had applied the input function u starting at t_0." If we look at the waveforms in Figures 1–20a and b, we see that $u(t_0) = z^{-\tau}u(t_0 + \tau)$ and that the new input function $z^{-\tau}u$ is obtained from the original input function u by shifting each value $u(t)$ *to the right* by τ units of time. We may thus think of z^{-1} as the **unit delay operator** which delays a function by one unit of time, so that $z^{-\tau}$ is the operator which delays the function by τ units of time; in Figure 1–20b we see that the output at time t is $u(t - \tau)$, the input that u would have given τ units earlier than at time t.

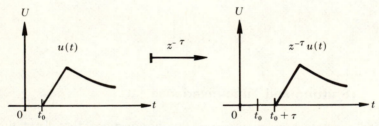

Figure 1–20   The operation $z^{-\tau}$ delays u by τ units. **(a)** The waveform u. **(b)** The waveform $z^{-\tau}u$.

The notion that a system changes state in a **stationary** or **time-invariant** fashion can now be expressed precisely by the following equation:

$$\phi(t_1, t_0, \tilde{x}, u) = \phi(t_1 + \tau, t_0 + \tau, \tilde{x}, z^{-\tau}u) \text{ for all times}$$
t_0, t_1, all (positive or negative) delays τ, all states $\tilde{x}$, and all   **(1)**
admissible input functions u.

This equation says that, for any delay τ, as long as we start in state $\tilde{x}$ and apply the same input value at corresponding times after the initial time, then the corresponding states will be the same, irrespective of whether our initial moment of time is t_0 or $t_0 + \tau$.

We need one more condition to specify stationarity or constancy of a system, and that is expressed by the following equation:

$$\eta(t_0, \hat{x}) = \eta(t_1, \hat{x}) \text{ for all times } t_0 \text{ and } t_1 \text{, and all states } \hat{x} \qquad \textbf{(2)}$$

which says that the output depends only on the present state $\hat{x}$, and not on the time at which the system is in that state.

Example 1

Consider the simple analog computer system shown in Figure 1–21, containing one integrator, two multipliers, and one summer as introduced in Example 1 of Section 1–1.

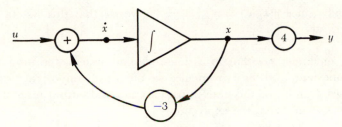

Figure 1–21   A simple one-integrator feedback system.

If we call the signal coming out of the integrator x, then the signal entering the integrator at time t is $\dot{x}(t)$ and is itself the output of the summer, where $-3x(t)$ and $u(t)$ are added. The output $y(t)$ is just 4 times $x(t)$. Thus, the system is described by the dynamic equations

$$\begin{cases} \dot{x}(t) = -3x(t) + u(t) \\ y(t) = 4x(t) \end{cases}$$

The first-order, linear differential equation

$$\dot{x}(t) + 3x(t) = u(t)$$

is easily solved for $x(t)$ in terms of an initial condition $x(t_0)$ and $u_{[t_0,t)}$ so the integrator output $x(t)$ is a good choice for the state variable. The solution yields

$$x(t_1) = \phi(t_1, t_0, \tilde{x}, u) = e^{-3(t_1-t_0)}\tilde{x} + \int_{t_0}^{t_1} e^{-3(t_1-\zeta)}u(\zeta)\, d\zeta$$

and from $y(t) = 4x(t)$, we have $\eta(t, \hat{x}) = 4\hat{x}$. Let us now test this system for time-invariance. Since

$$\phi(t_1 + \tau, t_0 + \tau, \tilde{x}, z^{-\tau}u) = e^{-3[(t_1+\tau)-(t_0+\tau)]}\tilde{x} + \int_{t_0+\tau}^{t_1+\tau} e^{-3(t_1+\tau-\zeta)}(z^{-\tau}u)(\zeta)\, d\zeta$$

$$= e^{-3(t_1-t_0)}\tilde{x} + \int_{t_0+\tau}^{t_1+\tau} e^{-3[t_1-(\zeta-\tau)]}u(\zeta - \tau)\, d\zeta$$

the change of variable $\sigma = \zeta - \tau$ in the last integration gives

$$\phi(t_1 + \tau, t_0 + \tau, \tilde{x}, z^{-\tau}u) = e^{-3(t_1-t_0)}\tilde{x} + \int_{t_0}^{t_1} e^{-3(t_1-\sigma)}u(\sigma)\, d\sigma$$

$$= \phi(t_1, t_0, \tilde{x}, u)$$

Finally, since $\eta(t_0, \hat{x}) = \eta(t_1, \hat{x}) = 4\hat{x}$, we see that this system is time-invariant. $\diamond$

We sometimes say that in a time-invariant system, the time origin t_0 is not important, and for convenience we often shift the origin to zero. To see this, let ϕ and η be the state transition map and output map of a time-invariant system. From the expression

$$\phi(t, t_0, \tilde{x}, u) = \phi(t + \tau, t_0 + \tau, \tilde{x}, z^{-\tau}u)$$

for all τ, if we let τ take the value $\tau = -t_0$, we get

$$\phi(t, t_0, \tilde{x}, u) = \phi(t - t_0, 0, \tilde{x}, z^{+t_0}u) \tag{3}$$

which shows how to shift the time origin from t_0 to zero and vice versa. Equation (3) reveals quite clearly that for time-invariant systems, ϕ depends upon the difference $(t - t_0)$ between observation time t and initial time t_0.

EXERCISES FOR SECTION 1–6

✔ **1.** Verify that the capacitor of Example 1 of Section 1–3 is a time-invariant system.

✔ **2.** Consider the analog computer system studied in Example 1 and Figure 1–21.

   (a) Replace the constant gain of 4 by the gain $h(t)$ and show that the system is time-invariant if and only if $h(t) =$ constant.

   (b) Leave the read-out gain at 4 but replace the state feedback gain -3 by the gain $f(t)$. Prove that the system is time-invariant if and only if the function $f(t) =$ constant.

   (c) Replace the constant gains -3 and 4 by $f(t)$ and $h(t)$ respectively, and derive conditions on $f(t)$ and $h(t)$ for time-invariance.

3. Verify that the circuit of Example 1–5–8 is a time-invariant system.

4. Suppose a resistive device has a resistance which increases with time: $R(t) = R_0 + \alpha t$. Taking current as input and voltage as output, prove that this device is *not* a time-invariant system. Compare it with the ordinary resistor R of Example 1–3–1.

✔ **5.** Another useful analog computer device is the **signal multiplier** shown in Figure 1–22a. Consider the analog system shown in Figure 1–22b, which uses the signal multiplier.

   (a) Write a differential equation describing the system. Show that x, the integrator output, is an acceptable choice of state variable, and find the maps ϕ and η.

   (b) Test this system for time invariance.

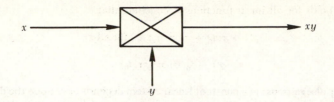

Figure 1–22(a) A signal multiplier.

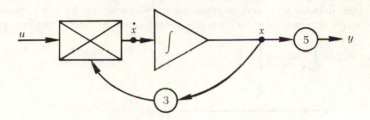

Figure 1–22(b) A feedback system using a multiplier.

6. Find out what sort of mathematical structure on the time set T is necessary to make sense out of the two expressions

$$(z^{-\tau_1 + (-\tau_2)})u$$

$$(z^{-a\tau})u$$

Prove that $(z^{-(\tau_1 + \tau_2)})u = z^{-\tau_1}(z^{-\tau_2}(u))$ if both sides are well-defined.

✔ **7.** Is the operator $z^{-\tau}$ **additive**; i.e., does $z^{-\tau}(u_1 + u_2) = z^{-\tau}u_1 + z^{-\tau}u_2$ for all u_1, $u_2 \in \Omega$?

Is $z^{-\tau}$ a **homogeneous** operator; i.e., does $z^{-\tau}(au) = a(z^{-\tau}u)$ for all constants a and all functions $u \in \Omega$?

8. Rewrite $z^{-\tau}((z^{-\tau}u)_{[t_0, t_1)})$ and $z^{+\tau}(z^{-\tau}u)$. Is $z^{-\tau}$ one-to-one? Onto?

✔ **9.** Use Equation (3) and part (c) of Exercise 4 of Section 1–5 to change the formula there for $\phi(t, 0, (i(0), v_c(0)), u)$ to the formula for $\phi(t, t_0, (a, b), u)$ where the initial state is $(i(t_0), v_c(t_0)) = (a, b)$ at t_0. Is this easier than the work you did there in parts (d) and (e)?

✔ **10.** In our discussion leading up to Example 9 of Section 1–5, we mentioned that for linear continuous-time systems the output $y(t)$ is often expressed by the superposition integral

$$y(t) = \int_{-\infty}^{\infty} g(t, \zeta)u(\zeta)\,d\zeta$$

(a) Let $u(t) = \delta(t - t_0)$, a **unit impulse occurring at** t_0, (a notion we make precise in Section 5–4) to show that $g(t, t_0)$ is the response at time t to an impulse applied at t_0.

(b) Let the system be time-invariant and set

$$\int_{-\infty}^{\infty} g(t + \tau, \zeta)(z^{-\tau}u)(\zeta)\,d\zeta$$

equal to $y(t)$ for all τ to get, after a change of variable,

$$\int_{-\infty}^{\infty} g(t + \tau, \sigma + \tau)u(\sigma)\,d\sigma = \int_{-\infty}^{\infty} g(t, \zeta)u(\zeta)\,d\zeta$$

Since this holds for all input functions u, conclude that

$$g(t + \tau, t_0 + \tau) = g(t, t_0) \text{ for all } \tau$$

Thus

$$g(t - t_0, 0) = g(t, t_0)$$

and the impulse response of a constant linear system depends only upon the difference or "age" variable $(t - t_0)$, and the superposition integral becomes the **convolution** integral.

EE   **11.**   Find the impulse response of the system in Figure 1–23 by letting $u(t) = \delta(t - \tau)$, an impulse at τ, and determining the response $y(t)$ for $t > \tau$. Show that this system is time-invariant. Why is this system called a "finite time integrator"?

$$\left[Answer: y(t) = \int_{t-T}^{t} u(\sigma)\, d\sigma \right]$$

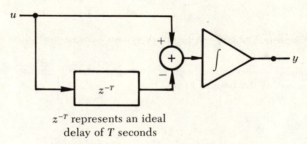

z^{-T} represents an ideal
delay of T seconds

Figure 1–23   A finite time integrator.

1–7   Summary

Our introductory discussion may be summarized in the following very general definitions, which, as we shall see later, embrace the familiar notions of classical linear systems theory.

A **deterministic mathematical system representation** is specified by five sets (T, U, Y, X, Ω) and two functions ϕ and η, where:

> T is the **time** set, a subset of the real numbers
> U is the **input** set
> Y is the **output** set
> X is the **state** set
> Ω is the set of **admissible input functions**, a subset of the set of all functions $T \rightarrow U$, which is closed under **splicing**; i.e., for all u, u' in Ω, all times $t_2 \in T$, there exists a u'' in Ω such that

$$u''(t) = \begin{cases} u(t) \text{ if } t \leq t_2 \\ u'(t) \text{ if } t_2 < t \end{cases}$$

$\phi: T \times T \times X \times \Omega \to X$ is the **state-transition function** and satisfies the **consistency** conditions:

$$\phi(t_0, t_0, \tilde{x}, u) = \tilde{x} \text{ for all times } t_0, \text{ all states } \tilde{x}, \text{ and}$$
$$\text{all admissible input functions } u,$$

$$\phi(t_2, t_1, \phi(t_1, t_0, \tilde{x}, u), u) = \phi(t_2, t_0, \tilde{x}, u);$$

and the **causality** condition:

$$\phi(t_1, t_0, \tilde{x}, u) = \phi(t_1, t_0, \tilde{x}, u') \text{ if } u(t) = u'(t) \text{ for } t_0 \leq t < t_1.$$

$\eta: T \times X \to Y$ is the **output function.**

The system is **discrete-time** if $T = \mathbf{Z}$, the set of integers; it is **continuous-time** if $T = \mathbf{R}$, the set of real numbers.

The system is **stationary (constant, time-invariant)** if T is closed under addition, Ω is closed under the shift operator $z^{-\tau}$ for every τ in T, and we have

$$\phi(t_1, t_0, \tilde{x}, u) = \phi(t_1 + \tau, t_0 + \tau, \tilde{x}, z^{-\tau}u) \text{ for all times } t_0, t_1,$$
$$\text{all delays } \tau, \text{ all}$$
$$\text{states } \tilde{x}, \text{ and all}$$
$$\text{admissible input}$$
$$\text{functions } u.$$

$$\eta(t_0, \hat{x}) = \eta(t_1, \hat{x}) \text{ for all times } t_0 \text{ and } t_1, \text{ and all states } \hat{x}.$$

Note that the above conditions allow us to remove the mention of one time variable, replacing ϕ and η by the simpler functions $\check{\phi}$ and $\bar{\eta}$ where

$$\check{\phi}: T \times X \times \Omega \to X \quad \text{by the rule} \quad (t_1, \tilde{x}, u) \mapsto \phi(t_1, 0, \tilde{x}, u)$$
$$\bar{\eta}: X \to Y \quad \text{by the rule} \quad \tilde{x} \mapsto \eta(0, \tilde{x})$$

since we know that for all choices of t_0

$$\phi(t_1, t_0, \tilde{x}, u) = \check{\phi}(t_1 - t_0, \tilde{x}, z^{+t_0} u)$$
$$\eta(t_1, \hat{x}) = \bar{\eta}(\hat{x})$$

In the future, we shall normally write η and ϕ whether we really mean $\bar{\eta}$ or η and $\check{\phi}$ or ϕ—leaving it to the context to determine which is intended.

CHAPTER 2

System Dynamics and
Local Transition Functions

This chapter falls into three parts:

In Sections 2–1 and 2–2 we consider ways in which the "global" transition function of Chapter 1 (which specifies for a given state at time t_0, and an input function applied from t_0 to some later time t_1, what the state will be at time t_1) may be reduced to a "local" transition function. In the case of a discrete-time system this is completely straightforward: we simply specify how the state and input at some time n determine the state at the next time $n + 1$ on the discrete-time scale. We then observe how induction can be used to recapture the "global" transition function from this local "one-step" description. However, the problem for continuous-time systems is more subtle, and we find that a "local" description exists only if the continuous-time system is sufficiently "smooth," in a sense which we shall make precise. If the system is in fact sufficiently smooth, the description takes the form of a differential equation $\dot{x}(t) = f(t, x(t), u(t))$ which reduces in the time-invariant case to the form $\dot{x}(t) = f(x(t), u(t))$. We shall postpone until Chapter 5 the seeking of conditions under which this sort of local description in terms of a differential equation allows us to reconstitute uniquely the global description by integrating the differential equations.

Sections 2–3 and 2–4 provide mathematical concepts which we shall need to apply in the further development of our study, and thus it is up to the teacher and the student whether to cover the material after Section 2–2, or to return to different parts of the material as they are called upon later in the book. Specifically, Section 2–3 introduces some basic notions of linear algebra. Thus, the reader who has already had a course in linear algebra may simply use this section as a reference for notation and some basic results. On the other hand, the reader who is completely new to linear algebra should work carefully through the section, doing the exercises, and perhaps doing supplementary reading in any of the standard texts on linear algebra. In Section 2–4 we introduce some useful mathematical spaces—metric spaces, normed linear spaces, and Banach spaces. While almost all of the book, save for the general integration theory of Chapter 5 and the optimization theory of Section 7–6, may be read without a knowledge of this section simply by pretending, for example, that all Banach spaces are finite dimensional real vector spaces, it turns out that these general ideas are both so powerful, so simple once the basic notation has been comprehended, and so widely adopted in the current research literature, that it seems that a moderate effort to master the unfamiliar notation of this section will be well repaid.

Finally, in Section 2–5, we show that the formalization of systems that we have presented does in fact embrace the usual formalizations of electrical and mechanical systems. Hopefully this section, coupled with the reader's own study of such topics as classical mechanics and circuit theory, will make it clear that the control theory which we develop in the ensuing chapters is applicable to the analysis of a wide variety of real systems.

2–1   Local Transition Functions For Discrete-Time Systems

Whereas, in a continuous-time system, any two distinct moments of time are separated by infinitely many other moments of time, in a discrete-time system we may talk of successive moments of time. Let us see how the input serves to update the state of a discrete-time system from one moment to the next. If at time n the state is $x(n)$ and the input is $u(n)$, then the state at time $n + 1$ will be $\phi(n + 1, n, x(n), u)$. Now we know that this last expression can depend only upon the values of time j for $n \leq j < n + 1$, and hence, only on the value of u at time n. The value of u at time $n + 1$ arrives simultaneously with the system settling into the state $x(n + 1)$ and so cannot affect that state, but only later states. Thus, we may replace $\phi(n + 1, n, x(n), u)$ by some simpler expression $f(n, x(n), u(n))$, where we have made explicit the fact that the transition from the state $x(n)$ at time n to the state $x(n + 1)$ at time $n + 1$ depends on the input only through the value u takes at time n. We call this "one-step" function f the **local transition function,** since it describes the state dynamics in the locality of time n, and we sometimes call the state transition function ϕ the **global transition function,** since it describes the dynamics for all times $n + k$.

Thus, for a discrete-time system, we may replace the specification of the complicated transition function $\phi: T \times T \times X \times \Omega \to X$, which is defined for arbitrary pairs of times and arbitrary admissible input *functions,* by the much simpler local or one-step transition function $f: T \times X \times U \to X$, which is defined for admissible input *values.* The crucial point, of course, is that given the one-step transition function f we may reconstitute the general transition function ϕ, as is made clear in the following diagram:

Time	n	$n + 1$	$n + 2$	$\ldots$
State	$x(n)$	$x(n + 1) =$ $f(n, x(n), u(n))$	$x(n + 2) =$ $f(n + 1, x(n + 1), u(n + 1))$	$\ldots$
Input	$u(n)$	$u(n + 1)$	$u(n + 2)$	$\ldots$

Since the output function was determined by the instantaneous value of the state, there is no need to modify it in giving a description of the system's behavior. We note that if the system is constant (time-invariant), then, of course, f will not depend explicitly upon the time value.

For a discrete-time system it will often be convenient to rewrite our global expression

$$x(n + k) = \phi(n + k, n, x(n), u)$$

in the form

$$x(n + k) = \phi(n, x(n), u_n, \ldots u_{n+k-1})$$

where we list explicitly only those input values which enter into the transition of the state from $x(n)$ to $x(n + k)$—namely $u_n, u_{n+1}, \ldots, u_{n+k-1}$ (with u_j occurring at time j). The value of u at time $n + k$ occurs simultaneously with the system settling into the state $x(n + k)$ and so cannot affect that state but only later states. Note that we did not write $n + k$ as an explicit variable in the last equation, since we may deduce it by adding to the initial time n the length k of the input sequence.

Thus, rather than specify the state-transition function for a discrete-time system in terms of the space Ω of all admissible input sequences, we shall often think of it as specified in terms of the space U^* of all *finite* input sequences†

$$\phi: T \times X \times U^* \to X$$

where $\phi(n, x(n), u_n, u_{n+1}, \ldots, u_{n+k-1})$ is the state at time $n + k$ if $x(n)$ is the state at time n, and u_j is the input at time j for $n \leq j < n + k$.

After obtaining a local transition function, it is always wise to check that its repeated application does indeed yield the desired overall, global behavior of the system. The appropriate method for such verification is

---

†The notation U^* to represent the set of all finite sequences constructible out of members of the set U is standard in the computer science and discrete-time systems literature and should not be confused with our later use (in Chapter 4) of the asterisk to denote the adjoint A^* of a matrix A.

mathematical induction, and so before proceeding further we remind the reader of some aspects of this method.

The principle of **mathematical induction** says that if a property of integers holds for $n = 0$, and if the truth of the property for any integer n allows us to guarantee the truth of the property for the integer $n + 1$, then in fact the property must hold for all non-negative integers. We call the truth of the statement for $n = 0$ the **basis for the induction,** and we call the proof that its truth for any integer n guarantees its truth for the integer $n + 1$ the **induction step.** The induction step lets us go from the basis $n = 0$ to infer the truth for $n = 1$. A second application of the induction step lets us then infer the truth for $n = 2$ from the truth for $n = 1$; continuing in this way, we may deduce the truth for any finite integer, as indicated in the following diagram:

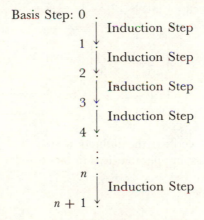

The reader should, of course, understand that the induction hypothesis may appear in other forms. For instance, the basis might tell us that a property holds for $n = 7$, and the induction step might tell us that if the property holds for any n, then it also holds for $n + 2$. We would then deduce, not that the property holds for all integers n, but rather that it holds for all odd integers greater than or equal to 7.

To make sure that the reader understands the use of induction in mathematical proofs, let us examine one very standard example.

Example 1

Consider the following array of equalities:

$$1 = 1 = 1^2$$
$$1 + 3 = 4 = 2^2$$
$$1 + 3 + 5 = 9 = 3^2$$
$$1 + 3 + 5 + 7 = 16 = 4^2$$

This suggests the hypothesis that the sum of the first n odd numbers equals n^2:

$$\sum_{j=1}^{n} (2j - 1) = n^2$$

We have already verified the basis step for $n = 1$ (and a few other cases as well). We know that to prove it true for all $n \geq 1$, it only remains to verify the induction step—in other words, to show that the truth of the hypothesis for any integer n implies its truth for the integer $n + 1$.

Assume for a given n that $\sum_{j=1}^{n} (2j - 1) = n^2$ is true. Then

$$\sum_{j=1}^{n+1} (2j - 1) = \left[\sum_{j=1}^{n} (2j - 1) \right] + 2n + 1$$

$$= [n^2] + 2n + 1 \quad \text{by the assumption on } n$$
$$= (n + 1)^2$$

and we have indeed verified the induction step. This, coupled with our verification of the basis, implies that it is in fact true for all positive integers n that $\sum_{j=1}^{n} (2j - 1) = n^2$. ◊

Note that we are using the word *basis* here in quite a different sense from that we shall use in vector space discussions.

Let us now study a simple system which crops up again and again as a subsystem or basic building block of general discrete-time systems.

Example 2

Consider the discrete-time feedback system of Figure 2–1a. The operational element represented in Figure 2–1b, with input function v and output function x, is a **unit-delay device** defined by $x = z^{-1}v$. Thus, at any time n we have $x(n) = (z^{-1}v)(n) = v(n - 1)$, consistent with our previous use of $z^{-\tau}$ for a delay of τ.

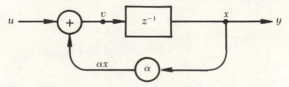

Figure 2–1(a)   A basic delay block with feedback.

The device of Figure 2–1c is a delayless multiplier or scaler whose present output equals α times its present input, and we call α the **gain** of the scaler. In Figure 2–1d we show a delayless summer or adder whose present output is the algebraic sum of its present inputs. Thus, the signal entering the delay box of Figure 2–1a is given by $v = u + \alpha x$, so that

$$x(n + 1) = (z^{-1}v)(n + 1) = (z^{-1}u)(n + 1) + \alpha(z^{-1}x)(n + 1)$$
$$= u(n) + \alpha x(n)$$

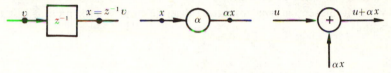

Figure 2–1(b)   The unit-delay device. **(c)** A constant multiplier. **(d)** A summer or adder.

As an exercise, the reader may easily check by induction that the solution $x(n + k)$ for any $k \geq 0$ is then determined by the forcing function $u_{[n,n+k)}$ and the "initial condition" term $x(n)$ by the formula

$$x(n + k) = \alpha^k x(n) + \alpha^{k-1}u(n) + \alpha^{k-2}u(n + 1) + \cdots + u(n + k - 1)$$

Thus the output, x, of the delay device is an acceptable choice for the state variable for this system, since knowing its value at time n and the inputs from n up through $n + k - 1$, we can determine its value at any subsequent time $n + k$.   ◊

For the system just discussed, the one-step transition function f is given by the formula $x(n + 1) = f(n, x(n), u(n)) = \alpha x(n) + u(n)$ and adequately describes the behavior of the system in the locality of time n. As we have just seen, f can be applied over and over to get the global transition function ϕ which gives the state $x(m)$ for *all* values of $m > n$, and not merely for $m = n + 1$. We have the formula

$$\phi(n, x(n), u_n, \ldots, u_{m-1}) = \alpha^{m-n}x(n) + \sum_{j=0}^{m-1-n} \alpha^j u_{m-1-j}$$

The second term is a "convolution summation" of the input sequence $u_n, u_{n+1}, \ldots, u_{m-1}$ with the sequence $\alpha^0, \alpha^1, \alpha^2, \ldots, \alpha^{m-1-n}$. We shall see a lot more of convolution summations in Chapter 4.

The reader may have noticed that we have not yet said what kind of values the functions x, u, and y can take. Clearly, if the product $\alpha x(n)$ is to make sense, the state function x must take on values compatible with the number α. For example, if α is a real number, then the set of states can be taken as $X = \mathbf{R}$, an infinite set. In Exercises 6 and 13 we discuss different kinds of number systems, called **fields,** in which we can do arithmetic such as multiplying and adding. If we felt like it, we could use a finite set such as the **binary number field** where the only numbers are 0 and 1, and do arithmetic *modulo* 2. Calling the set of field elements K, we can take $U = X = Y = K$, and as long as we only let the gain α take on values in K, the system of Example 1 and all the operations shown in Figure 2–1 make perfect sense under this kind of arithmetic.

We now present a non-numerical discrete-time system having just two state values.

Example 3

Suppose we have an input "alphabet" U consisting of just two letters a and b; i.e., $U = \{a, b\}$. Then U^* will be the set of all possible "words" we can form using letters from U. Let us design a deterministic system which will tell us whether there is an even or an odd number of occurrences of the letter a in each word. Our output set Y then consists of the symbols $Y = \{E, O\}$ where E is for even and O for odd. We can call our desired system a **parity checker,** since parity describes the evenness or oddness of a word with respect to the number of a's it contains. We illustrate our wishes with the "black box" of Figure 2–2a.

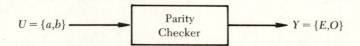

$$U = \{a,b\} \longrightarrow \boxed{\begin{array}{c} \text{Parity} \\ \text{Checker} \end{array}} \longrightarrow Y = \{E,O\}$$

Figure 2–2(a)   A system to compute the parity of a sequence of letters from $\{a, b\}$.

All we have so far is a global description of what we want our system to do. If we want to really get down to business and stop speaking in such generalities, we must start thinking about how *we* would decide the parity of an input string before designing a machine to do it. Given a string $w = (u_1, u_2, \ldots, u_k)$ of length k, we could just take the whole word and count the number of a's in it. A more enlightened approach would be to do the counting bit-by-bit: given a string w, first look at u_1; if it's an a we say "so far, w is odd" and think O. Next we look at u_2; if it's an a we lose oddness, but if it's a b we keep on thinking O. In this manner, we have defined the *local* dynamics of an effective procedure. Considering our *mind-set* to be a 2-state system, we may take $X = \{q_e, q_o\}$; at any instant n if we are in state q_o, we look at the input letter at that instant to decide whether our next state will be q_o or q_e. It is now easy to see that the 2-state system shown in Figure 2–2b is a realization of our local dynamic description.

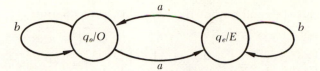

Figure 2–2(b)   A bit-by-bit parity checker.

We have labeled each circle (node) with the name of the state, q_o or q_e, which it represents and, after the slash mark, the value of the output, y, associated with each state; $\eta(q_o) = O$, and $\eta(q_e) = E$. The arrows show how from a current state $x(n)$ we proceed to the next state $x(n + 1)$ depending upon the input letter u_n at time n. Since there is no explicit dependence on n, we have

$$x(n + 1) = f(x(n), u(n)) = \begin{cases} q_e \text{ if } (x(n), u_n) = (q_o, a) \\ q_o \text{ if } (x(n), u_n) = (q_o, b) \\ q_o \text{ if } (x(n), u_n) = (q_e, a) \\ q_e \text{ if } (x(n), u_n) = (q_e, b) \end{cases} \qquad \Diamond$$

The crucial point to note about the last example is that we started out with a "global" description—we wanted the system to find the parity of arbitrarily long sequences. We used "intuition" to find an appropriate set of state-variables, and then we were able to realize our global desires by a concrete "local" transition function. Once we have the local function f, we can use it to find an explicit formula for the global transition map ϕ (Exercise 5). In the rest of this chapter we concentrate on finding local descriptions of systems, and postpone until Chapters 5 and 6 a systematic study of solving the local equations for the global transition function. In Chapters 6 and 8 we present several methods of realizing linear global equations.

As a final topic in this section, we illustrate by an example a procedure for drawing the block diagram, consisting of the basic unit-delay blocks, multipliers, and summers introduced in Example 2, which corresponds to local state dynamic equations of the form

$$x(n + 1) = F(n)x(n) + G(n)u(n)$$
$$y(n) = H(n)x(n)$$

where $F(n)$, $G(n)$, and $H(n)$ are *matrices*.

Example 4

Consider a two-dimensional, discrete-time system whose state $x(n) = \begin{bmatrix} x_1(n) \\ x_2(n) \end{bmatrix}$, input $u(n)$, and output $y(n)$ obey the local equations

$$\begin{bmatrix} x_1(n+1) \\ x_2(n+1) \end{bmatrix} = \begin{bmatrix} 0 & 1 \\ -a_1(n) & -a_2(n) \end{bmatrix} \begin{bmatrix} x_1(n) \\ x_2(n) \end{bmatrix} + \begin{bmatrix} b_1(n) \\ b_2(n) \end{bmatrix} u(n) \tag{1}$$

$$y(n) = \begin{bmatrix} c_1(n) & c_2(n) \end{bmatrix} \begin{bmatrix} x_1(n) \\ x_2(n) \end{bmatrix} \tag{2}$$

If we perform the simple matrix multiplications called for in the two matrix equations above, we get the three scalar equations:

$$x_1(n+1) = 0x_1(n) + 1x_2(n) + b_1(n)u(n) \tag{3}$$

$$x_2(n+1) = -a_1(n)x_1(n) - a_2(n)x_2(n) + b_2(n)u(n) \tag{4}$$

$$y(n) = c_1(n)x_1(n) + c_2(n)x_2(n) \tag{5}$$

Now we draw two unit-delay blocks, leaving lots of room between them, and label their outputs as $x_1(n)$ and $x_2(n)$, respectively, as shown in Figure 2–3a.

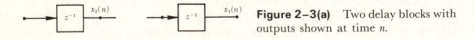

Figure 2–3(a)   Two delay blocks with outputs shown at time n.

With outputs at time n given by $x_1(n)$ and $x_2(n)$, the delay blocks in Figure 2–3a must have appearing at their input lines at time n the signals $x_1(n+1)$ and $x_2(n+1)$, respectively, as shown in Figure 2–3b.

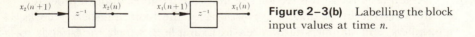

Figure 2–3(b)   Labelling the block input values at time n.

Now from equations (3) and (4) we see how to construct the signals $x_1(n+1)$ and $x_2(n+1)$ from the signals $x_1(n)$, $x_2(n)$, and $u(n)$. For example, if we take $-a_1(n)$ times $x_1(n)$, $-a_2(n)$ times $x_2(n)$, and $b_2(n)$ times $u(n)$ and add these resulting signals together then by equation (4) we get $x_2(n+1)$; so we construct paths corresponding to these operations on $x_1(n)$, $x_2(n)$, and $u(n)$ and connect the end result to the point already labelled with the value $x_2(n+1)$, as shown in Figure 2–3c.

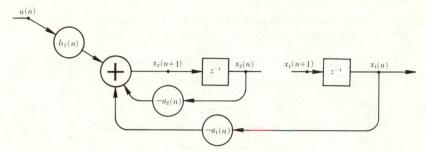

Figure 2–3(c) The construction of $x_2(n + 1)$ from $x_1(n)$, $x_2(n)$, and $u(n)$.

Similarly, "closing the loop" with the construction of signal $x_1(n + 1)$ using equation (3), we get the diagram of Figure 2–3d.

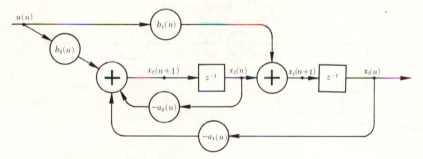

Figure 2–3(d) A realization of equations (3) and (4).

Finally, we use equation (5) to construct the system output $y(n)$ from the signals $x_1(n)$ and $x_2(n)$ to end up with the complete block diagram, which we draw neatly in Figure 2–3e.   ◊

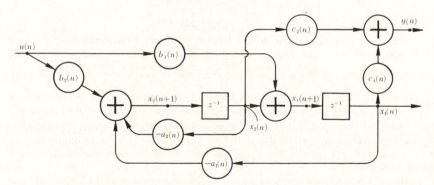

Figure 2–3(e) The complete block diagram for equations (1) and (2).

EXERCISES FOR SECTION 2–1

✔ **1.** **(a)**   Show that the system of Example 2 is time-invariant.

   (b)   Prove by induction that the formula given for $\phi(n, x(n), u_n, \ldots, u_{n+k-1})$ in Example 2 is true for all $k \in \mathbf{N}$.

   (c)   Find f, ϕ, and η for the discrete-time system of Figure 2–4a.
[*Answer:* $\phi(n + k, n, x(n), u_n, \ldots, u_{n+k-1}) = u_{n+k-1}.$]

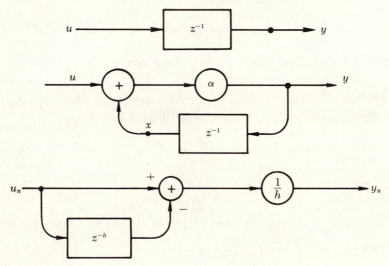

Figure 2–4

   (d)   Show, for the discrete-time system of Figure 2–4b, that

$$\phi(n + k, n, x, u_n, \ldots, u_{n+k-1}) = \alpha^k x + \sum_{j=0}^{k-1} \alpha^{1+j} u(n + k - 1 - j)$$

and

$$y(n + k) = \alpha^{1+k} x(n) + \sum_{j=0}^{k} \alpha^{1+j} u(n + k - j)$$

Give the formula for f. Compare with Example 2.

2.   Let a continuous-time function u be sampled every h seconds, where h is very small compared with the rate of variation of u. At the nth sampling instant, $t_n = nh$, the derivative of u is approximated by $y(n) = \dfrac{u(nh) - u((n-1)h)}{h} \doteq \dot{u}(nh)$.

   Thus, the discrete-time system of Figure 2–4c, employing the delay operator z^{-h}, generates approximate samples $y_n \doteq \dot{u}_n$. Find f and ϕ for this "differentiator."

3.   Suppose you open a savings account with an initial deposit of P_o, thereafter depositing d_j dollars on the last day of the jth month. The bank compounds interest at the rate of $r\%$ per month. Let P_n be the amount of money accumulated at the end of n months.

   (i)   Choose the amount of money, P_n, in your account as the "state" of your investment system and write the one-step transition map f for the system.

(ii)   Draw a realization of the difference equation $P_{n+1} = f(n, P_n, u_n)$ using delay boxes, multipliers, and summers.

(iii)   Solve the "local" equation for the total amount of money in your account at the end of k months.

$$\left[Answer: P_k = \left(1 + \frac{r}{100}\right)^k P_o + \sum_{j=0}^{k-1} \left(1 + \frac{r}{100}\right)^j d_{k-j}, \text{ for } k \geq 1.\right]$$

4.   In economic analysis, a fundamental concept is the notion of equilibrium. We illustrate this notion by writing dynamic equations to determine the price of a commodity, say bicycles. At time n let P_n be the price of a bike, $D(n)$ the demand for bikes, and $S(n)$ supply of bikes. It is found empirically that $D(n) = a + bP_n$ and $S(n) = c + dP_{n-1}$ (note that present supply was manufactured based on last price); a, b, c, and d are suitable numerical constants, where b is negative and d is positive (why?). Equilibrium occurs when supply equals demand.

Choose the price of bikes as the state variable of the bike market, and set up the difference equation which describes the price of bikes. Draw a block diagram. Interpret the signs of b and d in the light of stability considerations.

5.   Find the global equation for the parity checker of Example 3 and use induction to prove that it really works for input sequences of any length k.

✔ **6.**   The binary valued arithmetic used in our discussion is one of the simplest examples of mathematical structures known as fields.

The general definition of a field is given as follows:

A **field** $(K, +, \cdot)$ is a set K together with two operations, $+$ and $\cdot$. These operations $+ : K \times K \to K$ such that $(a, b) \mapsto a + b$ called "addition", and $\cdot : K \times K \to K$ such that $(a, b) \mapsto a \cdot b$ called "multiplication", together satisfy the following nine conditions:

(i)	$a \cdot (b + c) = a \cdot b + a \cdot c$	for all a, b, c in K
(ii)	$a + (b + c) = (a + b) + c$	for all a, b, c in K
(iii)	$a \cdot (b \cdot c) = (a \cdot b) \cdot c$	for all a, b, c in K
(iv)	$a + b = b + a$	for all a, b in K
(v)	$a \cdot b = b \cdot a$	for all a, b in K
(vi)	There exists an element 0 in K such that $a + 0 = a$	for all a in K
(vii)	There exists an element 1 in K such that $a \cdot 1 = a$	for all a in K
(viii)	To each a in K there corresponds an element $-a$ in K such that $a + (-a) = 0$. [We write $b - a$ as a shorthand for $b + (-a)$.]	
(ix)	To each $a \neq 0$ in K there corresponds an a^{-1} in K such that $a \cdot a^{-1} = 1$.	

The reader may note that **R** and **C** are indeed fields with the usual choice of addition and multiplication and of 0 and 1. However, there are other examples of fields, as the following exercises show:

(a)   Show that the "addition" and "multiplication" $+_2$ and $\cdot_2$ defined by the tables

$+_2$	0	1
0	0	1
1	1	0

$\cdot_2$	0	1
0	0	0
1	0	1

make the system $(\{0, 1\}, +_2, \cdot_2)$ a field. Identify the element -3, defined by $-(1 + 1 + 1)$, in this field? We call this field the **binary field** and denote it by $(\mathbf{Z}_2, +_2, \cdot_2)$

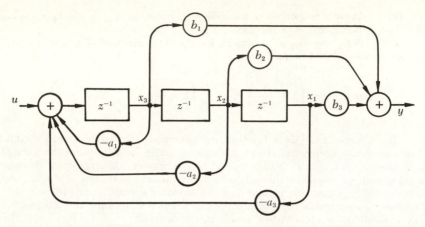

Figure 2–5 A canonical form realization of a third-order difference equation.

(b)   Give the tables for $+_3$ and $\cdot_3$ to make $(\{0, 1, 2\}, +_3, \cdot_3)$ a field. What is -3 in this **ternary field** $(Z_3, +_3, \cdot_2)$?

(c)   Repeat part (b) for the system $(\{0, 1, 2, 3\}, +_4, \cdot_4)$. What is -3 in this system? Notice that the "natural" definitions for $+_4$ and $\cdot_4$ do *not* give you a field!

✔ **7.**   This example has a lot of meat in it and is worth spending several hours on. Consider the discrete-time system of Figure 2–5 with state variables x_1, x_2, x_3 labeled as shown.

(a)   Find the one-step transition map f and the map η. Write them in the matrix form:

$$\begin{bmatrix} x_1(n+1) \\ x_2(n+1) \\ x_3(n+1) \end{bmatrix} = F \begin{bmatrix} x_1(n) \\ x_2(n) \\ x_3(n) \end{bmatrix} + G[u(n)]$$

$$y(n) = H \begin{bmatrix} x_1(n) \\ x_2(n) \\ x_3(n) \end{bmatrix}$$

Notice how the "state feedback" gains a_1, a_2, a_3 appear (in order) in your F matrix. Make similar observations about the "read in" matrix G and the "read out" matrix H. Because of this prominent display of the "feedback controls," this choice of state variables gives what we will call in Chapter 7 the "controllable canonical form" matrix representation of the system.

(b)   Relabel the state variables so that $x_1 \mapsto q_3$, $x_2 \mapsto q_2$, and $x_3 \mapsto q_1$, and find the new matrix representation for f for this choice of state variables. Compare the new matrices F', G', H' with F, G, H of part (a). Express your relabeling of the state variables by a constant transformation matrix T so that $x = Tq$. Express F', G', H' in terms of F, G, H, and T. Show that T is invertible so we can go from x to q and back without losing information.

(c)   Show that the system of part (a) has its output y related to its input u by the third-order scalar difference equation

$$y(n+3) + a_1 y(n+2) + a_2 y(n+1) + a_3 y(n) = b_1 u(n+2) + b_2 u(n+1) + b_3 u(n)$$

[*Hint:* If you get stuck, see Exercise 3 of Section 2–5 and Exercise 13 of Section 6–3. You should be able to manipulate the four equations

$$zx_1 = x_2$$

$$zx_2 = x_3$$

$$zx_3 = -a_1x_3 - a_2x_2 - a_3x_1 + u$$

$$y = b_3x_1 + b_2x_2 + b_1x_3$$

to get $(z^3 + a_1z^2 + a_2z + a_3)y = (b_1z^2 + b_2z + b_3)u$.]

(d)   Draw the block diagram for your representation F', G', H' in part (b), labeling the states and gains appropriately. Find the third order scalar difference equation relating input u and output y. Compare with part (c).

(e)   Find the global transition map ϕ in terms of F and G for the representation of part (a).

(f)   Suppose the gains shown in Figure 2–5 are time-varying $a_1(n), a_2(n), a_3(n)$; $b_1(n), b_2(n), b_3(n)$ instead of constants. Give the matrix representation $F(n), G(n), H(n)$ for this system with the states x_1, x_2, x_3 as shown in Figure 2–5.

Repeat part (b) now and find $F'(n)$, $G'(n)$, $H'(n)$ for the states q_1, q_2, q_3. Can you still relate x to q by a *constant* transformation matrix T?

✔ **8.**   In Chapter 3 we shall see that a finite-dimensional, linear, discrete-time system is always describable by a one-step transition function of the form

$$x(n + 1) = F(n)x(n) + G(n)u(n)$$

(a)   Plug away one step at a time and show that

$$x(n + 2) = F(n + 1)F(n)x(n) + F(n + 1)G(n)u(n) + G(n + 1)u(n + 1),$$

$$x(n + 3) = F(n + 2)F(n + 1)F(n)x(n) + F(n + 2)F(n + 1)G(n)u(n)$$
$$+ F(n + 2)G(n + 1)u(n + 1) + G(n + 2)u(n + 2)$$

(b)   Define the notation $\Phi(p, q) = \begin{cases} F(p - 1)F(p - 2)\cdots F(q) & \text{for } p > q \\ I \text{ (the identity matrix)} & \text{if } p = q \end{cases}$

so that

$$x(n + 3) = \Phi(n + 3, n)x(n) + \Phi(n + 3, n + 1)G(n)u(n)$$

$$+ \Phi(n + 3, n + 2)G(n + 1)u(n + 1) + \Phi(n + 3, n + 3)G(n + 2)u(n + 2)$$

$$= \Phi(n + 3, n)x(n) + \sum_{j=n}^{n+3-1} \Phi(n + 3, j + 1)G(j)u(j)$$

(c)   Guess the result for $x(n + k)$, and prove it by induction for all $k \geq 1$.

$$\left[\text{Answer: } x(n + k) = \Phi(n + k, n)x(n) + \sum_{j=n}^{n+k-1} \Phi(n + k, j + 1)G(j)u(j) \right]$$

(d)   Check your own work by regarding the state equation for Example 2 as a 1×1 matrix equation and verifying the result given there for $x(n + k)$.

(e)   Suppose the matrices $F(n)$ and $G(n)$ are constants F and G. Find $\Phi(p, q)$ and give the state transition function

$$x(n + k) = \Phi(n + k, n, x(n), u(n), \ldots, u(n + k - 1))$$

(f)   Show that the Φ of part (e), with F and G constant, is time-invariant, but that the Φ of part (c), with $F(n)$ and $G(n)$ depending on time n, is not time-invariant.

9.   **(a)**   Find the local (one-step) transition map for the single-input, single-output time-varying system of Figure 2–6.

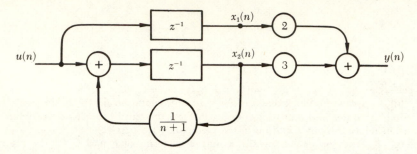

Figure 2–6   A time-varying system.

(b)   Use your general result from Exercise 8(c) to give the formula for $x(n + k)$ for the cases where u is zero, and where u is the step:

$$u(j) = \begin{cases} 1 & j \geq n \\ 0 & j < n \end{cases}$$

(c)   We shall develop the **z-transform** from scratch in Section 4–6. If you know already how to use it, solve the two scalar difference equations of parts (a) and (b) by the z-transform and check your results.

(d)   Notice that since

$$F(n) = \begin{bmatrix} 0 & 0 \\ 0 & \dfrac{1}{n+1} \end{bmatrix}$$

is singular for all n, the state transition matrix $\Phi(p, q)$ is *singular* for all $p > q$. We will see in Chapter 5 that this cannot happen in continuous-time systems, nor in discrete-time systems which are obtained by sampling continuous-time systems.

✔ **10.**   This exercise develops practice with many of the central ideas of linear system theory, and should be attempted at this stage by readers with at least an elementary understanding of matrix theory, even if all parts of it cannot yet be solved completely. Just attacking these ideas in this setting will develop a "mind-set" conducive to greater understanding of the material when we cover it later on in the text. In Figure 2–7, we present three different discrete-time systems, (F, G, H), (F', G', H'), and (F'', G'', H''), called **realizations,** which have the same input-output behavior.

(a)   Write the one-step matrices for the choices of state variables in Figures 2–7(a), (b), and (c).

(b)   For each of the three realizations, "eliminate" the state variables to arrive at a third order scalar difference equation relating input u to output y. [*Answer:* $y(n + 3) + 6y(n + 2) + 11y(n + 1) + 6y(n) = u(n)$.]

(c)   Compute $\Phi(n + k, n)$ for each realization. Which realization is the nicest for this calculation?

(d)   Verify that realization 2–7(c) could have been gotten analytically from realization 2–7(a) by making the matrix transformation $x = Tx''$, where T is the constant, invertible matrix

$$T = \begin{bmatrix} 1 & 1 & 1 \\ -1 & -2 & -3 \\ 1 & 4 & 9 \end{bmatrix}$$

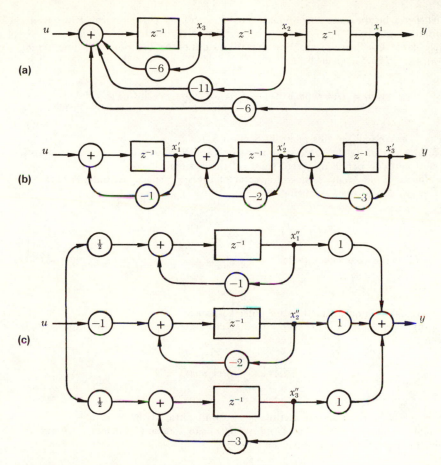

Figure 2-7(a) A "controllable canonical" representation.   **(b)** A triangular representation.   **(c)** A diagonal representation.

Express F'', G'', H'' in terms of F, G, H, T. Can you find an analogous matrix Q which yields realization 2-7(b) from 2-7(a) by setting $x = Qx'$?

    (e)   For the realization (F, G, H) of Figure 2-7(a), suppose that u is zero and the initial state $x(n)$ is given by

$$x(n) = \begin{bmatrix} 1 \\ -3 \\ 9 \end{bmatrix}$$

What is $x(n + 1)$? How about $x(n + k)$? Notice that this initial condition vector is just the third column of the matrix T in part (d). Try each of the first two columns as initial condition vectors and compute $x(n + 1)$ and $x(n + k)$. This shows that the columns of T are **eigenvectors** of the matrix F, a concept we develop in great detail in Section 6-2.

    (f)   Prove that the input-output (external) description of a constant linear system (F, G, H) is invariant with respect to invertible transformations of the state. [*Hint:* Let x be the state corresponding to (F, G, H), let Q be any invertible transformation, and let $x = Qx'$ where x' is the state of realization (F', G', H').]

[*Hint:* Let x be the state corresponding to (F, G, H), let Q be any invertible transformation, and let $x = Qx'$ where x' is the state of realization (F', G', H').]

✔ **11.** **(a)** Use Exercise 8 to show that the output of the linear discrete-time system $(F(n), G(n), H(n))$ is given by

$$y(m) = H(m)\Phi(m, n)x(n) + H(m)\sum_{j=n}^{m-1} \Phi(m, j+1)G(j)u(j)$$

for $m > n$. We will see this again in Section 5–1.

(b) Identify the "zero state response" and the "zero input response" terms. How would you define the "zero state" of this system?

(c) If $x(n) = 0_X$ is the zero vector of the state space X, find the output g at time m, due to the special input

$$u(j) = \delta_{ij} = \begin{cases} 0 & i \neq j \\ 1 & i = j \end{cases} \quad \text{where } n \leq i$$

δ_{ij} is called the **Kronecker delta**.

$$\left[Answer: g(m, i) = \begin{cases} H(m)\Phi(m, i+1)G(i), & n \leq i < m \\ 0 & m \leq i \end{cases} \right].$$

We call this output g the **unit pulse response**.

(d) Show that with $x(n) = 0$, the zero state response to any input $u(\cdot)$ is given by

$$y_0(m) = \sum_{i=n}^{m-1} g(m, i)u(i)$$

where $g(m, i)$ is the response at time m to a unit pulse at time $i < m$.

(e) Show that, if $F(n)$, $G(n)$ and $H(n)$ are constant matrices F, G, and H, then the unit pulse response can be written

$$g(m, i) = g(m - i, 0)$$

and the zero state response is given by a **convolution summation:**

$$y_0(m) = \sum_{j=n}^{m-1} g(m - j, 0)u(j),$$

which we will study in Sections 4–5 and 4–6.

(f) Illustrate the usefulness of these ideas by letting $u(\cdot)$ be a unit pulse occurring at time i, with the output set to zero at that instant, in the system of Example 1, and computing the output at time $m > i$ *by inspection* of Figure 2–1.

12. The national income of each of two countries equals domestic investment plus domestic consumers' goods production plus exports. The consumption and imports of each country are known functions of income in that country in the previous period.

Set up the system of difference equations describing the behavior of each national income over time. Specialize these by assuming constant marginal propensities to consume and import. [See Samuelson, P., *Foundations of Economic Analysis,* Atheneum Publishers, 1965.]

13.   This exercise illustrates the analysis of linear sequential circuits using symbols from a finite field, which are employed in coding theory and computer circuits.

Let K be the ternary field $(\mathbf{Z}_3, +_3, \cdot_3)$ studied in Exercise 6 (the "integers modulo 3") and let $X = K^2$, the set of 2-tuples whose entries are members of K. Consider the system of Figure 2–8; where the "summer" adds mod 3 and the gains —①— and —②— denote multiplication mod 3 by the numbers 1 and 2.

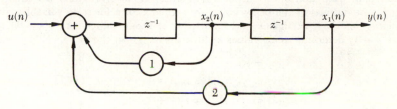

Figure 2–8   A system with two state variables and nine states.

(a)   Write the one-step state transition function in matrix form

$$x(n + 1) = Fx(n) + Gu(n)$$

(b)   Use Exercise 8(c), with each $u(j) = 0$, to find the zero input response (i.e., free of non-zero inputs) $y(n + k)$.

(c)   Show that the determinant $\det F = -2 = 1 \bmod 3$ and compute F^2, F^3, and so on up to F^7. Is $\Phi(n + k, n)$ singular?

(d)   Let $\chi(s) = \det[sI - F]$. Show that $\chi(s) = s^2 - s + 1$. Verify that $F^2 - F + I = 0$ (Cayley-Hamilton!). [The reader who is not familiar with determinants is referred to Chapter 1 of *Matrix Theory* by J. N. Franklin, Prentice-Hall, Inc., Englewood Cliffs, N.J., 1968.]

✔ **14.**   To find a discrete-time system which will have its input-output relationship satisfying a given nth order difference equation of the form

$$y(k + n) + a_1 y(k + n - 1) + a_2 y(k + n - 2) + \cdots$$
$$+ a_{n-1} y(k + 1) + a_n y(k) = u(k)$$

we have a simple algorithm:

(a)   Solve the equation for the "highest difference" term $y(k + n)$.

(b)   Assume $y(k + n)$ to be available as the input to a unit-delay device as in Figure 2–9, so $y(k + n - 1)$ will come out.

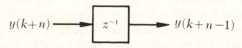

Figure 2–9

(c)   Feed $y(k + n - 1)$ into another unit delay to generate $y(k + n - 2)$. Repeat this procedure until all of the difference terms $y(k + n - 1), y(k + n - 2)$, $\ldots, (k + 1), y(k)$ appear as outputs of delay boxes.

(d)   Build $y(k + n)$ by "feeding back" appropriate amounts of $y(k + n - 1)$, $\ldots, y(k)$ and summing them with input $u(k)$ using an adder $\oplus$. (The appropriate amounts are, of course, given by the expression of part (a).)

(e)   Observe that you now have a system involving n delay boxes, whose overall input is $u(k)$ and whose output is $y(k)$, satisfying the given difference equation. Clearly, the outputs of the delay boxes are a good choice for state variables, since knowledge of them at time k and knowledge of input values $u(k), u(k + 1), \ldots,$ $u(k + n - 1)$ determine their values at time $k + n$. Thus, choose

$$x_1(k) = y(k)$$
$$x_2(k) = y(k + 1) = x_1(k + 1)$$
$$x_3(k) = y(k + 2) = x_2(k + 1)$$
$$\vdots \qquad\qquad \vdots$$
$$x_n(k) = y(k + n - 1) = x_{n-1}(k + 1)$$

Finally, using $x_n(k + 1) = y(k + n)$, which by the above equations and part (a) is also expressible in terms of $x_1(k), \ldots, x_n(k)$ and $u(k)$, you can write the matrix difference equation

$$x(k + 1) = Fx(k) + Gu(k)$$
$$y(k) = Hx(k)$$

The system (F, G, H) is a *realization* or *simulation* of the given scalar difference equation.

(f)   Illustrate this method by finding a realization for the difference equation

$$y(n + 3) + 6y(n + 2) + 11y(n + 1) + 6y(n) = u(n)$$

Compare your block diagram with that of Figure 2–7a.

(g)   Try to use this approach to realize the difference equation

$$y(n + 3) + 6y(n + 2) + 11y(n + 1) + 6y(n) = 3u(n + 2) + 4u(n + 1) + 5u(n)$$

Comment on how you provided $u(n + 1)$ and $u(n + 2)$ from $u(n)$.

15.   The famous Fibonacci numbers, $\{0, 1, 1, 2, 3, 5, 8, 13, \ldots\}$, can be generated as the output of an unforced discrete system. Find a second order scalar difference equation whose solution $y(k)$ is the kth Fibonacci number and a realization (F, G, H). Give an explicit expression for $y(k)$.

2–2   Local Descriptions for Continuous-Time Systems

It will be a much-used ploy in this book to use our understanding of discrete-time systems to ask questions about continuous-time systems, where the mathematics of the calculus lends a certain familiarity, but much of the analysis is more complicated and less intuitive. How can we extend our step-by-step analysis of the behavior of a discrete-time system over arbitrarily long finite periods of time—where each successive input causes a state transition which in turn yields a new output—to a continuous-time system? Here, we cannot take a system in a given state, apply a single input, and look at the next state—for there is no immediately successive time to any given time in a continuum. Rather, we concern ourselves with the behavior of the system over some finite interval of time.

In the continuous-time case we employ the notion of a derivative. If we have a curve in the plane, and we wish to indicate how it changes about a point, then we specify that change locally by specifying the derivative or "instantaneous chord."

In Figure 2–10, we have taken the simple case in which the state space is just the space of real numbers, and we have graphed a particular **trajectory*** of the state against time. If we look at a particular time t and the state $x(t)$, then we see that various chords may be drawn which specify how the state changes over some finite interval. By taking the limit of those chords as the interval tends to zero, we obtain the tangent to the curve which gives us the "instantaneous rate of change of state." Here, the crucial point is that there exists the theory of differential equations which tells us that, if we can specify the derivative to a curve at every point, and some one point on that curve, then, under very general conditions, we may reconstitute the curve uniquely. [We shall explicitly consider such general conditions in Section 5–2.] In other words, if we know the initial state, and if at every subsequent moment of time we may use the state and input at that time to compute the derivative of the state trajectory, then we are assured that we may integrate the resultant differential equation to obtain the actual behavior of the system in its state space. For a continuous-time system, then, we want a local description which takes the form of a differential equation

$$\frac{dx(t)}{dt} = f(t, x(t), u(t))$$

for some suitable function $f: T \times X \times U \to X$.

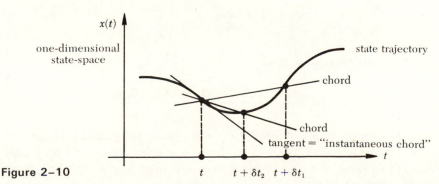

$x(t)$

one-dimensional
state-space

state trajectory

chord

chord

tangent = "instantaneous chord"

t

Figure 2–10

t $t + \delta t_2$ $t + \delta t_1$

---

*We call a mapping $x: T \to X$ a trajectory or motion in the space X. If $X = \mathbf{R}^n$ and T is an interval, $T = [a, b] \subset \mathbf{R}$, the trajectory is a curve in n-space and may be regarded as having been traced out by the tip of the state vector $x(t)$ as t increases from a to b.

Note that a given continuous-time system may not be describable by such a differential equation. For example, we might want to consider a continuous-time system which has a "reset to 0" input, such that when it is applied at time t, the subsequent behavior of the system will be as if the system had actually been in its 0 state at time t. Such behavior cannot be modelled by a differential equation, and yet the analysis of certain systems may be eased by assuming the use of such an input. Of course, any real system has inertia, and an *instantaneous* reset to zero would be impossible, so we could always model such a reset by a differential equation in which such an input, if applied over a finite (but very very small) interval of time, does indeed reset the system in a *continuous* fashion to 0. However, the study of such a differential equation would only be worthwhile if the settling transients were relevant to the control problem we were studying.

Given a description of a continuous-time system, it is thus a genuine question as to whether or not it can be modelled by a differential equation such as $\dfrac{dx(t)}{dt} = f(t, x(t), u(t))$.

Now, the derivative of the state at time t is given by $\dfrac{dx(t)}{dt} = \lim\limits_{\Delta t \to 0} \dfrac{x(t + \Delta t) - x(t)}{\Delta t}$, so that if the applied input is u, we have

$$\frac{dx(t)}{dt} = \lim_{\Delta t \to 0} \frac{\phi(t + \Delta t, t, x(t), u) - \phi(t, t, x(t), u)}{\Delta t}$$

Since in $\phi(t_1, t, x(t), u)$ we are holding all variables but the first fixed, $\dfrac{dx(t)}{dt}$ is thus defined if and only if the partial derivative of $\phi(t_1, t, x(t), u)$ is defined with respect to t_1, when t_1 actually equals t. If this partial derivative is well defined, then we may write

$$\frac{dx(t)}{dt} = \frac{\partial}{\partial t_1} \phi(t_1, t, x(t), u) \bigg|_{t_1 = t}$$

Now, the expression

$$\lim_{\Delta t \to 0} \frac{\phi(t + \Delta t, t, x(t), u) - \phi(t, t, x(t), u)}{\Delta t}$$

only depends on values of $u(\tau)$ for $t \leq \tau < t + \Delta t$. Since Δt is tending to 0, it is not unreasonable to expect the limit, if it exists, to depend on u only through its value at t, and *if* this is so (but it need not be—see below) we are justified in writing

$$\frac{dx(t)}{dt} = f(t, x(t), u(t))$$

where

$$f(t, x(t), u(t)) = \frac{\partial}{\partial t_1} \phi(t_1, t, x(t), u) \bigg|_{t_1=t}$$

The requirement that the system be described by a differential equation does *not* force us to restrict the input functions of the system to be differentiable—piecewise differentiable or piecewise continuous functions are still satisfactory, so long as plugging them into the differential equation still yields an integrable differential equation—and this will normally be the case.

Example 1

If we have the extremely simple system of Figure 2–11, the ideal integrator, governed by the differential equation $\dfrac{dx(t)}{dt} = u(t)$, and we apply the **Heaviside step function,**

$$1(t) = \begin{cases} 0 \text{ if } t \leq 0 \\ 1 \text{ if } t > 0 \end{cases},$$

then there is no trouble in integrating the differential equation even though this input function is not differentiable. We get the state trajectory

$$x(t) = \begin{cases} 0 \text{ if } t \leq 0 \\ t \text{ if } t > 0 \end{cases}$$

where the initial condition is $x(t_0) = 0$ for some $t_0 \leq 0$.   ◊

$$u(t) = \dot{x}(t) \qquad \qquad x(t)$$

Figure 2–11   An ideal integrator.

Let us examine more carefully the form of f. We have seen that it can only depend on values of $u(\tau)$ in the interval $[t, t + \Delta t)$ where Δt is *arbitrarily small*. But this does *not* mean that it is necessary that *only* the value $u(t)$ is relevant, for the formula

$$\dot{u}(t) = \lim_{\Delta t \to 0} \frac{u(t + \Delta t) - u(t)}{\Delta t}$$

shows that $\dot{u}(t)$—and in fact any derivative $u^{(n)}(t)$ of a suitably smooth function u—is determined by the values of u on $[t, t + \Delta t)$ for Δt arbitrarily small. Thus, there will be applications in which the local transition function depends not only on the value of u at t, but also on the values of a finite number of derivatives of u at t, say

$$\dot{x}(t) = \widehat{f}(t, x(t), u(t), \dot{u}(t), \ldots, u^{(n)}(t))$$

Mathematical aside. One might even consider systems in which $\dot{x}(t)$ depends on all derivatives of u, but we shall not do so in this book. Mathematicians speak of the **germ** of u at t to denote the totality of local information about u, i.e., that which can be obtained from $u_{(t-\varepsilon, t+\varepsilon)}$ no matter how small ε may be. We know that for certain functions, Taylor's theorem lets us "grow" the whole function from all its derivatives at a point:

$$u(t_1) = \sum_{j=0}^{\infty} u^{(j)}(t_0) \frac{(t_1 - t_0)^n}{n!}$$

and since these are all *part* of the germ (though e^{-1/x^2} shows that the germ contains even more—look at its derivatives at $x = 0$), we are thus using "germ" in the same sense in which the biologist talks of "germ plasm"—a localized concentration which contains all the information required to specify the complete function. But let us now return to the finite derivative case:

Example 2

Consider the capacitor of Figure 2–12a. If we take the current as the input and the voltage across it as the state, then we have the system equation

$$\dot{x}(t) = C^{-1}u(t)$$

which requires *no* derivatives of the input. However, if we take the voltage as input and current as state, as in Figure 2–12b, we have the system equation

$$\dot{x}(t) = C\ddot{u}(t)$$

which requires the *second* derivative of the input. ◊

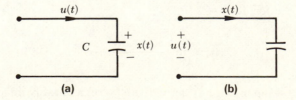

(a) **(b)**

Figure 2–12(a) The capacitor with voltage as state. **(b)** The capacitor with current as state.

In real systems, exact differentiation cannot, in fact, occur—although we can build systems that will differentiate slowly-varying functions with great accuracy. [For an example of such a system, see Exercise 2 of Section 2–1.] Note, however, that if the system equations are well-approximated by a function explicitly involving n derivatives

$$\dot{x}(t) = \hat{f}(t, x(t), u(t), \dot{u}(t), \ldots, u^{(n)}(t)) \tag{1}$$

we may interpose n differentiators between the input and a system of the form

$$\dot{x}(t) = f(t, x(t), v(t)) \tag{2}$$

for which the input is the $(n + 1)$-tuple

$$v(t) = (u(t), \dot{u}(t), \ldots, u^{(n)}(t))$$

as shown in Figure 2–13.

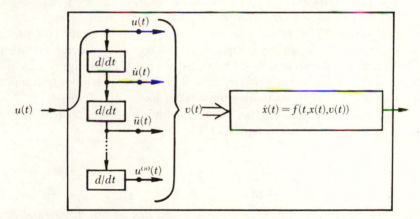

Figure 2–13   An implementation of $\dot{x}(t) = f(t, x(t), u(t), \dot{u}(t), \ldots, u^{(n)}(t))$ using n differentiators.

Thus, although we shall develop our general theory to handle continuous-time systems of the form (2), we shall also be able to apply the theory to systems of the form (1).

Even if we could build exact differentiators, there is an important non-deterministic reason for avoiding their use in actual systems.

Example 3

Suppose we have a signal s which we need to differentiate. If we apply s as the input to an ideal differentiator as in Figure 2–14, we get $\dot{s}$ as the output.

Now, in practice we won't ever really have a signal s; we will always have some corrupted version of s, say $\hat{s} = s + e$, where e is an error term

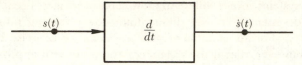

Figure 2–14   An ideal differentiator.

called "noise." When we try to differentiate s, we are really putting the signal $\hat{s}$ into the ideal differentiator and we get as output not $\dot{s}$ but $\hat{\dot{s}} = \dot{s} + \dot{e}$, producing an error of $\dot{e}$ in the result. Even when the error e in s is very small, the error $\dot{e}$ in $\dot{s}$ can be very large. To see this, let $e(t) = 10^{-3} \sin \omega t$ so that $\dot{e}(t) = 10^{-3}\omega \cos \omega t$ which has an amplitude of $10^{-3}\omega$. If ω is very large, say $\omega = 10^6$, the perfect differentiation gives us a very great relative error in the derivative. ◊

We say that differentiation amplifies the noise, which is often a small but rapidly varying function. On the other hand, integration reduces the relative effect of noise terms, and we say that integration is a "smoothing" process [Exercise 1]. Whenever possible in realizing or simulating continuous time systems, say on an analog computer, we will try to use integrators instead of differentiators. In Section 2–5, when we discuss different ways of selecting state variables for dynamic systems, we will see several ingenious techniques for using integrators to do differentiations for us.

So far we have seen that an arbitrary continuous-time system can be represented by a differential equation

$$\frac{dx(t)}{dt} = f(t, x(t), u(t))$$

only if the function ϕ in its original description has certain differentiability properties.

However, there is a deeper problem. We have been talking as if the state space were always the real line, so that we have no difficulty in talking about derivatives. But when we come to discuss more general state spaces, we cannot so blithely talk of taking derivatives without imposing certain special structures on the state space. For instance, two crucial operations in forming a derivative are that of forming the difference between two states, and that of performing a limiting operation upon the result. Thus, if we are to analyze continuous-time systems, using differential equations, then we shall have to restrict our spaces U, X, and Y, so that it is possible to perform on them the basic arithmetic operations, and also to take limits. The most commonly imposed restrictions require U, X, and Y to be $\mathbf{R}^k$, $\mathbf{C}^n$ or other **vector spaces** as discussed in Section 2–3 which have certain additional properties of the type we shall study in Section 2–4.

We should stress immediately that such restrictions were *not* necessary for our discussion of discrete-time systems. The local description of a discrete-time system in terms of its one-step transition function holds for arbitrary spaces U, X, and Y. If we are analyzing a computer, and we think of an input as being either a single instruction or a single datum, then we certainly would not demand that the sum of two instructions be well-defined as an instruction or datum; and, with a finite collection of instructions, we would expect the notion of "the limit of a set of instructions as $\Delta t \to 0$" to be essentially meaningless. We thus gain some immediate appreciation of the extra mathematical structure and subtlety which continuous-time system theory demands of us.

EXERCISES FOR SECTION 2-2

1.   Analogously to Example 3, study the effect of pure integration on a signal corrupted by noise.
2.   This exercise gives practice with several varieties of state selection.
     (i)   For the systems of Example 2, find the global state transition function ϕ for the two different state and input selections of Figure 2-12a and Figure 2-12b respectively.
     (ii)   For the state-input selection of Figure 2-12a, show that if $r(t) \triangleq 57x(t)$, then r is a perfectly acceptable state with which to describe the system. Give the local and global state equations using this r as the state. Is there anything special about the number 57? Generalize.

2-3   Vector Spaces and the Notion of a Basis

The purpose of this section is to introduce the reader to some general notions about vector spaces. Thus, the reader with a reasonable background in linear algebra may omit this section, though he may find it worthwhile to skim through it to see how the algebra is related to our concerns in system theory.

     Let us recall the basic properties of vectors in the plane, and then abstract from these to come up with the general notion of a vector space.

     Looking at Figure 2-15, we see that for vectors in the plane we have a well-defined notion of addition and of multiplication by a scalar. Addition is defined by the parallelogram law; multiplication by a scalar consists of leaving the direction of the vector unchanged, but multiplying its length by

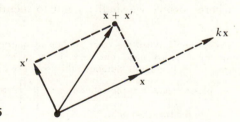

Figure 2-15

the given scalar—it being understood that a negative distance in a given direction is a positive distance in exactly the opposite direction. As we see in Figure 2–16, it is customary to refer a given vector x to a coordinate system. In this case, the coordinate system is specified by giving the vectors e_1 and e_2, and x can then be uniquely expressed as a combination $x_1 e_1 + x_2 e_2$, and so, given the coordinate system, we may denote x unambiguously by the ordered pair $\begin{bmatrix} x_1 \\ x_2 \end{bmatrix}$.

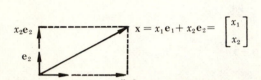

$$x = x_1 e_1 + x_2 e_2 = \begin{bmatrix} x_1 \\ x_2 \end{bmatrix}$$

Figure 2–16

Henceforth, we leave it to the judgment of the reader to determine whether x is a scalar (in this case a real number) or a vector, and we stop writing a vector x as $\mathbf{x}$. Generalizing the above discussion, we may see that, once we have chosen a coordinate system, any n-dimensional vector may be represented by an n-tuple† $x = (x_1, \ldots, x_n)^T$. The reader may then verify that the "coordinate-free" definitions of addition and multiplication by a scalar which we gave above in the two-dimensional case become the following coordinate-based definitions: If $x = (x_1, \ldots, x_n)^T$ and $x' = (x_1', \ldots, x_n')^T$ and k is any real number, then we define (read $\triangleq$ as "equals by definition")

$$x + x' \triangleq (x_1 + x_1', \ldots, x_n + x_n')^T \tag{1}$$

$$kx \triangleq (kx_1, \ldots, kx_n)^T \tag{2}$$

Historical note: In the nineteenth century, people used to label the components of a three-dimensional vector as x, y, and z; we prefer to label them x_1, x_2, and x_3 because we can then write compact statements like $\sum_{j=1}^{3} k_j x_j = k_1 x_1 + k_2 x_2 + k_3 x_3$ rather than expressions like $ax + by + cz$, which puts great strain on our imagination when it comes to inventing new variables. In fact, where possible we shall try not to mention the coordinates at all, and use

---

†Why the ^{T}? When we come to talk of matrices, it will be convenient to regard these n-tuples as columns, but here it is easier to write them as rows. So we write $(x_1, \ldots, x_n)^T$ to mean "although this looks like a row vector it should really be **Transposed**, so please regard it as the column vector $\begin{bmatrix} x_1 \\ \vdots \\ x_n \end{bmatrix}$." Some authors use a prime, x', instead of x^T to denote the transpose of x.

a compact notation which specifies the vector x rather than its components $(x_1, \ldots, x_n)^T$. This is the most modern approach.

Note that if we pick one coordinate vector, say e_1, and choose to identify the real number x_1 with the vector $x_1 e_1$, then in fact the notions of addition and of multiplication by a scalar are consistent with our definitions for real numbers—for instance, it is true that

$$x_1 e_1 + x_2 e_1 = (x_1 + x_2) e_1$$

Let us stress again that when we talk about n-dimensional space here, we are not trying to contradict a statement that the world is spatially 3-dimensional or—if you are a relativity fan—spatio-temporally 4-dimensional. Instead, we assert that if we are looking at a system whose state, say, can be characterized by the result of n measurements, then it is mathematically convenient to treat such collections of n measurements as a set of "all measurements with n components." Since position in ordinary Euclidean space can be characterized by triples of measurements, and since we refer to such a set as a three-dimensional space, so it is convenient to refer to an n-measurement set as an n-dimensional space. It refers to n degrees of freedom in carrying out measurement, not to n spatial dimensions in the space that we move and live in. If we think of each of our measurements as yielding real numbers, then we are led to consider Euclidean n-space (or real n-space) $\mathbf{R}^n$, the set of all n-tuples $(x_1, \ldots, x_n)^T$ of real numbers x_j. If we are to think of the measurements as yielding complex numbers, then we are led to consider complex n-space $\mathbf{C}^n$, the set of all n-tuples $(x_1, \ldots, x_n)^T$ of *complex* numbers. The reader may reconcile the notation $\mathbf{R}^n$ with our notion of a direct product, or Cartesian product, by identifying $\mathbf{R}^n$ with an n-fold Cartesian product

$$\underbrace{\mathbf{R} \times \cdots \times \mathbf{R}.}_{n \text{ times}}$$

Note that for both $\mathbf{R}^n$ and $\mathbf{C}^n$, addition is well defined by the formula (1). However, for $\mathbf{R}^n$ the definition of multiplication by a scalar, (2), can only be used when k is a *real* number, whereas the formula applies for $\mathbf{C}^n$ for any *complex* number k. If we refer to an element of $\mathbf{R}^n$ or $\mathbf{C}^n$ as a vector, and thus refer to the spaces themselves as vector spaces, we see that the specification of the vector space carries with it the specification of the "field" of scalars by which we may scale up vectors, as well as actually specifying the operations whereby vectors are added together and scaled up. Without being too precise about what we mean by a field [the reader can find a formal definition in Exercise 6 of Section 2–1], let us just comment that this is a term used in algebra to describe mathematical objects such as $\mathbf{R}$ and $\mathbf{C}$ in which addition and multiplication are defined and behave in certain convenient ways. It may be of interest, though, to note here that there are many other fields besides $\mathbf{R}$ and $\mathbf{C}$—for instance, finite fields [such as those in the cited exercise] which play a very important role in coding theory (Bobrow and Arbib [1973], Chapters 7 and 8).

In any case, we have seen that in specifying a vector space, we must specify the field K of scalars, that is, the "numbers" by which we wish to multiply our vectors. For instance, $\mathbf{R}^n$ and $\mathbf{C}^n$ are two examples of the vector space K^n, which is a vector space over the field K comprising n-tuples of scalars.

If $x = (x_1, \ldots, x_n)^T, x' = (x'_1, \ldots, x'_n)^T$ and $x'' = (x''_1, \ldots, x''_n)^T$ are vectors, and k and k' are scalars, then using the properties of numbers in a field, such as $\mathbf{R}$ or $\mathbf{C}$, we may note the following equalities:

$$x + x' = (x_1 + x'_1, \ldots, x_n + x'_n)^T$$
$$= (x'_1 + x_1, \ldots, x'_n + x_n)^T = x' + x$$

We note that **commutativity** of addition in the field of scalars (i.e., the fact that $x_j + x'_j = x'_j + x_j$ for any choice of the scalars x_j and x'_j) implies commutativity of addition in the vector space.

Similarly, the reader may verify that **associativity** in the field of scalars (i.e., the fact that $(x_j + x'_j) + x''_j = x_j + (x'_j + x''_j)$ for any choice of the scalars x_j, x'_j, and x''_j) implies associativity of addition in the vector space:

$$(x + x') + x'' = x + (x' + x'')$$

[It is worth noting that there are arithmetic ways of combining pairs of numbers which are *not* associative—for instance, subtraction is not associative: $(a - b) - c = a - (b + c) \neq a - (b - c)$].

Another property is that

$$k(k'x) = k(k'x_1, \ldots, k'x_n)^T$$
$$= (kk'x_1, \ldots, kk'x_n)^T = (kk')x$$

and that

$$1 \cdot x = (1 \cdot x_1, \ldots, 1 \cdot x_n)^T = (x_1, \ldots, x_n)^T = x$$

Similarly, the reader may verify that

$$k(x + x') = kx + kx'$$
$$(k + k')x = kx + k'x$$

If we call $\mathbf{0} = (0, \ldots, 0)^T$ the **zero vector,** we also have

$$\mathbf{0} + x = x$$
$$0 \cdot x = \mathbf{0}$$

[We shall seldom continue to use boldface type to denote the zero vector, since context will usually make it clear whether a given occurrence of 0 is a scalar or a vector].

At this stage, let us discuss the axiomatic method which plays such a big role in mathematics. Quite often, we use a number of examples to get the ideas which allow us to prove that certain properties hold; when we have finished, we may discover that we did not need to use everything about those examples, but only used certain particular properties of those examples. In particular, we shall find that much of the theory that we create for linear systems will not depend upon taking the state space to be, for instance, $\mathbf{R}^n$. Rather, most of the properties of linear systems will depend upon the fact that addition and multiplication by scalars are defined for our vectors in a way consistent with the properties we have just derived above for vectors in $\mathbf{R}^n$. For instance, we might have a system in which the input at any time is the pattern of light-wave amplitude and intensity across a whole lens. Such a wave pattern cannot be represented in $\mathbf{R}^n$ for any finite n—rather, we need to specify a complex number for every point of the lens—and we might thus speak of an infinite-dimensional complex vector space.

It is thus imperative that we have a notion of a vector space which is not bound to notions of finite dimensionality. In fact, the properties of vector addition and of multiplication by a scalar given above also remain true for patterns displayed upon a two-dimensional surface. Thus, in proving theorems about vector spaces, we must make sure that the steps in our proofs depend only upon the properties we have listed above (and not upon some special properties of $\mathbf{R}^n$). This will ensure that, later in our study of systems when we come to look at some new input or state space, if we can assure ourselves that this space satisfies all the above properties, then our theorems will remain true for this space.

Our strategy in proving theorems about vector spaces will normally be to use the considerable intuition we have for two-dimensional or three-dimensional vectors, to write out proofs using that intuition, and then to go back over the proofs and rewrite them more formally. We will do this in a way which ensures that we have used only those properties listed above, to assure ourselves that the results really will be true of all possible vector spaces, and not simply of those we have had in mind while making the proof. For instance, in our proof above that $x + x' = x' + x$, we might have written the proof thinking that it only applied to n-tuples of real numbers—but when we go back and read over the proof, we see that the only thing we used was commutativity of addition of real numbers. Thus, we realize that if we consider vectors that are n-tuples of scalars from any field, then in fact we shall have commutativity of addition of vectors, because it is a property of all fields that addition in them is commutative. There is thus a crucial interplay between the use of intuition to derive the sketch of a proof that something is true, and a process of formalization in which we go back over the proof and make it as generally applicable as we can.

We shall now formalize† the notion of a vector space which we have less formally given above. In reading the following definition, the reader should bear in mind that the notation $X \times X \to X$ such that $(x, x') \mapsto x + x'$ means that we have a function which takes elements of the Cartesian product $X \times X$ to yield an element of X (i.e., the function takes an ordered pair (x, x') and yields a specific vector which we label $x + x'$). Thus, when X is the vector space $\mathbf{R}^n$, then this function $\mathbf{R}^n \times \mathbf{R}^n \to \mathbf{R}^n$ would be the function which takes two n-tuples of real numbers and adds them together component by component.

DEFINTION 1

A **vector space over the field** K is a set X, the elements of K being called **scalars** and the elements of X being called **vectors,** together with two functions
   addition of vectors: $X \times X \to X$ such that $(x, x') \mapsto x + x'$
   multiplication by a scalar: $K \times X \to X$ such that $(k, x) \mapsto kx$ (or $k \cdot x$)
which satisfy the following conditions for all scalars k, k', and all vectors x, x' and x'':

$$x + x' = x' + x \qquad \text{(commutativity)}$$
$$(x + x') + x'' = x + (x' + x'') \text{ (associativity)}$$
$$k(x + x') = kx + kx'$$
$$(k + k')x = kx + k'x$$
$$k(k'x) = (kk')x$$
$$1 \cdot x = x$$

and there exists a unique vector **0,** called the **zero vector** (or **null vector**) such that for all vectors x we have

$$0 \cdot x = \mathbf{0}$$

where 0 is the zero element of the field K.   ○

Note that we can deduce from the above axioms that

$$\mathbf{0} + x = 0 \cdot x + 1 \cdot x = (0 + 1) \cdot x = 1 \cdot x = x$$

for all vectors x.

Given any vector space, let us use the shorthand

$$-x \text{ for } (-1) \cdot x$$
$$x - x' \text{ for } x + (-x')$$

Then for any vector x we have

$$x - x = 1 \cdot x + (-1) \cdot x = (1 + (-1)) \cdot x = 0 \cdot x = \mathbf{0}$$

---

† Our definition of a vector space is not the usual one, but is entirely equivalent to the usual list of axioms [see Nering, E. D., *Linear Algebra and Matrix Theory,* John Wiley & Sons, New York, 1970].

We have now seen how to abstract from our intuition on two-dimensional and three-dimensional spaces, or our arithmetic discussion of any K^n, a list of properties which we have gathered together to form the criteria which determine when we should consider a set X together with two functions $X \times X \to X$ and $K \times X \to X$ as a vector space over the field K. We have also seen a couple of very simple examples of how we may proceed from these axioms to properties which hold for all vector spaces. We now want to prove a very powerful result about vector spaces. Its interest to the system theorist should be immediate. If we have placed a vector space structure on the input space or the output space of a system, we might well wish to know whether we can similarly place a vector space structure on the set of all input functions or on the set of all state functions. In fact, this is always possible, and we shall see that we need not restrict our consideration to functions of time, but have the following very general result:

THEOREM 1

Let X be a vector space over some field K, and let A be *any* set. Let X^A stand for the set of all functions from A to X. Then, if we define addition of functions "pointwise" by

$$(f + f')(a) \overset{\Delta}{=} f(a) + f'(a) \text{ for all } a \text{ in } A$$

and multiplication of functions by scalars by

$$[kf](a) \overset{\Delta}{=} k[f(a)] \text{ for all } a \text{ in } A$$

then X^A, equipped with these operations, is itself a vector space over K.

Remark: When the set A is the time set for a system and (rather than X) we consider a vector space U of inputs, then our above definitions correspond to the natural notion of adding two input functions by adding together the inputs at corresponding times. Similarly, for instance, we would double an input function by doubling the input at each time.

We may now prove our theorem:

Proof

To prove that $f + f' = f' + f$ we must prove that $(f + f')(a) = (f' + f)(a)$ for every a in A:

$$\begin{aligned}
(f + f')(a) &= f(a) + f'(a) \text{ by the definition of} \\
&\qquad\qquad\qquad \text{addition of functions} \\
&= f'(a) + f(a) \text{ by commutativity in } X \\
&= (f' + f)(a) \text{ by definition.}
\end{aligned}$$

Similarly, the reader may verify the next five axioms in our list. To see the truth of the last axiom, let us introduce the zero function

$$\widehat{0}:A \to X \text{ such that } a \mapsto 0$$

whose value is precisely the null vector 0 of X, no matter what a we apply it to. But clearly, for any $f:X \to A$,

$$(0 \cdot f)(a) = 0 \cdot f(a) \text{ by definition of multiplication}$$
$$\text{by a scalar}$$
$$= 0 \qquad \text{by the property of the null vector in } X$$
$$= \widehat{0}(a) \qquad \text{by definition of } \widehat{0}$$

Thus, we have verified that

$$0 \cdot f = \widehat{0} \text{ for all } f \text{ in } X^A$$

Thus, X^A is in fact a vector space over K with the specified addition and multiplication by a scalar. Its null vector is the "zero function" $\widehat{0}$.   □

The reader is invited to supply the missing details. He should note that in the above proof we made use only of the vector space properties of X. It is the "pointwise" nature of our operations that allows the vector space structure of X to impose a vector space structure on X^A even when there is no special structure on the underlying set A.

To show how nicely things tie together, we might note that our earlier verification that K^n is a vector space is, in fact, a special case of the above theorem. If the reader glances back through the axioms, he will note that it is trivial that K is a vector space over K (1-dimensional) under the usual addition and multiplication, and the zero vector of K is just the scalar zero. Now let us introduce the set $[n] \overset{\Delta}{=} \{1, 2, 3, \ldots, n\}$ and note that any n-tuple of scalars $(x_1, \ldots, x_n)^T$ may be considered as a function $x:[n] \to K$ with $x(j)$ being just the jth component x_j. We may thus identify K^n with the space $K^{[n]}$ of all functions from $[n]$ to K. (Taking $X = K$ and $A = [n]$, the notation X^A at last makes sense!) Further, the definitions of addition and of multiplication by a scalar in K^n agree exactly with those in $K^{[n]}$ (check this) and so we may conclude from the fact that K is a vector space, and from the truth of our theorem above, that K^n is a vector space—quite independently of the direct verification that we made earlier. The reader should in fact convince himself that that direct verification was just a particular case of the steps we undertook in proving our theorem.

Example 1

An *infinite*-dimensional vector space of interest to the system theorist is an oscillating ideal string, as shown in Figure 2–17. The state of this system is

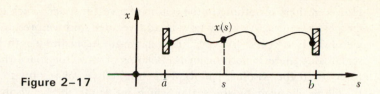

Figure 2–17

given by the function $x:[a, b] \rightarrow \mathbf{R}$ where $x(s)$ is the displacement of the string above point s at the given time. These states form an infinite-dimensional vector space, with addition given by

$$(x + x')(s) = x(s) + x'(s)$$

and multiplication by a scalar given by

$$(kx)(s) = kx(s) \qquad\qquad \lozenge$$

The Notion of a Basis.   We turn now to the notion of a basis, which lets us express vectors in terms of their coordinates: Let us consider once again the example of vectors in the plane. The reader will recall that we were able to define the sum of two vectors as the result of starting one vector at the origin, placing the tail of the second vector at the head of the first vector, and then letting the sum be the vector that went from the tail of the first vector to the head of the second vector. Similarly, we multiplied a vector by a scalar simply by leaving its direction unchanged but multiplying its length by that scalar. The important thing to note is that these definitions of vector addition and of multiplication by a scalar had nothing to do with any choice of coordinates. It was only when we chose "unit" vectors e_1 and e_2 along some coordinate system, and noted that a vector x could be uniquely expressed as the sum $x_1e_1 + x_2e_2$ that we were able to identify this vector in the plane with ordered pairs $\begin{bmatrix} x_1 \\ x_2 \end{bmatrix}$.

Now, when doing analytic geometry we usually choose our vectors e_1 and e_2 to be of unit length and at right angles as they are in Figure 2–18.

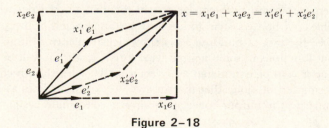

Figure 2–18

However, there is nothing in the notion of a vector space which tells us about lengths or angles. We know how to add vectors, and we presume that when we multiply a vector by a scalar, then we multiply the length by the same scalar. But there is nothing in the axioms of a vector space that allows us to say what the absolute length of the original vector is, only to suspect that it will be changed by the scalar which we apply to the vector. In fact, if we are to talk about the length of a vector and the angle between two vectors, we must place extra conditions upon our vector space. The resultant concept is that of an inner product space. This will be another example of a mathematical structure which may look rather forbidding if approached by the unwary; but we shall in fact see, in Section 4-4, that inner product space is as natural an expression of the properties of length and angles which we associate with vectors in $\mathbf{R}^n$ as were the properties of addition and multiplication by a scalar that we used as the basis of our abstraction of the notion of a vector space.

In any case, returning to Figure 2-18, we note that the pair of vectors e_1' and e_2' would do just as well—the point remains that, given a vector x, the numbers x_1' and x_2' such that $x = x_1'e_1' + x_2'e_2'$ are uniquely determined by x, and thus allow us to assign to the vector x as coordinates the pair $(x_1', x_2')^T$. In other words, there is nothing magical about the pair of numbers that we associate with a two-dimensional vector. However, once we have chosen a pair of "basis" vectors—such as the pair e_1 and e_2 or such as the pair e_1' and e_2'—then we have fixed the way in which coordinates are to be associated with vectors, because there will be unique numbers which tell us what multiples of the basis vectors must be added together to obtain x. To see why these numbers are unique in the above example, in Figure 2-18 we may start from the origin along a multiple of e_1 and note that we cannot get off that line. However, if we now add multiples of e_2, we can move off the line—but only parallel to e_2 along the line of departure. Thus, we draw a line along e_1 from the origin, and a line through x parallel to e_2; and from the point of intersection we can compute x_1 and x_2. (Can you make precise the notion of two lines being parallel in a vector space, using only vector space axioms, and making no use of the concept of distance or angle?)

We thus conclude that, given a vector space—in other words, a collection of vectors together with rules for combining vectors and multiplying them by scalars—there may be many ways of placing coordinates upon those vectors so that addition of vectors can be represented by component-by-component addition of the corresponding coordinate vectors, etc. It turns out to be a very important problem for system theorists to consider what is the best possible coordinate system to use for the state space. For instance, often we are given the task of building a system which transforms input functions into output functions in some desired way. Further, it may well be specified that the input is to be considered as a vector in $\mathbf{R}^p$, with the jth component actually being fed in along the jth of a given set of input lines as in Figure 2-19. Similarly, the output space may have a coordinate system imposed

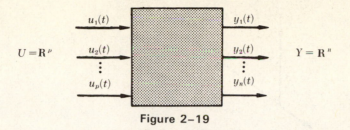

Figure 2–19

by restrictions as to where n output lines, each of them carrying a signal coded by a real number, are to go. The state space, however, may be given more abstractly. All that matters about the state space is that it can be updated in such a way as to preserve the appropriate relationships between input and output. In particular, our earlier mention of integration of differential equations suggests that we might wish to place coordinates upon the state space in such a way as to decouple the differential equations governing the change of the individual state components as much as possible. Thus, a firm understanding of the different coordinate systems which may be placed upon a given space, and how operators acting upon the space change under changes of coordinates, will be crucial to the system theorist. We shall discuss this matter very carefully in Section 3–3 and in Chapters 6 and 7.

For instance, if our state-space were two-dimensional, and the action of updating the state consisted of adding on a given vector every second as shown in Figure 2–20, then it would be well to make that vector one of the "basis" vectors, because then we would have to modify only one coordinate of the state vector rather than modifying several components.

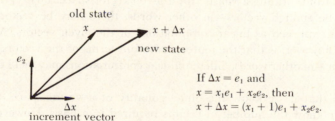

If $\Delta x = e_1$ and
$x = x_1 e_1 + x_2 e_2$, then
$x + \Delta x = (x_1 + 1)e_1 + x_2 e_2$.

Figure 2–20

Let us now analyze more carefully what it is about a pair of vectors that allows them to form the basis for a coordinate system in a two-dimensional vector space. If we had only one vector, it would not be enough—we could only give coordinates to vectors which were multiples of that vector—in other words, we could only cover one line of the whole plane. If we have two vectors, it is clear from Figure 2–21 that these will always serve to form the "basis" of a coordinate system so long as they are not collinear. We see

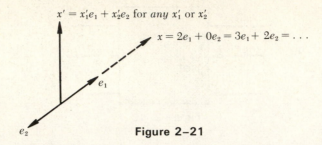

Figure 2-21

that, when e_1 and e_2 are collinear, one problem is that there are vectors off that line which cannot be represented as linear combinations of e_1 and e_2; but there is also the problem that vectors which are collinear with these two have many different representations as linear combinations of e_1 and e_2.

Finally, if we take three or more vectors, we see that they fail to form a basis because there are many different ways of representing a given vector in the plane as a linear combination of these three or more vectors as illustrated in Figure 2-22.

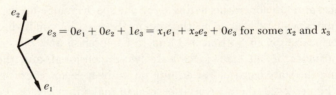

Figure 2-22

We thus see two distinct problems which may disqualify a collection of vectors from forming a "basis" for a given vector space. Firstly, there just may not be enough vectors—in other words, there may be vectors which cannot be expressed as linear combinations of the given vectors. A second problem, however, is that the representation in terms of the vectors may be redundant—in other words, different linear combinations may yield the same vector.

Note that if we know the dimensionality of a vector space, then we believe we know the number of vectors required in a basis. However, if we are only given a vector space abstractly, in terms of the elements and the operations of addition and multiplication by a scalar, then we may have no *a priori* idea of the dimensionality of the space. For instance, if we are interested in a state space for a system purely in terms of how it mediates the relationship between inputs and outputs, then considerable investigation may be required to find the minimum dimension state space which will work. Another question of some interest is to determine from the structure of a vector space how many elements are required to form a basis, without knowing that number *a priori* but by simply using the basic vector-space structure.

Let us note that if any one vector can be given two representations as linear combinations of a given set of vectors, for instance $x = x_1 e_1 + x_2 e_2 + \cdots + x_n e_n = x_1' e_1 + \cdots + x_n' e_n$ with $x_j \neq x_j'$ for some j, then so can any other vector. In particular, the zero vector has two different representations in this case:

$$
\begin{aligned}
0 &= 0e_1 + 0e_2 + \cdots + 0e_n \\
&= (x_1' - x_1)e_1 + \cdots + (x_n' - x_n)e_n
\end{aligned}
$$

with $x_j \neq x_j'$ for some j.

We are now ready to formalize the above discussion in a way that applies to any vector space which satisfies our general definition. In other words, we shall use the intuition gained above while discussing two-dimensional vectors to analyze arbitrary vectors. First, a word as to notation. By $\{x_\alpha\}_{\alpha \in A}$ we shall mean a collection of vectors, such that for each label α in the set A there is given a vector x_α in the collection.

We shall say that a collection of vectors is linearly independent if there are no relations between them which are such as to allow any vector, and in particular the zero vector, to have separate representations as linear combinations of the given vectors:

DEFINITION 2

A set $\{x_\alpha\}_{\alpha \in A}$ of vectors in a vector space X over the field K is said to be **linearly independent** if for any *finite* subset $\{x_\alpha\}_{\alpha \in A'}$, where $A' \subset A$, the only scalars $\{k_\alpha\}_{\alpha \in A'}$ such that

$$
\sum_{\alpha \in A'} k_\alpha x_\alpha = 0
$$

are all zero. The set is **linearly dependent** if it is not linearly independent. ○

To make more clear why we use the notation "linearly dependent," we prove the following:

PROPERTY 2

A set $\{x_\alpha\}_{\alpha \in A}$ is linearly dependent iff at least one x_γ depends linearly on a finite set of other vectors in the set; i.e., iff it can be written as a linear combination

$$
x_\gamma = \sum_{\beta \in B} \lambda_\beta x_\beta \quad \text{where } \gamma \notin B \subset A
$$

Proof

(The reader should check that the following argument only uses properties common to all vector spaces):

(i)   If $\{x_\alpha\}_{\alpha \in A}$ is linearly dependent, then some finite subset $\{x_\alpha\}_{\alpha \in A'}$ has

$$\sum_{\alpha \in A'} k_\alpha x_\alpha = 0$$

where at least one of the scalars, say k_γ, is *not* 0. But then, setting $B = A' - \{\gamma\}$, we have

$$k_\gamma x_\gamma + \sum_{\beta \in B} k_\beta x_\beta = 0$$

and so

$$x_\gamma = \sum_{\beta \in B} \lambda_\beta x_\beta$$

where each $\lambda_\beta = -\dfrac{k_\beta}{k_\gamma}$ (using the fact that, in a field, we may divide by any scalar so long as it is not zero); i.e., x_γ depends linearly on the set $\{x_\beta\}_{\beta \in B}$.

Conversely, if we do have a linear dependence of a vector x_γ on some finite set $\{x_\beta\}_{\beta \in A}$, i.e.,

$$x_\gamma = \sum_{\beta \in B} \lambda_\beta x_\beta \text{ where } \gamma \notin B \subset A$$

then we see that

$$\sum_{\alpha \in B \cup \{\gamma\}} k_\alpha x_a = 0 \text{ where } k_\alpha = \begin{cases} \lambda_\alpha & \text{if } \alpha \neq \gamma \\ -1 & \text{if } \alpha = \gamma \end{cases}$$

and thus the set $\{x_\alpha\}_{\alpha \in A}$ is not linearly independent.   □

Going back to our discussion of what qualifies a collection of vectors to be a "basis" for a linear space, we see that our notion of "linear independence" formalizes our requirement that a vector be uniquely represented as a linear combination of elements of the basis. We must now formalize the second requirement—that there be enough vectors in the basis so that every vector may be expressed as a linear combination of the basis vectors. For instance, in our two-dimensional example, a set consisting of a single non-zero vector is certainly independent, but it does not suffice to give us the whole space in terms of linear combinations of it. We may see that a pair of vectors in two-dimensional space meet our criterion when they are linearly independent—but they satisfy one extra condition not met by a single vector, namely that when we add one more vector the result can never be a linearly independent set. In two-dimensional space, three vectors are always linearly dependent.

DEFINITION 3

A set $\{x_\alpha\}_{\alpha \in A}$ of vectors in a vector space X is called a **basis** if it is linearly independent, and is maximal in that $\{x\} \cup \{x_a\}_{a \in A}$ is *not* linearly independent for any distinct vector x of X which does not belong to the collection $\{x_\alpha\}_{\alpha \in A}$.   ○

Let us now check that this definition allows us to prove formally that which our intuition has led us to suspect—namely, that a collection of vectors forms a basis for a linear space if and only if every vector in that space can be uniquely represented as a linear combination of a finite number of vectors from the basis.

THEOREM 3

A set $\{x_\alpha\}_{\alpha \in A}$ of vectors in a vector space X is a basis iff every vector x in X is representable in one and only one way as a linear combination $\sum_{\beta \in B} k_\beta x_\beta$ of a finite subset $\{x_\beta\}_{\beta \in B}$. (Of course, we do not regard $k_1 x_1 + k_2 x_2$ and $k_1 x_1 + k_2 x_2 + 0 x_3$, for example, as different representations.)

Proof

First let us show that every vector x has at least one such representation. Then we shall verify that the representation is unique.

If x actually belongs to $\{x_\alpha\}_{\alpha \in A}$, say $x = x_\beta$, then we immediately have the representation

$$x_\beta = 1 \cdot x_\beta$$

More subtly, consider $x \notin \{x_\alpha\}_{\alpha \in A}$. Since $\{x_\alpha\}_{\alpha \in A}$ is a basis, we know that there exists a finite subset of $\{x\} \cup \{x_\alpha\}_{\alpha \in A}$ which is linearly dependent, and so we may write

$$kx + \sum_{\beta \in B} k_\beta x_\beta = 0 \text{ for some finite subset } B \text{ of } A$$

where at least one of k or the k_β's is non-zero. In fact, k must be non-zero, for if k were zero, we would have $\sum_{\beta \in B} k_\beta x_\beta = 0$ with at least one k_β non-zero—and this would contradict the linear independence of the basis $\{x_\alpha\}_{\alpha \in A}$. But since k is non-zero, we may write

$$x = \sum_{\beta \in B} \left(-\frac{k_\beta}{k}\right) x_\beta$$

to obtain our desired representation of $\{x_\alpha\}$ as a linear combination of vectors in a finite subset of $\{x_\alpha\}_{\alpha \in A}$.

The proof of uniqueness is an easy consequence of the definition of linear independence: Suppose we can write some vector x as both $\sum_{\beta \in B} k_\beta x_\beta$ and $\sum_{\beta \in B} k'_\beta x_\beta$ for some finite set B. (Why is it permissible to use the same B in both cases?) Then we may deduce

$$\sum_{\beta \in B} (k_\beta - k'_\beta) x_\beta = 0$$

and so each $k_\beta - k'_\beta = 0$ by the linear independence of $\{x_\alpha\}_{\alpha \in A}$; i.e., the coefficients are uniquely determined by x.   □

Example 2

The vectors $e_1 = (1, 0, \ldots, 0)^T, \ldots, e_j = (0, \ldots, 1, \ldots, 0)^T$ with the 1 in the jth place, $\ldots, e_n = (0, \ldots, 0, 1)^T$ form a basis for $\mathbf{R}^n$ and we may write *uniquely*

$$(x_1, \ldots, x_n)^T = \sum_{j=1}^{n} x_j e_j \qquad \Diamond$$

We shall say that a space is n-**dimensional** if, like $\mathbf{R}^n$, it has a basis of n elements. It is in fact a theorem that any two bases for the same space must have the same number of elements, and so the dimension of a vector space is well-defined in this fashion. Of course, we then say a space is **finite-dimensional** if it is n-dimensional for some finite number n. We shall see in Example 3 that there are infinite-dimensional vector spaces which do have bases.

Whenever each vector of a space V can be expressed as a linear combination of a set of vectors $\{v_1, \ldots, v_n\}$ of V, we say that V is **spanned** by $\{v_1, \ldots, v_n\}$ and sometimes write

$$V = \text{span } \{v_1, \ldots, v_n\}$$

Often when dealing with a vector space X our attention is drawn to some subset, say W of X. If it turns out that the objects in W act like vectors under the addition and multiplication by scalars of the parent space X, then we say that W is a **subspace** of X. To check whether or not a given subset W is a subspace of X, we don't need to show that the elements of W satisfy all the axioms of a vector space given in Definition 1. Most of the axioms will automatically be satisfied by the members of W, since they are members of X, for which the axioms are known to hold. For example, for any x and x' in W we know

$$x + x' = x' + x$$

because the operation $+$ is commutative for all members of X. The only issues we have to check in deciding whether a set is a subspace are closure under addition and closure under multiplication by scalars. Thus, a nonempty subset W of a vector space X is a subspace of X if:

(i) whenever x and y belong to W, then $x + y$ belongs to W, and
(ii) if x belongs to W and k is any scalar, then kx belongs to W.

In Exercise 4 of Section 1-2 we introduced the symbol $C^k[a, b]$ to stand for the set of all real-valued functions having a continuous kth derivative on the closed interval $[a, b]$. We may now quite easily show that $C^k[a, b]$ is a vector space by making use of Theorem 1 and letting $X = K = \mathbf{R}$ and $A = [a, b]$, so that X^A is the set of all functions mapping $[a, b]$ into $\mathbf{R}$. Knowing that X^A is a vector space over $\mathbf{R}$ from Theorem 1, it is then a simple matter to show that the subset $C^k[a, b]$ of those functions which map $[a, b]$ into $\mathbf{R}$ and which also have a continuous kth derivative is a subspace of X^A. To check the closure properties (i) and (ii), we need only recall from calculus that if f and g have derivatives $\dot{f}$ and $\dot{g}$, then their sum $(f + g)$ has a derivative and $(f + g)^{\cdot} = \dot{f} + \dot{g}$; also, that if α is any real number, the function (αf) has a derivative and $(\alpha f)^{\cdot} = \alpha \dot{f}$.

This last discussion should illustrate that often we can tell that a set of objects is a vector space by showing it to be a subspace of some larger, known vector space. In Exercise 5 we provide some practice drill on recognizing subspaces.

Example 3

Consider the collection of all functions from $\mathbf{Z}$ to $\mathbf{R}$ which have the property that they are non-zero for only a finite set of integers, called the **support** of the function:

$$\text{supp}(f) = \{n | f(n) \neq 0\}.$$

Then if we define the function $\delta_j : \mathbf{Z} \to \mathbf{R}$ to be the function having as its support the single point j, where it takes the value 1:

$$\delta_j(n) = \begin{cases} 1 & \text{if } j = n \\ 0 & \text{if } j \neq n \end{cases}$$

we see that the set $\{\delta_j\}_{j \in \mathbf{Z}}$ forms a basis for this collection, since for any f we may write

$$f = \sum_{j \in \text{supp}(f)} f(j)\,\delta_j$$

since

$$\sum_{j \in \text{supp}(f)} f(j)\,\delta_j(k) = \begin{cases} f(k) & \text{if } k \in \text{supp}(f) \\ 0 & \text{if not—but then } f(k) \text{ is } 0. \end{cases} \qquad \Diamond$$

We close by noting that the collection of functions δ_j is *not* a basis for the vector space $\mathbf{R}^{\mathbf{Z}}$ of *all* functions from $\mathbf{Z}$ to $\mathbf{R}$. Why not?

To remove this problem, the notion of basis is changed somewhat when we talk of Banach spaces (to which we shall turn in the next section). Essentially, a Banach space is a vector space in which we may talk of limits. When we have limits available, we can talk of a set of linearly independent vectors being a basis, not when every vector in the space can be expressed as a finite linear combination of vectors in the basis, but when every vector in the space can be expressed as a *limit* of finite linear combinations of vectors in the basis—as in the theory of Fourier series, where we represent periodic functions as the limit of a weighted sum of suitable sines and cosines. In some cases, the limit is an infinite sum (sum of denumerably many terms), and in other cases it is an integral (sum of nondenumerably many terms).

EXERCISES FOR SECTION 2-3

1.   Here we define a "vector addition," $\oplus$, and multiplication by a scalar for the set $\mathbf{C}^2$ by the rules

$$\begin{bmatrix} x_1 \\ x_2 \end{bmatrix} \oplus \begin{bmatrix} y_1 \\ y_2 \end{bmatrix} = \begin{bmatrix} x_1 + y_1 \\ x_2 + y_2 \end{bmatrix}$$

and

$$c \begin{bmatrix} x_1 \\ x_2 \end{bmatrix} = \begin{bmatrix} cx_1 \\ 0 \end{bmatrix}$$

Check that these defined operations *almost* satisfy all the properties of a vector space, but violate one of the properties.

✔ **2.** Let K be the ternary field $(\{0, 1, 2\}, +_3, \cdot_3)$ as in Exercise 6(b) of Section 2-1, and let $X = \{a, b, c\}$. Define addition $\oplus$ on X by the following table:

$\oplus$	a	b	c
a	a	b	c
b	b	c	a
c	c	a	b

Example: $b \oplus c = a$

(a) What is the zero vector **0** for this addition? Compute $4b$, $3c$, and $3b$.

(b) What are the additive inverses $-a$, $-b$, and $-c$ for this addition?

(c) If you now try to define a multiplication of elements of X by scalars, you are trying to specify a map $f: K \times X \to X$ which is equivalent to filling in the following table:

$K \diagdown X$	a	b	c
0			
1			
2			

Fill in the table so that your definition makes X a vector space over K.

✔ **3.** The function space $C^k[a, b]$ is defined by

$$C^k[a, b] = \{f: [a, b] \to \mathbf{R} \mid f \text{ has a continuous } k\text{th derivative on } [a, b]\}$$

If f and g are two members of $C^k[a, b]$, define their sum $f + g$ as the map given by

$$(f + g)(x) = \underset{\text{(ordinary addition)}}{f(x) + g(x)} \qquad x \in [a, b]$$

and the multiplication of f by the scalar α as the map given by

$$(\alpha f)(x) = \underset{\text{(ordinary multiplication)}}{(\alpha)(f(x))} \qquad x \in [a, b]$$

We have already seen that these definitions make $C^k[a, b]$ a vector space over $\mathbf{R}$.

Show that the set $\{e^{0t}, e^{1t}, e^{2t}, \ldots, e^{nt}, \ldots\}$ of elements of $C^k[a, b]$ is *linearly independent* for all $n \in \mathbf{N}$ and hence that $C^k[a, b]$ is **infinite-dimensional** (not of finite dimension).

✔ **4.**   Check that $\{x_\alpha\}_{\alpha \in A}$ is always linearly dependent if one of the x_α's is the null vector 0.

✔ **5.**   **(a)**   Which of the following subsets are subspaces of $\mathbf{R}^3$?

(i)   All vectors of the form $\begin{bmatrix} a \\ b \\ 0 \end{bmatrix}$

(ii)   All vectors of the form $\begin{bmatrix} a \\ b \\ 2 \end{bmatrix}$

(iii)   All vectors which are perpendicular to the vector $\begin{bmatrix} 1 \\ 2 \\ 3 \end{bmatrix}$

(iv)   All vectors which are parallel to the vector $\begin{bmatrix} 1 \\ 2 \\ 3 \end{bmatrix}$

(v)   The set $\left\{ \begin{bmatrix} 0 \\ 0 \\ 0 \end{bmatrix} \right\}$ containing just the zero vector.

(vi)   The set $\left\{ x \,|\, Ax = \begin{bmatrix} 0 \\ 0 \\ 0 \end{bmatrix} \right\}$ where A is a 3×3 matrix.

(vii)   The set $\left\{ x \,|\, Ax = \begin{bmatrix} 1 \\ 2 \\ 3 \end{bmatrix} \right\}$ where A is a 3×3 matrix.

(b)   Which of the following sets are subspaces of $C^2[0, 1]$?

(i)   The set of all solutions of the homogeneous differential equation $\ddot{y} + 2\dot{y} + 3y = 0$

(ii)   The set $\{ y \,|\, \ddot{y} + 2\dot{y} + 3y = \cos(t) \}$

(iii)   The set $\{ f \,|\, f \in C^1[0, 1] \}$

(c)   Prove that the intersection of two subspaces of X is a subspace of X. Give an example showing that the union of two subspaces is not necessarily a subspace.

(d)   Let W be a subspace of X and suppose that V is a subspace of W. Prove that V is a subspace of X.

2–4   Some Useful Mathematical Spaces

In this section, we briefly survey some mathematical notions useful in the analysis of continuous-time systems. We have already seen how the general notion of a *vector space* captures the way in which the familiar two-dimensional vectors can be combined by addition or scaled up by multiplication by a scalar. In this section, we wish to provide similarly general frameworks for the study of distance. First we do this in complete generality with the idea of a *metric space*, for which we have a notion of distance between points, but do not have such operations as vector addition. We then move on to the idea of *normed linear space* in which the norm—or size measure—is fitted naturally to vector ideas. Finally, we shall see that *Banach spaces* arise naturally as normed linear spaces in which we can talk of limits and derivatives—and we shall then have the vocabulary for formalizing (in Chapter 3) the discussion of continuous-time systems sketched out in Section 2-2.

If we look at four points in the plane—a, b, c, and e as in Figure 2-23—and use $d(a, b)$ to denote the distance between a and b, we may observe relations such as the following:

$$d(a, a) = 0, \; d(a, b) > 0, \; d(a, c) = d(c, a)$$
$$d(a, b) = d(a, e) + d(e, b)$$
$$d(a, b) < d(a, c) + d(c, b)$$

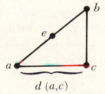

Figure 2-23 Four points and the distances between them.

$d(a,c)$

Generalizing from such considerations, we make the following definition, confident that $\mathbf{R}^2$ with the usual distance function

$$d\left(\begin{bmatrix} x_1 \\ x_2 \end{bmatrix}, \begin{bmatrix} x_1' \\ x_2' \end{bmatrix}\right) = \sqrt{(x_1 - x_1')^2 + (x_2 - x_2')^2}$$

is a special case:

DEFINITION 1

A **metric space** (X, d) is a set X with a function $d: X \times X \to \mathbf{R}$ which assigns to every pair of elements a and b of X a number $d(a, b) \geq 0$ which has the properties:

(i) $d(a, b) = 0$ iff $a = b$

(ii) $d(a, b) = d(b, a)$ (symmetry)

(iii) $d(a, b) \leq d(a, c) + d(c, b)$ (triangle inequality)

We call $d(a, b)$ the **distance between a and b** and we call d a **metric** or **distance function**. ○

It is always possible to define some (very uninteresting!) metric for a given set X:

Example 1

For any non-empty set X, define

$$d(a, b) = \begin{cases} 1 \text{ if } b \neq a \\ 0 \text{ if } b = a \end{cases}$$

Clearly this d, called the "trivial" or the **"discrete" metric,** satisfies (i), (ii), and (iii).   ◊

Since, if d is some metric on X, so is the function $3d$, we see that there are infinitely many metrics to choose from for any given set X. Often a clever choice of the metric will greatly facilitate analysis.

Once a metric or distance is defined on a set X, it is easy to talk about spheres; $S(x_0, r) = \{x \in X \mid d(x, x_0) < r\}$ is the **open sphere** of radius r centered at $x_0 \in X$, since it consists of all points of X closer than r units from x_0.

A sequence $x_1, x_2, \ldots, x_n, \ldots$ of points of X, written $\{x_n\}_1^\infty$, is said to **converge** to the limit $x \in X$ if we can get arbitrarily close to x by taking n to be sufficiently large; i.e.,

$$\lim_{n \to \infty} d(x_n, x) = 0$$

in that for every $\varepsilon > 0$ there exists $M \in \mathbf{N}$ such that $d(x_n, x) < \varepsilon$ for all $n > M$.

A sequence $\{x_n\}_1^\infty$ is a **Cauchy sequence** if we can force any pair of elements to be arbitrarily close together by taking them both far enough out in the sequence, i.e., if for every $\varepsilon > 0$, there exists $M \in \mathbf{N}$ such that whenever $n, m > M$ we have $d(x_n, x_m) < \varepsilon$.

Example 2

Let X be the *open* subinterval $(0, 2)$ of $\mathbf{R}$ and let us use the usual metric of $\mathbf{R}$, namely

$$d(x, y) = |x - y|$$

The sequence whose nth term is given by $x_n = \dfrac{1}{n}$ is easily shown to be a Cauchy sequence, since x_n and x_m get arbitrarily close for sufficiently large n and m. However, this sequence is *not* convergent since there is no point in $(0, 2)$ which x_n approaches (note that the point 0 is *not* included in $(0, 2)$. However, it would be convergent if we had taken X to be the *closed* interval $[0, 2]$ (or even $[0, 2)$).   ◊

While the last example shows that not all Cauchy sequences converge in a given metric space, it is easy to prove by the triangle inequality that all convergent sequences are Cauchy sequences [see Exercise 3].

Every now and then we come across a metric space (X, d) where every Cauchy sequence does converge, and we call such a space **complete** relative to the metric d. Example 2 showed a space which is not complete, but Exercise 4 shows how a different choice of metric makes $(0, 2)$ complete.

Just as in the calculus of real-valued functions, as soon as we have the notions of distance between points and of open sets [Exercise 2], we can define what we mean by a continuous mapping from one metric space to another. If (X, d) and (Y, σ) are two spaces and if $f : X \rightarrow Y$, we say f is **continuous at the point** x_0 in X if for all $\varepsilon > 0$ there exists $\delta > 0$ such that

$$d(x, x_0) < \delta \Longrightarrow \sigma(f(x), f(x_0)) < \varepsilon$$

i.e., if

$$\sigma(f(x), f(x_0)) \rightarrow 0 \text{ as } d(x, x_0) \rightarrow 0$$

Let us apply these ideas to prove a lemma, which at first sight seems to have nothing whatsoever to do with differential equations, but which will in fact allow us to prove existence and uniqueness theorems for these equations swiftly when, in Section 5–2, we see how to apply it. This lemma is called *the contraction lemma,* and it says that if we take any mapping of a space into itself which brings points closer together, then the mapping leaves one and only one point fixed—we say that the mapping has a unique fixed point. To think of this in rather more picturesque terms, imagine a metal disc on which we have placed a sheet of rubber which exactly covers the disc. Suppose now that we cool the disc so that the rubber sheet contracts, with any two points on the rubber sheet closer together after the cooling than before. Then the theorem we are about to prove asserts that one and only one point of the rubber sheet will be in the same place after the cooling as it was before the cooling.

Let us be given, then, a complete metric space M with metric d. We say that a function $f : M \rightarrow M$ of M into itself is a contraction if it "contracts" M in the sense that it brings points closer together—more formally, we say that f is a **contraction** if there exists some constant $K < 1$ such that for all points x and y in M we have $d(f(x), f(y)) \leq Kd(x, y)$. It is easy to show [Exercise 12] that if f is a contraction on M, then f is continuous.

With this definition, our above claim may be generalized to give the following formal statement, which we will prove shortly:

THEOREM 1

THE CONTRACTION LEMMA (*Fixed-Point Theorem for Contractions*):
    Let M be a complete metric space, with distance function d. Let $f : M \rightarrow M$ be a contraction of M with "contraction constant" $K < 1$; i.e.,

$$d(f(x), f(y)) \leq Kd(x, y) \text{ for all } x, y \text{ in } M.$$

Then f has a unique **fixed point** x_0; i.e., there is a unique point x_0 in M for which $f(x_0) = x_0$.

Moreover, for any point x in M, the sequence $\{x, f(x), f(f(x)), \ldots, f^n(x), \ldots\}$ is convergent and has as its limit the unique fixed point x_0.

Example 3

Consider radial contraction on a disc of radius R in which each point moves along its radius towards the center a fraction $K < 1$. More precisely, we use polar coordinates, so that M is the disc $\{(r, \theta) | 0 \leq r \leq R, 0 \leq \theta < 2\pi\}$ and $f: M \rightarrow M$ is defined by the equation

$$f(r, \theta) = (Kr, \theta)$$

Thus, if we apply f to the disc n times, we move (r, θ) to $(K^n r, \theta)$ and, since K is less than 1, the sequence $(K^n r, \theta)$ has the origin as its limit. Clearly, the origin is indeed the only fixed point for this contraction mapping. $\Diamond$

Proof of Theorem 1

We use the notation $f(x)$ to denote the result of a single application of the function f to x—let us then use the notation $f^n(x)$ to denote the result of n applications of f to x, so that in particular $f^{n+m}(x) = f^n(f^m(x))$ for all integers m and n greater than zero. We shall first prove that for every point x the sequence $\{f^n(x)\}$ is a Cauchy sequence, and thus, by the completeness of M, that it has a limit $\lim_{n \to \infty} f^n(x)$. Since it is quite clear [see Exercise 12] that

$$f\left[\lim_{n \to \infty} f^n(x)\right] = \lim_{n \to \infty} f[f^n(x)] = \lim_{n \to \infty} f^{n+1}(x) = \lim_{n \to \infty} f^n(x)$$

(removing the first term from a sequence does not change its limit) we see that $\lim_{n \to \infty} f^n(x)$ is in fact a fixed point of our contraction f. To conclude that the fixed point of the contraction is *unique*, we must simply show that for all points x and y of our space M we have that $\lim_{\varrho \to \infty} f^n(x) = \lim_{n \to \infty} f^n(y)$. (Consider, in particular, the case in which x itself is a fixed point.) But this latter equality is obvious, since we have for all n

$$d(f^n(x), f^n(y)) < Kd(f^{n-1}(x), f^{n-1}(y)) < \cdots < K^n d(x, y) < (1)^n d(x, y)$$

and so letting n tend to ∞ we see that the distance between $f^n(x)$ and $f^n(y)$ tends to 0, proving that $\lim_{n \to \infty} f^n(x) = \lim_{n \to \infty} f^n(y)$, as desired.

Let us then turn to the more subtle proof that $\{f^n(x)\}_1^\infty$ is indeed a Cauchy sequence, to complete the proof of our theorem. We must show that we can make $d(f^m(x), f^n(x))$ as small as desired by making m sufficiently large and then letting n take *any* value greater than m.

[Note that it is not enough to show this for some fixed value of $n > m$. We must show that all fluctuations after m are smaller than the desired amount—it might well be that after some fixed value of m the fluctuations in the next million or so terms were very small indeed, but checking this would not rule out that thereafter the fluctuations might become quite enormous, as is indicated in the sketch of Figure 2–24. In particular, it would not be enough to show that by taking m large the distance $d(f^m(x), f^{m+1}(x))$ would be small because, as n becomes much larger than m, the sum of these small "one-step" fluctuations could become very large indeed.]

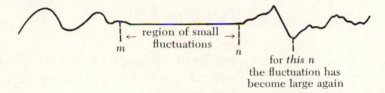

Figure 2–24

For $m < n$,

$$d(f^m(x), f^n(x)) = d(f^m(x), f^m[f^{n-m}(x)])$$
$$\leq K^m d(x, f^{n-m}(x))$$

since the result of applying the contraction f to a pair of points m times is to decrease their distance by a factor less than K^m.

Now, by repeated application of the triangle inequality, we know that it takes, if anything, less distance to go from x to $f^{n-m}(x)$ directly than it does to detour via the points

$$f(x), f^2(x), \ldots, f^{n-m-1}(x),$$

in other words,

$$d(x, f^{n-m}(x)) \leq d(x, f(x)) + d(f(x), f^2(x)) + \cdots + d(f^{n-m-1}(x), f^{n-m}(x))$$

In looking at the right-hand side of the above inequality, we see the wisdom of our detour—the two points in each term are obtained by applying f zero or more times to the pair $(x, f(x))$—and so we may use the fact that f is a contraction to rewrite our inequality in the form

$$d(x, f^{n-m}(x)) \leq d(x, f(x))[1 + K + K^2 + \cdots + K^{n-m-1}]$$

But since $K < 1$, we know that $\sum_{j=0}^{\infty} K^j$ converges, and in fact has the limit $\frac{1}{1-K}$. Since adding positive terms does not decrease the value of a series, we may substitute the infinite series for the finite one in the last inequality to get

$$d(x, f^{n-m}(x)) \leq d(x, f(x))[1 + K + K^2 + \cdots] = \frac{1}{1-k} d(x, f(x))$$

Combining this with our earlier observation that

$$d(f^m(x), f^n(x)) \leq K^m d(x, f^{n-m}(x)) \quad \text{for } m < n$$

we see that in fact

$$d(f^m(x), f^n(x)) \leq K^m \frac{d(x, f(x))}{1-K}$$

and since $K < 1$, this can be made as small as desired by making m sufficiently large, no matter what value n may take, so long as it exceeds m. $\{f^n(x)\}_1^{\infty}$ is thus a Cauchy sequence, and we are done. $\qquad\square$

Thus, as a fairly elegant exercise in our understanding of the notion of a Cauchy sequence, we have proved the contraction lemma—a lemma which has a graphic application to the cooling of rubber sheets, and which will prove invaluable in our study of differential equations in Section 5–2. For some immediate practical applications of the contraction lemma, see Exercise 14.

The fact that different distances may be defined on a given set may at first seem to be of merely academic interest. However, it actually is of very practical interest to the system theorist. For instance, a natural measure of the distance between two states of a system might be the amount of energy required to move from one to the other. The energy required to go from A to C will not exceed the energy required to go from A to B and then from B to C; the energy required will be non-negative; and under certain restrictions (corresponding to A and B having equal energy levels) the energy requirements may even be symmetric. Again, whereas in one type of control problem we may wish to analyze the system's behavior in terms of energy requirements, in another case we might wish to analyze it in terms of the minimal time in which we can bring about a certain change in the system, or in terms of the minimal amount of fuel which must be expended in bringing about a change. In each of these cases there will be systems for which these measures of time or fuel provide suitable distance measures on the state space. Thus, the choice of a distance for the state space of a system may well be dictated by physical considerations.

Let us now tie our metric ideas in with the idea of a *vector space,* or *linear space,* over a field of scalars.

Just as, in $\mathbf{R}^n$, we have a space with a "length" or "norm" $\|x\|$ for each vector x, so is any vector space X over the real or complex fields called a **normed linear space** if there is a map $\|\cdot\| : X \rightarrow \mathbf{R}_+$, called the **norm,** satisfying the three conditions:

(i) $\|x\| = 0$ iff $x = 0$, the zero vector

(ii) $\|\alpha x\| = |\alpha| \, \|x\|$ for all $x \in X$ and for all $\alpha \in K$

(iii) $\|x + y\| \leq \|x\| + \|y\|$ for all $x, y \in X$

The norm of a vector, $\|x\| \geq 0$, is a measure of the "size" or length of the vector, and properties (ii) and (iii) remind us that we can only define norms on sets X where scalar multiplication αx and addition $x + y$ make sense and for fields like $\mathbf{R}$ and $\mathbf{C}$ where the absolute value $|\alpha|$ makes sense.

In Example 4 we list several of the most commonly used norms in systems analysis for the finite dimensional vector space $\mathbf{R}^n$ and for the infinite dimensional vector space $C[a, b]$ of all real-valued continuous functions defined on the interval $[a, b]$. [See Exercise 3 of Section 2–3.]

Example 4   Some Useful Norms

Finite Dimensional

$$x = (x_1, \ldots, x_n)^T \in \mathbf{R}^n$$

$$\|x\|_\infty = \max_{1 \leq i \leq n} |x_i|$$

(the "uniform" norm)

$$\|x\|_1 = \sum_{i=1}^n |x_i|$$

(the ℓ_1 norm)

$$\|x\|_2 = \left(\sum_{i=1}^n |x_i|^2 \right)^{1/2}$$

(the ℓ_2 norm)

$$\|x\|_p = \left(\sum_{i=1}^n |x_i|^p \right)^{1/p}$$

(the ℓ_p norm)

Infinite Dimensional

$$f \in C[a, b]$$

$$\|f\|_\infty = \max_{a \leq t \leq b} |f(t)|$$

$$\|f\|_1 = \int_a^b |f(t)| \, dt$$

$$\|f\|_2 = \left(\int_a^b |f(t)|^2 \, dt \right)^{1/2}$$

$$\|f\|_p = \left(\int_a^b |f(t)|^p \, dt \right)^{1/p}$$

Since there are many choices of norm available for a space, it is often a critical problem of system analysis to decide which norm to use. For example, on $C[a, b]$ the ℓ_2 norm can be thought of as a measure of the energy of a signal

$$\|f\|_2 = \left(\int_a^b |f(t)|^2 \, dt \right)^{1/2}$$

and the uniform norm,

$$\|f\|_\infty = \max_{a \le t \le b} |f(t)|$$

as the peak or greatest excursion of a signal vector f. Clearly, whether a vector f is "larger" or "smaller" than the vector g depends on the norm being used. To see this, consider the two vectors f and g in $C[0, 1]$ whose graphs are shown in Figure 2–25. Then $\|f\|_\infty = \|g\|_\infty$ but $\|f\|_2 < \|g\|_2$.   ◊

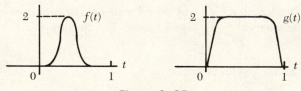

Figure 2–25

Thus, a sequence which converges when one norm is used may have a different limit or no limit at all under a different norm. It is a happy and remarkable fact of mathematics*, however, that in *finite-dimensional* spaces, all norms yield the same limits. The reader should verify for example that if any norm on $\mathbf{R}^n$ in Example 4 is small, then all are small.

We have now seen how a given set X may be endowed with a notion of distance by properly defining a metric $d: X \times X \to \mathbf{R}_+$ or with a notion of size by making X a vector space and suitably defining a norm $\|\cdot\| : X \to \mathbf{R}_+$. Thus, if we start out with X already a vector space, it may be possible to define a metric and a norm completely independently of each other [see Exercise 6].

However, once a norm is specified we can always define a metric, the **metric induced by the norm,** by the rule

$$d(x, y) = \|x - y\|$$

which sets the distance between two vectors equal to the length of their difference as in Figure 2–26. [See Exercises 7, 8, and 9.]

---

*See Epstein, B., Linear Functional Analysis, W. B. Saunders Co., 1970, page 83.

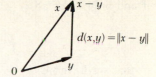

Figure 2–26   The metric induced by the norm
$\|\cdot\|: X \to \mathbf{R}_+$.

Thus, we can extend to normed linear spaces the notions of spheres, open sets, Cauchy sequences, convergence, completeness, continuity of mappings, and so forth, as defined earlier for metric spaces.

A normed vector space which is complete with respect to the metric induced by the norm is called a **Banach space.** If our state space X is a Banach space we will then be able to add vectors, multiply vectors by scalars, talk about lengths of vectors and distances between vectors, and take limits of sequences in X and of functions mapping X into or from other Banach spaces. This is a great deal of mathematical structure to impose on X, but it enables us to do calculus and to use differential equations to describe the local behavior of dynamic systems.

Thus, if $x: T \to \mathbf{R}^n$, we may specify x in terms of n real-valued time functions by

$$x(t) = (x_1(t), x_2(t), \dots, x_n(t))^T$$

We will now show that limits, derivatives, and integrals of $x(t)$ may all be reduced to performing these operations on the components $x_1(t), x_2(t), \dots, x_n(t)$.

Example 5

We show here that the limit

$$\lim_{t \to t_0} x(t) = \left(\lim_{t \to t_0} x_1(t), \dots, \lim_{t \to t_0} x_n(t) \right)^T$$

and leave differentiation and integration as exercises [see Exercise 10]. If we let $a = (a_1, a_2, \dots, a_n)^T$ we have that

$$\|x(t) - a\| = \sqrt{|x_1(t) - a_1|^2 + \cdots + |x_n(t) - a_n|^2} \text{ for all } t \in T$$

so that $x(t)$ close to a implies $x_i(t)$ close to a_i and vice versa. Thus,

$$\lim_{t \to t_0} (x_1(t), \dots, x_n(t))^T = (a_1, \dots, a_n)^T$$

if and only if for all i

$$\lim_{t \to t_0} x_i(t) = a_i$$

so that we have

$$\lim_{t \to t_0} (x_1(t), \ldots, x_n(t))^T = \left(\lim_{t \to t_0} x_1(t), \ldots, \lim_{t \to t_0} x_n(t) \right)^T$$

Similarly,

$$\frac{d}{dt} (x_1(t), \ldots, x_n(t))^T = \left(\frac{dx_1}{dt}(t), \ldots, \frac{dx_n}{dt}(t) \right)^T$$

and

$$\int_{t_0}^{t_1} (x_1(\xi), \ldots, x_n(\xi))^T \, d\xi = \left(\int_{t_0}^{t_1} x_1(\xi) \, d\xi, \ldots, \int_{t_0}^{t_1} x_n(\xi) \, d\xi \right)^T \qquad \lozenge$$

At this point the reader should work Exercise 11, which extends our component-by-component analysis to include Laplace transforms of vector functions. (The reader not yet acquainted with Laplace transforms can find them in Section 6–3.)

We have gone off on quite a tangent, cataloging the mathematical notions sufficient to define derivatives of state trajectories, and we will now return to the business of obtaining local descriptions of continuous-time systems.

In Section 4–4, when we need it, we will introduce the notion of inner products on vector spaces and still another type of space, the Hilbert space.

From here on, then, we shall adopt the following strategy. Whenever a result is true for *any* Banach space, we shall state it as a property of Banach spaces; we will do the same for vector spaces and Hilbert spaces. Moreover, we shall take care to formulate proofs which use only properties common to all spaces of the type we are considering. However, the proof will be completely intelligible to the reader who does *not* know the general theory, for he will recognize at every step of the proof that the properties we use hold in $\mathbf{R}^n$ or $\mathbf{C}^n$.

Thus, the reader may choose either of two strategies:

(1) In reading the rest of the book, mentally substitute $\mathbf{R}^n$ or $\mathbf{C}^n$ for any of the three phrases "vector space," "Banach space," or "Hilbert space."

(2) In reading the rest of the book, notice that many of the results hold true for infinite-dimensional systems. When we prove a result only for finite-dimensional systems, ask yourself if it is possible to generalize it.

Many readers may find it useful to adopt strategy (1) on a first reading, and then use strategy (2) on a later reading in preparation for more advanced study of mathematical system theory, for much of which Hilbert spaces and Banach spaces provide the appropriate setting. In the optimization theory of Section 7–6, the Banach space approach is decisive.

We close this section by noting the distinction between the integration of a given vector function which we described above, and which proceeds in a component-by-component fashion, and the "integration" of a vector differential equation in which, rather than having the function given to us for all time and then simply integrating its components, we now have only past values of the function, and must use these data to find a continuation of the function whose derivative will in fact meet the conditions stipulated by $\dfrac{dx(t)}{dt} = f(t, x(t), u(t))$ (Figure 2–27).

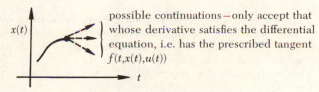

possible continuations – only accept that whose derivative satisfies the differential equation, i.e. has the prescribed tangent $f(t,x(t),u(t))$

Figure 2–27

In Example 5 it proved so easy to transfer the concepts of limit, derivative, and integral from one dimension to n dimensions that the reader might expect that solving differential equations in $\mathbf{R}^n$ might require an equally simple generalization. Unfortunately, it is *not* so easy. Let us write down, again, the differential equation we expect to use to describe continuous-time systems—it now being the case that the state-space of the system is $\mathbf{R}^n$.

$$\frac{dx(t)}{dt} = f(t, x(t), u(t))$$

Just as in any electrical circuit we would expect wires to interconnect the different subsystems which yield the state variables, we expect the change of any one component to depend on all the other components. In other words, if we decompose the above equation component by component to obtain

$$\frac{dx_j(t)}{dt} = f_j(t, x(t), u(t))$$

we must expect the function f_j, which governs the change of x_j, to depend on all the components $x_1(t), \ldots, x_n(t)$ and not simply on $x_j(t)$ alone. Thus, we cannot integrate the n differential equations for the different components independently. If we use our knowledge of x at t to extend the solution to some slightly later time $t + \tau$, then to extend the solution in any one component requires knowledge of the new values of all n components. It is this interdependence which will compel us to study, in Section 5–2, the theory of differential equations for the multivariable case, since we cannot employ a simple component-by-component analysis.

EXERCISES FOR SECTION 2–4

✔ **1.**   **(a)**   Using the discrete metric of Example 1 for the set $X = \mathbf{R}$, find $S(3, \frac{1}{2})$, $S(3, 1)$, and $S(3, 2)$

(b)   For $X = \mathbf{R}$, define a metric d by $d(x, y) = |x - y|$, the absolute value of the difference of x and y, the so-called **usual metric** for $\mathbf{R}$. Check that d satisfies conditions (i), (ii), and (iii) of Definition 1 and hence is a metric for $\mathbf{R}$.

(c)   Find for the metric d the open spheres $S(3, \frac{1}{2})$, $S(3, 1)$, and $S(3, 2)$ and compare with those of part (a).

2.   A set M in a metric space (X, d) is **open** if every point in M is the center of some open sphere (perhaps very small) which lies entirely in M. Show that $S(x_0, r)$ is itself an open set. Show that $[0, 2)$ is not an open subset of $\mathbf{R}$ but that $(0, 2)$ is open in $\mathbf{R}$ under the usual metric (Exercise 1(b)) for $\mathbf{R}$.

✔ **3.**   Prove that every convergent sequence $\{x_n\}_1^\infty$ in a metric space (X, d) is a Cauchy sequence.

4.   For any set X with the discrete metric of Example 1, show that every Cauchy sequence is of the form

$$a_1, \ldots, a_n, a, a, a, \ldots$$

for some $a \in X$ and hence is convergent to some point (namely a) of X. Thus (X, d) is complete relative to this metric d.

5.   For (X, d) as in Exercise 4, let (Y, σ) be any metric space and let $f : X \to Y$ be any map from X to Y. Prove that f is continuous.

6.   In Exercise 2 of Section 2–3, you constructed a vector space X over the ternary field K. To make X into a normed linear space by suitably defining $\| \cdot \| : X \to \mathbf{R}_+$ you must be able to fill in the following table with suitable non-negative numbers.

x	$\|x\|$
a	
b	
c	

(a)   Choose some numbers and check whether the norm axioms are satisfied.

(b)   Repeat for the choice $\|a\| = 0$, $\|b\| = \frac{1}{2}$, $\|c\| = 1$.

(c)   Let d be the discrete metric for X and notice that this d has nothing at all to do with your choices of norms in (a) and (b).

(d)   Now define a metric $\widehat{d}$ induced by the norm of part (a), $\widehat{d}(x, y) = \|x - y\|$, and fill in the table for $\widehat{d}$:

y x	a	b	c
a			
b			
c			

Compare with part (c).

✔ **7.**   **(a)**   Verify that the "metric" induced by the norm does indeed satisfy the three conditions for a metric.

   (b)   Give the metric induced by the norm for each entry in the table of Example 4.

8.   Suppose a metric d is specified for a vector space X.

   (a)   Can we induce a norm by the formula $\|x\| = d(x, 0)$ which takes the size of x as the distance of x away from the zero vector (the "origin")?

   (b)   Try to construct such a norm on the vector space of Exercise 6 using the discrete metric of 6(c).

   (c)   Repeat part (b) using the metric you found in Exercise 6(d). Can you generalize?

✔ **9.**   Sketch the unit sphere centered at the origin, $S(\begin{bmatrix}0\\0\end{bmatrix}, 1)$, for $\mathbf{R}^2$ for the metrics induced by the uniform norm, the ℓ_1 norm, and the ℓ_2 norm.

10.   Using the result for limits in Example 5, prove that

$$\frac{d}{dt}(\ldots, x_i(t), \ldots)^T = \left(\ldots, \frac{dx_i}{dt}(t), \ldots\right)^T$$

and

$$\int_{t_0}^{t_1}(\ldots, x_i(\xi), \ldots)^T \, d\xi = \left(\ldots, \int_{t_0}^{t_1} x_i(\xi) \, d\xi, \ldots\right)^T$$

✔ **11.**   Let $\mathcal{L}\{f(t)\}$ denote the Laplace Transform of f, and show that

$$\mathcal{L}\{(x_1(t), \ldots, x_n(t))^T\} = (\mathcal{L}\{x_1(t)\}, \ldots, \mathcal{L}\{x_n(t)\})^T$$

Use this result to Laplace Transform the system

$$\dot{x}(t) = \frac{d}{dt}\begin{bmatrix}x_1(t)\\x_2(t)\end{bmatrix} = \begin{bmatrix}-2 & -1\\1 & 0\end{bmatrix}\begin{bmatrix}x_1(t)\\x_2(t)\end{bmatrix} + \begin{bmatrix}1\\0\end{bmatrix}u(t)$$

and solve the resulting transform equation for $X(s)$. Use the inverse transform to get the solution $x(t)$. Compare with Exercise 1 of Section 1–3.

✔ **12.**   Let $f: M \to M$ be a contraction map on the metric space M.

   (a)   Prove that f is continuous on M.

   (b)   Let $x_0 = \lim_{n \to \infty} x_n$ where $\{x_n\}_1^\infty$ is any convergent sequence in M. Show that $f(x_0) = f\left(\lim_{n \to \infty} x_n\right) = \lim_{n \to \infty} f(x_n)$.

✔ **13.**   We have seen that for a normed vector space $(X, \|\cdot\|)$ over the field of complex (or real) numbers, the norm $\|\cdot\|$ is a mapping from X into $\mathbf{R}$

$$\|\cdot\| : X \to \mathbf{R}$$

(a)   Prove that $\|\cdot\|$ is continuous on X with respect to the metric $d(x, y) = \|x - y\|$.

(b)   Show that $\lim_{n \to \infty} \|x_n\| = \left\| \lim_{n \to \infty} x_n \right\|$ for any convergent sequence $\{x_n\}_1^\infty$ in X.

✔ **14.**   This exercise illustrates how to turn two familiar numerical problems into "fixed point" problems.

(a)   Suppose we want to find $\sqrt{3}$ numerically. We must solve $x^2 = 3$ for x. Adding x^2 to both sides, we get $2x^2 = x^2 + 3$ which, upon division by $2x$, yields:

$$x = \frac{1}{2}\left(x + \frac{3}{x}\right)$$

Now our answer x can be thought of as a fixed point of the map $f(t) = \frac{1}{2}\left(t + \frac{3}{t}\right)$. Using $d(a, b) = |a - b|$, show that

$$d(f(a), f(b)) = \frac{1}{2}\left|1 - \frac{3}{ab}\right| d(a, b) \text{ for all } a, b \in \mathbf{R}$$

and that f is contractive on the interval $[1, 2]$. Starting with the initial guess $\xi = 1$, generate the sequence $\xi, f(\xi), f(f(\xi)), \ldots, f^n(\xi), \ldots$ to find an approximation to $\sqrt{3}$.

(b)   Suppose that, for $x \in \mathbf{R}, f(x) = Fx + G$ where F is a known n by n matrix and G is a known n by 1 matrix. Use the ℓ_2 norm on $\mathbf{R}^n$ and the metric $d(x, y) = \|x - y\|_2$ and show that

$$d(f(x), f(y)) \leq K d(x, y)$$

where

$$K = \sqrt{\sum_{i=1}^n \|F_i\|_2^2} = \sqrt{\sum_{i=1}^n \sum_{j=1}^n (F_{ij})^2}$$

[*Hint:* Let F_i, $[F(x - y)]_i$, x_i, and y_i denote the ith rows of F, $F(x - y)$, x and y respectively and let $F_i \cdot x$ be the ordinary dot product of the n-tuples F_i and x. Since $|F_i \cdot x| \leq \|F_i\|_2 \|x\|_2$ (Cauchy-Schwartz inequality),

$$\|f(x) - f(y)\|_2 = \|(Fx + G) - (Fy + G)\|_2 = \|F(x - y)\|_2$$

$$= \sqrt{\sum_{i=1}^n [F(x - y)]_i^2} = \sqrt{\sum_{i=1}^n [F_i \cdot (x - y)]^2}$$

$$\leq \sqrt{\sum_{i=1}^n (\|F_i\|_2^2 \|x - y\|_2^2)} = \sqrt{\|x - y\|_2^2 \sum_{i=1}^n \|F_i\|_2^2}$$

$$\leq \left(\sqrt{\sum_{i=1}^n \|F_i\|_2^2}\right) \|x - y\|_2 \right]$$

(c)   Solving the matrix equation $Ax = b$ for $x \in \mathbf{R}^n$ can be posed as a fixed point problem of the function $f : \mathbf{R}^n \to \mathbf{R}^n$ by $f(x) = Fx + G$. To see how matrices F and G are found from A and b, write $Ax = b$ as

$$a_{11}x_1 + a_{12}x_2 + \cdots + a_{1n}x_n = b_1$$
$$a_{21}x_1 + a_{22}x_2 + \cdots + a_{2n}x_n = b_2$$
$$\vdots \qquad \vdots \qquad\qquad \vdots \qquad \vdots$$
$$a_{n1}x_1 + a_{n2}x_2 + \cdots + a_{nn}x_n = b_n$$

Dividing the ith equation by a_{ii} and solving it for x_i, show that $Ax = b$ can be rewritten in matrix form:

$$
\begin{bmatrix} x_1 \\ x_2 \\ \cdot \\ \cdot \\ \cdot \\ x_n \end{bmatrix}
=
\begin{bmatrix}
0 & -\dfrac{a_{12}}{a_{11}} & \cdots & -\dfrac{a_{1n}}{a_{11}} \\
-\dfrac{a_{21}}{a_{22}} & 0 & \cdots & -\dfrac{a_{2n}}{a_{22}} \\
\cdot & \cdot & \cdot & \cdot \\
\cdot & \cdot & \cdot & \cdot \\
\cdot & \cdot & \cdot & \cdot \\
-\dfrac{a_{n1}}{a_{nn}} & -\dfrac{a_{n2}}{a_{nn}} & \cdots & 0
\end{bmatrix}
\begin{bmatrix} x_1 \\ x_2 \\ \cdot \\ \cdot \\ \cdot \\ x_n \end{bmatrix}
+
\begin{bmatrix} \dfrac{b_1}{a_{11}} \\ \dfrac{b_2}{a_{22}} \\ \cdot \\ \cdot \\ \cdot \\ \dfrac{b_n}{a_{nn}} \end{bmatrix}
$$

where a_{ij} is the ijth entry of A and $b = (b_1, \ldots, b_n)^T$. This gives $x = Fx + G$. Observe that if the A matrix has a_{ii} as the largest entry in the ith row (**diagonal dominance**), the rows F_i of F will be small and $f(x) = Fx + G$, as in part (b), may be contractive.

Illustrate by solving $\begin{bmatrix} 5 & 2 \\ 1 & 3 \end{bmatrix}\begin{bmatrix} x_1 \\ x_2 \end{bmatrix} = \begin{bmatrix} 12 \\ 5 \end{bmatrix}$ by starting with $\xi = \begin{bmatrix} 1 \\ 1 \end{bmatrix}$ and gen-

erating $\xi, f(\xi), f(f(\xi)), \ldots$. Repeat with $\xi = \begin{bmatrix} 0 \\ 0 \end{bmatrix}$. What happens if you try this

on $\begin{bmatrix} 1 & 3 \\ 5 & 2 \end{bmatrix}\begin{bmatrix} x_1 \\ x_2 \end{bmatrix} = \begin{bmatrix} 5 \\ 12 \end{bmatrix}$, which you know is an *equivalent* system?

15.   A subset P of a metric space (X, d) is said to be **closed** if its complement $X - P$ is open as defined in Exercise 2. A point $b \in X$ is called a **limit point** of set P if there is some sequence $\{p_n\}_1^\infty$ of points of P such that $b = \lim_{n \to \infty} p_n$.

(a)   Prove that P is closed iff P contains all of its limit points.
(b)   Under the usual metric for $\mathbf{R}$, show that $[0, 2)$ is not a closed subset of $\mathbf{R}$ but that $[0, 2]$ is closed in $\mathbf{R}$.

2-5   Physical Systems and State Assignment

To make the ideas of Section 2-2 more explicit, and to introduce the reader to systems with multi-dimensional state-spaces, we shall show how the systems of Newtonian mechanics may be placed in a state-variable framework.

    This demonstration has another important purpose. The system theory we develop in this book enables us to study properties such as controllability, observability, and stability for a system represented in mathematical form.

It also shows us how to replace one mathematical representation by another more suited to analysis. But—apart from this section—we pay little attention to the question, "Given a real-world system—be it chemical, electrical, economic, ecological, mechanical, or biological—how do we obtain a mathematical representation to which we may apply the techniques of system theory?" Thus, the reader aiming for applicability rather than mathematical diversion is urged to ensure that he complements his study of system theory with a study of the modelling techniques which will enable him to render the real-world systems of interest to him into a form amenable to our methods of analysis.

In this section we make use of matrix notation. Presumably, most readers are familiar with this, especially those who have been working through the exercises; but for the novice let us just remark that a matrix equation of the form

$$Ax = b$$

where A is the matrix $\begin{bmatrix} a_{11} & a_{12} & \cdots & a_{1n} \\ \vdots & \vdots & \cdots & \vdots \\ a_{m1} & a_{m2} & \cdots & a_{mn} \end{bmatrix}$, x is the vector $\begin{bmatrix} x_1 \\ \vdots \\ x_n \end{bmatrix}$, and b is the vector $\begin{bmatrix} b_1 \\ \vdots \\ b_m \end{bmatrix}$, is simply a compact way of writing the m ordinary equations:

$$a_{11}x_1 + \cdots + a_{1n}x_n = b_1$$
$$\vdots \qquad\qquad \vdots$$
$$a_{m1}x_1 + \cdots + a_{mn}x_n = b_m$$

which have the alternative representation

$$\sum_{j=1}^{n} a_{ij}x_j = b_i \text{ for } i = 1, \ldots, m$$

We shall return to the explicit study of various properties of matrices in Section 3–3; for now, let us just use them. For example, when we write

$$\begin{bmatrix} 0 & 1 \\ 0 & 0 \end{bmatrix}\begin{bmatrix} x_1 \\ x_2 \end{bmatrix} + \begin{bmatrix} 0 \\ -g \end{bmatrix}$$

we multiply the matrix by the x vector to rewrite this as

$$\begin{bmatrix} 0 \cdot x_1 + 1 \cdot x_2 \\ 0 \cdot x_1 + 0 \cdot x_2 \end{bmatrix} + \begin{bmatrix} 0 \\ -g \end{bmatrix}$$

and then use our component-by-component definition of vector addition to obtain the result

$$\begin{bmatrix} x_2 \\ -g \end{bmatrix}$$

In this case, the matrix notation may seem to complicate rather than simplify; but the virtue of this representation is that it allows powerful generalizations, as we shall also see in Section 3-4.

With this as background, then, let us turn our attention to physical systems. The reader with a background in circuit theory will find that the exercises to this section will provide him with many opportunities to practice rendering circuits into state-variable representations.

Example 1

One of the simplest Newtonian systems is a particle falling under constant gravity (Figure 2-28).

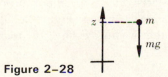

Figure 2-28

If z is the vertical coordinate, we know that the motion obeys the law $m\ddot{z} = -mg$ which has the general solution

$$z(t) = \frac{1}{2} g t^2 + c_1 t + c_2$$

where the constants c_1 and c_2 are determined by the initial conditions. This is a "free" system in that there is no externally controllable input. To put it in state-variable form, we need a state-vector x for which we can find a dynamic description

$$\dot{x}(t) = f(t, x(t), u(t))$$

which here reduces to

$$\dot{x}(t) = f(x(t))$$

since the system is input-free and is clearly time-invariant.

Certainly, we cannot express the velocity $\dot{z}$ as a function of z alone, but if we take both z and $\dot{z}$ as state-components

$$x = \begin{bmatrix} z \\ \dot{z} \end{bmatrix}$$

then we may write

$$\dot{x} = \begin{bmatrix} \dot{z} \\ \ddot{z} \end{bmatrix} = \begin{bmatrix} \dot{z} \\ -g \end{bmatrix}$$

so that we may express our system-dynamics in matrix form by

$$f\left(\begin{bmatrix} x_1 \\ x_2 \end{bmatrix}\right) = \begin{bmatrix} 0 & 1 \\ 0 & 0 \end{bmatrix}\begin{bmatrix} x_1 \\ x_2 \end{bmatrix} + \begin{bmatrix} 0 \\ -g \end{bmatrix}$$

where in our stationary f there is no variable input that rates explicit mention.
◊

This device of augmenting an output variable by a number of its derivatives to obtain a state-vector is an important one. To see this in a very general form, let us consider a system (Figure 2–29) specified by fixing two functions g and h such that we always have the relation between input and output

$$g(u(t), \dot{u}(t), \ldots, u^{(n)}(t), t) = h(y(t), \dot{y}(t), \ldots, y^{(m)}(t), t) \tag{1}$$

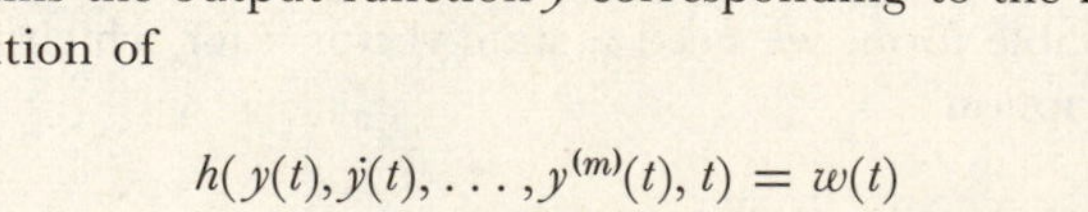

$$U \xrightarrow{u(t)} \boxed{} \xrightarrow{y(t)} Y$$

Figure 2–29

which constrains the output function y corresponding to the input function u to be a solution of

$$h(y(t), \dot{y}(t), \ldots, y^{(m)}(t), t) = w(t)$$

where the function w is obtained from a specified input u by applying g.†
Note that if h were the function which is identically zero, any y could be associated with any u. To restore some notion of causality, we demand

---

† As usual, $y^{(m)}(t)$ denotes the mth derivative of y with respect to t.

that, given $a_1, a_2, \ldots, a_m$, b, and t, the equation

$$h(a_1, a_2, \ldots, a_m, a_{m+1}, t) = b$$

be uniquely solvable for a_{m+1}—say, as the function $\phi_m(a_1, a_2, \ldots, a_m, b, t)$ of the specified values. This is a causality condition in that, if it were not to hold, different values of $y^{(m)}(t)$ would be consistent with given values of $y(t), \ldots, y^{(m-1)}(t)$ and the input.

In the light of the above discussion, it now seems plausible to introduce m components of a state vector by the assignments

$$x_1 = y$$
$$x_2 = \dot{y}$$
$$\vdots$$
$$x_m = y^{(m-1)}$$

In terms of this choice of state variables we can provide an appropriate state-transition equation:

$$
\begin{bmatrix} \dot{x}_1 \\ \dot{x}_2 \\ \vdots \\ \dot{x}_m \end{bmatrix}
=
\begin{bmatrix} x_2 \\ x_3 \\ \vdots \\ \phi_m(x(t), w(t), t) \end{bmatrix}
$$

Thus we see that all we need to know about the history of the input to a system specified by equation (1) can be summed up in a state vector comprising the present output and its first $(m - 1)$ derivatives.

Note well that if the output y is itself a vector, then each component x_j is a vector of the same dimension as y, so that our construction applied to a system with output space of dimension p will yield a state space of dimension mp.

Example 2

Let us apply this technique to the lossless mass-spring oscillator of Figure 2–30. Here the displacement coordinate y obeys the differential equation

$$M\ddot{y} = u - ky$$

where $k \geq 0$ is the spring constant, M is the mass of the cart, and u is the applied force. This equation is of the form

$$g(u) = h(y, \dot{y}, \ddot{y})$$

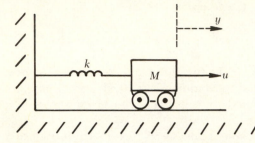

Figure 2–30 A mechanical oscillator.

where

$$g(u) = u$$

and

$$h(y, \dot{y}, \ddot{y}) = M\ddot{y} + ky$$

and h clearly satisfies our causality condition. Thus, we may write

$$x_1 = y$$
$$x_2 = \dot{y}$$

to obtain

$$\dot{x}_1 = x_2 \quad \text{and} \quad \dot{x}_2 = -\omega^2 x_1 + u/M$$

where

$$\omega^2 = k/M$$

Re-writing this in matrix form,

$$\begin{bmatrix} \dot{x}_1 \\ \dot{x}_2 \end{bmatrix} = \begin{bmatrix} 0 & 1 \\ -\omega^2 & 0 \end{bmatrix} \begin{bmatrix} x_1 \\ x_2 \end{bmatrix} + \begin{bmatrix} 0 \\ 1/M \end{bmatrix} u \qquad \Diamond$$

To the reader experienced with analog computers and systems simulation, the "solving" of a differential equation

$$h(y, \dot{y}, \ldots, y^{(m)}) = g(u, \dot{u}, \ldots, u^{(n)})$$

for the highest derivative $y^{(m)}$ in terms of u and lower derivatives of y is a familiar one. We illustrate with an example.

Example 3

Suppose a system has its input u and output y related by the third order linear differential equation.

$$y^{(3)}(t) + 2y^{(2)}(t) + 3\dot{y}(t) + 4y(t) = u(t)$$

Solving for the highest derivative of y,

$$y^{(3)}(t) = -2y^{(2)}(t) - 3\dot{y}(t) - 4y(t) + u(t)$$

Now if we had $y^{(3)}(t)$ available, we could put it into an integrator and get $y^{(2)}(t)$ at the output of the integrator of Figure 2–31.

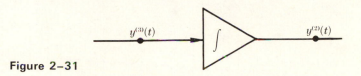

Figure 2–31

Connecting $y^{(2)}(t)$ to the input of another integrator yields $\dot{y}(t)$, and feeding this to a third integrator, as in Figure 2–32, yields $y(t)$, the desired solution of the differential equation.

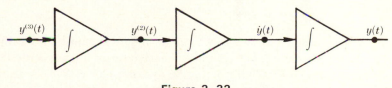

Figure 2–32

But if $y^{(2)}(t)$, $\dot{y}(t)$, $y(t)$ are now available as integrator outputs, we can use them along with u to construct $y^{(3)}(t)$ from the expression

$$y^{(3)}(t) = -2y^{(2)}(t) - 3\dot{y}(t) - 4y(t) + u(t)$$

This construction of $y^{(3)}(t)$ is easily accomplished on the analog computer, using three scalar multipliers and one summer to yield the circuit shown in Figure 2–33.

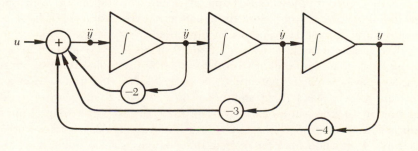

Figure 2–33

If we set the integrator outputs to prescribed initial values $y(0)$, $\dot{y}(0)$, and $\ddot{y}(0)$ at $t = 0$, these values get fed back through the various gains and added to the input, the integrators integrate, and the system perks away, actually generating the solution y as its output. Notice that besides just the solution y, we also get its derivatives $\dot{y}$, $\ddot{y}$, and $\dddot{y}$ free of charge by merely picking off the appropriate signals from the diagram.

The local description of the analog system is given by the matrix equations

$$\begin{bmatrix} \dot{x}_1 \\ \dot{x}_2 \\ \dot{x}_3 \end{bmatrix} = \begin{bmatrix} 0 & 1 & 0 \\ 0 & 0 & 1 \\ -4 & -3 & -2 \end{bmatrix} \begin{bmatrix} x_1 \\ x_2 \\ x_3 \end{bmatrix} + \begin{bmatrix} 0 \\ 0 \\ 1 \end{bmatrix} u$$

$$y = \begin{bmatrix} 1 & 0 & 0 \end{bmatrix} \begin{bmatrix} x_1 \\ x_2 \\ x_3 \end{bmatrix}$$

which may be written by inspection from the wiring diagram of Figure 2–33, taking $x_1 = y$, $x_2 = \dot{y}$, and $x_3 = \ddot{y}$. $\Diamond$

In the last example we saw how to go from a given input-output differential equation to an analog computer realization of that equation. By choosing the outputs of the integrators (the dynamic elements of the realization) as the state variables and labelling them x_1, x_2, x_3, and so on, we are then able to write (by looking at the signals present at the input of each integrator) a set of first-order scalar differential equations giving the derivative of each state variable in terms of the state variables and inputs. We then have no difficulty in rewriting the set of scalar differential equations as an equivalent pair of matrix differential equations of the form

$$\dot{x}(t) = F(t)x(t) + G(t)u(t)$$

$$y(t) = H(t)x(t)$$

which we shall denote by the three matrices (F, G, H).

This procedure provides a simple algorithm for coming up with both an analog realization and a matrix dynamic representation (F, G, H) of an input-output differential equation. By now, the reader who has been working the Exercises (especially Exercises 7 and 10 of Section 2–1 and Exercise 3 of this section) is also familiar with the problem of reversing this procedure; namely, given a matrix dynamic equation (F, G, H), first finding a circuit diagram which obeys those equations and, second, finding an equation relating input and output directly. The first part is rather simple and is outlined in Exercise 6 of Section 1–1 for continuous-time systems and in Example 4 of Section 2–1 for discrete-time systems, so that going back and forth between matrix equations and block diagrams is straightforward. The second part is a bit more involved, and different approaches to the problem of deriving input-output differential (or difference) equations are presented in Exercise 7 of Section 2–1, Example 7 of Section 3–2, and Exercise 13 of Section 6–3; later on, when we discuss *transfer functions* in Section 4–6 and Section 6–3 and *block diagram manipulations* and *Mason's Rule* in Section 6–4, we shall see still other simple methods of accomplishing this.

When the input-output differential equation involves derivatives of the input, it is not always possible to choose $y, \dot{y}, \ddot{y}, \ldots, y^{(m-1)}$ as state variables. The reader should work Exercises 4 and 6 to see two possible choices of state variables and their analog computer realizations which circumvent the difficulties presented by the presence of $u, \dot{u}, \ddot{u}, \ldots u^{(m-1)}$.

Newton's postulation of $F = ma$ as the fundamental equation of motion leads to the idea that any system with a finite number n of degrees of freedom may be specified by a vector

$$q = \begin{bmatrix} q_1 \\ q_2 \\ \vdots \\ q_n \end{bmatrix}$$

of generalized coordinates in such a way that the generalized acceleration $\ddot{q}$ depends only on the q's, on their derivatives, and on an externally applied forcing term $F(t)$:

$$\ddot{q} = \alpha(q, \dot{q}, F(t))$$

Applying our procedure to this system, we obtain a $2n$-dimensional state vector

$$x = \begin{bmatrix} x_1 \\ x_2 \end{bmatrix} = \begin{bmatrix} q \\ \dot{q} \end{bmatrix}$$

whose evolution is specified by the equation

$$\dot{x} = \begin{bmatrix} x_2 \\ \alpha(x_1, x_2, F(t)) \end{bmatrix}$$

where we may regard F as an input term. Thus, we specify the state of a Newtonian system by specifying the positions and velocities (or, equivalently, the momenta) of its constituent parts.

It is rather useful to be able to write equations which are valid irrespective of whether the components are scalars or vectors, just as it is better to write

$$Ax = b \text{ has solution } x = A^{-1}b \text{ if } A \text{ is invertible}$$

than to try to juggle piecemeal with the solution of such notational conglomerations as

$$ax + by + cz = d$$
$$cx + fy + gz = h$$
$$ix + jy + kz = l$$

In fact, as the following example illustrates, the state space X may not even be finite-dimensional, so that "component-by-component" analysis may not even be possible.

Example 4

Often, because of delays in transmission, a delayed version of a signal may be fed back to a subsystem, as in Figure 2–34.

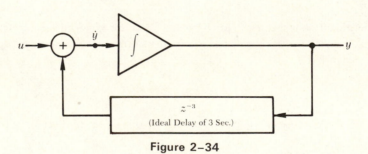

Figure 2–34

The input to the integrator is given by

$$\dot{y} = u + z^{-3}y$$

so that the system is described by the linear *differential-difference* equation*

$$\dot{y}(t) - y(t - 3) = u(t)$$

Reasoning intuitively, we see that the condition of the delay line, which contains the previous three seconds worth of $y(t)$, determines $\dot{y}(t)$, the rate

*See Bellman, R., and Cooke, K. L., *Differential-Difference Equations,* Academic Press, New York, 1963 or Norkin, S. B., *Differential Equations of the Second Order with Retarded Argument,* American Mathematical Society, Translations of Mathematical Monographs, Vol. 31, 1972 for a discussion of such equations.

of change of y for $t > t_0$. Thus, our choice for the state of the system at time t_0 is

$$x(t_0) = y_{[t_0-3, t_0)}$$

and the state space X is just $C^1[0, 3]$, the space of all differentiable real-valued functions defined on the real interval $[0, 3]$ [see Exercise 3 of Section 2–3], which is an infinite dimensional space.   ◊

STATE TRAJECTORIES IN PHASE SPACE

Let us consider in more detail the systems of Examples 1 and 2. In each case the state space X is two-dimensional, with position x_1 as one coordinate and velocity x_2 as the other, and $x(t)$ may be plotted versus t in the 3-space, $X \times T$, as illustrated in Figure 2–35.

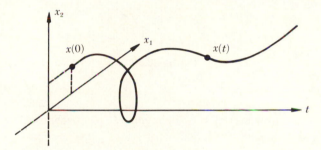

Figure 2–35

By "looking in" along the t axis, we may project the curve $x(t)$ onto the X plane to get a **phase portrait** [phase *profile* would be a more descriptive term] or **trajectory** of the motion $x(t)$. We call X the **phase plane,** and we plot trajectories as curves using t as a parameter.

Example 5

For the falling particle of Example 1, we can eliminate t to get x_1 as a function of x_2 by recalling that

$$\frac{dx_2}{dx_1} = \frac{dx_2}{dt} \bigg/ \frac{dx_1}{dt} = \frac{\dot{x}_2}{\dot{x}_1} = -\frac{g}{x_2}$$

This *separable* differential equation is easily solved by "multiplying" and "dividing" to get

$$x_2 dx_2 = -g dx_1$$

and integrating from t_0 to t so that

$$x_1 = -\frac{1}{2g}x_2^2 + c, \text{ where } c = x_1(t_0) + \frac{x_2^2(t_0)}{2g}$$

is determined by the initial conditions $x_1(t_0)$ and $x_2(t_0)$.

Thus, the state trajectories are parabolas with the orientation shown in Figure 2–36 for several different values of the integration constant c.   ◊

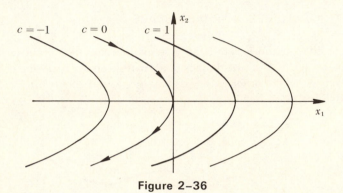

Figure 2–36

Example 6

To study the natural or unforced behavior of the mechanical oscillator of Example 2, we set the forcing function (input) $u = 0$. The slope of a trajectory at the point (x_1, x_2) is

$$\frac{dx_2}{dx_1} = \frac{\dot{x}_2}{\dot{x}_1} = -\omega^2 \frac{x_1}{x_2}$$

Integrating and simplifying, we get

$$x_2^2 = -\omega^2 x_1^2 + \alpha, \text{ where } \alpha = x_2^2(t_0) + \omega^2 x_1^2(t_0)$$

or

$$\frac{x_1^2}{c^2} + \frac{x_2^2}{\omega^2 c^2} = 1$$

where we set α (which, of course, is positive) equal to $\omega^2 c^2$. The trajectories are ellipses, as shown in Figure 2–37.   ◊

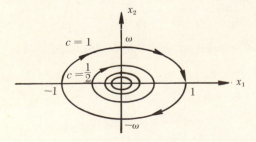

Figure 2–37

Observe that in both Figure 2-36 and Figure 2-37 we indicated by arrowheads the direction in which a particular trajectory is traversed as time increases. We accomplish this orientation by noting that $\dot{x}_1 = x_2$. Thus, wherever $x_2 > 0$, we know that x_1 will increase with time, while below the x_1 axis (where $x_2 < 0$), $\dot{x}_1 < 0$ and hence x_1 decreases with time. Since both examples were of the form

$$\frac{dx_2}{dx_1} = \frac{-f(x_1, x_2)}{x_2}$$

we see that the trajectories have infinite slope where they cross the x_1 axis ($x_2 = 0$).

Each of the preceding examples illustrates the motion of **free** or **unforced** (zero input) systems whose state spaces were two-dimensional, and which were therefore amenable to study in the phase plane. In general, for higher dimensional state spaces such graphical analyses are not applicable. Nevertheless, there are so many useful systems which do turn out to be two-dimensional, and there is such a volume of control theory literature making use of the phase plane, that it is worthwhile to become familiar with the concept here. It is often informative when dealing with analyses of n-dimensional systems to check the results by setting $n = 2$ and studying the motion, say by a digital computer simulation, in the phase plane.

We next make a phase-plane analysis of a very simple mechanical system which has a controlling input:

Example 7

Consider a cart of unit mass with wheels which are inertia-free and frictionless. Let the horizontal displacement be measured by the variable $y(t)$, and let $u(t)$ be a force acting upon the cart in the y-direction, as shown in Figure 2-38(a). Taking as state vector $= \begin{bmatrix} y \\ \dot{y} \end{bmatrix}$, the equation of motion $\ddot{y} = u$ may be recast in the form

$$\begin{bmatrix} \dot{x}_1 \\ \dot{x}_2 \end{bmatrix} = \begin{bmatrix} 0 & 1 \\ 0 & 0 \end{bmatrix} \begin{bmatrix} x_1 \\ x_2 \end{bmatrix} + \begin{bmatrix} 0 \\ 1 \end{bmatrix} u$$

$$y = \begin{bmatrix} 1 & 0 \end{bmatrix} \begin{bmatrix} x_1 \\ x_2 \end{bmatrix}$$

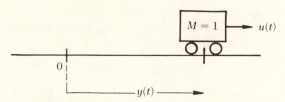

Figure 2-38(a)   Cart on frictionless wheels.

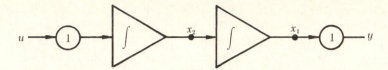

Figure 2-38(b)   Analog computer realization of Figure 2-38(a).

The analog computer realization of these equations is shown in Figure 2-38(b), and it should be clear why control engineers might call this system a "2-integrator plant."

Suppose that at time $t = 0$, the cart has position x_{10} and velocity x_{20}, and that we thereafter apply the constant force $u(t) = -1$.

Then $\ddot{y} = -1$ yields

$$y = -\frac{1}{2} t^2 + x_{20}t + x_{10}$$

In the phase-plane, we have $\dfrac{dx_2}{dx_1} = -\dfrac{1}{x_2}$, the solution of which yields the family of trajectories determined by

$$x_1 = c_1 - \frac{1}{2} x_2^2$$

where the parameter c_1 is determined by the initial conditions $x_1(t_0)$ and $x_2(t_0)$ to be $c_1 = x_{10} + \dfrac{1}{2} x_{20}^2$. Similarly, a constant force of $u = +1$ yields the family

$$x_1 = c_2 + \frac{1}{2} x_2^2, \text{ with } c_2 = x_{10} - \frac{1}{2} x_{20}^2$$

Consideration of the initial conditions

$$
\begin{array}{lll}
(\alpha) & x_{10} = x_{20} = 1 \\
(\beta) & x_{10} = x_{20} = 0
\end{array} \Big\} u = -1 \quad \Rightarrow
\begin{cases}
(\alpha) & c_1 = 3/2 \\
(\beta) & c_1 = 0
\end{cases}
$$

$$
\begin{array}{lll}
(\gamma) & x_{10} = x_{20} = -1 \\
(\delta) & x_{10} = x_{20} = 0
\end{array} \Big\} u = +1 \quad \Rightarrow
\begin{cases}
(\gamma) & c_2 = -(3/2) \\
(\delta) & c_2 = 0
\end{cases}
$$

yields the plots in the phase plane of Figure 2-39.

For a general constant force F and a (not necessarily unit) mass m, our equation of motion becomes

$$M\ddot{y} = F$$

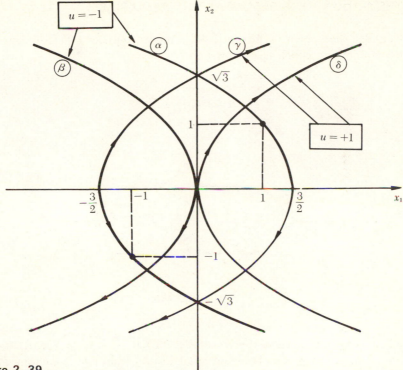

Figure 2-39

which yields the phase-plane trajectory

$$x_1 = \left[x_{10} - \frac{M}{2F} x_{20}^2 \right] + \frac{M}{2F} x_2^2$$

Thus, increasing the mass and decreasing the force both have the same effect on the parabolic trajectory—namely, to "shrink" the parabola and shift it horizontally. ◊

The reader should note from the last three examples the power of the phase plane approach. Given local dynamic equations of a system, *without actually solving* those differential equations for $x_1(t)$ and $x_2(t)$, we instead suppress the time variable and determine how x_2 varies with x_1. The resulting curves, while not revealing the actual solutions $x_1(t)$ and $x_2(t)$, nevertheless reveal many interesting facts about the solutions. For example, in Example 6, Figure 2-37 clearly shows that the solutions $x_1(t)$ and $x_2(t)$ are bounded, periodic oscillations with x_2 (speed) smallest when x_1 (displacement) is greatest; hence, we dubbed that system a mechanical oscillator. In Chapter 5 we shall see how to find the explicit solutions to differential equations describing the local state dynamics of continuous-time systems.

THE NOTION OF A REDUNDANT STATE VARIABLE

Consider the system of Figure 2–40, which comprises two interconnected integrators and which has the system equations

$$\dot{x}_1(t) = u(t)$$
$$\dot{x}_2(t) = 2u(t) \tag{1}$$
$$y(t) = x_1(t) + x_2(t)$$

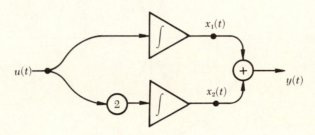

Figure 2–40

Then the second state variable is *redundant,* for $x_2(t) - x_2(t_0) = 2(x_1(t) - x_1(t_0))$ and so all information about the input history from t to t_1 contained in x_2 is already contained in x_1.

Thus, we can just as effectively describe the behavior of the system using one state variable:

$$\left. \begin{array}{l} \dot{x}_1(t) = u(t) \\ y(t) = 3x_1(t) \end{array} \right\}$$

since (after a little juggling with initial conditions) we can always recapture $x_2(t)$ as $2x_1(t)$. It is only when the system "malfunctions" so that (1) fails to describe the system and strict dependence between x_1 and x_2 breaks down that we need two state variables.

Electrical networks may be constructed so that an element which appears to be "storing" information, and hence adding a dimension to the state space, is really contributing nothing to information stored elsewhere. [See Exercises 8 and 15.]

THE HAMILTONIAN AND LAGRANGIAN FORMULATIONS OF DYNAMICS

At this stage, it will be useful to state two general principles, due to Hamilton and Lagrange, for writing equations of motion for dynamic systems. For a fuller discussion of these principles, the reader is referred to any good textbook on mechanics; here we shall simply attest to the plausibility of these principles by showing that they do yield the equations of motion for the falling particle and the simple harmonic oscillator. We shall return to the historical role of these principles in the theory of optimization in Section 7–6.

Consider a system with generalized coordinates (i.e., positions) $q_1, \ldots, q_n$. Let the force acting to change q_j be F_j. We say that the system is **conservative** if there exists a function $V = V(q)$ such that

$$F_j = -\frac{\partial V}{\partial q_j}$$

Such a V is called the **potential energy** of the system.

Let T denote the **kinetic energy** of the system—in the Hamiltonian formulation we let T be a function of q and the generalized momenta p, whereas in the Lagrangian formulation we let T be a function of q and the generalized velocities $\dot{q}$.

We may now set forth the two formulations and then "check" them with examples:

The Hamiltonian Formulation   Let the **Hamiltonian function** be the total energy

$$H(p, q) = T(p, q) + V(q)$$

Then the equations of motion of the system are given by

$$\frac{\partial H}{\partial p_j} = \dot{q}_j, \quad \text{for all } j$$

$$\frac{\partial H}{\partial q_j} = -\dot{p}_j, \text{ for all } j$$

The Lagrangian Formulation   Let the **Lagrangian function** be the difference

$$L(q, \dot{q}) = T(\dot{q}, q) - V(q)$$

Then the equations of motion of the system are given by

$$\frac{d}{dt}\left(\frac{\partial L}{\partial \dot{q}_j}\right) - \frac{\partial L}{\partial q_j} = 0 \text{ for all } j$$

Note that we only write half as many equations in the Lagrangian formulation, since $\dot{q}_j = \dot{q}_j$ is implicit! Note, too, that the derivative $\dfrac{\partial L}{\partial \dot{q}_j}$ is a true formal partial derivative of L with respect to $\dot{q}_j$ as a variable explicitly occurring in the formula for L, and makes no use of the dynamical interpretation of $\dot{q}_j$ as the time derivative of q_j.

Example 8

For the falling particle

$$m\ddot{z} = -mg$$

the force is $-mg$. Setting $V(z) = mgz$, we do indeed have

$$-mg = -\frac{\partial V}{\partial z}$$

Kinetic energy is just $\frac{1}{2}m\dot{z}^2$. We then have $q_1 = z$, $\dot{q}_1 = \dot{z}$, and $p_1 = m\dot{z}$. From the definition of the Hamiltonian function,

$$H(p_1, q_1) = \frac{1}{2m}p_1^2 + mgq_1$$

Thus

$$\frac{\partial H}{\partial p_1} = \frac{1}{m}p_1 \text{ and } \frac{\partial H}{\partial q_1} = mg$$

and this yields the actual equations of motion

$$\frac{1}{m}p_1 = \dot{q}_1; \ \dot{p}_1 = -mg$$

Now let us analyze this problem by setting up the Lagrangian function,

$$L(q_1, \dot{q}_1) = \frac{1}{2}m\dot{q}_1^2 - mgq_1$$

Then $\frac{\partial L}{\partial \dot{q}_1} = m\dot{q}_1$ and $\frac{\partial L}{\partial q_1} = -mg$, yielding the equation of motion

$$m\ddot{q}_1 + mg = 0 \qquad\qquad \Diamond$$

We close this section with an example using Lagrange's equations of motion to set up a state-variable description of a useful mechanical system.

Example 9

Consider the mechanical arm of Figure 2–41 moving in a horizontal plane so that we may ignore gravitational forces. We take the two bars to be identical, each of mass m and length ℓ. Assume that the arm is controlled by two motors, mounted at the joints, one applying a torque u_1 about pin P_1, and the other yielding a torque u_2 about P_2.

Appropriate generalized coordinates for the system are clearly the angles θ_1 and θ_2. We shall use the Lagrangian formulation and must thus compute T and V in terms of θ_1, $\dot{\theta}_1$, θ_2, $\dot{\theta}_2$, and externally applied forces.

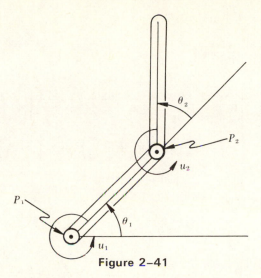

Figure 2-41

Since V must satisfy

$$\frac{\partial V}{\partial \theta_1} = -u_1; \; \frac{\partial V}{\partial \theta_2} = -u_2$$

we may take $V = -[u_1\theta_1 + u_2\theta_2]$.

The kinetic energy of each limb may be written as the sum of a term derived from the motion of the mass center, and a term derived from the rod's rotation. We thus have

$$T = \left[\frac{1}{2}mv_1^2 + \frac{1}{2}J\Omega_1^2 + \frac{1}{2}mv_2^2 + \frac{1}{2}J\Omega_2^2\right]$$

where

v_1 = velocity of the center of mass of the rod pinned at $P_1 = \dfrac{\ell}{2}\dot{\theta}_1$

J = moment of inertia of either rod about its center of mass
$\quad = \frac{1}{12}m\ell^2$

Ω_1 = angular velocity of rod 1 = $\dot{\theta}_1$

v_2 = velocity of the center of mass of rod 2, so that

$v_2^2 = \ell^2[\dot{\theta}_1^2 + \frac{1}{4}(\dot{\theta}_1 + \dot{\theta}_2)^2 + \dot{\theta}_1(\dot{\theta}_1 + \dot{\theta}_2)\cos\theta_2]$

Ω_2 = angular velocity of rod 2 = $(\dot{\theta}_1 + \dot{\theta}_2)$

We substitute these expressions for T and V into $L = T - V$, to obtain Lagrange's equations

$$\frac{d}{dt}\frac{\partial L}{\partial \dot{\theta}_1} - \frac{\partial L}{\partial \theta_1} = 0 \tag{2}$$

$$\frac{d}{dt}\frac{\partial L}{\partial \dot{\theta}_2} - \frac{\partial L}{\partial \theta_2} = 0 \tag{3}$$

We see that, since V is independent of $\dot{\theta}_1$,

$$\frac{\partial L}{\partial \dot{\theta}_1} = \frac{\partial T}{\partial \dot{\theta}_1} = \frac{1}{4} m \ell^2 \dot{\theta}_1 + \frac{1}{12} m \ell^2 \dot{\theta}_1$$

$$+ m \ell^2 \left[\dot{\theta}_1 + \frac{1}{4} (\dot{\theta}_1 + \dot{\theta}_2) + \frac{2\dot{\theta}_1 + \dot{\theta}_2}{2} \cos \theta_2 \right] + \frac{1}{12} m \ell^2 (\dot{\theta}_1 + \dot{\theta}_2)$$

$$= m \ell^2 \left[\frac{5}{3} \dot{\theta}_1 + \frac{1}{3} \dot{\theta}_2 + \left(\dot{\theta}_1 + \frac{1}{2} \dot{\theta}_2 \right) \cos \theta_2 \right]$$

so that

$$\frac{d}{dt} \left(\frac{\partial L}{\partial \dot{\theta}_1} \right)$$

$$= m \ell^2 \left[\frac{5}{3} \ddot{\theta}_1 + \frac{1}{3} \ddot{\theta}_2 + \left(\ddot{\theta}_1 + \frac{1}{2} \ddot{\theta}_2 \right) \cos \theta_2 - \left(\dot{\theta}_1 + \frac{1}{2} \dot{\theta}_2 \right) \dot{\theta}_2 \sin \theta_2 \right]$$

Since $\dfrac{\partial L}{\partial \theta_1} = u_1$, equation (2) takes the form

$$m \ell^2 \left[\frac{5}{3} \ddot{\theta}_1 + \frac{1}{3} \ddot{\theta}_2 + \left(\ddot{\theta}_1 + \frac{1}{2} \ddot{\theta}_2 \right) \cos \theta_2 - \left(\dot{\theta}_1 + \frac{1}{2} \dot{\theta}_2 \right) \dot{\theta}_2 \sin \theta_2 \right] = u_1 \quad \textbf{(2′)}$$

The reader should carry out the analogous manipulations to verify that equation (3) reduces to:

$$m \ell^2 \left[\frac{1}{3} (\ddot{\theta}_1 + \ddot{\theta}_2) + \frac{1}{2} \ddot{\theta}_1 \cos \theta_2 + \frac{1}{2} \dot{\theta}_1^2 \sin \theta_2 \right] = u_2 \quad \textbf{(3′)}$$

Let us now transform our equations into state-variable form by introducing the state vector x with components

$$x_1 = \theta_1; \; x_2 = \theta_2; \; x_3 = \dot{\theta}_1; \; x_4 = \dot{\theta}_2$$

so that we have two immediate equations of motion

$$\dot{x}_1 = x_3$$

$$\dot{x}_2 = x_4$$

The harder task is to manipulate our equations (2′) and (3′) to get explicit expressions for $\dot{x}_3 = \ddot{\theta}_1$ and $\dot{x}_4 = \ddot{\theta}_2$.

Setting $\widehat{u}_1 = \dfrac{u_1}{m\ell^2}$ and $\widehat{u}_2 = \dfrac{u_2}{m\ell^2}$, our equations may be rewritten as

$$\left(\frac{5}{3} + \cos x_2\right)\dot{x}_3 + \left(\frac{1}{3} + \frac{1}{2}\cos x_2\right)\dot{x}_4 = \widehat{u}_1 + \frac{2x_3 + x_4}{2}x_4 \sin x_2 \quad (2'')$$

$$\left(\frac{1}{3} + \frac{1}{2}\cos x_2\right)\dot{x}_3 + \frac{1}{3}\dot{x}_4 = \widehat{u}_2 - \frac{1}{2}x_3^2 \sin x_2 \qquad (3'')$$

which clearly may be solved explicitly for rather unpleasant equations of the form

$$\dot{x}_3 = f_2(x_2, x_3, x_4, \widehat{u}_1, \widehat{u}_2)$$
$$\dot{x}_4 = f_4(x_2, x_3, x_4, \widehat{u}_1, \widehat{u}_2)$$

Rather than explicitly solving for f_2 and f_4, let us note that for certain restricted values of the state variables we may easily find approximations to them.

First, take the special case where $\dot{\theta}_1$, $\dot{\theta}_2$, and θ_2 (i.e., x_2, x_3, and x_4) are small. Then we have $\cos x_2 \cong 1$ and $\sin x_2 \cong x_2$ so that $(2'')$ and $(3'')$ reduce, on omitting terms of higher than first order, to the two *linear* equations in the two variables $\dot{x}_3$ and $\dot{x}_4$:

$$\frac{8}{3}\dot{x}_3 + \frac{5}{6}\dot{x}_4 = \widehat{u}_1$$

$$\frac{5}{6}\dot{x}_3 + \frac{1}{3}\dot{x}_4 = \widehat{u}_2$$

Solving for $\dot{x}_2$ and $\dot{x}_3$ and substituting for $\widehat{u}_1$ and $\widehat{u}_2$ yields

$$\dot{x}_3 = \frac{1}{7m\ell^2}(12u_1 - 30u_2)$$

$$\dot{x}_4 = \frac{1}{7m\ell^2}(-30u_1 + 96u_2)$$

so that we can use matrix notation for the approximate state-transition equation for small $\dot{\theta}_1$, $\dot{\theta}_2$, and θ_2:

$$\dot{x} = \begin{bmatrix} 0 & 0 & 1 & 0 \\ 0 & 0 & 0 & 1 \\ 0 & 0 & 0 & 0 \\ 0 & 0 & 0 & 0 \end{bmatrix} x + \frac{1}{7m\ell^2}\begin{bmatrix} 0 & 0 \\ 0 & 0 \\ 12 & -30 \\ -30 & 96 \end{bmatrix}\begin{bmatrix} u_1 \\ u_2 \end{bmatrix}$$

The reader should now verify for another special case that if $\dot{\theta}_1$, $\dot{\theta}_2$, and $\left(\theta_2 - \dfrac{\pi}{2}\right)$ are small, then the state-transition equations may be approximated by the different matrix equations

$$\dot{x} = \begin{bmatrix} 0 & 0 & 1 & 0 \\ 0 & 0 & 0 & 1 \\ 0 & 0 & 0 & 0 \\ 0 & 0 & 0 & 0 \end{bmatrix} x + \frac{3}{4m\ell^2} \begin{bmatrix} 0 & 0 \\ 0 & 0 \\ 1 & -1 \\ -1 & 5 \end{bmatrix} \begin{bmatrix} u_1 \\ u_2 \end{bmatrix}$$

One might note here that even if we compute with linear equations in certain ranges, it is still useful to have a nonlinear "master" equation from which linear approximations may be read off at different stages of our control of a real system.   ◊

This last example illustrates the important notions of linearity and linearization, which we will study in depth in the next chapter.

EXERCISES FOR SECTION 2–5

1.   Using the analog computer elements defined in Example 1 of Section 1–1, show that the system of Figure 2–42 with states labeled as shown, is described by the same f as the mechanical system of Example 1.

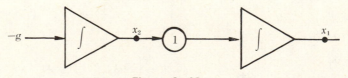

Figure 2–42

✔ **2.**   Exhibit an analog computer system which has the same f as the mechanical oscillator system of Example 2.

✔ **3.**   Consider the system with state variables x_1, x_2, x_3 as in Figure 2–43.

   (a)   Find a third order scalar differential equation relating output y to input u for this system. [*Answer:* $y^{(3)}(t) + a_1 y^{(2)}(t) + a_2 \dot{y}(t) + a_3 y(t) = b_1 u^{(2)}(t) + b_2 \dot{u}(t) + b_3 u(t)$]

   (b)   Directly from the wiring diagram, write the dynamic equations in the form

$$\dot{x}(t) = f(t, x(t), u(t)) \text{ and } y(t) = \eta(t, x(t))$$

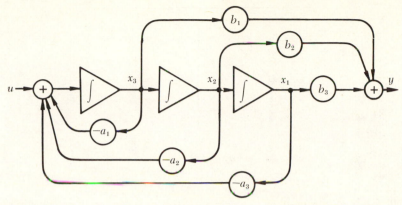

Figure 2–43   The so-called controllable canonical realization of input-output behavior.

using matrix notation

$$\dot{x}(t) = Fx(t) + Gu(t)$$
$$y(t) = Hx(t)$$

(c)   Relabel the state variables so that

$$x_1 \mapsto q_3, \; x_2 \mapsto q_2, \; x_3 \mapsto q_1$$

and write the new matrix dynamic equations for this choice. Find the differential equation relating y to u for this choice of state variables.

(d)   Compare this problem with Exercise 7 of Section 2–1, and repeat those parts of that exercise which apply to this continuous-time analogue of it.

✔ **4.**   Suppose a system has its output y related to its input u by the equation

$$y^{(3)}(t) + 2y^{(2)}(t) + 3\dot{y}(t) + 4y(t) = 5u^{(3)}(t) + 6u^{(2)}(t) + 7\dot{u}(t) + 8u(t)$$

(a)   Choose state variables $x_1 = y, \ldots, x_n = y^{(n-1)}$ as in Example 3 and draw an analog computer realization of this system. Notice that you have had to use differentiators to generate $\dot{u}$, $\ddot{u}$, and $\dddot{u}$ with this choice of state variables.

(b)   Try the choice $x_1 = y^{(n-1)}, \ldots, x_n = y$ and repeat the work of part (a). Is this an improvement?

(c)   Modify the analog diagram of Exercise 3 by adding a path with gain b_0 from input to output, so that that system will be a realization of our differential equation in this problem. From the diagram, write equations showing how the state variables x_1, x_2, and x_3 are related to y, $\dot{y}$, $\ddot{y}$, u, and $\dot{u}$. Notice that this realization based on Exercise 3 involves only integrators—no differentiators are required. Do the signals $\dot{u}, \ddot{u}$, and $\dddot{u}$ appear anywhere? How would you find $\dot{u}$?

✔ **5.**   Consider the simple pendulum of mass m suspended by a string of length ℓ. Let θ be the angular displacement as shown in Figure 2–44.

(a)   Write the differential equation of motion for this free system.

(b)   Define state variables by

$$x_1 = \theta \text{ and } x_2 = \dot{\theta}$$

and write the dynamic equations in the form

$$\dot{x} = f(t, x(t), u(t))$$

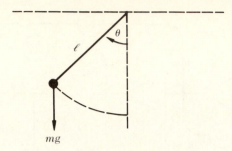

Figure 2–44

(c)   Find the possible trajectories in the phase plane as in Examples 5 and 6.

(d)   Sketch the trajectory corresponding to the initial condition $x(0) = \begin{bmatrix} -\pi/2 \\ 0 \end{bmatrix}$. Repeat for $x(0) = \begin{bmatrix} 0 \\ 1 \end{bmatrix}$.

(e)   Give an analog computer representation for this system using the state variables of part (b). Comment?

✔ **6.**   Here a continuous-time system is governed by a third order (linear) differential equation

$$y^{(3)} + 2y^{(2)} + 3\dot{y} + 4y = 5u^{(3)} + 6u^{(2)} + 7\dot{u} + 8u$$

We want the dynamic representation:

$$f\left(t, \begin{bmatrix} x_1 \\ x_2 \\ x_3 \end{bmatrix}, u\right) = \begin{bmatrix} 0 & 1 & 0 \\ 0 & 0 & 1 \\ -4 & -3 & -2 \end{bmatrix} \begin{bmatrix} x_1 \\ x_2 \\ x_3 \end{bmatrix} + \begin{bmatrix} b_1 \\ b_2 \\ b_3 \end{bmatrix} u$$

$$\eta\left(t, \begin{bmatrix} x_1 \\ x_2 \\ x_3 \end{bmatrix}, u\right) = \begin{bmatrix} 1 & 0 & 0 \end{bmatrix} \begin{bmatrix} x_1 \\ x_2 \\ x_3 \end{bmatrix} + [b_0]u$$

(a)   Choose state variables x_1, x_2, and x_3 to accomplish this and determine the values b_0, b_1, b_2, and b_3 that are necessary. [*Answer:* $b_0 = 5, b_1 = -4, b_2 = b_3 = 0$]

(b)   Draw the analog computer state diagram for your choice.

7.   Write state equations for the network shown in Figure 2–45. Give an analog computer representation.

EE 8.   A rule of thumb often used in writing state equations for electrical networks is, "Choose voltages across capacitors and currents through inductors as the state variables." In the circuit shown in Figure 2–46, this choice would yield $x_1 = v_1$, $x_2 = v_2$, and $x_3 = v_3$, where v_j is the voltage across capacitor C_j with the polarity shown by the + and − signs.

Let the element values be

$$R_1 = 2 \text{ ohms}, R_2 = 1 \text{ ohm}, C_1 = C_3 = 1 \text{ farad}, C_2 = 2 \text{ farads}$$

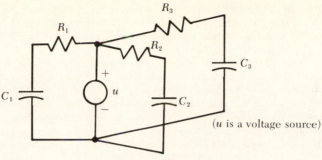

Figure 2-45

(u is a voltage source)

(a)   Write node equations for this circuit using Kirchoff's current law, and give the dynamic state equation which describes the system.

$$
[Answer: \quad \dot{x} = \begin{bmatrix} -\dfrac{1}{2} & 0 & \dfrac{1}{2} \\[2mm] 0 & -\dfrac{1}{2} & \dfrac{1}{2} \\[2mm] \dfrac{1}{2} & 1 & -\dfrac{3}{2} \end{bmatrix} x + \begin{bmatrix} 1 \\[2mm] -\dfrac{1}{2} \\[2mm] 0 \end{bmatrix} i \; ; \quad v = [1 \;-1 \;\; 0]x.]
$$

(b)   Solve the matrix-differential equation by Laplace transforming both sides, solving for $X(s)$, and inverse-Laplace transforming.

(c)   Suppose that, instead of voltages across capacitors, you used "node-pair voltages," V_A, V_B, and V_D, taking node E as the reference or ground and letting V_A be the voltage at node A with respect to node E.

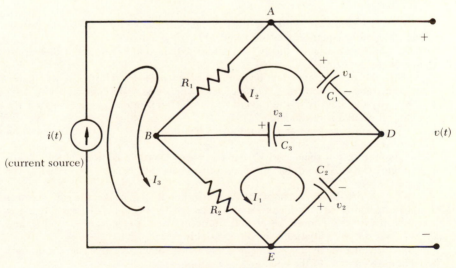

Figure 2-46

Write nodal equations in terms of V_A, V_B, and V_D, and choose $\bar{x}_1 = V_A$, $\bar{x}_2 = V_B$, and $\bar{x}_3 = V_D$ as state variables. Does this choice give a legitimate state? Try to express this new choice as a linear transformation of the first choice: $x = P\bar{x}$. Is the matrix P invertible?

(d)   Analyze the network using the method of "loop-currents" and Kirchoff's voltage law. Try I_1, I_2, and I_3 as state variables.

✔ **9.**   In Exercise 10 of Section 2–1, we saw three different realizations of a given input-output equation for discrete-time systems. Replace each unit-delay device (Figure 2–47(a)) with an integrator (Figure 2–47(b)) in the three discrete-time realizations shown in Figure 2–7 to get three corresponding continuous-time realizations of the input-output differential equation

$$y^{(3)}(t) + 6y^{(2)}(t) + 11\dot{y}(t) + 6y(t) = u(t)$$

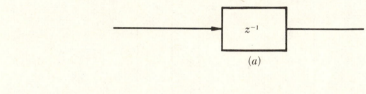

(a)

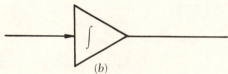

(b)

Figure 2–47

(a)   In each case, write the state equations and verify that we do get the above differential equation relating y and u.

(b)   Study parts (d) and (f) of Exercise 10, Section 2–1, to see if they apply to our continuous-time analogues.

(c)   Solve the state equations you found in part (a). [Hint: Use the easiest form, and transform from that choice of state variables to the others.]

10.   Verify the validity of the Lagrangian and Hamiltonian formulations for the mechanical oscillator of Example 2.

11.   Use the Lagrangian and Hamiltonian formulations on the simple pendulum of Exercise 5.

12.   Assume that a slender stick of mass m and length ℓ is mounted on a cart, as in Figure 2–48. Assume that the mass of the cart, M_c, is large compared to the mass of the stick.

(a)   Write the equations of motion, including a force $u(t)$ on the cart. Assume that no external force is applied to the stick except by motion of the cart. The cart wheels generate a friction force of magnitude $h\dot{y}$. Let $x_1 = y$, $x_2 = \dot{y}$, $x_3 = \theta$, and $x_4 = \dot{\theta}$, and put your equations in state variable form.

(b)   Make the assumption that $\dot{\theta}$ and θ are small, and give the small-angle linear approximation to the equations of motion.

13.   Consider the simple inverted pendulum shown in Figure 2–49, pinned at a joint without friction. The mass of 4 kg may be considered to be concentrated at

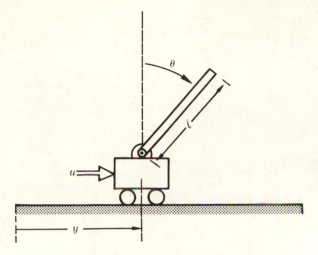

Figure 2–48   Stick-balancing cart.

a point 2 meters from the pin. Torque $u(t)$ in Newton-meters can be applied about the support. Take the acceleration of gravity to be 10 meters/sec² along $y = \pi$ (down).

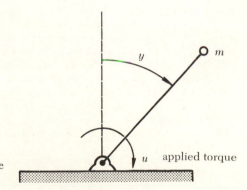

Figure 2–49   Inverted simple pendulum.

    (a)   Write the model equations of motion in terms of the angle, $y(t)$, from a vertical reference as shown.

    (b)   Give the *linear* equation of motion when y is small.

    (c)   Let $x_1 = y$ and $x_2 = \dot{y}$, and write the linear equations in first order (state) form.

14.   Write state equations for the mechanical system shown in Figure 2–50. Under no mechanical constraint, the length of spring i is ℓ_i, and the force needed to compress spring i by $\Delta \ell_i$ is $F_i = k_i \Delta \ell_i$.

    (a)   Assume first that $\ell_1 + \ell_2 + \ell_3 = \ell$. Indicate clearly the state variables.

    (b)   Now, let $\ell_1 + \ell_2 + \ell_3 \neq \ell$. What difference does this make?

EE 15.   This exercise illustrates the notion of redundant state variables.

    (a)   Use the three inductor currents i_1, i_2, and i_3 as state variables, and write the state equations for the electrical network shown in Figure 2–51.

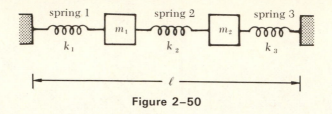

Figure 2–50

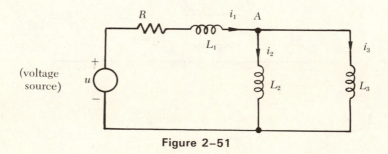

Figure 2–51

(b)   Use Kirchoff's current law at node A to show that if any two of i_1, i_2, and i_3 are known, the third is also known. Give a two-dimensional state variable formulation for this system.

(c)   Draw analog computer realizations for both the three-dimensional state selection and the reduced two-dimensional selection.

(d)   Using the **duality theorem** of circuit theory, draw the electrical dual of the circuit of part (a) and write both three- and two-dimensional state equations for it. Draw analog computer diagrams.

(e)   Choose all capacitor voltages and inductor currents to get a five-dimensional state equation for the circuit shown in Figure 2–52. Eliminate redundancies and choose state variables to give a minimal-dimension state dynamic equation for the circuit.

Note: There is a systematic way to eliminate redundancy and obtain a state dynamic equation description of any connected electrical network containing resis-

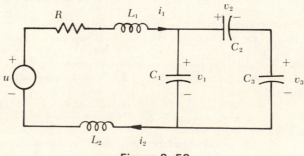

Figure 2–52

tors, capacitors, inductors, and independent current and voltage sources. The method uses "cut-sets," "tie-sets," "trees," and "branches," all of which are concepts of graph theory and network topology. See *Introduction to the State Variable Approach to Network Theory* by R. A. Rohrer, McGraw-Hill Book Co., 1970, Chapters 11 and 14, especially page 198.

✔ **16.** Consider the analog computer system with state variables x_1 and x_2, shown in Figure 2–53.

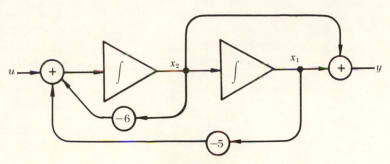

Figure 2–53

(a) Write the matrix state dynamic equations.

(b) Find the scalar differential equation relating input u to output y. Notice anything?

(c) Give a different choice of a state variable which will yield another analog computer realization of the input-output equation you found in part (b). Draw the new realization and write the dynamic state equation.

CHAPTER 3

Linear Systems and Linearization

Whereas, in Chapter 2, we concentrated upon the way in which inputs caused the system to change state, we now pay attention to the equally important way in which inputs cause the system to emit outputs. A response function for a system is one which specifies, for a given initial state at some time t_0 and the system input over the interval from t_0 to some later time t_1, what the *output* will be at time t_1. We develop the basic properties of such response functions, and provide appropriate examples in Section 3–1. In Section 3–2 we discuss the general concept of a linear transformation, and note what happens when the response function of a system is zero-state linear or zero-input linear. In Section 3–3, we then pause once again to provide further background material for the reader who is weak in linear algebra, by stressing the ways in which matrices arise as *numerical representations* of the linear transformations introduced in Section 3–2. Finally, we unify these ideas in Section 3–4 by showing that linear systems may be thought of as approximations to actual systems, if the nonlinearities of such systems are sufficiently smooth. Approximating the dynamic system functions f and η by the first derivative term of their Taylor Series, we shall derive the characterizations

$$\dot{x}(t) = F(t)x(t) + G(t)u(t)$$
$$y(t) = H(t)x(t)$$

and

$$x(t + 1) = F(t)x(t) + G(t)u(t)$$
$$y(t) = H(t)x(t)$$

134

where F, G, and H are linear operators, to model systems over their range of linear operation.

3–1 Input-Output Behavior and Response Functions

There are many situations in which we do not take explicit note of the way in which a system changes state, but rather observe its *input/output behavior*. This is the way in which the system responds to inputs by producing outputs. Of course, the response of the system to a succession of inputs depends on the state of the system at the beginning of our period of observation and—in the case of a time-varying system—on the time at which our observations commence.

Thus, given a system Σ described by the septuple $(T, U, X, Y, \Omega, \phi, \eta)$, it will be convenient to introduce a function

$$\mathscr{S}_{t_0, \hat{x}} : T \times \Omega \to Y$$

which is **the response function of Σ started at time t_0 in state $\hat{x}$**, and is defined by the equation

$$\mathscr{S}_{t_0, \hat{x}}(t, u) = \eta(t, \phi(t, t_0, \hat{x}, u))$$

i.e., we start Σ in state $\hat{x}$ at time t_0, apply an admissible input function $u \in \Omega$ to obtain state $\phi(t, t_0, \hat{x}, u)$ at time t, and then apply η to determine the corresponding output at time t.

Example 1

For the system of Example 1–5–8 we had

$$\phi(t, t_0, (v, u(t_0)), u) = \left[e^{-(t-t_0)/RC} v + \frac{1}{RC} \int_{t_0}^{t} e^{-(t-\zeta)/RC} u(\zeta) \, d\zeta, u(t) \right]$$

$$\eta(t, (v_c, u(t))) = \frac{R_1}{R_1 + R_2} [u(t) - v_c]$$

and we thus deduce that for this system

$$\mathscr{S}_{t_0, \left[\begin{smallmatrix} v_0 \\ u(t_0) \end{smallmatrix}\right]}(t, u) = \frac{R_1}{R_1 + R_2} \left[u(t) - e^{-(t-t_0)/RC} v_0 - \frac{1}{RC} \int_{t_0}^{t} e^{-(t-\zeta)/RC} u(\zeta) \, d\zeta \right]$$

$\Diamond$

It is easy to show [see Exercise 2] that for time-invariant systems

$$\mathcal{S}_{t_0+\tau,\hat{x}}(t + \tau, z^{-\tau}u) = \mathcal{S}_{t_0,\hat{x}}(t, u) \tag{1}$$

for all times t, t_0, all delays τ, all states $\hat{x}$, and all admissible input functions u. Thus, for time-invariant systems we may omit mention of the initial time variable and simply make use of

$$\mathcal{S}_{\hat{x}}: T \times \Omega \to Y$$

which is **the response function of Σ started in state $\hat{x}$** (at time 0). This is defined by the equation

$$\mathcal{S}_{\hat{x}}(t, u) = \mathcal{S}_{0,\hat{x}}(t, u) = \eta(\phi(t, \hat{x}, u))$$

where (recall Section 1–6) we have used the simplified η and ϕ for time-invariant systems. Thanks to (1), we may reconstitute $\mathcal{S}_{t_0,\hat{x}}$ from $\mathcal{S}_{\hat{x}}$ for any t_0 by the equation

$$\mathcal{S}_{t_0,\hat{x}}(t, u) = \mathcal{S}_{\hat{x}}(t - t_0, z^{t_0}u)$$

obtained from (1) by setting $\tau = -t_0$.

Often, in talking about a system we may make such comments as "we did nothing to the system at time t" or "the system made no response at time t." Since our formalization of a system requires that for every time $t \in T$ an input $u(t)$ and an output $y(t)$ be specified, it will be convenient to include a **zero input** 0_U in U and a **zero output** 0_Y in Y so that we may paraphrase our above comments as "$u(t) = 0_U$" and "$y(t) = 0_Y$," respectively. In cases where U or Y already has an additive structure with a zero element—such as the role played by 0 in **R** or **Z**—it is usually worth a little extra care to ensure that 0_U or 0_Y does in fact coincide with this 0. We drop the subscripts on our 0s when no confusion should arise.

Example 2

Consider the set of three letters $\{a, b, c\}$ with an addition $\oplus$ defined by the following table.

$\oplus$	a	b	c
a	c	a	b
b	a	b	c
c	b	c	a

We see that the letter b serves as the "additive identity" for the space $(\{a, b, c\}, \oplus)$. Thus, if we were labelling an output set Y with the three elements $\{a, b, c\}$ and hoped to make use of the additive structure of $\oplus$, it would be most convenient if we could take $0_Y = b$.   ◇

Having chosen a zero input 0_U in U, we may define the **zero input function** 0_Ω in Ω by the equation

$$0_\Omega(t) = 0_U \text{ for all } t \text{ in } T$$

Then, for a system Σ with zero input 0_U and zero output 0_Y, we say that a state $\hat{x}$ is a **zero state** if, whenever we start the system in state $\hat{x}$ and "do nothing" to the system, then the system "does nothing" in return. In terms of our above formalism, this says that $\hat{x}$ is a zero state of the system Σ whenever

$$\mathcal{S}_{t_0, \hat{x}}(t, 0_\Omega) = 0_Y \text{ for all times } t \geq t_0$$

We now give an example which shows that a system may have many different zero states.

Example 3

Consider the circuit of Figure 3–1, where the input i is from an ideal current source and the output y is the voltage across the series-connected capacitors.

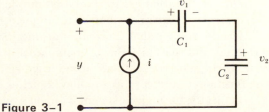

Figure 3–1

Choosing the capacitor voltages v_1 and v_2 as state variables, we get the equations

$$\frac{d}{dt}\begin{bmatrix} v_1(t) \\ v_2(t) \end{bmatrix} = \begin{bmatrix} 0 & 0 \\ 0 & 0 \end{bmatrix} \begin{bmatrix} v_1(t) \\ v_2(t) \end{bmatrix} + \begin{bmatrix} 1/C_1 \\ 1/C_2 \end{bmatrix} i(t) \tag{2}$$

$$y = \begin{bmatrix} 1 & 1 \end{bmatrix} \begin{bmatrix} v_1(t) \\ v_2(t) \end{bmatrix}$$

Here we take $T = \mathbf{R}$, $Y = \mathbf{R}$, and $X = \mathbf{R}^2$. We set $0_Y = 0$ and $0_X = \begin{bmatrix} 0 \\ 0 \end{bmatrix}$.

We take for Ω the set of integrable functions $i : T \to \mathbf{R}$, and then the zero input function 0_Ω is the "zero current" waveform with $i(t) \equiv 0$.

The solution of equations (2) yields

$$\begin{bmatrix} v_1(t) \\ v_2(t) \end{bmatrix} = \begin{bmatrix} v_1(t_0) \\ v_2(t_0) \end{bmatrix} + \begin{bmatrix} 1/C_1 \\ 1/C_2 \end{bmatrix} \left(\int_{t_0}^{t} i(\xi)\, d\xi \right)$$

while

$$y(t) = v_1(t) + v_2(t)$$

Thus,

$$\mathcal{S}_{t_0, \begin{bmatrix} v_1(t_0) \\ v_2(t_0) \end{bmatrix}} (t, i) = v_1(t_0) + v_2(t_0) + \frac{C_1 + C_2}{C_1 C_2} \int_{t_0}^{t} i(\xi)\, d\xi \tag{3}$$

Hence, the response of Σ started in any state $\begin{bmatrix} a \\ b \end{bmatrix}$ to the zero input function is

$$\mathcal{S}_{t_0, \begin{bmatrix} a \\ b \end{bmatrix}} (t, 0_\Omega) = a + b \tag{4}$$

since the integral in (3) is then zero.

The expression (4) is zero for all t only when $a + b = 0$. Hence, any state of the form

$$x_\theta = \begin{bmatrix} a \\ -a \end{bmatrix} \text{ for } a \in \mathbf{R}$$

satisfies our definition of a zero state.   ◊

Notice, then, that a zero state need not be the additive zero of the state space, and that there need not be a unique zero state.

Henceforth, we shall normally assume that a system has at least one zero state, and shall pick one of them, label it x_θ, and refer to it as *the* zero state. If the state space X has an additive zero, 0, we shall usually set up our functions η and ϕ so as to allow the choice $x_\theta = 0$.

Let us fix, for a given system Σ, upon a suitable choice of 0_U, and thus of 0_Ω, of 0_Y, and of x_θ. We then refer to the function

$$\mathcal{S}_{t_0, x_\theta} : T \times \Omega \to Y$$

as **the zero-state response** of Σ started at time t_0, and we refer to the function

$$\mathcal{S}_{t_0} : T \times X \to Y \text{ such that } (t, \widehat{x}) \mapsto \mathcal{S}_{t_0, \widehat{x}} (t, 0_\Omega)$$

as **the zero-input response** of Σ started at time t_0, telling us what outputs different states will yield under the action of the zero input function. The zero-input response is also referred to as the **free, natural** or **unforced** response.

Example 4

For the system of Example 3 we have the zero-state response,

$$\mathcal{S}_{t_0, \begin{bmatrix} a \\ -a \end{bmatrix}}(t, u) = \frac{C_1 + C_2}{C_1 C_2} \int_{t_0}^{t} u(\zeta)\, d\zeta$$

and zero-input response

$$\mathcal{S}_{t_0}\left(t, \begin{bmatrix} a \\ b \end{bmatrix}\right) = a + b \qquad \diamond$$

Whenever we have a function $\mu : A \times B \times C \to D$ of, say, three variables, we can obtain a function of fewer variables by fixing the values of one or more variables. Such functions are

$$\mu(a, \cdot, \cdot) : B \times C \to D \text{ such that } (b, c) \mapsto \mu(a, b, c)$$

and $\qquad \mu(a, \cdot, c) : B \to D$ such that $b \mapsto \mu(a, b, c)$

If we now define the **overall response function of a system** Σ

$$\mathcal{S} : T \times T \times X \times \Omega \to Y$$

by the equation

$$\mathcal{S}(t, t_0, \widehat{x}, u) = \eta(t, \phi(t, t_0, \widehat{x}, u)) \tag{5}$$

then we see that our above definitions of various particular response functions may be summed up in the formulae:

$$\mathcal{S}_{t_0, \widehat{x}} = \mathcal{S}(\cdot, t_0, \widehat{x}, \cdot) : T \times \Omega \to Y \tag{6}$$

$$\mathcal{S}_{t_0, x_\theta} = \mathcal{S}(\cdot, t_0, x_\theta, \cdot) : T \times \Omega \to Y \tag{7}$$

and

$$\mathcal{S}_{t_0} = \mathcal{S}(\cdot, t_0, \cdot, 0_\Omega) : T \times X \to Y \tag{8}$$

Again, if we fix upon two times t_1 and t_0 in T, then we see that the function

$$\mathcal{S}(t_1, t_0, \cdot, \cdot) : X \times \Omega \to Y,$$

which specifies how the system responds at time t_1 when started at time t_0, may be expressed as the composition of the function

$$\phi(t_1, t_0, \cdot, \cdot) : X \times \Omega \to X$$

which specifies how the system changes state from time t_0 to time t_1, and the function

$$\eta(t_1, \cdot):X \to Y$$

which specifies output at time t_1. The composition is

$$\mathcal{S}(t_1, t_0, \cdot, \cdot) = \eta(t_1, \cdot) \circ \phi(t_1, t_0, \cdot, \cdot)$$

Note that, as is required for a legal composition, the codomain of $\phi(t_1, t_0, \cdot, \cdot)$ is indeed the domain X of $\eta(t_1, \cdot)$.

It is possible to build on this notation to show how we may use the response functions of two systems to compute the response functions of the systems formed by "connecting" them in some way.

We consider three ways (illustrated in Figure 3–2) of combining two systems with the same time base:

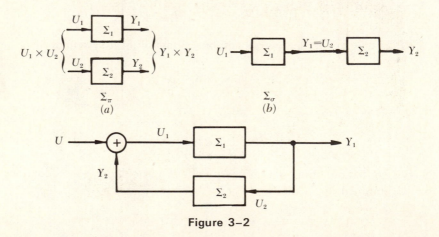

Figure 3–2

$\Sigma_1 = (T, U_1, Y_1, X_1, \Omega_1, \phi_1, \eta_1)$ with overall response function $\mathcal{S}_1$; and $\Sigma_2 = (T, U_2, Y_2, X_2, \Omega_2, \phi_2, \eta_2)$ with overall response function $\mathcal{S}_2$ and the same time base T. We consider three ways of combining these two systems. In each case, we know the state of the overall system when we know the state of each of the two subsystems. Thus, each system of Figure 3–2 has state space $X_1 \times X_2$.

In the **parallel "connection"** Σ_π of Figure 3–2(a), the two systems are simply run side-by-side. Thus, the input set is $U_1 \times U_2$ and the output set is $Y_1 \times Y_2$; the overall response function is obviously given by

$$\mathcal{S}_\pi:T \times T \times (X_1 \times X_2) \times (\Omega_1 \times \Omega_2) \to (Y_1 \times Y_2)$$

such that

$$(t_1, t_0, (\widehat{x}_1, \widehat{x}_2), (u_1, u_2)) \mapsto (\mathcal{S}_1(t_1, t_0, \widehat{x}_1, u_1), \mathcal{S}_2(t_1, t_0, \widehat{x}_2, u_2))$$

In the **series connection**—or **cascade**—Σ_σ of Figure 3–2(b), the output of the first system provides the input to the second system. We thus have to restrict Σ_2 so that it accepts the output of Σ_1 as an admissible input—so that not only must we have $Y_1 \subset U_2$, but we must also have that all possible output functions from Σ_1 be in Ω_2. We now assume that such conditions have been met. Then Σ_σ has input set U_1 and output set U_2, and the response

$$\mathcal{S}_\sigma(t_1, t_0, (\widehat{x}_1, \widehat{x}_2), u_1)$$

of Σ_σ to the input function u_1 in Ω, when started in state $(\widehat{x}_1, \widehat{x}_2)$, must clearly be the response of Σ_2 to the input function u_2 (when Σ_2 is started in state $\widehat{x}_2$) during that period; and u_2 is, in turn, just the response $\mathcal{S}_1(\cdot, t_0, \widehat{x}_1, u_1)$ of Σ_1 when started in state $\widehat{x}_1$. Thus, the overall response function is given by

$$\mathcal{S}_\sigma : T \times T \times (X_1 \times X_2) \times \Omega_1 \to Y_2$$

such that

$$(t_1, t_0, (\widehat{x}_1, \widehat{x}_2), u_1) \mapsto \mathcal{S}_2(t_1, t_0, \widehat{x}_2, \mathcal{S}_1(\cdot, t_0, \widehat{x}_1, u_1))$$

In the **feedback connection** Σ_f of Figure 3–2(c), we take any input set U, we take Y_1 as the output set, and we require the same conditions as for Σ_σ to allow the output of Σ_1 to serve as the input to Σ_2. We then provide a function $\alpha : U \times Y_2 \to U_1$, and specify that the input $u_1(t)$ to Σ_1 at time t must be expressed in terms of the system input $u(t)$ and the output $y_2(t)$ of Σ_2 at that time by the equation

$$u_1(t) = \alpha(u(t), y_2(t))$$

Perhaps the most familiar choice of α is that commonly used in error-controlled negative feedback systems, in which

$$U = Y_2 = U_1$$

are all vector spaces and

$$\alpha(u(t), y_2(t)) = u(t) - y_2(t)$$

In any case, the overall response function $\mathcal{S}_f$ of Σ_f is given rather subtly as the solution of an equation, rather than by explicit formulas as was the case for $\mathcal{S}_\pi$ and $\mathcal{S}_\sigma$. [The reader may wish to skip the details and move on to Example 5 at a first reading.]

When Σ_f is started in state $(\widehat{x}_1, \widehat{x}_2)$ at time t_0 and fed input u, its response is by definition given by the function

$$\mathcal{S}_f(\cdot, t_0, (\widehat{x}_1, \widehat{x}_2), u)$$

It is this response which provides the input to Σ_2, whose output in these circumstances is thus given by the function

$$\mathcal{S}_2(\cdot, t_0, \widehat{x}_2, \mathcal{S}_f(\cdot, t_0, (\widehat{x}_1, \widehat{x}_2), u))$$

Hence, the input to Σ_1 is the function $\alpha(u, \mathcal{S}_2(\cdot, t_0, \widehat{x}_2, \mathcal{S}_f(\cdot, t_0, (\widehat{x}_1, \widehat{x}_2), u)))$ for which

$$\alpha(u, \mathcal{S}_2(\cdot, t_0, \widehat{x}_2, \mathcal{S}_f(\cdot, t_0, (\widehat{x}_1, \widehat{x}_2), u)))(t) = \alpha(u(t), \mathcal{S}_2(t, t_0, \widehat{x}_2, \mathcal{S}_f(\cdot, t_0, (\widehat{x}_1, \widehat{x}_2), u)))$$

$\mathcal{S}_f$ is then simply (!) the response of Σ_1 to this function, when started in state $\widehat{x}_1$ at time t_0, so that we have

$$\mathcal{S}_f(t, t_0, (\widehat{x}_1, \widehat{x}_2), u) = \mathcal{S}_1(t, t_0, \widehat{x}_1, \alpha(u, \mathcal{S}_2(\cdot, t_0, \widehat{x}_2, \mathcal{S}_f(\cdot, t_0, (\widehat{x}_1, \widehat{x}_2), u))))$$

However, since $\mathcal{S}_f$ appears on both sides, it is very difficult indeed to deduce the form of $\mathcal{S}_f$. This difficulty, in fact, provides a powerful argument for the local transition functions introduced in Chapter 2:

Example 5

Suppose that $T = \mathbf{R}$ and that Σ_1 and Σ_2 have the local descriptions

$$\Sigma_1 \begin{cases} \dot{x}_1(t) = f_1(t, x_1(t), u_1(t)) \\ y_1(t) = \eta_1(t, x_1(t)) \end{cases}$$

and

$$\Sigma_2 \begin{cases} \dot{x}_2(t) = f_2(t, x_2(t), u_2(t)) \\ y_2(t) = \eta_2(t, x_2(t)) \end{cases}$$

Then when Σ_1 and Σ_2 are connected in Σ_f, we see that

$$u_1(t) = \alpha(u(t), \eta_2(t, x_2(t)))$$
$$u_2(t) = \eta_1(t, x_1(t))$$

so that we can immediately write down for Σ_f the local description *explicitly* as

$$\Sigma_f \begin{cases} \begin{bmatrix} \dot{x}_1(t) \\ \dot{x}_2(t) \end{bmatrix} = \begin{bmatrix} f_1(t, x_1(t), \alpha(u(t), \eta_2(t, x_2(t)))) \\ f_2(t, x_2(t), \eta_1(t, x_1(t))) \end{bmatrix} \\ u_2(t) = \eta_1(t, x_1(t)) \end{cases}$$

which we may then integrate to find ϕ_f and thus $\mathcal{S}_f$.   ◊

We close this section by explicitly noting that, apart from Example 5, our above discussion of system composition is completely general and has nothing to do with the vector spaces or linearity which will occupy us in the ensuing section of this chapter. This point will be of great importance in our very general approach to observability and controllability in Chapter 4. We emphasize this by the following example of a discrete-time non-linear finite-state system.

Example 6

Consider the time-invariant system which has

$$T = \mathbf{Z}$$
$$U = \{0, 1\} = Y$$
$$X = \{a, b, c\}$$

and whose state-transition and output functions are represented in the graph of Figure 3–3, where the notation $a/0$ within a circle indicates that $\eta(a) = 0$, while an arrow marked 1 going from the circle corresponding to state a to that for state b indicates that the one-step transition (as in Section 2–1) is given by $f(a, 1) = b$.

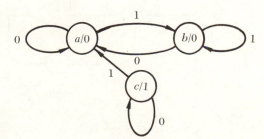

Figure 3–3

We take $0_U = 0_Y = 0$. The reader may wish to compute sample values of response functions for himself. Here let us simply verify that both a and b are zero states for this system. If we apply 0_Ω to the system started in state

a, the system stays in state a; while if we apply 0_Ω to the system started in state b, the first 0 sends the system to state a, and all subsequent 0 inputs keep it in state a. Since both $\eta(a)$ and $\eta(b)$ are zero, both the initial output and all subsequent outputs are zero, and so a and b are indeed zero states.

◇

EXERCISES FOR SECTION 3-1

✔ **1.** Find the overall response function $\mathcal{S}$ for the systems of
   (a) Example 1-3-1.
   (b) Example 1-6-1.
   (c) Exercise 10 of Section 2-1.
   (d) Example 2-5-3.
   (e) Example 2-5-7.

✔ **2.** Prove that for time-invariant systems,

$$\mathcal{S}(t_1 + \tau, t_0 + \tau, \widehat{x}, z^{-\tau}u) = \mathcal{S}(t_1, t_0, \widehat{x}, u)$$

for all

$$t_1, t_0, \tau \in T, \widehat{x} \in X, \text{ and } u \in \Omega.$$

3. **(a)** Find $0_Y, 0_U, 0_\Omega$, and x_θ for the systems of
   (i) Example 1-6-1
   (ii) Exercise 5 of Section 1-6
   (iii) Example 2-5-7
   (iv) Exercise 10 of Section 2-1
   (b) Rework Example 3 using a *one*-dimensional state vector.

✔ **4.** Consider the system of Figure 3-4, where the right-hand box denotes a "squarer."
   (a) Find ϕ and η.
   (b) Give $0_Y, 0_U, 0_\Omega$ and x_θ.
   (c) Does $\mathcal{S}(t_1, t_0, \widehat{x}, u) = \mathcal{S}(t_1, t_0, x_\theta, u) + \mathcal{S}(t_1, t_0, \widehat{x}, 0_\Omega)$?
Compare with Example 3.

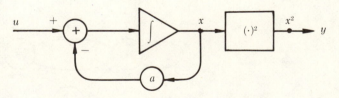

Figure 3-4

✔ **5.** Let V and W be vector spaces over the *same* field K. Let $+_V$ and $+_W$ denote vector addition on V and W, and let kv and kw stand for the respective definitions of multiplication of vectors $v \in V$ and $w \in W$ by scalars $k \in K$. Consider the set $V \times W = \{(v, w) | v \in V, w \in W\}$. For $V \times W$ define addition $+_{V\times W}$ and multiplication by scalars by

$$(v, w) +_{V\times W} (v', w') = (v +_V v', w +_W w')$$

and

$$k(v, w) = (kv, kw)$$

Show that these definitions make $V \times W$ a vector space over K.

6.   Define $f: \mathbf{R}^2 \to \mathbf{R}$ and $g: \mathbf{R} \to \mathbf{R}^2$ by the formulas

$$f\left(\begin{bmatrix} x_1 \\ x_2 \end{bmatrix}\right) = x_1 + 2x_2 \text{ for all } \begin{bmatrix} x_1 \\ x_2 \end{bmatrix} \in \mathbf{R}^2$$

$$g(x) = \begin{bmatrix} 3x \\ -x \end{bmatrix} \text{ for all } x \in \mathbf{R}$$

(a)   Compute the formulas for $f \circ g$ and $g \circ f$.

(b)   Check that $f \circ g$ is onto even though g is not onto; also, check that $f \circ g$ is one-to-one even though f is not one-to-one.

✔ **7.**   Consider the systems Σ_1 and Σ_2 cascaded in Figure 3–5.

(a)   Compute $\mathcal{S}_1$ and $\mathcal{S}_2$, the overall response functions for each component system.

(b)   Compute $\mathcal{S}$, the overall response function for the cascaded system.

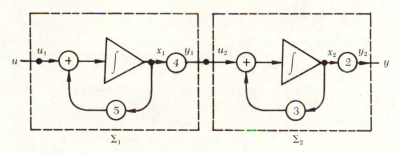

Figure 3–5

8. (A)   Compute the overall (i) local transition function f, (ii) state transition function ϕ (in both discrete- and continuous-time cases) and (iii) output function η in terms of those of Σ_1 and Σ_2 for the three "connections" of Figure 3–2:

(a) Σ_π, parallel

(b) Σ_σ, series

(c) Σ_f, feedback

(B)   Illustrate your work in Part A using the two one-dimensional analog systems Σ_1 and Σ_2 of Figure 3–5.

✔ **9.**   Suppose that a certain continuous-time system has its steady-state response function given by

$$\mathcal{S}(t, -\infty, 0, u) = \int_{-\infty}^{t} g(t - \xi) u(\xi) \, d\xi$$

which is a convolution integral. We may think of this expression as defining an operator L into which we plug a function u and get out a new function Lu. The new function Lu has the value at time t given by

$$(Lu)(t) = \int_{-\infty}^{t} g(t - \xi) u(\xi) \, d\xi$$

(a)   Show that $L(ku) = k(Lu)$ for all scalars k and that $L(u_1 + u_2) = (Lu_1) + (Lu_2)$ for all functions u_1 and u_2. Combine these properties to get $L(k_1u_1 + k_2u_2) = k_1(Lu_1) + k_2(Lu_2)$.

(b)   For any time τ, show that the delay operator $z^{-\tau}$ commutes with the operator L, so that

$$z^{-\tau} \circ L = L \circ z^{-\tau}$$

That is, for all u, $(z^{-\tau}(Lu))(t) = (L(z^{-\tau}u))(t)$, for all t.

3–2   Linearity

A map L of one vector space V into another vector space W is said to be **additive** if

$$L(v_1 + v_2) = L(v_1) + L(v_2)$$

for all vectors $v_1, v_2 \in V$, where the addition on the left is in V, while that on the right is in W.

We say that $L: V \to W$ is **homogeneous** iff $L(kv) = kL(v)$ for all scalars $k \in K$ and all vectors $v \in V$.

It is clear that for homogeneity both V and W must be vector spaces over the same field K of scalars: kv must make sense in V, and $kL(v)$ must make sense in W. Normally, in this book, we shall assume that the same field K—usually $\mathbf{R}$ or $\mathbf{C}$—is used throughout any connected discussion, and we will not explicitly state that all vector spaces considered are over the same field.

Example 1

Consider the map $\mathbf{R}^2 \to \mathbf{R}$ with $L\left(\begin{bmatrix} a \\ b \end{bmatrix}\right) = a + b$. Since

$$L\left(\begin{bmatrix} a_1 \\ b_1 \end{bmatrix} + \begin{bmatrix} a_2 \\ b_2 \end{bmatrix}\right) = L\left(\begin{bmatrix} a_1 + a_2 \\ b_1 + b_2 \end{bmatrix}\right)$$

$$= (a_1 + a_2) + (b_1 + b_2)$$
$$= (a_1 + b_1) + (a_2 + b_2)$$

$$= L\left(\begin{bmatrix} a_1 \\ b_1 \end{bmatrix}\right) + L\left(\begin{bmatrix} a_2 \\ b_2 \end{bmatrix}\right)$$

we see that L is additive.

Since for all scalars $k \in$ **R,**

$$L\left(k\begin{bmatrix} a \\ b \end{bmatrix}\right) = L\left(\begin{bmatrix} ka \\ kb \end{bmatrix}\right)$$

$$= (ka) + (kb)$$
$$= k(a + b)$$

$$= kL\left(\begin{bmatrix} a \\ b \end{bmatrix}\right)$$

we see that L is also homogeneous.   ◊

A map may be neither additive nor homogeneous.

Example 2

For the simple unforced pendulum of Figure 3–6 (see Exercise 2–5–5) with $x = \begin{bmatrix} \theta \\ \dot{\theta} \end{bmatrix}$ and $\dot{x} = f(t, x(t), u(t))$ we have

$$f\left(t, \begin{bmatrix} x_1(t) \\ x_2(t) \end{bmatrix}, 0\right) = \begin{bmatrix} x_2(t) \\ -\omega^2 \sin x_1(t) \end{bmatrix}$$

Figure 3–6 The simple, non-
linear pendulum. mg

where

$$\omega^2 = g/\ell$$

Thus for each fixed choice of $t, f(t, \cdot, 0): \mathbf{R}^2 \to \mathbf{R}^2$ but clearly

$$f\left(t, \begin{bmatrix} a_1 \\ b_1 \end{bmatrix} + \begin{bmatrix} a_2 \\ b_2 \end{bmatrix}, 0\right) \neq f\left(t, \begin{bmatrix} a_1 \\ b_1 \end{bmatrix}, 0\right) + f\left(t, \begin{bmatrix} a_2 \\ b_2 \end{bmatrix}, 0\right)$$

and

$$f\left(t, k\begin{bmatrix} a_1 \\ b_1 \end{bmatrix}, 0\right) \neq kf\left(t, \begin{bmatrix} a_1 \\ b_1 \end{bmatrix}, 0\right)$$

for all

$$\begin{bmatrix} a_1 \\ b_1 \end{bmatrix}, \begin{bmatrix} a_2 \\ b_2 \end{bmatrix} \in \mathbf{R}^2$$

and all $k \in \mathbf{R}$, so that this map $f(t, \cdot, 0)$ is neither additive nor homogeneous.

◊

In Exercises 8 and 9 the reader will find a map which is additive but not homogeneous, and one which is homogeneous but not additive.

DEFINITION 1

If $L : V \to W$ is both additive and homogeneous, we say that L is **linear** or that it is a **linear transformation** from V to W. ○

The map $L : \mathbf{R}^2 \to \mathbf{R}$ of Example 1 is thus linear, while $f(t, \cdot, 0)$ of Example 2 is *non-linear*.

There appear in the literature two common equivalent definitions of linearity for a map $L : V \to W$:

(i) L is linear iff

$$L(av_1 + bv_2) = aL(v_1) + bL(v_2)$$

for all $v_1, v_2 \in V$ and all scalars a, b.

(ii) L is linear iff

$$L(k(v_1 - v_2)) = kL(v_1) - kL(v_2)$$

for all $v_1, v_2 \in V$ and all scalars k.

We shall make especially heavy use of (i) in what follows, so let us verify that it is indeed equivalent to our definition—the reader will be asked to check the validity of (ii) in Exercise 11.

Example 3

If $L(av_1 + bv_2) = aL(v_1) + bL(v_2)$

then choosing $a = b = 1$ implies $L(v_1 + v_2) = L(v_1) + L(v_2)$
and choosing $a = k, b = 0$ implies $L(kv_1) = kL(v_1)$

so that L is additive and homogeneous. Conversely, if we take an L which is additive and homogeneous, then

$$L(av_1 + bv_2) = L(av_1) + L(bv_2) \text{ by additivity}$$
$$= aL(v_1) + bL(v_2) \text{ by homogeneity.} \qquad \square$$

Example 4

Consider the simple unforced pendulum of Example 2, where

$$f\left(t, \begin{bmatrix} x_1(t) \\ x_2(t) \end{bmatrix}, 0\right) = \begin{bmatrix} x_2(t) \\ -\omega^2 \sin x_1(t) \end{bmatrix}$$

Suppose we define the angular velocity $\dot{\theta}$ as the output y of this system, so that $y(t) = \eta\left(t, \begin{bmatrix} x_1(t) \\ x_2(t) \end{bmatrix}\right) = x_2(t)$.

The output map $\eta(t, \cdot): \mathbf{R}^2 \to \mathbf{R}^1$ is easily seen to be linear, even though the local transition map $f(t, \cdot, 0): \mathbf{R}^2 \to \mathbf{R}^2$ is non-linear. ◊

We saw in Section 1–2 that for *any* map $f: A \to B$ we may define the **range** or **image** of f to be the subset $\mathcal{R}(f) = f(A) = \{f(a) | a \in A\}$ of B. If B is a vector space, we define the **null space** or **kernel** of f to be the subset $\mathcal{N}(f) = \ker(f) = \{a | f(a) = 0\}$ comprising all points of A whose image is 0.

Let us now see what extra properties inhere in the range and the kernel when our map f is in fact a linear map $L: V \to W$ of one vector space into another.

Firstly (see Exercise 1), we know that if L is linear, then $L(0) = 0$, so that $\mathcal{N}(L)$ is nonempty, for it certainly contains $0 \in V$.

Next, note that if v_1 and v_2 are in $\mathcal{N}(L)$, and k_1 and k_2 are any scalars, then

$$L(k_1 v_1 + k_2 v_2) = k_1 L(v_1) + k_2 L(v_2) \text{ by linearity}$$
$$= k_1 \cdot 0 + k_2 \cdot 0$$
$$= 0$$

so that $k_1 v_1 + k_2 v_2$ is also in $\mathcal{N}(L)$.

Thus we have verified that $\mathcal{N}(L)$ is a subspace of V for a linear L, where by **subspace** we simply mean a set which is closed under addition and multiplication by scalars. The reader may similarly check, as an exercise, that $\mathcal{R}(L)$ is a subspace of W for a linear L:

LEMMA 1

For any linear transformation $L: V \to W$, we have that $\mathfrak{N}(L)$ is a subspace of V, while $\mathfrak{R}(L)$ is a subspace of W.   □

To explore linearity further, we must make a few comments on the status of our functions when U, X, and Y are vector spaces—be they our familiar examples of $\mathbf{R}$, $\mathbf{R}^2$, $\mathbf{R}^n$, or $\mathbf{C}^n$ for finite n, or some of the more exotic, infinite-dimensional, examples considered in Chapter 2. First we note that if U is a vector space, then so is the set of all functions from T into U—whether or *not* T is a vector space—using the "pointwise" definitions of addition $(u_1, u_2) \mapsto u_1 + u_2$ and of multiplication by a scalar $(k, u_1) \mapsto ku_1$ by the equations

$$\left.\begin{array}{r}[u_1 + u_2](t) = u_1(t) + u_2(t)\\[u_1](t) = k[u_1(t)]\end{array}\right\} \tag{1}$$

Now, if u_1 and u_2 are both piecewise integrable, or piecewise continuous, or piecewise differentiable, then so too are $u_1 + u_2$ and ku_1. Thus, in almost all cases in which U has a vector space structure, the space Ω of admissible input functions will also have a vector space structure as induced by the equations (1) above.

Henceforth in this chapter, we shall only consider systems for which X, Y, U, and Ω are vector spaces over the same field K.

We denote vectors in the product space $X \times \Omega$ by $(\hat{x}, u)$, and then define (see Exercise 5 of Section 3–1) addition on $X \times \Omega$ by

$$(\hat{x}_1, u_1) + (\hat{x}_2, u_2) = (\hat{x}_1 + \hat{x}_2, u_1 + u_2)$$

where, as is our custom, we abuse notation and just use one plus sign $+$ for the vector additions $+_X$, $+_\Omega$, and $+_{X \times \Omega}$ on each of the spaces X, Ω, and $X \times \Omega$ when no ambiguity can (or should) arise.

Similarly, we define the multiplication of a vector $(\hat{x}, u) \in X \times \Omega$ by a scalar k from the common field K by $k(\hat{x}, u) = (k\hat{x}, ku)$, since both $k\hat{x}$ and ku are well-defined for the spaces X and Ω, respectively. With these definitions, $X \times \Omega$ becomes a vector space over K.

With this as background, we may now say that a *system* is linear if and only if it changes state and emits outputs in a linear way. More precisely, we have the definition:

DEFINITION 2

The system $(T, U, Y, X, \Omega, \phi, \eta)$ is said to be a **linear system** if

(i) U, Y, X, and Ω are vector spaces over the same field K; and

(ii) for each choice of times t_0 and t_1 in T, the maps

$$\phi(t_1, t_0, \cdot, \cdot): X \times \Omega \to Y$$

and $$\eta(t_1, \cdot): X \to Y$$

are linear.   ○

Example 5

We will show now that the system of Example 3–1–3 is linear. There we had

$$\phi\left(t_1, t_0, \begin{bmatrix} x_1(t_0) \\ x_2(t_0) \end{bmatrix}, u\right) = \begin{bmatrix} x_1(t_0) + \dfrac{1}{C_1} \displaystyle\int_{t_0}^{t_1} u(\xi)\, d\xi \\[2ex] x_2(t_0) + \dfrac{1}{C_2} \displaystyle\int_{t_0}^{t_1} u(\xi)\, d\xi \end{bmatrix}$$

and

$$\eta\left(t_1, \begin{bmatrix} x_1(t_1) \\ x_2(t_1) \end{bmatrix}\right) = x_1(t_1) + x_2(t_1)$$

Since $X = \mathbf{R}^2$, Ω is the set of piecewise integrable functions $\mathbf{R} \to \mathbf{R}$, and $Y = \mathbf{R}$, where these sets are all vector spaces over the real field $\mathbf{R}$, we need only show that $\phi(t_1, t_0, \cdot, \cdot)$ and $\eta(t_1, \cdot)$ are linear maps.

For any $t_1 \in \mathbf{R}$, $\eta(t_1, \cdot)$ is just the map L of Example 1, and so is linear. To see that $\phi(t_1, t_0, \cdot, \cdot)$ is linear on $X \times \Omega$, we compute as follows:

$$\phi(t_1, t_0, k(x(t_0), u) + k'(x'(t_0), u'))$$
$$= \phi(t_1, t_0, kx(t_0) + k'x'(t_0), ku + k'u')$$

$$= \phi\left(t_1, t_0, \begin{bmatrix} kx_1(t_0) + k'x'_1(t_0) \\ kx_2(t_0) + k'x'_2(t_0) \end{bmatrix}, ku + k'u'\right)$$

$$= \begin{bmatrix} [kx_1(t_0) + k'x'_1(t_0)] + \dfrac{1}{C_1} \displaystyle\int_{t_0}^{t_1} [ku(\xi) + k'u'(\xi)]\, d\xi \\[2ex] [kx_2(t_0) + k'x'_2(t_0)] + \dfrac{1}{C_2} \displaystyle\int_{t_0}^{t_1} [ku(\xi) + k'u'(\xi)]\, d\xi \end{bmatrix}$$

$$= \begin{bmatrix} kx_1(t_0) + \dfrac{1}{C_1} \displaystyle\int_{t_0}^{t_1} ku(\xi)\, d\xi \\[2ex] kx_2(t_0) + \dfrac{1}{C_2} \displaystyle\int_{t_0}^{t_1} ku(\xi)\, d\xi \end{bmatrix} + \begin{bmatrix} k'x'_1(t_0) + \dfrac{1}{C_1} \displaystyle\int_{t_0}^{t_1} k'u'(\xi)\, d\xi \\[2ex] k'x'_2(t_0) + \dfrac{1}{C_2} \displaystyle\int_{t_0}^{t_1} k'u'(\xi)\, d\xi \end{bmatrix}$$

$$= k\phi\left(t_1, t_0, \begin{bmatrix} x_1(t_0) \\ x_2(t_0) \end{bmatrix}, u\right) + k'\phi\left(t_1, t_0, \begin{bmatrix} x'_1(t_0) \\ x'_2(t_0) \end{bmatrix}, u'\right)$$

so we do indeed have that

$$\phi(t_1, t_0, k(x(t_0), u) + k'(x'(t_0), u'))$$
$$= k\phi(t_1, t_0, (x(t_0), u)) + k'\phi(t_1, t_0, (x'(t_0), u')) \quad \Diamond$$

We say that the system is **zero-state linear** if the map $\mathcal{S}(t_1, t_0, 0, \cdot):\Omega \to Y$ is a linear map for each $t_1, t_0 \in T$, and we call the system **zero-input linear** if the map $\mathcal{S}(t_1, t_0, \cdot, 0):X \to Y$ is linear for each $t_1, t_0 \in T$. Note that we are now identifying the zero state and input with the 0 elements of the vector spaces X and Ω. (See Exercise 15.)

THEOREM 2

The response function

$$\mathcal{S}(t_1, t_0, \cdot, \cdot):X \times \Omega \to Y$$

of a linear system $(T, U, Y, X, \Omega, \phi, \eta)$ is a linear map for each $t_1, t_0 \in T$. Moreover, every linear system is also zero-state linear.

Proof

$$\mathcal{S}(t_1, t_0, \cdot, \cdot) = \eta(t_1, \cdot) \circ \phi(t_1, t_0, \cdot, \cdot)$$

and so, being the composition of two linear maps, is also linear (Exercise 4b).
In particular, we can decompose the zero-state response as

$$\mathcal{S}(t_1, t_0, 0, \cdot) = \eta(t_1, \cdot) \circ \phi(t_1, t_0, 0, \cdot)$$

and $\phi(t_1, t_0, \cdot, \cdot)$ being linear implies that $\phi(t_1, t_0, 0, \cdot)$ is linear (by Exercise 4a). □

As an exercise (Exercise 10) the reader should verify that each linear system is also zero-input linear.
The following property of linear systems is fundamental.

THEOREM 3

The response function of a linear system is the sum of its zero-state response and the zero-input response:

$$\mathcal{S}(t_1, t_0, (x, u)) = \mathcal{S}(t_1, t_0, (0, u) + (x, 0))$$
$$= \mathcal{S}(t_1, t_0, 0, u) + \mathcal{S}(t_1, t_0, x, 0)$$

Proof

By additivity of $\mathcal{S}(t_1, t_0, \cdot, \cdot)$.   □

Example 6

We saw in Example 3-1-3 that the complete response of the given system was

$$\mathcal{S}\left(t_1, t_0, \begin{bmatrix} v_1(t_0) \\ v_2(t_0) \end{bmatrix}, i\right) = v_1(t_0) + v_2(t_0) + \frac{C_1 + C_2}{C_1 C_2} \int_{t_0}^{t_1} i(\xi)\, d\xi$$

and, in Example 3-1-4, that

$$\mathcal{S}\left(t_1, t_0, \begin{bmatrix} 0 \\ 0 \end{bmatrix}, i\right) = \frac{C_1 + C_2}{C_1 C_2} \int_{t_0}^{t_1} i(\xi)\, d\xi$$

and

$$\mathcal{S}\left(t_1, t_0, \begin{bmatrix} v_1(t_0) \\ v_2(t_0) \end{bmatrix}, 0\right) = v_1(t_0) + v_2(t_0)$$

Thus

$$\mathcal{S}\left(t_1, t_0, \begin{bmatrix} v_1(t_0) \\ v_2(t_0) \end{bmatrix}, i\right) = \mathcal{S}\left(t_1, t_0, \begin{bmatrix} v_1(t_0) \\ v_2(t_0) \end{bmatrix}, 0\right) + \mathcal{S}\left(t_1, t_0, \begin{bmatrix} 0 \\ 0 \end{bmatrix}, i\right)$$

for this linear system.

For the non-linear system of Exercise 4 of Section 3-1, however, this decomposition property of the response function does not hold.   ◊

The reader who has studied circuit theory may observe that the **superposition principle** of that theory is just the statement that *the zero-state response is linear:*

$$\mathcal{S}(t_1, t_0, 0, au_1 + bu_2) = a\mathcal{S}(t_1, t_0, 0, u_1) + b\mathcal{S}(t_1, t_0, 0, u_2)$$

We mention in passing that there exist systems which are zero-state linear but not linear, as illustrated in Exercise 14 and Exercise 1-1-5. The interested reader should be able to construct other examples from the following idea: the state changes in a nonlinear way but the output function discards these nonlinearities. When, in Chapter 8, we come to study realization theory, we shall see that whenever a zero-state linear response function may be derived from a nonlinear system, we can always find a linear system (perhaps with an infinite-dimensional state-space) which has the given response function.

Let us now tie together several of our previous observations about discrete-time systems. In Exercises 8 and 11 of Section 2-1 we studied discrete-time, finite-dimensional systems described by dynamic equations of the form

$$x(n + 1) = F(n)x(n) + G(n)u(n)$$
$$y(n) = H(n)x(n)$$

(1)

where $F(\cdot)$, $G(\cdot)$, and $H(\cdot)$ are $N \times N$, $N \times M$ and $Q \times N$ matrices, and we found that the state transition map ϕ and the output map η were given in terms of the matrix $\Phi(p, q)$ as

$$x(m) = \phi(m, n, \widehat{x}, u_n, \ldots, u_{m-1}) = \Phi(m, n)\widehat{x} + \sum_{j=n}^{m-1} \Phi(m, j + 1)G(j)u(j) \quad (2)$$

$$y(m) = \eta(m, x(m)) = H(m)\Phi(m, n)\widehat{x} + H(m) \sum_{j=n}^{m-1} \Phi(m, j + 1)G(j)u(j) \quad (3)$$

where $x(n) = \widehat{x}$ and $m > n$.

Using simple properties of matrix multiplication, such as distributivity over addition, the reader should have no trouble showing that the maps ϕ and η of equations (2) and (3) are linear [Exercise 3a] and hence that systems having representations as in equation (1) are linear systems. Later, in Section 3–4, we shall prove that every discrete-time linear system can be represented by matrix difference equations of the form of equation (1). Thus, our solutions of equation (1) given by the formulas of equations (2) and (3) have completely solved the problem of finding the state and output behavior of all linear discrete-time systems. If, in addition, a system is represented by constant matrices (F, G, H) having no dependence upon n, our solution of equation (1) covers the important subclass of time-invariant, linear, discrete-time systems [Exercise 3b.]

We close this section with an example illustrating how linearity enables us to derive a useful canonical realization of input-output behavior described by linear difference equations.

Example 7

Suppose that a system has its input u and output y related by the difference equation

$$y(n + 3) + 6y(n + 2) + 11y(n + 1) + 7y(n) = 3u(n + 2) + 4u(n + 1) + 5u(n)$$

In Exercise 14 of Section 2–1 we presented an algorithm for constructing a system whose input u and output $\widehat{y}$ are related by the simpler equation

$$\widehat{y}(n + 3) + 6\widehat{y}(n + 2) + 11\widehat{y}(n + 1) + 7\widehat{y}(n) = u(n)$$

This simpler system, realized by solving for

$$\hat{y}(n + 3) = -6\hat{y}(n + 2) - 11\hat{y}(n + 1) - 7\hat{y}(n) + u(n)$$

is shown in Figure 3–7(a). The output $\hat{y}(n)$ of Figure 3–7(a) is the response of the system shown to input $u(n)$, and from the block diagram we can see

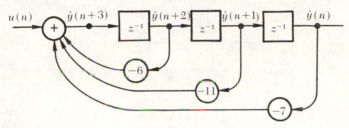

Figure 3–7(a)   A realization of $\hat{y}(n + 3) + 6\hat{y}(n + 2) + 11\hat{y}(n + 1) + 7\hat{y}(n) = u(n)$.

that this system can be described by the matrix difference equation of equation (1) with the particular matrices $F = \begin{bmatrix} 0 & 1 & 0 \\ 0 & 0 & 1 \\ -7 & -11 & -6 \end{bmatrix}$, $G = \begin{bmatrix} 0 \\ 0 \\ 1 \end{bmatrix}$,

and $H = \begin{bmatrix} 1 & 0 & 0 \end{bmatrix}$. Since these matrices are constant, we know that the system of Figure 3–7a is linear and time-invariant [Exercise 3].

By linearity and time-invariance, we may then say that if $u(n)$ produces $\hat{y}(n)$, then $u(n + 1)$ produces $\hat{y}(n + 1)$ and $u(n + 2)$ produces $\hat{y}(n + 2)$, so that the right-hand side forcing function

$$3u(n + 2) + 4u(n + 1) + 5u(n)$$

produces a response of

$$y(n) = 3\hat{y}(n + 2) + 4\hat{y}(n + 1) + 5\hat{y}(n)$$

Since each of the signals $\hat{y}(n + 2), \hat{y}(n + 1), \hat{y}(n)$ is available on the diagram in Figure 3–7(a), we can pick them off and multiply them respectively by the gains 3, 4, and 5 to arrive at the desired realization of

$$y(n + 3) + 6y(n + 2) + 11y(n + 1) + 7y(n) = 3u(n + 2) + 4u(n + 1) + 5u(n)$$

as shown in Figure 3–7(b).

If we choose the outputs of the delay boxes as state variables with

$$x_1(n) = \hat{y}(n)$$
$$x_2(n) = \hat{y}(n + 1)$$
$$x_3(n) = \hat{y}(n + 2)$$

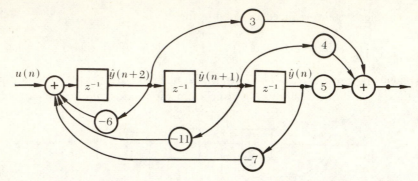

Figure 3–7(b) A realization of $y(n + 3) + 6y(n + 2) + 11y(n + 1) + 7y(n) = 3u(n + 2) + 4u(n + 1) + 5u(n)$.

we can write the matrix difference equations

$$\begin{bmatrix} x_1(n + 1) \\ x_2(n + 1) \\ x_3(n + 1) \end{bmatrix} = \begin{bmatrix} 0 & 1 & 0 \\ 0 & 0 & 1 \\ -7 & -11 & -6 \end{bmatrix} \begin{bmatrix} x_1(n) \\ x_2(n) \\ x_3(n) \end{bmatrix} + \begin{bmatrix} 0 \\ 0 \\ 1 \end{bmatrix} u(n)$$

$$y(n) = \begin{bmatrix} 5 & 4 & 3 \end{bmatrix} \begin{bmatrix} x_1(n) \\ x_2(n) \\ x_3(n) \end{bmatrix}$$

directly from the block diagram. Notice how the coefficients of the given scalar difference equation appear in the matrices and in the block diagram of this realization. ◊

The realization in the last example of a given input-output equation is called the *controllable canonical* realization, and we shall see more of it in Chapter 7 when we study controllability. (See Exercise 16.)

EXERCISES FOR SECTION 3–2

✔ **1.** Let $f: V \to W$ with V and W vector spaces.
  (a) Show that f additive $\Rightarrow f(0) = 0$.
  (b) Show that f homogeneous $\Rightarrow f(0) = 0$.
Thus, linear maps send zero elements to zero elements.
  (c) Prove that $-v = (-1)v$ for all $v \in V$.
  (d) Prove that f additive or homogeneous $\Rightarrow f(-v) = -f(v)$.
✔ **2.** Both Exercises 7 and 8 of Section 2–1 and Example 7 of this section were special cases of systems where the local transition map f could be written as a matrix equation:

$$f(t, x(t), u(t)) = A(t)x(t) + B(t)u(t)$$

where $A(t)$ and $B(t)$ are $N \times N$ and $N \times M$ matrices, $x(t) \in \mathbf{R}^N$ and $u(t) \in \mathbf{R}^M$. Show that $f(t, \cdot, \cdot): X \times U \to X$ is linear for each $t \in T$.

✔ **3.**   **(a)**   Show that for all times n and m with $m > n$, the maps $\phi(m, n, \cdot, \cdot)$ and $\eta(m, \cdot)$ defined by equations (2) and (3) are linear.

    (b)   Show that if the matrices $F(n)$, $G(n)$, and $H(n)$ are constant matrices (F, G, H), then a system described by

$$x(n + 1) = Fx(n) + Gu(n)$$
$$y(n) = Hx(n)$$

is a time-invariant linear system.

[*Note:* The matrix $\Phi(p, q)$ of equations (2) and (3) was defined in Exercise 8 of Section 2–1 and was studied further in Exercise 11 of that section.]

✔ **4.**   **(a)**   Prove that if $g: V \times W \to X$ is linear than $g(0, \cdot): W \to X$ and $g(\cdot, 0): V \to X$ are also linear maps.

    (b)   Prove that the composition of two linear maps is a linear map.

✔ **5.**   Let $C^k[a, b]$ be the set of all real valued functions having a continuous kth derivative on the interval $[a, b]$. Let

$$f: C^k[a, b] \to C^{k-1}[a, b]$$

and

$$g: C^k[a, b] \to C^k[a, b]$$

be defined by

$$f(u) = \dot{u}, \text{ for each } u \in C^k[a, b]$$
$$g(u) = \alpha u + \beta, \text{ for each } u \in C^k[a, b]$$

where α and β are two nonzero real numbers.

    (a)   Give the rules defining

$$g \circ f: C^k[a, b] \to C^{k-1}[a, b]$$

and

$$f \circ g: C^k[a, b] \to C^{k-1}[a, b]$$

(Note that $C^k[a, b] \subset C^{k-1}[a, b]$.)

    (b)   Show that $f \circ g$ is linear but that g is not linear. (Note the contrast with Exercise 4.)

✔ **6.**   Let $L: V \to W$, and let L be linear. Prove that L is one-to-one $\Leftrightarrow \mathfrak{N}(L) = \{0\}$.

7.   Let $L: C[0, 1] \to \mathbf{R}$ be defined by the rule

$$L(u) = \int_0^1 u(\xi) \, d\xi$$

Prove that L is linear and onto but not one-to-one.

8.   Let $f: \mathbf{C}^2 \to \mathbf{C}$ be defined by

$$f\left(\begin{bmatrix} x_1 \\ x_2 \end{bmatrix}\right) = \overline{x_1 + x_2}$$

where the bar denotes the complex conjugate, and $\mathbf{C}^2$ and $\mathbf{C}$ are both vector spaces over the field $\mathbf{C}$. Show that f is additive but not homogeneous.

9.   Define

$$F: C^k[a, b] \rightarrow C^{k+1}[a, b]$$

by the rule

$$F(u) = \left(\int_a^b [u(\xi)]^3 \, d\xi \right)^{1/3}$$

Show that F is homogeneous but not additive.

✔ 10.   Prove that a linear system is zero-input linear.

11.   Let $L: V \rightarrow W$, where V and W are vector spaces over the same field K. Prove that L is linear iff

$$L(k(v_1 - v_2)) = kL(v_1) - kL(v_2)$$

for all $k \in K$ and all $v_1, v_2 \in V$.

12.   Let $L: V \rightarrow W$ be an additive map of vector spaces over $K = \mathbf{R}$.

   (a)   Show that $L(nv) = nL(v)$ for all $n \in \mathbf{N}$.

   (b)   Show that $L\left(\dfrac{n}{m} v\right) = \dfrac{n}{m} L(v)$ for all n, $m \in \mathbf{N}$ and $m \neq 0$, so that any *rational number* can be pulled outside of $L(\cdot)$.

   (c)   Now let $v_1, v_2, \ldots, v_n, \ldots$ be any convergent sequence in V and let $\lim_{n \to \infty} v_n = v$. We say that L is **continuous** on V iff $\lim_{n \to \infty} L(v_n) = L\left(\lim_{n \to \infty} v_n\right) = L(v)$. Prove that L additive and continuous $\Rightarrow$ L homogeneous.

13.   An **inner product** $\langle \cdot, \cdot \rangle$ is defined for a vector space V over the complex field $\mathbf{C}$ as a mapping $\langle \cdot, \cdot \rangle : V \times V \rightarrow \mathbf{C}$ which satisfies the properties
   (i) $\langle v_1, v_2 \rangle = \langle \overline{v_2, v_1} \rangle$
   (ii) $\langle v_1, v_2 + v_3 \rangle = \langle v_1, v_2 \rangle + \langle v_1, v_3 \rangle$
   (iii) $\langle \lambda v_1, v_2 \rangle = \lambda \langle v_1, v_2 \rangle$
   (iv) $\langle v, v \rangle \geq 0$ for all $v \in V$, while $\langle v, v \rangle = 0$ only if $v = 0$.

   (a)   Show that $\langle v_0, \cdot \rangle : V \rightarrow \mathbf{C}$ is not linear but that $\langle \cdot, v_0 \rangle : V \rightarrow \mathbf{C}$ is linear for each fixed $v_0 \in V$.

   (b)   Is the map $\langle \cdot, \cdot \rangle : V \times V \rightarrow \mathbf{C}$ linear?

✔ 14.   Consider the analog computer system of Figure 3–8(a), where the subsystem of Figure 3–8(b) is a multiplier.

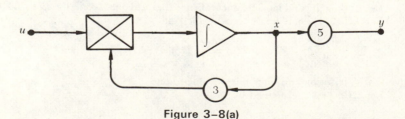

Figure 3–8(a)

   (a)   Give X, U, Ω, Y, the local state transition map f, and η.
   (b)   Find ϕ, η and $\mathcal{S}$ by solving the differential equation $\dot{x}(t) = f(t, x(t), u(t))$.
   (c)   Give 0_Y, 0_U, 0_Ω, 0_X and x_θ.

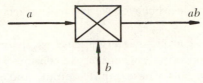

Figure 3–8(b)

(d)   Does

$$\mathcal{S}(t_1, t_0, \hat{x}, u) = \mathcal{S}(t_1, t_0, x_\theta, u) + \mathcal{S}(t_1, t_0, \hat{x}, 0_\Omega)?$$

(e)   Show that this system is both zero-state linear and zero-input linear, but *not* linear.

(f)   Prove that this system is time-invariant.

✔ **15.**   **(a)**   Show that if a system is zero-input linear, it has a zero state x_θ and the zero vector 0_X is always a zero state. [*Hint:* Use Exercise 1 on the given linear map $\mathcal{S}(t, t_0, \cdot, 0_\Omega):X \rightarrow Y$ to show $\mathcal{S}(t, t_0, 0_X, 0_\Omega) = 0_y.$]

(b)   Show that 0_X is a zero state for every linear system.

✔ **16.**   Use the technique of Example 7 to get the controllable canonical realization of the input-output difference equation

$$y(n + 3) + a_1 y(n + 2) + a_2 y(n + 1) + a_3 y(n) = b_1 u(n + 2) + b_2 u(n + 1) + b_3 u(n)$$

(a)   Draw the block diagram and write the matrix dynamic equations describing your realization. Notice the order in which the given coefficients appear.

(b)   Compare with Exercise 7 of Section 2–1.

(c)   Generalize this realization method to the continuous-time case, where input and output are related by the linear differential equation

$$y^{(3)}(t) + a_1 y^{(2)}(t) + a_2 \dot{y}(t) + a_3 y(t) = b_1 \ddot{u}(t) + b_2 \dot{u}(t) + b_3 u(t)$$

[*Hint:* First realize

$$\hat{y}^{(3)}(t) + a_1 \hat{y}^{(2)}(t) + a_2 \dot{\hat{y}}(t) + a_3 \hat{y}(t) = u(t)$$

by solving for highest derivative; then use time-invariance and linearity. State clearly any other assumptions needed.] Compare your block diagram and matrix differential equations with Exercise 3 of Section 2–5.

3–3   Matrices—Tools for Linear System Characterization

In our study of a mechanical arm in Section 2–5, we saw how a non-linear system could be approximated in certain regions of its state space by linear systems, and we used matrices to write down the description of the approximations. In the next section we shall present the general theory of *linearization*—of approximating a system by a linear system—but first we should make explicit how matrices arise as *numerical representations* of linear transformations, and so must play a role whenever we try to characterize a linear system numerically.

To avoid tedious convergence arguments, we shall only consider finite matrices; these are appropriate whenever we consider *finite-dimensional linear systems*, i.e., linear systems whose input, output, and state spaces are all finite-dimensional. By way of contrast, the reader may recall that the system of Example 2–5–4 had an infinite-dimensional state space. If dim $X = n$ (i.e., it takes n scalar quantities to characterize the state), we say that our system is *n-dimensional*. Thus, the dimension of the system depends upon the scalars we are prepared to use, as the following example makes clear:

Example 1

If $X = \mathbf{C}^2$ over the field **C**, then the two vectors $\left\{\begin{bmatrix} 1 \\ 0 \end{bmatrix}, \begin{bmatrix} 0 \\ 1 \end{bmatrix}\right\}$ are a basis for X and dim $(X) = 2$. However, if we let $X = \mathbf{C}^2$ *over the real field* **R**, the four vectors $\left\{\begin{bmatrix} 1 \\ 0 \end{bmatrix}, \begin{bmatrix} i \\ 0 \end{bmatrix}, \begin{bmatrix} 0 \\ 1 \end{bmatrix}, \begin{bmatrix} 0 \\ i \end{bmatrix}\right\}$ are a basis for X, and dim $(X) = 4$.   ◊

Suppose that V and W are finite-dimensional vector spaces over the field K, with dim $V = n$ and dim $W = m$. Let us now see how any linear map $L : V \to W$ may be naturally represented by an $m \times n$ matrix whose entries are elements of the field K, *as soon as we pick bases for V and W:*

Let $\mathcal{B} = \{v_1, \ldots, v_n\}$ be an ordered basis for V, and let $\mathcal{B}' = \{w_1, \ldots, w_m\}$ be an ordered basis for W. Then every vector v in V may be uniquely represented by the column vector $[v]_{\mathcal{B}}$ (the notation is short for **the representation of v with respect to the basis** $\mathcal{B}$) whose components $k_1, \ldots, k_n$ are uniquely defined by the equation

$$v = k_1 v_1 + k_2 v_2 + \cdots + k_n v_n$$

We also call the k's the **coordinates** of the vector v with respect to the basis $\mathcal{B}$.

Similarly, a vector w in W may be represented as $[w]_{\mathcal{B}'} \in K^m$ in terms of its coordinates with respect to the basis $\mathcal{B}'$.

Having represented vectors in terms of their coordinates with respect to fixed bases $\mathcal{B}$ in V and $\mathcal{B}'$ in W, we naturally ask: "How may the linear transformation $L : V \to W$ be represented in terms of *scalar* operations upon the coordinates in $[v]_{\mathcal{B}}$ which yield the coordinates in $[Lv]_{\mathcal{B}'}$?"

The answer follows simply by linearity: If

$$[v]_{\mathcal{B}} = \begin{bmatrix} k_1 \\ \vdots \\ k_n \end{bmatrix} \text{ we have that}$$

$$[Lv]_{\mathcal{B}'} = [L(k_1 v_1 + \cdots + k_n v_n)]_{\mathcal{B}'} \text{ by definition of the } k\text{'s}$$
$$= [k_1 L v_1 + \cdots + k_n L v_n]_{\mathcal{B}'} \text{ by linearity of } L$$

$$= k_1[Lv_1]_{\mathcal{B}'} + \cdots + k_n[Lv_n]_{\mathcal{B}'} \text{ by linearity of the}$$
$$\text{coordinate representation}$$

Thus all the numbers we need to characterize L are contained in the array of n column vectors

$$^{\mathcal{B}'}L^{\mathcal{B}} = [[Lv_1]_{\mathcal{B}'} \cdots [Lv_n]_{\mathcal{B}'}]$$

whose jth column is just $[L(v_j)]_{\mathcal{B}'}$, the coordinate vector of $L(v_j)$ with respect to the basis $\mathcal{B}'$. Since $^{\mathcal{B}'}L^{\mathcal{B}}$ has n columns, each with m components, it is an $m \times n$ matrix, and we call it **the matrix of L with respect to the bases $\mathcal{B}'$ for W and $\mathcal{B}$ for V.** Further, the rule for obtaining the vector $[Lv]_{\mathcal{B}'}$ from the matrix $^{\mathcal{B}'}L^{\mathcal{B}}$ and the vector $[v]_{\mathcal{B}}$ is precisely the one we have already given for multiplying a vector and a matrix:

$$[Lv]_{\mathcal{B}'} = {^{\mathcal{B}'}L^{\mathcal{B}}} \cdot [v]_{\mathcal{B}}$$

Example 2

Let $V = \mathbf{R}^3$ and $W = \mathbf{R}^4$. Let $\mathcal{B} = \{v_1, v_2, v_3\}$ and $\mathcal{B}' = \{w_1, w_2, w_3, w_4\}$ where

$$v_1 = \begin{bmatrix} 1 \\ 1 \\ 0 \end{bmatrix}, v_2 = \begin{bmatrix} 0 \\ 1 \\ 1 \end{bmatrix}, v_3 = \begin{bmatrix} 1 \\ 0 \\ 1 \end{bmatrix}$$

and

$$w_1 = \begin{bmatrix} 1 \\ 2 \\ 0 \\ 3 \end{bmatrix}, w_2 = \begin{bmatrix} 0 \\ 1 \\ -1 \\ 0 \end{bmatrix}, w_3 = \begin{bmatrix} 0 \\ 0 \\ 1 \\ 2 \end{bmatrix}, w_4 = \begin{bmatrix} 0 \\ 0 \\ 0 \\ 1 \end{bmatrix}$$

Define $L : \mathbf{R}^3 \to \mathbf{R}^4$ by the rule

$$L\left(\begin{bmatrix} x_1 \\ x_2 \\ x_3 \end{bmatrix}\right) = \begin{bmatrix} x_1 + x_2 + x_3 \\ x_2 - x_3 \\ 3x_1 - 5x_2 \\ x_1 \end{bmatrix}$$

Thus, $Lv_1 = \begin{bmatrix} 2 \\ 1 \\ -2 \\ 1 \end{bmatrix}$ and, to find $[Lv_1]_{\mathcal{B}'}$, we write

$$Lv_1 = k_1w_1 + k_2w_2 + k_3w_3 + k_4w_4$$

and solve for the coordinates $k_1, \ldots, k_4$:

From

$$
\begin{bmatrix} 2 \\ 1 \\ -2 \\ 1 \end{bmatrix} = k_1 \begin{bmatrix} 1 \\ 2 \\ 0 \\ 3 \end{bmatrix} + k_2 \begin{bmatrix} 0 \\ 1 \\ -1 \\ 0 \end{bmatrix} + k_3 \begin{bmatrix} 0 \\ 0 \\ 1 \\ 2 \end{bmatrix} + k_4 \begin{bmatrix} 0 \\ 0 \\ 0 \\ 1 \end{bmatrix}
$$

we get

$$
\begin{bmatrix} 2 \\ 1 \\ -2 \\ 1 \end{bmatrix} = \begin{bmatrix} k_1 \\ 2k_1 + k_2 \\ -k_2 + k_3 \\ 3k_1 + 2k_3 + k_4 \end{bmatrix}
$$

Equating components and solving the resulting four equations in four unknowns, we get

$$
\begin{bmatrix} k_1 \\ k_2 \\ k_3 \\ k_4 \end{bmatrix} = \begin{bmatrix} 2 \\ -3 \\ -5 \\ 5 \end{bmatrix}
$$

so that

$$
[Lv_1]_{\mathcal{B}'} = \begin{bmatrix} 2 \\ -3 \\ -5 \\ 5 \end{bmatrix}
$$

Similarly,

$$
[Lv_2]_{\mathcal{B}'} = \begin{bmatrix} 2 \\ -4 \\ -9 \\ 12 \end{bmatrix} \quad \text{and} \quad [Lv_3]_{\mathcal{B}'} = \begin{bmatrix} 2 \\ -5 \\ -2 \\ -1 \end{bmatrix}
$$

so that

$$
{}^{\mathcal{B}'}L^{\mathcal{B}} = [[Lv_1]_{\mathcal{B}'} \quad [Lv_2]_{\mathcal{B}'} \quad [Lv_3]_{\mathcal{B}'}] = \begin{bmatrix} 2 & 2 & 2 \\ -3 & -4 & -5 \\ -5 & -9 & -2 \\ 5 & 12 & -1 \end{bmatrix} \quad \lozenge
$$

The nice thing about the matrix of a linear map is contained in the
following statement:

THEOREM 1

$w = Lv$ and only if $[w]_{\mathcal{B}'} = {}^{\mathcal{B}'}L^{\mathcal{B}}[v]_{\mathcal{B}}$.   □

Thus, to see what map L does to $v \in V$, we can take the coordinates of v and multiply this $n \times 1$ matrix $[v]_{\mathcal{B}}$ by the $m \times n$ matrix $^{\mathcal{B}'}L^{\mathcal{B}}$ to get the m coordinates of Lv with respect to the basis $\mathcal{B}'$ of W.

Example 3

For the map L and bases $\mathcal{B}$ and $\mathcal{B}'$ of Example 2, let

$$v = 1v_1 - 2v_2 + 3v_3 = \begin{bmatrix} 4 \\ -1 \\ 1 \end{bmatrix}$$

Then $[v]_{\mathcal{B}} = \begin{bmatrix} 1 \\ -2 \\ 3 \end{bmatrix}$, so

$$^{\mathcal{B}'}L^{\mathcal{B}}[v]_{\mathcal{B}} = \begin{bmatrix} 2 & 2 & 2 \\ -3 & -4 & -5 \\ -5 & -9 & -2 \\ 5 & 12 & -1 \end{bmatrix} \begin{bmatrix} 1 \\ -2 \\ 3 \end{bmatrix}$$

$$= \begin{bmatrix} 4 \\ -10 \\ 7 \\ -22 \end{bmatrix}$$

Thus

$$[Lv]_{\mathcal{B}'} = \begin{bmatrix} 4 \\ -10 \\ 7 \\ -22 \end{bmatrix}$$

so that

$$Lv = 4w_1 - 10w_2 + 7w_3 - 22w_4$$

$$= \begin{bmatrix} 4 \\ 8 \\ 0 \\ 12 \end{bmatrix} + \begin{bmatrix} 0 \\ -10 \\ 10 \\ 0 \end{bmatrix} + \begin{bmatrix} 0 \\ 0 \\ 7 \\ 14 \end{bmatrix} + \begin{bmatrix} 0 \\ 0 \\ 0 \\ -22 \end{bmatrix}$$

$$= \begin{bmatrix} 4 \\ -2 \\ 17 \\ 4 \end{bmatrix}$$

Checking, we compute Lv directly by plugging $v = \begin{bmatrix} 4 \\ -1 \\ 1 \end{bmatrix}$ into the

formula for L given in Example 2 and get the same result. At this point, the reader should work Exercise 1 to observe the effect on the above calculations of changing one or both of the bases.   ◊

We emphasize that as we change bases in V and W, the column vectors that represent v and Lv will change, and so will the matrix that represents L; nonetheless, a particular vector v in V will change under a fixed linear transformation into the same vector in W despite these changes in numerical representation. How then are all the different representations related?

Suppose that on the space X, for which we have basis $\mathcal{B} = \{u_1, u_2, \ldots, u_n\}$ and for which $x = x_1 u_1 + x_2 u_2 + \cdots + x_n u_n$, we now impose a new basis $\mathcal{B}' = \{v_1, v_2, \ldots, v_n\}$ for which $x = x_1' v_1 + x_2' v_2 + \cdots + x_n' v_n$. Let us now make explicit the ways in which the two representations of x are related. First, let us introduce the column vectors $[v_1]_{\mathcal{B}}, [v_2]_{\mathcal{B}}, \ldots, [v_n]_{\mathcal{B}}$ which we write as

$$[v_j]_{\mathcal{B}} = \begin{bmatrix} a_{1j} \\ a_{2j} \\ \vdots \\ a_{nj} \end{bmatrix}$$

so that $v_j = \sum_i a_{ij} u_i$. We then have

$$x = \sum_j x_j' v_j = \sum_j x_j' \left(\sum_i a_{ij} u_i \right) = \sum_i \left(\sum_j a_{ij} x_j' \right) u_i$$

But we had $x = \sum_i x_i u_i$, and thus the coefficients of our given vector with respect to the basis $\mathcal{B}' = \{v_1, v_2, \ldots, v_n\}$ are related to those with respect to the basis $\mathcal{B} = \{u_1, u_2, \ldots u_n\}$ by the simple formula

$$x_i = \sum_j a_{ij} x_j' \tag{1}$$

If we thus denote by A the matrix whose ij element is a_{ij}, our change of basis reduces to the simple formula

$$[x]_{\mathcal{B}} = A[x]_{\mathcal{B}'} \tag{2}$$

This representation is completely intuitive. The above equation is to be thought of as working in the space X when using the basis $\mathcal{B} = \{u_1, u_2, \ldots, u_n\}$. The columns of the matrix A are simply the representations of the vectors v_j with respect to the basis, and our equation just tells us that our vector x is that linear combination of the vectors $\{v_1, v_2, \ldots, v_n\}$ of $\mathcal{B}'$ indicated by the column vector $[x]_{\mathcal{B}'}$.

Conversely, we may introduce another matrix B, whose columns give us the representations of the vectors $\{u_1, u_2, \ldots, u_n\}$ of $\mathcal{B}$ with respect to the basis $\mathcal{B}'$. The above argument applies, *mutatis mutandis,* to tell us that $[x]_{\mathcal{B}'} = B[x]_{\mathcal{B}}$. We thus deduce that for all x in X, we have $[x]_{\mathcal{B}'} = BA[x]_{\mathcal{B}'}$ and $[x]_{\mathcal{B}} = AB[x]_{\mathcal{B}}$. From this, we further deduce that $AB = BA = I$, the identity matrix, so that A and B are inverse matrices; that is, $A = B^{-1}$ and $B = A^{-1}$, and each "undoes" the effect of the other.

Next, let us be given a linear transformation $F: X \to X$, and let the bases $\mathcal{B}$ and $\mathcal{B}'$ be related by

$$[x]_{\mathcal{B}} = A[x]_{\mathcal{B}'}; \quad [x]_{\mathcal{B}'} = A^{-1}[x]_{\mathcal{B}}$$

for a suitable invertible matrix A. If we now consider the action of the linear transformation F upon the vector x, we obtain the result ${}^{\mathcal{B}}F^{\mathcal{B}}[x]$ with respect to the first basis and the result ${}^{\mathcal{B}'}F^{\mathcal{B}'}[x]$ with respect to the second basis. Putting these together, we have

$$ {}^{\mathcal{B}}F^{\mathcal{B}}[x]_{\mathcal{B}} = A \ {}^{\mathcal{B}'}F^{\mathcal{B}'}[x]_{\mathcal{B}'} = A \ {}^{\mathcal{B}'}F^{\mathcal{B}'} A^{-1}[x]_{\mathcal{B}} $$

Since this equality must hold for every choice of x, we conclude that

$$ {}^{\mathcal{B}}F^{\mathcal{B}} = A \ {}^{\mathcal{B}'}F^{\mathcal{B}'} A^{-1} \tag{3} $$

and we thus have the following:

THEOREM 2

Two matrices $\widehat{F}$ and $\widetilde{F}$ correspond to the same linear transformation

$$F: X \to X$$

for suitable choices of basis on X if and only if there exists an invertible matrix A such that

$$\widehat{F} = A^{-1}\widetilde{F}A$$

We say that $\widehat{F}$ and $\widetilde{F}$ are **similar** if such an A exists. $\square$

In Section 6–1, we shall study how, given some matrix $\widetilde{F}$, to find A in such a way as to reduce $\widetilde{F}$ to a form $\widehat{F}$ that is more convenient for the analysis of the system behavior.

Example 4

Even though the vectors v_j which form a basis of V are linearly independent, there is no reason to expect that the Lv_j will be linearly independent for an arbitrary linear transformation $L: V \to W$. If $V = W$ and our transformation is the identity I, then $Lv_j = v_j$ and the Lv_j certainly are linearly independent; however, if L is the zero transformation, then for each j we have $Lv_j = 0$ and the Lv_j certainly are *not* linearly independent. $\Diamond$

As the reader may check by working Exercise 3, the dimensions of the range $\mathfrak{R}(L) = L(V)$ and null space $\mathfrak{N}(L)$ of a linear map $L: V \to W$ are related by the equality

$$\dim \mathfrak{R}(L) + \dim \mathfrak{N}(L) = \dim V$$

when the dimension of V is *finite*. Since a proper subspace of an infinite-dimensional vector space may have the same dimension as the whole space, this relationship may be somewhat confusing in general (see Exercises 4 and 5):

Example 5

If $L: C[0, 1] \to \mathbf{R}$ is defined by $L(u) = \displaystyle\int_0^1 u(\xi)\, d\xi$ for each $u \in C[0, 1]$, then

$\dim V = \infty$ and $\dim L(V) = \dim \mathbf{R} = 1$, since L is onto (by Exercise 7 of Section 3–2). In this case $\mathfrak{N}(L) \neq V$ even though $\dim \mathfrak{N}(L) = \infty$, since $\mathfrak{N}(L)$ is the set of all continuous functions having zero area on $[0, 1]$. $\Diamond$

It is customary to call $\dim \mathfrak{R}(L)$ the **rank** of L, or of the matrix of L, and to call $\dim \mathfrak{N}(L)$ the **nullity** of L. For an $m \times n$ matrix A of field elements $a_{ij} \in K$, let $[A]_{*j}$ denote the jth *column* $\begin{bmatrix} a_{1j} \\ a_{2j} \\ \vdots \\ a_{mj} \end{bmatrix}$; let $[A]_{i*} = [a_{i1} \ \ a_{i2} \ldots a_{in}]$ be

the ith *row* of A; and let $[A]_{ij} = a_{ij}$ be the ijth *entry* of A. Then we may define the **column space** of A, denoted by $CS(A)$, by

$$CS(A) = \{k_1[A]_{*1} + \cdots + k_n[A]_{*n} \,|\, k_j \in K\}$$

which is the set of all $v \in K^m$ which can be built by linear combinations of the n vectors $[A]_{*1}, \ldots, [A]_{*n}$ of K^m. The **solution space** of a matrix equation $Ax = 0$ is just the set

$$\{x \in K^n \,|\, Ax = 0 \in K^m\}$$

and is thus just the kernel or nullspace of A.

THEOREM 3

The rank of L is just the dimension of the *column space* of the matrix ${}^{\mathfrak{B}'}L^{\mathfrak{B}}$, and the nullity of L is the dimension of the *solution space* of the matrix equation

$$ {}^{\mathfrak{B}'}L^{\mathfrak{B}} \begin{bmatrix} x_1 \\ \vdots \\ x_n \end{bmatrix} = \begin{bmatrix} 0 \\ \vdots \\ 0 \end{bmatrix} $$

Proof

Exercises 6 and 7.   □

As the reader may verify in Exercises 9 and 10, multiplication of matrices corresponds to composition of the linear transformations they represent; addition of two matrices corresponds to the addition of the two linear transformations they represent; and multiplication of a matrix by a scalar corresponds to multiplying the linear transformation that the matrix represents by the given scalar.

In the next section we discuss how to approximate a given system by a linear system, and how (for finite-dimensional systems) we can compute the matrices which describe the linear approximations to f and η.

EXERCISES FOR SECTION 3-3

✔ **1.**   In Examples 2 and 3, suppose that we decided not to use the given basis $\mathfrak{B}' = \{w_1, w_2, w_3, w_4\}$ but the "usual basis"

$$ \mathfrak{B}'' = \left\{ \begin{bmatrix} 1 \\ 0 \\ 0 \\ 0 \end{bmatrix}, \begin{bmatrix} 0 \\ 1 \\ 0 \\ 0 \end{bmatrix}, \begin{bmatrix} 0 \\ 0 \\ 1 \\ 0 \end{bmatrix}, \begin{bmatrix} 0 \\ 0 \\ 0 \\ 1 \end{bmatrix} \right\} $$

for $\mathbf{R}^4$ instead.

   (a)   Compute the new matrix ${}^{\mathfrak{B}''}L^{\mathfrak{B}}$ for L.

(b) Compute $^{\mathfrak{B}''}L^{\mathfrak{B}}[v]_{\mathfrak{B}}$ and compare with $[Lv]_{\mathfrak{B}''}$ for $v = \begin{bmatrix} 4 \\ -1 \\ 1 \end{bmatrix}$.

(c) Repeat (a) and (b) using $\mathfrak{B}''' = \left\{ \begin{bmatrix} 1 \\ 0 \\ 0 \end{bmatrix}, \begin{bmatrix} 0 \\ 1 \\ 0 \end{bmatrix}, \begin{bmatrix} 0 \\ 0 \\ 1 \end{bmatrix} \right\}$, the usual basis for $\mathbf{R}^3$.

✔ **2.** Let $L(v) = \dfrac{dv}{dx}$ on $V = \{ f | f(x) = \alpha_2 x^2 + \alpha_1 x + \alpha_0$, for all $\alpha_i \in \mathbf{R}$ and for all $x \in \mathbf{R}\}$

    (a) Show that if $v_1(x) = -1 + x^2$
$$v_2(x) = -1 + x$$
$$v_3(x) = 3 - x - x^2 \text{ for } x \in \mathbf{R},$$
then $\{v_1, v_2, v_3\}$ is a basis $\mathfrak{B}$ for V.

    (b) Find $^{\mathfrak{B}}L^{\mathfrak{B}}$, the matrix of $L: V \to V$ relative to the basis $\mathfrak{B}$. Check your work by computing the action of L on $v(x) = x^2 - 2x$.

✔ **3.** Prove $\dim \mathfrak{R}(L) + \dim \mathfrak{N}(L) = \dim V$, where $\dim V = n$ and L maps V linearly into W. [*Hint:* Let $v_1, \ldots, v_k$ be a basis for the subspace $\mathfrak{N}(L)$, and adjoin linearly independent vectors $v_{k+1}, \ldots, v_n$ which make up a basis $v_1, \ldots, v_k, v_{k+1}, \ldots, v_n$ for all of V. Then show that $L(v_{k+1}), \ldots, L(v_n)$ span $\mathfrak{R}(V)$ and are linearly independent and hence a basis for $\mathfrak{R}(V)$.]

4. Let V be the set of all polynomials in t with real coefficients, and let $L: V \to V$ be defined by $L(u) = \dfrac{du}{dt}$ for each $u \in V$. Show that $\dim L(V) = \dim V = \infty$ and $\dim \mathfrak{N}(L) = 1$.

5. For $L: V \to W$ as in Exercise 3, let $\dim W = \dim V = n$. Prove that L is onto if and only if L is one-to-one. Compare this result with the infinite-dimensional cases of Exercise 4 and Example 5.

✔ **6.** Prove that $CS(^{\mathfrak{B}'}L^{\mathfrak{B}})$ is the set of all coordinates for the vectors in the range of L, to deduce that the rank of L equals the dimension of the column space of $^{\mathfrak{B}'}L^{\mathfrak{B}}$.

✔ **7.** Prove (yes, it *is* trivial) that the nullity of L is just the dimension of the solution space of the matrix equation

$$^{\mathfrak{B}'}L^{\mathfrak{B}} \begin{bmatrix} x_1 \\ \vdots \\ x_n \end{bmatrix} = \begin{bmatrix} 0 \\ \vdots \\ 0 \end{bmatrix}$$

8. **(a)** Let $L: \mathbf{R} \to \mathbf{R}$, and let $\mathfrak{B} = \mathfrak{B}' = \{1\}$, the usual basis for $\mathbf{R}^1$. Show that if L is linear it must have the formula $L(x) = ax$ for some real number a. Find the matrix $^{\mathfrak{B}}L^{\mathfrak{B}}$ from the formula $L(x) = ax$.

    (b) Let $V = \mathbf{R}$ and let W be any vector space. Let $L: V \to W$ be any linear map. Choose bases $\mathfrak{B}$ and $\mathfrak{B}'$ for V and W and compute the matrix $^{\mathfrak{B}'}L^{\mathfrak{B}}$ for L.

    (c) Repeat part (b) with $V = \mathbf{R}^n$.

    (d) Let $L: \mathbf{R}^n \to \mathbf{R}^m$ be a linear map, and let $\mathfrak{B}$ and $\mathfrak{B}'$ be the usual bases for $\mathbf{R}^n$ and $\mathbf{R}^m$, respectively. Show that L must have the formula $L(x) = Ax$ (that is, ordinary matrix multiplication), where A is an $m \times n$ matrix. From the above formula for L, compute $^{\mathfrak{B}'}L^{\mathfrak{B}}$.

✔ **9.** This exercise is designed to show that the usual definition of matrix multiplication arises by considering composition of linear maps. Let $A: V \to W$ and $B: U \to V$ be two linear transformations. Then we define their **composite** $AB: U \to W$ such that $u \mapsto A(Bu)$.

(a)   Check that, if A and B are linear, then so too is AB.

(b)   Let us choose bases $\mathcal{B}$, $\mathcal{B}'$, and $\mathcal{B}''$ on U, V, and W and let $^{\mathcal{B}''}A^{\mathcal{B}'}$ be the $n \times p$ matrix $[a_{ij}]$, let $^{\mathcal{B}'}B^{\mathcal{B}}$ be the $m \times p$ matrix $[b_{ij}]$, and let $^{\mathcal{B}''}(AB)^{\mathcal{B}}$ be the $m \times n$ matrix $[c_{ij}]$. Verify, by computing $[ABu]$ in two distinct ways, that

$$c_{ij} = \sum_{k=1}^{p} a_{ik}b_{kj}$$

That is, show that matrix multiplication does indeed correspond to composition of linear maps.

✔ **10.**   Given any two functions f and f' from one vector space V to another vector space W over the same field K, we define (by point-to-point operations) the sum of the two functions $f + f' : V \to W$ such that $v \mapsto f(v) + f'(v)$; and we may multiply any such function by any scalar k according to $kf : V \to W$ such that $v \mapsto k \cdot f(v)$. Since these definitions work for *arbitrary* functions f and f', we know that we may add together any pair of linear transformations with the same domain and range, and we may multiply any linear transformation by a scalar.

(a)   Prove that if L and M are *linear* maps $V \to W$ then so is $L + M$, and so is kL for any scalar k in K.

(b)   Let $\mathcal{B}$ and $\mathcal{B}'$ be bases for V and W, respectively. Then verify that, with the conventions of part (a), if $^{\mathcal{B}'}L^{\mathcal{B}} = [l_{ij}]$ and $^{\mathcal{B}'}M^{\mathcal{B}} = [m_{ij}]$, then

$$^{\mathcal{B}'}(L + M)^{\mathcal{B}} = [l_{ij} + m_{ij}]$$

and

$$^{\mathcal{B}'}(kL)^{\mathcal{B}} = [kl_{ij}]$$

(c)   Give the argument that for fixed bases $\mathcal{B}$ and $\mathcal{B}'$, the matrix $^{\mathcal{B}'}L^{\mathcal{B}}$ of the linear map $L : V \to W$ is *unique*.

(d)   Suppose that L is one-to-one and onto, so that the inverse map $L^{-1} : W \to V$ exists. Prove that L^{-1} is linear, and also that

$$^{\mathcal{B}}(L^{-1})^{\mathcal{B}'} = (^{\mathcal{B}'}L^{\mathcal{B}})^{-1}$$

3–4   System Linearization and System Approximation

The task of this section is to develop the general theory of *linearization*, i.e., of approximating the behavior of a system in some region of phase space by a *linear* system, that is, one whose state-transition equation can be written

$$x(t + 1) = F(t)x(t) + G(t)u(t)$$

in the discrete-time case, or

$$\dot{x}(t) = F(t)x(t) + G(t)u(t)$$

in the continuous-time case, where $F(t)$ and $G(t)$ are suitable linear maps (which will vary with time t if the system is time-varying).

To motivate our general theory of linearization, let us see how in ordinary calculus we consider the derivative as providing a linear approximation to a function:

Given a differentiable function $f: \mathbf{R} \to \mathbf{R}$, its derivative at a point x is often thought of as the slope of the tangent to the graph of f at the abscissa x; it is thus determined by considering the limiting slopes of nearby chords, as in Figure 3–9, so that the derivative is given by the formula

$$f'(x) = \lim_{h \to 0} \frac{f(x + h) - f(x)}{h} \tag{1}$$

and is defined whenever this limit exists.

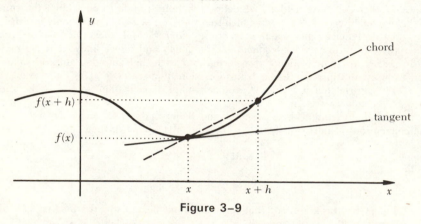

Figure 3–9

Rephrasing the above definition of the derivative, we may write

$$f(x + h) = f(x) + f'(x) \cdot h + \psi(h) \tag{2}$$

where $\psi(h)$ is a term that tends to 0 faster than h tends to 0, i.e., a term such that $\dfrac{\psi(h)}{h} \to 0$. We say that "$\psi(h)$ **is of smaller order than** h," and abbreviate this as "$\psi(h)$ is $o(h)$" where o stands for order, a small o indicating a small order.

What the above equation really says is that near to the point $(x, f(x))$, the tangent $h \mapsto f'(x)h$ yields that line which best approximates the curve, considered as a function of the increment h. We may refer to the tangent as the *best linear approximation* to the curve at the point $(x, f(x))$. Of course, as the graph shows, as the increment h becomes large, the linear approximation may become increasingly bad. We should note, though, that if the curve is itself a straight line, then the linear approximation is perfect, no matter how big the increment. That the extent of the local description may be fairly large is evidenced by the wide applicability of Ohm's law, which gives us a linear relationship between the current through a resistor and the voltage across the resistor. This linear relationship holds over a wide range, but we realize that if we increase the current to the melting point, then the linear relationship will no longer hold.

Turning now to system theory, it will be important for us to know that we may generalize our one-dimensional notion of a linear approximation to obtain a general notion of linear approximation which holds in arbitrary Banach spaces. [We remind the reader that he should not feel uncomfortable about the phrase "arbitrary Banach space" but may interpret the phrase as, for example, "$\mathbf{R}^n$ or $\mathbf{C}^n$ for arbitrary n."] Thus, given any system whose input, output, and state spaces are all Banach spaces, we may, in a wide range of circumstances, set up linear approximations to the system of interest.

THE DERIVATIVE AS A LINEAR MAP

Let $f: X \to W$ where X and W are Banach spaces, and let us *fix* $x_0 \in X$. Note that f need *not* be linear. We say that f is **differentiable at** x_0 if there exists a continuous linear map $L: X \to W$ and a map ψ such that for all h

$$f(x_0 + h) = f(x_0) + L(h) + \psi(h) \tag{3}$$

where the map $\psi: X \to W$ is such that $\psi(h)$ is $o(h)$, in the sense that

$$\lim_{\|h\| \to 0} \frac{\|\psi(h)\|}{\|h\|} = 0 \tag{4}$$

The linear map L is regarded as a linear approximation to f near x_0 and the map $\psi(h)$ is an "error" term, which goes to zero faster than h. We consider h to be a vector increment, usually small ($\|h\|$ small), in the argument of f; and we call L, when it exists, the **derivative of** f **at** x_0. This does indeed correspond to our discussion of scalar functions:

Example 1

Let $f: \mathbf{R} \to \mathbf{R}$ be a differentiable function at $x_0 \in \mathbf{R}$ and call the number $f'(x_0)$ the derivative of f at x_0 just as in elementary calculus. To show that this previous notion is compatible with our new definitions, we must find a linear map $L: \mathbf{R} \to \mathbf{R}$ and a map $\psi: \mathbf{R} \to \mathbf{R}$ having the appropriate properties.

If we try the maps $L(h) = f'(x_0) \cdot h$ (ordinary multiplication in $\mathbf{R}$) and $\psi(h) = (f(x_0 + h) - f(x_0)) - (f'(x_0)h)$ we see that L is linear and

$$\lim_{\|h\| \to 0} \frac{\|\psi(h)\|}{\|h\|} = \lim_{\|h\| \to 0} \left(\frac{|(f(x_0 + h) - f(x_0)) - f'(x_0)h|}{|h|} \right)$$

$$= \lim_{|h| \to 0} \left(\left| \frac{f(x_0 + h) - f(x_0)}{h} - f'(x_0) \right| \right)$$

since the norm $\| \cdot \|$ on $\mathbf{R}$ is just $|\cdot|$, the absolute value. But from calculus, $f'(x_0)$ is defined as

$$f'(x_0) = \lim_{h \to 0} \frac{f(x_0 + h) - f(x_0)}{h}$$

so

$$\lim_{\|h\| \to 0} \frac{\|\psi(h)\|}{\|h\|} = 0$$

as hoped. [See Exercise 1.]

Thus, the two maps we tried have the required properties and we conclude that differentiable functions from calculus have derivatives in our new sense, too. ◊

LEMMA 1

The derivative of a map at a point is unique, if it exists.

Proof Outline

Let $f: X \to W$ be an arbitrary function, and choose $x_0 \in X$. Suppose that there exist maps L_1 and ψ_1 such that $f(x_0 + h) = f(x_0) + L_1(h) + \psi_1(h)$, where L_1 is linear and continuous and $\psi_1(h)$ is $o(h)$.

Suppose that there also exist maps L_2 and ψ_2 such that L_2 is linear and continuous and $\psi_2(h)$ is also $o(h)$, for which $f(x_0 + h) = f(x_0) + L_2(h) + \psi_2(h)$. Equating the two expressions for $f(x_0 + h)$, we get $L(h) = \psi(h)$ for all increments, where $L = L_1 - L_2$ and $\psi = \psi_2 - \psi_1$. It is a simple matter to show that ψ is $o(h)$ and that L is linear (Exercise 12).

Let x be any nonzero vector in X, and let $e = \dfrac{1}{\|x\|} x$. Let $\alpha \neq 0$ be any real number, and let $h = \alpha e$. Then $L(e) = \dfrac{\psi(\alpha e)}{\alpha}$ for all $\alpha \neq 0$, and we may then show from $\lim\limits_{\|h\| \to 0} \dfrac{\|\psi(h)\|}{\|h\|} = 0$ that $\lim\limits_{\alpha \to 0} \dfrac{\psi(\alpha e)}{\alpha} = 0$ and hence that $L(e) = 0$. Thus, $L(x) = 0$ for all $x \in X$; therefore, $L_1 = L_2$, the unique derivative. □

Thus, we may speak of *the* derivative of a differentiable map.

We saw in the last example that the matrix of our linear map L (our derivative) was just $[f'(x_0)]$. (See Exercise 8 of Section 3–3.) This was no accident, and we are motivated to write $L(\cdot)$ as $f'(x_0, \cdot)$ for our derivative map, where the notation reminds us that the particular linear map required depends upon x_0, the point where we want the derivative, as well as upon f.

Let us now consider what happens when we try to find the derivative $f'(x_0, \cdot)$ of a linear map.

Example 2

Let $f: X \to W$ be linear and continuous. To see if f is differentiable at $x_0 \in X$, we seek maps L and ψ such that L is linear and ψ is $o(h)$ as $\|h\| \to 0$. Suppose we try L equal to f itself as our linear map and try $\psi(h) \equiv 0$, the identically zero map. Can we write

$$f(x_0 + h) = f(x_0) + L(h) + \psi(h)$$

with this choice of L and ψ? Yes, since f is linear. Substituting L and ψ into this equation, we obtain

$$
\begin{aligned}
f(x_0 + h) &= f(x_0) + f(h) \\
&= f(x_0) + L(h) + \psi(h)
\end{aligned}
$$

Thus, $f'(x_0, \cdot) = f(\cdot)$, so the derivative of a linear continuous map is the map itself.   ◊

This last result is not all that surprising, in that it states that the linear approximation to a linear function is the function itself.

When the Banach spaces X and W are $\mathbf{R}^n$ and $\mathbf{R}^m$, the derivative turns out to have a simple representation as a matrix of certain ordinary partial derivatives.

Example 3

Let $f: \mathbf{R}^n \to \mathbf{R}^m$ be a differentiable function

$$f(x + h) = f(x) + f'(x, h) + \psi(h) \quad \text{with} \quad \psi(h) = o(h)$$

Representing vectors in the two spaces as column matrices, we may write the action of f in terms of components as

$$
\begin{bmatrix} x_1 \\ x_2 \\ \vdots \\ x_n \end{bmatrix}
\mapsto
\begin{bmatrix} f_1(x_1, \ldots, x_n) \\ f_2(x_1, \ldots, x_n) \\ \vdots \\ f_m(x_1, \ldots, x_n) \end{bmatrix}
$$

where $f_j : \mathbf{R}^n \to \mathbf{R}$, for $1 \leq j \leq m$, is the jth coordinate function of f. Now let e_j be the jth unit vector in the usual basis for $\mathbf{R}^n$,

$$e_j = \begin{bmatrix} 0 \\ 0 \\ \vdots \\ 1 \\ \vdots \\ 0 \end{bmatrix} \leftarrow j\text{th of } n \text{ entries}$$

and consider increments constrained to lie along the jth coordinate direction

$$h = \alpha e_j \text{ with } \alpha \in \mathbf{R}$$

so that $h \in \mathbf{R}^n$. Plugging such an h into the equation

$$f(x + h) = f(x) + f'(x, h) + \psi(h)$$

we get

$$f(x + \alpha e_j) = f(x) + f'(x, \alpha e_j) + \psi(\alpha e_j)$$

Looking at the ith components of the terms of this equation, we have

$$f_i(x + \alpha e_j) = f_i(x) + [f'(x, \alpha e_j)]_i + \psi_i(\alpha e_j)$$
$$= f_i(x) + \alpha [f'(x, e_j)]_i + \psi_i(\alpha e_j)$$

Thus,

$$[f'(x, e_j)]_i = \frac{f_i(x + \alpha e_j) - f_i(x)}{\alpha} - \frac{\psi_i(\alpha e_j)}{\alpha}$$

and since

$$\|h\| = \|\alpha e_j\| = |\alpha| \, \|e_j\| = |\alpha|$$

and

$$|\psi_i(h)| \leq \|\psi(h)\| = \sqrt{\sum_{k=1}^{m} [\psi_k(h)]^2}$$

we arrive at

$$\lim_{\|h\| \to 0} \frac{\psi(h)}{\|h\|} = 0 \implies \lim_{\alpha \to 0} \frac{\psi_i(\alpha e_j)}{\alpha} = 0$$

Using these relationships,

$$[f'(x, e_j)]_i = \lim_{\alpha \to 0} \frac{f_i(x + \alpha e_j) - f_i(x)}{\alpha}$$

$$= \frac{\partial f_i}{\partial x_j}(x)$$

Thus, the ith component of our derivative map when evaluated at e_j is just the jth partial derivative of f_i at x.

What, then, is the matrix representation of our linear map $f'(x, \cdot)$? Using the standard bases $\mathfrak{B}$ and $\mathfrak{B}'$ for $\mathbf{R}^n$ and $\mathbf{R}^m$ and denoting by J_f the matrix of $f'(x, \cdot)$, the jth column of J_f is just the $m \times 1$ matrix $f'(x, e_j)$. The ith entry of the jth column, $[J_f]_{ij}$, is $[f'(x, e_j)]_i$, which we just showed to be $\dfrac{\partial f_i}{\partial x_j}(x)$.

The resulting matrix of $f'(x, \cdot)$,

$$J_f(x) = {}^{\mathfrak{B}'} f'(x, \cdot)^{\mathfrak{B}} = \begin{bmatrix} \dfrac{\partial f_1}{\partial x_1}(x) & \dfrac{\partial f_1}{\partial x_2}(x) & \cdots & \dfrac{\partial f_1}{\partial x_n}(x) \\ \vdots & \vdots & & \vdots \\ \dfrac{\partial f_m}{\partial x_1}(x) & \dfrac{\partial f_m}{\partial x_2}(x) & \cdots & \dfrac{\partial f_m}{\partial x_n}(x) \end{bmatrix} \tag{5}$$

is called the **Jacobian Matrix of f at x** and appears extensively in the literature. ◊

Notice in this last example that when $f: \mathbf{R}^n \to \mathbf{R}^1$, the matrix of the derivative is just the **gradient** of f,

$$\left[\dfrac{\partial f}{\partial x_1}(x), \dfrac{\partial f}{\partial x_2}(x), \ldots, \dfrac{\partial f}{\partial x_n}(x) \right] \tag{6}$$

Example 4

Let $f: \mathbf{R}^3 \to \mathbf{R}^2$ where

$$\begin{bmatrix} x_1 \\ x_2 \\ x_3 \end{bmatrix} \mapsto \begin{bmatrix} x_1 x_2 \\ x_1 + x_2 + x_3 \end{bmatrix}$$

i.e.,

$$f_1(x_1, x_2, x_3) = x_1 x_2$$

and

$$f_2(x_1, x_2, x_3) = x_1 + x_2 + x_3$$

Proceeding directly,

$$f(x + h) = f \begin{bmatrix} x_1 + h_1 \\ x_2 + h_2 \\ x_3 + h_3 \end{bmatrix} = \begin{bmatrix} (x_1 + h_1)(x_2 + h_2) \\ (x_1 + h_1) + (x_2 + h_2) + (x_3 + h_3) \end{bmatrix}$$

$$= \begin{bmatrix} x_1 x_2 \\ x_1 + x_2 + x_3 \end{bmatrix} + \begin{bmatrix} x_1 h_2 + x_2 h_1 \\ h_1 + h_2 + h_3 \end{bmatrix} + \begin{bmatrix} h_1 h_2 \\ 0 \end{bmatrix}$$

$$= f(x) + \begin{bmatrix} x_2 & x_1 & 0 \\ 1 & 1 & 1 \end{bmatrix} \begin{bmatrix} h_1 \\ h_2 \\ h_3 \end{bmatrix} + \begin{bmatrix} h_1 h_2 \\ 0 \end{bmatrix}$$

Thus

$$f'(x, h) = \begin{bmatrix} x_2 & x_1 & 0 \\ 1 & 1 & 1 \end{bmatrix} \begin{bmatrix} h_1 \\ h_2 \\ h_3 \end{bmatrix}$$

and

$$\psi(h) = \begin{bmatrix} h_1 h_2 \\ 0 \end{bmatrix}$$

and

$$\frac{\psi(h)}{\|h\|} = \frac{1}{\sqrt{h_1^2 + h_2^2}} \begin{bmatrix} h_1 h_2 \\ 0 \end{bmatrix} \rightarrow 0 \text{ as } \|h\| \rightarrow 0 \quad [\text{See Exercise 1}]$$

The reader may check, by computing the partial derivatives, that $J_f(x)$ computed as $\left[\dfrac{\partial f_i}{\partial x_j} (x) \right]$ does indeed equal $\begin{bmatrix} x_2 & x_1 & 0 \\ 1 & 1 & 1 \end{bmatrix}$. ◊

Thus, not only does our Banach space definition yield a compact description of derivative well-suited for the development of theory, but it is one that "unpacks" easily for computational purposes, whenever we wish to work with respect to given bases.

We now illustrate how our definition of derivative also applies to maps on the infinite-dimensional spaces frequently encountered in modern system theory.

Example 5

Let $f: C[0, 2\pi] \rightarrow \mathbf{R}$ be defined by the rule $f(x) = \displaystyle\int_0^{2\pi} x^2(\xi) \, d\xi$ for $x \in C[0, 2\pi]$. If we consider $f(x + h)$, we see that

$$f(x + h) = \int_0^{2\pi} x^2(\xi) \, d\xi + 2 \int_0^{2\pi} x(\xi) h(\xi) \, d\xi + \int_0^{2\pi} h^2(\xi) \, d\xi$$

$$= f(x) + 2 \int_0^{2\pi} x(\xi) h(\xi) \, d\xi + \int_0^{2\pi} h^2(\xi) \, d\xi$$

Thus, logical candidates for our two maps are

$$f'(x, h) = 2 \int_0^{2\pi} x(\xi) h(\xi) \, d\xi$$

and

$$\psi(h) = \int_0^{2\pi} h^2(\xi) \, d\xi$$

since $f'(x, \cdot)$ is linear and $\psi(h)$ looks like it might vanish as h gets small.

To carry out the details, we use the uniform norm

$$\|x\| = \max_{0 \leq \xi \leq 2\pi} |x(\xi)|$$

on $C[0, 2\pi]$. With this norm,

$$|h(\xi)| \leq \|h\| \text{ for } 0 \leq \xi \leq 2\pi$$

so

$$0 \leq \int_0^{2\pi} h^2(\xi) \, d\xi \leq \int_0^{2\pi} \|h\|^2 \, d\xi$$

$$\leq \|h\|^2 \int_0^{2\pi} d\xi$$

$$= \|h\|^2 \, 2\pi$$

Thus,

$$0 \leq \frac{\psi(h)}{\|h\|} \leq 2\pi \frac{\|h\|^2}{\|h\|} = \|h\| \, (2\pi)$$

so as $\|h\| \to 0$, $\dfrac{\psi(h)}{\|h\|}$ is trapped and crowded to approach zero as we require.

This completes the proof that the derivative of f at x is given by

$$f'(x, h) = 2 \int_0^{2\pi} x(\xi)h(\xi) \, d\xi \quad \text{[See Exercises 2, 3, and 4.]} \qquad \lozenge$$

Returning to a comment on our definition of the derivative of a map, note that in writing down the term $o(h)$ we make no commitment about its size as h becomes large. As we can see from our ordinary one-dimensional use of the tangent, it gives a linear approximation which may be very bad indeed as we move away from the tangent point—all we guarantee is that our approximation is good locally, and we express this by saying that the term $o(h)$ is guaranteed to go to 0 faster than h itself as the increment h tends to 0.

THE LINEARIZATION OF DYNAMIC SYSTEMS

Let us now use our notion of the derivative of a function as its linear approximation to introduce the idea of a linear system as an approximation to more general systems.

First, let us consider a local state transition map, $f: T \times X \times U \to X$, for a deterministic system. In so doing, we shall be accomplishing two tasks at once, for our theory will hold irrespective of whether f enters into the dynamics

$$x(t + 1) = f(t, x(t), u(t))$$

of a discrete-time system, or into the dynamics

$$\dot{x}(t) = f(t, x(t), u(t))$$

of a continuous-time system. For our theory of linearization, we shall consider derivatives not in the time variable, but rather in the state and input variables. Specifically, for each fixed $t_0 \in T$, we shall require $f(t_0, \cdot, \cdot)$ to be a differentiable function defined on the Banach space $X \times U$ [see Exercise 5: the norm is $\|(\hat{x}, \hat{u})\| = \|\hat{x}\| + \|\hat{u}\|$] and taking values in the Banach space X.

In our general theory, we had—changing the notation to avoid confusion below—that $g: V \to W$ has a derivative $g'(v_0, \cdot)$ at $v_0 \in V$ if there exists a linear map $g'(v_0, \cdot): V \to W$ and a function $\psi(h)$ which is $o(h)$ such that

$$g(v_0 + h) = g(v_0) + g'(v_0, h) + \psi(h)$$

Now, we take g to be the function $f(t_0, \cdot, \cdot)$ so that V becomes $X \times U$ and W becomes X. Let us now carry out our expansion about the origin of $X \times U$, so that v_0 becomes $(0_X, 0_U)$, and let us take h to be any vector $(\hat{x}, \hat{u})$ in $X \times U$. Then our general rule becomes

$$\begin{aligned} f(t_0, \hat{x}, \hat{u}) &= f(t_0, (0_X, 0_U) + (\hat{x}, \hat{u})) \\ &= f(t_0, 0_X, 0_U) + f'(t_0, (0_X, 0_U), (\hat{x}, \hat{u})) + \psi(\hat{x}, \hat{u}) \end{aligned}$$

where $f'(t_0, (0_X, 0_U), \cdot)$ is a linear map of $X \times U \to X$, and $\dfrac{\psi(\hat{x}, \hat{u})}{\|(\hat{x}, \hat{u})\|} \to 0$

as $\|(\hat{x}, \hat{u})\| \to 0$, i.e., as both $\|\hat{x}\|$ and $\|\hat{u}\|$ tend to 0, since $\|(\hat{x}, \hat{u})\| = \|\hat{x}\| + \|\hat{u}\|$.

To simplify our equations, let us abbreviate $f'(t_0, (0_X, 0_U), \cdot)$ to $f_0'(t_0, \cdot)$. Since $f_0'(t_0, \cdot)$ is linear on $X \times U$, we write

$$\begin{aligned} f_0'(t_0, \hat{x}, \hat{u}) &= f_0'(t_0, (\hat{x}, 0_U) + (0_X, \hat{u})) \\ &= f_0'(t_0, \hat{x}, 0_U) + f_0'(t_0, 0_X, \hat{u}) \end{aligned}$$

Thus, we have two maps

$$f_1(t_0): X \to X \text{ such that } \hat{x} \mapsto f_0'(t_0, \hat{x}, 0_U)$$

and

$$f_2(t_0): U \to X \text{ such that } \hat{u} \mapsto f_0'(t_0, 0_X, \hat{u})$$

which are both linear, and such that we may write

$$f(t_0, \hat{x}, \hat{u}) = f(t_0, 0_X, 0_U) + f_1(t_0)(\hat{x}) + f_2(t_0)(\hat{u}) + \psi(\hat{x}, \hat{u})$$

Now let us learn directly who f_1 and f_2 *really* are. If we set $\hat{u} = 0$, the last equation reduces to

$$f(t_0, \hat{x}, 0) = f(t_0, 0, 0) + f_1(t_0)(\hat{x}) + \psi(\hat{x}, 0)$$

where $\psi(\hat{x}, 0)$ is $o(\hat{x})$. [See Exercise 5(c).] Thus, $f_1(t_0)$ is simply the derivative of $f(t_0, \hat{x}, 0)$ *with respect to* $\hat{x}$ at $(0, 0)$. Similarly, $f_2(t_0)$ is the derivative of $f(t_0, 0, \hat{u})$ *with respect to* $\hat{u}$ at $(0, 0)$. There was forethought in our notation, then: $f_1(t_0)$ is the partial derivative with respect to the *first* (vector) variable $\hat{x}$ of the pair $(\hat{x}, \hat{u})$, while $f_2(t_0)$ is the partial derivative with respect to the *second* (vector) variable $\hat{u}$ of the pair $(\hat{x}, \hat{u})$.

When we interpret the function f as the next-state function of a discrete-time system, it is natural to demand that if the system is in the zero state, and receives the zero input, then it will remain in the zero state. If we are considering continuous-time systems, then it is natural to demand that when the system is in the zero state, and receives the zero input, then its rate of change will remain zero. We call such points, where $\dot{x}(t)$ is zero, **equilibrium points;** we shall see more of them in Section 7–4, when we study stability theory.

Example 6

For the simple unforced pendulum of Example 3–2–2, we have

$$\dot{x}(t_0) = f\left(t_0, \begin{bmatrix} x_1(t_0) \\ x_2(t_0) \end{bmatrix}, 0\right) = \begin{bmatrix} x_2(t_0) \\ -\omega^2 \sin x_1(t_0) \end{bmatrix}$$

and clearly,

$$f\left(t_0, \begin{bmatrix} 0 \\ 0 \end{bmatrix}, 0\right) = \begin{bmatrix} 0 \\ 0 \end{bmatrix}$$

so $\left(\begin{bmatrix} 0 \\ 0 \end{bmatrix}, 0\right)$ is an equilibrium point. So is $\left(\begin{bmatrix} \pi \\ 0 \end{bmatrix}, 0\right)$ since $\dot{x}(t_0) = f\left(t_0, \begin{bmatrix} \pi \\ 0 \end{bmatrix}, 0\right) = \begin{bmatrix} 0 \\ 0 \end{bmatrix}$. We call the latter an unstable equilibrium point; can you guess why?   ◊

Thus, in system theory it will be natural to regard the local function f as satisfying the condition

$$f(t_0, 0_X, 0_U) = 0_X$$

so that

$$f(t_0, \hat{x}, \hat{u}) = f_1(t_0)(\hat{x}) + f_2(t_0)(\hat{u}) + \psi(\hat{x}, \hat{u})$$

This means that our expression for the local state transition map takes the form

$$f(t_0, \widehat{x}, \widehat{u}) = F_{t_0}\widehat{x} + G_{t_0}\widehat{u} + \psi(\widehat{x}, \widehat{u}) \tag{7}$$

where $F_{t_0}: X \to X$ and $G_{t_0}: U \to X$ are both linear transformations, depending, of course, on the time $t_0 \in T$.

Throwing away the "higher order" error term $\psi(\widehat{x}, \widehat{u})$, we get the **linear approximation**

$$f(t, \widehat{x}, \widehat{u}) \approx F_t\widehat{x} + G_t\widehat{u}$$

If the spaces X and U are finite dimensional, say $\mathbf{R}^n$ and $\mathbf{R}^m$, we can use the matrix representations $F(t)$ and $G(t)$, say, of the linear maps F_t and G_t, respectively, to get the familiar-looking [see Exercise 2 of Section 3–2] equation

$$f(t, x(t), u(t)) \approx F(t)x(t) + G(t)u(t) \tag{8}$$

A similar line of reasoning yields the linear approximation $\eta(t, x(t)) = H_t x(t)$ to the output function $\eta(t, \cdot): X \to Y$, with X and Y being Banach spaces. In the finite-dimensional case, say $X = \mathbf{R}^n$ and $Y = \mathbf{R}^p$, letting $H(t)$ be the $p \times n$ matrix of H_t, we get

$$y(t) = H(t)x(t) \tag{9}$$

The pair of matrix equations

$$\left.\begin{array}{l} \dot{x}(t) = F(t)x(t) + G(t)u(t) \\ y(t) = H(t)x(t) \end{array}\right\} \tag{10}$$

is taken as the starting point in most texts on continuous-time, linear systems. We have taken some care to point out the philosophy and underlying mathematical structure of this formulation, so that the reader understands that *linear systems arise as approximate descriptions of real-world systems, and must be treated accordingly.*

We shall consider linear systems to be time-varying if the linear transformations F_t, G_t, and H_t introduced earlier are themselves functions of time. We are thus led to the following working definitions of linear systems:

DEFINITION 1

A discrete-time system is **linear** if it has *vector* spaces for its input, output, and state spaces U, Y, and X, and if there exist, for each time t in T, three linear transformations

$$F(t):X \to X$$
$$G(t):U \to X$$
$$H(t):X \to Y$$

such that state transitions and outputs are given by the equations

$$x(t+1) = F(t)x(t) + G(t)u(t)$$
$$y(t) = H(t)x(t) \qquad \bigcirc$$

DEFINITION 2

A continuous-time system is **linear** if it has *Banach* spaces for its input, output, and state spaces U, Y, and X, and if there exist, for each time t in T, three linear transformations

$$F(t):X \to X$$
$$G(t):U \to X$$
$$H(t):X \to Y$$

such that state transitions and outputs are given by the equations

$$\dot{x}(t) = F(t)x(t) + G(t)u(t)$$
$$y(t) = H(t)x(t) \qquad \bigcirc$$

In either case, of course, the system will be constant or time-invariant if the linear transformations F, G, and H are themselves constant, i.e., are independent of the time at which we use them. It is on linear systems that much of our ensuing development will focus.

Example 7

To ensure that the reader has followed the linearization argument, let us restate it in terms of partial derivatives for a time-invariant continuous-time system with input space $U = \mathbf{R}^m$, state space $X = \mathbf{R}^n$, and output space $Y = \mathbf{R}^p$ which is described by the $n + p$ equations:

$$\dot{x} = f(x, u) \quad \begin{cases} \dot{x}_1 = f_1(x_1, \ldots, x_n, u_1, \ldots, u_m) \\ \vdots \qquad \vdots \\ \dot{x}_n = f_n(x_1, \ldots, x_n, u_1, \ldots, u_m) \end{cases}$$

$$y = \eta(x) \begin{cases} y_1 = \eta_1(x_1, \ldots, x_n) \\ \vdots \qquad \vdots \\ y_p = \eta_p(x_1, \ldots, x_n) \end{cases}$$

and where we assume that $f(0, 0) = 0$ and $\eta(0) = 0$.

We may then use Taylor's theorem to expand these equations about $x = 0$, $u = 0$ to obtain

$$\begin{cases} \dot{x}_1 = f_1(0, 0) + \dfrac{\partial f_1}{\partial x_1} x_1 + \cdots + \dfrac{\partial f_1}{\partial x_n} x_n + \dfrac{\partial f_1}{\partial u_1} u_1 + \cdots + \dfrac{\partial f_1}{\partial u_m} u_m \\ \qquad\qquad\qquad\qquad\qquad \vdots \qquad\qquad\qquad + \{\text{higher order terms}\} \\ \dot{x}_n = f_n(0, 0) + \dfrac{\partial f_n}{\partial x_1} x_1 + \cdots + \dfrac{\partial f_n}{\partial x_n} x_n + \dfrac{\partial f_n}{\partial u_1} u_1 + \cdots + \dfrac{\partial f_n}{\partial u_m} u_m \\ \qquad\qquad\qquad\qquad\qquad\qquad\qquad\qquad\qquad + \{\text{higher order terms}\} \end{cases}$$

$$\begin{cases} y_1 = \eta_1(0) + \dfrac{\partial \eta_1}{\partial x_1} x_1 + \cdots + \dfrac{\partial \eta_1}{\partial x_n} x_n + \{\text{higher order terms}\} \\ \qquad\qquad\qquad\qquad \vdots \\ y_1 = \eta_p(0) + \dfrac{\partial \eta_p}{\partial x_1} x_1 + \cdots + \dfrac{\partial \eta_p}{\partial x_n} x_n + \{\text{higher order terms}\} \end{cases}$$

which may be written compactly in matrix notation as

$$\dot{x} = Fx + Gu + \{\text{higher order terms}\}$$
$$y = Hx \qquad\ + \{\text{higher order terms}\}$$

where the matrices F, G, and H are given by the following arrays of partial derivatives:

$$F = \frac{\partial f}{\partial x} = \begin{bmatrix} \dfrac{\partial f_1}{\partial x_1} & \cdots & \dfrac{\partial f_1}{\partial x_n} \\ & \cdots & \\ \dfrac{\partial f_n}{\partial x_1} & \cdots & \dfrac{\partial f_n}{\partial x_n} \end{bmatrix}; \quad G = \frac{\partial f}{\partial u} = \begin{bmatrix} \dfrac{\partial f_1}{\partial u_1} & \cdots & \dfrac{\partial f_1}{\partial u_m} \\ & \cdots & \\ \dfrac{\partial f_n}{\partial u_1} & \cdots & \dfrac{\partial f_n}{\partial u_m} \end{bmatrix};$$

$$H = \frac{d\eta}{dx} = \begin{bmatrix} \dfrac{\partial \eta_1}{\partial x_1} & \cdots & \dfrac{\partial \eta_1}{\partial x_n} \\ & \cdots & \\ \dfrac{\partial \eta_p}{\partial x_1} & \cdots & \dfrac{\partial \eta_p}{\partial x_n} \end{bmatrix}$$

In the display, the partial derivatives of f are evaluated at $(x, u) = (0, 0)$, and the partial derivatives of η are evaluated at $x = 0$.   ◊

In Example 2–5–9 we saw that the dynamics of a two-jointed mechanical arm could be approximated by two different linear systems, with the choice of the system depending upon the point of the state space in which we are using the approximation. We leave it as Exercise 6 for the reader to take the non-linear equations of Example 2–5–9 and linearize them by the methods of this section.

Let us now turn to an example from electronics and analyze a circuit containing a non-linear component.

Example 8

A tunnel diode has an operating characteristic

$$i_D = f(v_D)$$

as exemplified by the current-voltage curve of Figure 3–10, where v_D is the potential difference across the diode and i_D is the current flowing through

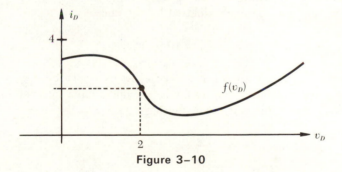

Figure 3–10

the diode. Observe that there is a region of "negative resistance," where the slope $\dfrac{di_D}{dv_D}$ is negative.

Suppose that the diode is employed in the circuit of Figure 3–11. From node and loop equations,

$$i_L = C \frac{dv_D}{dt} + i_D$$

$$u = i_L R + L \frac{di_L}{dt} + v_D$$

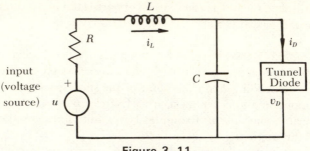

Figure 3–11

and, since $i_D = f(v_D)$, these equations can be written

$$\frac{di_L}{dt} = -\frac{R}{L}i_L - \frac{1}{L}v_D + \frac{1}{L}u$$

$$\frac{dv_D}{dt} = \frac{1}{C}i_L - \frac{1}{C}f(v_D)$$

These are the local state equations for the system, with i_L and v_D as the state variables. To linearize them, assume that over the region of negative resistance $f(v_D)$ may be approximated by the line

$$i_D = 10 - 3v_D$$

We shall compute the bias voltage u_0 necessary to hold $v_D = 2$ volts, in the negative resistance zone, and write the linear equations valid for small fluctuations about this bias point. Setting

$$\frac{di_L}{dt} = \frac{dv_D}{dt} = 0 \text{ at } u = u_0$$

we get

$$0 = -Ri_{L0} - v_{D0} + u_0$$
$$0 = i_{L0} + 3v_{D0} - 10$$

and for $v_{D0} = 2$, we need

$$u_0 = 2 + 4R$$

as the bias.

Now, this problem is simple enough to be linearized by inspection. Instead, we will carry out all the steps in detail to illustrate the general procedures.

Letting $x = \begin{bmatrix} x_1 \\ x_2 \end{bmatrix} = \begin{bmatrix} i_L \\ v_D \end{bmatrix}$, our equations become

$$x = f(t, x_0 + h, u_0 + k) = \begin{bmatrix} -\dfrac{R}{L}(x_{10} + h_1) - \dfrac{1}{L}(x_{20} + h_2) + \dfrac{1}{L}(u_0 + k_0) \\[2mm] -\dfrac{1}{C}(x_{10} + h_1) - \dfrac{1}{C}f(x_{20} + h_2) \end{bmatrix}$$

$$= \begin{bmatrix} -\dfrac{R}{L}h_1 - \dfrac{1}{L}h_2 + \dfrac{1}{L}k_0 \\[2mm] \dfrac{1}{C}h_1 - \dfrac{1}{C}(f(x_{20} + h_2) - f(x_{20})) \end{bmatrix}$$

for our choices of x_{10}, x_{20}, and u_0. Then

$$f_1(t) = \frac{\partial f}{\partial h}(t, x_0, u_0) = \begin{bmatrix} -\dfrac{R}{L} & -\dfrac{1}{L} \\[2mm] \dfrac{1}{C} & -\dfrac{1}{C}f'(x_{20}) \end{bmatrix}$$

while

$$f_2(t) = \frac{\partial f}{\partial k}(t, x_0, u_0) = \begin{bmatrix} \dfrac{1}{L} \\[2mm] 0 \end{bmatrix}$$

But $f'(x_{20}) = -3$. Thus, letting $x = x_0 + h$ and $u = u_0 + k$, we get

$$\dot{x} = \begin{bmatrix} -\dfrac{R}{L} & -\dfrac{1}{L} \\[2mm] \dfrac{1}{C} & \dfrac{3}{C} \end{bmatrix} \begin{bmatrix} h_1 \\[2mm] h_2 \end{bmatrix} + \begin{bmatrix} \dfrac{1}{L} \\[2mm] 0 \end{bmatrix} k$$

as our "small signal" linear model of the circuit's behavior near (x_0, u_0).

Electronic engineers frequently express this linearization by an *equivalent linear circuit*, as shown in Figure 3–12, where it is understood that $k = (u - u_0)$,

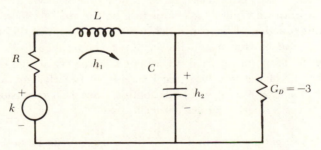

Figure 3–12 A "small signal" equivalent circuit.

a small perturbation about the bias voltage u_0, that $h_1 = (i_L - i_{L0})$ and $h_2 = (v_D - v_{D0})$, and that the diode acts like a conductance of $G_D = \dfrac{df(v_D)}{dv_D}$ $= -3$ mhos. $\diamond$

We are interested in such linear approximations because it may be impossible to analyze the differential equation

$$\dot{x}(t) = f(t, x(t), u(t))$$

for arbitrary local functions f, and so our attention turns to simpler forms of f which we can handle with relative ease, yet which still offer useful approximations to real situations.

However, we must stress that *it is often possible, and sometimes necessary, to treat nonlinearities directly, without resorting to linearization.* Too many students are taught control theory for linear systems only. However, we shall take care to show that many concepts are valid in general. Even though many computational techniques will require linearization, nonetheless much of our later theory will actually be *simplified* by considering the general case.

For completeness, we now present an example from optimal control theory where linearization is neither possible nor desirable.

Example 9   A non-removable nonlinearity in optimal control:

Suppose that we wish to find a control law for an elevator which will determine, as a function of the elevator's current position and velocity, the accelerating force we must apply to the elevator in order to make it go from the first to the top floor in the shortest time possible. Practical considerations will usually determine that there is a maximal acceleration $u_1 > 0$ and a maximal deceleration $u_2 < 0$ which can be applied. We are thus considering a problem of time-optimal control with bounded inputs. It can in fact be shown (see Appendix A–3) that, even if we linearize the elevator's dynamics, the time-optimal control law is to maximally accelerate until the elevator reaches some critical height, at which time we switch to maximal deceleration until we bring the elevator to a halt at the top floor. Thus, the input-output function of the controller, which has for input the height $\hat{x}$ of the elevator, and has for output the force u to be applied to the elevator, is given by the graph of Figure 3–13, where x_C is the "switching point" and x_1 is the height of the top floor [see Exercise 8]. Clearly, any attempt to linearize the *controller* near x_C would be foolish, since any "excursions" about x_C will by necessity be large ones: either u_1 or u_2. $\diamond$

We shall sometimes find that a non-linear system can be approximated by *linear dynamics* over a wide range so long as we encode the input to an appropriate linear system, and decode the output from this system, by

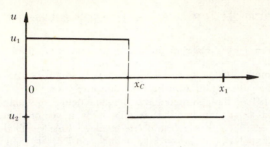

Figure 3-13

appropriate non-linear "memoryless" maps, as in Figure 3-14. Thus, if the dynamics are in some sense more slowly changing than the input and output specifications, we can use a model of this form which will let us apply most of our knowledge of linear system theory, to be built up in subsequent chapters, in some useful cases in which there is no direct linear relationship between input and output. [See Exercise 9.]

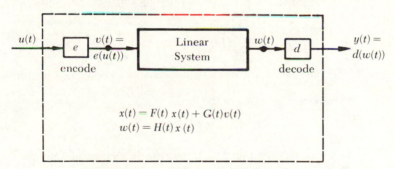

Figure 3-14   Overall Nonlinear System

EXERCISES FOR SECTION 3-4

1.   **(a)**   Show that $\lim\limits_{\|h\|\to 0} \dfrac{\|\psi(h)\|}{\|h\|} = 0$ iff $\lim\limits_{\|h\|\to 0} \dfrac{\psi(h)}{\|h\|} = 0$.

  (b)   For $\psi(h) = \begin{bmatrix} h_1 h_2 \\ 0 \end{bmatrix}$, show that $\dfrac{\psi(h)}{\|h\|} \to 0$ as $\|h\| \to 0$.

✔ 2.   For $f : C[0, 2\pi] \to \mathbf{R}$ defined by $f(x) = \displaystyle\int_0^{2\pi} x^2(\xi)\, d\xi$, find $f'(x_a, \cdot), f'(x_b, \cdot)$ and $f'(x_c, \cdot)$, where

$$x_a(\xi) \equiv 0$$
$$x_b(\xi) = \sin n\,\xi$$
$$x_c(\xi) = \cos n\,\xi, \quad \text{for } 0 \le \xi \le 2\pi, n \in \mathbf{N}$$

3.   Define $F: C[0, 2\pi] \to \mathbf{R}$ by $F(x) = \int_0^{2\pi} [x(\xi)]^n \, d\xi$. Compute $F'(x, \cdot)$.

✔ **4.**   Let $G_t(x) = \int_{-\infty}^{\infty} g(t, \tau) x(\tau) \, d\tau$ for all $x \in C(-\infty, \infty)$ and $t \in (-\infty, \infty)$, where g is some fixed map in $C[(-\infty, \infty) \times (-\infty, \infty)]$. [$g$ might be the impulse response of a time varying system and x an input]. Compute $G_t'(x, \cdot)$.

5.   Let V and W be Banach spaces over the same field K of scalars, with respective norms $\|\cdot\|_V$ and $\|\cdot\|_W$. We know that $V \times W$ is a vector space over K, with $(v, w) + (v', w') = (v + v', w + w')$ and $k(v, w) = (kv, kw)$. Verify that if we impose the norm $\|\cdot\|_{V \times W}$ such that $(v, w) \mapsto \|v\|_V + \|w\|_W$ on $V \times W$, then $V \times W$ becomes a Banach space; i.e., check that

    (a) $\|\cdot\|_{V \times W}$ is indeed a norm on $V \times W$.

    (b) $V \times W$ is complete with respect to this norm.

    (c)   If $\dfrac{\|\psi(v, w)\|}{\|(v, w)\|} \to 0$ as $\|(v, w)\| \to 0$, then $\dfrac{\|\psi(v, 0)\|}{\|v\|} \to 0$ as $\|v\| \to 0$ and $\psi(0, w)$ is $o(w)$.

✔ **6.**   In Example 2–5–9 we found a mechanical system to be described by the nonlinear equations $\dot{x}_1 = x_3$, $\dot{x}_2 = x_4$,

$$\left(\frac{5}{3} + \cos x_2\right) \dot{x}_3 + \left(\frac{1}{3} + \frac{1}{2}\cos x_2\right)\dot{x}_4 = \hat{u}_1 + \frac{2x_3 + x_4}{2} x_4 \sin x_2$$

and

$$\left(\frac{1}{3} + \frac{1}{2}\cos x_2\right)\dot{x}_3 + \frac{1}{3}\dot{x}_4 = \hat{u}_2 - \frac{1}{2} x_3^2 \sin x_2$$

There we made some "small excursion" approximations and came up with linear equations.

    (a)   Write these equations in the form

$$\dot{x}(t) = f(t, x(t), u(t))$$

    (b)   Linearize this function f by the derivative method of Example 4, employing the Jacobian matrix, about the points

$$x_a = [s, 0, 0, 0]^T$$

and

$$x_b = \left[s, \frac{\pi}{2}, 0, 0\right]^T$$

where s is any real number.

✔ **7.**   Linearize the simple pendulum of Exercise 5, Section 2–5, about the points $x_a = \left[-\frac{\pi}{2}, 0\right]^T$, $x_b = [0, 1]^T$, and $x_c = [0, 0]^T$. Solve the linearized equations for the expansion about x_c and sketch the trajectory in the phase plane corresponding to the initial condition $x(0) = [\pi/6, 0]^T$.

8.   For the elevator of Example 9, determine x_c as a function of x_1, u_1, u_2, and the mass m of the elevator.

9.   Consider the system shown in Figure 3–15a, where the actual input v is "encoded" by the device of Figure 3–15b to become the input of a linear system followed by a "decoder" d. Devise a decoder to make this stupid system do something sensible.

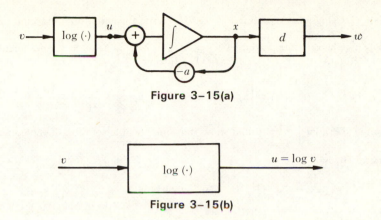

Figure 3-15(a)

Figure 3-15(b)

10. Consider a device in which we work with the set of decimal digits $\{0, 1, 2, \ldots, 9\}$ rather than the real line for each component of our input, state, and output spaces. Let it have a one-dimensional input, and let the output corresponding to an input sequence

$$\ldots . 000\, v_1 v_2 v_3 \ldots v_n$$

be simply the remainder after dividing $\sum_{L=1}^{n} v_L$ by 1000. Draw a simulation diagram for the system, using components which do arithmetic modulo 10. Write down the state and input-output equations for the system. What is the dimensionality of the state space? Is the system linear?

✔ **11.** Consider the problem of trying to control the orbit of a satellite to keep it circular. As a simplified mathematical model, consider a point mass m acted upon by an inverse square law force (gravity) $F_r = \dfrac{k}{r^2}$ radially inward, and assume that the satellite has thrust jets enabling it to apply a radial force u_1 and a tangential force u_2 as illustrated in Figure 3-16.

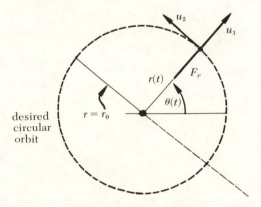

desired
circular
orbit

Figure 3-16 A satellite.

(a)   Using polar coordinates $(r(t), \theta(t))$, write the equations of motion for the satellite.

(b)   Choose state variables $x_1 = r$, $x_2 = \dot{r}$, $x_3 = \theta$, $x_4 = \dot{\theta}$ and write the dynamic equations $\dot{x}(t) = f(t, x(t), u(t))$. Is f linear?

$$\text{Answer: } f(t, x(t), u(t)) = \left[x_2, \frac{-k}{mx_1^2} + x_1 x_4^2 + \frac{1}{m} u_1, x_4, \frac{-2x_2 x_4}{x_1} + \frac{1}{mx_1} u_2 \right]^T.$$

(c)   Let $u = \begin{bmatrix} u_1 \\ u_2 \end{bmatrix} = 0$ and show that one possible solution of the equations

of (a) and (b) is

$$\begin{cases} r(t) = r_0 & \text{constant} \\ \theta_0(t) = \omega_0 t, & \omega_0 \text{ constant} \end{cases}$$

a **circular** orbit. $\left[\textit{Answer: } \text{The radius is related to } \omega_0 \text{ by } r_0^3 = \frac{k}{m\omega_0^2}. \right]$

(d)   Linearize the dynamic equations of (b) about the desired circular orbit. Take the reference vectors as

$$x_0 = \begin{bmatrix} r_0 \\ 0 \\ \omega_0 t \\ \omega_0 \end{bmatrix}, \quad u_0 = \begin{bmatrix} 0 \\ 0 \end{bmatrix}$$

so that the incremental excursions in x_1, x_2, x_3, x_4 are given by h_1, h_2, h_3, h_4 where $h_1 = r - r_0$, $h_2 = \dot{r} - 0$, $h_3 = \theta - \omega_0 t$, $h_4 = \dot{\theta} - \omega_0$ and get the linear equation $\dot{h}(t) = Fh(t) + Gu(t)$. [*Answer:*

$$\begin{bmatrix} \dot{h}_1 \\ \dot{h}_2 \\ \dot{h}_3 \\ \dot{h}_4 \end{bmatrix} = \begin{bmatrix} 0 & 1 & 0 & 0 \\ 3\omega_0^2 & 0 & 0 & 2\left(\frac{\omega_0 k}{m}\right)^{1/3} \\ 0 & 0 & 0 & 1 \\ 0 & -2\left(\frac{\omega_0^5 m}{k}\right)^{1/3} & 0 & 0 \end{bmatrix} \begin{bmatrix} h_1 \\ h_2 \\ h_3 \\ h_4 \end{bmatrix} + \begin{bmatrix} 0 & 0 \\ \frac{1}{m} & 0 \\ 0 & 0 \\ 0 & \left(\frac{\omega_0^2}{m^2 k}\right)^{1/3} \end{bmatrix} \begin{bmatrix} u_1 \\ u_2 \end{bmatrix}$$

follows easily from

$$\dot{h} = \left[\frac{\partial f_i}{\partial x_j} \right]_{x_0, u_0} h + \left[\frac{\partial f_i}{\partial u_j} \right]_{x_0, u_0} u \quad \text{and} \quad r_0^3 = \frac{k}{m\omega_0^2}.]$$

✔ **12.**   Fill in the details of the proof of Lemma 1.

13.   Look up Taylor's theorem for higher dimensions in any calculus or advanced calculus book and verify that it can be written:

$$g(P + H) = g(P) + \frac{(H \cdot \nabla)}{1!} g(P) + \frac{(H \cdot \nabla)^2}{2!} g(P) +$$

$$+ \cdots + \frac{(H \cdot \nabla)^{r-1}}{(r-1)!} g(P) + \frac{(H \cdot \nabla)^r}{r!} (P + \tau H)$$

for some $\tau \in (0, 1)$ provided that $g : \mathbf{R}^n \to \mathbf{R}$ has continuous partial derivatives of order r.

Here ∇ is the operator

$$\begin{bmatrix} \dfrac{\partial}{\partial x_1} \\ \vdots \\ \dfrac{\partial}{\partial x_n} \end{bmatrix}, \quad \text{the increment } H = \begin{bmatrix} h_1 \\ \vdots \\ h_n \end{bmatrix},$$

and $H \cdot \nabla$ (the dot product) is given by

$$(H \cdot \nabla) g(P) = \begin{bmatrix} h_1 \\ \vdots \\ h_n \end{bmatrix} \cdot \begin{bmatrix} \dfrac{\partial g}{\partial x_1}(P) \\ \vdots \\ \dfrac{\partial g}{\partial x_n}(P) \end{bmatrix} = h_1 \frac{\partial g}{\partial x_1}(P) + \cdots + h_n \frac{\partial g}{\partial x_2}(P)$$

Similarly, $(H \cdot \nabla)^2 = (H \cdot \nabla) \circ (H \cdot \nabla)$ (composition of the operators $H \cdot \nabla$ and $H \cdot \nabla$), and

$$(H \cdot \nabla)^2 g(P) = (H \cdot \nabla)((H \cdot \nabla) g) \Big|_{\text{at } P}$$

$$= h_1 \frac{\partial}{\partial x_1}(H \cdot \nabla) g + \cdots + h_n \frac{\partial}{\partial x_n}(H \cdot \nabla) g \Big|_{\text{at } P}$$

$$= h_1^2 \frac{\partial^2 g}{\partial^2 x_1}(P) + h_1 h_2 \frac{\partial^2 g}{\partial x_1 \partial x_2}(P) + \cdots + h_n^2 \frac{\partial^2 g}{\partial^2 x_n}(P)$$

(a) Use Taylor's theorem on each coordinate function $f_i : \mathbf{R}^n \to \mathbf{R}$ of the map $f : \mathbf{R}^n \to \mathbf{R}^m$ and show that

$$\begin{bmatrix} f_1(P + H) \\ \vdots \\ f_m(P + H) \end{bmatrix} = \begin{bmatrix} f_1(P) \\ \vdots \\ f_m(P) \end{bmatrix} + \begin{bmatrix} \dfrac{\partial f_1}{\partial x_1} & \dfrac{\partial f_1}{\partial x_2} & \cdots & \dfrac{\partial f_1}{\partial x_n} \\ \vdots & \vdots & & \vdots \\ \dfrac{\partial f_m}{\partial x_1} & \dfrac{\partial f_m}{\partial x_2} & \cdots & \dfrac{\partial f_m}{\partial x_n} \end{bmatrix} \begin{bmatrix} h_1 \\ \vdots \\ h_n \end{bmatrix}$$

$$+ \{\text{terms involving higher order partial derivatives}\}$$

(b) Illustrate Taylor's theorem on the map of Example 4. What is

$$((H \cdot \nabla)^2 f) \left(\begin{bmatrix} x_1 \\ x_2 \\ x_3 \end{bmatrix} \right) ?$$

Can you find a $\tau \in (0, 1)$ to express the "higher order term" in terms of

$$(H \cdot \nabla)^2 f \left(\begin{bmatrix} x_1 \\ x_2 \\ x_3 \end{bmatrix} + \tau \begin{bmatrix} h_1 \\ h_2 \\ h_3 \end{bmatrix} \right) ?$$

CHAPTER 4

Reachability and Observability

In the first three chapters of this book we have developed a general language for talking about systems. In Chapter 1 we introduced the basic notion of a deterministic system characterized by spaces of inputs, states, and outputs, by an output function, and by a state transition function. In Chapter 2 we saw how, in many cases, the "global" transition function of Chapter 1 could be replaced by a "local" transition function; we also developed some mathematical concepts related to vector spaces and Banach spaces, among others, and related our formalization of systems to the study of electrical and mechanical systems. Then, in Chapter 3, we studied the response function of a system, and related linearity to these functions. After briefly reviewing some of the basic notions of matrices, we closed by analyzing conditions under which we could approximate a system by a linear system. Section 4–1, *The Story So Far,* is designed for the reader for whom most of this material is either familiar, or too far removed from the particular applications of system theory he had in mind. It gives sufficient formal details, supplementing the above quick account of the preceding chapters, to allow the reader to continue through the book with relatively little reference to what has gone before.

The remainder of the chapter is devoted to the crucial system-theoretic concepts of reachability, controllability, and observability. These are concepts at the heart of modern system theory, and for many readers, this will be where the story really begins.

In our study of reachability and controllability we analyze the extent to which a system can be transferred from one state to another, while we say that a state is observable if we can deduce that the system was in that state from observations of its input-output behavior. In Section 4–2, we provide the formal statements of these informal definitions, and observe that there are a number of key results about reachability, controllability, and observability that hold for the completely general systems of Section 1–5, without any reference to whether the system be discrete-time or continuous-time, time-invariant or time-variant, linear or nonlinear. This is a particularly important point, for it shows that a number of properties which many people are only aware of as the result of tedious manipulations of the matrices of linear systems are in fact accessible, and more easily understandable, in our general setting. In Section 4–3 we specialize our study of reachability and controllability to the case in which our systems are discrete-time, with special attention to linear systems. Although the results are couched in algebraic formalization, we repeat our assurance, made in the introduction to Chapter 1, that with practice the results here will come to seem intuitive. This will turn out to be particularly useful when, in Chapters 6 and 7, we find that many results for continuous-time linear systems are remarkably similar to the discrete-time results—so much so, in fact, that one may often work out an algorithm for the analysis of a continuous-time system by looking at the discrete-time system described by the same matrices F, G, and H. In Section 4–5 we study observability for linear discrete-time systems, using the notion of inner-product spaces and adjoints of Section 4–4 to show that observability is, in an interesting sense, dual to reachability in the discrete-time case. As we shall see in Chapter 7, a continuous-time linear system is controllable if and only if it is reachable, and thus we have a foretaste of the well-known duality of controllability and observability for continuous-time linear systems. Finally, in Section 4–6 we present a self-contained introduction to the useful Z-transform, which lets us associate with every discrete-time function an infinite power series in the variable z, and vice versa. This transform permits us to go back and forth between the "time domain" and the "z-domain" in the analysis of discrete-time, constant linear systems very much like the Laplace transform does for continuous-time systems; a table of Z-transforms is given in Appendix A-2. We develop the important concept of the discrete-time transfer function $\mathfrak{H}(z)$ and show how it is related to the unit-pulse response and the Z-transform. We close the chapter with discussions of the resolvent matrix $(zI - F)^{-1}$, the characteristic polynomial $\chi_F(z)$ of F, and a proof of the powerful Cayley-Hamilton theorem, all of which we shall see again in more detail in Chapter 6.

4–1 The Story So Far

To aid the reader who wishes to commence his study of the volume at this point, we now recapitulate the way in which we have narrowed down our description of deterministic systems. We started with a description which specified a function ϕ which, given the state $\tilde{x}$ in the state space X at any time t_0, and the input u over any ensuing finite interval $[t_0, t_1)$, would determine the state at the end of that interval as $x(t_1) = \phi(t_1, t_0, \tilde{x}, u)$. We stress that u denotes an input *function* from a collection Ω of admissible functions.

More formally, then, our introductory discussion in Chapter 1 may be summarized in the following very general definitions, which we shall see later embrace the familiar notions of classical linear systems theory.

A **deterministic mathematical system representation** is specified by five sets (T, U, Y, X, Ω) and two functions ϕ and η, where:

T is the *time* set, a subset of the real numbers
U is the *input* set
Y is the *output* set
X is the *state* set
Ω is the set of *admissible input functions*
$\phi: T \times T \times X \times \Omega \to X$ is the *state-transition function*
$\eta: T \times X \to Y$ is the *output function*

Ω is required to be a subset of the set of all functions $T \to U$, which is closed under *splicing;* i.e., for all u, u' in Ω and all times $t_2 \in T$, there exists a u'' in Ω such that

$$u''(t) = \begin{cases} u(t) & \text{if } t \le t_2 \\ u'(t) & \text{if } t_2 < t \end{cases}$$

ϕ is required to satisfy the following consistency conditions:

$$\phi(t_0, t_0, \tilde{x}, u) = \tilde{x}$$

$$\phi(t_2, t_1, \phi(t_1, t_0, \tilde{x}, u), u) = \phi(t_2, t_0, \tilde{x}, u)$$

for all times t_0, t_1, t_2 all states $\tilde{x}$, and all admissible input functions u

and the causality condition:

$$\phi(t_1, t_0, \tilde{x}, u) = \phi(t_1, t_0, \tilde{x}, u')$$
$$\text{if } u(t) = u'(t) \text{ for } t_0 \le t < t_1$$

The system is **discrete-time** if $T = \mathbf{Z}$, the set of integers; it is **continuous-time** if $T = \mathbf{R}$, the set of real numbers. The system is **time-invariant** (stationary, constant) if T is closed under addition, Ω is closed under the shift operator $z^{-\tau}$ for every τ in T (where $[z^{-\tau}u](t) = u(t + \tau)$ for all t in T), and we have

$\phi(t_1, t_0, \tilde{x}, u) = \phi(t_1 + \tau, t_0 + \tau, \tilde{x}, z^{-\tau}u)$ for all times t_0, t_1, all delays τ, all states $\tilde{x}$, and all admissible input functions u

$\eta(t_0, \hat{x}) = \eta(t_1, \hat{x})$ for all times t_0 and t_1, and all states $\hat{x}$

Note that the above conditions allow us to remove the mention of one time variable, replacing ϕ and η by $\breve{\phi}$ and $\breve{\eta}$ where

$$\breve{\phi}: T \times X \times \Omega \to X \text{ by the rule } (t_1, \tilde{x}, u) \mapsto \phi(t_1, 0, \tilde{x}, u)$$

$$\breve{\eta}: X \to Y \text{ by the rule } \tilde{x} \mapsto \eta(0, \tilde{x})$$

since we know that for all choices of t_0

$$\phi(t_1, t_0, \tilde{x}, u) = \breve{\phi}(t_1 - t_0, \tilde{x}, z^{+t_0}u)$$

$$\eta(t_1, \hat{x}) = \breve{\eta}(\hat{x}).$$

In the future, we shall normally write η and ϕ whether we really mean $\breve{\eta}$ or η and $\breve{\phi}$ or ϕ, leaving it to the context to determine which is intended.

In Chapter 2, we found that for *discrete-time* systems we could reduce the state transition function to a compact form specifying how the present state and present input determined the next state:

$$x(t + 1) = f(t, x(t), u(t))$$

We saw that the general description in terms of ϕ could be built up by repeated application of the one-step transition function f:

$$x(t + n) = f(t + n - 1, x(t + n - 1), u(t + n - 1))$$
$$= f(t + n - 1, f(t + n - 2, x(t + n - 2), u(t + n - 2)),$$
$$u(t + n - 1)) = \cdots$$

In the *continuous-time* case, our search for a compact local description led us to consider the time derivative $\dfrac{\partial \phi}{\partial t_1}$ of the state transition function ϕ. We realized that to get a local description

$$\dot{x}(t) = f(t, x(t), u(t))$$

for the rate of change of the state in a continuous-time system, we had to make certain smoothness restrictions upon the function ϕ.

We then saw in Section 3–4 that we could approximate the resulting local transition function $f(t, x(t), u(t))$, if it was smooth enough, by a linear function $F(t)x(t) + G(t)u(t)$. The measure of accuracy for that linear approximation is the discrepancy term $\psi(x, u)$ where

$$\psi(x(t), u(t)) = f(t, x(t), u(t)) - [F(t)x(t) + G(t)u(t)]$$

and we may consider a system to be linear over the range for which this discrepancy is negligibly small, and for which the output function η has a satisfactory approximation of the form

$$\eta(t, x(t)) = H(t)x(t)$$

A linear system, then, is one for which the discrepancy term is negligible over the whole range of operation in which we expect the system to be employed. For other systems, the range over which the discrepancy is negligible will be rather small, and we will have the problem of piecing together local linear approximations to get a useful global nonlinear analysis of the system. We are thus led to the following working definitions of linear systems:

DEFINITION 1

A discrete-time system is **linear** if it has **vector** spaces for its input, output, and state spaces U, Y and X, and if there exist, for each time t in T, three linear transformations

$$F(t): X \to X$$
$$G(t): U \to X$$
$$H(t): X \to Y$$

such that state-transitions and outputs are given by the equations

$$x(t + 1) = F(t)x(t) + G(t)u(t)$$
$$y(t) = H(t)x(t) \qquad\qquad \bigcirc$$

DEFINITION 2

A continuous-time system is **linear** if it has **Banach** spaces† for its input, output, and state spaces U, Y and X, and if there exist, for each time t in

---

† The reader will find an account of Banach spaces in Section 2–4. However, for many readers it will suffice to consider the case in which $U = \mathbf{R}^m$, $X = \mathbf{R}^n$, **and** $Y = \mathbf{R}^q$ for suitable integers m, n, and q.

T, three linear transformations

$$F(t):X \to X$$
$$G(t):U \to X$$
$$H(t):X \to Y$$

such that state-transitions and outputs are given by the equations

$$\dot{x}(t) = F(t)x(t) + G(t)u(t)$$
$$y(t) = H(t)x(t)$$

In either case the system will be constant (time-invariant) if the linear transformations F, G, and H are themselves constant, i.e., are independent of the time at which we use them.

We should comment on the fact that in discrete-time systems we have only called for vector spaces, whereas in the continuous-time systems we require Banach spaces. We used Banach spaces in Section 3–4 to set up the approximation theory which suggested that linear systems will often provide good approximations to other systems. In the case of continuous-time systems, we shall continue to use Banach spaces, because differential and integral *calculus* is necessary to go from local to global descriptions in continuous time. However, we do not need calculus for the analysis of discrete-time systems, and thus we prefer to state our definition for discrete-time systems in the utmost generality—in other words, in terms of vector spaces. For example, in coding theory (Bobrow and Arbib, *Discrete Mathematics,* Chapters 7 and 8) it is quite usual to consider coding devices, which are linear discrete-time systems, for which the state space is a vector space over a finite field. Given a finite state space, it is not meaningful to talk of limits, and so the restriction of discrete-time systems to having Banach spaces for their state spaces would unduly limit the applicability of our theory.

The reader should note that the transformation F plays somewhat different roles in discrete-time and continuous-time systems. In the discrete-time case, it tells us *to what* the system changes state over one unit of time under the zero input. However, in the continuous-time case, F tells us the *rate* at which the system changes state under the zero input: If $u(t) = 0$, we have

$$x(t + 1) = F(t)x(t) \text{ in discrete-time systems}$$
$$\dot{x}(t) = F(t)x(t) \text{ in continuous-time systems}$$

To give the reader more feel for this, consider the case in which $X = \mathbf{R}^1$, and $F(t)$ is constant, being represented by multiplication by the scalar k. Then in the discrete-time case, we simply have

$$x(t + 1) = kx(t)$$

whereas in the continuous-time case we integrate

$$\dot{x}(t) = kx(t)$$

to obtain

$$x(t + \tau) = e^{k\tau}x(t)$$

so that, in particular,

$$x(t + 1) = e^{k}x(t)$$

In Chapter 6, we shall find that it is possible to define for each linear transformation F a new linear transformation e^{Ft} for each time t with the property that, for a constant linear system with zero input, the state at time t can be expressed in terms of the state at time 0 by the relation $x(t) = e^{Ft}x(0)$. In particular, we see that the F of the discrete-time system corresponds to e^{F}, the exponential of the F of the continuous-time system.

4–2   General Definitions

Having studied state-selection and the writing of state equations for deterministic systems, we now turn to questions of the dynamic possibilities of our systems: Can we *control* a system in a given state, applying inputs to bring it to its resting state? Given a system in its resting state, can we apply inputs which will force it to *reach* a desired state? Given a system in an unknown state, can we *observe* its behavior in such a way as to determine what that state was?

In this section, we shall set up the terminology required to discuss these questions for general (nonlinear, time-varying, discrete- or continuous-time) deterministic systems. In the remainder of the chapter we shall pursue the answers to these questions for *discrete-time* systems, introducing the necessary mathematical apparatus of inner-product spaces and adjoint operators in Section 4–4 so that we may provide algorithms for the study of reachability and observability of discrete-time *linear* systems in Section 4–5. The corresponding theory for continuous-time linear systems will be taken up in Chapter 7.

We shall call a pair $(\tau, \tilde{x})$, consisting of a time τ and a state $\tilde{x}$, an **event.**

We use this terminology because in time-varying systems, in order to determine the effect of an input upon the system, we must know not only the state $\tilde{x}$ but also the time τ at which the system was in the state $\tilde{x}$.†

An event $(\tau, \tilde{x})$ is **reachable** from the zero state 0_X iff‡ there exists some time $s \leq \tau$ and some input $u \in \Omega$ such that

$$\tilde{x} = \phi(\tau, s, 0_X, u)$$

In other words, an event is reachable if it is possible for the system to be in the zero state at some earlier time, and be transferred by an appropriate choice of input to the desired state at the desired time.

Example 1

The system of Figure 4–1 has as its state space $X = \mathbf{R}$, with the voltage across the capacitor as the state variable and the voltage $y = v$ as output. Recall

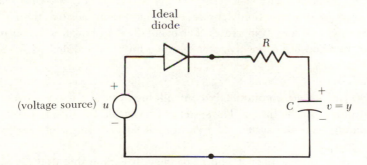

Figure 4–1   A system with some unreachable states.

from Figure 1–4 that the ideal diode permits current to flow freely in its reference direction as long as the voltage across it is positive, but acts like an open circuit when the voltage is negative. Here there is a unique zero state $0_X = 0$. The event $(\tau, \tilde{x}) = (5, 4)$ is reachable from the zero state because we can take $s = 0 < 5$ and apply a positive step of voltage of just the right amplitude to make the ideal diode conduct and charge the capacitor up to $v = 4$ volts at time $\tau = 5$ seconds.

The event $(\tau, \tilde{x}) = (5, -4)$ is not reachable from the zero state, since

† In what follows, the reader will be able to obtain simpler definitions for time-invariant systems by replacing the event $(\tau, \tilde{x})$ by the state $\tilde{x}$.

‡ It is universally agreed that in mathematical *definitions* the word **if** always means **if and only if** (abbreviated **iff.**).

no matter how far back in time we go, the diode will not allow the capacitor to be charged to any negative voltage, let alone $v = -4$ volts.   ◊

We say that a system is **completely reachable** at time τ if every event $(\tau, \tilde{x})$ is reachable at time τ. The system of Example 1 is thus *not* completely reachable; for an example of a system which is completely reachable, see Exercise 1. Having said that an event is reachable if we can be in the zero state in the past and apply an input to get to the desired event, we may consider interchanging past and future, and ask whether there is some input which will carry us from the given event to the zero state at some future time. We say an event $(\tau, \bar{x})$ is **controllable** to the zero state 0_X if there exists some time $t \geq \tau$ and some input $u \in \Omega$ such that

$$0_X = \phi(t, \tau, \bar{x}, u)$$

Example 2

For the system of Figure 4–1, the event $(\tau, \bar{x}) = (5, 4)$ is *not* controllable to zero, since the diode will only conduct when u is more positive than 4; hence v either stays at 4 volts or increases. The event $(5, -4)$, however, is controllable to zero by choosing a positive step voltage for u as detailed in Exercises 2 and 3.   ◊

We call a system **completely controllable** at time τ if every event $(\tau, \bar{x})$ is controllable at the time τ. The system of Example 1 is thus not completely controllable, but the system of Example 1–6–1 is completely controllable [Exercise 4].

From a control theorist's point of view, it is desirable that a system be both completely reachable and completely controllable, since [see Exercise 5] he may then drive an event $(\tau, \bar{x})$ to zero at time t, and then drive $(t, 0_X)$ to any desired event $(\hat{\tau}, \hat{x})$ if $\hat{\tau}$ is sufficiently larger than τ.

In the language of computer scientists, we call a system **strongly connected** [Exercise 6] if for every event $(\tau, \tilde{x})$ and state $\hat{x}$ there exists a time $\hat{\tau} \geq \tau$ and an input function u such that

$$\hat{x} = \phi(\hat{\tau}, \tau, \tilde{x}, u)$$

i.e., we can control the system to change it to any desired final state.

In later sections we will develop methods for deciding directly from the local functions f and η whether or not a *linear* system is reachable or controllable. We pause now to present an example of a theorem which is easier to prove in general than by special formulas for linear systems, to remind the reader that skillful use of intuition is always preferable to the mindless plugging-in of standard formulas.

Suppose we have two systems Σ_1 and Σ_2 such that the output space

Y_1 of the first system equals the input space U_2 of the second system. Then we may connect the output lines of Σ_1 to the input lines of Σ_2 to obtain the series connection Σ_σ as shown in Figure 4–2. To specify the state of Σ_σ we simply specify the state of each of Σ_1 and Σ_2; thus, $X_\sigma = X_1 \times X_2$.

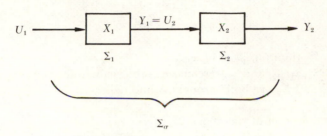

Figure 4–2   Two systems in cascade.

If Σ_1 is in state x_1 and Σ_2 is in state x_2 we shall denote the state of Σ_σ by $X_\sigma = \begin{bmatrix} x_1 \\ \hdashline x_2 \end{bmatrix}$ where the dashed line indicates how we have stacked the column x_1 on top of the column x_2 to get the resulting partitioned matrix.†

   Note, in particular, that if $X_1 = \mathbf{R}^{n_1}$ and $X_2 = \mathbf{R}^{n_2}$, then X_σ will be $\mathbf{R}^{n_1+n_2}$ and

$$x_\sigma = \begin{bmatrix} x_1 \\ \hdashline x_2 \end{bmatrix} \begin{matrix} \} \ n_1 \\ \} \ n_2 \end{matrix}$$

will be a vector with $n_1 + n_2$ real components, the first n_1 of which form $x_1 \in \mathbf{R}^{n_1}$ and the last n_2 of which form $x_2 \in \mathbf{R}^{n_2}$.

   We now prove, in complete generality, the following theorem:

THEOREM 1

If the series connection Σ_σ of Σ_1 and Σ_2 is strongly connected, then each of Σ_1 and Σ_2 is itself strongly connected.

Proof

To say that Σ_σ is strongly connected is to say that, given any event $(\tau, \bar{x}_1)$ and state $\hat{x}_1$ of Σ_1 and any event $(\tau, \bar{x}_2)$ and state $\hat{x}_2$ of Σ_2, there exists a

---

†Throughout the rest of this book we shall make frequent use of this idea of partitioning a matrix into various, convenient sub-matrices or blocks. The reader who needs it will find in Exercise 13 a quick survey of block partitioned matrices.

time $\hat{\tau} \geq \tau$ and an input u which, when applied to Σ_σ on $[\tau, \hat{\tau})$, will drive it from $\left(\tau, \left[\frac{\bar{x}_1}{\bar{x}_2}\right]\right)$ to $\left(\hat{\tau}, \left[\frac{\hat{x}_1}{\hat{x}_2}\right]\right)$. But then it is obvious that u applied to Σ_1 alone over the interval $[\tau, \hat{\tau})$ will drive Σ_1 from $(\tau, \bar{x}_1)$ to $(\hat{\tau}, \hat{x}_1)$. The corresponding output function for Σ_1

$$\eta_1(\,\cdot\,, \phi_1(\,\cdot\,, \tau, \bar{x}_1, u))$$

will drive Σ_2 from $(\tau, \bar{x}_2)$ to $(\hat{\tau}, \hat{x}_2)$.

Since $\bar{x}_1, \hat{x}_1, \bar{x}_2$, and $\hat{x}_2$ were chosen arbitrarily in their appropriate state spaces, Σ_1 and Σ_2 are both strongly connected. $\square$

A modification of the above proof yields:

COROLLARY 2

If the series connection Σ_σ of Σ_1 and Σ_2 is reachable (controllable), then each of Σ_1 and Σ_2 is itself reachable (controllable). $\square$

Can you construct an example to show that the converse of Theorem 1 and Corollary 2 is not true in general? The proof of Theorem 1 is very simple, yet it holds for non-linear as well as for linear systems. If the man on the street had been asked to prove Corollary 2 for time-invariant linear discrete-time systems, he might well have called upon the famous theorem which we will prove in Section 4–3:

"If the state space of the linear system whose state transitions are described by $x(t + 1) = Fx(t) + Gu(t)$ has dimension n, then the system is reachable if and only if the block-partitioned matrix $[F^{n-1}G \mid \cdots \mid FG \mid G]$ has rank n." For example, by this theorem the system described by

$$F = \begin{bmatrix} 2 & 3 \\ 0 & 1 \end{bmatrix}, \; G = \begin{bmatrix} 1 \\ 0 \end{bmatrix}$$

is not reachable since $[FG \mid G] = \begin{bmatrix} 2 & 1 \\ 0 & 0 \end{bmatrix}$ has rank $1 < 2$.

In Exercise 14 the reader is asked to determine F_σ and G_σ, the matrices for Σ_σ in terms of F_1, G_1, F_2, and G_2 for Σ_1 and Σ_2. We could, thus, plug F_σ and G_σ into the theorem just quoted to determine the conditions under which Σ_σ is reachable. We could then manipulate the resulting matrices to deduce that the reachability conditions (in terms of rank) for F_1, G_1 and F_2, G_2 respectively do indeed hold, so that Σ_1 and Σ_2 are reachable. The reader need only attempt this [Exercise 15] to convince himself that this is not the best approach! In Exercise 16, the reader is asked to attempt a similar analysis to decide whether or not the series connection of two reachable systems must also be reachable.

The "plug-into-a-standard-formula" proof is in this case tedious; the "let's-use-a-little-intuition-about-what-the-system-is-doing" proof not only is shorter, but also yields insights which get lost in the matrix manipulations of the longer proof. In other cases, we may have to resort to manipulation to get any answer at all. The good system designer, then, is one with enough experience to proceed by a judicious blend of "plugging-in" of formulas and of using physical insights to take a "short-cut." We know of no recipe to achieve this fortunate skill. Our sermonizing over for the time being, we return now to our general development.

It may well be the case in a given system that two events are **indistin-guishable,** in that we cannot tell from the input-output behavior of the system which of the two states $\bar{x}$ and $\hat{x}$ the system was in or will be in.

We say that two events $(\tau, \bar{x})$ and $(\tau, \hat{x})$ are **indistinguishable in the future** if for all inputs $u \in \Omega$ and all times $t \geq \tau$ we have

$$\eta(t, \phi(t, \tau, \bar{x}, u)) = \eta(t, \phi(t, \tau, \hat{x}, u))$$

That is, for each choice of input u applied on $[\tau, t)$, $\bar{x}$ and $\hat{x}$ produce the same output at time t.

Similarly, $(\tau, \bar{x})$ and $(\tau, \hat{x})$ are **indistinguishable in the past** if for all inputs $u \in \Omega$ and all times $t \leq \tau$ we have

$$\eta(t, \phi(t, \tau, \bar{x}, u)) = \eta(t, \phi(t, \tau, \hat{x}, u))$$

Note that we are extending our notation $\phi(t, \tau, \bar{x}, u)$ to mean not only the state that we are carried to from state $\bar{x}$ at time τ if we apply the input u until time t for $t \geq \tau$, but also the state that we would have been in at time $t \leq \tau$ if input u will carry us to state $\bar{x}$ at time τ. Although in certain systems there may be many states which satisfy the latter criterion, we shall see in Chapter 5 that for most continuous-time systems of interest to us, $\phi(t, \tau, \bar{x}, u)$ is uniquely defined not only for $t \geq \tau$ but also for $t \leq \tau$.

Example 3

In the system of Figure 4–3, the states $\bar{x} = \begin{bmatrix} 1 \\ 1 \end{bmatrix}$ and $\hat{x} = \begin{bmatrix} 1 \\ 0 \end{bmatrix}$ are indistin-guishable in the future and in the past, since the output y is in no way affected by x_2, the second component of x.

The states $\begin{bmatrix} 1 \\ 1 \end{bmatrix}$ and $\begin{bmatrix} 2 \\ 0 \end{bmatrix}$ are, however, easily distinguishable in the future and in the past [Exercise 7].   ◊

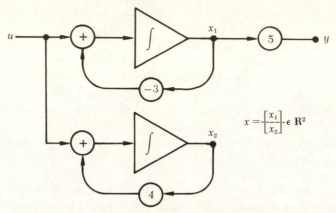

Figure 4–3 A system with some indistinguishable states.

In linear systems, two states $\bar{x}$ and $\hat{x}$ are indistinguishable if and only if their difference $x - \hat{x}$ is indistinguishable from the zero state [Exercise 9]. Thus, in much of our analysis, rather than look at the general problem of indistinguishability of two arbitrary states, we shall instead consider the notions of *observability* and *identifiability*:

An event $(\tau, \bar{x})$ is **unobservable** if it is indistinguishable in the future from the zero event $(\tau, 0_X)$. An event $(\tau, \bar{x})$ is **unidentifiable** if it is indistinguishable in the past from the zero event $(\tau, 0_X)$. In Example 3, the event $\left(\tau, \begin{bmatrix} 0 \\ 1 \end{bmatrix}\right)$ is indistinguishable from $\left(\tau, \begin{bmatrix} 0 \\ 0 \end{bmatrix}\right)$, and hence is unobservable, while the event $\left(\tau, \begin{bmatrix} 1 \\ 0 \end{bmatrix}\right)$ is distinguishable in the future from $\left(\tau, \begin{bmatrix} 0 \\ 0 \end{bmatrix}\right)$, and hence is observable.

We call a system **observable** at time τ if every event $(\tau, \bar{x})$ for the given τ is observable. Similarly, the system is said to be **identifiable** at time τ if every event $(\tau, \bar{x})$ for the given τ is identifiable.

Let us return for a moment to the idea of the indistinguishability in the future of two states. Suppose we are asked to transfer the system from a state $\bar{x}$ to an indistinguishable state $\hat{x}$. The answer is that we cannot if we are unable to tell which of the states we are in. But the reader might well ask, "Since you cannot tell which of the two states you are in, why should you ever care to transfer the system from one state to the other?" The reason for this is not a mathematical one, but a practical one. We must remember that the mathematical system we are looking at is an abstraction from a real system, and we may only expect this abstraction to model the behavior of the system accurately over a certain range. Thus, while we may expect two states to be indistinguishable as long as the model is applicable, they may be very different in that one state may lead to a behavior which takes us out of the domain of applicability, whereas the other does not. Clearly, then, we would prefer to be in the latter state.

For example, in the system of Example 3 the state $x(t) = \begin{bmatrix} x_1(t) \\ x_2(t) \end{bmatrix}$ is given by

$$x_1(t) = e^{-3(t-t_0)}x_1(t_0) + 3 \int_{t_0}^{t} e^{-3(t-\xi)}u(\xi)\, d\xi$$

and

$$x_2(t) = e^{+4(t-t_0)}x_2(t_0) - 4 \int_{t_0}^{t} e^{+4(t-\xi)}u(\xi)\, d\xi$$

for $t \geq t_0$, so that, with $u = 0$, the indistinguishable initial states $x(t_0) = \begin{bmatrix} 1 \\ 1 \end{bmatrix}$ and $x'(t_0) = \begin{bmatrix} 1 \\ 0 \end{bmatrix}$ lead to quite different behavior. The former state, where $x_2(t_0) = 1$, causes $x_2(t)$ to increase without bound, perhaps until wires melt and fuses blow and our system model no longer applies. It is because of such considerations that we shall often wish to take careful notice of differences between two states even when they are indistinguishable with respect to a given model of our system.

In control theory, the **regulator problem** amounts to trying to force a system (the "plant") to follow some pre-assigned state trajectory. The solution of the regulator problem consists of giving a **control law** which prescribes the values of the controlling inputs as a function of the measured deviation from the pre-assigned path.

Example 4

In Example 7 of Section 2-5 we made a phase-plane analysis of a cart on frictionless wheels. Suppose we want to design a controller to make that two-integrator plant follow the state trajectory of Figure 4-4; that is, to behave like a mechanical oscillator.

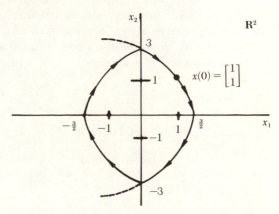

Figure 4-4 A specified trajectory in the phase plane.

A glance at the phase plane portraits of Figure 2–39 and a little cunning lead to the control law

$$u\left(\begin{bmatrix} x_1 \\ x_2 \end{bmatrix}\right) = -\mathrm{sgn}\,(x_1)$$

where sgn $(\cdot)$ is the "signum" function defined by

$$\mathrm{sgn}\,(\zeta) = \begin{cases} +1 \text{ if } \zeta > 0 \\ -1 \text{ if } \zeta < 0 \end{cases}$$

This type of abrupt switching in the control law is called **bang-bang** control and is easily realized by relays or electronic "flip-flop" devices. ◊

Since the design of a controller to solve an arbitrary regulator problem requires that we know or else can estimate the state arbitrarily well, we see that a necessary condition for the solution of the regulator problem is that the system be both observable and identifiable. Moreover, in some control problems, we may not even know what the appropriate equations are for describing the system we wish to control. Thus, the controller not only has the task of estimating the state of a given system, but must also estimate what the actual equations of that system are. This problem of generating a state-variable description of the system under study is called **the realization problem,** and we shall study it with great care in Chapter 8.

We shall see that the realization problem (i.e., that of going from the input-output behavior of a system to its state equations) is solvable if and only if the system is both reachable and observable. We shall also see that given any system, we may use its input-output behavior to identify a reachable and observable system with the same input-output behavior. If the system under study is itself reachable and observable, then the system that we identify with identical input-output behavior will have essentially the same state behavior.

Let us now turn from our general development of reachability and observability and study these notions in detail for discrete-time systems, where the necessary mathematics appears in its most simple, uncluttered form. With the insights and intuition developed by our study of discrete-time systems, the more complicated manipulations of continuous-time systems will seem less formidable.

EXERCISES FOR SECTION 4–2

✔ **1.**   Show that the system in Example 1 of Section 1–6 is completely reachable, and find a simple input u which will cause the event $(9, 3)$ to be reached in exactly two seconds. Repeat for the event $(5, 4)$.

EE 2.   For the system of Example 1, find a step voltage $u(t) = k\mathbf{1}(t)$ which results in the event $(5, 4)$ being reached in 5 seconds.

EE 3.   Show that the event $(5, -4)$ is controllable to zero in the system of Example 1 by finding a u which will drive v to 0 in precisely 3 seconds.

Observe that if $u(t) \equiv 0$ for $t \geq 5$, then the capacitor will take forever to discharge to the zero state at $t = \infty$.

✔ **4.**   Show the system of Example 1–6–1 to be completely controllable, and find a simple control input u which "controls" the event $(5, 4)$ to be driven to $(7, 0)$.

Compare this control u with that needed in Exercise 1 to get to $(5, 4)$ from $(3, 0)$.

✔ **5.**   Find a control signal u which will drive the system of Example 1–6–1 from the event $(5, 4)$ to $(9, 3)$. Do this directly, and compare with the result of splicing the controls you found in Exercises 1 and 4.

✔ **6.**   In automata theory, computer scientists call a time-invariant system *strongly connected* if for any pair of states there is an input sequence which will cause the system to change from the first state to the second. Does this property imply that every event $(\tau, \bar{x})$ is both reachable and controllable? The converse is certainly not true.

What conditions must be met besides reachability and controllability of each event in order to guarantee that the system is strongly connected? [*Hint:* As in Exercise 4, consider trajectories passing through the zero state.]

✔ **7.**   In the system of Example 3, show that $\bar{x} = \begin{bmatrix} 1 \\ 1 \end{bmatrix}$ and $\hat{x} = \begin{bmatrix} 2 \\ 7 \end{bmatrix}$ are distinguishable in the future by exhibiting a simple input u which when applied on $[\tau, t)$ leads to different outputs $\eta(t, \bar{x})$ and $\eta(t, \hat{x})$ for any $t \geq \tau$.

Make a similar analysis to show that $(\tau, \bar{x})$ and $(\tau, \hat{x})$ are distinguishable in the past.

8.   Consider the discrete-time system of Figure 4–5. Show that $\bar{x} = \begin{bmatrix} 1 \\ 1 \end{bmatrix}$ and $\hat{x} = \begin{bmatrix} 2 \\ 7 \end{bmatrix}$ are distinguishable in the future. What about distinguishability in the past?

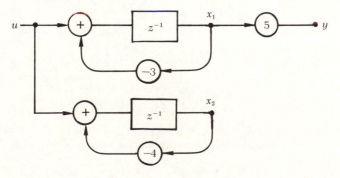

Figure 4–5

✔ **9.**   Prove that in a linear system the states $\bar{x}$ and $\hat{x}$ are indistinguishable if and only if their difference $(\bar{x} - \hat{x})$ is indistinguishable from 0_X.

✔ **10.**   In the system discussed in Example 4, suppose we consider the control law

$$u\left(\begin{bmatrix} x_1 \\ x_2 \end{bmatrix}\right) = -\text{sgn}\,(x_2 + 3x_1)$$

(a) Sketch the trajectory followed by the system starting in the initial state $x(0) = \begin{bmatrix} 1 \\ 1 \end{bmatrix}$. [Notice that the phase plane is divided into two regions by the switching line $x_2 = -3x_1$ and that this control law shows that $\bar{x} = \begin{bmatrix} 1 \\ 1 \end{bmatrix}$ can be controlled to the zero state $0_X = \begin{bmatrix} 0 \\ 0 \end{bmatrix}$ but that this control takes an infinite number of switchings.]

(b) Find a straight line switching curve which controls $x(0) = \begin{bmatrix} 1 \\ 1 \end{bmatrix}$ to $0_X = \begin{bmatrix} 0 \\ 0 \end{bmatrix}$ such that the control changes sign only once. This will prove that $\bar{x} = \begin{bmatrix} 1 \\ 1 \end{bmatrix}$ is controllable to the origin in *finite time* and hence is controllable.

(c) Does the switching line you found in (b) also cause the initial state $\hat{x} = \begin{bmatrix} -1 \\ 2 \end{bmatrix}$ to be driven to the origin in finite time? Can you find a switching *curve* (not a straight line) such that any initial state will be driven to the origin in finite time, with at most one change of sign in the control signal?

11. Determine whether or not the system of Exercise 14 of Section 3–2 is reachable. Is it controllable? Observable?

✔ **12.** Consider the systems Σ_1 and Σ_2 and their connection in cascade as shown in Exercise 7 of Section 3–1.

(a) Show that Σ_1 and Σ_2 are separately reachable, controllable, and observable.

(b) Show that the composite system of Σ_1 and Σ_2 in cascade is reachable, controllable, and observable.

(c) Give a different realization $\hat{\Sigma}$ for the cascaded system.

✔ **13.** This exercise provides practice with block-partitioned matrices, which will be used in Exercise 14 and extensively used in Sections 4–3, 4–5, and 6–6 and in all of Chapters 7 and 8. **Make up some simple matrices of your own and numerically verify each formula presented in parts (a) through (g).**

Let

$$A = \begin{bmatrix} A_{11} & A_{12} \\ A_{21} & A_{22} \end{bmatrix} \qquad B = \begin{bmatrix} B_{11} & B_{12} \\ B_{21} & B_{22} \end{bmatrix}$$

be two matrices partitioned into blocks as shown.

(a) Show that

$$A + B = \begin{bmatrix} A_{11} + B_{11} & A_{12} + B_{12} \\ A_{21} + B_{21} & A_{22} + B_{22} \end{bmatrix}$$

if A_{ij} is the same size as B_{ij} [A and B identically partitioned].

(b) In order that we can do multiplication by blocks and write

$$AB = \begin{bmatrix} (A_{11}B_{11} + A_{12}B_{21}) & (A_{11}B_{12} + A_{12}B_{22}) \\ (A_{21}B_{11} + A_{22}B_{21}) & (A_{21}B_{12} + A_{22}B_{22}) \end{bmatrix}$$

what must be true about the column partitioning of A and the row partitioning of B? (We call this *conformable* or *compatible* partitioning.)

(c) Let A be $n \times n$ and partitioned so that A_{11}, A_{22} are square (not necessarily of the same size). If $A_{21} = 0$, show that $\det A = \det(A_{11}) \det(A_{22})$. [This looks like the product of diagonal entries of a triangular matrix.] What happens if $A_{12} = 0$?

(d) With A as in part (c) and $A_{21} = 0$, find $A^2, A^3, \dots, A^k$ [powers of block triangular matrices].

(e)   Show that if A_{11} and A_{22} are invertible, then

$$\left[\begin{array}{c|c} A_{11} & A_{12} \\ \hline 0 & A_{22} \end{array}\right]^{-1} = \left[\begin{array}{c|c} A_{11}^{-1} & -A_{11}^{-1}A_{12}A_{22}^{-1} \\ \hline 0 & A_{22}^{-1} \end{array}\right]$$

(f)   Study the invertibility of the partitioned matrix

$$\left[\begin{array}{c|c|c} A_{11} & A_{12} & A_{13} \\ \hline 0 & A_{22} & A_{23} \\ \hline 0 & 0 & A_{33} \end{array}\right]$$

where the diagonal blocks A_{11}, A_{22}, and A_{33} are each square and invertible.

(g)   Show that

$$\left[\begin{array}{c|c} A_{11} & 0 \\ \hline 0 & A_{22} \end{array}\right]^{k} = \left[\begin{array}{c|c} A_{11}^{k} & 0 \\ \hline 0 & A_{22}^{k} \end{array}\right]$$

for all integers k (positive and negative) for which $(A_{ij})^{k}$ makes sense; in particular, let $k = -1$.

✔ **14.**   Let Σ_1 and Σ_2 be linear, discrete-time, time-invariant systems described by

$$\Sigma_1 \begin{cases} x_1(t+1) = F_1 x_1(t) + G_1 u_1(t) \\ \quad\; y_1(t) = H_1 x_1(t) \end{cases}$$

$$\Sigma_2 \begin{cases} x_2(t+1) = F_2 x_2(t) + G_2 y_1(t) \\ \quad\; y_2(t) = H_2 x_2(t) \end{cases}$$

Prove that their series connection Σ_σ (Σ_1 followed by Σ_2) is linear, and find the block-partitioned matrices F_σ, G_σ, H_σ which yield

$$\Sigma_\sigma \begin{cases} \left[\begin{array}{c} x_1(t+1) \\ \hline x_2(t+1) \end{array}\right] = F_\sigma \left[\begin{array}{c} x_1(t) \\ \hline x_2(t) \end{array}\right] + G_\sigma u_1(t) \\[20pt] \qquad\qquad y_2(t) = H_\sigma \left[\begin{array}{c} x_1(t) \\ \hline x_2(t) \end{array}\right] \end{cases}$$

[See Exercise 13 for practice with block-partitioning if you need it.]

15.   Write out the "matrix plug-in proof" sketched in the text following Corollary 2.

✔ **16.**   Is it the case for any pair of "compatible" systems Σ_1 and Σ_2, each of which is reachable, that their series connection Σ_σ must also be reachable? If not, indicate conditions on Σ_1 and Σ_2 under which Σ_σ will be reachable.

Repeat, substituting controllable for reachable. When is Σ_σ strongly connected?

17.   A subset B of a real vector space is said to be **convex** if for any $x, y \in B$ the point $\lambda x + (1 - \lambda)y$ remains in B for all $\lambda \in [0, 1]$ (we call the locus of all such points the *line segment* between x and y).

(a)   Show that the set Ω of admissible input functions of a linear system is convex.

(b)   Let $S^{(\theta)}$ be the set of all states reachable from the zero state of a linear system. Show that $S^{(\theta)}$ is convex.

(c)   Let $S^{(x_0)} = \{\bar{x} \,|\, \bar{x} \text{ is reachable from } x_0\}$. Show that for a linear system $S^{(x_0)}$ is convex.

(d)   Let $S_t^{(x_0)} = \{\bar{x} \,|\, \bar{x} \text{ is reachable from } x_0, \text{ at time } t\}$. Is this set convex for a linear system? How about for a time-invariant linear system?

4-3   Reachability and Controllability for Discrete Time

We have seen that linear systems can be used to approximate real systems. Yet we have also stressed that much more of system theory applies to non-linear systems than appears from most textbook treatments. Thus, before we develop in Chapter 5 the mathematical machinery which allows us to compute readily the various properties of continuous-time linear systems, it seems well to gain what insight we can into such notions as reachability, observability, and controllability for discrete-time systems, *even if they are non-linear.* As the reader may be prepared to believe from our discussion of controllability of a series connection of systems in the previous section, the resulting treatment will actually yield insights that tend to be obscured in the matrix manipulations used so often in linear system theory.

Let us focus, then, on a discrete-time, time-invariant system Σ with input set U, state set X and output set Y. Since we shall be interested in the way in which a *sequence* of inputs affects the system, let us introduce the special notation U^* to denote the set of all input sequences:

$$U^* \triangleq \{u_0 u_1 \ldots u_{n-1} | n \geq 0 \text{ and each } u_j \in U\}$$

Note that we allow $n = 0$; we use Λ to denote the empty string and use "apply Λ to Σ" as a synonym for "do nothing to Σ" just as, in arithmetic, "add 0 to a" is a synonym for "don't add anything to a." One point of caution, though—applying an input whose value is 0 is *not* the same as applying no input at all. For instance, consider the constant linear system described by

$$x(t + 1) = Fx(t) + Gu(t) \tag{1}$$

Then if we write $\phi(\hat{x}, \Lambda)$ to indicate that we have not yet applied an input to the system in state $\hat{x}$, we have

$$\begin{array}{ccc} \phi(\hat{x}, & \Lambda) & = & \hat{x} \\ \uparrow & \uparrow & & \uparrow \\ x(t) & \text{no input} & & \text{still } x(t) \\ & \text{applied} & & \end{array}$$

whereas if we write $\phi(\hat{x}, 0)$ to indicate that we have applied a single input which happened to take the value 0, we have

$$\begin{array}{ccc} \phi(\hat{x}, & 0) & = & F\hat{x} \\ \uparrow & \uparrow & & \uparrow \\ x(t) & u(t) & & x(t + 1) \end{array}$$

We shall sometimes find it convenient to use the notation 0^n, not as in arithmetic as a number equal to 0, but rather to denote a string of n consecutive inputs, each taking the value 0. We would then have the scheme

$$\widehat{x} \xrightarrow{0} \phi(\widehat{x}, 0) = F\widehat{x} \xrightarrow{0} \phi(\widehat{x}, 0^2) = F^2\widehat{x} \xrightarrow{0} \cdots \xrightarrow{0} \phi(\widehat{x}, 0^n) = F^n\widehat{x}$$

for our linear system (1).

Anyway, the point of introducing all this notation is to stress that every discrete-time, time-invariant system can be described by a pair of functions

$$\phi: X \times U^* \to X$$

and

$$\eta: X \to Y$$

such that if we have

$$x(t) = \widehat{x} \quad \text{and} \quad u(t + j) = u_j \quad 0 \le j < n$$

then we may deduce that

$$x(t + n) = \phi(\widehat{x}, u_0 u_1 \ldots u_{n-1}) \text{ while } y(t) = \eta(x(t))$$

Let us now restate the notions of reachability and observability for discrete-time, time-invariant systems:

A system is *reachable* from state $x_0 \in X$ if every state of the system is reachable from x_0, i.e., if for every state $\widehat{x} \in X$ there exists at least one input sequence $w \in U^*$ such that

$$\phi(x_0, w) = \widehat{x}$$

In modern algebraic terminology, this definition may be simply rephrased as:

A system is *reachable* from state $x_0 \in X$ if and only if the function

$$\phi(x_0, \cdot): U^* \to X$$

is *onto*.

Let us review the notation $\phi(x_0, \cdot)$ and the word "onto." We see that ϕ is a function of *two* arguments, which assigns to any pair $(\widehat{x}, w) \in X \times U^*$ a state $\phi(\widehat{x}, w)$ in X. We then use $\phi(x_0, \cdot)$ to denote the function of *one* argument (obtained by fixing the first argument of ϕ to be the state x_0) which assigns to any sequence w of U^* the state $\phi(x_0, w)$ in X:

$$\phi(x_0, \cdot): U^* \to X \text{ such that } w \mapsto \phi(x_0, w)$$

A map from U^* into X is *onto* iff for every state $\widehat{x} \in X$ there is some input sequence $w \in U^*$ which the map sends to $\widehat{x}$. That is, $\phi(x_0, \cdot): U^* \to X$ is onto iff $\phi(x_0, U^*) = X$. [See Exercise 7 of Section 1–2.]

Let us now turn to *observability*. Suppose we are given a "black box" Σ in state $\hat{x}$ as in Figure 4–6. How can we verify that Σ is in state $\hat{x}$ without "opening" the box? It is done simply by applying inputs to the system and checking that the resulting outputs are indeed appropriate to Σ started in state $\hat{x}$.

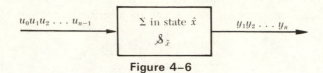

$$u_0 u_1 u_2 \ldots u_{n-1} \longrightarrow \boxed{\begin{array}{c} \Sigma \text{ in state } \hat{x} \\ \mathcal{S}_{\hat{x}} \end{array}} \longrightarrow y_1 y_2 \ldots y_n$$

Figure 4–6

We are thus interested in the *input-output* function, or *response* function, of Σ started in state $\hat{x}$, i.e., the function $\mathcal{S}_{\hat{x}}$ which tells us, for any input sequence w, the output $\mathcal{S}_{\hat{x}}(w)$ that will be emitted by Σ, if started in state $\hat{x}$, after it has processed w. We clearly have

$$\mathcal{S}_{\hat{x}} : U^* \to Y \quad \text{by the formula} \quad w \mapsto \mathcal{S}_{\hat{x}}(w) = \eta[\phi(\hat{x}, w)]$$

since w sends Σ from state $\hat{x}$ to state $\phi(\hat{x}, w)$, for which the output is $\eta[\phi(\hat{x}, w)]$.†

Note that although $\mathcal{S}_{\hat{x}}(w)$ only tells us the final output resulting from an input sequence $w = u_0 u_1 \ldots u_{n-1}$, we can reconstruct the whole output sequence as

$$\mathcal{S}_{\hat{x}}(u_0), \mathcal{S}_{\hat{x}}(u_0 u_1), \ldots, \mathcal{S}_{\hat{x}}(u_0 u_1 \ldots u_{n-1})$$

Now we have said that we can only check the state of a system "from the outside" by applying input sequences w and checking the responses $\mathcal{S}_{\hat{x}}(w)$. If two states $\hat{x}$ and $\tilde{x}$ have identical response functions, then there is no way of telling "from the outside" whether Σ is in state $\hat{x}$ or state $\tilde{x}$. This suggests the definition:

A system is **observable** if for every pair of distinct states $\hat{x}$ and $\tilde{x}$ there exists at least one input sequence to which they respond differently; i.e., there exists w in U^* such that

$$\mathcal{S}_{\hat{x}}(w) \neq \mathcal{S}_{\tilde{x}}(w)$$

Note that this definition only says of an observable system that when we have two hypotheses, "Σ is in state $\hat{x}$" and "Σ is in state $\tilde{x}$," one of which is correct, we can tell which one is indeed correct by applying a suitable input sequence w. Later we shall see that the definition sometimes implies observa-

---

†The reader will recall our use of $\mathcal{S}_{\hat{x}} : T \times \Omega \to Y$, for the response of a time-invariant system Σ, in Section 3–1. For a discrete-time system, our use of U^* instead of Ω makes the T-variable redundant, and so what we might denote by $\mathcal{S}_{\hat{x}} : T \times \Omega \to Y$ reduces to our $\mathcal{S}_{\hat{x}} : U^* \to Y$.

bility in a stronger sense; e.g., for linear systems we can provide an algorithm for computing the initial state of an observable system on the basis of its response to a string of zero inputs.

Let us use the symbol $[U^*, Y]$ to denote the set of *all* functions mapping U^* into Y. We can then use $\rho : X \rightarrow [U^*, Y]$ (ρ for Response assignment) to denote the function which assigns to each state $\hat{x}$ its response function $\mathcal{S}_{\hat{x}}$. Since two functions differ whenever they assign a different image to at least one point, we have the algebraic reformulation of our definition of observability:

A system is **observable** if and only if the function

$$\rho : X \rightarrow [U^*, Y] \text{ by the rule } \hat{x} \mapsto \mathcal{S}_{\hat{x}}$$

is one-to-one.

Finally, let us define the **realization problem:** Suppose that we are given a function $\mathcal{F} : U^* \rightarrow Y$, possibly obtained as the result of measurements upon some system $\tilde{\Sigma}$ of interest to us, but of which the dynamic equations are not known. To control $\tilde{\Sigma}$, we need a "model," that is, a system Σ whose equations may not describe the "internal behavior" of $\tilde{\Sigma}$ but which will have the input-output behavior summarized in the function $\mathcal{F}$. This motivates the formal definition:

The Realization Problem.   Given a function $\mathcal{F} : U^* \rightarrow Y$, find a system Σ with an input set U, an output set Y, a state $\hat{x}$, and a response function $\mathcal{S}_{\hat{x}} : U^* \rightarrow Y$ such that

$$\mathcal{S}_{\hat{x}} = \mathcal{F}$$

We call Σ a **realization** of $\mathcal{F}$.

We stress that a given $\mathcal{F}$ might have been found by measuring the response function $\mathcal{S}_{\tilde{x}}$ of some system $\tilde{\Sigma}$ started in state $\tilde{x}$. However, we must usually expect our computations, based on values of $\mathcal{F}$ rather than knowledge of $\tilde{\Sigma}$, to yield a realization Σ different from $\tilde{\Sigma}$. We give a continuous-time example:

Example 1

If $\tilde{\Sigma}$ were the system of Figure 4–7(a), whose response at time t to an input function u is the integral $2 \int_{t_0}^{t} u(\tau) \, d\tau$, we would expect any reasonable identification procedure to construct a realization Σ with one integrator, as shown in Figure 4–7(b).   ◊

Let us now give a finite-state example:

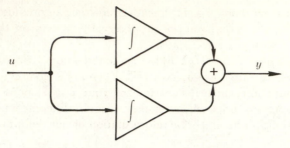

Figure 4–7(a)   A two-dimensional system $\tilde{\Sigma}$.

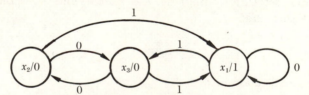

Figure 4–7(b)   A one-dimensional realization Σ.

Example 2

Consider a system $\tilde{\Sigma}$ with binary inputs and outputs, in other words, with $U = Y = \{0, 1\}$, and three states, $X = \{x_1, x_2, x_3\}$. We can represent it by the graph in Figure 4–8(a). There are three nodes, where, for instance,

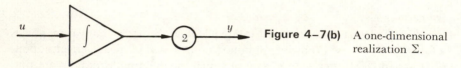

Figure 4–8(a)   A three-state system $\tilde{\Sigma}$.

the label $x_2/0$ on a node indicates that $\eta(x_2) = 0$. One arrow leaves a node for each distinct input, where, for instance, the label 1 on the arrow leading from x_2 to x_1 indicates that $\phi(x_2, 1) = x_1$.

Noting that $\mathcal{S}_{x_j}(\Lambda) = \eta(x_j)$ for any system and any state x_j and recalling that, in general, $\mathcal{S}_{\hat{x}}(w) = \eta(\phi(\hat{x}, w))$, we can compute the responses to various input sequences w. For example, $\mathcal{S}_{x_1}(\Lambda) = 1$, while $\mathcal{S}_{x_2}(11) = 0$, since 11 takes us from x_2 to x_1 and thence to x_3, and x_3 has output 0. In this way we can fill in the table of Figure 4–8(b), where the value $\mathcal{S}_{\hat{x}}(w)$ appears in the intersection of row $\mathcal{S}_{\hat{x}}$ and column w:

This table suggests the following result, which the reader is asked to prove by induction† (i.e., prove it for $w = \Lambda$, and then prove that truth for sequences of length n implies truth for sequences of length $n + 1$):

---

†Readers who are rusty on proofs by mathematical induction should review Section 2–1.

$S_x \backslash^w$	Λ	0	1	00	01	10	11	000	...
S_{x_1}	1	1	0	1	0	0	1	1	...
S_{x_2}	0	0	1	0	1	1	0	0	...
S_{x_3}	0	0	1	0	1	1	0	0	...

Figure 4–8(b)   Outputs corresponding to different input strings and initial states.

For all sequences w,

$$S_{x_2}(w) = S_{x_3}(w) = \text{``parity''} \text{ of } w = \begin{cases} 1 \text{ if } w \text{ has an odd number of 1's} \\ 0 \text{ if } w \text{ has an even number of 1's} \end{cases}$$

$$S_{x_1}(w) = 1 - S_{x_2}(w)$$

[See Exercise 1 of Section 4–5 for another way to express $S_{x_2}(w)$.]

Now, if we were asked to come up with a realization of the parity function, we would probably come up with the Σ of Figure 4–8(c). In fact,

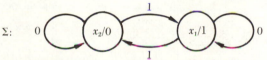

$$\Sigma:$$

Figure 4–8(c)   A two-state realization Σ.

we may regard Σ as being obtained from $\tilde{\Sigma}$ by "merging" the state x_3 into the state x_2, from which it is indistinguishable. Σ and $\tilde{\Sigma}$ are both reachable, while only Σ is observable. Interestingly, although $\tilde{\Sigma}$ is not observable—there is no way of telling *initial* state x_2 from initial state x_3—we can apply an input 1 following an output 0 or an input 0 following an output 1 to ensure that its *final* state is x_1! Of course, we can only make this determination because we know the dynamics of the system. However, once we have this "lids-off" knowledge of how the system works, we only use "black-box" information to determine its final state. This asymmetry between initial and final states can only occur for discrete-time systems, for reasons that will become clear in Chapters 5 and 7, where we shall discuss the "reversibility" of continuous-time systems.   ◊

Now that we are familiar with the basic algebraic terminology, we may briefly summarize the section so far, it being understood that from here on every system Σ in this chapter is *discrete-time* and *time-invariant* unless explicit notice is given to the contrary:

DEFINITION 1

Recalling that a system Σ is described by two functions

$$\phi : X \times U^* \to X$$
$$\eta : X \to Y$$

(subject to the usual conditions on ϕ), we say that Σ is **reachable** from state x_0 if $\phi(x_0, \cdot) : U^* \to X$ such that $w \mapsto \phi(x_0, w)$ is onto; and that Σ is **observable** if the function $\rho : X \to [U^*, Y]$ such that $\hat{x} \mapsto \mathcal{S}_{\hat{x}}$ is one-to-one, where $\mathcal{S}_{\hat{x}} : U^* \to Y$ by $w \mapsto \eta(\phi(\hat{x}, w))$ is the response function of Σ started in state $\hat{x}$. The **realization problem** for a given function $\mathcal{F} : U^* \to Y$ is to find a **realization** of $\mathcal{F}$, that is, a system Σ with input set U and output set Y which has a state $\hat{x}$ whose response function $\mathcal{S}_{\hat{x}}$ equals $\mathcal{F}$. $\bigcirc$

Our task will now be to study reachability, observability, and controllability in more detail, and in particular to see what extra information is available when Σ is further restricted to be a *linear* system.

THE REACHABLE STATES

DEFINITION 2

We introduce the following notations for a fixed system Σ:

$S^{(x_0)} =$ the set of all states reachable from x_0
   $= \{ \hat{x} \mid \text{there exists } w \in U^* \text{ with } \phi(x_0, w) = \hat{x} \}$
   $= \phi(x_0, U^*)$, the range of the map $\phi(x_0, \cdot)$

$\overset{e}{S}_j^{(x_0)} =$ the set of all states reachable from x_0 by applying a sequence of *exactly* j inputs
   $= \{ \hat{x} \mid \text{there exists } u_0 u_1 \ldots u_{j-1}, \text{ each } u_i \in U, \text{ with}$
       $\phi(x_0, u_1 u_2 \ldots u_j) = \hat{x} \}$
   $= \phi(x_0, U^j)$ where we use U^j to denote the j-fold cartesian product of U with itself

$S_k^{(x_0)} =$ the set of all states reachable from x_0 in *at most* k steps
   $= \{ \hat{x} \mid \text{there exists } u_0 \ldots u_{j-1}, 0 \le j \le k, \text{ with}$
       $\phi(x_0, u_0 \ldots u_{j-1}) = \hat{x} \}$
   $= \overset{e}{S}_0^{(x_0)} \cup \overset{e}{S}_1^{(x_0)} \cup \ldots \cup \overset{e}{S}_k^{(x_0)}$

The last union of $k + 1$ sets may be written as $\displaystyle\bigcup_{j=0}^{k} \overset{e}{S}_j^{(x_0)} = \bigcup_{j=0}^{k} \phi(x_0, U^j)$. Note that the only input string of length 0 is Λ, so that $U^0 = \{\Lambda\}$, whence $S_0^{(x_0)} = \{x_0\}$ for any x_0.   ○

Example 3

Consider the system with two states, two output symbols, and one input symbol in Figure 4–9. For this system,

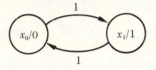

Figure 4–9   A two-state, two-output, one-input system.

$$S^{(x_0)} = \{x_0, x_1\}$$

$$S_0^{(x_0)} = \{x_0\}, \quad S_k^{(x_0)} = \{x_0, x_1\} \text{ for } k \geq 1$$

while

$$\overset{e}{S}_j^{(x_0)} = \begin{cases} \{x_0\} \text{ if } j \text{ is even} \\ \{x_1\} \text{ if } j \text{ is odd} \end{cases} \qquad ◇$$

The following simple result is an immediate consequence of our definitions:

PROPERTY 1

$$S^{(x_0)} = \bigcup_{j=0}^{\infty} \overset{e}{S}_j^{(x_0)} = \bigcup_{k=0}^{\infty} S_k^{(x_0)}$$

Furthermore,

$$S_0^{(x_0)} \subset S_1^{(x_0)} \subset S_2^{(x_0)} \subset S_3^{(x_0)} \subset \cdots \subset S^{(x_0)} \qquad □$$

Note: We write $A \subset B$ to say that A is a subset of B, and do *not* rule out the case $A = B$. For instance, we saw in the above example that $S_k^{(x_0)} = S^{(x_0)}$ for all $k \geq 1$. By contrast, let us see a simple example of a system which is reachable but is such that $S_k^{(x_0)} \neq S^{(x_0)}$ for any finite integer k:

Example 4

Consider the system of Figure 4–10 with two inputs $U = \{0, 1\}$ and with an infinite set of states $X = \mathbf{N}$, the set $\{0, 1, 2, 3, \ldots\}$ of all non-negative integers. (We omit outputs, since our interest here is restricted to reachability.)

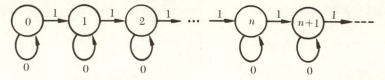

Figure 4–10   An infinite-state, two-input system.

We have
$$\phi(n, 0) = n$$
$$\phi(n, 1) = n + 1$$

and, in fact,
$$\phi(n, u_0 \ldots u_{j-1}) = n + \sum_{i=0}^{j-1} u_i$$

Then $S_k^{(0)} = \{0, 1, \ldots, k\} \overset{e}{=} \bar{S}_k^{(0)}$ and we have $S_k^{(0)} \neq S_{k'}^{(0)}$ for every $k \neq k'$. Yet the system is reachable from 0, since for every state n we have

$$n = \phi(0, 1^n)$$

i.e., we can reach n from 0 by applying the 1-input n times in a row. Note that this system is *not* reachable from any state $n \neq 0$, since no input sequence w can cause the state to decrease from n to $n - 1$.   ◊

In Example 3, the $S_k^{(x_0)}$'s increased with k until an equality was reached, at which time we had obtained all reachable states. In Example 4, however, the $S_k^{(x_0)}$'s continued to increase as k increased, and no equality was ever reached. In fact, the situation can be characterized by the following theorem:

THEOREM 2

Let Σ be a system with a state x_0 for which there exists an integer k such that $S_k^{(x_0)} = S_{k+1}^{(x_0)}$. Then

$$S_k^{(x_0)} = S^{(x_0)}$$

Remark: In Example 3 we can take $k = 1$; in Example 4, no such k exists. Although we shall give a formal proof below, this theorem is quite obvious as suggested by the paraphrase: If you reach a place [namely, S_k] from which one step will carry you no further [i.e., $S_k = S_{k+1}$], then no finite number of steps can carry you any further [i.e., $S_k = S$]. In the formal proof of Theorem 2 we shall use a fact which we present here as a lemma ("little" sub-theorem), whose proof is outlined in Exercise 3.

LEMMA 3

If there exists an integer $k \in \mathbf{N}$ such that $S_{k'}^{(x_0)} = S_k^{(x_0)}$ for all $k' \geq k$, then

$$S^{(x_0)} = \bigcup_{j=0}^{\infty} S_j^{(x_0)} = S_k^{(x_0)} \qquad \square$$

Proof of Theorem 2

By Lemma 3, it suffices to prove that $S_k^{(x_0)} = S_{k'}^{(x_0)}$ for all $k' > k$. We do this by induction on j for $k' = k + j$:

Basis Step: $k' = k + 1$: $S_{k+1}^{(x_0)} = S_k^{(x_0)}$ by hypothesis.

Induction Step: Given that $S_k^{(x_0)} = S_{k'}^{(x_0)}$ for $k' = \ell$, prove that

$$S_k^{(x_0)} = S_{\ell+1}^{(x_0)}$$

But this is immediate:

$$S_{\ell+1}^{(x_0)} = \phi(x_0, U^{\ell+1}) = \phi(x_0, U^{k+1} U^{\ell-k})$$

since a sequence of length $\ell + 1$ can always be decomposed into a sequence of length $k + 1$ followed by a sequence of length $\ell - k$. Thus,

$$S_{\ell+1}^{(x_0)} = \phi(S_{k+1}^{(x_0)}, U^{\ell-k})$$

Also, $\phi(x_0, w_1 w_2) = \phi(\phi(x_0, w_1), w_2)$ for all $w_1 \in U^{k+1}$ and $w_2 \in U^{\ell-k}$, so that

$$\begin{aligned} S_{\ell+1}^{(x_0)} &= \phi(S_k^{(x_0)}, U^{\ell-k}) \text{ by hypothesis} \\ &= S_\ell^{(x_0)} \\ &= S_k^{(x_0)}, \text{ by induction hypothesis.} \qquad \square \end{aligned}$$

REACHABILITY FOR LINEAR SYSTEMS

Let us now see what this general theory means for the linear system

$$x(t + 1) = Fx(t) + Gu(t)$$
$$y(t) = Hx(t)$$

which we shall henceforth refer to as (F, G, H).

Recall that for such a system, U, X, and Y must be vector spaces over the same field, while F, G, and H are linear transformations

$F: X \to X$: updating the state

$G: U \to X$: reading in the input

$H: X \to Y$: reading out the output

If we restrict ourselves to finite-dimensional real vector spaces, say $X = \mathbf{R}^n$, $U = \mathbf{R}^m$, and $Y = \mathbf{R}^q$, this just says that

F is an $n \times n$ matrix

G is an $n \times m$ matrix

H is a $q \times n$ matrix

Now, for the system (F, G, H), it is clear that $\phi(\hat{x}, u(t)) = F\hat{x} + Gu(t)$ and $\eta(x) = Hx$. But what is the correct formulation for $\phi: X \times U^* \to X$? Let us look at ϕ for sequences of length ≤ 3:

If we let $x_i = x(t_0 + i)$ and $u_i = u(t_0 + i)$ for $i = 0, 1, 2, \ldots$ then

$\phi(x_0, \Lambda) = x_0$

$\phi(x_0, u_0) = Fx_0 + Gu_0 = x_1$

$\phi(x_0, u_0 u_1) = Fx_1 + Gu_1 = F^2 x_0 + FGu_0 + Gu_1 = x_2$

$\phi(x_0, u_0 u_1 u_2) = Fx_2 + Gu_2 = F^3 x_0 + F^2 Gu_0 + FGu_1 + Gu_2 = x_3$

The reader should by now have perceived the pattern, and be prepared for the general result:

$$x_k = \phi(x_0, u_0 u_1 \ldots u_{k-1}) = F^k x_0 + \sum_{j=0}^{k-1} F^{k-j-1} Gu_j, \text{ for } k > 0 \qquad (2)$$

The formula is completely intuitive; at each instant in time we update the old state by F and add in the contribution Gu of the new input u. Thus, $\phi(x_0, u_0 \ldots u_{k-1})$ has $k + 1$ summands; the first, $F^k x_0$, is obtained by updating the initial state every time an input is read in; while the term $F^{k-j-1} Gu_j$ is obtained by reading in u_j with G and then updating this contribution every time a subsequent input is read in.

For linear systems, our interest centers on states reachable from the zero state 0_X. Thus, for simplicity, let us use S_k to denote $S_k^{(0_X)}$, S for $S^{(0_X)}$, and $\overset{e}{S_k}$ for $\overset{e}{S}_k^{(0_X)}$. Let us now derive an explicit formula for S_k for the system (F, G, H).

First we note that a zero input 0 leaves the zero state 0_X unchanged:

$$0_X = F0_X + G0$$

Thus, if we can reach a state $\hat{x}$ from the zero state by applying a sequence of $j < k$ inputs $u_0 u_1 \ldots u_{j-1}$, we can also reach $\hat{x}$ by applying the zero input $k - j$ times in succession, after which we are still in the zero state, and *then* applying $u_0 u_1 \ldots u_{j-1}$. Thus, for all $u_0 u_1 \ldots u_{j-1} \in U^j$ and all $k > j$, our linear system (F, G, H) satisfies

$$\phi(0_X, u_0 u_1 \ldots u_{j-1}) = \phi(0_X, 0^{k-j} u_0 u_1 \ldots u_{j-1})$$

where we remind the reader that 0^{k-j} denotes a string of $k - j$ successive inputs, each with value 0. In other words, for our linear system (F, G, H), we have $S_k = \overset{e}{S}_k$.

Thus, $\hat{x}$ is in S_k iff there exists a sequence $u_0 u_1 \ldots u_{k-1}$ of length *exactly* k (though some of the initial inputs may be 0) such that

$$\hat{x} = F^k \cdot 0_X + \sum_{j=0}^{k-1} F^{k-j-1} G u_j$$

$$= [F^{k-1}G \mid F^{k-2}G \mid \ldots \mid G] \begin{bmatrix} u_0 \\ \hline u_1 \\ \hline \vdots \\ \hline u_{k-1} \end{bmatrix}$$

where we have made use of the fact that $F^k \cdot 0_X = 0_X$, and have stacked up the sequence of vectors $u_0 u_1 \ldots u_{k-1}$ as a big vector. Note, in the finite-dimensional case $U = \mathbf{R}^m$, that $\begin{bmatrix} u_0 \\ \hline u_1 \\ \hline \vdots \\ \hline u_{k-1} \end{bmatrix}$ is a vector in $\mathbf{R}^{mk}$. Thus, our state

$\phi(0_X, u_0 u_1 \ldots u_{k-1})$ may be expressed as the product of the block-partitioned†

matrix $[F^{k-1}G, F^{k-2}G, \ldots, G]$ and the block-partitioned vector $\begin{bmatrix} u_0 \\ u_1 \\ \hline \vdots \\ \hline u_{k-1} \end{bmatrix}$, where

for aesthetic reasons we have left out some of the dashed partitioning lines, replacing them with commas where necessary to avoid confusion. Of course, we could just as well have written this product as

$$[G, FG, F^2G, \ldots, F^{k-1}G] \begin{bmatrix} u_{k-1} \\ \hline u_{k-2} \\ \vdots \\ \hline u_0 \end{bmatrix}$$

---

† See Exercise 13 of Section 4–2 if you need practice with block-partitioned matrices.

in which case our subsequent discussion would involve the matrix $Q = [G, FG, \ldots, F^{k-1}G]$ instead of $\mathcal{C} = [F^{k-1}G, \ldots, FG, G]$. We have thus proved:

LEMMA 4

For the linear system (F, G, H), the set of states reachable from the zero state 0_X in at most k steps is the range of the linear transformation $[F^{k-1}G, F^{k-2}G, \ldots, G] : U^k \to X$. That is,

$$S_k = \mathcal{R}([F^{k-1}G, F^{k-2}G, \ldots, G]) \qquad \square$$

Now remember that we proved for any state x_0 of any (linear or non-linear) system Σ that if there exists a k with $S_k^{(x_0)} = S_{k+1}^{(x_0)}$, then $S_k^{(x_0)} = S$. Let us apply this to deduce immediately the following important result for linear systems:

THEOREM 5

Let (F, G, H) have dimension n (which is both the dimension of the state space X and the size of F). Then every state reachable from 0_X can be reached in at most n steps; that is,

$$S_n = S$$

Proof

First the reader should verify [Exercise 5] that for each k, S_k is a *subspace* of S_{k+1} and of X. Thus, if for any k we have $S_{k+1} \neq S_k$, we must have $\dim S_{k+1} \geq 1 + \dim S_k$. Now consider the series of set inclusions

$$S_0 \subset S_1 \subset S_2 \subset \cdots \subset S_{n-1} \subset S_n \subset \cdots \subset S$$

If $S_n = S_{n-1}$, then $S = S_{n-1}$ by our general result in Theorem 2, and certainly $S_n = S$. If $S_n \neq S_{n-1}$ then we must have $S_0 \neq S_1 \neq S_2 \neq S_3 \neq \cdots \neq S_{n-1} \neq S_n$. But the dimension of S_j increases by at least one at each step of this series. Thus $\dim S_n \geq n$. But S_n is a subspace of X, which is itself of dimension n. Thus $S_n = X$, and since $S_n \subset S \subset X$ we deduce that $S_n = S$.

Note that if $S_n = S_{n-1}$, we may or may not have $S = X$, while if $S_n \neq S_{n-1}$ we must have $S = X$. In either case, we do have $S = S_n$, as claimed. $\qquad \square$

COROLLARY 6

A linear system (F, G, H) of dimension n is reachable (i.e., reachable from 0_X) if and only if

$$\text{rank } [F^{n-1}G, F^{n-2}G, \ldots, G] = n$$

Proof

The system is reachable if and only if $X = S$ or, since $S = S_n$, if and only if

$$\begin{aligned}
\dim X &= \dim S_n \\
&= \dim (\mathcal{R}[F^{n-1}G, F^{n-2}G, \ldots, G]) \\
&= \text{rank } [F^{n-1}G, \ldots, G].
\end{aligned}$$ □

The following example should help the reader develop an intuitive feeling for these results.

Example 5

Consider the system with state variables chosen as the outputs of the unit-delay elements, as shown in Figure 4–11. Here we have $U = \mathbf{R}^2$, $X = \mathbf{R}^2$,

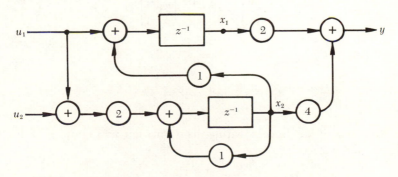

Figure 4–11   A two-dimensional linear system.

and $Y = \mathbf{R}$. From the diagram we may write, by inspection, the equations

$$\begin{aligned}
x_1(t + 1) &= x_2(t) + u_1(t) \\
x_2(t + 1) &= x_2(t) + 2[u_1(t) + u_2(t)] \\
y(t) &= 2x_1(t) + 4x_2(t)
\end{aligned}$$

which are of the form $x(t + 1) = Fx(t) + Gu(t)$ and

$$y(t) = Hx(t)$$

where

$$F = \begin{bmatrix} 0 & 1 \\ 0 & 1 \end{bmatrix}, G = \begin{bmatrix} 1 & 0 \\ 2 & 2 \end{bmatrix}, \text{ and } H = \begin{bmatrix} 2 & 4 \end{bmatrix}$$

We see that G is of rank 2, since $\begin{bmatrix} 1 \\ 2 \end{bmatrix}$ and $\begin{bmatrix} 0 \\ 2 \end{bmatrix}$ are linearly independent, and thus

$$S_1 = \Re(G) = X$$

i.e., all states are reachable from 0_X in exactly one step. Since rank $G = 2$ it is, of course, also true that

$$\text{rank } [F^{2-1}G, G] = 2$$

so that our reachability criterion is indeed satisfied.

The reader should be pleased to see that here, at least, theory agrees with intuition. If we have the system in the zero state at time t, and want it in state $\begin{bmatrix} a \\ b \end{bmatrix}$ at time $t + 1$, we simply set

$$\begin{bmatrix} a \\ b \end{bmatrix} = \begin{bmatrix} 1 & 0 \\ 2 & 2 \end{bmatrix} \begin{bmatrix} u_1(t) \\ u_2(t) \end{bmatrix}$$

and solve to get

$$u_1(t) = a \text{ and } u_2(t) = \frac{1}{2}b - a \qquad \Diamond$$

Note that if we have a wiring-diagram such as Figure 4–11 for the discrete-time system

$$x(t + 1) = Fx(t) + Gu(t) \qquad (3)$$

then by simply replacing each delay element as in Figure 4–12(a) by an integrator as in Figure 4–12(b), we obtain a wiring-diagram for the continuous-time system

$$\dot{x}(t) = Fx(t) + Gu(t) \qquad (4)$$

Figure 4–12(a) A unit delay block.   **(b)**   A unit integrator block.

Thus the reader may not be too surprised to find that in Chapter 7 we shall prove that system (4) is also reachable iff

$$\text{rank } [F^{n-1}G, \ldots, FG, G] = n$$

when F is an $n \times n$ constant matrix.

Let us now study controllability for the system $x(t + 1) = Fx(t) + Gu(t)$. We say that (F, G, H) is **controllable** if we can bring it to the zero state from any other state; that is, if for every state $x_0 \in X$, there exists a sequence $u_0 u_1 \ldots u_{k-1}$ of inputs such that

$$0_X = F^k x_0 + \sum_{j=0}^{k-1} F^{k-j-1} Gu_j$$

Thus x_0 is *controllable* if and only if there exists a sequence $v_0 v_1 \ldots v_{k-1}$ (with $v_j = -u_j$) such that

$$F^k x_0 = \sum_{j=0}^{k-1} F^{k-j-1} Gv_j$$

i.e., if and only if the state $F^k x_0$ is *reachable* in k steps:

$$F^k x_0 \in \mathcal{R}([F^{k-1}G, \ldots, FG, G])$$

Since $\mathcal{R}(F^k) = F^k(X) = \{F^k x_0 \mid x_0 \in X\}$, we may say quite succinctly that (F, G, H) is controllable if and only if there exists a $k \in \mathbf{N}$ such that

$$\mathcal{R}(F^k) \subset \mathcal{R}([F^{k-1}G, \ldots, FG, G]).$$

Now, since dim $X = n$ and $F:X \to X$ is linear, we have [Exercise 11] the chain:

$$\{0_X\} \subset \cdots \subset \mathcal{R}(F^{i+1}) \subset \mathcal{R}(F^i) \subset \cdots \subset \mathcal{R}(F^2) \subset \mathcal{R}(F) \subset X$$

where $\mathcal{R}(F^i) = F^i(X)$, the range or image of the map F^i.

Since $1 + \dim \, \Re(F^{i+1}) \leq \dim \, \Re(F^i)$ whenever the inclusion $\Re(F^{i+1}) \subset \Re(F^i)$ is proper, there must be some integer k, where $0 \leq k \leq n$, such that $\dim \, \Re(F^i) = \dim \, \Re(F^k)$ for all $i \geq k$. By Exercise 6(b), since $\Re(F^i) \subset \Re(F^k)$ for all $i \geq k$, this implies $\Re(F^i) = \Re(F^k)$ for all $i \geq k$, and the range of F^i stops decreasing at some step $i = k$ before or equal to n. With the help of Exercise 13 we have thus proved the following lemma:

LEMMA 7

For $F\!:\!X \to X$ and $\dim X = n$, $\Re(F^n) \subset \Re(F^i)$ for all $i \in \mathbf{N}$, and $\Re(F^n) = \Re(F^i)$ when $i \geq n$. $\qquad\square$

Now the statements

"there exists $k \in \mathbf{N}$ such that $\Re(F^k) \subset \Re[F^{k-1}G, \ldots, FG, G]$"

and

"$\Re(F^n) \subset \Re[F^{n-1}G, \ldots, FG, G]$"

are easily [Exercise 13] shown to be equivalent in light of Lemma 7 and the fact that $\Re[F^iG, \ldots, FG, G]$ attains its maximum when $i = n - 1$, and stays there thereafter [Theorem 5].

We have then deduced the following:

THEOREM 8

The n-dimensional system $(F, G)\!:\!x(t + 1) = Fx(t) + Gu(t)$ is controllable if and only if

$$\Re(F^n) \subset \Re[F^{n-1}G, \ldots, FG, G] \qquad\square$$

It is "easier" for a discrete-time system to be controllable than it is to be reachable:

COROLLARY 9

(i) (F, G) reachable $\Rightarrow (F, G)$ controllable

(ii) $\left.\begin{array}{c} (F, G) \text{ controllable} \\ \text{and} \\ F \text{ invertible} \end{array}\right\} \Rightarrow (F, G)$ reachable

Proof

(i) (F, G) reachable implies $\Re[F^{n-1}G, \ldots, FG, G] = X$ by Corollary 6, so $\Re(F^n) \subset \Re[F^{n-1}G, \ldots, FG, G]$ and (F, G) is controllable by Theorem 8.

(ii) F invertible implies $\mathfrak{R}(F^n) = \mathfrak{R}(F) = X$ so (F, G) controllable implies $X \subset \mathfrak{R}[F^{n-1}G, \ldots, FG, G]$, whence $X = \mathfrak{R}[F^{n-1}G, \ldots, FG, G]$. By Corollary 6, $\dim \mathfrak{R}[F^{n-1}G, \ldots, FG, G] = n$ implies (F, G) reachable. $\square$

We say that $F: X \to X$ is **nilpotent** if some power of F is zero. An example of a nonzero nilpotent F is

$$F = \begin{bmatrix} 0 & 0 & 0 \\ 1 & 0 & 0 \\ 0 & 1 & 0 \end{bmatrix} \text{ where } F^2 = \begin{bmatrix} 0 & 0 & 0 \\ 0 & 0 & 0 \\ 1 & 0 & 0 \end{bmatrix} \text{ and } F^3 = \begin{bmatrix} 0 & 0 & 0 \\ 0 & 0 & 0 \\ 0 & 0 & 0 \end{bmatrix}$$

By Lemma 7, if F is nilpotent, then $F^n = 0$ and we may deduce the following:

COROLLARY 10

F nilpotent implies (F, G) controllable.

Proof

Since $F^n = 0$ we have $\mathfrak{R}(F^n) = \{0_X\}$. Since $\{0_X\}$ is certainly contained in $\mathfrak{R}[F^{n-1}G, \ldots, G]$, then (F, G) is controllable by Theorem 8. $\square$

This is in fact trivially true, since any state x_0 will move to zero $(0_X = F^n x_0)$ freely, under the action of a sequence of zero inputs.

As an exercise, the reader should verify that if the block diagram for $x(t + 1) = Fx(t) + Gu(t)$ contains no feedback loops, then F is nilpotent, and relate this observation to Corollary 10. What about the converse?

We now present some single-input, two-dimensional examples to provide an interpretation of reachability in more intuitive terms.

Example 6

Consider the system shown in Figure 4-13. By inspection we have

$$F = \begin{bmatrix} \alpha & 0 \\ 0 & \beta \end{bmatrix}, G = \begin{bmatrix} a \\ b \end{bmatrix}, \text{ and } H = [c \quad d]$$

The matrix $\mathcal{C} = [FG, G]$ is $\mathcal{C} = \begin{bmatrix} \alpha a & a \\ \beta b & b \end{bmatrix}$ and $\det \mathcal{C} = ab(\alpha - \beta)$.

We see the following facts immediately:

(i) If either a or b is zero, $\mathcal{C}$ has a zero row and is singular. Thus, from Corollary 6 the system is not reachable. This is pretty obvious in this example where the state variables are uncoupled: if the input is not connected to one of the state variables, that variable cannot be controlled.

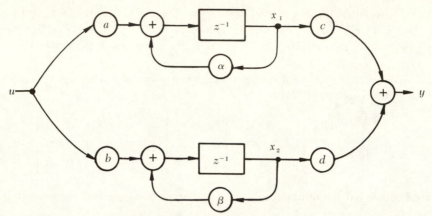

Figure 4–13   Two blocks in parallel.

(ii) If $\alpha = \beta$, then one column of $\mathcal{C}$ is a multiple of the other, and again control is lost. This is a bit more subtle. As will become clearer in Section 7–3, two systems with *identical time constants*† driven in parallel cannot be independently controlled.   ◊

The significance of the phrase "driven in parallel" becomes clearer when we examine two systems with identical time constants α connected in a series structure rather than in parallel.

Example 7

Consider the system of Figure 4–14. For this system the coefficient matrices are $F = \begin{bmatrix} \alpha & 1 \\ 0 & \alpha \end{bmatrix}$, $G = \begin{bmatrix} a \\ b \end{bmatrix}$, and $H = [c \quad d]$; the matrix $\mathcal{C} = [FG, G]$ is

$$\mathcal{C} = \begin{bmatrix} \alpha a + b & a \\ \alpha b & b \end{bmatrix}$$

and $\det \mathcal{C} = b^2$.

The system is reachable if and only if b is nonzero. Because the component blocks of this system are in *series*, the coefficient a may be zero without destroying reachability.   ◊

The situations for the n-dimensional systems corresponding to these

---

†"Time constant" is a term used in engineering and physics to measure the amount of time it takes for an exponential or geometric function to grow or decay a prescribed fraction of its total growth. In Example 6 α and β determine the natural, zero-input rates of growth (decay) of the system; in Section 6–3 we will see that they are "eigenvalues" or natural frequencies of the system.

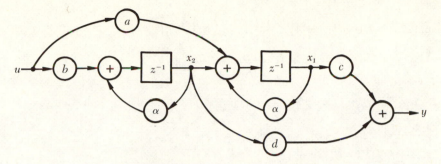

Figure 4–14   A chain of blocks.

examples are simple generalizations of these instances. If the state variables are so defined that the coefficient matrix F is in *Jordan normal form* (to be discussed in Section 6–6), inspection of F and G will reveal the truth about reachability. The control must enter the *beginning* of every chain such as in Figure 4–14, and no two separate chains can be associated with the same time constant or eigenvalue. Thus, we see that a reachable system is, intuitively speaking, *strongly connected:* each dynamic mode may be individually controlled. We shall return to this topic and elucidate it in Section 7–3.

EXERCISES FOR SECTION 4–3

✓ **1.**   The assertion of Theorem 2 about the sets $\{S_k^{(x_0)}\}_1^\infty$ is not true for an arbitrary sequence of sets $\{A_i\}_1^\infty$. Consider the sequence of real intervals

$$A_1 = [0, 1], \qquad A_2 = [0, 1], \qquad A_3 = [0, 1],$$
$$A_n = [0, n] \text{ for } n \geq 4$$

Observe that there exists a $k \in \mathbf{N}$ such that $A_k = A_{k+1}$ but it is not true that $A_k = A_{k+l}$ for all $l \in \mathbf{N}$.

2.   **(a)**   Show that for any sets A and B

$$A \cup B = A \Leftrightarrow B \subset A$$

Let V and W be two sets, let $f : V \to W$, and let $\{A_i\}_1^\infty$ be any sequence of subsets of V.

(b)   Prove $f\left(\bigcup_{i=1}^{\infty} A_i\right) = \bigcup_{i=1}^{\infty} f(A_i)$

(c)   Prove $A_i \subset A_j \Rightarrow f(A_i) \subset f(A_j)$.

✓ **3.**   Let $\{A_i\}_1^\infty$ be any sequence of sets. Suppose that there exists $k \in \mathbf{N}$ such that $A_i = A_k$ for all $i \geq k$. Prove $\bigcup_{j=0}^{\infty} A_j = \bigcup_{j=0}^{k} A_j$.

[*Hint:* $s \in \bigcup_{j=0}^{\infty} A_j \Rightarrow s \in A_m$ for some $m \in \mathbf{N}$. If $m \leq k$ then $A_m \subset \bigcup_{j=0}^{k} A_j$ so $s \in \bigcup_{j=0}^{k} A_j$

and if $m > k$ then $A_m = A_k \subset \bigcup_{j=0}^{k} A_j$, so $s \in \bigcup_{j=0}^{k} A_j$]

4.   For the system $x(t + 1) = Fx(t) + Gu(t)$
$$y(t) = Hx(t)$$

with $x_i = x(t_0 + i)$, $u_i = u(t_0 + i)$ for $i = 0, 1, 2 \ldots$, prove by induction that

$$\phi(x_0, u_0 u_1 \ldots u_{k-1}) = F^k x_0 + \sum_{j=0}^{k-1} F^{k-j-1} G u_j$$

✔ **5.**   As in Theorem 5, for the system (F, G, H) let $S_k = S_k^{(0x)}$. Prove that for each k, S_k is a subspace of X and $S = S^{(0x)}$ is a subspace of X.

✔ **6.**   In the proof of Theorem 5 we used a fundamental fact about finite-dimensional vector spaces. Let dim $V = n$ and let W be a subspace of V.

   (a)   Prove dim $W \leq$ dim V.
   (b)   Prove dim $W =$ dim $V \Rightarrow W = V$.

7.   In what follows, I is the *identity* $x \mapsto x$ and 0 is the zero matrix or *nullor* $x \mapsto 0_X$.

   (a)   Under what circumstances is $(F, 0, H)$ reachable?
   (b)   Prove that $S_1 = X$ for (F, I, H). Check that rank $[F^{n-1}G, F^{n-2}G, \ldots, G]$
$= n$ the easy way. What assumptions have you made about X, U, and n?

✔ **8.**   Write down F, G and H for the systems Σ_1 and Σ_2 of Figures 4–15(a) and 4–15(b).

   (a)   What is the subspace of X reachable from the 0 state of Σ_1 in one step? Two steps? Three steps? Do the same for Σ_2. In each case, do *not* use F, G, and H, but give a verbal argument based on propagation of signals around the network.
   (b)   For each system, compute

$$\mathcal{C}_k = [F^{k-1}G, \ldots, FG, G] \text{ for } k = 1, 2, 3$$

and find a basis for the range of these six matrices.

   (c)   Comment on the relation between your answers to (a) and (b).

9.   Let $\phi: X \times U^* \to X$ be the state-transition function of a non-linear system with a finite number n of states. Prove that for every state x_0, all states reachable from x_0 can be reached in at most $n - 1$ steps: $S^{(x_0)} = S_{n-1}^{(x_0)}$.

10.   Compute $S_j^{(\hat{x})}$ for $j = 0, 1, 2, 3, 4, 5, 6, 7$ for the finite-state system of Figure 4–16 for $\hat{x} = x_0$ and for $\hat{x} = x_7$.

✔ **11.**   Let $F: X \to X$ where F is any mapping (linear or not) and X is any set (maybe a vector space, maybe not). Prove, by induction, that for every $k \in \mathbf{N}$, $F^{k+i}(X) \subset F^k(X)$ where $F^i(X) = \mathcal{R}(F^i)$, the range or image of the map $F^i = F \circ F \circ \cdots \circ F$ (i-fold composition).

✔ **12.**   With F and X general, as in Exercise 11:

   (a)   Prove that if $F^{k+1}(X) = F^k(X)$ for some $k \in \mathbf{N}$, then

$$F^{k+i}(X) = F^k(X) \text{ for all } i \in \mathbf{N}$$

   (b)   Argue that, if equality occurs at the kth step in the infinite chain $\cdots \subset$ $F^{k+1}(X) \subset F^k(X) \subset \cdots \subset F^2(X) \subset F(X) \subset X$, then the chain terminates in $F^k(X)$ and

$$\bigcap_{i=0}^{\infty} F^i(X) = F^k(X) = \mathcal{R}(F^k)$$

Σ_1:

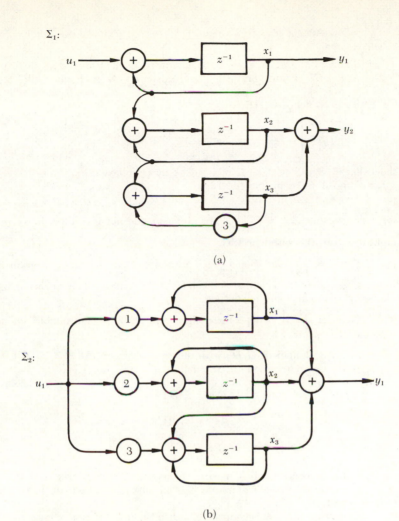

(a)

Σ_2:

(b)

Figure 4–15(a) and (b)

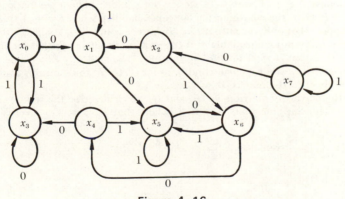

Figure 4–16

231

✔ **13.**   Let $\dim X = n$ and $F: X \to X$ be a linear map. Prove that $\Re(F^n) \subset \Re[F^{n-1}G, \ldots, FG, G]$ if and only if there exists $k \in \mathbf{N}$ such that $\Re(F^k) \subset \Re[F^{k-1}G, \ldots, FG, G]$.
[*Hint:* For all $i \in \mathbf{N}$, define $\mathcal{C}_i : U^i \to X$ by

$$\mathcal{C}_i \begin{bmatrix} u_0 \\ \vdots \\ u_{i-1} \end{bmatrix} = \sum_{j=0}^{i-1} F^{i-j-1}Gu_j = [F^{i-1}G, \ldots, FG, G] \begin{bmatrix} u_0 \\ \vdots \\ u_{i-1} \end{bmatrix}$$

and show that $\Re(\mathcal{C}_i) \subset \Re(\mathcal{C}_{i+1}) \subset \Re(\mathcal{C}_n) \subset X$ by Corollary 6.]

✔ **14.**   Let F and X be as in Exercise 13.

(a)   Prove that F invertible $\Rightarrow F$ not nilpotent.

(b)   Show that $F = \begin{bmatrix} 1 & 0 \\ 1 & 0 \end{bmatrix}$ is not nilpotent and not invertible, to see that the implication in (a) is only one way.

(c)   Let $G = \begin{bmatrix} 1 \\ 0 \end{bmatrix}$ and $F = \begin{bmatrix} 1 & 0 \\ 1 & 0 \end{bmatrix}$ as above. Give the block diagram for the discrete-time system (F, G) and notice that (F, G) is reachable (hence controllable) even though F is singular and not nilpotent.

(d)   Now let $G = \begin{bmatrix} 1 \\ 1 \end{bmatrix}$ in (c) and show that (F, G) is controllable but not reachable.

✔ **15.**   For the linear system (F, G, H) with $\dim X = n$, where $F(n \times n)$, $G(n \times m)$, and $H(q \times n)$ are *real* matrices, define the matrix

$$\mathcal{C}_k \overset{\Delta}{=} [F^{k-1}G, \ldots, FG, G]$$

so that

$$\mathcal{C}_k : \mathbf{R}^{n \times km} \to \mathbf{R}^n$$

Then

$$\mathcal{C} \overset{\Delta}{=} \mathcal{C}_n = [F^{n-1}G, \ldots, FG, G]$$

is the **reachability** or **controllability** matrix. Define the $(n \times n)$ matrix $W = \mathcal{C}_n \mathcal{C}_n^T$.

(a)   Prove that the $(n \times nm)$ matrix $\mathcal{C}_n$ is of rank n if and only if the $(n \times n)$ matrix W is invertible.

(b)   Prove that if (F, G) is controllable, then the input sequence $u_0, \ldots u_{nm-1}$ given by $\begin{bmatrix} u_0 \\ \vdots \\ u_{nm-1} \end{bmatrix} = \mathcal{C}_n^T W^{-1} x_0$ will drive x_0 to zero.

✔ **16.**   Suppose that in Figure 4–14 we let the x_2 block have a feedback gain of β where $\beta \neq \alpha$. Give the condition for noncontrollability when $b \neq 0$. Explain.

17.   Suppose that only the output $y = Hx$ is to be controlled. We define the system (F, G, H) to be **output controllable** iff there exists an input u such that for any initial state x_0, the output is zero at some finite time $k: y(k) = 0$.

Give necessary and sufficient conditions for (F, G, H) to be output controllable, and relate these to the matrix W of Exercise 15.

18.   Write down the matrices F_s, G_s, and H_s which describe the series connection of Figure 4–17. Suppose that F_1 is $n_1 \times n_1$ and that F_2 is $n_2 \times n_2$. Verify directly that

$$\operatorname{rank} [G_s, F_s G_s, \ldots, F_s^{n_1 + n_2 - 1} G_s] = n_1 + n_2$$

implies not only

$$\operatorname{rank} [G_1, F_1 G_1, \ldots, F_1^{n_1 - 1} G_1] = n_1$$

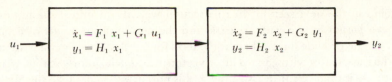

Figure 4–17

but also

$$\text{rank } [G_2, F_2 G_2, \ldots, F_2^{n_2-1} G_2] = n_2$$

by matrix manipulation and judicious use of the properties of the rank of various combinations of matrices. [See Corollary 4–2–2 and Exercises 13 and 14 of Section 4–2.]

✔ **19.**   Let $A : V \to W$ and $B : W \to U$ both be linear. Let BA stand for $B \circ A : V \to U$.

    (a)   Prove that $\mathcal{R}(BA) \subset \mathcal{R}(B)$.

    (b)   Prove that rank $(BA) \leq \min \{\text{rank } (B), \text{rank } (A)\}$.

20.   Show that $W(n, 0)$ as given by $\mathcal{CC}^T$ can be generated by the recurrence relation

$$W(k, 0) = F(k) W(k - 1, 0) F^T(k) + G(k) G^T(k)$$

with the initial condition

$$W(0, 0) = G(0) G^T(0)$$

This may be a convenient way to construct W for numerical work.

21.   Let F, G, and S_k be as in Lemma 4.

    (a)   Is S_k an *invariant subspace of* F; that is, does $F(S_k) \subset S_k$?

    (b)   Prove that $S_n = S = \mathcal{R}[F^{n-1}G, \ldots, FG, G]$ is an invariant subspace of F. Try not to use the Cayley-Hamilton theorem, even if you know it. [*Hint:* $b \in F(S) \Rightarrow b = Fa$ for some $a \in S$. Write a as a linear combination of columns of $[F^{n-1}G, \ldots, FG, G]$ and show that Fa is also spanned by these columns.]

    (c)   Suppose that there exists a polynomial

$$P(s) = s^q + a_1 s^{q-1} + \cdots + a_{q-1} s + a_q$$

such that $P(F) = F^q + a_1 F^{q-1} + \cdots + a_q I = 0$ and that q is the *lowest degree* of any polynomial of which F is a zero; then we call $P(s)$ a **minimal polynomial** of F.

Prove that $S_q = S_{q+1}$ for all $i \geq 0$.

    (d)   Let F be a given $(n \times n)$ matrix and choose any $(n \times 1)$ matrix G. Suppose that $S_n = X$ (that is, rank$[F^{n-1}G, \ldots, FG, G] = n$). Prove that $F^n + a_1 F^{n-1} + \cdots + a_{n-1} F + a_n I = 0$ where $s^n + a_1 s^{n-1} + \cdots + a_{n-1} s + a_n = \det(sI - F)$, the characteristic equation of F. [You have just proved the CAYLEY-HAMILTON theorem, which we shall study in Section 4–6, for this case.]

4–4   Inner Product Spaces and Adjoints

The elegant results of the last section on reachability of linear systems will yield results on observability, thanks to a duality principle which we shall develop in the next section. In this section we will introduce the notion of an *inner product* for a vector space V and the notion of an *adjoint* of a linear operator, to provide the mathematical tools for our duality principle. An inner product associates with each pair of vectors v and w a unique scalar, written $\langle v|w \rangle$. As is made clear in Example 3, this notion will generalize the familiar "dot product" of ordinary vectors in real 3-space and will enable us to speak quite geometrically about the "*angle*" between two vectors in V, their *lengths,* and their *distance* apart.

While it is possible to define an inner product operation for an abstract vector space V over any field K [Exercise 1], we shall most often be concerned with vector spaces over the field $\mathbf{C}$ of complex numbers and we shall use the bar over a number λ to denote its complex conjugate $\bar{\lambda}$.

DEFINITION 1

We call the operation $\langle \cdot | \cdot \rangle : V \times V \to \mathbf{C}$ an **inner product** if the following properties hold for all u, v, $w \in V$ and all $\lambda \in \mathbf{C}$:

(i) $\langle u|v \rangle = \overline{\langle v|u \rangle}$
(ii) $\langle u + v|w \rangle = \langle u|w \rangle + \langle v|w \rangle$
(iii) $\langle \lambda u|v \rangle = \bar{\lambda}\langle u|v \rangle$
(iv) $\langle u|u \rangle \geq 0$, and $\langle u|u \rangle = 0$ if and only if $u = 0_V$.

A vector space together with an inner product defined on it is called an **inner-product space** or a **pre-Hilbert space,** because one additional property will make it into a **Hilbert space** (as defined in Definition 2). ○

Example 1

In the space $V = \mathbf{C}^n$ over the field $\mathbf{C}$, the *standard*† *inner product* of two vectors $x = [x_1, \ldots, x_n]^T$ and $y = [y_1, \ldots, y_n]^T$ is $\langle x|y \rangle = \sum_{i=1}^{n} \bar{x}_i y_i$, which can be written in terms of matrices as $\langle x|y \rangle = \bar{x}^T y$, where $\bar{x}^T = \overline{(x^T)}$ is the conjugate of the transpose of the $(n \times 1)$ matrix x.

That this definition satisfies properties (i), (ii), (iii), and (iv) is easily shown. We will verify (iv) and leave the others to the reader:
For all $x \in \mathbf{C}^n$,

$$\langle x|x \rangle = \sum_{i=1}^{n} \bar{x}_i x_i = \sum_{i=1}^{n} |x_i|^2$$

---

† *Caution:* Some authors use $\sum_{i=1}^{n} x_i \bar{y}_i = x^T \bar{y}$ as the standard inner product on $\mathbf{C}^n$; for some reason, many systems authors do it our way.

which is a sum of non-negative terms and hence is non-negative. Then

$$\langle x | x \rangle = \sum_{i=1}^{n} |x_i|^2 = 0$$

iff each $|x_i|^2 = 0$, whence each $x_i = 0$. ◊

Of course, if V is a vector space over the real numbers **R**, as is the case in most of this book, the conjugates may be omitted in properties (i) to (iv). The *standard inner product* for $V = \mathbf{R}^n$ over the field **R** is given by

$$\langle x | y \rangle = \sum_{i=1}^{n} x_i y_i$$

See Exercise 2(a) for a different, less standard, but perfectly valid, inner product for $\mathbf{R}^n$.

If V is a complex inner product space, it is a fundamental fact, known as the **Cauchy-Buniakovsky-Schwartz (CBS) Inequality**, that

$$|\langle u | v \rangle|^2 \leq \langle u | u \rangle \langle v | v \rangle$$

for all $u, v \in V$, where equality holds if and only if $u = \lambda v$ or $v = 0_V$. A proof of this inequality is outlined in Exercise 4, and from it, it is easy to show that the inner product induces a norm† on V.

LEMMA 1

If V is a complex or real inner-product space with inner product $\langle \cdot | \cdot \rangle$, it is also a normed vector space with norm $\| \cdot \| : V \to \mathbf{R}$ defined by $\| v \| = \langle v | v \rangle^{1/2}$

Proof

Exercise 5.   □

Thus, every inner product space is also a normed vector space and a metric space, so we may use notions of "size" and "distance" meaningfully.

---

†See Section 2-4 for a discussion of normed vector spaces and metric spaces.

Example 2

Let $V = C[a, b]$, the space of all continuous functions $[a, b] \to \mathbf{R}$, and let the field in which we wish to define our scalar product be $\mathbf{R}$. The definition

$$\langle f \mid g \rangle \overset{\Delta}{=} \int_a^b f(\xi) g(\xi) \, d\xi$$

for all $f, g \in C[a, b]$ is easily shown [Exercise 6] to be an inner product for V. In fact, it may be viewed as an infinite-dimensional generalization of the ordinary inner product on $\mathbf{R}^n$ as follows. Suppose that f and g are sampled at a large number n of points. The resulting lists of sample values are points of $\mathbf{R}^n$, and their ordinary inner product is given by

$$\langle \, [f(t_1), f(t_2), \dots, f(t_n)]^T \mid [g(t_1), g(t_2), \dots, g(t_n)]^T \, \rangle = \sum_{i=1}^n f(t_i) g(t_i)$$

which, as $n \to \infty$, starts looking more and more like an integral of $f(t)g(t)$, especially if we weight each sample $f(t_j)$ and $g(t_j)$ by $\sqrt{t_{j+1} - t_j}$.

The size of a function $f \in C[a, b]$ is just the norm of f, and using the norm induced by the inner product we have

$$\|f\| = \left(\int_a^b |f(\xi)|^2 \, d\xi \right)^{1/2}$$

The distance between two functions $f, g \in C[a, b]$, measured in the metric induced by the inner-product norm, is

$$d(f, g) = \|f - g\| = \left(\int_a^b |f(\xi) - g(\xi)|^2 \, d\xi \right)^{1/2} \qquad \Diamond$$

DEFINITION 2

Let the complex inner product (pre-Hilbert) space V be normed by defining

$$\|v\| = \langle v \mid v \rangle^{1/2}$$

and metrized by

$$d(v, w) = \|v - w\| = \langle v - w \mid v - w \rangle^{1/2}$$

If all Cauchy sequences in V have limits in V, so that V is complete with respect to the metric induced by the inner product, we say that V is a **Hilbert space.** $\bigcirc$

Every Hilbert space is thus a Banach space, too, and we can do both geometry and calculus in Hilbert spaces, which accounts for their popularity in science.

The most familiar examples of Hilbert spaces are $\mathbf{R}^n$ and $\mathbf{C}^n$, which are easily shown to be complete. Infinite-dimensional Hilbert spaces arise naturally in the study of quantum mechanics and Fourier optics, as well as in modern system theory. In communication theory, where we try to extract a signal from noise, we consider a Hilbert space of functions as in Example 2, except that we usually use "Lebesgue square-integrable" functions, rather than just continuous functions. If we let the function v stand for a message signal and n for a noise signal, the *detection problem* is to extract from $n + v$ the best† "least squares" approximation to v. Since the norm in this space is given by the integral of the square of an element, "least squares" approximation reduces to "projecting" $n + v$ onto an appropriate subspace of square-integrable messages.

In the next example, we show how our axioms for inner products generalize the familiar "dot product" of calculus and physics fame. For convenience, we look at ordinary vectors in real 2-space, $\mathbf{R}^2$.

Example 3

Suppose that $x, \hat{x} \in \mathbf{R}^2$ have respective lengths r and $\hat{r}$. We define their **dot product** by $x \cdot \hat{x} = \hat{r}r \cos \theta$, where θ is the angle between x and $\hat{x}$, as shown in Figure 4–18(a).

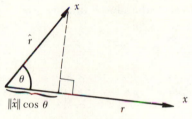

Figure 4–18(a)

We want to prove that the dot product $x \cdot \hat{x}$ is the same as $\langle x | \hat{x} \rangle$, the inner product $x^T \hat{x}$ of Example 1. Suppose that x makes an angle ϕ with the unit vector $e_1 = \begin{bmatrix} 1 \\ 0 \end{bmatrix}$, as in Figure 4–18(b).

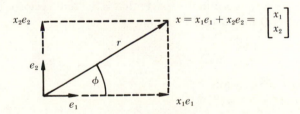

Figure 4–18(b)

Then $x_1 = r \cos \phi$ and $x_2 = r \sin \phi$. Similarly, if $\hat{x}$ makes angle $\hat{\phi}$ with e_1, we have $\hat{x}_1 = \hat{r} \cos \hat{\phi}$ and $\hat{x}_2 = \hat{r} \sin \hat{\phi}$. Since $\theta = \phi - \hat{\phi}$ so that

$$\cos \theta = \cos \phi \cos \hat{\phi} + \sin \phi \sin \hat{\phi}$$

---

†In Section 7–6 we formalize the notion of *best* in our study of optimization.

we have

$$x \cdot \hat{x} = r\hat{r}(\cos \phi \cos \hat{\phi} + \sin \phi \sin \hat{\phi})$$
$$= x_1 \hat{x}_1 + x_2 \hat{x}_2$$

which is just what our standard definition of inner product on $\mathbf{R}^2$ yields. Note that for $e_1 = \begin{bmatrix} 1 \\ 0 \end{bmatrix}$ and $e_2 = \begin{bmatrix} 0 \\ 1 \end{bmatrix}$ we have

$$\|e_1\| = \sqrt{\langle e_1 | e_1 \rangle} = \sqrt{1 \cdot 1 + 0 \cdot 0} = 1$$
$$\|e_2\| = \sqrt{\langle e_2 | e_2 \rangle} = \sqrt{0 \cdot 0 + 1 \cdot 1} = 1$$

which is why we call e_1 and e_2 "unit" vectors.

When $\|x\| \neq 0$ and $\|\hat{x}\| \neq 0$, then

$$0 = \langle x | \hat{x} \rangle = \|x\| \, \|\hat{x}\| \cos \theta \quad \text{if and only if} \quad \cos \theta = 0$$

that is, if and only if x and $\hat{x}$ are at "right angles" (*orthogonal*). Since $\langle e_1 | e_2 \rangle = 0$, we call $\{e_1, e_2\}$ an *orthonormal* set—they are mutually *ortho*gonal and are "*normal*ized" in that they have unit length.   ◊

DEFINITION 3

In any inner product space V, we say that two vectors v, w are **orthogonal** if $\langle v | w \rangle = 0$.   ○

In Exercise 7 we make use of the CBS inequality to define the angle between two nonzero vectors in terms of the inner product, and we point out the distinction between perpendicularity and orthogonality in complex inner product spaces.

There is precisely one vector in any complex inner product space which is orthogonal to every vector in the space; that is the zero vector.† This follows because $\langle v | w \rangle = 0$ for all $w \in V$ implies

$$\langle v | v \rangle = \|v\|^2 = 0$$

which implies $v = 0_V$, while

$$\langle 0_V | w \rangle = \langle 0 0_V | w \rangle = \bar{0} \langle 0_V | w \rangle$$
$$= 0$$

---

†Any inner product space in which this statement holds is said to be *non-degenerate*. See Exercise 1.

DEFINITION 4

If S is a set of vectors in V, we call S an **orthogonal set** if every pair of distinct vectors in S is orthogonal. An **orthonormal** set is just an orthogonal set S with the additional property that

$$\|s\| = 1 \text{ for all } s \in S \qquad \bigcirc$$

The standard bases for $\mathbf{R}^n$ and $\mathbf{C}^n$ are orthonormal with respect to the standard inner product on $\mathbf{R}^n$ or $\mathbf{C}^n$. [Verify!]

Any orthogonal set of nonzero vectors is linearly independent, and from any list $\{v_1, v_2, v_3, \ldots\}$ of linearly independent vectors we can construct an orthonormal set $\{e_1, e_2, e_3, \ldots\}$ by the **Gram-Schmidt procedure:**

$$e_1 = x_1/\|x_1\|$$
$$e_2 = [x_2 - \langle e_1 | x_1 \rangle e_1]/\|x_2 - \langle e_1 | x_1 \rangle e_1\|$$
$$\vdots$$

$$e_n = \left[x_n - \sum_{i=1}^{n-1} \langle e_1 | x_n \rangle e_i \right] \Big/ \left\| x_n - \sum_{i=1}^{n-1} \langle e_i | x_n \rangle e_i \right\|$$

$$\vdots$$

for $n \geq 2$. This forms the rationale for the following theorem:

THEOREM 2

Every finite-dimensional inner product space V has an orthonormal basis. Such a basis can be formed from any known basis for the space by the Gram-Schmidt procedure.

Proof

Exercises 8 through 10.   □

DEFINITION 5

If S is any subset of V, we define the orthogonal complement of S, denoted $S^\perp$ and pronounced "S-perp," as the set of all vectors of V which are orthogonal to each element of S; that is,

$$S^\perp = \{v \in V | \langle v | s \rangle = 0 \text{ for all } s \in S\} \qquad \bigcirc$$

It is easy to show that, even if S is just a set (maybe not a subspace), $S^{\perp}$ is a subspace of V; we also have the following extremely useful decomposition:

THEOREM 3

If S is a finite-dimensional subspace of the inner product space V, then every vector $v \in V$ can be decomposed uniquely as $v = s_1 + s_2$, where $s_1 \in S$ and $s_2 \in S^{\perp}$.

Proof

Exercise 12(a).   □

For the situation of Theorem 3, we say that V is the **direct sum** of S and $S^{\perp}$, and we write $V = S \oplus S^{\perp}$.

We now make use of the inner product to define the *adjoint* of an operator.

DEFINITION 6

Let L be any mapping of V into W, where V and W are inner product spaces. We say that L **has an adjoint** if there exists a map L^* of W into V such that

$$\langle Lv \,|\, w \rangle_W = \langle v \,|\, L^* w \rangle_V$$

for all $v \in V$ and all $w \in W$. When the map L^* exists, we call it **the adjoint of L** and pronounce it "L-adjoint."   ○

Example 4

Let $L : \mathbf{R}^3 \to \mathbf{R}^1$ be such that $L \begin{bmatrix} x_1 \\ x_2 \\ x_3 \end{bmatrix} = 2x_1 + x_2 - x_3$. Consider the map

$F : \mathbf{R}^1 \to \mathbf{R}^3$ such that $Fy = \begin{bmatrix} 2y \\ y \\ -y \end{bmatrix}$. Since

$$\langle L(x) \,|\, y \rangle_{\mathbf{R}^1} = (2x_1 + x_2 - x_3) y$$

$$\langle x \,|\, Fy \rangle_{\mathbf{R}^3} = [x_1 \; x_2 \; x_3] \begin{bmatrix} 2y \\ y \\ -y \end{bmatrix} = 2yx_1 + yx_2 - yx_3$$

we have

$$\langle Lx\,|\,y\rangle = \langle x\,|\,Fy\rangle$$

so that F can be used as the map L^* whose existence proves that L has an adjoint.   ◊

The action of the adjoint of an operator may be conveniently expressed by saying:

"Taking the adjoint corresponds to 'flipping' an operator from one side of the inner product to the other."

With the following simple lemmas, we can deduce many interesting facts about maps and their adjoints.

LEMMA 4

Let $L: V \to W$ be a mapping, linear or not, with V and W perhaps infinite-dimensional. Then $\langle Lv\,|\,w\rangle = 0$ for all $v \in V$ and for all $w \in W \Leftrightarrow L = 0$, the zero map.

Proof

($\Longrightarrow$) Choose any v in V and then set $w = Lv$ to obtain

$$\langle Lv\,|\,Lv\rangle = 0$$

which implies $\|Lv\| = 0$, so that $Lv = 0_W$ for all $v \in V$.

($\Longleftarrow$) If $L = 0$, then for all $v \in V$ and for all $w \in W$, we have

$$\langle Lv\,|\,w\rangle = \langle 0_W\,|\,w\rangle = 0$$ □

LEMMA 5

Let $L_1: V \to W$ and let $L_2: V \to W$ be linear or not. Then we have

$$\langle L_1v\,|\,w\rangle = \langle L_2v\,|\,w\rangle$$

for all $v \in V$ and for all $w \in W$, if and only if $L_1 = L_2$.

Proof

Let $L = L_1 - L_2$ so that $\langle Lv\,|\,w\rangle = \langle L_1v\,|\,w\rangle - \langle L_2v\,|\,w\rangle$, and apply Lemma 4 to get $L = 0$, whence $L_1 = L_2$.   □

In Exercise 16, we outline a proof that for any inner product spaces V and W, finite-dimensional or not, if a map $L: V \to W$ has an adjoint $L^*: W \to V$, then it has only one (uniqueness), and

$$L^{**} = L$$

Also, if L^* exists for a linear map L, then L^* is linear [Exercise 19].

Now let us present some useful facts about adjoints:

THEOREM 6

Let A_1 and A_2 be linear maps on V which have adjoints; let I be the identity map on V, and let θ be the zero map or nullor on V. Then

(i) $I^* = I$ and $\theta^* = \theta$

(ii) $(A_1 + A_2)^* = A_1^* + A_2^*$ and $(cA_1)^* = \bar{c} A_1^*$ for $c \in \mathbf{C}$

(iii) $(A_2 A_1)^* = A_1^* A_2^*$ (note order)

(iv) A_1 is invertible if and only if A_1^* is invertible and $(A_1^*)^{-1} = (A^{-1})^*$

(v) $A_1^{**} = A_1$

Proof

We will prove (iii) and leave the other parts to the reader as Exercise 20. Consider $A_1: V \to V$. Since $(A_2 A_1)v = A_2(A_1 v)$, we have $\langle (A_2 A_1)v \mid w \rangle = \langle A_2(A_1 v) \mid w \rangle = \langle A_1 v \mid A_2^* w \rangle$ by the property of A_2^*. By the property of A_1^*,

$$\langle A_1 v \mid A_2^* w \rangle = \langle v \mid A_1^*(A_2^* w) \rangle$$
$$= \langle v \mid (A_1^* A_2^*)w \rangle$$

Also $\langle (A_2 A_1)v \mid w \rangle = \langle v \mid (A_2 A_1)^* w \rangle$

so $\langle v \mid (A_2 A_1)^* w \rangle = \langle v \mid (A_1^* A_2^*)w \rangle$

for all $v \in V$ and all $w \in W$. By Lemma 5 $(A_2 A_1)^* = A_1^* A_2^*$.   □

Not all maps on inner product spaces have adjoints,† but all linear maps on *finite-dimensional* inner product spaces *do*, as we now show.

---

† Not even all *linear* maps do. See K. Hoffman and R. Kunze, *Linear Algebra* (Prentice-Hall, 1971), p. 240, Example 16.

THEOREM 7

If $\dim V = n$ and $\dim W = m$, then every linear map $L: V \to W$ has a unique adjoint L^* which is itself a linear map. Moreover, if $\mathfrak{B} = \{v_1, \ldots, v_n\}$ and $\mathfrak{B}' = \{w_1, \ldots, w_m\}$ are orthonormal bases for V and W, respectively, then $^{\mathfrak{B}}(L^*)^{\mathfrak{B}'} = (^{\mathfrak{B}'}L^{\mathfrak{B}})^T$, where $^{\mathfrak{B}'}L^{\mathfrak{B}}$ is the $m \times n$ matrix† of L with respect to the bases $\mathfrak{B}$ and $\mathfrak{B}'$.

Proof

Call $A = {}^{\mathfrak{B}'}L^{\mathfrak{B}}$ and let $[A]_{*j}$, $[A]_{i*}$ and $[A]_{ij}$ denote the jth column, ith row and i,jth entry, respectively, of A.

By the definition of A, we have $[A]_{*j} = [Lv_j]_{\mathfrak{B}'}$ so that $Lv_j = \sum_{k=1}^{m} [A]_{kj}w_k$.

Since $\mathfrak{B}'$ is an orthonormal basis for W,

$$\langle Lv_j | w_l \rangle = \sum_{k=1}^{m} [\bar{A}]_{kj} \langle w_k | w_l \rangle = [\bar{A}]_{lj}$$

Thus, in terms of the projections of Lv_j on the unit vectors $\{w_1, \ldots, w_m\}$, we get

$$[A]_{kj} = \overline{\langle Lv_j | w_k \rangle}$$

Now, if L has an adjoint $L^*: W \to V$, with $C = {}^{\mathfrak{B}}(L^*)^{\mathfrak{B}'}$, we have by the preceding result that

$$[C]_{kj} = \overline{\langle L^* w_j | v_k \rangle}$$

But $\quad [A]_{kj} = \overline{\langle Lv_j | w_k \rangle}$

$\qquad\qquad = \overline{\langle v_j | L^* w_k \rangle}$ by definition of L^*

$\qquad\qquad = \langle L^* w_k | v_j \rangle$ by the conjugate symmetry of $\langle \cdot | \cdot \rangle$

Thus $\quad [A]_{kj} = [\bar{C}]_{jk}$, the conjugate of the j, kth entry of C

so $C = \bar{A}^T$ as claimed, and the matrix of L^*, if it exists at all, is uniquely determined as the transpose of the conjugate of A.

To show that an adjoint L^* really does exist under the conditions of the theorem, we simply define L^* as the linear map of W into V whose matrix

---

†This notation and the relationship between matrices and linear maps are discussed in Section 3–3.

is $(\overline{{}^{\circledR}L^{\circledR}})^T$ and show [Exercise 15] that this map L^* satisfies

$$\langle Lv \mid w \rangle = \langle v \mid L^*w \rangle \text{ for all } (v, w) \in V \times W \qquad \square$$

As outlined in Exercise 15, this last result may be expressed as

$$\left\langle A \begin{bmatrix} x_1 \\ \vdots \\ x_n \end{bmatrix} \middle| \begin{bmatrix} y_1 \\ \vdots \\ y_m \end{bmatrix} \right\rangle_{\mathbf{C}^m} = \left\langle \begin{bmatrix} x_1 \\ \vdots \\ x_n \end{bmatrix} \middle| (\overline{A})^T \begin{bmatrix} y_1 \\ \vdots \\ y_m \end{bmatrix} \right\rangle_{\mathbf{C}^n}$$

for an $(m \times n)$ complex matrix A, so we call the matrix $(\overline{A})^T$ the **adjoint** of A and write $A^* = (\overline{A})^T$. Using this notation, we may rewrite the standard inner product for $\mathbf{C}^n$, given in Example 1, as $\langle x \mid y \rangle = x^*y$.

Caution.   Some older linear algebra books use the words "adjoint of a matrix" to denote the matrix of cofactors used in computing the inverse of a square matrix by Cramer's rule. Today, to avoid confusion, most mathematicians call that particular matrix the *adjugate*.

At this point the reader should work Exercises 13 and 14 to enhance his facility with the operations and notation involved.

We showed in Theorem 7 that every linear map on a finite-dimensional inner product space has an adjoint, and we mentioned that this is not true in general for infinite-dimensional spaces. The most useful generalization of Theorem 7 that can be made for infinite-dimensional spaces is the following statement given here without proof:[†]

THEOREM 8

For any complex inner product spaces V and W, every bounded linear mapping $L: V \to W$ has a *unique* bounded linear adjoint L^*, where L is said to be **bounded** if there is a real number M such that, for all $v \in V$, $\|Lv\| \leq M\|v\|$. $\qquad \square$

Exercise 18 shows that all linear maps on finite-dimensional spaces are bounded and thus, by Exercise 17(b), are continuous.

We are now ready to present the primary reason for introducing the adjoint. Let us recall the following notions for any *linear* transformation

---

[†] See D. G. Luenberger, *Optimization by Vector Space Methods* (Wiley, 1969), p. 151.

$A : V \rightarrow W$:

$$\mathcal{R}(A), \text{ the } \textit{range} \text{ of } A = \{Av \mid v \in V\} \subset W$$
$$\mathfrak{N}(A), \text{ the } \textit{nullspace} \text{ of } A = \{v \mid Av = 0_W\} \subset V$$

$\mathcal{R}(A)$ is the space that Av ranges over as v ranges over V. It is a *subspace* of W, since

$$c_1 A v_1 + c_2 A v_2 = A(c_1 v_1 + c_2 v_2)$$

$\mathfrak{N}(A)$ is the space of all vectors which are nulled by A. It is a subspace of V, since

$$c_1 A v_1 + c_2 A v_2 = 0_W \text{ if } A v_1 = 0_W \text{ and } A v_2 = 0_W$$

Example 5

Let us compute $\mathcal{R}(A)$ and $\mathfrak{N}(A^*)$ for

$$A = \begin{bmatrix} 1 & 2 \\ 2-i & 4-2i \end{bmatrix}$$

Remembering that the range of a matrix is just the space of linear combinations of its columns, we have

$$\mathcal{R}(A) = \left\{ \begin{bmatrix} 1 & 2 \\ 2-i & 4-2i \end{bmatrix} \begin{bmatrix} x_1 \\ x_2 \end{bmatrix} \middle| x_1, x_2 \in \mathbf{C} \right\}$$

$$= \left\{ (x_1 + 2x_2) \begin{bmatrix} 1 \\ 2-i \end{bmatrix} \middle| x_1, x_3 \in \mathbf{C} \right\}$$

$$= \text{ the set of all scalar multiples of } \begin{bmatrix} 1 \\ 2-i \end{bmatrix}$$

Now, $A^* = \begin{bmatrix} 1 & 2+i \\ 2 & 4+2i \end{bmatrix}$. Thus, $\begin{bmatrix} x_1 \\ x_2 \end{bmatrix} \in \mathfrak{N}(A^*)$ iff $[1 \quad 2+i]\begin{bmatrix} x_1 \\ x_2 \end{bmatrix} = 0$ (since

$[2 \quad 4+2i] = 2[1 \quad 2+i]$); that is, if and only if

$$\left\langle \begin{bmatrix} 1 \\ 2-i \end{bmatrix} \middle| \begin{bmatrix} x_1 \\ x_2 \end{bmatrix} \right\rangle = 0$$

Thus, every vector in $\mathfrak{N}(A^*)$ is orthogonal to every vector in $\mathfrak{R}(A)$. We will see that this is indeed true for every A. ◊

THEOREM 9

For every linear transformation $A: V \to W$ with adjoint A^*, we have $[\mathfrak{R}(A)]^\perp = \mathfrak{N}(A^*)$.

Proof

$w \in [\mathfrak{R}(A)]^\perp \Leftrightarrow \langle w \mid y \rangle = 0$, all $y \in \mathfrak{R}(A)$

$\Leftrightarrow \langle w \mid Av \rangle = 0$, all $v \in V$

$\Leftrightarrow \langle A^*w \mid v \rangle = 0$, all $v \in V$ since $A^{**} = A$

$\Leftrightarrow A^*w = 0_V$ since A^*w is orthogonal to every vector in V

$\Leftrightarrow w \in \mathfrak{N}(A^*)$ □

Notice that since $A^{**} = A$ we immediately get the following corollary:

COROLLARY 10

$$[\mathfrak{R}(A^*)]^\perp = \mathfrak{N}(A). \qquad \square$$

Now, if we "take the perp of both sides" (as in Exercise 11(c)) of this last result, we get

$$[\mathfrak{R}(A^*)]^{\perp\perp} = [\mathfrak{N}(A)]^\perp$$

Unfortunately, we *cannot* conclude from this that $\mathfrak{R}(A^*) = [\mathfrak{N}(A)]^\perp$, because in general† $[\mathfrak{R}(A^*)]^{\perp\perp} \neq \mathfrak{R}(A^*)$, as we indicate in Exercise 11(a) and (b). In finite-dimensional spaces, however, it *is* true [See Exercise 21] that $[\mathfrak{R}(A^*)]^{\perp\perp} = \mathfrak{R}(A^*)$, so we have the following result:

---

†In general, $S^{\perp\perp}$ = the smallest *closed* subspace containing S, and the closure of $\mathfrak{R}(A^*) = \mathfrak{N}(A)^\perp$. See Luenberger (1969), p. 52 and p. 157.

THEOREM 11

If $A : V \to W$ is linear, and V is finite-dimensional, then $\mathfrak{R}(A) = [\mathfrak{N}(A^*)]^\perp$.

□

This last result is very important, since it enables us to decide whether or not the operator equation $Ax = y$ has a solution by studying the solvability of the simpler equation

$$A^* w = 0$$

That is, $Ax = y$ has a solution if and only if y is in the range of A, $y \in \mathfrak{R}(A)$, and this happens if and only if y is orthogonal to every vector in $\mathfrak{N}(A^*)$; that is, $\langle y | w \rangle = 0$ for all solutions w of $A^* w = 0_V$. [See Exercises 22 and 23.]

Example 6

The row rank of a matrix A is equal to the column rank of A. The row rank, the maximum number of linearly independent rows of A, is the dimension of the row space of A:

$$\text{row rank } (A) = \dim \{x_1[A]_{1*} + \cdots + x_m[A]_{m*} \,|\, x_1, \ldots, x_m \in \mathbf{C}\}$$

where $[A]_{i*}$ is the ith row of the $(m \times n)$ complex matrix $A : \mathbf{C}^n \to \mathbf{C}^m$.

The reader should convince himself that linear independence of vectors is not affected by taking complex conjugates. Since the rows of A^* are just the conjugates of the columns of A^T,

$$
\begin{aligned}
\text{row rank } (A) &= \text{column rank of } A^T \\
&= \text{rank } (A^*) \\
&= \dim \mathfrak{R}(A^*) \\
&= \dim [\mathfrak{N}(A)]^\perp \quad [\text{Theorem 11}] \\
&= n - \dim \mathfrak{N}(A) \quad [\text{Exercise 21}]
\end{aligned}
$$

and, since $\dim \mathfrak{N}(A) + \dim \mathfrak{R}(A) = n$, we have row rank $(A) = \dim \mathfrak{R}(A)$, which is the column rank of A.   ◊

Another immediate consequence of Theorem 11 is:

COROLLARY 13

Let $A : V \to W$ be linear, and let V be finite-dimensional. Then A is one-to-one if and only if A^* is onto.

Proof

A is one-to-one $\Leftrightarrow \mathfrak{N}(A) = \{0_V\}$

$\qquad\qquad \Leftrightarrow [\mathfrak{N}(A)]^\perp = \{0_V\}^\perp = V$ (Exercise 21)

$\qquad\qquad \Leftrightarrow \mathfrak{R}(A^*) = V$ (using $A^{**} = A$ in Theorem 11)

$\qquad\qquad \Leftrightarrow A^*$ is onto $\qquad\qquad\qquad\qquad\qquad\qquad$ $\square$

We are now ready to show, in the next section, how the observability and reachability of the system (F, G, H) are linked together by the useful properties of the adjoint developed in this section.

EXERCISES FOR SECTION 4–4

1.   Let V be a vector space over a general field K. Define $\langle \cdot | \cdot \rangle : V \times V \to K$ such that
   (a) $\langle v | w \rangle = \langle w | v \rangle$
   (b) $\langle u | v + w \rangle = \langle u | v \rangle + \langle u | w \rangle$
   (c) $\langle u | kv \rangle = k \langle u | v \rangle$
are satisfied for all $u, v, w \in V$ and all $k \in K$. Then $\langle v | w \rangle$ is called the scalar or inner product of v and w. If $\langle \cdot | \cdot \rangle$ also satisfies
   (d) $v \in V$ and $\langle v | w \rangle = 0$ for all $w \in V$ implies $v = 0_V$, we say that the inner product is **non-degenerate**. Let $V = (\{a, b, c\}, +)$ and let $K = (\{0, 1, 2\}, +_3, \cdot_3)$ as in Exercise 2 of Section 2–3.
   Fill in the following table to define an inner product $\langle \cdot | \cdot \rangle$ for the space V over K.

v \ v	a	b	c
a		0	
b			
c			

Example:
$\langle a | b \rangle = 0$

Is your $\langle \cdot | \cdot \rangle$ non-degenerate? How many ways could you have filled in the table to have gotten different inner products?
2.   Let $V = \mathbf{R}^n$ and $K = \mathbf{R}$.
   (a)   Determine which of the following definitions of $\langle \cdot | \cdot \rangle : \mathbf{R}^n \times \mathbf{R}^n \to \mathbf{R}$ satisfy the properties of an inner product.

   (i) $\langle x | y \rangle = \sum_{j=1}^{n} \frac{1}{j} x_j y_j$

   (ii) $\langle x | y \rangle = \sum_{j=1}^{n} (-1)^j x_j y_j$

(iii) $\langle x | y \rangle = \sum_{j=1}^{n} (x_j^2 y_j^2)^{1/2}$

(b)   Which are non-degenerate?

3.   Any vector space over the real field **R** with an inner product $\langle \cdot | \cdot \rangle$ defined for it is said to have a **positive definite** inner product if $\langle v|v \rangle \geq 0$ for all $v \in V$ and if $\langle v|v \rangle = 0$ implies $v = 0_V$. Which of the $\langle \cdot | \cdot \rangle$'s of Exercise 2 are positive definite? What about the standard inner product for $\mathbf{R}^n$ over **R**?

✔ **4.**   Let $v = 0_V$ and show that $|\langle u|v \rangle|^2 \leq \langle u|u \rangle \langle v|v \rangle$. Then let $v \neq 0_V$. Since for all scalars $\lambda \in \mathbf{C}$

$$0 \leq \langle u - \lambda v | u - \lambda v \rangle$$

it also holds for the particular value $\lambda = \dfrac{\langle v|u \rangle}{\langle v|v \rangle}$. The CBS inequality follows.

✔ **5.   (a)**   Define $\|v\| = \langle v|v \rangle^{1/2}$ and show that the triangle inequality $\|v + u\| \leq \|v\| + \|u\|$ follows from $0 \leq \langle u + v|u + v \rangle$. [*Hint:* $\langle u|v \rangle + \langle v|u \rangle = 2\,\mathfrak{Re}\langle u|v \rangle \leq 2|\langle u|v \rangle|$.]

(b)   Prove that this choice of $\|\cdot\|$ satisfies all the properties of a norm for the space V.

(c)   Prove that $\|v + w\|^2 + \|v - w\|^2 = 2\|v\|^2 + 2\|w\|^2$, and illustrate with a sketch.

✔ **6.**   Let $C[a, b]$ denote the space of continuous complex-valued functions on the interval $[a, b]$ and define an inner product $\langle \cdot | \cdot \rangle$ by $\langle f | g \rangle = \int_a^b \overline{f(\zeta)} g(\zeta)\, d\zeta$. Show that all the properties of the inner product hold.

7.   Let V be an inner product space over **C**.

(a)   Using $|\mathfrak{Re}z| \leq |z|$ for any complex number z, show that for any non-zero vectors v and w,

$$-1 \leq \frac{\mathfrak{Re}\langle v|w \rangle}{\|v\|\,\|w\|} \leq 1$$

(b)   This permits us to define the angle θ between v and w as the unique angle between 0 and π radians such that

$$\mathfrak{Re}\langle v|w \rangle = \|v\|\,\|w\|\cos\theta$$

If we define two nonzero vectors to be **perpendicular** if they form a right angle ($\cos\theta = 0$, or $\theta = \pi/2$), show that u, v orthogonal implies u, v perpendicular.

(c)   Give an example in $\mathbf{C}^2$, using the standard inner product of two nonzero vectors which are perpendicular but *not* orthogonal.

(d)   Show that in *real* inner product spaces the notions of perpendicularity and orthogonality are equivalent.

✔ **8.   (a)**   Prove that an orthogonal set of nonzero vectors is linearly independent.

[*Hint:* Use $\left\langle v_j \left| \sum_{i=1}^{k} c_i v_i \right. \right\rangle = \sum_{i=1}^{k} c_i \langle v_j | v_i \rangle$ and $\langle v|0_V \rangle = 0$ for all $v \in V$.]

(b)   If $v = c_1 v_1 + \cdots + c_n v_n$ and $\{v_1, \ldots, v_n\}$ is *orthonormal*, prove that $c_i = \overline{\langle v|v_i \rangle} = \langle v_i|v \rangle$; that is, $v = \sum_{i=1}^{n} \langle v_i|v \rangle v_i$. (Note order.)

✔ **9.**   (Gram-Schmidt) Let $\{x_1, x_2, \ldots\}$ be a list of linearly independent vectors in an inner product space V. Define $e_1 = x_1/\|x_1\|$ and, for $n \geq 2$, $e_n = z_n/\|z_n\|$, where

$$z_n = x_n - \sum_{i=1}^{n-1} \langle e_i \,|\, x_n \rangle \, e_i$$

(a)   Prove that for each n, the set $\{e_1, e_2, \ldots, e_n\}$ spans the same space that the set $\{x_1, x_2, \ldots, x_n\}$ spans.

(b)   Prove that $z_n \neq 0_V$ and, by direct computation, that $\langle z_n \,|\, e_i \rangle = 0$ for $i = 1, 2, \ldots, n-1$.

✔ **10.**   Let V be the set of all real polynomials of degree at most 3, and define

$$\langle f \,|\, g \rangle = \int_0^1 f(t)g(t)\, dt.$$ Let $f_1(t) = 1$, $f_2(t) = t$, $f_3(t) = t^2$, and $f_4(t) = t^3$.

(a)   Show that $\{f_1, f_2, f_3, f_4\}$ is a basis for V.

(b)   Construct an orthonormal basis $\{e_1, e_2, e_3, e_4\}$ for V.

✔ **11.**   Let V be a pre-Hilbert space, and let $S \subset V$. S is just a sub*set*, and perhaps not a sub*space*.

(a)   Prove that $S^\perp$ is a subspace of V. In $\mathbf{R}^2$, let $S = \left\{ \begin{bmatrix} 1 \\ 1 \end{bmatrix} \right\}$, which of course is not a subspace. Find $S^\perp$. Find $S^{\perp\perp} \overset{\Delta}{=} (S^\perp)^\perp$.

(b)   Prove that $S \subset S^{\perp\perp}$. Is $S \subset S^\perp$ or $S^\perp \subset S$? Is $S^{\perp\perp} \subset S$? [*Hint:* If so, then $S = S^{\perp\perp}$ and S would have to be a sub*space*.]

(c)   Prove that $S \subset T \subset V \Rightarrow T^\perp \subset S^\perp$.

✔ **12.**   Let S be a finite-dimensional subspace of the pre-Hilbert space V.

(a)   Prove that if $v \in V$, then v has the unique representation $v = s_1 + s_2$ for some $s_1 \in S$ and $s_2 \in S^\perp$. We write this symbolically as $V = S \oplus S^\perp$ and say "V is the direct sum of S and $S^\perp$." [*Hint:* Let $\{e_1, \ldots, e_m\}$ be an orthonormal basis of

S, and take $s_1 = \sum_{i=1}^{m} \langle e_i \,|\, v \rangle e_i$ and $s_2 = v - s_1$. Then show that $\langle s_2 \,|\, e_j \rangle = 0$ for

$j = 1, \ldots, m$ whence $s_2 \in S^\perp$. Finally show that $S \cap S^\perp = \{0_V\}$ from which unique-ness follows.]

(b)   Prove that $\displaystyle\sum_{i=1}^{m} |\langle v \,|\, e_i \rangle|^2 \leq \|v\|^2$.

(c)   Generalize (b) to get Bessel's Inequality:

$$\sum_{i=1}^{m} \frac{|\langle v \,|\, v_i \rangle|^2}{\|v_i\|^2} \leq \|v\|^2$$

where $\{v_1, \ldots, v_m\}$ is any orthogonal set of nonzero vectors of V. Show that equality occurs iff v is in the set spanned by $\{v_1, \ldots, v_m\}$.

13.   Let $A = \begin{bmatrix} 2 - i & 1 + i & 4 - i \\ 3 & -i & 1 \end{bmatrix}$. Using the standard inner products on $\mathbf{C}^3$

and $\mathbf{C}^2$. Compute $\langle Ax \,|\, y \rangle$ and $\langle x \,|\, A^*y \rangle$ for the vectors $y = \begin{bmatrix} -1 + i \\ -2 \end{bmatrix}$ and

$x = \begin{bmatrix} 1 + i \\ 2 \\ -i \end{bmatrix}$

14.   Let $A = \begin{bmatrix} 1+i & -7i \\ 3+i & 4-i \end{bmatrix}$ and $B = [3+i \quad 1-6i]$.

    (a)   Compute A^*, B^*, $(BA)^*$, and A^*B^*.

    (b)   Compute BB^*, B^*B, and $(BB^*)^*$.

✔ **15.**   With V, W, L, $\mathfrak{B}$, $\mathfrak{B}'$, and A all defined as in Theorem 7, define $F: W \to V$ as the linear map whose matrix $C = {}^{\mathfrak{B}}F^{\mathfrak{B}'} = (\overline{A})^T$. That is,

$$F(w) = F(y_1 w_1 + \cdots + y_m w_m) = \sum_{i=1}^{m} y_i F(w_i)$$

$$= \sum_{i=1}^{m} y_i \left(\sum_{j=1}^{n} [C]_{ji} v_j \right)$$

$$= \sum_{i=1}^{m} y_i \sum_{j=1}^{n} [\overline{A}]_{ij} v_j$$

    (a)   If $v = x_1 v_1 + \cdots + x_n v_n$, and $w = y_1 w_1 + \cdots + y_m w_m$, show that

$$\langle L(v) \,|\, w \rangle_W = \sum_{j=1}^{n} x_j \sum_{i=1}^{m} \overline{y_i} [\overline{A}]_{ij}$$

$$= \left\langle A \begin{bmatrix} x_1 \\ \vdots \\ x_n \end{bmatrix} \,\Bigg|\, \begin{bmatrix} y_1 \\ \vdots \\ y_m \end{bmatrix} \right\rangle_{\mathbf{C}^m}$$

    (b)   Prove that $\langle L(v) \,|\, w \rangle = \langle v \,|\, F(w) \rangle$ for all $v \in V$ and all $w \in W$, so that $F = L^*$.

✔ **16.**   Let $L: V \to W$ be any map, where V and W are perhaps infinite-dimensional inner product spaces, where not even all linear maps have adjoints.

    (a)   Prove that if $F: W \to V$ and $G: W \to V$ both serve as adjoint maps for L, then $F = G$. Thus, the adjoint is unique if it exists.

    (b)   Prove that if L has an adjoint L^*, then L^* has an adjoint and, in fact, $L^{**} = L$. [*Hint:* Use Lemma 5.]

17.   If $L: V \to W$ where V and W are any normed linear spaces, we say that L is *bounded* if there exists $M \in \mathbf{R}$ such that $\|L(v)\| \leq M \|v\|$ for all $v \in V$. The "smallest" such M which satisfies this condition is defined to be the **norm of L** and is denoted by $\|L\|$.

    (a)   Show that

$$\|L\| = \sup\{\|L(v)\| \mid \|v\| \leq 1\}$$

and

$$\|L\| = \sup \left\{ \frac{\|L(v)\|}{\|v\|} \,\Big|\, v \neq 0_V \right\}$$

where sup stands for **supremum,** the **least upper bound** of a set of real numbers.

    (b)   Prove that a linear operator is bounded iff it is continuous.

18. Let $A : \mathbf{C}^n \to \mathbf{C}^m$, where A is an $m \times n$ matrix. We show that A is bounded.
    (a) Prove that $\|Ax\| \leq k \|x\|$ for all $x \in \mathbf{C}^n$, where

$$k^2 = \sum_{i=1}^{m} \sum_{j=1}^{n} |[A]_{ij}|^2$$

[*Hint:* Let $\|Ax\|^2 = \sum_{i=1}^{m} |[Ax]_i|^2$, where $[Ax]_i = \sum_{i=1}^{n} [A]_{ij}x_j$ is the ith component of the vector Ax, and use the CBS inequality.]
    (b) Prove that $\langle Ax \,|\, y \rangle \leq k \|x\| \|y\|$ for all $x \in \mathbf{C}^n$, $y \in \mathbf{C}^m$.

✔ **19.** Let $L : V \to W$ be linear and have adjoint L^*. Prove that L^* is linear. [*Outline:* For any $v \in V$ and any $w_1, w_2 \in W$,

$$\begin{aligned}
\langle v \,|\, L^*(w_1 + w_2) \rangle &= \langle Lv \,|\, w_1 + w_2 \rangle \\
&= \langle Lv \,|\, w_1 \rangle + \langle Lv \,|\, w_2 \rangle \\
&= \langle v \,|\, L^*w_1 \rangle + \langle v \,|\, L^*w_2 \rangle \\
&= \langle v \,|\, L^*w_1 + L^*w_2 \rangle
\end{aligned}$$

and use Lemma 5.]

✔ **20.** A map $A : V \to W$ is said to be **invertible** if there exists a map $F : W \to V$ such that $FA = I_V$ and $AF = I_W$, where I_V and I_W are the identity maps on V and W, respectively. Such an F is called **the inverse** of A and is written A^{-1}.
    (a) Prove that A is invertible iff A is both one-to-one and onto.
    (b) Prove that A is invertible iff A^* is invertible, and that $(A^*)^{-1} = (A^{-1})^*$.
[*Hint:* Use Theorem 6 (iii) on $AA^{-1} = I$.]

✔ **21.** In Exercise 11, we saw that $S \subset S^{\perp\perp}$ but that $S^{\perp\perp} \neq S$ in general. In Exercise 12, we saw that if S was a finite-dimensional subspace of V, then $V = S \oplus S^{\perp}$. Let us now restrict our attention to the case where V is finite-dimensional and S is a *subspace* of V.
    (a) Prove that $V = S \oplus S^{\perp}$ and dim $S^{\perp} = $ dim $V - $ dim S.
    (b) Prove $S^{\perp\perp} = S$. [*Hint:* Use (a) to establish $S^{\perp}$ finite-dimensional, whence $V = S^{\perp} \oplus (S^{\perp})^{\perp}$, so dim $S^{\perp\perp} = $ dim $V - $ dim $S^{\perp} = $ dim S.]
    (c) Prove that, in finite dimensional spaces, for subspaces S and T we have $S = T \Leftrightarrow S^{\perp} = T^{\perp}$, so we can "take the perp" of both sides.

22. **(a)** Show that the linear system of equations (with more equations than "unknowns"):

$$\begin{aligned}
2x_1 + 3x_2 &= y_1 \\
x_1 - x_2 &= y_2 \\
x_1 - 2x_2 &= y_3
\end{aligned}$$

has a solution iff $y = [y_1 \, y_2 \, y_3]^T$ satisfies $3y_1 - 11y_2 - 5y_3 = 0$.
    (b) Consider the system described by $Ax = y$ where

$$A = \begin{bmatrix} 2 & -3 & 1 \\ -3 & 2 & -4 \\ 1 & -4 & -2 \end{bmatrix} \text{ and } y = [1 \quad a \quad b]^T$$

Find a basis for $\mathfrak{N}(A^*)$. For what numbers a and b can you solve the given system of equations? Find a solution if $a = 2$ and $b = 4$. [*Hint:* If $\{v_1, \ldots, v_k\}$ span $\mathfrak{N}(A^*)$, then the inhomogeneous equation $Ax = y$ has a solution iff $\langle v_j | y \rangle = 0$ for $j = 1$, $2, \ldots, k$; that is, iff y is orthogonal to the solutions of the homogeneous adjoint equation.]

(c)   If A is a square matrix, prove that dim $\mathfrak{N}(A) = $ dim $\mathfrak{N}(A^*)$.

23.   Let $V = \{u \in C^2[0, \pi] \,|\, u(0) = 0$ and $u(\pi) = 0\}$. Define $L: V \to V$ by $Lu = \ddot{u} + u$ for $u \in V$. Use the customary definition $\langle u | v \rangle = \displaystyle\int_0^{\pi} u(\zeta)\, v(\zeta)\, d\zeta$.

(a)   Show that $\langle Lu \,|\, v \rangle = \langle u \,|\, Lv \rangle$ for all u, $v \in V$. [This means that L is self-adjoint, $L = L^*$.]

(b)   Show that $\mathfrak{N}(L^*) = \{ag \,|\, a \in \mathbf{R}^1\}$ where $g(t) = \sin(t)$, $0 \le t \le \pi$ and find $\mathfrak{R}(L)$. Is L one to one? Onto? Linear?

(c)   Let $f \in C[0, \pi]$, and consider the equation $Lu = f$. Show that a solution exists iff $\displaystyle\int_0^{\pi} f(t) \sin(t)\, dt = 0$. [*Note:* This exercise illustrates how to set up the "boundary value problem" $\ddot{u} + u = f$ for $u(0) = u(\pi) = 0$ in terms of linear operator theory.]

24.   Use the methods of this section on Exercise 19 of Section 4–3 to show that rank $AB \le$ min $\{$rank A, rank $B\}$.

25.   If $(Lu)(t) = \displaystyle\int_0^1 K(t, s)\, u(s)\, ds$, $(Fv)(s) = \displaystyle\int_0^1 K(s, t)\, v(t)\, dt$, and $\langle u | v \rangle = \displaystyle\int_0^1 u(\xi)\, v(\xi)\, d\xi$, verify that $F = L^*$.

26.   Let $V = W = \{$square-integrable real functions on $[0, 1]\}$ denoted by $L_2[0, 1]$. Let $K(t, s)$ satisfy $\displaystyle\int_0^1 \int_0^1 |K(t, s)|^2\, ds\, dt < \infty$.

Let $\langle v | w \rangle = \displaystyle\int_0^1 v(\xi) w(\xi)\, d\xi$ and $A: V \to W$ by

$$(Av)(t) = \int_0^t K(t, s)\, v(s)\, ds \text{ for } t \in [0, 1]$$

Prove that A^* is given by

$$(A^*w)(t) = \int_t^1 K(\xi, t) w(\xi)\, d\xi \text{ for } t \in [0, 1]$$

[*Hint:* Evaluate $\langle Av | w \rangle$ and change the order of the resulting double integration.]

27.   Let V be a complex inner product space, and let $L: V \to V$. We say that L is **self-adjoint** or **Hermitian** if $L^* = L$.

(a)   Show that $\langle v | Lv \rangle$ is a *real* number for all $v \in V$, if L is self-adjoint.

(b)   Let A be an $n \times m$ complex matrix and let $W = AA^*$. Show that W is self-adjoint. Are there any restrictions on m and n?

✔ **28.**   A self-adjoint (Hermitian) operator L is said to be **positive definite** if $\langle v | Lv \rangle > 0$ for all $v \ne 0$, and **positive semi-definite** if $\langle v | Lv \rangle \ge 0$ for all $v \in V$.

(a)   Use the CBS inequality to prove that if L is positive definite, then L is invertible.

(b)   As in Exercise 27(b), let $W = AA^*$. Prove that rank $A = n$ if and only if W is positive definite.

(c)   Illustrate the concepts of these two exercises by letting A first be the 2×1 matrix $\begin{bmatrix} 1 \\ 3 \end{bmatrix}$ and then the 1×2 matrix $[1 \quad 3]$.

4-5 Observability for Linear Discrete-Time Systems

Let us recall that any time-invariant system is *observable* if every pair of distinct states has distinct response functions; i.e., for any pair of distinct $\widehat{x}$ and $\widetilde{x}$ of Σ we have $\mathcal{S}_{\widehat{x}} \neq \mathcal{S}_{\widetilde{x}}$. $\mathcal{S}_{\widehat{x}} : U^* \to Y$ is defined by $w \mapsto \eta(\phi(\widehat{x}, w))$, since the response to input sequence w of Σ started in state $\widehat{x}$ is the output from the state $\phi(\widehat{x}, w)$ to which Σ moves from $\widehat{x}$ after reading in the whole sequence w.

Let us now discuss observability for the linear system (F, G, H):

$$\begin{cases} x(t+1) = Fx(t) + Gu(t) \\ y(t) = Hx(t) \end{cases}$$

where F, G, and H are $n \times n$, $n \times m$, and $q \times n$ matrices respectively. The first thing we must do is to compute the response function of a state $\widehat{x}$ of (F, G, H):

We recall from equation (2) of Section 4–3 that, for $\ell > 0$,

$$x(\ell) = \phi(\widehat{x}, u_0 \ldots u_{\ell-1}) = F^\ell \widehat{x} + \sum_{j=0}^{\ell-1} F^{\ell-j-1} Gu_j \tag{1}$$

which is the sum of the contribution $F^\ell \widehat{x}$ of $\widehat{x}$ updated ℓ times by F, and of the contributions of the inputs, with u_j being read in by G and then updated $(\ell - j - 1)$ times by F to yield the contribution $F^{\ell-j-1} Gu_j$ after u_ℓ has been read in.

THE UNIT-PULSE RESPONSE AND CONVOLUTION

Since the output map of (F, G, H) is $\widehat{x} \mapsto H\widehat{x}$, we immediately deduce, using the linearity of H to pass it inside the summation of equation (1), that the total response is given by

$$y(\ell) = \mathcal{S}_{\widehat{x}}(u_0 \ldots u_{\ell-1}) = H\phi(\widehat{x}, u_0 \ldots u_{\ell-1}), = HF^\ell \widehat{x} + \sum_{j=0}^{\ell-1} HF^{\ell-j-1} Gu_j \tag{2}$$

This last equation is very significant. It holds for any state $\widehat{x}$ of our linear system (F, G, H) and for any sequence $u_0 \ldots u_{\ell-1}$ of inputs. In particular, it will hold for the case $\widehat{x} = 0_X$ to yield the

zero-state response $\mathcal{S}_{0_X}(u_0 \ldots u_{\ell-1}) = \sum_{j=0}^{\ell-1} HF^{\ell-j-1} Gu_j \tag{3}$

since the term $HF^l\widehat{x}$ vanishes when $\widehat{x} = 0_X$ (noting that any linear transformation maps 0 to 0).

Secondly, we may take $\widehat{x}$ to be *any* state (which may or may not be 0_X) and apply the sequence 0^l of l consecutive 0 inputs, $\underbrace{0 \ldots 0}_{l \text{ times}}$ to get the

zero-input response $\qquad S_{\widehat{x}}(0^l) = HF^l\widehat{x}$ $\hfill$ (4)

By identifying these terms, we recognize that equation (2) expresses, for the discrete-time (F, G, H), the crucial property of all linear systems which we proved in Theorem 3–3–2:

$$S_{\widehat{x}}(u_0 \ldots u_{l-1}) = \underbrace{S_{\widehat{x}}(0^l)}_{\substack{\text{zero-input} \\ \text{response}}} + \underbrace{S_{0_X}(u_0 \ldots u_{l-1})}_{\substack{\text{zero-state} \\ \text{response}}}$$

In other words, the response of a *linear* system started in state $\widehat{x}$ to an input sequence $u_0 \ldots u_{l-1}$ can be *decomposed* into two parts—one which is independent of the inputs, $S_{\widehat{x}}(0^l)$, and one which is independent of the initial state $\widehat{x}$, $S_{0_X}(u_0 \ldots u_{l-1})$.

The zero-state response S_{0_X} of equation (3) admits an interesting interpretation. Suppose that, starting in the zero state 0_X, we drive the system (F, G, H) with the input sequence $w = (1, 0, 0, 0, \ldots.)$ which we may denote by δ. What will be the zero state output $y_0(l)$ at the lth time instant? From equation (4), all but the $j = 0$ term drop out and we have

$$y_0(l) = S_{0_X}(1, \underbrace{0, 0, \ldots, 0}_{l - 1 \text{ zeros}}) = \begin{cases} HF^{l-1}G, & \text{for } l \geq 1 \\ 0, & \text{for } l = 0 \end{cases} \hfill (5)$$

Thus, the expression $HF^{l-1}G$ can be thought of as the zero-state response at time $l > 0$ of (F, G, H) to $\delta \overset{\Delta}{=} (1, 0, 0, \ldots.)$, a **"unit-pulse"** input at time 0. If we call the **unit-pulse response** function g, then

$$g(l) \overset{\Delta}{=} S_{0_X}(1, \underbrace{0, 0, \ldots, 0}_{l - 1 \text{ zeros}}) = \begin{cases} HF^{l-1}G, & l > 0 \\ 0, & l = 0 \end{cases} \hfill (6)$$

and we see that

$$g(l - j) = HF^{l-j-1}G, \text{ for } l > j$$

and that the zero-state response formula in (4) may be rewritten as

$$S_{0_X}(u_0, u_1, \ldots, u_{l-1}) = \sum_{j=0}^{l-1} g(l - j)u_j \hfill (7)$$

The operation on the right-hand side of (7) is called the **convolution summation**† of the two sequences

$$(g(1), g(2), \ldots, g(\ell)) \text{ and } (u_0, u_1, \ldots, u_{\ell-1})$$

because it looks so much like the famous **convolution integral**

$$y_0(t) = \int_0^t g(t - \zeta)u(\zeta)\, d\zeta$$

which, as we will see in Section 5–4, gives the zero-state response of a continuous-time linear system in terms of the input function u and the "unit impulse response" function g.

In Section 4–6 we will return to the discussion of the unit-pulse response g and convolution summations, and we will show how they are related to the Z-transform, a useful tool for analyzing discrete-time linear systems. The rest of this section is devoted to the observability of such systems.

THE INDISTINGUISHABLE STATES

Just as we approached reachability of a system by seeing how many states can be reached in at most k steps, as k gets larger and larger, so shall we approach observability by seeing what pairs of states can or cannot be distinguished by input sequences of length at most k, as k gets larger and larger:

DEFINITION 1

Two states $\bar{x}$ and $\hat{x}$ of a system Σ are said to be k-**indistinguishable**, and we write $\bar{x} \sim_k \hat{x}$, if and only if $\mathcal{S}_{\bar{x}}(w) = \mathcal{S}_{\hat{x}}(w)$ for all input sequences w of length at most k.

Two states $\bar{x}$ and $\hat{x}$ of a system Σ are said to be **indistinguishable**, and we write $\bar{x} \sim \hat{x}$, if and only if $\mathcal{S}_{\hat{x}}(w) = \mathcal{S}_{\bar{x}}(w)$ for *all* input sequences w. ○

Of course, when two states $\bar{x}$ and $\hat{x}$ are *not* indistinguishable we say they are **distinguishable** and sometimes we write $\bar{x} \not\sim \hat{x}$ or $\bar{x} \not\sim_k \hat{x}$ to denote their distinguishability.

Our definitions immediately yield the following facts:

LEMMA 1

For any discrete-time, time-invariant system Σ we have

(i) For all $k \leq p$, $\bar{x} \sim_p \hat{x} \Longrightarrow \bar{x} \sim_k \hat{x}$

---

†Also called the *serial product* or *Cauchy product* of the two sequences. See R. M. Bracewell, *The Fourier Transform*, McGraw-Hill, 1965, pp. 30–38.

(ii) $\bar{x} \sim \hat{x} \Leftrightarrow [\bar{x} \sim_k \hat{x}$ for all k.]

(iii) Σ is observable $\Leftrightarrow$ no two distinct states are indistinguishable

$$\Leftrightarrow [\bar{x} \sim \hat{x} \Rightarrow \bar{x} = \hat{x}] \qquad \square$$

Let us see what we can deduce when Σ is a *linear* system. Pick any input sequence $u_0 \ldots u_{l-1}$ and any two states $\bar{x}$ and $\hat{x}$ which may or may not be 0_X. Then, if Σ is linear, we have:

$$\mathcal{S}_{\bar{x}}(u_0 \ldots u_{l-1}) = \mathcal{S}_{\bar{x}}(0^l) + \mathcal{S}_{0_X}(u_0 \ldots u_{l-1})$$
$$\mathcal{S}_{\hat{x}}(u_0 \ldots u_{l-1}) = \mathcal{S}_{\hat{x}}(0^l) + \mathcal{S}_{0_X}(u_0 \ldots u_{l-1})$$

Note that the second term is the same in both cases, and so the sums are equal if and only if the first terms are equal:

$$\mathcal{S}_{\bar{x}}(u_0 \cdots u_{l-1}) = \mathcal{S}_{\hat{x}}(u_0 \cdots u_{l-1}) \Leftrightarrow \mathcal{S}_{\bar{x}}(0^l) = \mathcal{S}_{\hat{x}}(0^l)$$

Hence, we deduce an important fact:

LEMMA 2

Two states of a given *linear* system are k-indistinguishable if and only if they yield the same response to all zero input sequences comprising at most k terms.

Proof

By Lemma 1, $\bar{x} \sim_k \hat{x} \Leftrightarrow \mathcal{S}_{\bar{x}}(u_0 \ldots u_{l-1}) = \mathcal{S}_{\hat{x}}(u_0 \ldots u_{l-1})$ for all $l \leq k$ $\Leftrightarrow \mathcal{S}_{\bar{x}}(0^l) = \mathcal{S}_{\hat{x}}(0^l)$ for all $l \leq k$ $\qquad \square$

Thus, to test k-indistinguishability for linear systems we only have to make $k + 1$ tests, namely with $\Lambda = 0^0$, $0 = 0^1$, $00 = 0^2$, $000 = 0^3, \ldots, 0^k$. If the system Σ is not linear, it is not necessarily true that zero input sequences alone suffice to test distinguishability. [See Exercises 1, 2, 3 and 4.]

Note that our discussion of distinguishability says that two states are distinguishable if there is at least one input sequence to which they respond differently. Since $\mathcal{S}_{\bar{x}}(0^l) = \mathcal{S}_{\hat{x}}(0^l)$ if and only if $HF^l\bar{x} = HF^l\hat{x}$, which is true if and only if $HF^l(\bar{x} - \hat{x}) = 0_Y = HF^l0_X$, we have the following:

LEMMA 3

In linear systems, two states $\bar{x}$ and $\hat{x}$ are distinguishable if and only if their difference $(\bar{x} - \hat{x})$ is distinguishable from 0_X. $\qquad \square$

The reader should, of course, realize that a given nonzero state $\hat{x}$ might be indistinguishable for a while from 0_X under the application of zero inputs. All we demand is that $\hat{x}$ eventually causes a nonzero output under a long enough string of zero inputs, in order that $\hat{x}$ be distinguishable from 0_X.

Example 1

Let Σ be the linear system (F, G, H) where

$$F = \begin{bmatrix} 0 & 0 & 1 \\ 1 & 0 & 0 \\ 0 & 1 & 0 \end{bmatrix}, G = \begin{bmatrix} 1 \\ 0 \\ 0 \end{bmatrix}, H = \begin{bmatrix} 0 & 0 & 1 \end{bmatrix}$$

We show that $\hat{x} = \begin{bmatrix} 1 \\ 0 \\ 0 \end{bmatrix}$ is distinguishable from $0_X = \begin{bmatrix} 0 \\ 0 \\ 0 \end{bmatrix}$.

Here the next-state matrix F simply permutes the components of the state vector, since

$$F \begin{bmatrix} x_1 \\ x_2 \\ x_3 \end{bmatrix} = \begin{bmatrix} x_3 \\ x_1 \\ x_2 \end{bmatrix}$$

and H just reads out the last component of the state vector

$$H \begin{bmatrix} x_1 \\ x_2 \\ x_3 \end{bmatrix} = x_3$$

Thus, if we start in state

$$\begin{bmatrix} 1 \\ 0 \\ 0 \end{bmatrix}$$

Σ will have output zero; applying a zero input will then send us to state

$$\begin{bmatrix} 0 \\ 1 \\ 0 \end{bmatrix}$$

which also yields a zero output, and it is not until we apply one further zero input to obtain state

$$\begin{bmatrix} 0 \\ 0 \\ 1 \end{bmatrix}$$

that we finally obtain a nonzero output, namely 1. Thus,

$$\mathcal{S}_{\hat{x}}(\Lambda) = \mathcal{S}_{0_X}(\Lambda)$$
$$\mathcal{S}_{\hat{x}}(0) = \mathcal{S}_{0_X}(0)$$

but

$$\mathcal{S}_{\hat{x}}(0^2) \neq \mathcal{S}_{0_X}(0^2)$$

Thus, $\hat{x}$ is distinguishable from 0_X even though $\hat{x} \sim_0 0_X$ and $\hat{x} \sim_1 0_X$.   ◊

In this last example, we have an output that only lets us look at one component of the state variable at a time. We may think of the next-state function as moving the components around, so that we may not see any nonzero components of a state until after it has been moved around several times. With this sort of terminology, it may seem plausible to expect that if we are ever going to see a component rotated into view, so to speak, then this must happen in a time less than the number of dimensions of the state space—at each time, we would expect to see one new dimension, or never to see any new dimensions again. This type of argument should seem reminiscent of our discussion of reachability in Section 4–3, and will now be formalized.

Recall the notion of a **block-partitioned** matrix.† Suppose that an n-vector x satisfies some k equalities

$$\left. \begin{array}{l} A_1 x = b_1 \\ A_2 x = b_2 \\ A_3 x = b_3 \\ \quad \vdots \\ A_k x = b_k \end{array} \right\} \tag{8}$$

where each A_j is an $m_j \times n$ matrix and b_j is an m_j-vector. Then we may rewrite these k equalities of m_j-vectors by one equality of $(m_1 + m_2 + \cdots + m_k)$-vectors:

$$\begin{bmatrix} A_1 x \\ A_2 x \\ \vdots \\ A_k x \end{bmatrix} = \begin{bmatrix} b_1 \\ b_2 \\ \vdots \\ b_k \end{bmatrix}$$

† Or else see Exercise 4–2–13.

and this equality may be simply rewritten [Exercise 5] as:

$$\begin{bmatrix} A_1 \\ A_2 \\ \vdots \\ A_k \end{bmatrix} x = \begin{bmatrix} b_1 \\ b_2 \\ \vdots \\ b_k \end{bmatrix} \tag{9}$$

Example 2

The 2-vector $x = \begin{bmatrix} x_1 \\ x_2 \end{bmatrix}$ satisfies the two equations

$$\begin{bmatrix} 2 & 1 \\ 0 & 3 \end{bmatrix} \begin{bmatrix} x_1 \\ x_2 \end{bmatrix} = \begin{bmatrix} 5 \\ 9 \end{bmatrix}$$

and

$$\begin{bmatrix} 1 & 6 \\ -2 & 3 \end{bmatrix} \begin{bmatrix} x_1 \\ x_2 \end{bmatrix} = \begin{bmatrix} 19 \\ 7 \end{bmatrix}$$

if and only if it satisfies the single equation

$$\begin{bmatrix} 2 & 1 \\ 0 & 3 \\ 1 & 6 \\ -2 & 3 \end{bmatrix} \begin{bmatrix} x_1 \\ x_2 \end{bmatrix} = \begin{bmatrix} 5 \\ 9 \\ 19 \\ 7 \end{bmatrix}$$

as may be verified by matrix multiplication.   ◊

Combining Lemma 2 with this last observation, we immediately have:

LEMMA 4

For the linear system (F, G, H) we have for *any* two states $\bar{x}$ and $\hat{x}$,

$$\bar{x} \sim_k \hat{x} \text{ if and only if } (\bar{x} - \hat{x}) \in \mathfrak{N}\left(\begin{bmatrix} H \\ HF \\ \vdots \\ HF^k \end{bmatrix} \right),$$

where the last expression is the null space of the matrix $\begin{bmatrix} H \\ HF \\ \vdots \\ HF^k \end{bmatrix} : \mathbf{C}^n \to \mathbf{C}^{kq}$ when F is $n \times n$ and H is $q \times n$.

Proof

$$\bar{x} \sim_k \hat{x} \Leftrightarrow \mathcal{S}_{\bar{x}}(0^l) = \mathcal{S}_{\hat{x}}(0^l) \quad \text{for } l \leq k$$
$$\Leftrightarrow HF^l\bar{x} = HF^l\hat{x} \quad \text{for } l \leq k$$
$$\Leftrightarrow \begin{bmatrix} H \\ HF \\ \vdots \\ HF^k \end{bmatrix} (\bar{x} - \hat{x}) = \begin{bmatrix} 0_Y \\ 0_Y \\ \vdots \\ 0_Y \end{bmatrix} = 0 \quad \text{(the zero vector of } \mathbf{C}^{kq}\text{)}$$

where we have used the linearity of HF^l to see that $HF^l\bar{x} = HF^l\hat{x}$ if and only if $HF^l(\bar{x} - \hat{x}) = 0_Y$. $\square$

Now recall Corollary 4-4-10, which states that for any linear transformation A, a vector y is in $\mathfrak{N}(A)$ if and only if y is orthogonal to $\mathfrak{R}(A^*)$. Now, given matrices $A_1, \ldots, A_k$, where A_j is $(n_j \times n)$, the adjoint of the

block-partitioned $(n_1 + n_2 + \cdots + n_k) \times n$ matrix $\begin{bmatrix} A_1 \\ A_2 \\ \vdots \\ A_k \end{bmatrix}$ is just

$[A_1^*, A_2^*, \ldots, A_k^*]$.

For example, if

$$\begin{bmatrix} A_1 \\ A_2 \end{bmatrix} = \begin{bmatrix} 1 - i & 2 + i \\ \hline 3 & -2i \end{bmatrix} \text{ then } \begin{bmatrix} A_1 \\ A_2 \end{bmatrix}^* = [A_1^* \ A_2^*] = \begin{bmatrix} 1 + i & 3 \\ 2 - i & 2i \end{bmatrix}$$

If we further recall that the adjoint of a product is the product of the adjoints in *reverse order*, we immediately deduce that

$$\begin{bmatrix} H \\ HF \\ \vdots \\ HF^k \end{bmatrix}^* = [H^*, F^*H^*, \ldots, (F^*)^k H^*]$$

so that we have the crucial result:

LEMMA 5

For the linear system (F, G, H),

$$\bar{x} \sim_k \hat{x} \iff (\bar{x} - \hat{x}) \perp \mathfrak{R}([H^*, F^*H^*, \ldots, (F^*)^k H^*])$$

where by $\perp$ we mean "is orthogonal† to." $\square$

DUALITY

Before reading the next paragraph, the reader is asked to consider how he would have interpreted the expression

$$\mathfrak{R}([H^*, F^*H^*, \ldots, (F^*)^k H^*])$$

if he had come upon it in our discussion of reachability in Section 4-3.

(Pause for Consideration)

---

† See Exercise 7 of Section 4-4 for a distinction between perpendicularity and orthogonality in complex vector spaces.

The answer is that it is precisely the set of states reachable in $(k + 1)$ steps or less from the 0-state of the linear system whose state-transitions are governed by the equation

$$p(t + 1) = F^*p(t) + H^*w(t) \tag{10}$$

If we let S_k^* denote the set of states reachable from the zero state in at most k steps, and let S^* be the set of all states reachable from the zero state for the system (F^*, H^*) in equation (10), then Lemma 5 may be written as:

LEMMA 5A

For the linear system (F, G, H),

$$\bar{x} \sim_k \hat{x} \iff (\bar{x} - \hat{x}) \in (S_{k+1}^*)^\perp \qquad \square$$

Of course, as in Theorem 4–3–5 on reachability, we have

$$S_k^* \subset S_{k+1}^* \subset S_n^* = S^*$$

since F and F^* are $(n \times n)$ matrices.

We thus have the following beautiful result:

THEOREM 6

The linear system (F, G, H) is observable if and only if the system (F^*, H^*), written as

$$p(t + 1) = F^*p(t) + H^*w(t)$$

is reachable.

Proof

(F, G, H) is observable

$\iff [\bar{x} \sim \hat{x} \Rightarrow \bar{x} = \hat{x}]$ (by definition)

$\iff [\bar{x} \sim_k \hat{x} \text{ for all } k \geq 0 \Rightarrow \bar{x} = \hat{x}]$ (by definition)

$\iff [(\bar{x} - \hat{x}) \in (S_{k+1}^*)^\perp \text{ for all } k \geq 0 \Rightarrow \bar{x} = \hat{x}]$ (by Lemma 5A)

$\iff [(\bar{x} - \hat{x}) \in (S_n^* = S^*)^\perp \Rightarrow \bar{x} = \hat{x}]$ (by Theorem 4–3–5

and Exercise 7)

$$\Leftrightarrow [(S^*)^\perp = \{0_X\}]$$

$\Leftrightarrow$ [The only vector orthogonal to all the reachable states of (F^*, H^*) is 0_X]

$\Leftrightarrow [S^* = \{0_X\}^\perp = X]$ (since F^* and F are $(n \times n)$)

$\Leftrightarrow$ all states of (F^*, H^*) are reachable. $\square$

COROLLARY 7

The linear system (F, G, H) of dimension n is observable if and only if

$$\text{rank} ([H^*, F^*H^*, \ldots, (F^*)^{n-1}H^*]) = n \qquad \square$$

We obtained the system (10) from (F, G, H) by replacing G by H and then taking adjoints. This suggests that the appropriate output map for (10) should be G^*. We thus make the definition:†

DEFINITION 2

The **dual** of the linear system (F, G, H) is the linear system (F^*, H^*, G^*) with state vector p, input w, and output v:

$$\begin{cases} p(t + 1) = F^*p(t) + H^*w(t) \\ v(t) = G^*p(t) \end{cases} \qquad \bigcirc$$

This definition proves to be a happy one, for we see that our criteria yield the famous Kalman Duality Theorem:

THEOREM 8 (KALMAN DUALITY THEOREM)

The system (F, G, H) is observable (respectively, reachable) if and only if its dual (F^*, H^*, G^*) is reachable (respectively, observable). $\square$

This is a very useful result, for it says that whenever we have algorithms to test reachability of any system, it requires no extra effort to handle observability—to test the observability of a system, we simply apply the reachability criterion to the dual system.

---

†In most books which deal primarily with continuous-time systems, it is traditional to define $(-F^*, H^*, G^*)$ as the dual of (F, G, H). When we study observability in continuous-time systems in Section 7–3, we shall embrace this tradition. For the present we proceed minus the minus.

Note that we have used the basic duality result of linear algebra, $\mathfrak{N}(A) = [\mathfrak{R}(A^*)]^\perp$, to prove the duality theorem for linear systems, so that we may then call on our study of reachability to deduce properties of observability "without further effort."

Example 3

Consider the system (F, G, H) of Example 1. The dual system (F^*, H^*, G^*) is described by matrices

$$F^* = \begin{bmatrix} 0 & 1 & 0 \\ 0 & 0 & 1 \\ 1 & 0 & 0 \end{bmatrix}, H^* = \begin{bmatrix} 0 \\ 0 \\ 1 \end{bmatrix}, G^* = [1 \quad 0 \quad 0]$$

The reachability matrix for (F^*, H^*, G^*) is

$$[(F^*)^2 H^*, F^* H^*, H^*] = \begin{bmatrix} 1 & 0 & 0 \\ 0 & 1 & 0 \\ 0 & 0 & 1 \end{bmatrix} = I$$

of rank 3, so (F^*, H^*, G^*) is reachable and (F, G, H) is observable.   ◊

It is often instructive to draw the "wiring-diagram" of the dual system. In Figure 4–19(a) we show a realization of the system (F, G, H) of Example 1. In Figure 4–19(b) we show the dual system (F^*, H^*, G^*), which has the same structure as (F, G, H). To further illustrate the use of duality and to get more practice with the notation of Lemma 5A and Theorem 6, we will

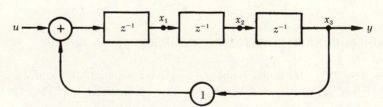

Figure 4–19(a)   The system (F, G, H) of Example 1.

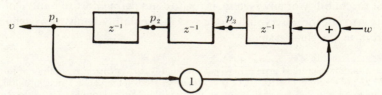

Figure 4–19(b)   The dual system (F^*, H^*, G^*) of Figure 4–19(a).

now rework Example 1 and show that $\hat{x} = \begin{bmatrix} 1 \\ 0 \\ 0 \end{bmatrix}$ is distinguishable from $0_X = \begin{bmatrix} 0 \\ 0 \\ 0 \end{bmatrix}$ by an easy analysis of the dual system.

Example 4

By Lemma 5A we know that $\hat{x} \sim_k 0_X$ iff $\hat{x} \in (S_{k+1}^*)^\perp$. From Figure 4–19(b) we can tell by inspection that

$$S_1^* = \left\{ \begin{bmatrix} 0 \\ 0 \\ a \end{bmatrix} \,\middle|\, a \in \mathbf{R} \right\}$$

$$S_2^* = \left\{ \begin{bmatrix} 0 \\ a \\ b \end{bmatrix} \,\middle|\, a \in \mathbf{R}, b \in \mathbf{R} \right\}$$

and

$$S_3^* = \left\{ \begin{bmatrix} a \\ b \\ c \end{bmatrix} \,\middle|\, a \in \mathbf{R}, b \in \mathbf{R}, c \in \mathbf{R} \right\}$$

Thus

$$(S_1^*)^\perp = \left\{ \begin{bmatrix} x_1 \\ x_2 \\ 0 \end{bmatrix} \,\middle|\, x_1 \in \mathbf{R}, x_2 \in \mathbf{R} \right\}$$

$$(S_2^*)^\perp = \left\{ \begin{bmatrix} x_1 \\ 0 \\ 0 \end{bmatrix} \,\middle|\, x_1 \in \mathbf{R} \right\}$$

and

$$(S_3^*)^\perp = \left\{ \begin{bmatrix} 0 \\ 0 \\ 0 \end{bmatrix} \right\}$$

Since $\hat{x} \in (S_1^*)^\perp$ and $\hat{x} \in (S_2^*)^\perp$, we have $\hat{x} \sim_0 0_X$ and $\hat{x} \sim_1 0_X$, but since $\hat{x} \notin (S_3^*)^\perp$, $\hat{x} \not\sim_2 0_X$, so $\hat{x}$ and 0_X are distinguishable.   ◊

For the n-dimensional system (F, G, H), the block-partitioned matrices

$$\mathcal{O} \triangleq \begin{bmatrix} H \\ HF \\ \vdots \\ HF^{n-1} \end{bmatrix} \quad \text{and} \quad \mathcal{C} \triangleq [F^{n-1}G, \dots, FG, G]$$

are known in the literature as the **observability** and **controllability (reachability) matrices** respectively.

The single-input, single-output, two-dimensional system of Example 4-3-6 provides a very intuitive interpretation of observability.

Example 5

For the system (F, G, H) shown in Figure 4-13, we have $F^* = \begin{bmatrix} \bar{\alpha} & 0 \\ 0 & \bar{\beta} \end{bmatrix}$ and $H^* = \begin{bmatrix} \bar{c} \\ \bar{d} \end{bmatrix}$, so that the reachability matrix of the dual system (F^*, H^*) is just the 2×2 matrix $\mathcal{C} = [F^*H^*, H^*] = \begin{bmatrix} \bar{\alpha}\bar{c} & \bar{c} \\ \bar{\beta}\bar{d} & \bar{d} \end{bmatrix}$. Since $\det \mathcal{C} = \overline{cd(\alpha - \beta)} = 0$

iff $\alpha = \beta$ or either one of c or d is zero, the system is not observable for precisely these conditions. When $\alpha = \beta$, we have the familiar case of placing in parallel two subsystems with the same time constant. When either c or d is zero, the corresponding state variable is decoupled from the output y. Since observability concerns finding out about behavior of the state variables from observations of the input-output behavior, if there is no path connecting a state variable to the output, that variable can't be observed.   ◊

Let us now show that if a *linear* system is observable in the sense that every pair of distinct states is distinguishable, then it is observable in the stronger sense that, given the system in *any* initial state, we can conduct a finite experiment to tell what that state *was:*

Suppose that we start our n-dimensional linear system (F, G, H) in an arbitrary state $\hat{x}$ and apply $(n - 1)$ inputs, each of which is zero, so that we observe a sequence of n outputs, which we label by $y_0, y_1, \ldots y_{n-1}$:

$$\mathcal{S}_{\hat{x}}(\Lambda) = H\hat{x} = y_0$$
$$\mathcal{S}_{\hat{x}}(0) = HF\hat{x} = y_1$$
$$\vdots$$
$$\mathcal{S}_{\hat{x}}(0^{n-1}) = HF^{n-1}\hat{x} = y_{n-1}$$

Then to say that Σ is observable is to say that the equation

$$\begin{bmatrix} H \\ HF \\ \vdots \\ HF^{n-1} \end{bmatrix} \hat{x} = \begin{bmatrix} y_0 \\ y_1 \\ \vdots \\ y_{n-1} \end{bmatrix}$$

has a unique solution. Thus, given the matrices H and F of a system and the n observations $y_0, \ldots, y_{n-1}$, we can compute the initial state.

Example 6

Consider the system

$$x(t + 1) = \begin{bmatrix} 1 & 2 \\ 0 & 1 \end{bmatrix} x(t) + \begin{bmatrix} 2 \\ 3 \end{bmatrix} u(t)$$

$$y(t) = [1 \quad 3]x(t)$$

If we start it in state $\begin{bmatrix} 1 \\ 0 \end{bmatrix}$, then the immediate output will be $y_0 = 1$. But

it is clear that any state $\begin{bmatrix} x_1 \\ x_2 \end{bmatrix}$ for which $x_1 + 3x_2 = 1$ would yield the same

output.

Now, under a single 0 input, state $\begin{bmatrix} x_1 \\ x_2 \end{bmatrix}$ changes to state

$\begin{bmatrix} 1 & 2 \\ 0 & 1 \end{bmatrix}\begin{bmatrix} x_1 \\ x_2 \end{bmatrix} = \begin{bmatrix} x_1 + 2x_2 \\ x_2 \end{bmatrix}$ for which the output is $[1 \quad 3]\begin{bmatrix} x_1 + 2x_2 \\ x_2 \end{bmatrix} =$

$x_1 + 5x_2$. In particular, $\begin{bmatrix} 1 \\ 0 \end{bmatrix}$ yields the new output $y_1 = 1$.

Suppose, then, that we are given the data $y_0 = y_1 = 1$. Can we determine

that the initial state was $\begin{bmatrix} 1 \\ 0 \end{bmatrix}$? We must solve the equation

$$\begin{bmatrix} H \\ HF \end{bmatrix}\begin{bmatrix} x_1 \\ x_2 \end{bmatrix} = \begin{bmatrix} 1 \\ 1 \end{bmatrix}$$

But $\begin{bmatrix} H \\ HF \end{bmatrix} = \begin{bmatrix} 1 & 3 \\ 1 & 5 \end{bmatrix}$, and it is clear that

$$x_1 + 3x_2 = 1$$
$$x_1 + 5x_2 = 1$$

implies first that $x_2 = 0$, and second that $x_1 = 1$, as desired.   ◇

We shall return to the study of reachability, controllability, and observability in Chapter 7, after we develop the theory for solving the local transition equations of continuous-time systems in Chapter 5 and study some basic properties of linear systems in Chapter 6.

EXERCISES FOR SECTION 4–5

1.   Consider the parity checker system Σ of Figure 4–20.

(a)   Verify that we can write $\mathcal{S}_{x_2}(u_0, u_1, \ldots, u_{l-1}) = u_0 +_2 u_1 \cdots +_2 u_{l-1}$

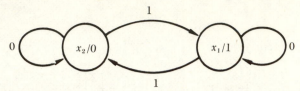

Figure 4–20 The parity checker of Example 4–3–2.

where $+_2$ denotes addition modulo 2, and that

$$\mathcal{S}_{x_1}(u_0, u_1, \ldots, u_{l-1}) = 1 +_2 \mathcal{S}_{x_2}(u_0, \ldots, u_{l-1})$$

(b) What is the zero state for this system? Notice that $X = \{x_1, x_2\}$ is not a vector space (unless we go out of our way to define addition and multiplication by scalars to make it one), so we cannot even speak about zero-input linearity. We can, however, consider zero-state linearity, since U^* and Y are vector spaces over the binary field $(\{0, 1\}, +_2, \cdot_2)$. Show that Σ is zero-state linear.

(c) Show that x_1 is not k-indistinguishable from x_2 for any k in **N**, so that Lemma 2 holds even though Σ is non-linear.

2. In the system of Figure 4–21, neither $U = \{1\}$ nor $X = \{x_0, x_1\}$ is a vector space (although $Y = \{0, 1\}$ can be), so linearity is not even a possible issue. Suppose

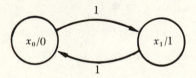

Figure 4–21 The system of Example 4–3–3.

that we make U into a vector space over $K = (\{0, 1\}, +_2, \cdot_2)$ by defining an addition $+'$ on U by the table

u \\ u	1
1	1

(that is, $1 +' 1 = 1$) and defining multiplication of elements of U by scalars from K by the table

k \\ u	1
0	1
1	1

(note that both operations are the only ones possible).

(a) What is 0_U? What is $\Omega = U^*$? What is 0_Ω? What is the zero state? What does $\mathcal{S}_{\hat{x}}(0^2)$ mean here?

(b) Give $\mathcal{S}_{0_X}(u_0, \ldots, u_{l-1})$, the zero-state response. Is the system zero-state linear?

(c) Show that x_0 and x_1 are not k-indistinguishable for any $k \in$ **N**.

✔ **3.** Consider the system of Example 4–3–4, modified in Figure 4–22 to show the

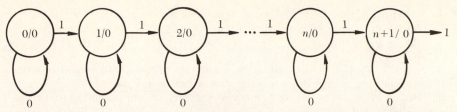

Figure 4–22 A non-linear, infinite-state system which is zero-state linear.

outputs. Here, $X = \mathbf{N}$ is not a vector space (unless really fancy footwork is employed to make it one), while $U = \{0, 1\}$ and $Y = \{0\}$.

(a)   Observe that every state is a zero state, and show that the system is zero-state linear even though it is not linear.

(b)   Show, for any pair of states $\bar{x}$ and $\hat{x}$, that $\mathcal{S}_{\bar{x}}(0^{\ell}) = \mathcal{S}_{\hat{x}}(0^{\ell})$ for all $\ell \in \mathbf{N}$, and thus that Lemma 2 may hold even when the system is not linear.

✔ **4.**   Consider the system Σ of Figure 4–23. Verify that $\mathcal{S}_{x_1}(0^{\ell}) = \mathcal{S}_{x_3}(0^{\ell})$ for all

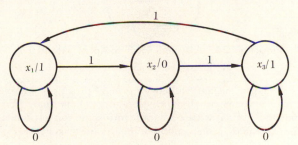

Figure 4–23 A non-linear system.

ℓ, but that x_1 *is* distinguishable from x_3, so that Lemma 2 might not hold if Σ is not linear.

5.   Prove that, if A_j is an $(m_j \times n)$ matrix, x is an $(n \times 1)$ matrix, and b_j is an $(m_j \times 1)$ matrix, then

$$\begin{matrix} A_1 x = b_1 \\ A_2 x = b_2 \\ \vdots \\ A_k x = b_k \end{matrix} \quad \Longleftrightarrow \quad \begin{bmatrix} A_1 \\ A_2 \\ \vdots \\ A_k \end{bmatrix} x = \begin{bmatrix} b_1 \\ b_2 \\ \vdots \\ b_k \end{bmatrix}$$

Can you give a vector space interpretation of this in terms of subspaces and direct sums?

[*Hint:* Both statements require the elements of x to satisfy the same linear equations.]

✔ **6.**   Let A_j be an $(n_j \times n)$ matrix for $j = 1, \ldots, k$, and let A_j^* be the adjoint of A_j. Prove that

$$\begin{bmatrix} A_1 \\ A_2 \\ \vdots \\ A_k \end{bmatrix}^{*} \quad = \quad [A_1^*, A_2^*, \ldots, A_k^*]$$

$$(n_1 + \cdots + n_k) \times n \qquad n \times (n_1 + \cdots + n_k)$$

✔ **7.** Let S_k^*, S_n^*, and S^* be defined as in Lemma 5A.

    (a) Prove that $\bar{x} \sim_k \hat{x}$ for all $k \geq 0 \Leftrightarrow (\bar{x} - \hat{x}) \in (S_{k+1}^*)^\perp$ for all $k \geq 0$.

    (b) Prove that $y \in (S_n^*)^\perp \Leftrightarrow y \in (S_k^*)^\perp$ for all $k \geq 0$. [*Hint:* $S_k^* \subset S_n^* \Rightarrow (S_n^*)^\perp \subset (S_k^*)^\perp$]

8.   **(a)** Prove that the dual of the connection of two linear systems in series is the connection of their duals in series. But in which order?

    (b) What is the wiring-diagram of the dual of the system given by the wiring-diagram of Figure 4–24?

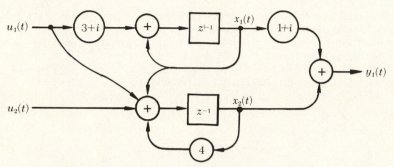

Figure 4–24 A system with complex parameters.

    (c) Can you provide, and validate, a general prescription for going from the wiring-diagram of a system to the wiring-diagram of its dual? [*Hint:* It may be necessary to put the original wiring-diagram into some sort of standard form.]

✔ **9.** Carry out the analysis of Example 6 for the linear systems (F, G, H), where:

    (a)
$$F = \begin{bmatrix} 1 & 2 & 0 \\ 3 & -1 & -2 \\ 1 & 0 & -1 \end{bmatrix}, H = \begin{bmatrix} 1 & 0 & 0 \\ 2 & 1 & 0 \end{bmatrix}, \text{and } \hat{x} = \begin{bmatrix} 1 \\ 0 \\ 2 \end{bmatrix}$$

    (b)
$$F = \begin{bmatrix} 1 & 2 & 0 \\ 3 & -2 & 0 \\ 1 & 2 & -1 \end{bmatrix}, H = \begin{bmatrix} 3 & 1 & 0 \end{bmatrix}, \text{and } \hat{x} = \begin{bmatrix} 1 \\ 0 \\ 2 \end{bmatrix}$$

Compute the *observability matrix* $\mathcal{O}$ in each case to confirm your results.

10. Is (F, G), where $F = \begin{bmatrix} 1 & 2 & 3 \\ 1 & 4 & 6 \\ 2 & 1 & 7 \end{bmatrix}$ and $G = \begin{bmatrix} 1 & 9 \\ 0 & 0 \\ 2 & 0 \end{bmatrix}$, controllable?

✔ **11.** Suppose that the constant linear systems (F, G, H) with state x and $(\hat{F}, \hat{G}, \hat{H})$ with state $\hat{x}$ are related by a constant similarity transformation matrix P, so that $x(t) = P\hat{x}(t)$.

    (a) Show that $\hat{F} = P^{-1}FP$, $\hat{G} = P^{-1}G$, and $\hat{H} = HP$.

    (b) Prove that (F, G, H) is reachable, controllable, or observable if and only if $(\hat{F}, \hat{G}, \hat{H})$ has the corresponding property.

    (c) How are the dual systems of (F, G, H) and $(\hat{F}, \hat{G}, \hat{H})$ related?

✔ **12.** Let $v_1, \ldots, v_n$ be any n vectors in the complex inner product space V. We outline a test for linear independence.

    (a) Take the inner product of $c_1 v_1 + c_2 v_2 + \cdots + c_n v_n = 0$ respectively with

$v_1, v_2, \ldots$, and v_n and write the resulting set of n equations as the matrix product

$$
W \begin{bmatrix} c_1 \\ \vdots \\ c_n \end{bmatrix} = \begin{bmatrix} 0 \\ \vdots \\ 0 \end{bmatrix}
$$

for an appropriate $n \times n$ matrix W (called the **Grammian**).

(b) State conditions on W for linear independence of $v_1, \ldots, v_n$.

(c) Note that this worked in any inner product space. Look now at a special case where $V = \mathbf{C}^m$ (each v_i is an m-tuple), and $\langle v_i | v_j \rangle = v_i^* v_j$. What does W look like? Can you write $W = AA^*$ for some $n \times m$ matrix A?

(d) Observe that if $n > m$ with $V = \mathbf{C}^m$, the vectors $v_1, \ldots, v_n$ are obviously dependent, so consider the more interesting situation where $n \le m$. Prove that $W = AA^*$ iff A is of rank n.

4–6   The Z-transform and Discrete-time Transfer Functions

In the last section, we saw that the zero-state response of the discrete-time linear system (F, G, H) is given by

$$
y_0(k) = \mathcal{S}_{0x}(u(0), u(1), \ldots, u(k-1)) = \begin{cases} \displaystyle\sum_{j=0}^{k-1} HF^{k-1-j}Gu(j) & \text{for } k > 0 \\ 0 & \text{for } k = 0 \end{cases} \tag{1a}
$$

This relationship shows how the infinite sequence of output values

$$
y_0 = (y_0(0), y_0(1), y_0(2), \ldots, y_0(k), \ldots) \tag{1b}
$$

is determined from the infinite input sequence

$$
u = (u(0), u(1), \ldots, u(k), \ldots) \tag{1c}
$$

by zero-state operation of our system.

The Z-transform is defined as an operation into which we plug a *sequence* f and get out an *infinite series*, which we will call $Z\{f\}$:

DEFINITION 1

If f is the sequence

$$
f = (f(0), f(1), f(2), \ldots, f(k), \ldots)
$$

then the **Z-transform** of f, denoted by $Z\{f\}$, is the formal† power series

---

† We say the power series is **formal** because we wish to emphasize that $Z\{f\}$ is the expression written out in (2a), and *not* a function obtained by substituting complex numbers for the indeterminate z. Thus questions of convergence of complex power series are *irrelevant* to the present discussion.

$$\mathbf{Z}\{f\} \overset{\Delta}{=} f(0) + f(1)z^{-1} + f(2)z^{-2} + \cdots \qquad \text{(2a)}$$

or, more compactly,

$$\mathbf{Z}\{f\} = \sum_{k=0}^{\infty} f(k)z^{-k} \qquad \text{(2b)}$$

We say that f is the **inverse Z-transform** of $\mathbf{Z}\{f\}$, and sometimes write
$f = \mathbf{Z}^{-1}\{\mathbf{Z}\{f\}\}$.   ○

Example 1

If

$$f = (1, 0, 0, 0, \ldots, 0, \ldots)$$
$$\underset{\text{0th place}}{\uparrow}$$

then $\mathbf{Z}\{f\} = 1 + 0z^{-1} + 0z^{-2} + \cdots = 1$. More generally, writing δ_m for
the function $k \mapsto \delta(k - m)$,† if

$$f = \delta_m \overset{\Delta}{=} (0, 0, \ldots, 0, 1, 0, 0, \ldots)$$
$$\qquad\quad \underset{\text{0th place}}{\uparrow} \qquad \underset{\text{mth place}}{\uparrow}$$

with the unit input occurring at time $k = m$, then

$$\mathbf{Z}\{\delta_m\} = 0 + 0z^{-1} + \cdots + 1z^{-m} + 0 + \cdots = z^{-m}. \qquad ◊$$

The reader who is familiar with the Laplace transform (for an exposition,
see Section 6–3) $\mathcal{L}\{f\}$ of a *continuous*-time function f given by the integral

$$\mathcal{L}\{f\} \overset{\Delta}{=} \int_0^{\infty} f(t)e^{-st}\,dt \qquad \text{(3)}$$

will be struck by the similarity of the two operations $\mathcal{L}$ and $\mathbf{Z}$, especially
if he replaces e^s by z.

Example 2

Let f be the sequence whose formula is $f(k) = r^k$ for $k \geq 0$. If we plug this
formula for $f(k)$ into the series (2b) and sum over the "dummy variable"
of summation k, we get

---

† In discrete-time theory, δ is used to denote the **Kronecker delta**, $\delta(k) = 0$ if $k \neq 0$, while
$\delta(0) = 1$. In continuous-time theory, δ is used to denote the unit impulse (also known as **Dirac's
delta "function,"** as described in Section 5–4).

$$Z\{f\} = \sum_{k=0}^{\infty} f(k)z^{-k} = \sum_{k=0}^{\infty} r^k z^{-k} = \sum_{k=0}^{\infty} \left(\frac{r}{z}\right)^k$$

This formal power series sums to the complex power series

$$\frac{1}{1 - \dfrac{r}{z}} \text{ for } \left|\frac{r}{z}\right| < 1$$

$$= \frac{z}{z - r} \text{ for } |r| < |z| \qquad \Diamond$$

The reader should have no trouble proving that the Z-transform is a linear operation:

THEOREM 1

If a and b are any scalars and v and w are any sequences, then

$$Z\{av + bw\} = aZ\{v\} + bZ\{w\} \qquad \Box$$

Let's clear up one pesky point about notation right away. Why do we use the variable z^{-1} in the defining series for the Z-transform? Might it not cause hopeless confusion with our previous use of z^{-1} to represent the unit-delay operator in block diagrams throughout this book?

Consider the following sequences and their transforms:

$$Z\{1, 0, 0, 0, \ldots)\} = 1 \tag{4a}$$

$$Z\{(0, 1, 0, 0, \ldots)\} = z^{-1} \tag{4b}$$

Clearly, the second sequence is exactly like the first sequence, only delayed by one unit of time, and its Z-transform is just z^{-1} times the Z-transform of the first sequence.

Example 3

The sequence $v = (0, 1, r, r^2, r^3, \ldots)$ has Z-transform

$$Z\{v\} = 0 + 1z^{-1} + rz^{-2} + r^2 z^{-3} + \cdots$$
$$= z^{-1}(1 + rz^{-1} + r^2 z^{-2} + \cdots)$$

From Example 2, we recognize that $1 + rz^{-1} + r^2 z^{-2} + \cdots = Z\{f\}$ where f is the sequence $f = (1, r, r^2, r^3, \ldots)$

Thus, $Z\{v\} = z^{-1}Z\{f\}$. But the sequence v is just f delayed by one time unit, so, sticking our necks out, we might use our delay operator notation to write $v = z^{-1}f$. Are we in trouble? If we just left the equation $v = z^{-1}f$ lying around, suppose that some novice found it and, not knowing that we meant z^{-1} to be an operation, mistook it for a number. If he then tried to take the Z-transform of v, he might use the linearity of Theorem 1 to "pull the number z^{-1} outside" and get

$$Z\{v\} = Z\{z^{-1}f\} = z^{-1}Z\{f\}$$

But no harm is done, since this is just what we got when we worked out $Z\{v\}$ and $Z\{f\}$ directly.   ◊

By the same token, we shall use z itself for the *left shift* operator (delay of -1 time units) for which $zf = (f(1), f(2), f(3), \ldots, f(m+1), \ldots)$.

$$\uparrow$$
$$m\text{th place}$$

It should now be clear that this coincidence is precisely the reason that we decided to use the symbol z^{-1} to stand for the unit-delay operator at the beginning of this book. We chose it so that it would fit with the vast body of literature on the Z-transform.

More generally, we have the translation theorem:

THEOREM 2

For any $m \geq 0$,

$$Z\{(0, 0, \ldots, 0, f(0), f(1), \ldots)\} = z^{-m}Z\{(f(0), f(1), \ldots)\}$$
$$\uparrow$$
$$m\text{th place}$$

that is, $Z\{z^{-m}f\} = z^{-m}Z\{f\}$, where the z on the left is the shift operator, and the z on the right is our transform variable.

Proof

$$Z\{z^{-m}f\} = 0 + 0z^{-1} + \cdots + 0z^{-(m-1)} + f(0)z^{-m} + f(1)z^{-(m+1)} + \cdots$$
$$= z^{-m}(f(0) + f(1)z^{-1} + \cdots)$$
$$= z^{-m}Z\{f\} \qquad \qquad \square$$

Analogous to the theorem on Laplace transforms that

$$\mathcal{L}\{\dot{f}(t)\} = s\mathcal{L}\{f(t)\} - f(0)$$

we have:

THEOREM 3

If $f = (f(0), f(1), f(2), \ldots)$ and $v = (f(1), f(2), f(3), \ldots) = zf$, then

$$\mathbb{Z}\{v\} = z\mathbb{Z}\{f\} - zf(0).$$

Proof

$$
\begin{aligned}
\mathbb{Z}\{v\} &= f(1) + f(2)z^{-1} + f(3)z^{-2} + \cdots \\
&= z^{+1}(f(0) + f(1)z^{-1} + f(2)z^{-2} + \cdots) - z^{+1}f(0) \\
&= z\mathbb{Z}\{f\} - zf(0)
\end{aligned}
$$

□

Just as we abuse notation in Laplace transforms letting $\dot{x}(t)$ and $x(t)$ stand for "the functions $\dot{x}$ and x whose values at time t are given by $\dot{x}(t)$ and $x(t)$," it is customary to let $f(k)$ stand for the sequence whose kth term is the value $f(k)$, and write Theorem 3 as

$$\mathbb{Z}\{f(k + 1)\} = z\mathbb{Z}\{f(k)\} - zf(0) \tag{5}$$

With this notation it is no trouble to prove the following result:

THEOREM 4

For any $m \geq 0$,

$$\mathbb{Z}\{f(k + m)\} = z^m\mathbb{Z}\{f(k)\} - (z^m f(0) + z^{m-1}f(1) + \cdots + zf(m - 1))$$

□

Together, Theorems 1, 3, and 4 are the reasons why the Z-transform is useful in solving linear difference equations with constant coefficients, such as our n-dimensional equations (F, G, H).

Let us now use the Z-transform to analyze discrete-time linear systems. Taking the Z-transforms of the sequence $y_0 = (y_0(0), y_0(1), \ldots)$ and using the formula for $y_0(k)$ given in equation (1), we get

$$\mathbb{Z}\{y_0\} = \sum_{k=0}^{\infty} y_0(k)z^{-k}$$

$$= \sum_{k=1}^{\infty} \left(\sum_{j=0}^{k-1} HF^{k-j-1}Gu(j) \right) z^{-k}$$

$$= \sum_{j=0}^{\infty} \sum_{k>j} z^{-1} \left[HF^{(k-j-1)}z^{-(k-j-1)}G \right] z^{-j}u(j)$$

$$= \sum_{j=0}^{\infty} z^{-1} \left(\sum_{l=0}^{\infty} HF^{l}z^{-l}G \right) u(j)z^{-j}$$

$$= z^{-1} \left(\sum_{l=0}^{\infty} HF^{l}Gz^{-l} \right) \sum_{j=0}^{\infty} u(j)z^{-j} \tag{6}$$

But $\sum_{j=0}^{\infty} u(j)z^{-j} \triangleq \mathbb{Z}\{u\}$ where $u = (u(0), u(1), \ldots.)$, and the expression

$\sum_{l=0}^{\infty} HF^{l}Gz^{-l}$ is an infinite series of matrices which can be written

$$\sum_{l=0}^{\infty} HF^{l}Gz^{-l} = H \left(\sum_{l=0}^{\infty} F^{l}z^{-l} \right) G = H(I + Fz^{-1} + F^{2}z^{-2} + \cdots)G$$

Reasoning by analogy with the scalar case in Example 2, where

$$1 + rz^{-1} + r^{2}z^{-2} + \cdots = z\frac{1}{z - r} = z(z - r)^{-1}$$

we might well guess that

$$I + Fz^{-1} + F^{2}z^{-2} + \cdots = z(zI - F)^{-1} \tag{7a}$$

holds for matrices.† Term-by-term multiplication shows that our guess is correct:

$$(zI - F)(I + Fz^{-1} + F^{2}z^{-2} + \cdots) = zI + F + F^{2}z^{-1} + \cdots$$
$$- F - F^{2}z^{-1} - \cdots$$
$$= zI \tag{7b}$$

and that

$$(zI - F)^{-1} = z^{-1}(I + Fz^{-1} + F^{2}z^{-2} + \cdots) \tag{7c}$$

Thus, we have explicitly summed the matrix series in equation 6, and we may write

$$\sum_{l=0}^{\infty} HF^{l}Gz^{-l} = zH(zI - F)^{-1}G \tag{7d}$$

If we now define the function $\mathfrak{H}(z)$ for any system (F, G, H) by the formula

$$\mathfrak{H}(z) \triangleq H(zI - F)^{-1}G \tag{8}$$

---

†We may regard this as true for *formal* matrix power series in the sense that (7b) holds purely formally. Of course, if we interpret z as a complex number, then just as the scalar series only made sense (converged) for large enough values of z (namely $|z| > |r|$), the infinite matrix series only converges for large enough values of z. In Section 5–3 we define, for a matrix F, its norm $\|F\|$ which measures the size of F, and with this notation we can show that the infinite matrix series (7a) converges for all z such that $|z| > \|F\|$.

equation 6 can be written in the form

$$Z\{y_0\} = \mathcal{H}(z)Z\{u\} \tag{9}$$

We shall give an alternate derivation of the validity of (7d) for $\mathcal{H}(z)$ below. In any case, we can say from equation (9) that the Z-transform of the output is equal to the Z-transform of the input times the function $\mathcal{H}(z)$. Symbolically, we may divide both sides of (9) by the series $Z\{u\}$ to get

$$\mathcal{H}(z) = H(zI - F)^{-1}G = \frac{Z\{y_0\}}{Z\{u\}} = \frac{\text{output transform}}{\text{input transform}} \tag{10}$$

and call $\mathcal{H}(z)$ the **discrete-time transfer function** of (F, G, H). From equation (8) we can compute $\mathcal{H}(z)$ once and for all for any given (F, G, H); then, if someone gives us an input sequence u, we can find the zero-state output sequence y_0 by taking the Z-transform $Z\{u\}$ of u and multiplying it by $\mathcal{H}(z)$ to get $Z\{y_0\}$ and then taking the inverse Z-transform.

Example 4

Consider a unit-delay device in the time domain, as in Figure 4–25(a). What is its transfer function? Recalling Theorem 3, we have

$$Z\{x(n + 1)\} = zZ\{x(n)\} - zx(0)$$

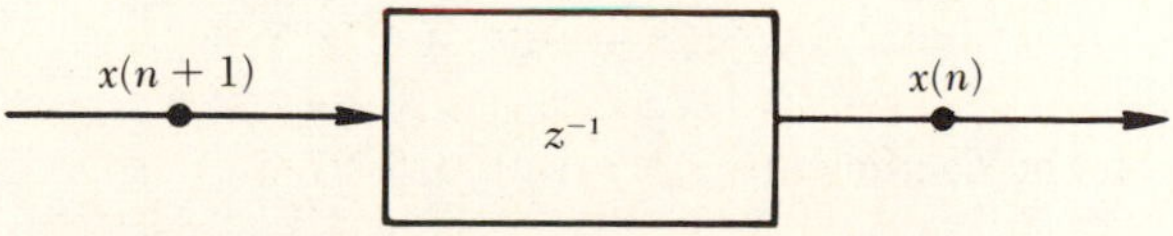

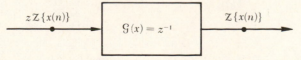

Figure 4–25(a)   A unit-delay in the time domain.

Setting $x(0) = 0$ for zero-state response and dividing output by input transforms gives

$$\mathcal{G}(z) = \frac{Z\{x(n)\}}{Z\{x(n + 1)\}} = \frac{1}{z} = z^{-1}$$

Thus, we say that a unit of delay in the time domain corresponds to multiplication by z^{-1} in the Z-transform domain, as shown in Figure 4–25(b),

Figure 4–25(b)   Block diagram showing transfer function of a unit-delay device.

and we see that we have been labelling the unit-delay block with its transfer function z^{-1} all along.   ◊

Later, in Section 6–4, we will draw the block diagram of a system, label each individual block with its transfer function, and, by simple manipulations, compute the overall transfer function of the whole system. The methods we shall present there are equally applicable to continuous-time and discrete-time systems.

Once we have calculated $Z\{y_0\}$ from the transfer function $\mathfrak{H}(z)$ and the input transform $Z\{u\}$, how do we "inverse-transform" to find the sequence y_0 from its Z-transform $Z\{y_0\}$? There are several ways, ranging from looking in tables of Z-transform pairs (such as the one in Appendix A-2) to equating infinite series term-by-term.

Example 5

Suppose that $Z\{y_0\} = \dfrac{1}{z - 3}$. Then

$$\frac{1}{z - 3} = \frac{1}{z\left[1 - \left(\dfrac{3}{z}\right)\right]} = z^{-1}\left[\frac{1}{1 - \left(\dfrac{3}{z}\right)}\right]$$

$$= z^{-1}\left[1 + \left(\frac{3}{z}\right) + \left(\frac{3}{z}\right)^2 + \left(\frac{3}{z}\right)^3 + \cdots\right]$$

$$= z^{-1} + 3z^{-2} + 9z^{-3} + 27z^{-4} + \cdots$$

So, reading off the coefficients, $y_0 = (0, 1, 3, 9, 27, \ldots)$.   ◊

We can also use partial fractions:

Example 6

If $Z\{y_0\} = \dfrac{3z - 7}{z^2 - 4z + 3}$, we may write

$$\frac{3z - 7}{z^2 - 4z + 3} = \frac{3z - 7}{(z - 1)(z - 3)} = \frac{2}{z - 1} + \frac{1}{z - 3}$$

Now $\dfrac{1}{z - 1} = z^{-1} + z^{-2} + z^{-3} + \cdots = Z\{(0, 1, 1, 1, \ldots.)\}$

so $\dfrac{2}{z - 1} = Z\{(0, 2, 2, 2, \ldots)\}$ (linearity) and $\dfrac{1}{z-3} = \mathbf{Z}\{(0, 1, 3, 9, 27, \ldots .)\}$ (by Example 5)

and, using linearity again, the inverse transform y_0 is given by

$$y_0 = (0, 2, 2, 2, \ldots) + (0, 1, 3, 9, \ldots .)$$
$$= (0, 3, 5, 11, 29, \ldots .)$$

as the reader may check in Exercise 2.   ◊

It is interesting to see what happens in equation (9) if we let the input sequence be the unit pulse $u = \delta = (1, 0, 0, 0, \ldots .)$. For this input, $\mathbf{Z}\{u\} = 1$ and $\mathbf{Z}\{y_0\} = \mathcal{B}(z) \cdot 1$.

Thus we have the characterization:

THEOREM 5

The discrete-time transfer function $\mathcal{B}(z)$ of a constant linear system is the Z-transform of the zero-state response to the unit-pulse input:

$$\mathcal{B}(z) = \mathbf{Z}\{\mathcal{S}_{0_x}(\delta)\}$$ □

This just says that if we know what the output of (F, G, H) is for a unit-pulse input, we can tell what it will be for any input $(u(0), u(1), u(2), \ldots .)$. This is not too surprising, since $(u(0), u(1), u(2), \ldots) = u(0)(1, 0, 0, \ldots) + u(1)(0, 1, 0, \ldots .) + u(2)(0, 0, 1, 0, \ldots) + \cdots$, and by linearity (Theorem 1) and shifting (Theorem 2) we can add up the pieces.

Back in Section 4–5, we used g to stand for the zero-state unit-pulse response

$$g(k) = \mathcal{S}_{0_x}(\delta) = \begin{cases} HF^{k-1}G, & k > 0 \\ 0 & , k = 0 \end{cases}$$ **(11)**

and the zero-state response was given by the convolution summation

$$y_0(k) = \mathcal{S}_{0_x}(u(0), u(1), u(2), \ldots .) = \sum_{j=0}^{k-1} g(k-j)u(j)$$ **(12)**

Using the expression of equation (12) for y_0 in equation (9), we can write

$$\mathbf{Z}\left\{\sum_{j=0}^{k-1} g(k-j)u(j)\right\} = \mathcal{B}(z)\mathbf{Z}\{u(k)\}.$$ But we can also express $\mathcal{B}(z)$ in terms of

$g(k)$ by substituting $g(k)$ for $\mathcal{S}_{0_x}(\delta)$ in Theorem 5 to get $\mathcal{B}(z) = \mathbf{Z}\{g(k)\}$. Combining these last two relationships, we have proved:

THEOREM 6

The discrete-time transfer function $\mathcal{G}(z)$ and the unit-pulse response $g(k)$ are related by

$$\mathcal{G}(z) = \mathbb{Z}\{g(k)\}$$

Moreover,

$$\mathbb{Z}\left\{\sum_{j=0}^{k-1} g(k-j)u(j)\right\} = \mathbb{Z}\{g(k)\} \cdot \mathbb{Z}\{u(k)\}$$

In other words, the $\mathbb{Z}$-transform of the convolution of two sequences is the product of their $\mathbb{Z}$-transforms. $\qquad\square$

As a check, suppose we use equation (11) to compute $\mathbb{Z}\{g(k)\}$ directly:

$$\mathbb{Z}\{g(k)\} = \sum_{j=0}^{\infty} g(j)z^{-j} = 0 + \sum_{j=1}^{\infty} HF^{j-1}Gz^{-j}$$

Changing the dummy variable of summation to ℓ by the transformation $\ell = j - 1$,

$$\mathbb{Z}\{g(k)\} = \sum_{\ell=0}^{\infty} HF^{\ell}Gz^{-(\ell+1)} = z^{-1}\left(\sum_{\ell=0}^{\infty} HF^{\ell}Gz^{-\ell}\right)$$

which is just what we used as the basis for defining $\mathcal{G}(z)$ in equation (8).

It only remains for us now to consider $\mathbb{Z}$-transforming the dynamic state equations

$$x(k+1) = Fx(k) + Gu(k)$$
$$y(k) = Hx(k)$$

directly. By Theorem 3 and Theorem 1,

$$\mathbb{Z}\{x(k+1)\} = \mathbb{Z}\{Fx(k) + Gu(k)\}$$

yields

$$z\mathbb{Z}\{x(k)\} - zx(0) = F\mathbb{Z}\{x(k)\} + G\mathbb{Z}\{u(k)\}$$

Solving for $\mathbb{Z}\{x(k)\}$,

$$(zI - F)\mathbb{Z}\{x(k)\} = zx(0) + G\mathbb{Z}\{u(k)\} \qquad\qquad \textbf{(13)}$$

whence

$$\mathbb{Z}\{x(k)\} = (zI - F)^{-1}zx(0) + (zI - F)^{-1}G\mathbb{Z}\{u(k)\} \tag{14}$$

Since $y(k) = \mathcal{S}_{x(0)}(u(0), u(1), \ldots) = Hx(k)$, we have

$$\underbrace{\mathbb{Z}\{y(k)\}}_{} = \underbrace{H(zI - F)^{-1}zx(0)}_{} + \underbrace{H(zI - F)^{-1}G\mathbb{Z}\{u(k)\}}_{} \tag{15}$$

$$\mathbb{Z}\left\{\begin{matrix}\text{complete}\\\text{response}\end{matrix}\right\} = \mathbb{Z}\left\{\begin{matrix}\text{zero-input}\\\text{response}\end{matrix}\right\} + \mathbb{Z}\left\{\begin{matrix}\text{zero-state}\\\text{response}\end{matrix}\right\} \tag{16}$$

Thus, equating equation (9) and the zero-state response term of equation (16), we get

$$\mathcal{G}(z)\mathbb{Z}\{u(k)\} = H(zI - F)^{-1}G\mathbb{Z}\{u(k)\}$$

and since this must hold for all input sequences $u(\cdot)$, we get an independent confirmation of our defining equality $\mathcal{G}(z) = H(zI - F)^{-1}G$.

The matrix $(zI - F)^{-1}$ is called the **resolvent matrix** of F and yields a well-defined matrix over $\mathbf{C}$ for any z in $\mathbf{C}$ such that $\det(zI - F) \neq 0$. We call the nth degree polynomial χ_F defined by the determinant,

$$\chi_F(z) \overset{\Delta}{=} \det(zI - F)$$

the **characteristic polynomial of** F, and there are n values of z for which it is zero. We call these roots of $\chi_F(z)$ for which $(zI - F)$ is singular the **characteristic values** or **eigenvalues** of F, and we call the set of all eigenvalues the **spectrum** of F.

We shall see a great deal more of eigenvalues and resolvent matrices when we study continuous-time linear systems in Chapter 6 and find that the continuous-time transfer function $\mathcal{G}(s)$ is the Laplace transform of the impulse response $g(t)$ and that $\mathcal{G}(s) = H(sI - F)^{-1}G$. Later, in Chapters 7 and 8, we will do quite a few tricks with the expression

$$H(sI - F)^{-1}G$$

and everything that we do there with an s for continuous-time systems, you can translate into a corresponding fact about discrete-time systems by replacing s by z.

Before launching our study of continuous-time systems, we pause for a moment to derive one of the most famous theorems of linear algebra.

THE CAYLEY-HAMILTON THEOREM

We recall from the theory of determinants that for an $(n \times n)$ matrix B,

$$B \cdot \text{adj } B = \det B \cdot I \tag{17a}$$

where adj B denotes the **adjugate matrix** of B, consisting of the various cofactors of B, and det B is the determinant. Equation (17a) is just *Cramer's rule* of linear algebra, since, when det $B \neq 0$, it can be written

$$B \cdot \frac{\text{adj } B}{\text{det } B} = I \tag{17b}$$

yielding the expression

$$B^{-1} = \frac{\text{adj } B}{\text{det } B} \tag{17c}$$

for the inverse matrix B^{-1} of B.

If we let B be the matrix $(zI - A)$, we get

$$(zI - A) \cdot \text{adj}(zI - A) = \det(zI - A) \cdot I$$

or

$$(zI - A) \cdot \text{adj}(zI - A) = \chi_A(z)I \tag{18}$$

since $\det(zI - A)$ is just the characteristic polynomial χ_A of A.

If we set

$$\text{adj}(zI - A) = \sum_{i=0}^{n-1} R_i z^i \tag{19}$$

where $R_0, R_1, \ldots, R_{n-1}$ are the constant matrix coefficients of the $(n-1)$ degree *matrix polynomial* adj$(zI - A)$, and note the commutativity

$$(zI - A)\,\text{adj}(zI - A) = \text{adj}(zI - A)\,(zI - A) \tag{20}$$

we can subtract $z\,\text{adj}(zI - A) = (zI)\,\text{adj}(zI - A) = \text{adj}(zI - A)(zI)$ from both sides to deduce that

$$A \left(\sum_{i=0}^{n-1} R_i z^i \right) = \left(\sum_{i=0}^{n-1} R_i z^i \right) A \tag{21a}$$

or

$$\sum_{i=0}^{n-1} (AR_i)z^i = \sum_{i=0}^{n-1} (R_i A)z^i \tag{21b}$$

Equating coefficients term by term in equation (21b), we see that

$$AR_i = R_i A, \text{ for } i = 0, 1, \ldots, n-1 \tag{21c}$$

and so A commutes with all the matrix coefficients of the polynomial adj$(zI - A)$. Thanks to this commutativity, we can substitute A for the variable z in the equality

$$(zI - A) \sum_{i=0}^{n-1} R_i z^i = \chi_A(z) \cdot I$$

rearranging terms with impunity to obtain

$$(AI - A) \sum_{i=0}^{n-1} R_i A^i = \chi_A(A) \cdot I$$

Since $AI = A$, this tells us that

$$0 = \chi_A(A)I$$

and we have proved the famous result:

THEOREM 7 (CAYLEY-HAMILTON)

Let A be any $n \times n$ matrix, with characteristic polynomial $\chi_A(z) = \det(zI - A) = z^n + \alpha_1 z^{n-1} + \cdots + \alpha_{n-1} z + \alpha_n$. Then

$$\chi_A(A) = A^n + \alpha_1 A^{n-1} + \cdots + \alpha_{n-1} A + \alpha_n I = 0$$

i.e., in a formal sense every matrix is a "zero" of its own characteristic polynomial.   □

In Section 6–5, Exercise 1, we will see a simpler proof of this theorem in a form which develops the coefficient matrices $R_0, R_1, \ldots, R_{n-1}$ and yields a recursive procedure for evaluating the matrix adj$(zI - A)$ without actually having to compute all those determinants for the cofactors; [See Exercise 7].

We could have used the Cayley-Hamilton theorem in Section 4–3 to show that the range of the matrix $[G, FG, F^2G, \ldots, F^{k-1}G]$ does not just keep on increasing as we add more columns to it by letting k increase; in fact, $F^l G$ can be written as a linear combination of $F^0G, FG, \ldots, F^{n-1}G$ for all integers l.

Example 7

For the system (F, G, H) shown in Figure 4–26, we have

$$F = \begin{bmatrix} 0 & 1 \\ 2 & 1 \end{bmatrix}, G = \begin{bmatrix} 0 \\ 1 \end{bmatrix}, H = \begin{bmatrix} 1 & a \end{bmatrix}$$

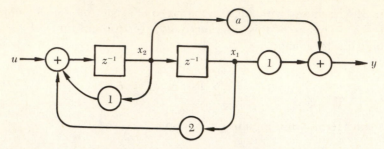

Figure 4-26   A discrete-time system.

The characteristic equation of F is

$$\chi_F(z) = \det \begin{bmatrix} z & -1 \\ -2 & z-1 \end{bmatrix} = z^2 - z - 2$$

so by Cayley-Hamilton, $F^2 - F - 2I = 0$, as the reader should verify by adding F^2, $-F$, and $-2I$ together. Thus,

$$F^2 = 2I + F$$

and so we can compute all powers of F in terms of I and F:

$$F^3 = FF^2 = F(2I + F) = 2F + F^2 = 2F + (2I + F) = 2I + 3F,$$

$$F^4 = FF^3 = F(2I + 3F) = 2F + 3F^2 = 2F + 3(2I + F) = 6I + 5F,$$

$$\dots$$

Thus, for all $k \geq 1$,

$$F^k = \alpha_0(k)I + \alpha_1(k)F$$

where $\alpha_0(k)$ and $\alpha_1(k)$ are numbers which depend upon k, so

$$F^k G = \alpha_0(k)G + \alpha_1(k)FG$$

We see that the columns of $F^k G$ are in the column space of $\mathcal{C} = [FG, G]$ even for $k \geq n = 2$.   ◊

In the next three chapters we shall use our experience with discrete-time systems to develop the theory of continuous-time systems. As you work through Chapters 5, 6, and 7 for continuous-time systems, where the mathematics is more complicated, try to ask yourself how each result developed there would look for the discrete-time case.

EXERCISES FOR SECTION 4–6

1.   Prove Theorem 4.

✔ **2.**   Check the inverse Z-transformation of Example 6 in two ways:

(a)   Write $\dfrac{3z-7}{z^2-4z+3} = 3z\left(\dfrac{1}{z^2-4z+3}\right) - 7\left(\dfrac{1}{z^2-4z+3}\right)$ and ex-

pand $\dfrac{1}{z^2-4z+3}$ by partial fractions.

(b)   Use long division

$$
\begin{array}{r}
3z^{-1} + 5z^{-2} + \cdots \\
z^2 - 4z + 3\overline{)\,3z - 7} \\
3z - 12z^0 + 9z^{-1} \\
\hline
5 \quad - \quad 9z^{-1} \\
5 \quad - 20z^{-1} + 15z^{-2} \\
\hline
11z^{-1} - 15z^{-2}
\end{array}
$$

to generate an infinite power series in (z^{-1}) and identify terms.

✔ **3.**   **(a)**   Find $\mathcal{B}(z)$, the transfer function, for the system of Figure 4–27(a). [*Hint:* Put a unit pulse δ in at $t = 0$ and chase signals around to get the unit pulse response sequence; then Z-transform to get

$$
\mathcal{B}(z) = \frac{1}{z-a}.
$$

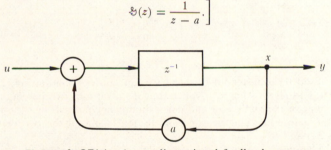

Figure 4–27(a)   A one-dimensional feedback system.

(b)   Repeat part (a) for the system of Figure 4–27(b).

$$
\left[\text{Answer: } \mathcal{B}(z) = \frac{bc}{z-a}.\right]
$$

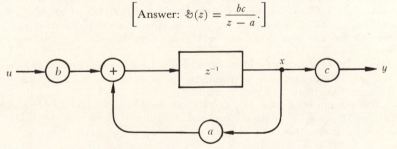

Figure 4–27(b)   A related system.

4. Find the unit-pulse response g and $\mathcal{L}(z)$ for the systems of Exercises 7 and 10 of Section 2–2.

✔ **5.** **(a)** Find the unit-pulse response g for Exercise 9 of Section 2–2 and compute the zero-state response to the input string $w = (0, 0, 2, 5, 2, 0, 0, \ldots)$.

(b) Compute the Z-transform $\mathbb{Z}\{g\}$ for the g of part (a), and also $\mathbb{Z}\{w\}$ and $\mathbb{Z}\{y_0\}$. Does

$$\mathbb{Z}\{y_0\} = \mathbb{Z}\{g\}\mathbb{Z}\{u\}?$$

Why not?

✔ **6.** Compute $\mathcal{L}(z)$ by equation (8) and by each of Theorems 5 and 6 for the systems (F, G, H) of:

(a) Examples 1 and 3 of Section 4–5. Find the values of z for which $(zI - F)$ is not invertible (the eigenvalues).

(b) Examples 6 and 7 of Section 4–3. Find the eigenvalues of F and relate them to "time constants."

Check your work by using the convolution equation (12).

✔ **7.** Let $F = \begin{bmatrix} 0 & 1 \\ 2 & 1 \end{bmatrix}$, $G = \begin{bmatrix} 0 \\ 1 \end{bmatrix}$, $H = [1 \quad a]$ as in Example 7 and Figure 4–26.

(a) From the characteristic polynomial $\chi_F(z)$, find the eigenvalues λ_1 and λ_2 for which

$$\chi_F(\lambda_1) = \chi_F(\lambda_2) = 0$$

(b) Compute the matrices $(zI - F)$, $\mathrm{adj}(zI - F)$, and

$$(zI - F)^{-1} = \frac{\mathrm{adj}(zI - F)}{\det(zI - F)}$$

for $z \neq \lambda_1, \lambda_2$.

(c) Find the transfer function $\mathcal{L}(z)$ by successively computing $(zI - F)^{-1}$, $(zI - F)^{-1}G$, and $H(zI - F)^{-1}G$. [Notice that it's a ratio of two polynomials in z.]

(d) Factor the numerator and denominator polynomials of $\mathcal{L}(z)$. For which values of the gain a can you cause a cancellation of one of the two denominator terms $(z - \lambda_1)$ and $(z - \lambda_2)$?

(e) Compute $\mathcal{C}$ and $\mathcal{O}$, the controllability and observability matrices of (F, G, H), and see which values of a affect controllability and observability. Compare with part (d).

(f) Suppose you let $u = Kx$ where $K = [k_1, k_2]$. (We say you have introduced **linear state feedback**). Draw the branches with gains k_1 and k_2 in on the block diagram of (F, G, H). Write the new matrices $(F + GK, G, H)$ which describe the "closed loop" feedback system, and compute its characteristic polynomial $\chi_{F+GK}(z)$. Choose K so that the closed loop system will have eigenvalues at $z = -1$ and $z = -1$. [*Answer:* $K = [-3 \quad -6]$.]

(g) Try to repeat part (f) on the system of Example 6 of Section 4–3, letting $\alpha = -1$ and $\beta = 2$. Can you determine the gains k_1 and k_2 by inspection of the block diagram for the closed loop system?

EE **8.** This exercise is for those who know about delta functions (unit impulses) and Laplace transforms. (The necessary material is contained in Sections 5–4 and 6–3.) For a given sequence $u = (u(0), u(1), u(2), \ldots)$, consider the time function

$$f(t) = u(0)\delta(t) + u(1)\delta(t - 1) + u(2)\delta(t - 2) + \cdots$$

where δ is now the continuous-time unit impulse, or Dirac delta "function." Thus, f is a "weighted train" of impulses, where the unit impulse occurring at time $t = a$ is multiplied by the value $u(a)$.

   (a)   Take the Laplace transform of $f(t)$ and get an infinite series of powers of the term e^{-s}.

   (b)   Replace e^s by z in the Laplace transform of the impulse-train f and get a series of powers of (z^{-1}). Compare it with $Z\{u\}$. This correspondence is why many sampled-data systems analysts consider the Z-transform to be a special case of the Laplace transform, and regard the sequence $(u(0), u(T), u(2T) \ldots)$ as having been gotten from some continuous-time waveform $u(t)$ by **modulating** (multiplying) the

impulse train $\displaystyle\sum_{i=0}^{\infty} \delta(t - iT)$ by $u(t)$. By the sampling property of the impulse,

$$u(t) \sum_{i=0}^{\infty} \delta(t - iT) = \sum_{i=0}^{\infty} u(t)\delta(t - iT) = \sum_{i=0}^{\infty} u(iT)\delta(t - iT)$$

   (c)   Illustrate parts (a) and (b) for the function $u(t) = e^{\lambda t}$ for $t \geq 0$. Find the Laplace transform of $u(t)$ and of the impulse train of period T modulated by $u(t)$.

   (d)   Do you see why systems engineers say that "poles in the left half of the s-plane correspond to poles inside the unit circle of the z-plane"?

✔ **9.**   **(a)**   For the two-dimensional system of Example 7 we saw that for all $k \geq 1$,

$$F^k = \alpha_0(k)I + \alpha_1(k)F$$

for some appropriate scalar functions $\alpha_0(\cdot)$ and $\alpha_1(\cdot)$. Find the particular functions $\alpha_0(\cdot)$ and $\alpha_1(\cdot)$ for that system.

   (b)   Generalize this result to $n \times n$ matrices F, and give an expression for F^k in terms of $I, F, F^2, \ldots, F^{n-1}$. Explain how you would actually determine the scalar functions $\alpha_0(\cdot), \ldots, \alpha_{n-1}(\cdot)$.

CHAPTER 5

Global Behavior of Continuous-Time Systems

Given the differential equations

$$\left.\begin{array}{l} \dot{x}(t) = F(t)x(t) + G(t)u(t) \\ y(t) = H(t)x(t) \end{array}\right\} \tag{1}$$

for a time-varying, continuous-time linear system, its state $x(t_0)$ at time t_0, and the admissible input function u in Ω which is applied to the system (so that $u(t)$ is the input at each time t in the time interval $[t_0, t_1]$ of interest), we wish to be able to compute both

$$\phi(t_1, t_0, x(t_0), u) = \text{the state at time } t_1 \tag{2}$$

and

$$\mathcal{S}(t_1, t_0, x(t_0), u) = \text{the output at time } t_1 \tag{3}$$

In other words, we wish to *integrate* the dynamic equations (1) to obtain an explicit formula for (2). It is then straightforward to obtain the response function $\mathcal{S}$ from the formula

$$\mathcal{S}(t_1, t_0, x(t_0), u) = H(t_1)\phi(t_1, t_0, x(t_0), u) \tag{4}$$

We start, in Section 5–1, by seeing how to integrate the discrete-time dynamics

$$x(t + 1) = F(t)x(t) + G(t)u(t)$$

288

to *motivate,* but not rigorously derive, the general solution

$$\phi(t_1, t_0, x(t_0), u) = \Phi(t_1, t_0)x(t_0) + \int_{t_0}^{t_1} \Phi(t_1, \tau)G(\tau)u(\tau)\,d\tau \tag{5}$$

where $\Phi(t, \tau)$ is a suitable matrix function of τ for each t_1.

Then, in Section 5–2, we forsake linearity for a while, and see how our study of Banach spaces, and in particular the Contraction Lemma of Section 2–4, lets us establish conditions under which a differential equation of the form

$$\dot{x}(t) = f(t, x(t), u(t)), \quad t_1 \leq t \leq t_2$$

with $x(t_0)$ fixed, $t_0 \in [t_1, t_2]$, has a unique solution.

In Section 5–3 we shall show that these conditions hold for linear systems as long as $F(t)$ is uniformly bounded, in a sense we shall make precise, over every finite interval. On this basis, we can define the *state transition matrix* $\Phi(t_1, \tau)$ appearing in equation (5), and then give a rigorous proof for the system (1), noting in particular that $\Phi(t_1, \tau) = e^{F(t_1 - \tau)}$, when F is constant, if we make an appropriate definition of the matrix exponential.

Combining equations (4) and (5), we note in Section 5–4 that the zero-state response of the linear system (1) is given by the equation

$$\mathcal{S}(t_1, t_0, 0, u) = \int_{t_0}^{t_1} g(t_1, \tau)u(\tau)\,d\tau \tag{6}$$

where

$$g(t_1, \tau) = H(t_1)\Phi(t_1, \tau)G(\tau) \tag{7}$$

We explain why g is called the *impulse response* of the system (1), briefly providing some insight into the theory of distributions on the way.

Finally, in Section 5–5 we analyze equation (7) to describe situations in which two linear systems Σ and Σ' are zero-state equivalent, in that

$$\mathcal{S}_\Sigma(t_1, t_0, 0, u) = \mathcal{S}_{\Sigma'}(t_1, t_0, 0, u)$$

for all admissible input functions u in Ω.

5–1   From Discrete-Time to Continuous-Time for Time-Varying Systems

In the equations for a discrete-time linear system

$$\begin{aligned} x(t + 1) &= F(t)x(t) + G(t)u(t) \\ y(t) &= H(t)x(t) \end{aligned} \tag{1}$$

we may view $F(t)$ as updating the old state $x(t)$ to give its contribution to the new state $x(t + 1)$, whereas $G(t)$ "reads in" an input contribution to be added to that made by the old state. $H(t)$ simply "reads out" the present state to yield the present output. In particular, if we set $u(t) = 0$, we see that F tells us *how much* the state changes over one moment of time when the input is held to zero. In the continuous-time case, $\dot{x}(t) = F(t)x(t) + G(t)u(t)$, we see that F tells us *the rate* at which the state changes when the input is held to zero.

Let us now see, by repeated application of the equations (1), how a state at time t changes under the application of a sequence of inputs $u(t)$, $u(t + 1), \ldots, u(t + n - 1)$ to yield the new state $x(t + n)$ at time $t + n$. This will extend the analysis for $F(t) = F$, $G(t) = G$ made in Section 4–3. Input $u(t)$ at time t changes state $x(t)$ to state

$$x(t + 1) = F(t)x(t) + G(t)u(t)$$

Successive inputs act on the system according to similar equations, until input $u(t + n - 1)$ at time $t + n - 1$ changes state $x(t + n - 1)$ to state

$$x(t + n) = F(t + n - 1)x(t + n - 1) + G(t + n - 1)u(t + n - 1)$$

To combine the above formulas into a pleasingly simple form, let us first consider what happens in the case in which all the inputs are zero; in other words, $u(t) = u(t + 1) = \cdots = u(t + n - 1) = 0$. Repeatedly applying the above equalities in this case, and using the notation $x^{\circ}(t + j)$ to denote the state of the system at time $t + j$ when only zero inputs have been applied starting at time t, we get the following equations:

$$x^{\circ}(t) = x(t)$$
$$x^{\circ}(t + 1) = F(t)x^{\circ}(t)$$
$$x^{\circ}(t + 2) = F(t + 1)x^{\circ}(t + 1) = F(t + 1)F(t)x^{\circ}(t)$$
$$\vdots \qquad \qquad \vdots$$
$$x^{\circ}(t + n) = F(t + n - 1)x^{\circ}(t + n - 1) = F(t + n - 1)\ldots F(t + 1)F(t)x^{\circ}(t)$$

To simplify our notation, let us introduce $\Phi(t_1, t_0): X \to X$ which, for each fixed pair of times $t_1 \geq t_0$, is a linear transformation of the state space into itself. We define Φ by the equations $\Phi(t_0, t_0) = I$, the identify transformation, and then, proceeding by induction, $\Phi(t_1 + 1, t_0) = F(t_1)\Phi(t_1, t_0)$ for $t_1 \geq t_0$.

Our equations above for the change of state of the system under zero input then have the very pleasing form

$$x^{\circ}(t_1) = \Phi(t_1, t_0)x^{\circ}(t_0) \tag{2}$$

and we thus call Φ the **state transition matrix** of the system; $\Phi(t_1, t_0): X \to X$

is the matrix which takes the state of a system at time t_0 and yields the state to which the system will move *under zero input* by time t_1.

By returning to

$$x(t + 1) = F(t)x(t) + G(t)u(t)$$
$$y(t) = H(t)x(t)$$

we see that the input contribution $u(t + j - 1)$ is read in by the matrix $G(t + j - 1)$, and first makes a contribution to the state at time $t + j$. If we now want to see what contribution it makes at time $t + k$, we must use the state transition matrix $\Phi(t + k, t + j)$ to update it from time $t + j$ to time $t + k$. This motivates the following general formula:

$$x(t + k) = \Phi(t + k, t)x(t) + \sum_{j=1}^{k} \Phi(t + k, t + j)G(t + j - 1)u(t + j - 1) \quad \text{(3)}$$

Its meaning is clear—it is the sum of $k + 1$ terms, the first of which reflects the updating of the original state from time t to time $t + k$, while each of the other terms reflects the contribution of an input, presented to the system at time $t + j - 1$, read in for the first time as a contribution to the state at time $t + j$, and then updated from time $t + j$ until time $t + k$.

To prove (3) formally, note first that it holds for $k = 0$, from

$$x(t) = \Phi(t, t)x(t) + \sum_{j=1}^{0} \Phi(t, t + j)G(t + j - 1)u(t + j - 1)$$

from $\Phi(t, t) = I$, and from the common computer programming convention that a summation whose lower index is higher than its upper index is zero.

The general validity of the formula follows by induction. Given that (3) holds for any t and $t + k$, we check that it holds for t and $t + k + 1$ as follows:

$$x(t + k + 1) = F(t + k)x(t + k) + G(t + k)u(t + k)$$
$$= F(t + k)\Phi(t + k, t)x(t)$$

$$+ \sum_{j=1}^{k} F(t + k)\Phi(t + k, t + j)G(t + j - 1)u(t + j - 1) + G(t + k)u(t + k)$$

$$= \Phi(t + k + 1, t)x(t) + \sum_{j=1}^{k+1} \Phi(t + k + 1, t + j)G(t + j - 1)u(t + j - 1)$$

With this, we have the basis and the induction step, and so we are assured of the validity of (3) for all $k \geq 0$.

Note that in the case of a time-invariant system in which $F(t)$ equals a fixed F for all t, the state transition matrix Φ reduces to the simple form

$$\Phi(t_1, t_0) = F^{(t_1 - t_0)} \text{ for all } t_1 \geq t_0$$

[where we adopt the usual convention, modelled on the case that the 0th power of a real number is always one, that $F^0 = I$, the identity transformation]. Thus, in the constant case our general equation (3) takes the simple form,

$$x(t + k) = F^k x(t) + \sum_{j=1}^{k} F^{k-j} Gu(t + j - 1) \tag{4}$$

which is familiar from Section 4–3. Whether our discrete-time *linear* system is time-varying or constant, we get the following interesting decomposition:

$$\phi(t_1, t_0, x, u) = \underbrace{\phi(t_1, t_0, x, 0_\Omega)}_{\substack{\text{the zero-input} \\ \text{state trajectory}}} + \underbrace{\phi(t_1, t_0, 0_X, u)}_{\substack{\text{the zero-initial-state} \\ \text{state trajectory}}}$$

In other words, the effect of the initial state and the effect of the input sequence are independent; to find the state to which a system is sent by an input sequence, starting in a given state, we simply add together two state vectors—one obtained by driving the given initial state x with zero inputs for the desired length of time, and the other obtained by starting in the zero state and driving it with the given input function u for the desired length of time.

Let us now indicate the appearance of the result corresponding to (3) for a time-varying *continuous-time* linear system:

$$\left. \begin{aligned} \dot{x}(t) &= F(t)x(t) + G(t)u(t) \\ y(t) &= H(t)x(t) \end{aligned} \right\} \tag{5}$$

The first equation tells us that if we take a large integer $n \geq 0$, so that the time interval $\frac{1}{n}$ is very small and $F(t)$ and $G(t)$ do not vary much in the time interval $\left[t, t + \frac{1}{n} \right]$, then we have, to a good approximation, that

$$\frac{x\left(t + \frac{1}{n} \right) - x(t)}{1/n} = F(t)x(t) + G(t)u(t)$$

or, in other words, that

$$x\left(t + \frac{1}{n}\right) = \left[I + \frac{1}{n}F(t)\right]x(t) + \left[\frac{1}{n}G(t)u(t)\right]$$

$$y(t) = H(t)x(t) \tag{6}$$

What this tells us is that any "smooth" continuous-time linear system (5) may be approximated by a discrete-time system (6)—but now with time-interval ("sampling period") $\frac{1}{n}$ rather than 1. Our general formulas (2) and (3) for the linear system (1) may then be immediately adapted to the linear system (6) to yield:

$$\left.\begin{aligned}
&\Phi_n(t_0, t_0) = I, \text{ the identity transformation} \\
&\Phi_n\left(t_1 + \frac{1}{n}, t_0\right) = \left[I + \frac{1}{n}F(t_1)\right]\Phi_n(t_1, t_0) \text{ for } t_1 - t_0 \text{ a positive} \\
&\qquad\qquad\qquad\qquad\qquad \text{integral multiple of } \frac{1}{n}
\end{aligned}\right\} \tag{7}$$

and

$$x(t_1) = \Phi_n(t_1, t_0)x(t) + \sum_{t=t_0\left[\frac{1}{n}\right]}^{t_1} \Phi_n\left(t_1, t + \frac{1}{n}\right)\left[\frac{1}{n}G(t)\right]u(t) \tag{8}$$

where the summation symbol indicates that we have replaced the index j in the summation by the quantity $t = t_0 + \frac{j}{n}$, so that t increases from t_0 to t_1 in $n(t_1 - t_0)$ steps, each of length $\frac{1}{n}$.

Rewriting the latter equation of (7) in the form

$$\frac{\Phi_n\left(t_1 + \frac{1}{n}, t_0\right) - \Phi_n(t_1, t_0)}{1/n} = F(t_1)\Phi_n(t_1, t_0)$$

and letting $n \to \infty$ $\left(\text{i.e., } \frac{1}{n} \to 0\right)$, we may introduce the function

$$\Phi(t_1, t_0) = \lim_{n\to\infty} \Phi_n(t_1, t_0)$$

and deduce that

$$\left.\begin{aligned}
&\Phi(t_0, t_0) = I, \text{ the identity transformation} \\
&\frac{\partial}{\partial t_1}\Phi(t_1, t_0) = F(t_1)\Phi(t_1, t_0) \text{ for } t_1 \geq t_0
\end{aligned}\right\} \tag{9}$$

The reason that the present treatment is not rigorous, and that we shall later have to go back and carry out a careful study of linear matrix differential equations, is that we are not taking any care now to ensure that the limit we have just defined really exists; all we are doing is deducing the properties of the limit *if* it exists. It will remain for a later section to verify that there does actually exist a matrix function $\Phi(t_1, t_0)$ of two variables which satisfies the matrix differential equation (9).

Let us now verify that, if in fact we do have a solution to equation (9), then it will be the state transition map for our continuous-time system (5). That is, if at time t_0 the system is in state $x(t_0) = \Phi(t_0, t_0)x(t_0)$, then we wish to verify that, should we continuously apply the zero input to the system, then at some later time t_1 it will be in state $x(t_1) = \Phi(t_1, t_0)x(t_0)$. In other words, we wish to verify that the function $x(t_1) = \Phi(t_1, t_0)x(t_0)$ of t_1 is a solution of the unforced differential equation $\dot{x}(t_1) = F(t_1)x(t_1)$. But this is almost immediate:

$$\dot{x}(t_1) = \frac{d}{dt_1} x(t_1)$$

$$= \frac{\partial}{\partial t_1} [\Phi(t_1, t_0)x(t_0)]$$

$$= \frac{\partial}{\partial t_1} \Phi(t_1, t_0)x(t_0)$$

$$= F(t_1)\Phi(t_1, t_0)x(t_0) \quad \text{by (9)}$$

$$= F(t_1)x(t_1) \quad \text{as desired}$$

Let us now turn our energies to equation (8). As $n \to \infty$, i.e. as the time interval $\frac{1}{n} \to 0$, and as our "sampled-data system" becomes a better and better approximation to our given continuous-time system, then we shall have that $\Phi_n\left(t_1, t + \frac{1}{n}\right) \to \Phi(t_1, t)$ and the summation

$$\sum_{t=t_0\left[\frac{1}{n}\right]}^{t_1} \Phi\left(t_1, t + \frac{1}{n}\right) G(t)u(t) \cdot \frac{1}{n}$$

will become a better and better approximation to the integral

$$\int_{t_0}^{t_1} \Phi(t_1, t)G(t)u(t)\, dt$$

where the increment $\frac{1}{n}$ has gone over into the infinitesimal dt. The reader should thus be prepared to believe that if we can solve the matrix differential equation (9), we shall obtain the linear transformation $\Phi(t_1, t_0)$ which takes a state at time t_0 to the state at time t_1 to which the system will move under zero input, and that the response of the system to an arbitrary continuous input function can then be written in the form

$$x(t_1) = \Phi(t_1, t_0)x(t_0) + \int_{t_0}^{t_1} \Phi(t_1, t)G(t)u(t) \, dt \tag{10}$$

Example 1

The system of Figure 5–1 is described by a scalar (one-dimensional differential equation of the form

$$\dot{x}(t) = Fx(t) + Gu(t)$$

where F and G are real constants.

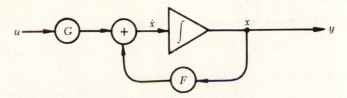

Figure 5–1   A constant, one-dimensional linear system.

Solving $\dfrac{\partial}{\partial t_1} \Phi(t_1, t_0) = F\Phi(t_1, t_0)$ and $\Phi(t_0, t_0) = 1$

gives $\Phi(t_1, t_0) = e^{F(t_1-t_0)}$.

Thus, the complete solution is

$$x(t_1) = e^{F(t_1-t_0)}x(t_0) + \int_{t_0}^{t_1} e^{F(t_1-\xi)}Gu(\xi) \, d\xi$$

just as we saw in Example 1–6–1 with $F = -3$ and $G = 1$.   ◊

Speaking rather loosely, we may say that the input $u(t)$ makes a contribution $G(t)u(t) \, dt$ to the state at time t, which is then updated from time t to time t_1 by the transition matrix $\Phi(t_1, t)$, to obtain precisely the same decomposition as we had in the discrete-time case:

$$\underbrace{\phi(t_1, t_0, x, u)}_{} = \underbrace{\phi(t_1, t_0, x, 0_\Omega)}_{\substack{\text{the zero-input} \\ \text{state trajectory}}} + \underbrace{\phi(t_1, t_0, 0_X, u)}_{\substack{\text{the zero-initial-state} \\ \text{state trajectory}}} \tag{11}$$

The reader should note that once we have derived our formulas for updating the state of a system, the problem of reading out the output is completely trivial, since it is always solved by the equation $y(t_1) = H(t_1)x(t_1)$, which is equally simple to apply whether the system be discrete-time or continuous-time, constant or time-varying.

Our derivation of equations (10) and (11) for the continuous-time linear system has been very heuristic. A more rigorous discussion is presented in the next section, where we study the solution of the differential equations

$$\dot{x}(t) = f(t, x(t), u(t))$$

which describe continuous-time systems. The discussion there will be extremely general, applying to non-linear systems as well as linear ones, and holding in any Banach space, not just $\mathbf{R}^n$.

EXERCISES FOR SECTION 5-1

1.   Using the **continued product** notation

$$\prod_{i=1}^{m} a_i = (a_1)(a_2) \dots (a_m)$$

carry through the step-by-step analysis of the linear time-varying discrete-time system

$$x(t + 1) = F(t)x(t) + G(t)u(t)$$

to derive an expression for the state transition matrix $\Phi(t + k, t)$ in terms of $F(\cdot)$.

Check your work by showing that when $F(t) = F$, constant, your expression reduces to $\Phi(t + k, t) = F^k$ for $k \geq 0$.

✔ 2.   In Example 1 we looked at the scalar differential system with constant coefficients F and G. Consider now the time-varying *scalar* equation $\dot{x}(t) = F(t)x(t) + G(t)u(t)$, where $F(\cdot)$ and $G(\cdot)$ are continuous real functions.

Solve

$$\begin{cases} \dfrac{\partial}{\partial t} \Phi(t, t_0) = F(t)\Phi(t, t_0) \\[2mm] \Phi(t_0, t_0) = I \end{cases} \tag{12}$$

for the (1×1) state transition matrix Φ, and plug this Φ into equation (10) to get the complete solution to the differential equation.

✔ 3.   **Leibnitz' rule** for differentiating an integral is given by:

$$\frac{d}{dt}\left(\int_{\alpha(t)}^{\beta(t)} f(t, \xi)\, d\xi\right) = \int_{\alpha(t)}^{\beta(t)} \left[\frac{\partial}{\partial t} f(t, \xi)\right] d\xi + f(t, \beta(t))\frac{d\beta(t)}{dt} - f(t, \alpha(t))\frac{d\alpha(t)}{dt}$$

Use this rule to differentiate equation (10) to get

$$\frac{d}{dt_1} x(t_1) = \frac{\partial}{\partial t_1} \Phi(t_1, t_0)x(t_0) + \int_{t_0}^{t_1} \frac{\partial}{\partial t_1} \Phi(t_1, t)G(t)u(t)\, dt + \Phi(t_1, t_1)G(t_1)u(t_1)$$

Then use the properties of Φ given by Equation (9) to show that this expression reduces to

$$\frac{d}{dt_1} x(t_1) = F(t_1)x(t_1) + G(t_1)u(t_1)$$

which shows that (10) really is a solution of this differential equation.

4.   Give the complete solution to the initial value problem

$$\begin{cases} \dot{x}(t) = -tx(t) + 2 & \text{for } t \geq 0 \\ x(0) = 1 \end{cases}$$

P 5.   For the continuous-time scalar system described in Exercise 4, carry out the steps of equations (6) through (8) to:

(a)   Find a discrete-time system which approximates the given system.

(b)   Determine $\Phi_n(t_1, t_0)$ and investigate its behavior as $n \to \infty$.

(c)   Solve the discrete-time system for an answer $x_n(t_1)$ and investigate its behavior as $n \to \infty$.

(d)   Let $n = 10$ and find

$$x_n(0), x_n\left(\frac{1}{10}\right), x_n\left(\frac{2}{10}\right), \ldots, x_n\left(\frac{9}{10}\right), x_n(1)$$

as an approximation to the actual solution you found in Exercise 4, on the interval $0 \leq t \leq 1$. Is this approximation a good one?

5-2   Differential Systems

In this section we present a rather general discussion of the solution of the differential equation

$$\dot{x}(t) = f(t, x(t), u(t)) \tag{1}$$

which describes the local behavior of continuous-time systems. The reader will recall† that the natural mathematical setting for the study of such equations is the Banach space—a vector space which has a norm $\|\cdot\|$ defined on it, and is complete with respect to the metric

$$d(x, y) = \|x - y\|$$

induced by the norm.

By developing the theory in the context of general Banach spaces, we ensure that our results will apply to non-linear systems whose state spaces may be infinite-dimensional, as well as to the more familiar finite-dimensional linear systems having $\mathbf{R}^n$ or $\mathbf{C}^n$ as their state spaces.

---

†Or reread Section 2–4.

The equation

$$\dot{x}(t) = f(t, x(t), u(t))$$

tells us how quickly our state changes as a function of the present time, the present state, and the present input. If a system is to really be deterministic, we must prove that this differential equation specifies the future behavior of the system uniquely, given the state at some fixed time, say t_0, and subsequent values of the input function u.

We shall in fact find that we can obtain these results by a simple application of a result which we proved quite easily in Section 2–4:

THE CONTRACTION LEMMA

Let $\mathfrak{M}$ be a complete metric space with distance function d. Let $S : \mathfrak{M} \to \mathfrak{M}$ be a contraction of $\mathfrak{M}$ with "contraction constant" $K < 1$; i.e.,

$$d(S(x), S(y)) \leq K d(x, y) \quad \text{for all } x, y \text{ in } \mathfrak{M}$$

Then S has a unique *fixed point* x_0; i.e., there is a unique point x_0 in $\mathfrak{M}$ for which $S(x_0) = x_0$. Further, for any x in $\mathfrak{M}$, the sequence $\{S^n(x)\}$ has x_0 for its limit. □

Before showing what the contraction lemma has to do with the solution of differential equations, we must make a few remarks about integration in Banach spaces.

In much of modern system theory we have to use the Lebesgue integral, a generalization of the ordinary Riemann integral studied in calculus courses. The Lebesgue integral is developed† for general Banach spaces, using measure theory, so that it "undoes" differentiation:

$$\int_{t_0}^{t} \dot{f}(\xi) \, d\xi = f(t) - f(t_0) \tag{2}$$

(*The Fundamental Theorem of Calculus*)

where $f : \mathbf{R} \to X$ is a differentiable map of the reals into a Banach space X and $\dot{f} : \mathbf{R} \to X$ is the derivative

$$\dot{f}(t) = \lim_{h \to 0} \frac{f(t + h) - f(t)}{h}$$

---

† See W. Rudin, *Real and Complex Analysis* (McGraw-Hill, 1966) or any graduate text on analysis. Essentially, all functions for which you have ever written an integral are Lebesgue integrable, but there exist other functions, too "bumpy" to have a Riemann integral, which do have a Lebesgue integral. For example, the theory of the Laplace transform needs the full power ("bump tolerance") of the Lebesgue integral to handle all the situations we run into in system theory when dealing with "limiting situations."

Similarly, if $g(t)$ is continuous from the right,

$$\frac{d}{dt}\left[\int_{t_0}^{t} g(\xi)\, d\xi\right] = g(t) \tag{3}$$

differentiation "undoes" integration, and more generally, Leibnitz' rule holds:

$$\frac{d}{dt}\int_{\alpha(t)}^{\beta(t)} g(\xi)\, d\xi = g(\beta(t))\frac{d\beta(t)}{dt} - g(\alpha(t))\frac{d\alpha(t)}{dt} \tag{4}$$

Without going into any more detail, we shall freely make use of the fact that we can integrate as well as differentiate functions in Banach spaces and that our Banach space calculus is just a generalization of ordinary freshman calculus.

Example 1

With the state space $X = \mathbf{R}^n$, the integral is defined by

$$\int_{t_0}^{t}\begin{bmatrix} x_1(\xi) \\ x_2(\xi) \\ \vdots \\ x_n(\xi) \end{bmatrix} d\xi = \begin{bmatrix} \int_{t_0}^{t} x_1(\xi)\, d\xi \\ \int_{t_0}^{t} x_2(\xi)\, d\xi \\ \vdots \\ \int_{t_0}^{t} x_n(\xi)\, d\xi \end{bmatrix}$$

i.e., the Lebesgue integral of a vector function is given by Lebesgue integration of the component functions. Thus

$$\int_{t_0}^{t}\begin{bmatrix} \cos(\xi) \\ \sin(\xi) \end{bmatrix} d\xi = \begin{bmatrix} \sin(t) - \sin(t_0) \\ -\cos(t) + \cos(t_0) \end{bmatrix}$$

and

$$\frac{d}{dt}\begin{bmatrix} \sin(t) - \sin(t_0) \\ -\cos(t) + \cos(t_0) \end{bmatrix} = \begin{bmatrix} \cos(t) \\ \sin(t) \end{bmatrix} \qquad \diamond$$

Let us now return to our differential equation

$$\dot{x}(t) = f(t, x(t), u(t))$$

and observe that as soon as we have specified what the applied input function

$u(\cdot)$ is, our equation may be written in the form

$$\dot{x}(t) = \bar{f}(t, x(t)) \tag{5}$$

The specification of the input function u has been "swallowed," along with the specification of the old function $f: T \times X \times U \to X$, into the specification of a new function $\bar{f}: T \times X \to X$. We are interested in finding a solution $x(t)$ of equation (5) which satisfies the specified initial condition $x(t_0) = x_0$.

We assume that $\bar{f}$ is continuous to ensure that the following integrals exist. Then, integrating both sides of equation (5) from t_0 to t yields

$$\int_{t_0}^{t} \dot{x}(\xi) \, d\xi = \int_{t_0}^{t} \bar{f}(\xi, x(\xi)) \, d\xi$$

or

$$x(t) - x(t_0) = \int_{t_0}^{t} \bar{f}(\xi, x(\xi)) \, d\xi$$

Using the specified value x_0 for $x(t_0)$, we get the equivalent **integral equation**

$$x(t) = x_0 + \int_{t_0}^{t} \bar{f}(\xi, x(\xi)) \, d\xi \tag{6}$$

corresponding to the **initial value problem**

$$\dot{x}(t) = \bar{f}(t, x(t)) \tag{7}$$
$$x(t_0) = x_0$$

in differential equations.

This type of integral equation, of the form $x(t) = A + \int_{t_0}^{t} \bar{f}(\xi, x(\xi)) \, d\xi$ where the solution x appears both on the left side and on the right under the integral, is called a "Volterra integral equation of the second kind."† If the function $\bar{f}(t, \cdot)$ is linear, we get a "linear Volterra equation of the second kind," but we shall continue our general discussion without assuming $\bar{f}$ to be linear.

Now, the operations called for by the right hand side of equation (6) may be thought of in their own right as defining a map S which takes a function x and transforms it to the function (Sx) whose value at time t is given by

$$(Sx)(t) = x_0 + \int_{t_0}^{t} \bar{f}(\xi, x(\xi)) \, d\xi \tag{8}$$

---

† See F. B. Hildebrand, *Methods of Applied Mathematics*, 2nd ed. (Prentice-Hall, 1965).

Because substituting the function x into S to get the function (Sx) involves first substituting values of x into $\bar{f}$ and then *integrating*, we call the map S an **integral operator.**†

Example 2

Let S map any integrable function g into the function Sg by the formula

$$(Sg)(t) = 1 + \int_0^t 2\xi\, g(\xi)\, d\xi$$

Let g_1 and g_2 be the two functions given by $g_1(t) = t^2$ and $g_2(t) = e^{t^2}$.

Then $(Sg_1)(t) = 1 + \displaystyle\int_{t_0}^t 2\xi\xi^2\, d\xi = 1 + \tfrac{1}{2}t^4$, and

$$(Sg_2)(t) = 1 + \int_0^t 2\xi e^{\xi^2}\, d\xi = 1 + \left. e^{\xi^2}\right|_0^t$$

$$= e^{t^2} = g_2(t) \qquad\qquad\qquad \lozenge$$

As a result of this discussion, we may now say that a function x is a solution of the differential equation $\dot{x}(t) = \bar{f}(t, x(t))$, with initial condition $x(t_0) = x_0$, if and only if the functions x and Sx are identical. Thus, x is a **solution** of our differential equation if and only if x is **a fixed point** of (i.e., a point left fixed by) the operator S.

Our strategy should now be clear to the reader. We want to prove that when we take our differential equation $\dot{x}(t) = \bar{f}(t, x(t))$ with specified initial condition $x(t_0) = x_0$, it has one and only one solution. But this amounts to showing that S has one and only one fixed point. Thus, if we can show that the collection of functions $\mathbf{E}$ upon which S acts may be considered as a complete metric space, and then show that (with respect to the metric on this collection of functions) S is a contraction which maps $\mathbf{E}$ into itself, we will have shown not only that our differential equation $\dot{x}(t) = \bar{f}(t, x(t))$ with initial condition $x(t_0) = x_0$ has a solution, but also that the solution is unique. The shrinking of rubber sheets certainly seems relevant to system theory after all!

We emphasize that our operator S may act upon a function x whether or not x satisfies the differential equation (7). For most functions, of course, x will differ from Sx. What we are saying is that the function x satisfies our differential equation precisely when the operation of S does not change x at all.

---

† This is a Volterra integral operator, since the upper limit of integration is the variable t; see Hildebrand, *op. cit.*

Example 3

Consider the scalar differential equation $\dot{x}(t) = 2tx(t)$ with specified initial condition $x(0) = 1$. Here $\bar{f}(t, x(t)) = 2tx(t)$ and the equivalent integral equation is

$$x(t) = 1 + \int_0^t 2\xi x(\xi)\, d\xi$$

as the reader should verify by differentiating.

The operator S for this problem acts upon any integrable function g by the rule

$$(Sg)(t) = 1 + \int_0^t 2\xi g(\xi)\, d\xi$$

and is just the operator studied in Example 2. We saw in that example two functions g_1 and g_2, where

$$(Sg_1)(t) = 1 + \frac{1}{2} t^4 \neq t^2 = g_1(t)$$

but

$$(Sg_2)(t) = e^{t^2} = g_2(t)$$

Thus, g_1 was not a fixed point of S, since $Sg_1 \neq g_1$. However, g_2 was left unchanged by S, and hence is a fixed point of S and a solution of the given differential equation.   ◊

Our task now is to specify suitable conditions upon the function $\bar{f}$ in equation (8) to guarantee that our operator S is a contraction taking a suitable complete metric space of functions into itself. [Actually, it will prove not to be S, but a close relative of S, that does the job.]

Clearly, the time has come to be specific about the Banach space in which S operates. Having fixed the Banach space X in which our vectors lie, and a finite time interval $[t_1, t_2]$ containing t_0 over which we shall study the behavior of our differential equation, we let E be the space of all *continuous* functions

$$x\colon [t_1, t_2] \to X$$

In other words, for any $\varepsilon > 0$, there exists $\delta > 0$ such that, no matter where τ_1 and τ_2 lie in $[t_1, t_2]$, we have

$$|\tau_1 - \tau_2| < \delta \Rightarrow \|x(\tau_1) - x(\tau_2)\| < \varepsilon$$

First note that this condition implies that $\|x(\cdot)\|$ is *bounded* for

$\tau \in [t_1, t_2]$, for if we pick ε and δ as above, it clearly follows that

$$\|x(\tau) - x(t_1)\| \le \left(\frac{t_2 - t_1}{\delta}\right)\varepsilon$$

so that

$$\|x(\tau)\| \le \|x(t_1)\| + \|x(\tau) - x(t_1)\| \le \|x(t_1)\| + \left(\frac{t_2 - t_1}{\delta}\right)\varepsilon$$

We call the smallest M such that

$$\|x(\tau)\| \le M \text{ for all } \tau \in [t_1, t_2]$$

the **norm** of x, and write it as $\|x\|$.

To verify that **E** is a **Banach space** with respect to this norm is straightforward, and so is left to the reader in Exercise 10.

Our task now is to set forth reasonable conditions under which $x \in$ **E** ensures that Sx is also in **E**. In other words, given that x is continuous on $[t_1, t_2]$ with least upper bound $\|x\|$, what conditions on $\bar{f}$ ensure that

$$(Sx)(\cdot) = x_0 + \int_{t_0}^{(\cdot)} \bar{f}(\xi, x(\xi))\, d\xi$$

is also continuous on $[t_1, t_2]$, so that, by making $|\tau_2 - \tau_1|$ sufficiently small we can ensure that

$$\|Sx(\tau_2) - Sx(\tau_1)\| = \left\|\int_{\tau_1}^{\tau_2} \bar{f}(\xi, x(\xi))\, d\xi\right\|$$

is as small as we require? If $\bar{f}$ satisfies the simple requirement that

$$\|\bar{f}(\xi, \hat{x})\| \le K\|\hat{x}\| \text{ for all } \xi \in [t_1, t_2],\ \hat{x} \in X \tag{9}$$

then we deduce that

$$\|Sx(\tau_2) - Sx(\tau_1)\| \le K\|x\|\,|\tau_2 - \tau_1|$$

which implies that Sx certainly belongs to **E** (and that $\|Sx\| \le K|t_1 - t_0|\,\|x\|$).

Notice that we routinely make use of the fact that our Banach space integrals are linear operations, so that the integral of a sum is the sum of the corresponding integrals, and so that multiplication by a scalar passes through the integral sign. We have used the familiar property of integrals that

$$\left\|\int_{t_0}^{t} g(\xi)\, d\xi\right\| \le \int_{t_0}^{t} \|g(\xi)\|\, d\xi \le |t - t_0|\,\|g\| \tag{10}$$

where $\|g\| = \sup\limits_{t_0 \leq \xi \leq t_1} \|g(\xi)\|$ denotes the *least upper bound* of the set of numbers $\{\|g(\xi)\|\}$ as ξ takes on values from t_0 to t_1.

To move on to the question of when S is a contraction, assume that α and β belong to **E,** and consider the following difference for any t in $[t_1, t_2]$:

$$(S\alpha)(t) - (S\beta)(t) = \left[x_0 + \int_{t_0}^t \bar{f}(\xi, \alpha(\xi)) \, d\xi \right] - \left[x_0 + \int_{t_0}^t \bar{f}(\xi, \beta(\xi)) \, d\xi \right]$$

$$= \int_{t_0}^t [\bar{f}(\xi, \alpha(\xi)) - \bar{f}(\xi, \beta(\xi))] \, d\xi \tag{11}$$

which yields

$$\|(S\alpha)(t) - (S\beta)(t)\| \leq |t - t_0| \sup_{t_0 \leq \xi \leq t} \|\bar{f}(\xi, \alpha(\xi)) - \bar{f}(\xi, \beta(\xi))\| \tag{12}$$

We may now formulate a condition on $\bar{f}$ which fits in nicely with (12):

DEFINITION 1

We say that $\bar{f}: [t_0, t_1] \times X \to X$ satisfies a **Lipschitz condition** on X uniformly with respect to t in $[t_0, t_1]$ if there exists a number $K > 0$ such that

$$\|\bar{f}(t, x) - \bar{f}(t, y)\| \leq K\|x - y\| \quad \text{for all } x, y \text{ in } X \text{ and } t \text{ in } [t_0, t_1]$$

Such a number K is called a **Lipschitz constant** for $\bar{f}$. [Note that this is a strengthening of condition (9).] $\bigcirc$

Thus, if our $\bar{f}$ in equation (8) satisfies a Lipschitz condition uniformly on X, with a Lipschitz constant of K, then

$$\|\bar{f}(\xi, \alpha(\xi)) - \bar{f}(\xi, \beta(\xi))\| \leq K\|\alpha(\xi) - \beta(\xi)\| \tag{13}$$

so that

$$\sup_{t_0 \leq \xi \leq t} \|\bar{f}(\xi, \alpha(\xi)) - \bar{f}(\xi, \beta(\xi))\| \leq K \sup_{t_0 \leq \xi \leq t} \|\alpha(\xi) - \beta(\xi)\| \tag{14}$$

and from the inequality (12),

$$\|(S\alpha)(t) - (S\beta)(t)\| \leq |t - t_0| \cdot K \cdot \|\alpha - \beta\|_{\mathbf{E}} \tag{15}$$

Hence, if we keep t within an interval $[t_0, t_1]$ for which $|t_1 - t_0| \cdot K$ is less than one, then the functions $S\alpha$ and $S\beta$ will be closer together than were α and β—that is, S is a contraction. We have thus shown that, given some initial time t_0, we can find a neighborhood of t_0, say $t_0 - \dfrac{1}{k} < t < t_0 + \dfrac{1}{k}$ which keeps $|t - t_0| \cdot K < 1$, over which we can show that a solution exists.

Let us now consider solving the differential equation $\dot{x}(t) = \overline{f}(t, x(t))$ with initial condition $x(t_0) = x_0$ over an *arbitrary* finite time interval $[t_1, t_2]$ with $t_1 < t_0 < t_2$. As before, we know that we have a solution of our equation with specified initial condition if and only if we have a fixed point of the operator $S: \mathbf{E} \to \mathbf{E}$, where for each $x \in \mathbf{E}$,

$$(Sx)(t) = x_0 + \int_{t_0}^{t} \overline{f}(\xi, x(\xi)) \, d\xi \text{ for } t_1 \le t \le t_2$$

As before, we assume a Lipschitz condition to ensure that S defined by

$$x \mapsto x_0 + \int_{t_0}^{(\cdot)} \overline{f}(\xi, x(\xi)) \, d\xi \tag{16}$$

maps $\mathbf{E}$ into $\mathbf{E}$, where $\mathbf{E}$ is the Banach space of all continuous functions on $[t_1, t_2]$ with the norm defined by

$$\|x\| = \sup_{t_1 \le t \le t_2} \|x(t)\|, \text{ for } x \in \mathbf{E}$$

Now let us see if S will be a contraction over the whole interval $[t_1, t_2]$:

$$\|S\alpha - S\beta\| = \sup_{t_1 \le t \le t_2} \|(S\alpha)(t) - (S\beta)(t)\|$$

$$\le \sup_{t_1 \le t \le t_2} [K|t - t_0| \, \|\alpha - \beta\|] \text{ if } \overline{f} \text{ is uniformly}$$
$$\text{Lipschitzian on}$$
$$[t_1, t_2], \text{ by (15)}$$

$$= \left(\sup_{t_1 \le t \le t_2} K|t - t_0| \right) \|\alpha - \beta\| \tag{17}$$

Now if $\left(\sup\limits_{t_1 \le t \le t_2} K|t - t_0| \right)$ were a number less than 1, we would have that S is contractive on $\mathbf{E}$.

But

$$\sup_{t_1 \le t \le t_2} K|t - t_0| = K \max \{|t_1 - t_0|, |t_2 - t_0|\} \tag{18}$$

[verify using Exercise 2], and there is no guarantee that this number is less than one.

In other words, in general, S is *not* a contraction. But all is not lost! The reader may readily verify by induction on n [Exercise 6] that, just as the n-fold integral of 1 is $\dfrac{t^n}{n!}$, so do we have

$$\|S^n\alpha - S^n\beta\| \leq \frac{[K\max(|t_1 - t_0|, |t_2 - t_0|)]^n}{n!}\|\alpha - \beta\| \qquad \text{(19)}$$

and since the powers of any fixed number grow more slowly than the factorial grows, we can choose n so large that

$$\frac{[K\max(|t_1 - t_0|, |t_2 - t_0|)]^n}{n!} < 1$$

and so S^n *is* a contraction. It thus has a unique fixed point

$$x:[t_1, t_2] \to X \text{ with } S^n x = x$$

But we are looking for a fixed point of S, not of S^n. Consider, then, the following:

$$S^n(Sx) = S(S^n x) = Sx$$

This tells us that Sx is also a fixed point of S^n. But we have said that S^n has a *unique* fixed point. We thus conclude that $x = Sx$ and x is in fact a fixed point of our original mapping S, and so is a solution of our differential equation $\dot{x}(t) = \bar{f}(t, x(t))$ defined for $t_1 \leq t \leq t_2$ and with initial condition $x(t_0) = x_0$. Conversely, if $\hat{x}$ is also a fixed point of S, then it is also a fixed point of S^n. But then, by uniqueness of fixed points of a contraction, we must have $x = \hat{x}$, so that x is indeed the *unique* fixed point of S. We may sum all this up in the following very useful theorem:

THEOREM 1 (Existence and Uniqueness)

Let $\bar{f}: T \times X \to X$ satisfy a Lipschitz condition

$$\|\bar{f}(t, x_1) - \bar{f}(t, x_2)\| \leq K\|x_1 - x_2\|$$

uniformly in t. Then the differential equation

$$\dot{x}(t) = \bar{f}(t, x(t))$$

with arbitrary initial condition

$$x(t_0) = x_0$$

has a unique solution $x : T \to X$ defined for *all* $t \in T$.   □

It should be emphasized that our discussion so far not only has given us conditions under which we can tell that there *is* a unique solution of the initial value problem

$$\begin{cases} \dot{x}(t) = \bar{f}(t, x(t)) \\ x(t_0) = x_0 \end{cases}$$

but has given us a procedure to construct the solution as the limit of a sequence of **successive approximations:**

From $(Sx)(t) = x_0 + \displaystyle\int_{t_0}^{t} \bar{f}(\xi, x(\xi)) \, d\xi$, the equivalent integral equation, we define the sequence

$$r_0(t) = x_0$$

$$r_1(t) = (Sr_0)(t) = x_0 + \int_{t_0}^{t} \bar{f}(\xi, r_0(\xi)) \, d\xi$$

$$\vdots$$

$$r_{m+1}(t) = (Sr_n)(t) = x_0 + \int_{t_0}^{t} \bar{f}(\xi, r_n(\xi)) \, d\xi \qquad (20)$$

and the solution $x(t) = \lim_{n \to \infty} r_n(t)$.

Moreover, each approximation $r_n(\cdot)$ automatically satisfies the initial condition, since $r_0(t_0) = x_0$ and $r_{n+1}(t_0) = (Sr_n)(t_0) = x_0$.

Example 4

Solve $\begin{bmatrix} \dot{x}_1(t) \\ \dot{x}_2(t) \end{bmatrix} = \begin{bmatrix} x_2(t) \\ -x_1(t) \end{bmatrix}$ subject to the initial condition $\begin{bmatrix} x_1(0) \\ x_2(0) \end{bmatrix} = \begin{bmatrix} 0 \\ 1 \end{bmatrix}$.

Here

$$\bar{f}\left(t, \begin{bmatrix} x_1(t) \\ x_2(t) \end{bmatrix}\right) = \begin{bmatrix} x_2(t) \\ -x_1(t) \end{bmatrix}, \quad x_0 = \begin{bmatrix} 0 \\ 1 \end{bmatrix}$$

and

$$\left(S\begin{bmatrix} x_1 \\ x_2 \end{bmatrix}\right)(t) = \begin{bmatrix} 0 \\ 1 \end{bmatrix} + \int_0^t \begin{bmatrix} x_2(\xi) \\ -x_1(\xi) \end{bmatrix} d\xi$$

Starting with the initial approximation $r_0(t) = x_0 = \begin{bmatrix} 0 \\ 1 \end{bmatrix}$, we get

$$r_1(t) = \begin{bmatrix} 0 \\ 1 \end{bmatrix} + \int_0^t \begin{bmatrix} 1 \\ 0 \end{bmatrix} d\xi = \begin{bmatrix} t \\ 1 \end{bmatrix}$$

$$r_2(t) = \begin{bmatrix} 0 \\ 1 \end{bmatrix} + \int_0^t \begin{bmatrix} 1 \\ -\xi \end{bmatrix} d\xi = \begin{bmatrix} t \\ 1 - \dfrac{t^2}{2} \end{bmatrix}$$

$$r_3(t) = \begin{bmatrix} 0 \\ 1 \end{bmatrix} + \int_0^t \begin{bmatrix} 1 - \dfrac{\xi^2}{2} \\ -\xi \end{bmatrix} d\xi = \begin{bmatrix} t - \dfrac{t^3}{3!} \\ 1 - \dfrac{t^2}{2!} \end{bmatrix}$$

$$\vdots$$

$$r_n(t) = \begin{bmatrix} t - \dfrac{t^3}{3!} + \cdots + \dfrac{t^n}{n!} \\ 1 - \dfrac{t^2}{2!} + \cdots + \dfrac{t^{n-1}}{(n-1)!} \end{bmatrix}, \ (n \text{ odd})$$

Since $\lim\limits_{n\to\infty} r_n(t) = \begin{bmatrix} \sin(t) \\ \cos(t) \end{bmatrix}$, the solution is $x(t) = \begin{bmatrix} \sin(t) \\ \cos(t) \end{bmatrix}$, and it is easily checked that $x(t)$ is the fixed point of S. ◊

Of course, it seldom happens in a real problem that we can get a closed-form expression for the limit of the sequence of successive approximations. Typically, we won't even be able to evaluate the integrals in closed form and will be forced to do the whole problem numerically.

The statement in our existence and uniqueness theorem that our solution is defined for all time is justified by the fact that we have shown that it is defined uniquely for any finite time interval. Thus, we can keep extending that interval further and further to get the unique continuation of the trajectory as far as we like. The reader should note well that in the above proof, when we wrote an integral of the form $\int_{t_0}^t (\cdot)$, we did *not* imply that $t > t_0$, so the uniqueness of the solution of our differential equation is guaranteed in the past as well as in the future. In other words, we are not simply saying that the evolution of the system is well defined in that if we know the present state, we determine uniquely what the state will be at any future time—we also say that it is *reversible,* in the sense that if we know the state now, then given any earlier time, there is a unique state at that time which could have led to the state that we observe at present.

It is interesting to recall that the feature of *reversibility* of continuous-time differential equations is *not* shared by discrete-time difference equations. For example, if we have the very simple difference equation written out in component form as

$$\begin{bmatrix} x_1(t + 1) \\ x_2(t + 1) \end{bmatrix} = \begin{bmatrix} x_1(t) - x_2(t) \\ 0 \end{bmatrix}$$

we see that the states of the system divide sharply into two classes—those which could not possibly arise from the action of the system upon a previous state (those states whose second component is not zero) and those which could have arisen from infinitely many immediately prior states (those in which the second component is zero). This system is not very interesting after the second time step, and the reader might wish to construct more subtle examples. In terms of our general equations for a linear discrete-time system $x(t + 1) = F(t)x(t)$, we see immediately that it shares the "time-reversal" properties of the continuous-time systems if and only if the matrix $F(t)$ is invertible for all values of the time parameter t. [Recall Exercise 2–1–9d.]

We may make this point even more forcefully by considering the absolutely trivial *scalar* equations with $F(t) \equiv 0$, for all t. In the discrete-time case we have $x(t + 1) = 0$, so that a state yields absolutely no information about the previous state, if we allow that the previous state could have been set by an experimenter, rather than being a result of the dynamics. However, in the continuous-time case, we have that $\dot{x}(t) = 0$, which implies $x(t) = x(t_0)$ for all $t \geq t_0$, and so the state at any time certainly gives us complete information about the earlier states as long as there was no outside intervention, and the dynamics of the system took its course. What we are saying is that in the continuous-time linear system, even if we start with singular matrix equations, when we integrate them, we obtain non-singular solution matrices. We do not have this process of integration in the discrete-time case, and so we must content ourselves with non-reversibility of time unless we can guarantee invertibility for every $F(t)$.

Summing up, we have seen that local descriptions of continuous-time systems by differential equations yield in a very large variety of cases— whether linear and finite-dimensional or neither—the global descriptions of the systems with which we started in Chapter 1. We also add with great emphasis that *whenever a system is described by a "smooth"† differential equation, then specifying an initial condition and an input regime determines not only how the system will behave in the future, but also how it must have behaved in the past.*

We close this section by distinguishing between necessary and sufficient conditions. We have shown that, in order to get uniqueness of solutions of differential equations, it is *sufficient* that our function $\bar{f}$ satisfy a Lipschitz condition. We have not proved that it is *necessary* for it to satisfy a Lipschitz condition—and in fact the reader may well find, in practical applications,

---

† Smoothness is assured by $\bar{f}$ being at least Lipschitzian. See Kalman, Falb, and Arbib, *Topics in Mathematical System Theory* (McGraw-Hill, 1969), pp. 10, 12, 28.

that he will encounter functions $\bar{f}$ for which there is a unique solution, but for which no Lipschitz condition is satisfied. Thus, although we have shown for a very broad class that we can get unique solutions to our differential equations, we have in no way ruled out the possibility that other useful systems will also yield unique solutions. However, we should also note that there are cases in which there is no uniqueness of solution—and of course, in such cases, the Lipschitz condition will not be met.

Example 5

The initial value scalar problem

$$\dot{x}(t) = x(t)^{1/3} \text{ with } x(0) = 0$$

has *two* distinct solutions:

$$x(t) \equiv 0 \quad \text{and} \quad x(t) = \begin{cases} 0 & \text{for } t < 0 \\ \dfrac{2}{3} t^{3/2} & \text{for } t \geq 0 \end{cases}$$

Here $\bar{f}(t, x(t)) = [x(t)]^{1/3}$ and since

$$\bar{f}(t, x(t)) - \bar{f}(t, 0) = x(t)^{1/3} - 0 = x(t)^{-2/3}[x(t) - 0]$$

and the norm is just the absolute value,

$$|\bar{f}(t, x(t)) - \bar{f}(t, 0)| = |x(t)^{-2/3}| \, |x(t) - 0|$$

Since $x(t)^{-2/3}$ is unbounded near $x = 0$, this $\bar{f}$ does *not* satisfy a Lipschitz condition.   ◊

The crucial point for uniqueness of solutions is simply that some smoothness condition be met—and we hope that we have said enough to convince the reader that neither linearity nor finite dimensionality is requisite for the meeting of this condition.

Having done what we can about the general case, we will next focus our attention on the solution of linear vector differential equations in finite dimensional spaces.

EXERCISES FOR SECTION 5–2

✔ 1.   The **supremum** or **least upper bound** of a set of real numbers A is often denoted by

$$\sup \{x \,|\, x \in A\}, \quad \sup_{x \in A} \{x\}, \quad \text{or} \quad \text{l.u.b. } (A)$$

the last being shorthand for *least upper bound*. The sup is a generalization of "largest" and is precisely defined. For a subset $A \subset \mathbf{R}$ which is bounded above, L is the sup of A if (i) L is *an* upper bound for A and (ii) no number smaller than L is an upper bound for A.

(a)   Give an analogous definition for the **infimum** (inf), the **greatest lower bound** (g.l.b.) of A. Prove that $\inf\{x \mid x \in A\} = -\sup\{-x \mid x \in A\}$.

(b)   List three different upper bounds for the set $A = [0, 1)$, the half-open interval. Show that A has a smallest element but does not have a largest element. Find $\sup\limits_{x \in A}\{x\}$.

(c)   Find $\sup\limits_{x \in A}\{f(x)\}$ and $\sup\limits_{1 \le x \le 3}\left\{\int_2^x f(s)\,ds\right\}$ if $f(t) = e^{-t}$. What happens if $f(t) = e^t$?

(d)   Let A and B be non-empty sets of real numbers and assume that no member of A exceeds any member of B. Prove that

$$\sup\limits_{x \in A}\{x\} \le \inf\limits_{x \in B}\{x\}$$

✔ **2.**   For the Banach space $\mathbf{E} = C[a, b]$, the set of continuous real-valued functions defined on the interval $[a, b]$, the "sup" norm reduces to the "max" norm discussed in Example 2–4–4 because of a theorem from calculus which states that a continuous real function f always has a maximum value at some point of the closed, bounded interval $[a, b]$.

(a)   Show that $\sup\limits_{a \le t \le b} |f(t)| = \max\limits_{a \le t \le b} |f(t)| = \|f\|_\infty$ for $f \in C[a, b]$.

(b)   Now consider $\mathbf{E} = C[0, 1) = \{$continuous real functions defined for $0 \le t < 1\}$. Let $f(t) = 2t$ for $t \in [0, 1)$ and observe that f does not attain a maximum on the interval $[0, 1)$; hence $\max\limits_{0 \le t < 1} |f(t)|$ makes no sense. Find $\sup\limits_{0 \le t < 1} |f(t)|$.

(c)   Show that $\sup\limits_{x \in A} |f(x)| = 0 \Rightarrow f(x) \equiv 0$ on A.

(d)   Prove the properties of sup used in deriving equations (17) and (18). Determine the corresponding properties for inf.

✔ **3.**   (a)   Show that the $\bar{f}$ in Example 3 satisfies a Lipschitz condition uniformly on $[-5, 7]$ and give a suitable Lipschitz constant K for $\bar{f}$.

(b)   In Example 5–1–1, let the input $u(t) = 10 \cos \omega_0 t$ and let F, G be arbitrary real constants. Find the equivalent integral equation. Give the operator S for this example and find a good Lipschitz constant for $f(t, x(t), u(t))$ on $[0, 5]$.

4.   Repeat Exercise 3(b), letting $u(t) = 1$ for $0 \le t \le 5$. Letting $F = -3$ and $G = 1$, and $x(0) = 4/3$, show that $x(t) = \dfrac{1}{3} + e^{-3t}$, for $0 \le t \le 5$, is a fixed point of the operator S. Verify that this $x(t)$ is a solution of the differential equation of Example 5–1–1 for the initial condition $x(0) = \dfrac{4}{3}$.

5.   We have developed the notion of a Banach space in Section 2–4 as a mathematical structure which lets us add objects, multiply them by scalars, and take limits. In Section 3–4 we introduced the concept of the derivative of a map from one Banach space into another. Using the derivative, we may formulate for Banach spaces a theorem which looks just like a familiar theorem of differential calculus. Let $g : \mathbf{E} \to \mathbf{F}$ be a differentiable function from one Banach space, $\mathbf{E}$, into another, $\mathbf{F}$. Then if the derivative $g'(x)$ is a bounded linear transformation with norm $< M$, we have

$$\|g(x) - g(y)\| \le M\|x - y\| \text{ for } x, y \text{ in } \mathbf{E}$$

Consider the function $\bar{f} : \mathbf{R} \times \mathbf{E} \to \mathbf{E}$ of equations (6) and (7). For each fixed ξ, the map $\bar{f}(\xi, \cdot)$ maps $\mathbf{E}$ into itself. Show that if $\bar{f}(\xi, \cdot)$ has a derivative bounded above by the constant M, then $\bar{f}$ satisfies a Lipschitz condition uniformly on $\mathbf{E}$ with respect to $\mathbf{R}$.

✔ **6.**   Let $(S\alpha)(t) = x_0 + \int_{t_0}^{t} \bar{f}(s, \alpha(s)) \, ds$, $a \leq t_0 \leq t \leq b$, with $\bar{f}$ uniformly Lipschitzian on $[a, b]$ with Lipschitz constant K.

   (a)   Show that $\|(S\alpha)(t) - (S\beta)(t)\| \leq K\|\alpha - \beta\|(t - t_0)$, $a \leq t \leq b$, where

$$\|x - y\| = \sup_{a \leq t \leq b} \|x(t) - y(t)\|$$

   (b)   Show that $\|(S^2\alpha)(t) - (S^2\beta)(t)\| \leq \int_{t_0}^{t} K\|(S\alpha)(s) - (S\beta)(s)\| \, ds$

$$\leq \int_{t_0}^{t} K \cdot K\|\alpha - \beta\|(s - t_0) \, ds$$

$$\leq K^2\|\alpha - \beta\| \frac{(t - t_0)^2}{2} \quad \text{on } [a, b]$$

   (c)   Show why it is all right to "take the sup of both sides" of the inequalities of parts (a) and (b) to get

$$\|S\alpha - S\beta\| \leq \sup_{a \leq t \leq b} K\|\alpha - \beta\|(t - t_0) \leq K\|\alpha - \beta\| \max\{|a - t_0|, |b - t_0|\}$$

and

$$\|S^2\alpha - S^2\beta\| \leq \sup_{a \leq t \leq b} K^2\|\alpha - \beta\| \frac{(t - t_0)^2}{2} \leq \frac{K^2\|\alpha - \beta\|}{2} \max\{|a - t_0|^2, |b - t_0|^2\}$$

[*Hint:* See Exercise 1(d).]

   (d)   Prove that $\|S^n\alpha - S^n\beta\| \leq \dfrac{K^n\|\alpha - \beta\|}{n!} \max\{|a - t_0|^n, |b - t_0|^n\}$.

✔ **7.**   Let $f : \mathbf{R} \to \mathbf{R}$ by $f(t) = t^2$.
   (a)   Is f a contraction on $\mathbf{R}$?
   (b)   Prove that f is a contraction on $\left[0, \dfrac{1}{3}\right]$ and find a fixed point of f.
   (c)   Study the behavior of the sequence

$$x_0 = \frac{1}{10}$$

$$x_1 = f(x_0)$$

$$\vdots$$

$$x_n = f(x_{n-1})$$

Repeat for the initial point $x_0 = \dfrac{1}{4}$.

Repeat for the initial point $x_0 = 2$, which is outside the region where f is contractive.

COMP   8.   **(a)**   Solve the second order differential equation $\ddot{y} - 4\dot{y} + 3y = 0$ subject to the initial data $y(0) = 2$, $\dot{y}(0) = 4$ and compute the values $y(0.5)$ and $\dot{y}(0.5)$.

(b)   Transform the equation to a first order matrix differential equation $\dot{x}(t) = \bar{f}(t, x(t))$, $x(0) = x_0$ by choosing state variables $x_1 = y$, $x_2 = \dot{y}$. Does it satisfy a Lipschitz condition? Generate the first four terms of the sequence of successive approximations $r_0(t), \ldots, r_3(t)$ and compare $r_3(0.5)$ with your answer from part (a).

MP **9.**   When we established the existence and uniqueness theorem for the initial value problem, we transformed an operator equation of the form

$$Tx = 0 \quad \left(\text{we had } Tx = \frac{d}{dt} x - \bar{f}(t, x) \right)$$

into an equivalent operator equation of the form

$$x = Sx \quad \left(\text{we had } Sx(\cdot) = x_0 + \int_{t_0}^{(\cdot)} \bar{f}(\xi, x(\xi)) \, d\xi \right)$$

so that our solution would be a fixed point of S. We found the solution as the limit of a sequence of successive approximations $x_{n+1} = S(x_n)$, which under appropriate conditions on T made S contractive. This method of *iterating* to get the solution is used in many different contexts.
(a)   In any elementary book on numerical analysis, look up the Newton-Raphson method for finding roots of equations and the Seidel or Gauss-Seidel methods of solving matrix equations.
(b)   Use the Newton-Raphson method to find a real root of the real polynomial $T(x) = x^3 - 0.35x^2 - 0.35x - 1.35$, starting with a first approximation of $x_0 = 1.00$. Repeat for $x_0 = 10.0$.

(c)   Use the Gauss-Seidel method to solve $Tx = b$ with $T = \begin{bmatrix} 40 & -2 \\ 1 & 20 \end{bmatrix}$ and $b = \begin{bmatrix} -42 \\ 19 \end{bmatrix}$ with initial guess $x_0 = \begin{bmatrix} 0 \\ 0 \end{bmatrix}$. Repeat with $T = \begin{bmatrix} 1 & 20 \\ 40 & -2 \end{bmatrix}$, $b = \begin{bmatrix} 19 \\ -42 \end{bmatrix}$
and compare.
10.   Let E be the set of all continuous maps from a finite interval $[t_1, t_2]$ of $\mathbf{R}$ into a Banach space X. Verify that E is a Banach space by the following steps:
(a)   Define linear combinations $r_1 x_1 + r_2 x_2$ for r_1, r_2 in $\mathbf{R}$ and x_1, x_2 in X by

$$(r_1 x_1 + r_2 x_2)(t) = r_1 x_1(t) + r_2 x_2(t) \text{ for all } t \text{ in } [t_1, t_2]$$

and check that $r_1 x_1 + r_2 x_2$ is continuous, and so does indeed belong to E.
(b)   Check (recalling the formal definition of a norm from Section 2–4) that

$$\|x\|_{\mathbf{E}} = \sup_{t_1 \leq t \leq t_2} \|x(t)\|_X$$

really does make E a normed linear space.
(c)   Prove that E is *complete*, i.e., that if $(x_1, x_2, \ldots, x_n, \ldots)$ is a Cauchy sequence, then $x = \lim_{n \to \infty} x_n$ (defined by $x(t) = \lim_{n \to \infty} x_n(t)$) really exists, and belongs to E. [*Hint:* First prove that $(x_1(t), x_2(t), \ldots)$ is a Cauchy sequence in X, and so has a limit, since X is complete, for each t. Then use the inequality

$$\|x(\tau_1) - x(\tau_2)\|_X \leq \|x(\tau_1) - x_n(\tau_1)\|_X + \|x_n(\tau_1) - x_n(\tau_2)\|_X + \|x_n(\tau_2) - x(\tau_2)\|_X$$

to prove that x is continuous.]

5–3   Linear Differential Systems

NORMS OF LINEAR OPERATORS AND MATRICES

Suppose, now, that we are studying the *linear* differential system described by

$$\dot{x}(t) = F(t)x(t) + G(t)u(t) \tag{1}$$

Having fixed upon a particular input function *u*, we are then considering the case in which

$$\bar{f}(t, x) = F(t)x + G(t)u(t) \tag{2}$$

To say that $\bar{f}$ satisfies a Lipschitz condition is thus simply to say that there exists a constant *K* for which

$$\|\bar{f}(t, x) - \bar{f}(t, \hat{x})\| \leq K\|x - \hat{x}\| \tag{3}$$

But $\bar{f}(t, x) - \bar{f}(t, \hat{x}) = F(t)x - F(t)\hat{x} = F(t)(x - \hat{x})$ by linearity of $F(t)$. We thus demand that

$$\|F(t)(x - \hat{x})\| \leq K\|x - \hat{x}\| \tag{4}$$

holds true for all values of *t*. Thus, we make the following definition:

DEFINITION 1

A linear transformation $A: X \to Y$ of a normed vector space *X* into the normed vector space *Y* is **bounded** if there exists a finite positive constant *M* such that

$$\|Ax\|_Y \leq M\|x\|_X \text{ for all } x \text{ in } X$$

We call the smallest such *M* the **norm** of *A*, and denote it by $\|A\|$.   ◯

Thus, equation (4) simply says that $F(t)$ must be *uniformly* bounded as a family of linear transformations:

$$\|F(t)\| \leq M \text{ for all } t$$

Using the notation *inf* to stand for *infimum* or *greatest lower bound* [see Exercise 1 of Section 5–2], we may write

$$\|A\| = \inf_{0 < M < \infty} \{M \mid \|Ax\| \leq M\|x\|, \text{ for all } x \in X\} \tag{5}$$

as a more precise definition for the norm of an operator. Thus A is bounded if and only if $\|A\|$ exists.

Example 1

Let $X = Y$ be any normed vector space and let $A = I$, the identity map on X. Then

$$\|Ix\| = \|x\| \leq M\|x\|$$

for all real $M \geq 1$. Thus, I is bounded and

$$\|I\| = \inf_{0 < M < \infty} \{M \mid \|Ix\| \leq M\|x\|, \text{ for all } x \in X\}$$
$$= \inf_{0 < M < \infty} \{M \mid 1 \leq M\} = 1 \qquad \diamond$$

There are several useful equivalent [see Exercise 1] definitions for the norm of a bounded linear operator:

$$\|A\| = \sup_{\|x\| = 1} \|Ax\| \qquad (6)$$

$$\|A\| = \sup_{\|x\| \leq 1} \|Ax\| \qquad (7)$$

$$\|A\| = \sup_{x \neq 0} \frac{\|Ax\|}{\|x\|} \qquad (8)$$

and, in particular, if X is finite-dimensional [Exercise 2],

$$\|A\| = \max_{\|x\| = 1} \|Ax\| \qquad (9)$$

the maximum value of $\|Ax\|$ as x ranges over the unit sphere $\{x \mid \|x\| = 1\}$.

To provide practice with these alternative expressions for $\|A\|$, we will use equation (6) to recapture the property of Definition 1 that

$$\|Ax\| \leq \|A\| \, \|x\| \text{ for all } x \qquad (10)$$

when A is bounded:

From (6), if $\|x\| = 1$, then $\|Ax\| \leq \|A\| = \|A\| \, \|x\|$. If $x = 0$, then $Ax = 0$ also, and $\|Ax\| = \|0\| = 0 \leq \|A\| \cdot 0 = \|A\| \, \|x\|$. Finally, if $x \neq 0$, so $\dfrac{x}{\|x\|}$ has unit length, then $\left\| A\left(\dfrac{x}{\|x\|}\right) \right\| \leq \|A\|$, by (6). But $\left\| A\left(\dfrac{x}{\|x\|}\right) \right\| = \left\| \dfrac{1}{\|x\|} A(x) \right\| = \dfrac{1}{\|x\|} \|Ax\|$, so $\|Ax\| \leq \|A\| \, \|x\|$ for all x.

Example 2

Let $B: X \to Y$ and $A: Y \to Z$ be bounded linear maps so that AB maps X into Z.

Then $\|(AB)(x)\| = \|A(Bx)\| \leq \|A\| \, \|Bx\|$, by (10) and $\|Bx\| \leq \|B\| \, \|x\|$, so $\|(AB)(x)\| \leq \|A\| \, \|B\| \, \|x\|$.

Thus the number $M = \|A\| \, \|B\|$ shows that the composition AB is bounded when A and B are. Moreover, since $\|AB\|$ is the infimum of all such M, we have

$$\|AB\| \leq \|A\| \, \|B\|$$

Similarly [Exercise 4], if $C: X \to Y$ is a bounded linear map, then

$$\|C + B\| \leq \|C\| + \|B\| \qquad \qquad \lozenge$$

Our remarks so far apply to linear operators on infinite-dimensional spaces as well as finite ones. When the dimension of the space X is finite, *every* linear transformation is bounded† [Exercises 5 and 7], and our conclusions thus apply to all linear finite-dimensional matrix-differential equations.

Since the norm of an $n \times n$ matrix A is given by

$$\|A\| = \sup_{\|x\|=1} \|Ax\|$$

or the equivalent expressions (5), (7), or (8), we will get different norms for our operators if we use different norms for our vectors. For example [see Exercise 6], if we use $\|x\|_\infty$, the uniform norm of Example 2-4-4, we get

$$\|A\|_\infty = \max_{1 \leq i \leq n} \left\{ \sum_{j=1}^{n} |a_{ij}| \right\}, \quad \binom{\text{row}}{\text{norm}}$$

but if we use $\|x\|_1$, the ℓ_1 norm for x [Example 2-4-4], we get

$$\|A\|_1 = \max_{1 \leq j \leq n} \left\{ \sum_{i=1}^{n} |a_{ij}| \right\}, \quad \binom{\text{column}}{\text{norm}}$$

where a_{ij} is the ij element of A.

It can be shown [Luenberger [1969], p. 146; also, see Exercise 7] that using the ℓ_2 (Euclidean) norm,

$$\|x\|_2 = \left(\sum_{i=1}^{n} |x_i|^2 \right)^{1/2}$$

---

†For two other proofs of this fact, see G. E. Shilov, *Introduction to the Theory of Linear Spaces* (Prentice-Hall, 1961), p. 145, and D. G. Luenberger, *Optimization by Vector Space Methods* (Wiley, 1969), p. 146.

yields the corresponding matrix norm

$$\|A\|_2 = \sqrt{\lambda_{max}}$$

where λ_{max} is the largest eigenvalue† of the matrix $Q = A^*A$.

Our proof, in Section 5–2, of the existence of a unique solution for the differential equation (1) requires that, given any continuous trajectory defined over a finite time period (a, b), we may form the integral

$$\int_a^b \overline{f}(t, x(t))\, dt = \int_a^b [F(t)x(t) + G(t)u(t)]\, dt$$

and to ensure this works, it will usually suffice to demand that $G(t)$ also forms a family of uniformly bounded linear operators.

In particular, we know that whenever we have a *finite-dimensional* linear differential equation of the form $\dot{x}(t) = F(t)x(t) + G(t)u(t)$, it has a uniquely defined solution for every initial condition, so long as the specified input function $u(\cdot)$ is piecewise continuous.

Example 3

Solve the homogeneous matrix differential equation

$$\dot{x}(t) = F(t)x(t)$$

subject to the initial condition $x(t_0) = 0$, where $F(t)$ is a real $n \times n$ matrix whose elements are known piecewise continuous real functions.

Here the input $u(t) \equiv 0$ (certainly piecewise continuous), and the matrix $F(t)$ is bounded on every closed interval, so we know that our initial value problem has a unique solution. Clearly $x(t) \equiv 0$, for $t \geq t_0$, is a solution of the differential equation and satisfies the specified initial condition at $t = t_0$. Thus, by uniqueness, $x(t) \equiv 0$ is the solution, and we have shown that an unforced, initially relaxed linear system remains relaxed.‡   ◊

TIME-VARYING, FINITE-DIMENSIONAL SYSTEMS

Consider the n-dimensional, time-varying linear system represented by the equations

$$\dot{x}(t) = F(t)x(t) + G(t)u(t) \tag{11}$$

$$y(t) = H(t)x(t)$$

---

†Eigenvalues and eigenvectors of linear systems will be discussed in Section 6–2.
‡This is hardly surprising since, from $x(t) = \phi(t, t_0, x_0, u)$, when $x_0 = 0$ and $u(t) = 0$, we have $x(t) = \phi(t, t_0, 0, 0)$, which is zero by linearity of $\phi(t, t_0, \cdot, \cdot)$ [Section 3–3].

where $F(\cdot)$, $G(\cdot)$, and $H(\cdot)$ are respectively $n \times n$, $n \times m$, and $q \times n$ matrices whose entries are real-valued, piecewise continuous, uniformly bounded functions of t. Then for any piecewise continuous uniformly bounded input u, and any initial state x_0, there is a unique solution for the state equation (11).

THE HOMOGENEOUS SOLUTION

Just as in Section 5–1 in our motivating study of discrete-time systems, we will first turn our attention to the solution of the homogeneous (unforced) portion of equation (11) obtained by setting $u(t) \equiv 0$:

For concreteness, we shall focus on the case in which X is of finite dimension, say n. The first thing we shall prove is that the set of solutions of the homogeneous equation is a vector space of the same dimension as the underlying state space X.

THEOREM 1

Let $F(t):X \to X$ be a family of linear transformations, uniformly bounded on any finite interval. Then the set X_F of solutions of the *l*inear *h*omogeneous (LH) equation

$$\text{LH:} \quad \dot{x}(t) = F(t)x(t)$$

form a vector space of the same dimension as X.

Proof

Let $B = \{v_1, v_2, \ldots, v_n\}$ be any basis for X. By our existence theorem, each initial value problem

$$\dot{x}(t) = F(t)x(t)$$
$$x(t_0) = v_j$$

has a unique solution $W_j(\cdot)$, for $1 \leq j \leq n$. Thus

$$\dot{W}_j(t) = F(t)W_j(t)$$

and

$$W_j(t_0) = v_j \tag{12}$$

and the set of solutions, X_F, is non-empty. To show that the set of functions comprising X_F is indeed a vector space under the usual definitions of addition and multiplication by scalars of functions, we show that X_F is closed under linear combinations:

Let w and $\hat{w}$ be solutions of LH and let k and k' be scalars. Then

$$\frac{d}{dt}(kw + k'\hat{w}) = k\frac{d}{dt}w + k'\frac{d}{dt}\hat{w} \quad \text{by linearity of } \frac{d}{dt}$$

$$= kFw + k'F\hat{w} \qquad \text{by LH}$$
$$= F(kw + k'\hat{w}) \qquad \text{by linearity of } F$$

Thus $kw + k'\hat{w}$ belongs to X_F when w and $\hat{w}$ do, and X_F is a vector space.

Now we shall show that the solutions $W_1(\cdot), \ldots, W_n(\cdot)$ are a basis for X_F. Suppose that some linear combination of the functions W_j is the zero function so that

$$c_1 W_1(t) + \cdots + c_n W_n(t) = 0, \quad \text{for all } t$$

Evaluation at t_0 yields

$$c_1 W_1(t_0) + \cdots + c_n W_n(t_0) = 0$$

and, from (12),

$$c_1 v_1 + \cdots + c_n v_n = 0$$

Since the v_j form a basis for X, we conclude that all the $c_j = 0$, and hence $W_1(\cdot), \ldots, W_n(\cdot)$ are linearly independent.

To show that the $W_j(\cdot)$ span X_F, let $\psi(\cdot)$ be any solution of LH. Since $\psi(t_0)$ belongs to X, it can be written in terms of the basis as

$$\psi(t_0) = a_1 v_1 + \cdots + a_n v_n$$

for suitable unique scalars $a_1, \ldots, a_n$.

Consider the function

$$w(\cdot) = a_1 W_1(\cdot) + \cdots + a_n W_n(\cdot)$$

Its value at t_0,

$$w(t_0) = a_1 W_1(t_0) + \cdots + a_n W_n(t_0)$$
$$= a_1 v_1 + \cdots + a_n v_n$$
$$= \psi(t_0)$$

is the value taken by $\psi(\cdot)$ at t_0. Now $w(\cdot)$, being a linear combination of the $W_j(\cdot)$ which are solutions of LH, is itself a solution of LH. But the solution of LH with a specified initial condition is unique, so

$$\psi(\cdot) = w(\cdot) = a_1 W_1(\cdot) + \cdots + a_n W_n(\cdot)$$

and the functions $W_1(\cdot), \ldots, W_n(\cdot)$ span X_F.

Since the set $W_j(\cdot)$, $1 \leq j \leq n$, is a basis for X_F we have dim X_F = dim X as claimed. □

We call the set of solutions $W_j(\cdot)$ of equations (12) a **fundamental system of solutions** of LH.

Example 4

Consider the LH equation

$$\begin{bmatrix} \dot{x}_1(t) \\ \dot{x}_2(t) \end{bmatrix} = \begin{bmatrix} 1 & 0 \\ 0 & 2t \end{bmatrix} \begin{bmatrix} x_1(t) \\ x_2(t) \end{bmatrix} \tag{13}$$

and let $v_1 = \begin{bmatrix} 1 \\ 0 \end{bmatrix}$, $v_2 = \begin{bmatrix} 1 \\ 1 \end{bmatrix}$ so that $\{v_1, v_2\}$ is a basis for $X = \mathbf{R}^2$.

The matrix equation (13) yields the two scalar equations

$$\dot{x}_1(t) = x_1(t)$$
$$\dot{x}_2(t) = 2t\, x_2(t)$$

whose general solutions (verify!) are

$$x_1(t) = x_1(t_0)e^{(t-t_0)}$$

and

$$x_2(t) = x_2(t_0)e^{(t^2 - t_0^2)}$$

Thus

$$x(t) = \begin{bmatrix} x_1(t_0)e^{(t-t_0)} \\ x_2(t_0)e^{(t^2 - t_0^2)} \end{bmatrix} \tag{14}$$

is the general solution of (13). For the fundamental system of solutions $W_1(\cdot)$, $W_2(\cdot)$ we require solutions (14) such that $W_1(t_0) = \begin{bmatrix} 1 \\ 0 \end{bmatrix}$ and $W_2(t_0) = \begin{bmatrix} 1 \\ 1 \end{bmatrix}$.

Setting

$$W_1(t_0) = x(t_0) = \begin{bmatrix} x_1(t_0) \\ x_2(t_0) \end{bmatrix} = \begin{bmatrix} 1 \\ 0 \end{bmatrix}$$

we get

$$W_1(t) = \begin{bmatrix} e^{(t-t_0)} \\ 0 \end{bmatrix}$$

and from

$$W_2(t_0) = x(t_0) = \begin{bmatrix} x_1(t_0) \\ x_2(t_0) \end{bmatrix} = \begin{bmatrix} 1 \\ 1 \end{bmatrix}$$

we get

$$W_2(t) = \begin{bmatrix} e^{(t-t_0)} \\ e^{(t^2-t_0^2)} \end{bmatrix}$$

It is easily checked that $W_1(\cdot)$ and $W_2(\cdot)$ form a basis for

$$X_F = \left\{ \begin{bmatrix} x_1(t_0)e^{(t-t_0)} \\ x_2(t_0)e^{(t^2-t_0^2)} \end{bmatrix} \;\middle|\; \begin{bmatrix} x_1(t_0) \\ x_2(t_0) \end{bmatrix} \in \mathbf{R}^2 \right\}$$

In fact, since $x(t_0) = a_1 v_1 + a_2 v_2$ yields

$$\begin{bmatrix} x_1(t_0) \\ x_2(t_0) \end{bmatrix} = a_1 \begin{bmatrix} 1 \\ 0 \end{bmatrix} + a_2 \begin{bmatrix} 1 \\ 1 \end{bmatrix} = \begin{bmatrix} a_1 + a_2 \\ a_2 \end{bmatrix}$$

whence

$$a_1 = x_1(t_0) - x_2(t_0) \text{ and } a_2 = x_2(t_0),$$

we have

$$x(t) = \begin{bmatrix} x_1(t_0)e^{(-t_0)} \\ x_2(t_0)e^{(t^2-t_0^2)} \end{bmatrix} = (x_1(t_0) - x_2(t_0)) \, W_1(t) + x_2(t_0) W_2(t) \qquad \Diamond$$

Note that the proof of Theorem 1 shows us that if we take any particular time t_0, there is a one-to-one correspondence between the basis $W_1(\cdot), \ldots,$ $W_n(\cdot)$ for the space of solutions, and the basis $v_1, \ldots, v_n$ for the state space, if the latter are considered as representing initial conditions at that time.

Since each fundamental solution $W_j(\cdot)$ is a solution of LH, if we construct the matrix $W(t)$ having $W_1(t), \ldots, W_n(t)$ as its columns,

$$W(t) = [W_1(t) \quad W_2(t) \ldots W_n(t)]$$

then $W(t)$ satisfies the matrix equation

$$\dot{W}(t) = F(t)W(t) \tag{15}$$

because

$$\begin{aligned} F(t)[W_1(t) \quad W_2(t) \ldots W_n(t)] &= [F(t)W_1(t) \quad F(t)W_2(t) \ldots F(t)W_n(t)] \\ &= [\dot{W}_1(t) \quad \dot{W}_2(t) \quad \ldots \dot{W}_n(t) \quad] \\ &= \frac{d}{dt} [W_1(t) \quad W_2(t) \quad \ldots W_n(t) \quad] \end{aligned}$$

We call such a matrix $W(t)$, whose columns consist of n linearly independent solutions of LH, a **fundamental matrix** or a **Wronskian matrix**.

Example 5

For the LH of Example 4 we found the fundamental solutions

$$W_1(t) = \begin{bmatrix} e^{(t-t_0)} \\ 0 \end{bmatrix}, \; W_2(t) = \begin{bmatrix} e^{(t-t_0)} \\ e^{(t^2-t_0^2)} \end{bmatrix}$$

Hence the matrix

$$W(t) = \begin{bmatrix} e^{(t-t_0)} & e^{(t-t_0)} \\ 0 & e^{(t^2-t_0^2)} \end{bmatrix}$$

is a fundamental matrix for that LH. ◊

Conversely, any matrix function $P(t)$ which satisfies $\dot{P}(t) = F(t)P(t)$ and is non-singular for some t_0, so that the columns of $P(t_0)$ constitute some basis for X, is a fundamental matrix for LH.

We will now show that because the fundamental matrix $W(t)$ is non-singular at some time t_0, it will be non-singular for all t. [Take a peek at Exercise 8.]

THEOREM 2

If $W(t) = [W_1(t) \ldots W_n(t)]$ is a fundamental matrix of LH, then $W(t)$ is non-singular for all t.

Proof

Let t_0 be the time at which $W(t)$ is known to be non-singular. Suppose, for contradiction, that $W(t)$ is singular at some time t_1. Then the columns $W_1(t_1), \ldots, W_n(t_1)$ are linearly dependent, so there exist scalars $c_1, \ldots c_n$, not all zero, such that

$$c_1 W_1(t_1) + \cdots + c_n W_n(t_1) = 0$$

Taking t_1 as an initial time, we thus have determined a function $c_1 W_1(\cdot) + \cdots + c_n W_n(\cdot)$ which is a solution of LH and is zero at t_1.

Since the zero function is a solution of LH which is zero at t_1, we have by uniqueness that

$$c_1 W_1(t) + \cdots + c_n W_n(t) = 0 \text{ for all } t$$

so that at t_0 we certainly have

$$c_1 W_1(t_0) + \cdots + c_n W_n(t_0) = 0$$

which contradicts the linear independence of $W_1(t), \ldots, W_n(t)$ at t_0. □

The determinant, det $W(t)$, of a fundamental (Wronskian) matrix $W(t)$ of LH is called the **Wronskian** of the functions $W_1(\cdot), \ldots, W_n(\cdot)$.

COROLLARY 3

The Wronskian of a set of fundamental solutions $W_1(\cdot), \ldots, W_n(\cdot)$ of LH is never zero. □

It can be shown [Exercise 15] after a little matrix calculus [Exercises 9 and 11] that the Wronskian is given by the so-called **Liouville formula:**

$$\det W(t) = \det W(t_0) \ e^{\int_{t_0}^{t} \text{trace } F(\tau) \, d\tau} \tag{16}$$

where trace $F(t) = \text{trace } [f_{ij}(t)] = \displaystyle\sum_{i=1}^{n} f_{ii}(t)$.

Now, let $W(\cdot)$ be any fundamental matrix of LH. Since $W(\cdot)$ is a solution of equation (15) and since $W(t_0)$ is a non-singular constant matrix for any t_0, the matrix $\Phi(\cdot, \cdot)$ defined by

$$\Phi(t, t_0) = W(t) W^{-1}(t_0) \text{ for all } t, t_0 \tag{17}$$

also satisfies (15). [Verify by differentiating Φ; see Exercise 14(d).] Thus,

$$\frac{\partial}{\partial t} \Phi(t, t_0) = F(t) \Phi(t, t_0) \tag{18}$$

and

$$\Phi(t_0, t_0) = I \text{ (the identity matrix)}$$

for all t_0.

We call this special fundamental matrix $\Phi(t, t_0)$ the **state transition matrix** of $\dot{x}(t) = F(t) x(t)$. Since each column Φ_j is a solution of the initial value problem

$$\frac{\partial}{\partial t} \Phi_j(t, t_0) = F(t)\Phi_j(t, t_0)$$

$$\Phi_j(t_0, t_0) = e_j$$

(where e_j is the jth unit vector of the standard basis), it is unique and $\Phi(t, t_0)$ does not depend on the particular $W(t)$ used in equation (17).

Example 6

The matrix $W(t) = \begin{bmatrix} e^t & e^t \\ 0 & e^{t^2} \end{bmatrix}$

is a fundamental matrix for the LH

$$\dot{x}(t) = \begin{bmatrix} 1 & 0 \\ 0 & 2t \end{bmatrix} x(t) \quad \text{[Example 4]}$$

since

$$\frac{d}{dt} W(t) = \frac{d}{dt} \begin{bmatrix} e^t & e^t \\ 0 & e^{t^2} \end{bmatrix} = \begin{bmatrix} e^t & e^t \\ 0 & 2te^{t^2} \end{bmatrix}$$

$$= \begin{bmatrix} 1 & 0 \\ 0 & 2t \end{bmatrix} \begin{bmatrix} e^t & e^t \\ 0 & e^{t^2} \end{bmatrix}$$

$$= F(t)W(t)$$

and since

$$W(t_0) = \begin{bmatrix} e^{t_0} & e^{t_0} \\ 0 & e^{t_0^2} \end{bmatrix}$$

is non-singular. Since $\det W(t_0) = e^{t_0 + t_0^2} \neq 0$, we have

$$W^{-1}(t_0) = \frac{1}{e^{t_0 + t_0^2}} \begin{bmatrix} e^{t_0^2} & -e^{t_0} \\ 0 & e^{t_0} \end{bmatrix}$$

$$= \begin{bmatrix} e^{-t_0} & -e^{-t_0^2} \\ 0 & e^{-t_0^2} \end{bmatrix}$$

Thus

$$\Phi(t, t_0) = W(t)W^{-1}(t_0) = \begin{bmatrix} e^t & e^t \\ 0 & e^{t^2} \end{bmatrix} \begin{bmatrix} e^{-t_0} & -e^{-t_0^2} \\ 0 & e^{-t_0^2} \end{bmatrix}$$

$$= \begin{bmatrix} e^{(t-t_0)} & 0 \\ 0 & e^{(t^2 - t_0^2)} \end{bmatrix}$$

is the state transition matrix for this system.

To check,

$$\Phi(t_0, t_0) = \begin{bmatrix} 1 & 0 \\ 0 & 1 \end{bmatrix} = I$$

and

$$\frac{\partial}{\partial t} \Phi(t, t_0) = \frac{\partial}{\partial t} \begin{bmatrix} e^{(t-t_0)} & 0 \\ 0 & e^{(t^2-t_0^2)} \end{bmatrix} = \begin{bmatrix} e^{(t-t_0)} & 0 \\ 0 & 2te^{(t^2-t_0^2)} \end{bmatrix}$$

$$= \begin{bmatrix} 1 & 0 \\ 0 & 2t \end{bmatrix} \begin{bmatrix} e^{(t-t_0)} & 0 \\ 0 & e^{(t^2-t_0^2)} \end{bmatrix} = F(t)\Phi(t, t_0) \qquad \Diamond$$

Because the state transition matrix satisfies the matrix differential equation (15) and reduces to the identity matrix at $t = t_0$, we have the following result:

THEOREM 4

The unforced finite-dimensional system

$$\dot{x}(t) = F(t)x(t)$$

starting from the initial state $x(t_0) = x_0$, has the unique state trajectory $x(t) = \phi(t, t_0, x_0, 0)$ given by

$$x(t) = \phi(t, t_0, x_0, 0) = \Phi(t, t_0)x_0 \qquad (19)$$

where the matrix $\Phi(t, t_0)$ is the unique matrix satisfying

$$\frac{\partial}{\partial t} \Phi(t, t_0) = F(t)\Phi(t, t_0)$$

and

$$\Phi(t_0, t_0) = I$$

Proof

Differentiating $x(t) = \Phi(t, t_0)x_0$ yields

$$\dot{x}(t) = (\dot{\Phi}(t, t_0))x_0 = (F(t)\Phi(t, t_0))x_0 = F(t)x(t)$$

and

$$x(t_0) = \Phi(t_0, t_0)x_0 = Ix_0 = x_0 \qquad \square$$

Equation (19) of Theorem 4 reveals the reason that $\Phi(t, t_0)$ is called the state transition matrix. It describes the free, unforced, "natural" motion of the state vector $x(t)$ of a linear system, and represents the linear transformation which maps the initial state x_0 at time t_0 into the state x at time t. This vindicates our heuristic derivation of equation (3) of Section 5–1.

The following theorem summarizes some of the most important properties of the state transition matrix.

THEOREM 5

If $\Phi(t, t_0)$ is the state transition matrix of the system $\dot{x}(t) = F(t)x(t)$, then for all t_0, t_1, t_2:

$$\Phi(t_2, t_0) = \Phi(t_2, t_1)\Phi(t_1, t_0) \tag{20}$$

$$\Phi^{-1}(t_2, t_0) = \Phi(t_0, t_2) \tag{21}$$

$$\det \Phi(t_1, t_0) = e^{\int_{t_0}^{t_1} \text{trace } F(\tau)\, d\tau} \tag{22}$$

and its "time reversed adjoint" $\Phi^*(t_0, t)$ is the state transition matrix of the system

$$\dot{q}(t) = -F^*(t)q(t) \tag{23}$$

Proof

If $W(t)$ is any fundamental matrix of $\dot{x} = Fx$, then

$$\Phi(t_2, t_0) = W(t_2)W^{-1}(t_0) = W(t_2)[W^{-1}(t_1)W(t_1)]W^{-1}(t_0)$$
$$= [W(t_2)W^{-1}(t_1)][W(t_1)W^{-1}(t_0)] = \Phi(t_2, t_1)\Phi(t_1, t_0)$$

which is (20).

Using (20), and letting $t_2 = t_0$,

$$I = \Phi(t_0, t_0) = \Phi(t_0, t_1)\Phi(t_1, t_0)$$

whence (21) follows.

Equation (22) follows from the Liouville formula (16), since Φ is itself a fundamental matrix of $\dot{x} = Fx$. Substituting Φ for W in (16):

$$\det \Phi(t, t_0) = \det \Phi(t_0, t_0)e^{\int_{t_0}^{t} \text{trace } F(\tau)\, d\tau}$$

and since $\det \Phi(t_0, t_0) = \det I = 1$, (22) results.

Finally, using Exercise 9(b) to find the derivative of the inverse of a matrix on $\Phi^{-1}(t, t_0)$,

$$\frac{\partial}{\partial t} \Phi(t_0, t) = \frac{\partial}{\partial t} \Phi^{-1}(t, t_0) = -\Phi^{-1}(t, t_0)\left[\frac{\partial}{\partial t} \Phi(t, t_0)\right]\Phi^{-1}(t, t_0)$$

$$= -\Phi^{-1}(t, t_0)[F(t)\Phi(t, t_0)]\Phi^{-1}(t, t_0)$$

$$= -\Phi^{-1}(t, t_0)f(t)$$

$$\frac{\partial}{\partial t} \Phi(t_0, t) = -\Phi(t_0, t)F(t)$$

Since the adjoint of a product is the product of the adjoints in reverse order [Section 4-4, Theorem 8], and since

$$\left(\frac{\partial}{\partial t} \Phi(t_0, t)\right)^* = \frac{\partial}{\partial t} \Phi^*(t_0, t)$$

we obtain

$$\frac{\partial}{\partial t} \Phi^*(t_0, t) = -F^*(t)\Phi^*(t_0, t) \text{ (note order of the arguments)}$$

as claimed in property (23).   □

Property (20) is often called the "semigroup property" or "composition rule" of the state transition matrix, and it is illustrated in Figure 5-2. Since

$$\begin{aligned}
x(t_2) &= \Phi(t_2, t_0)x(t_0) \\
&= [\Phi(t_2, t_1)\Phi(t_1, t_0)]x(t_0) \\
&= \Phi(t_2, t_1)[\Phi(t_1, t_0)x(t_0)] \\
&= \Phi(t_2, t_1)x(t_1)
\end{aligned}$$

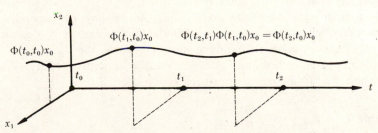

Figure 5-2   Two-dimensional state space illustration of the semigroup property of $\Phi(t, t_0)$.

and since $x(t) = \phi(t, t_0, x(t_0), 0)$ by equation (19), property (20) is just a verification that our state transition equation satisfies the consistency equation discussed in Section 1-5:

$$\phi(t_2, t_0, x(t_0), u) = \phi(t_2, t_1, \phi(t_1, t_0, x(t_0), u))$$

Property (21) reveals the time reversal property of continuous-time systems and shows that, since $\Phi(t_2, t_0)$ transforms the state at time t_0 into the state at time t_2, then the inverse operation $\Phi^{-1}(t_2, t_0)$ transforming the state at time t_2 back to that at t_0 is accomplished by simply reversing the roles of t_0 and t_2 in the transition matrix. This property is illustrated in Figure 5–3.

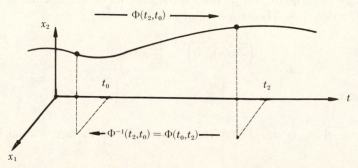

Figure 5–3   Two-dimensional state space illustration of the time reversal property.

The system $\dot{q}(t) = -F^*(t)q(t)$ of property (23) is called the **adjoint system** of the system $\dot{x} = Fx$. Since

$$\Phi^*(t_0, t) = [\Phi^{-1}(t, t_0)]^*$$

equation (23) shows that the adjoint equation determines the rate of change of Φ with respect to its second argument while equation (18) specifies the rate of change of Φ with respect to its first argument. The most important property of the adjoint equation is that it shows how to propagate the solution of the original equation backwards in time. As we show in Appendix A–3, this property is used extensively in optimization theory. It is also used in simulation and in solving boundary-value problems. We shall see more of the adjoint system in Section 7–3, when we discuss controllability, observability, and duality for continuous-time systems.

Example 7

Consider the homogeneous nth order scalar differential equation

$$y^{(n)} + a_1(t)y^{(n-1)} + \cdots + a_n(t)y = 0 \tag{24}$$

By choosing state variables

$$x_1 = y$$
$$x_2 = \dot{y}$$
$$\vdots$$
$$x_n = y^{(n-1)}$$

we get the equivalent first order, n-dimensional system

$$\dot{x}(t) = \begin{bmatrix} 0 & 1 & 0 & \cdots & 0 \\ 0 & 0 & 1 & & 0 \\ \vdots & \vdots & & & \\ 0 & 0 & & & 1 \\ -a_n & -a_{n-1} & & \cdots & -a_1 \end{bmatrix} x(t)$$

Notice that the coefficients of the scalar equation appear in reverse order and with minus signs in the bottom row of the matrix F.

The adjoint system is

$$\dot{q}(t) = \begin{bmatrix} 0 & 0 & & 0 & \bar{a}_n \\ -1 & 0 & \cdots & 0 & \bar{a}_{n-1} \\ \vdots & \vdots & & & \vdots \\ 0 & 0 & \cdots & -1 & \bar{a}_1 \end{bmatrix} q(t)$$

and corresponds to the component equations

$$\dot{q}_1 = \bar{a}_n q_n$$
$$\dot{q}_2 = -q_1 + \bar{a}_{n-1} q_n$$
$$\dot{q}_{n-1} = -q_{n-2} + \bar{a}_2 q_n$$
$$\dot{q}_n = -q_{n-1} + \bar{a}_1 q_n$$

To see what scalar differential equation this system corresponds to, differentiate the last equation to get

$$\ddot{q}_n = -\dot{q}_{n-1} + (\bar{a}_1 \dot{q}_n)$$

and substitute

$$\dot{q}_{n-1} = -q_{n-2} + \bar{a}_2 q_n$$

from the next to last equation to get

$$\ddot{q}_n = q_{n-2} - \bar{a}_2 q_n + (\bar{a}_1 \dot{q}_n)$$

Differentiate this expression again and substitute for $\dot{q}_{n-2}$, and continue in this way until all that remains is the equation in q_n and its derivatives:

$$(-1)^n q_n^{(n)} + (-1)^{n-1}(\bar{a}_1 q_n)^{(n-1)} + \cdots + (-1)(\bar{a}_{n-1} \dot{q}_n) + \bar{a}_n q_n = 0 \quad (25)$$

The scalar equation (25) is called the adjoint equation of (24).   ◊

Since the state transition matrix Φ of the system $\dot{x} = Fx$ satisfies the matrix differential equation

$$\frac{\partial}{\partial t} \Phi(t, t_0) = F(t)\Phi(t, t_0)$$

we get, upon integrating both sides from t_0 to t,

$$\Phi(t, t_0) - \Phi(t_0, t_0) = \int_{t_0}^{t} F(\tau)\Phi(\tau, t_0)\, d\tau$$

Using $\Phi(t_0, t_0) = I$, we see that $\Phi(t, t_0)$ also satisfies the equivalent integral equation

$$\Phi(t, t_0) = I + \int_{t_0}^{t} F(\tau)\Phi(\tau, t_0)\, d\tau$$

and thus can be evaluated by the sequence of successive approximations:

$$\Phi_{(k+1)}(t, t_0) = I + \int_{t_0}^{t} F(\tau)\Phi_{(k)}(\tau, t_0)\, d\tau, \quad k \geq 0 \tag{26}$$

starting with the initial approximation

$$\Phi_{(0)}(t, t_0) = I$$

so that

$$\Phi(t, t_0) = \lim_{k \to \infty} \Phi_{(k+1)}(t, t_0) \tag{27}$$

Example 8

For the system $\dot{x}(t) = \begin{bmatrix} 0 & t \\ 0 & 0 \end{bmatrix} x(t)$ the sequence of successive approximations yields

$$\Phi_{(0)}(t, t_0) = I$$

$$\Phi_{(1)}(t, t_0) = I + \int_{t_0}^{t} \begin{bmatrix} 0 & \tau \\ 0 & 0 \end{bmatrix} \Phi_{(0)}(\tau, t_0)\, d\tau$$

$$= I + \int_{t_0}^{t} \begin{bmatrix} 0 & \tau \\ 0 & 0 \end{bmatrix} d\tau = I + \begin{bmatrix} 0 & \frac{1}{2}(t^2 - t_0^2) \\ 0 & 0 \end{bmatrix}$$

$$= \begin{bmatrix} 1 & \frac{1}{2}(t^2 - t_0^2) \\ 0 & 1 \end{bmatrix}$$

$$\Phi_{(2)}(t, t_0) = I + \int_{t_0}^t \begin{bmatrix} 0 & \tau \\ 0 & 0 \end{bmatrix} \Phi_{(1)}(\tau, t_0) \, d\tau$$

$$= I + \int_{t_0}^t \begin{bmatrix} 0 & \tau \\ 0 & 0 \end{bmatrix} \begin{bmatrix} 1 & \frac{1}{2}(\tau^2 - t_0^2) \\ 0 & 1 \end{bmatrix} d\tau$$

$$= I + \int_{t_0}^t \begin{bmatrix} 0 & \tau \\ 0 & 0 \end{bmatrix} d\tau$$

$$= \begin{bmatrix} 1 & \frac{1}{2}(t^2 - t_0^2) \\ 0 & 1 \end{bmatrix} = \Phi_{(1)}(t, t_0) \text{ again.}$$

Since $\Phi_{(1)}(t, t_0)$ is a fixed point of the operator $S(A)(t) = I + \int_{t_0}^t F(\tau)A(\tau) \, d\tau$, the sequence of successive approximations has converged after only two iterations to the result

$$\Phi(t, t_0) = \begin{bmatrix} 1 & \frac{1}{2}(t^2 - t_0^2) \\ 0 & 1 \end{bmatrix} \qquad \Diamond$$

As we observed in Section 5-2 after Example 4, it is a rare problem indeed in which we can get a closed-form expression for Φ from the sequence of successive approximations. In practical problems either we solve the differential equation $\dot{\Phi} = F\Phi$ numerically, perhaps by "predictor-corrector" or "Runge-Kutta" methods,† or we use some approximate integration such as the trapezoidal or Simpson rules on the integral equation

$$\Phi_{(k+1)}(t, t_0) = I + \int_{t_0}^t F(\tau)\Phi_{(k)}(\tau, t_0) \, d\tau$$

to convert it into an approximate matrix equation. Both of these approaches are tantamount to approximating a continuous-time system by a discrete-time system, as discussed in Section 5-1 [see Exercises 21 and 22].

The method of successive approximations, of course, converges for many different choices of the initial guess. If we start the process (26) with initial guess $\Phi_{(0)}(t, t_0) = I$, we get

$$\Phi_{(1)}(t, t_0) = I + \int_{t_0}^t F(\tau)I \, d\tau = I + \int_{t_0}^t F(\tau) \, d\tau$$

---

† See Todd, *Survey of Numerical Analysis*, (McGraw-Hill, 1962), Sections 2.24–2.38, 9.1.

$$\Phi_{(2)}(t, t_0) = I + \int_{t_0}^{t} F(\tau)\Phi_{(1)}(\tau, t_0)\, d\tau = I + \int_{t_0}^{t} F(\tau) \left[I + \int_{t_0}^{\tau} F(\tau_1)\, d\tau_1 \right] d\tau$$

$$= I + \int_{t_0}^{t} F(\tau)\, d\tau + \int_{t_0}^{t} \int_{t_0}^{\tau} F(\tau)F(\tau_1)\, d\tau_1\, d\tau$$

so that for $k \geq 2$,

$$\Phi_{(k+1)}(t, t_0) = I + \int_{t_0}^{t} F(\tau)\, d\tau + \cdots$$

$$+ \int_{t_0}^{t} F(\tau) \int_{t_0}^{\tau} F(\tau_1) \cdots \int_{t_0}^{\tau_{k-2}} F(\tau_{k-1}) \int_{t_0}^{\tau_{k-1}} F(\tau_k)\, d\tau_k \cdots d\tau_1\, d\tau$$

Thus

$$\Phi(t, t_0) = \lim_{k \to \infty} \Phi_{k+1}(t, t_0) \tag{28}$$

$$= I + \int_{t_0}^{t} F(\tau)\, d\tau + \int_{t_0}^{t} F(\tau) \int_{t_0}^{\tau} F(\tau_1)\, d\tau_1\, d\tau + \cdots$$

which is an infinite series of nested integrals of the matrix F. Such a series is called the **Neuman series,** and its limit is called the **resolvent** or **matrizant**† for the integral operator S. The expression (28) for $\Phi(t, t_0)$ is often called the **Peano-Baker** formula [see Exercise 23].

Example 9

If $F(t) = F$, constant, then the Peano-Baker formula yields

$$\Phi(t, t_0) = I + F \int_{t_0}^{t} 1\, d\tau + F^2 \int_{t_0}^{t} \int_{t_0}^{\tau} 1\, d\tau_1\, d\tau + \cdots$$

$$+ F^3 \int_{t_0}^{t} \int_{t_0}^{\tau} \int_{t_0}^{\tau_1} 1\, d\tau_2\, d\tau_1\, d\tau + \cdots$$

so

$$\Phi(t, t_0) = I + (t - t_0)F + \frac{1}{2!}(t - t_0)^2 F^2 + \frac{1}{3!}(t - t_0)^3 F^3 + \cdots \tag{29}$$

a function of the difference $(t - t_0)$. Thus $\Phi(t + \tau, t_0 + \tau) = \Phi(t, t_0)$ for all τ, and

$$\Phi(t, t_0) = \Phi(t - t_0, 0) \tag{30}$$

---

†See F. B. Hildebrand, *Methods of Applied Mathematics* (Prentice-Hall, 1965).

Moreover, using (29),

$$\Phi(t, 0)\Phi(s, 0) = \Phi(t + s, s)\Phi(s, 0) = \Phi(t + s, 0)$$

when the matrix F is constant.   ◊

Since the scalar exponential function has the Taylor series expansion

$$e^x = 1 + x + \frac{1}{2!} x^2 + \frac{1}{3!} x^3 + \cdots$$

the matrix series we found in equation (29) and in Exercise 23 motivates us to define the **matrix exponential** by

$$e^A = I + A + \frac{1}{2!} A^2 + \frac{1}{3!} A^3 + \cdots \tag{31}$$

for any square matrix A.

This series converges because

$$\left\| \sum_{i=m}^{m+k} \frac{A^i}{i!} \right\| \le \sum_{i=m}^{m+k} \frac{\|A^i\|}{i!} \le \sum_{i=m}^{m+k} \frac{\|A\|^i}{i!}$$

for all non-negative integers m and k, and the number on the right can be made as small as we like by choosing m sufficiently large.

With this notation we may summarize the results of Exercises 23 and 25 in the following theorem.

THEOREM 6

The state transition matrix of the system $\dot{x}(t) = F(t)x(t)$ is given by

$$\Phi(t, t_0) = e^{\int_{t_0}^{t} F(\tau)\, d\tau}$$

if and only if $F(t)$ and $\int_{t_0}^{t} F(\tau)\, d\tau$ commute.   □

THE COMPLETE SOLUTION

Having discussed the linear homogeneous equation in some detail, let us now turn to the question that is really of interest to us as systems theorists—namely, the inhomogeneous equation in which we specify the action of some input $\dot{x}(t) = F(t)x(t) + G(t)u(t)$. The reader will recall from the study of discrete-time systems [equation (4) of Section 5–1] that we motivated the formula

$$x(t) = \Phi(t, t_0)x(t_0) + \int_{t_0}^{t} \Phi(t, \tau)G(\tau)u(\tau)\, d\tau$$

for the continuous-time case. We now have the machinery to check very easily that this is true:

THEOREM 7

Let $F(t)$ be a linear operator on $\mathbf{R}^n$ for each t, which is uniformly bounded on each interval. Then the unique solution of

$$\dot{x}(t) = F(t)x(t) + G(t)u(t)$$

with initial condition $x(t_0) = x_0$ is given by

$$x(t) = \Phi(t, t_0)x_0 + \int_{t_0}^{t} \Phi(t, \xi)G(\xi)u(\xi)\, d\xi \tag{32}$$

where $\Phi(t, \xi)$ is the state-transition matrix for F.

Proof

It suffices to check that (32) yields a solution—uniqueness follows from the general theory.

Since $\Phi(t_0, t_0) = I$, we see in (32) that $x(t_0) = x_0$ as desired. Now let us check the derivative of $x(t)$ [using the Leibnitz rule:

$$\frac{d}{dt}\int_{t_0}^{t} a(t, \xi)\, d\xi = \int_{t_0}^{t} \frac{\partial}{\partial t} a(t, \xi)\, d\xi + a(t, t)$$

see Exercise 3, Section 5–1]

$$\frac{d}{dt} x(t) = F(t)\Phi(t, t_0)x_0 + \int_{t_0}^{t} F(t)\Phi(t, \xi)G(\xi)u(\xi) \, d\xi + \Phi(t, t)G(t)u(t)$$

$$= F(t)x(t) + G(t)u(t)$$

as desired. $\square$

As usual, we have made use of elementary operations of calculus which are valid in sufficient generality for our purposes, and are not restricted to the one-dimensional case.

The important thing to note here is that our intuition for discrete-time systems immediately gave us the formula (32)—and it was then a completely straightforward procedure for us to check that this equation really was valid. In many textbooks on differential equations, which discuss the inhomogeneous linear equation without the intuition from system theory, the formula (32) either is produced by magic out of the air, or is gotten by a procedure called variation of parameters [Exercise 27], which not only is non-intuitive, but also itself seems to have been pulled out of the air! We thus see that our stress on the interplay between discrete-time and continuous-time systems pays off in a deepened intuition and understanding of what is going on in the mathematics.

The general formula (32) immediately yields the following description of the input-output behavior of a system:

COROLLARY 8

The input-output behavior of the linear system

$$\begin{cases} \dot{x}(t) = F(t)x(t) + G(t)u(t) \\ y(t) = H(t)x(t) \end{cases}$$

is given by the equation

$$y(t) = H(t)\Phi(t, t_0)x(t_0) + \int_{t_0}^{t} H(t)\Phi(t, \xi)G(\xi)u(\xi) \, d\xi \qquad (33)$$

where for each fixed ξ, $\Phi(t, \xi)$ is the solution of the matrix differential equation

$$\frac{d}{dt} \Phi(t, \xi) = F(t)\Phi(t, \xi)$$

with initial condition $\Phi(\xi, \xi) = I$.

Example 10

Find the response to a unit step input at t_0 of the system

$$\dot{x}(t) = \begin{bmatrix} 1 & 0 \\ 0 & 2t \end{bmatrix} x(t) + \begin{bmatrix} 1 \\ t \end{bmatrix} u(t)$$

$$y(t) = [e^{-t} \quad te^{-t}]x(t)$$

Here our input is $u(t) = 1(t - t_0)$, unit step, and we already know from Example 4 that

$$\Phi(t, \xi) = \begin{bmatrix} e^{(t-\xi)} & 0 \\ 0 & e^{t^2 - \xi^2} \end{bmatrix}$$

Thus

$$H(t)\Phi(t, \xi) = [e^{-t} \quad te^{-t}] \begin{bmatrix} e^{(t-\xi)} & 0 \\ 0 & e^{t^2 - \xi^2} \end{bmatrix}$$

$$= [e^{-\xi} \quad te^{t^2 - t - \xi^2}]$$

$$H(t)\Phi(t, \xi)G(\xi) = [e^{-\xi} \quad te^{t^2 - t - \xi^2}] \begin{bmatrix} 1 \\ \xi \end{bmatrix}$$

$$= e^{-\xi} + t\xi e^{t^2 - t - \xi^2}$$

and

$$\int_{t_0}^{t} H(t)\Phi(t, \xi)G(\xi)u(\xi)\, d\xi = \int_{t_0}^{t} (e^{-\xi} + t\xi e^{t^2 - t - \xi^2})1(\xi - t_0)\, d\xi$$

$$= \int_{t_0}^{t} e^{-\xi}\, d\xi + (te^{t^2 - t}) \int_{t_0}^{t} e^{-\xi^2}\xi\, d\xi$$

$$= -(e^{-t} - e^{-t_0}) - \frac{1}{2} te^{t^2 - t}(e^{-t^2} - e^{-t_0^2})$$

If $x_0 = x(t_0)$, the state of the system when the unit step was applied, we have from (33)

$$y(t) = [e^{-t_0} \quad te^{t^2 - t - t_0^2}] \begin{bmatrix} x_{10} \\ x_{20} \end{bmatrix} - (e^{-t} - e^{-t_0}) - \frac{te^{t^2 - t}}{2}(e^{-t^2} - e^{-t_0^2}) \qquad \Diamond$$

Equations (32) and (33), and this last example, clearly illustrate the decomposition property of linear systems. From the state equation (32):

$$\phi(t, t_0, x_0, 0) = \Phi(t, t_0)x_0$$

$$\phi(t, t_0, 0, u) = \int_{t_0}^{t} \Phi(t, \xi)G(\xi)u(\xi)\, d\xi$$

so that

$$\phi(t, t_0, x_0, u) = \phi(t, t_0, x_0, 0) + \phi(t, t_0, 0, u)$$

as we discussed in Section 3–3.

EXERCISES FOR SECTION 5-3

✔ **1.** In this exercise we outline the proofs of the equivalence of expressions (5), (6), (7), and (8) for the norm of a bounded linear operator A. Give the full proofs.

(a) Call $l = \sup \{\|Ax\| \mid \|x\| = 1\}$. Let M be any positive number such that $\|Ax\| \le M\|x\|$ for all $x \in X$. If $\|x\| = 1$, then $\|Ax\| \le M$, so M is an upper bound for the set $\{\|Ax\| \mid \|x\| = 1\}$. Since l is the *least* upper bound for this set, we have $l \le M$. Thus l is *a* lower bound for all such M's. But $\|A\|$, by (5), is the *greatest* lower bound for all such M. Conclude that $l \le \|A\|$.

Conversely, if $l < \infty$, for every $x \ne 0$,

$$\|Ax\| = \left\| A\frac{x}{\|x\|} \right\| \|x\| \le l\|x\|$$

and hence $\|A\| \le l$. Conclude that $\|A\| = l$.

(b) Use the trick that for any $x \ne 0$, $\hat{x} = \dfrac{x}{\|x\|}$ satisfies $\|\hat{x}\| = 1$ and $A\hat{x} = \dfrac{1}{\|x\|}Ax$ to prove that

$$\|A\| = \sup\{\|A\hat{x}\| \mid \|\hat{x}\| = 1\} = \sup\left\{ \frac{\|Ax\|}{\|x\|} \,\middle|\, x \ne 0 \right\}$$

(c) Use $\|A\| = \sup\{\|Ax\| \mid \|x\| = 1\} \le \sup\{\|Ax\| \mid \|x\| \le 1\}$ and the fact that, if $\|x\| \le 1$, and $x \ne 0$, then $\dfrac{\|Ax\|}{\|x\|} \ge \|Ax\|$, to get the inequality

$$\|A\| = \sup\left\{ \frac{\|Ax\|}{\|x\|} \,\middle|\, x \ne 0 \right\} \ge \sup\left\{ \frac{\|Ax\|}{\|x\|} \,\middle|\, \|x\| \le 1 \right\} \ge \sup\{\|Ax\| \mid \|x\| \le 1\}$$

from which equality (7) follows.

2. Use the fact that if X is finite dimensional, then the closed, bounded set $\{x \mid \|x\| = 1\}$ is compact, and the property that $\|A(\cdot)\|$ is a *continuous* function on this set and hence attains its maximum on it, to prove equation (9).

3. Let X and Y be Banach spaces and let $A : X \to Y$ be a linear map. Prove that A is bounded if and only if it maps the unit sphere $S(0; 1) = \{x \mid \|x\| = 1\}$ onto a bounded subset of Y. [*Hint:* A bounded, $\|x\| < 1 \Rightarrow \|Ax\| \le \|A\|$. Conversely, let $\|A\hat{x}\| \le M$ for $\hat{x} \in S(0; 1)$; for any $x \ne 0$, take $\hat{x} = \dfrac{x}{\|x\|}$ so $\left\| A\dfrac{x}{\|x\|} \right\| \le M$ and $\|Ax\| \le M\|x\|$.]

✔ **4.**   Use the triangle inequality and Definition 1 to prove that $\|C + B\| \leq \|C\| + \|B\|$ for C, B bounded linear mappings of X into Y.

5.   Let $\mathcal{B} = \{v_1, \ldots, v_n\}$ and $\mathcal{B}' = \{w_1, \ldots, w_m\}$ be bases for the respective normed spaces V and W, and let $L: V \to W$ be linear. Let $A = [a_{ij}]$ be the matrix of L, $^{\mathcal{B}'}L^{\mathcal{B}} = A = [a_{ij}]$, and let $[x_1, \ldots, x_n]^T$ and $[y_1, \ldots, y_m]^T$ be the coordinates of x and y with respect to $\mathcal{B}$ and $\mathcal{B}'$.

(a)   Show that $\|x\|_\infty = \max\limits_{1 \leq i \leq n} |x_i|$ is **a** norm for the space V. [This shows that every finite-dimensional vector space over **R** (or **C**) can be normed by using one of the norms of the coordinates in $\mathbf{R}^n$ (or $\mathbf{C}^n$).]

(b)   Call $y = Lx = \sum\limits_{j=1}^{m} y_j w_j$ so that $y_j = A[x_1, \ldots, x_n]^T = \sum\limits_{i=1}^{n} a_{ji} x_i$, and show that $\|Lx\|_\infty \leq M\|x\|_\infty$ for all $x \in V$ where

$$M = \max_{i \leq j \leq m} \left\{ \sum_{i=1}^{n} |a_{ji}| \right\}$$

[*Outline:* $\|Lx\|_\infty = \|y\|_\infty = \max\limits_{1 \leq j \leq m} |y_j| = \max\limits_{1 \leq j \leq m} \left| \sum\limits_{i=1}^{n} a_{ji} x_i \right|$

$$\leq \max_{1 \leq j \leq m} \sum_{i=1}^{n} |a_{ji}| \, |x_i| \leq \left(\max_{1 \leq i \leq n} |x_i| \right) \left(\max_{1 \leq j \leq m} \sum_{i=1}^{n} |a_{ji}| \right)$$

Thus, L is bounded using this coordinate norm $\|\cdot\|_\infty$.]

(c)   Look in B. Epstein, *Linear Functional Analysis* (W. B. Saunders Co., 1970, p. 83) or any book on functional analysis, and find out that all norms on finite-dimensional spaces are *equivalent*. Thus, since L is bounded with respect to $\|\cdot\|_\infty$, it is bounded with respect to any other norm $\|\cdot\|$.

6.   **(a)**   For a square complex matrix A, show that if $\|x\|_\infty = \max\limits_{1 \leq i \leq n} |x_i|$ is used for the norm of x, then the **matrix norm** will be

$$\|A\|_\infty = \max_{1 \leq i \leq n} \left\{ \sum_{j=1}^{n} |a_{ij}| \right\}$$

[*Outline:* $\|A\|_\infty = \max\limits_{\|x\|_\infty = 1} \|Ax\|_\infty$ and

$$\|Ax\|_\infty = \max_{1 \leq i \leq n} \left(\left| \sum_{j=1}^{n} a_{ij} x_j \right| \right) \leq \max_{1 \leq i \leq n} \left(\sum_{j=1}^{n} |a_{ij}| \, |x_j| \right)$$

$$\leq \left(\max_{1 \leq j \leq n} |x_j| \right) \left(\max_{1 \leq i \leq n} \sum_{j=1}^{n} |a_{ij}| \right) = \|x\|_\infty \max_{1 \leq i \leq n} \sum_{j=1}^{n} |a_{ij}|$$

To get the inequality in the other direction, let

$$\sum_{j=1}^{n} |a_{kj}| = \max_i \left\{ \sum_{j=1}^{n} |a_{ij}| \right\}$$

and define

$$\tilde{x} = (\tilde{x}_1, \ldots, \tilde{x}_n) \text{ by } \tilde{x}_j = \begin{cases} 1 \text{ if } a_{kj} = 0 \\ \dfrac{|a_{kj}|}{a_{kj}} \text{ if } a_{kj} \neq 0 \end{cases}$$

so that $\|\tilde{x}\|_\infty = 1$ and $\|A\tilde{x}\|_\infty \leq \|A\|_\infty \|\tilde{x}\|_\infty = \|A\|_\infty$. Thus

$$\|A\|_\infty \geq \|A\tilde{x}\|_\infty = \max_i \left| \sum_{j=1}^n a_{ij}\tilde{x}_j \right| \geq \sum_{j=1}^n a_{kj}\tilde{x}_j = \sum_{j=1}^n |a_{kj}|. \bigg]$$

(b)   Prove that if $\|x\|_1 = \displaystyle\sum_{i=1}^n |x_i|$, the ℓ_1 norm, is used for x, then the matrix norm is

$$\|A\|_1 = \max_{1 \leq j \leq n} \sum_{i=1}^n |a_{ij}|$$

the maximum of the sum of the moduli of the column elements. Note that $\|A\|_1 = \|A^*\|_\infty$ where $A^* = (\bar{A})^T$, the adjoint of A.

✔ **7.**   Let $A = (a_{ij})$ be any $m \times n$ complex matrix, so $A : \mathbf{C}^n \to \mathbf{C}^m$ using the standard basis.

(a)   Show that for any x in $\mathbf{C}^n$

$$\|Ax\|_2 \leq K\|x\|_2$$

where

$$K^2 = \sum_{i=1}^m \sum_{j=1}^n |a_{ij}|^2$$

$$\left[Hint: (\|Ax\|_2)^2 = \langle Ax, Ax \rangle = \sum_{i=1}^m |[Ax]_i|^2 = \sum_{i=1}^m \left| \sum_{j=1}^n a_{ij}x_j \right|^2 \right.$$

where $[Ax]_i$ is the ith component of Ax. By the C.B.S. inequality,

$$\left. \left| \sum_{j=1}^n a_{ij}x_j \right|^2 \leq \sum_{j=1}^n |a_{ij}|^2 \sum_{j=1}^n |x_j|^2 = (\|x\|_2)^2 \sum_{j=1}^n |a_{ij}|^2 \right]$$

(b)   Show that for any $x \in \mathbf{C}^n$ and any $y \in \mathbf{C}^m$

$$|\langle y, Ax \rangle| \leq K\|x\|_2 \|y\|_2$$

with K as in part (a).

(c)   Show that $A = A^* \Rightarrow \|A^2\| = \|A\|^2$. [Such an A is called *bounded Hermitian*]. *Note:* The number K^2 can also be written $K^2 = \text{trace}(A^*A)$.

✔ **8.**   Let $A(t) = \begin{bmatrix} t^2 & t & 1 \\ t & 1 & 0 \\ 0 & 0 & 0 \end{bmatrix}$ and show that the columns $A_1(\cdot), A_2(\cdot), A_3(\cdot)$ are linearly independent on $(0, \infty)$ even though $\det A(t) \equiv 0$ on $(0, \infty)$. Is this surprising? Corollary 3 says that this can not happen when $A(t)$ is a fundamental matrix of LH.

The next few exercises provide a review of matrix calculus.

✔ **9.** Let $A = \begin{bmatrix} a_{11}(t) & \cdots & a_{1n}(t) \\ \vdots & & \vdots \\ a_{m1}(t) & \cdots & a_{mn}(t) \end{bmatrix} = [a_{ij}(t)]$ and define

$$\frac{dA}{dt} = \frac{d}{dt}[a_{ij}(t)] = \left[\frac{d}{dt}a_{ij}(t)\right]$$

(a) Show that $\frac{d}{dt}(AB) = \frac{dA}{dt}B + A\frac{dB}{dt}$. (Note order!)

(b) Use (a) to show that

$$\frac{d}{dt}A^{-1} = -A^{-1}\frac{dA}{dt}A^{-1}$$

(c) Give an expression for

$$\frac{\partial}{\partial\tau}\Phi(t,\tau)$$

where Φ is the state transition matrix of the system $\dot{x}(t) = F(t)x(t)$.

✔ **10.** For $x: T \to \mathbf{R}^n$ and $y: T \to \mathbf{R}^n$ show that

$$\frac{d}{dt}\left\langle x(t)\,|\,y(t)\right\rangle = \left\langle\frac{dx(t)}{dt}\,|\,y(t)\right\rangle + \left\langle x(t)\,|\,\frac{dy(t)}{dt}\right\rangle$$

11. For $W(t) = [w_{ij}(t)]$, let $q(t) = \det W(t)$.

(a) Show that $\frac{dq}{dt}(t) = \sum_{i=1}^{m}\sum_{j=1}^{n}\frac{\partial q}{\partial w_{ij}}\cdot\frac{dw_{ij}}{dt}$

[*Hint:* The determinant q depends on all the entries $w_{11}, \ldots, w_{mn}$, so use the chain rule to compute $\frac{dq}{dt}$.]

(b) Now expand $q = \det W$ about the ith row of W to get $q = \sum_{k=1}^{n} w_{ik}\cdot\det q_{ik}$ where $\det q_{ik}$ is the i, kth cofactor of q. Since q_{ik} does not contain w_{ik} [to get the cofactor we strike the i, kth element], differentiate this last expression to get

$$\frac{\partial q}{\partial w_{ij}} = \det q_{ij}$$

(c) Define a matrix ${}^iW = \begin{bmatrix} w_{11} & \cdots & w_{1n} \\ \vdots & & \\ \frac{d}{dt}w_{i1} & \cdots & \frac{d}{dt}w_{in} \\ \vdots & & \\ w_{m1} & \cdots & w_{mn} \end{bmatrix}$

which is the same as W except that the ith row is replaced by its time derivative.

Expand $\det{}^iW$ about the ith row to get

$$\det{}^iW = \sum_{k=1}^{n} \frac{d}{dt}\,w_{ik}\cdot\det q_{ik}$$

Use the result from (b) to get

$$\det{}^iW = \sum_{k=1}^{n} \frac{\partial q}{\partial w_{ik}}\cdot\frac{d}{dt}\,w_{ik}$$

and the result from (a) to get

$$\frac{dq}{dt} = \sum_{i=1}^{m} \det{}^iW$$

or

$$\frac{d}{dt}\det\begin{bmatrix} w_{11} & \cdots & w_{1n} \\ \vdots & & \vdots \\ w_{i1} & \cdots & w_{in} \\ \vdots & & \vdots \\ w_{m1} & \cdots & w_{mn} \end{bmatrix} =$$

$$\det\begin{bmatrix} \dfrac{d}{dt}w_{11} & \cdots & \dfrac{d}{dt}w_{1n} \\ w_{21} & \cdots & w_{2n} \\ w_{m1} & \cdots & w_{mn} \end{bmatrix} + \cdots + \det\begin{bmatrix} w_{11} & \cdots & w_{1n} \\ w_{21} & \cdots & w_{2n} \\ \dfrac{d}{dt}w_{n1} & \cdots & \dfrac{d}{dt}w_{mn} \end{bmatrix}$$

which is the rule for differentiating determinants.

✔ **12.** For $W(t) = \begin{bmatrix} \cos(t) & \sin(t) \\ -\sin(t) & \cos(t) \end{bmatrix}$, use the formula of Exercise 11(c) to compute $\dfrac{d}{dt}\det W$ and check it by computing $\det W$ and differentiating. Does $\det\left(\dfrac{dW}{dt}\right) = \dfrac{d}{dt}(\det W)$?

✔ **13.** Let $A(t) = \begin{bmatrix} 1 & t \\ 0 & 3 \end{bmatrix}$ and find

$$B(t) = \frac{d}{dt}A(t) \text{ and } C(t) = \int_0^t A(\tau)\,d\tau$$

Do A and B commute? Do A and C?

✔ **14.** For $A(t) = [a_{ij}(t)]$, define

$$\int_{t_0}^t A(\tau)\,d\tau = \left[\int_{t_0}^t a_{ij}(\tau)\,d\tau\right] \text{ (integrate each component)}$$

and

$$\alpha(t)A(t) = [\alpha(t)a_{ij}(t)]$$

for any scalar function of t, α.

(a) Use these definitions to define $\mathcal{L}\{A(t)\}$, the Laplace transform of a matrix $A(t)$.

(b) Find $\mathcal{L}\left\{\begin{bmatrix} \sin(t) & te^{-t} \\ -e^{-t} & (1+t)e^{-t} \end{bmatrix}\right\}$.

(c) Define the inverse Laplace transform $\mathcal{L}^{-1}\{A(s)\}$ and find

$$\mathcal{L}^{-1}\left\{\begin{bmatrix} \dfrac{1}{s+2} & \dfrac{1}{(s+2)^2} \\ \dfrac{s}{s+2} & \dfrac{s}{(s+2)^2} \end{bmatrix}\right\}$$

(d) If M and K are constant matrices, show that:

$$\det(MA(t)K) = \det M \cdot \det A(t) \cdot \det K$$

$$\frac{d}{dt}(MA(t)K) = M\frac{dA}{dt}K \quad [\text{Use Exercise 9}]$$

$$\mathcal{L}\{MA(t)K\} = M\mathcal{L}\{A(t)\}K$$

(e) Derive the Laplace transform of the derivative of a matrix

$$\mathcal{L}\left\{\frac{d}{dt}A(t)\right\} = s\mathcal{L}\{A(t)\} - A(0)$$

and check it for the matrix $A(t)$ of part (b).

15. Use the formula of Exercise 11(c) for differentiating determinants to derive equation (16) (Liouville's formula).
[*Outline:*

$$\frac{d}{dt}\det W(t) = \det\begin{bmatrix} \dot{w}_{11} & \cdots & \dot{w}_{1n} \\ w_{21} & \cdots & w_{2n} \\ & \vdots & \\ w_{n1} & \cdots & w_{nn} \end{bmatrix} + \cdots + \det\begin{bmatrix} w_{11} & \cdots & w_{1n} \\ w_{21} & \cdots & w_{2n} \\ & \vdots & \\ \dot{w}_{n1} & \cdots & \dot{w}_{nn} \end{bmatrix} \quad \textbf{(34)}$$

Each column $W_j(t) = \begin{bmatrix} w_{1j}(t) \\ w_{2j}(t) \\ \vdots \\ w_{nj}(t) \end{bmatrix}$ is a solution of LH

so

$$\dot{w}_{ij}(t) = \sum_{k=1}^{n} f_{ik}(t)w_{kj}(t)$$

where

$$F(t) = [f_{ij}(t)]$$

Consider the ith term of the sum of determinants on the right side of (34). Multiply the kth row, for $k \neq i$, of this matrix by $-f_{ik}(t)$ and then add it to the ith row. This gives

$$\det \begin{bmatrix} w_{11} & \cdots & w_{1n} \\ \vdots & & \\ f_{ii}w_{i1} & \cdots & f_{ii}w_{im} \\ \vdots & & \\ w_{n1} & \cdots & w_{nn} \end{bmatrix} = f_{ii}(t) \det \begin{bmatrix} w_{11} & \cdots & w_{1n} \\ \vdots & & \\ \dot{w}_{i1} & \cdots & \dot{w}_{in} \\ \vdots & & \\ w_{n1} & \cdots & w_{nn} \end{bmatrix}$$

of (34). Adding all the terms on the right of (34) gives

$$\frac{d}{dt} \det W(t) = \sum_{i=1}^{n} f_{ii}(t) \det W(t)$$

$$= (\text{trace } F(t)) \det W(t)$$

whose solution is (16).]

✔ **16.** Find the state transition matrix $\Phi(t, t_0)$ for the LH

$$\dot{x}(t) = \begin{bmatrix} t & 0 \\ 0 & 0 \end{bmatrix} x(t)$$

and compute $\Phi^{-1}(t, t_0)$ by inverting your result. Compare with $\Phi(t_0, t)$.

17. Consider the scalar time-varying differential equation

$$\ddot{y} + \frac{4}{t}\dot{y} + \frac{2}{t^2}y = u$$

(a) Define state variables $x_1 = y$ and $x_2 = \dot{y}$, and derive the representation $(F(\cdot), G(\cdot), H(\cdot))$ given by

$$\begin{cases} \dot{x}(t) = \begin{bmatrix} 0 & 1 \\ -\dfrac{2}{t^2} & -\dfrac{4}{t} \end{bmatrix} x(t) + \begin{bmatrix} 0 \\ 1 \end{bmatrix} u(t) \\ y(t) = [1 \quad 0]x(t) \end{cases}$$

(b) Verify that $\psi_1(t) = \dfrac{1}{t}$ and $\psi_2(t) = \dfrac{1}{t^2}$ are two linearly independent solutions of the homogeneous scalar equation.

(c) Construct the matrix $W(t) = \begin{bmatrix} \psi_1(t) & \psi_2(t) \\ \dot{\psi}_1(t) & \dot{\psi}_2(t) \end{bmatrix} = \begin{bmatrix} \dfrac{1}{t} & \dfrac{1}{t^2} \\ -\dfrac{1}{t^2} & -\dfrac{2}{t^3} \end{bmatrix}$ and verify

that it is a fundamental matrix for $\dot{x} = Fx$.

(d) Find the state transition matrix $\Phi(t, t_0)$ and check it.

✔ **18.**   Suppose we are given that $\Phi(t, 0) = \begin{bmatrix} \cos\theta(t) & \sin\theta(t) \\ -\sin\theta(t) & \cos\theta(t) \end{bmatrix}$, where $\theta(t) = \int_0^t w(\tau)\, d\tau$, is the state transition matrix of an LH system $\dot{x} = Fx$.

(a)   Find $F(t)$. Can you always find F from Φ?

(b)   Set $y = x_1$ and $\dot{y} = x_2$, and find the scalar differential equation in y corresponding to $\dot{x} = Fx$. Give the solution $y(t)$ of this equation for the special case when the angular frequency $w(t) = w_0$, a constant.

(c)   Find the state transition matrix $\psi(t, 0)$ of the system

$$\dot{q} = -F^*q$$

19.   Let $\Phi(t, t_0)$ and $\Psi(t, t_0)$ be the state transition matrices of the systems $\dot{x} = Fx$ and $\dot{q} = -F^*q$ respectively, so $\Psi^*(t, t_0) = \Phi^{-1}(t, t_0)$. Let $x(t)$ and $q(t)$ be respective solutions of $\dot{x} = Fx$ and $\dot{q} = -F^*q$. Prove that the inner product of $x(t)$ and $q(t)$ is a constant. [*Hint:* Use Exercise 10 on $\dfrac{d}{dt}\langle x(t)\,|\,q(t)\rangle$.]

20.   **(a)**   Find the state transition matrix $\Phi(t, t_0)$ for the system

$$\dot{x}(t) = \begin{bmatrix} 0 & \dfrac{2}{t^2} \\ -1 & \dfrac{4}{t} \end{bmatrix} x(t)$$

using your results from Exercise 17.

(b)   Carry out the steps of Example 7 to find a second order scalar differential equation for which this matrix equation is a representation. Compare your result with the scalar equation of Exercise 17.

COMP 21.   Consider the system $\dot{x}(t) = F(t)x(t)$, and suppose that we wish to know its behavior as t ranges over the observation interval $[t_s, t_f]$ (we use t_s for start and t_f for finish).

(a)   Using the approximation $\dfrac{d}{dt}\Phi(t, t_0) \approx \dfrac{\Phi(t + h, t_0) - \Phi(t, t_0)}{h}$ for small h, show that

$$\Phi(t + h, t_0) \approx [I + hF(t)]\Phi(t, t_0) \quad \textbf{(Euler's method)}$$

(b)   Partition the time interval $[0, 1]$ into N equal parts by points

$$t_i = t_s + ih, \, i = 0, 1, 2, \ldots, N$$

with $h = \dfrac{(t_f - t_s)}{N}$, to get the discrete system

$$\Phi_h(t_{i+1}, t_s) \approx B(t_i)\Phi_h(t_i, t_s), \, i = 0, 1, \ldots, N$$

where

$$B(t_i) = I + hF(t_i), \, \Phi_h(t_s, t_s) = I$$

(c)   More generally, get the discrete approximation

$$\Phi_h(t_{i+1}, t_j) \approx B(t_i)\Phi_h(t_i, t_j)$$

with

$$B(t_i) = I + hF(t_i) \text{ and } \Phi_h(t_j, t_j) = I$$

for $i, j \in \{0, 1, \ldots, N\}$.

(d)   Illustrate this "discretization" for the case where

$$F(t) = \begin{bmatrix} 1 & e^{-5t} \\ 0 & 1 \end{bmatrix} \text{ for } t, t_0 \in [t_s, t_f] = [0, 1]$$

choosing $N = 10$ (rather crude), and comparing your generated samples $\Phi_h(t_i, t_s)$, $i \in \{0, 1, \ldots, N\}$, with the exact values gotten from

$$\Phi(t, \tau) = e^{(t-\tau)} \begin{bmatrix} 1 & \dfrac{1}{5}(e^{-5\tau} - e^{-5t}) \\ 0 & 1 \end{bmatrix}$$

(e)   How would you find the N^2 samples $\Phi_h(t_i, t_j)$ for $i, j \in \{0, 1, \ldots, N\}$? Note that $\Phi_h(t_6, t_3)$ is easy enough, but what about $\Phi_h(t_3, t_6)$? Is it equal to $\Phi_h^{-1}(t_6, t_3)$? Do the calculations, and compare with $\Phi(t_6, t_3)$ and $\Phi(t_3, t_6)$.

(f)   Will $\Phi_h(t_i, t_j)$ be non-singular in general for an arbitrary system $\dot{x}(t) = F(t)x(t)$?

(g)   Use the Euler method of approximating the derivative to solve the simple pendulum problem studied in Exercise 11 of Section 5–2. Do this for the same parameters, initial conditions, and step sizes h you used there in parts (c), (d), and (e). Compare.

MP **22.**   The **"fourth-order Runge-Kutta"** algorithm for solving differential equations of the form

$$\begin{cases} \dot{x}(t) = f(t, x(t)) \text{ [equation (5) of Section 5–2]} \\ x(t_0) = C \text{ (given)} \end{cases}$$

may be shown to have a truncation error of order h^5, whereas the Euler method of Exercise 21 is only of order h^2. Using the notation of Exercise 21, the algorithm is:

(i)   Start with $x_0 = C$, the given initial condition vector.
(ii)   Define, for each integer i, the vectors

$$B_1 = hf(t_i, x(t_i))$$

$$B_2 = hf\left(t_i + \frac{h}{2}, x(t_i) + \frac{1}{2}B_1\right)$$

$$B_3 = hf\left(t_i + \frac{h}{2}, x(t_i) + \frac{1}{2}B_2\right)$$

$$B_4 = hf(t_i + h, x(t_i) + B_3)$$

and compute $x(t_{i+1})$ from the recursion formula

$$x(t_{i+1}) = x(t_i) + \frac{1}{6}(B_1 + 2B_2 + 2B_3 + B_4).$$

(a)   Use the Runge-Kutta method to solve the two-dimensional equation $\dot{\Phi} = F\Phi$ for the F of Exercise 21(d) and compare with your results using the Euler method.

(b)   Solve the non-linear simple pendulum of Exercise 11, Section 5–2, parts (c), (d), and (e) by Runge-Kutta and compare with the solution by successive approximations and trapezoidal rule integration you used there. Compare also with the Euler method of Exercise 21(g).

23.   Investigate what happens in the Peano-Baker formula (28) for the special class of matrices which commute with their integral:

$$F(\tau) \left(\int_{t_0}^{\tau} F(\tau_1) \, d\tau_1 \right) = \left(\int_{t_0}^{\tau} F(\tau_1) \, d\tau_1 \right) F(\tau)$$

Outline:

(a)   Show that if two matrices A and B commute, then so do A^n and B^m for any powers n and m. Thus,

$$[F(\tau)]^n \left[\int_{t_0}^{\tau} F(\tau_1) \, d\tau_1 \right]^m = \left[\int_{t_0}^{\tau} F(\tau_1) \, d\tau_1 \right]^m [F(\tau)]^n$$

(b)   Since A and $\dot{A}$ do not in general commute,

$$\frac{d}{dt} (A)^2 = \frac{d}{dt} (AA) = \dot{A}A + A\dot{A} \neq 2A \frac{d}{dt} A$$

in general. Show that, for our F's,

$$\frac{d}{dt} \left[\int_{t_0}^{t} F(\sigma) \, d\sigma \right]^2 = 2F(t) \left[\int_{t_0}^{t} F(\sigma) \, d\sigma \right]$$

and that

$$\frac{d}{dt} \left[\int_{t_0}^{t} F(\sigma) \, d\sigma \right]^k = kF(t) \left[\int_{t_0}^{t} F(\sigma) \, d\sigma \right]^{(k-1)}$$

(c)   Since $\frac{d}{dt} (UV) = \dot{U}V + U\dot{V}$, we have the matrix version of integration by parts:

$$\int_{t_0}^{t} U(\tau)\dot{V}(\tau) \, d\tau = U(\tau)V(\tau) \Big|_{t_0}^{t} - \int_{t_0}^{t} \dot{U}(\tau)V(\tau) \, d\tau$$

Use this, with $U(\tau) = \int_{t_0}^{\tau} F(\tau_1) \, d\tau_1$ and $\dot{V}(\tau) = F(\tau)$, to show that

$$\int_{t_0}^{t} F(\tau) \int_{t_0}^{\tau} F(\tau_1) \, d\tau_1 \, d\tau = \frac{1}{2} \left[\int_{t_0}^{t} F(\sigma) \, d\sigma \right]^2$$

Show also that

$$\int_{t_0}^{t} F(\tau) \left[\int_{t_0}^{\tau} F(\sigma) \, d\sigma \right]^k d\tau = \frac{1}{k+1} \left[\int_{t_0}^{t} F(\sigma) \, d\sigma \right]^{k+1}$$

$$\left[\text{Thanks to commutativity, the integral is of the form } \int [U(\tau)]^k \frac{d}{d\tau} U(\tau) \, d\tau. \right]$$

(d) Use part (c) on the kth term of the Peano-Baker series, working outward from the middle,

$$\int_{t_0}^{t} F(\tau) \int_{t_0}^{\tau} F(\tau_1) \cdots \left[\int_{t_0}^{\tau_{k-2}} F(\tau_{k-1}) \int_{t_0}^{\tau_{k-1}} F(\tau_k) \, d\tau_k \, d\tau_{k-1} \right] d\tau_{k-2} \cdots d\tau_1 \, d\tau$$

$$= \int_{t_0}^{t} F(\tau) \int_{t_0}^{\tau} F(\tau_1) \cdots \left\{ \int_{t_0}^{\tau_{k-3}} F(\tau_{k-2}) \left[\frac{1}{2} \left(\int_{t_0}^{\tau_{k-2}} F(\sigma) \, d\sigma \right)^2 \right] d\tau_{k-2} \right\} \cdots d\tau_1 \, d\tau$$

$$= \int_{t_0}^{t} \int_{t_0}^{\tau} F(\tau_1) \cdots \left\{ \frac{1}{2} \cdot \frac{1}{3} \left[\int_{t_0}^{\tau_{k-3}} F(\sigma) \, d\sigma \right]^3 \right\} d\tau_{k-3} \cdots d\tau_1 \, d\tau$$

$$= \frac{1}{k!} \left[\int_{t_0}^{t} F(\sigma) \, d\sigma \right]^k$$

(e) Thus, if we call $B(t) = \int_{t_0}^{t} F(\sigma) \, d\sigma$, we have, when $B\dot{B} = \dot{B}B$, that

$$\Phi(t, t_0) = I + B(t) + \frac{1}{2!} B^2(t) + \frac{1}{3!} B^3(t) + \cdots + \frac{1}{k!} B^k(t) + \cdots$$

(f) To show that there is a large class of F's which do commute with their integrals, consider $F(t) = \sum_{i=1}^{m} \alpha_i(t) A_i$, where the $\alpha_i(\cdot)$ are scalar functions of t and the A_i are constant matrices such that $A_i A_j = A_j A_i$.

24. Let $A(t)$ be a time-varying square matrix. Prove that $\dfrac{dA}{dt}$ commutes with A if and only if

$$\frac{d}{dt} A^n = nA^{n-1} \frac{dA}{dt} = n \frac{dA}{dt} A^{n-1} \text{ for all } n = 1, 2, \ldots$$

[*Note:* The "if" part was used in Exercise 23(b).]

25. Consider the system $\dot{x}(t) = F(t)x(t)$. Let $B(t) = \int_{t_0}^{t} F(\tau) \, d\tau$ and let

$$P(t) = I + B(t) + \frac{1}{2!} B^2(t) + \frac{1}{3!} B^3(t) + \cdots$$

Prove that if P is the state transition matrix of $\dot{x} = Fx$, then $\dot{B}B = B\dot{B}$, so that $F(t)$ commutes with $\int_{t_0}^{t} F(\tau) \, d\tau$.

[*Hint:* Write the series for $e^{B(t)s}$; use $\dot{P} = FP$ to get

$$\frac{d}{dt} e^{B(t)} = F(t)e^{B(t)} = \dot{B}(t)e^{B(t)}$$

and thus, that $\dfrac{d}{dt} e^{B(t)s} = s \dfrac{dB}{dt} e^{B(t)s}$; equate coefficients of powers of s to get $\dfrac{dB^n}{dt} = n \dfrac{dB}{dt} B^{n-1}$ and use Exercise 24.]

26.   Prove that $\dfrac{d}{dt} e^{B(t)} = \dfrac{dB}{dt} e^{B(t)}$, where $B(t) = \displaystyle\int_{t_0}^{t} F(\tau)\, d\tau$, if and only if $F(t_1)F(t_2) = F(t_2)F(t_1)$ for all $t_1,\, t_2$.

[*Hint:* See Kinariwala, "Analysis of Time-Varying Networks," *IRE Inter. Conv. Record,* 1961, Part 4, pp. 268–276.]

27.   **Lagrange's variation of parameters:** To find the forced response of the system

$$\begin{cases} \dot{x}(t) = F(t)x(t) + G(t)u(t) \\ x(t_0) = x_0 \end{cases}$$

from knowledge of the homogeneous solution $x_0(t) = \Phi(t, t_0)x_0$, assume a solution of the form

$$x(t) = \Phi(t, t_0)v(t)$$

where $v(t)$ is the vector of "parameters" to be determined and Φ is the state transition matrix of LH.

(a)   Substitute this expression for $x(t)$ in the inhomogeneous differential equation and deduce that

$$\frac{dv}{dt}(t) = \Phi^{-1}(t, t_0)G(t)u(t)$$

whence

$$v(t) = v(t_0) + \int_{t_0}^{t} \Phi^{-1}(\xi, t_0)G(\xi)u(\xi)\, d\xi$$

(b)   Using this value of $v(t)$, show that $x(t) = \Phi(t, t_0)v(t)$ reduces to equation (32).

✔ **28.**   (a)   Show that the complete solution given by Equation (32) can also be expressed by the equivalent equations:

$$x(t) = \Phi(t, t_0)\left[x(t_0) + \int_{t_0}^{t} \Phi(t_0, \xi)G(\xi)u(\xi)\, d\xi \right]$$

$$x(t) = \Phi(t, t_0)\left[x(t_0) + \int_{t_0}^{t} \Phi^{-1}(\xi, t_0)G(\xi)u(\xi)\, d\xi \right]$$

and

$$x(t) = \Phi(t, t_0)\left[x(t_0) + \int_{t_0}^{t} \Psi^{*}(\xi, t_0)G(\xi)u(\xi)\, d\xi \right]$$

where $\Psi(t, t_0)$ is the state transition matrix of the adjoint system $\dot{q}(t) = -F^{*}(t)q(t)$.

(b)   Use the above equation to solve equation (32) for $x(t_0)$:

$$x(t_0) = \Phi^{-1}(t, t_0)x(t) - \int_{t_0}^{t} \Phi^{-1}(\xi, t_0)G(\xi)u(\xi)\, d\xi$$

$$x(t_0) = \Phi^{-1}(t, t_0)x(t) + \int_{t}^{t_0} \Phi^{-1}(\xi, t_0)G(\xi)u(\xi)\, d\xi$$

$$x(t_0) = \Psi^{*}(t, t_0)x(t) + \int_{t}^{t_0} \Psi^{*}(\xi, t_0)G(\xi)u(\xi)\, d\xi$$

This last equation illustrates how the solution of the adjoint system lets us "run backward in time" from t to t_0 to find what state $x(t_0)$ we should start in, to end up in $x(t)$ with the input $u(\cdot)$.

29.   Let $\Phi(t, t_0)$ be the state transition matrix of the system $\dot{x}(t) = F(t)x(t)$.

(a)   Show that if $F(\cdot)$ is *periodic* with period T_0 (i.e., $F(t + T_0) = F(t)$, for all t) then $\Phi(\cdot, t_0)$ is also periodic with the same period.

$$[Hint: \frac{d}{dt}\Phi(t + T_0, t_0) = \frac{d}{d(t + T_0)}\Phi(t + T_0, t_0)\left[\frac{d(t + T_0)}{dt}\right]$$

$$= F(t + T_0)\Phi(t + T_0, t_0) = F(t)\Phi(t + T_0, t_0)$$

so $\Phi(t + T_0, t)$ is *a fundamental matrix of* $\dot{x} = Fx$.]

(b)   Show that there exists a nonsingular matrix C_0 such that $\Phi(t + T_0, t_0) = \Phi(t, t_0)C_0$, for all t. What is C_0 if $F(t) = F$, constant?

(c)   Determine the conditions under which the output y of the inhomogeneous system (F, G, H), with $F(t + T_0) = F(t)$ for all t, will be periodic.

MP **30.**   Here we numerically evaluate the state transition matrix $\Phi(t, t_0)$ for $t \in [t_0, t_f]$ by using the trapezoidal rule on the integral in the sequence of successive approximations (26):

$$\Phi_{(k+1)}(t, t_0) = I + \int_{t_0}^{t} F(\tau)\Phi_{(k)}(\tau, t_0)\, d\tau$$

We get

$$\Phi_{(k+1)}(t_j, t_0) \doteq I + h\left\{\frac{1}{2}F(t_0)\Phi_{(k)}(t_0, t_0) + F(t_1)\Phi_{(k)}(t_1, t_0)\right.$$

$$\left. + \cdots + F(t_{j-1})\Phi_{(k)}(t_{j-1}, t_0) + \frac{1}{2}F(t_j)\Phi_{(k)}(t_j, t_0)\right\}$$

where $t_j = t_0 + jh$ for $j = 0, 1, 2, \ldots N$, and the sampling interval is $h = \dfrac{t_f - t_0}{N}$ for some suitably large N to make the error tolerable. Note that this formula shows us how to take a table of approximate samples $\Phi_{(k)}$:

$$\{\Phi_{(k)}(t_0, t_0), \Phi_{(k)}(t_1, t_0), \Phi_{(k)}(t_2, t_0), \ldots, \Phi_{(k)}(t_f, t_0)\}$$

and from them construct a more accurate table:

$$\{\Phi_{(k+1)}(t_0, t_0), \Phi_{(k+1)}(t_1, t_0), \ldots, \Phi_{(k+1)}(t_f, t_0)\}$$

(a)   Let $F(t) = \begin{bmatrix} 1 & e^{-5t} \\ 0 & 1 \end{bmatrix}$, $[t_0, t_f] = [0, 1]$,   and $N = 10$;   start   with $\Phi_{(1)}(t_i, t_0) = I$, $i = 0, 1, \ldots, N$ and generate the successive approximation tables $\{\Phi_{(2)}(t_i, t_0)\}_{i=1}^{N}, \ldots \{\Phi_{(7)}(t_i, t_0)\}_{i=1}^{N}$. At each iteration compare the entries of $\{\Phi_{(k+1)}(t_i, t_0)\}_{i=1}^{N}$ with the preceding table $\{\Phi_{(k)}(t_i, t_0)\}_{i=1}^{N}$. Did you need to do seven iterations for this problem?

(b)   Compare your tables for $\Phi(t_i, t_0)$ with the values you found in Exercise 21(d). [Note that in Exercise 21, by approximating the derivative as we did, we effectively used the "rectangle rule" of integration instead of the trapezoidal rule as in this exercise.]

(c)   Repeat part (a) with a finer mesh, $N = 100$, and monitor the change in the tables:

$$\|\Phi_{(k+1)}(t_i, t_0) - \Phi_{(k)}(t_i, t_0)\|$$

as you go from each iteration to the next. Use this "measure of error" to determine when to stop iterating. Can you stop before seven iterations?

   (d)   Write out the iterative matrix equations which result from using Simpson's rule (parabolic) on the integral in equation (26).

5–4   Input-Output Descriptions

THE IDEA OF THE IMPULSE RESPONSE

We saw in the last section that if we consider the linear continuous-time system

$$\dot{x}(t) = F(t)x(t) + G(t)u(t)$$
$$y(t) = H(t)x(t)$$

then we can find the zero-input state-transition matrix $\Phi(t, t_0)$ such that

$$\frac{\partial}{\partial t} \Phi(t, t_0) = F(t)\Phi(t, t_0)$$

$$\Phi(t_0, t_0) = I$$

This allows us to express the behavior of our system in terms of the following equations:

$$x(t) = \Phi(t, t_0) + \int_{t_0}^{t} \Phi(t, \tau)G(\tau)u(\tau)\,d\tau$$

$$y(t) = H(t)x(t) = H(t)\Phi(t, t_0)x(t_0) + \int_{t_0}^{t} g(t, \tau)u(\tau)\,d\tau \qquad (1)$$

where the linear map

$$g(t, \tau) = H(t)\Phi(t, \tau)G(\tau)$$

is called the **impulse response** of the system. Let us now justify this appellation.

In the study of classical mechanics or classical electricity, we have become quite used to the idea of considering a point mass or a point charge. Although we know that in physical reality, any mass or charge must be distributed over a nonzero volume in space, we agree that it is a useful approximation in many problems to consider a mass distributed over a small volume to be actually concentrated at a single point. The reason for this is that one can show that as soon as one is at a distance from the mass that is large compared with its actual diameter, the effects of the approximation are negligible. Analogously, then, to concentrating a mass distributed over a small volume into a single point, in system theory we find it useful to approximate an input distributed over a small interval of time by a hypothetical input delivered in one moment of time. (The student who is familiar with Lebesgue integration, and who is thus used to thinking of the behavior of a function at any one point as being negligible, will thus realize that such an input cannot be a function in the usual sense of the word. In fact, it is what is known as a *distribution*, as we shall briefly sketch below.)

Consider, then, a system with one-dimensional input space. Let us denote by $\delta_{(\varepsilon)}$ the ε-impulse, which is the input function shown in Figure 5–4 and

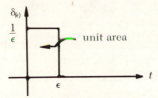

Figure 5–4   An approximation to the unit impulse.

given by the equation

$$\delta_{(\varepsilon)}(t) = \begin{cases} \dfrac{1}{\varepsilon} & \text{if } 0 \leq t \leq \varepsilon \\ 0 & \text{otherwise} \end{cases}$$

We are to imagine ε as being a small number. Thus, we apply a nonzero input to the system only during the short interval of length ε starting at time zero, in such a way that the total integral of the applied pulse is 1.

Now suppose that at some negative time, we start our system in the zero state. If we examine the output at some time t after we have applied the impulse, then that output will in fact equal, by equation (1) with $u = \delta_{(\varepsilon)}$,

$$y(t) = \int_{t_0}^{t} g(t, \tau)\, \delta_{(\varepsilon)}(\tau)\, d\tau$$

$$= \frac{1}{\varepsilon} \int_{0}^{\varepsilon} g(t, \tau)\, d\tau$$

which is just the value taken by $g(t, \tau)$ averaged over the short period of time $\tau \in [0, \varepsilon]$.

Now, for a function $g(t, \tau)$ that is continuous in τ, as ε gets smaller and smaller, this average will approximate better and better the value taken by $g(t, \tau)$ at $\tau = 0$:

$$\lim_{\varepsilon \to 0} \frac{1}{\varepsilon} \int_{0}^{\varepsilon} g(t, \tau)\, d\tau = g(t, 0)$$

We may thus say "$g(t, 0)$ is the limit, as $\varepsilon \to 0$, of the response of a linear system at time t to an ε-impulse delivered at time 0." If (naughtily) we thus define an **impulse** as being the "limit" as $\varepsilon \to 0$ of the ε-impulse, we may then say that "$g(t, 0)$ is the response of a linear system at time t to an impulse delivered at time 0." The only trouble, of course, is that there is no function δ such that $\delta(t) = \lim_{\varepsilon \to 0} \delta_{(\varepsilon)}(t)$ is true. This is because we have

$$\lim_{\varepsilon \to 0} \delta_{(\varepsilon)}(t) = 0, t \neq 0$$

while

$$\lim_{\varepsilon \to 0} \delta_{(\varepsilon)}(0) = \infty$$

and a function is only well defined when it takes a finite value at every point. The foolishness of expecting to find a suitable δ-function is heightened when we note the following limit:

$$\lim_{\varepsilon \to 0} \int_{t_1}^{t_2} \delta_{(\varepsilon)}(t) \, dt = 1 \text{ when } t_1 < 0 < t_2$$

and so we are not only asking our δ-function to be infinity at the origin, but are asking it to be infinite in such a way that its integral over any interval near the origin is 1—distinguishing this infinity from a value of infinity at the origin which gives us an integral of 2, or what have you! But, just as the unit point mass and unit point charge have proved to be serviceable approximations in gravitation and electromagnetism, so the unit impulse is a valuable concept in system theory—and in quantum mechanics, too, for that matter.

The solution of this dilemma is to consider a real-valued function *not* as a map $f : \mathbf{R} \to \mathbf{R}$ which assigns a real number to every real number, but rather as a linear operator $L_f : \mathfrak{D}(\mathbf{R}) \to \mathbf{R}$ which assigns a real number to every function in some collection $\mathfrak{D}(\mathbf{R})$ of *sufficiently smooth* real-valued functions (in the old mapping sense!) by the formula

$$g \mapsto L_f(g) = \int_{-\infty}^{\infty} f(t)g(t) \, dt$$

Now, if we associate with each ε-impulse $\delta_{(\varepsilon)}$ its linear operator

$$L_{\delta_{(\varepsilon)}} : g \mapsto \frac{1}{\varepsilon} \int_0^{\varepsilon} g(t) \, dt$$

it is clear that we can define a linear operator $L_\delta = \lim_{\varepsilon \to 0} L_{\delta_{(\varepsilon)}}$ since

$$\lim_{\varepsilon \to 0} \frac{1}{\varepsilon} \int_0^{\varepsilon} g(t) \, dt = g(0)$$

for any continuous function g. Note that the assignment

$$\mathfrak{D}(\mathbf{R}) \to \mathbf{R} \text{ such that } g \mapsto g(0)$$

is in fact linear, since in $\mathfrak{D}(\mathbf{R})$ the vector space operations are pointwise, as in our usual discussion of functions on vector spaces. However, there exists no function $f:\mathbf{R} \to \mathbf{R}$, in the old point-by-point sense, for which we have $L_f = (g \mapsto g(0))$. Roughly speaking, then, we see that, with any function defined in the old pointwise sense, we can associate a linear operator $L_f:\mathfrak{D}(\mathbf{R}) \to \mathbf{R}$. If we then take limits, we may well end up by having no function, in the old sense, which is the limit of the sequence of functions—and yet there may well be a well-defined linear operator which is the limit of the linear operators attached to the functions in the sequence. Thus, our ploy is to introduce the idea of a **distribution,** which is simply any sufficiently smooth linear transformation $\mathfrak{D}(\mathbf{R}) \to \mathbf{R}$ which attaches real numbers to all the functions in some collection of suitably smooth functions in the old pointwise sense. We then see that, even though there is no function which has the properties of an impulse, there is an impulse *distribution*—and this is in fact well-defined as the limit of the ε-impulses, when they are considered as distributions rather than as functions. It is a common abuse of language in the literature of engineering and physics to refer to the distribution δ as the δ-function—the reader may permit such lapses, as long as he realizes that it is nonsensical to consider δ to be a function in the old pointwise sense, but quite permissible to think of it in the new sense of a distribution, which is defined in terms of the way it attaches numbers to functions, rather than to points of the real line.

Let us just check that, with this definition of the impulse as the distribution with the property $\mathfrak{D}(\mathbf{R}) \to \mathbf{R}:g \mapsto g(0)$, the impulse response of a system is indeed the response to an impulse! We consider the formula

$$y(t) = \int_0^t g(t, \tau)u(\tau)\, d$$

which gives the response of a system when started in the zero state at time zero. We "apply" (the reader should note the quotes) the δ "function" to get the response

$$y(t) = \int_0^t g(t, \tau)\, \delta(\tau)\, d\tau = g(t, 0)$$

exactly as desired.

We stress, again, that the δ "function" is not an input which we can actually apply to a real system, but it does describe, in the limit, the effect of applying ε-impulses, which can be considered as applicable inputs.

To determine the ij component of the impulse response **matrix** of a multidimensional system, we simply apply an impulse on the ith input line alone at time 0, and then measure the output on the jth output line at the desired time t.

Note that when we attempted to regard δ as a function in the pointwise sense we had the absurd requirement that it be infinite at zero but that its integral be one. This would further imply that by multiplying δ by 2 we could produce an infinity, in some sense twice as large, which would yield an integral of two. However, such bizarre requirements become reasonable and useful when we turn to looking at δ as a distribution, for then we simply have

$$2\delta : g \mapsto 2g(0)$$

Clearly 2δ is simply the operator which assigns to every smooth function g twice its value at the origin.

When we actually write out the theory of distributions rigorously, we see that the passage from functions to distributions is very much like the passage from rational numbers to real numbers. We start with a collection of entities for which we can define Cauchy sequences, and yet which does not contain limits for all of these sequences. We then adjoin to the old collection a new collection of "limits"—and we then have to check carefully that the overall collection includes the original collection in a respectable way, and that all the operations that were of use to us in the original domain go through in the new domain. For instance, we can (but not in this volume!) study how to differentiate distributions, how to add them together, how to tell whether two sequences tend to the same limiting distribution, how to take Fourier transforms of distributions, and many other operations that are useful for functions. Just as we have come to think of real numbers as real (!) rather than artificial constructs made from rational numbers, so may the reader eventually come to view distributions as completely natural things, of which functions are felt to be an artificial limitation, rather than an underlying reality. Recall that a rational number is precisely one that can be expressed by two integers—and thus in terms of a finite expression—whereas an irrational number cannot be expressed as such a ratio; the best we can hope for is an infinite decimal expansion. Similarly, we may often have to think of a distribution in terms of the limiting behavior of an infinite sequence of functions. On the other hand, as in our simple prescription $\mathfrak{D}(\mathbf{R}) \to \mathbf{R} : g \mapsto g(0)$ for a δ function, we may sometimes actually consider the distributions to be simpler than the general notion of a function. If one wishes to be profoundly philosophical, one might suggest that it is in fact impossible to apply an arbitrary continuous function, in the old pointwise sense, to a system—at most one can manipulate a finite number of finitely specifiable parameters. In this case, one might actually consider the application of the δ function to be a more realistic operation than the application of some rather twisty old-fashioned function, which requires an infinite series for its specification. The theory of distributions and generalized functions goes back to the electrical engineer Oliver Heaviside, who introduced his operational calculus in the 1880's. Because he did not have a rigorous theory which made quite clear the distinction between an ordinary function and the generalized, or distribution, notion of function, he had a great deal of trouble with the pure mathematicians of his day—it is to his credit that he is quoted as saying, "Even Cambridge mathematicians deserve justice." It is the talent of a real mathematician to avoid rejecting something simply because in its first formulation it contains some logical errors—the challenge is to see through these errors to the underlying mathematical structure, and bring that structure out in such a way that the errors can then be eliminated. This is what we have done in our transition from functions to distributions. The idea of applying a unit impulse, or of using a point mass or point charge, is a completely intuitive and useful one. If one makes a mistake, it is not for the mathematician to deny its validity because it does not fit into his framework—rather it is a challenge to discover the framework into which it can fit.

THE IMPULSE RESPONSE MATRIX

Having provided some motivation for the idea of a δ-"function," we can discuss in more detail the response of the system to impulses "applied" to the system at different times.

From equation (31) of Section 5–3, the output of our linear system is seen to be the sum of two terms

$$\mathcal{S}(t, t_0, x_0, u) = \mathcal{S}(t, t_0, x_0, 0) + \mathcal{S}(t, t_0, 0, u)$$

the zero-input (natural) response

$$\mathcal{S}(t, t_0, x_0, 0) = H(t)\Phi(t, t_0)x_0 \tag{2a}$$

and the zero-state (forced) response

$$\mathcal{S}(t, t_0, 0, u) = \int_{t_0}^{t} H(t)\Phi(t, \xi)G(\xi)u(\xi) \, d\xi \tag{2b}$$

When we compute the zero-state response of a linear system to a unit impulse δ_τ at time τ [where $\delta_\tau(t) = \delta(t - \tau)$ is not to be confused with the ε-impulse $\delta_{(\varepsilon)}$] occurring at time $\tau \in (t_0, t)$, we get

$$\mathcal{S}(t, t_0, 0, \delta_\tau) = \int_{t_0}^{t} H(t)\Phi(t, \xi)G(\xi) \, \delta_\tau(\xi) \, d\xi$$

Using the "sampling" property of the δ-function, namely that for continuous functions f,

$$\int_{t_0}^{t} f(\xi)\,\delta_\tau(\xi)\,d\xi = \int_{t_0-\tau}^{t-\tau} f(\xi + \tau)\,\delta(\xi)\,d\xi$$

$$= f(\xi + \tau)|_{\xi=0} = f(\tau) \qquad \text{if } t_0 < \tau < t$$

we get

$$\mathcal{S}(t, t_0, 0, \delta_\tau) = H(t)\Phi(t, \tau)G(\tau) \text{ for } t_0 < \tau < t$$

so that $H(t)\Phi(t, \tau)G(\tau)$ is just the response of the system at time t to the impulse δ_τ.

Thus, we are indeed justified in defining the **impulse response**† of a system as the response at time t of the relaxed (zero initial conditions) system to a unit impulse "applied" at time τ and denoting it by the matrix $g(t, \tau)$. Thus

$$g(t, \tau) = \mathcal{S}(t, t_0, 0, \delta_\tau) \text{ for } t_0 < \tau < t$$

and for the linear system $(F(\cdot), G(\cdot), H(\cdot))$

$$g(t, \tau) = \begin{cases} H(t)\Phi(t, \tau)G(\tau) & t > \tau \\ 0 & t < \tau \end{cases} \tag{3}$$

We may now rewrite equation (31) of Section 5–3 as

$$y(t) = H(t)\Phi(t, t_0)x(t_0) + \int_{t_0}^{t} g(t, \tau)u(\tau)\,d\tau$$

For the *relaxed* system, $x(t_0) = 0$, the input-output relation is given by

$$y(t) = \int_{t_0}^{t} g(t, \tau)u(\tau)\,d\tau \tag{4}$$

which is called the **superposition** integral. Often we write

$$y(t) = \int_{-\infty}^{\infty} g(t, \tau)u(\tau)\,d\tau$$

assuming the system to have been relaxed at $t = -\infty$, and relying on causality, i.e., $g(t, \tau) = 0$ for $t < \tau$ [see Exercise 3 of Section 1–5] to help us determine the actual limits of integration corresponding to inputs which are nonzero on the interval $[t_0, t)$.

---

† Some authors use the term "weighting pattern" or "Green's function."

SUPERPOSITION AND SMOOTHNESS

We will now present an informal rederivation of the superposition integral (4) to emphasize its underlying structure and limitations.

If we let $P_\varepsilon^{t_i}$ be the rectangular pulse of width ε and height $\dfrac{1}{\varepsilon}$ as shown in Figure 5-5, then

$$P_\varepsilon^{t_i}(t) = \begin{cases} 0 \text{ for } t \le t_i \\[2mm] \dfrac{1}{\varepsilon} \text{ for } t_i < t \le t_i + \varepsilon \\[2mm] 0 \text{ for } t_i + \varepsilon < t \end{cases}$$

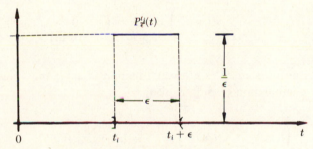

Figure 5-5   A rectangular pulse approximation to an impulse at t_i.

Since $P_\varepsilon^{t_i}$ has unit area for all ε, as ε approaches zero we see that $P_\varepsilon^{t_i}$ approaches the δ_{t_i}.

Now let u be any piecewise continuous input function. As shown in Figure 5-6, we can approximate u by a series of n weighted rectangular pulses $P_\varepsilon^{t_i}$

$$u(t) \approx \sum_{i=1}^{n} u(t_i)\varepsilon P_\varepsilon^{t_i}(t)$$

From the zero-state response function $\mathcal{S}$ we get

$$y(t) = \mathcal{S}(t, t_0, 0, u)$$

$$= \mathcal{S}\left(t, t_0, 0, \lim_{\substack{\varepsilon \to 0 \\ n \to \infty}} \sum_{i=1}^{n} u(t_i)\varepsilon P_\varepsilon^{t_i}\right)$$

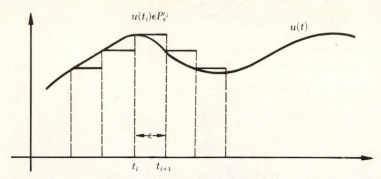

Figure 5–6 Rectangular pulse train approximation to an input function u.

Now if the system is *smooth*, then $\mathcal{S}$ is continuous in u, and the limit may be safely brought outside [see Section 2–4, Exercise 12]:

$$\mathcal{S}\left(t, t_0, 0, \lim_{\substack{\varepsilon \to 0 \\ n \to \infty}} \sum_{i=1}^{n} u(t_i)\varepsilon P_\varepsilon^{t_i}\right) = \lim_{\substack{\varepsilon \to 0 \\ n \to \infty}} \mathcal{S}\left(t, t_0, 0, \sum_{i=1}^{n} u(t_i)\varepsilon P_\varepsilon^{t_i}\right)$$

Since, for a linear system, $\mathcal{S}$ is a linear function of u, we may now pull the finite linear combination outside $\mathcal{S}$ also:

$$\mathcal{S}\left(t, t_0, 0, \sum_{i=1}^{n} u(t_i)\varepsilon P_\varepsilon^{t_i}\right) = \sum_{i=1}^{n} u(t_i)\varepsilon \mathcal{S}(t, t_0, 0, P_\varepsilon^{t_i})$$

Thus

$$y(t) = \lim_{\substack{\varepsilon \to 0 \\ n \to \infty}} \sum_{i=1}^{n} \mathcal{S}(t, t_0, 0, P_\varepsilon^{t_i}) u(t_i)\varepsilon$$

and in this limit $P_\varepsilon^{t_i}$ tends toward an impulse; the response $\mathcal{S}(t, t_0, 0, P_\varepsilon^{t_i})$ tends toward the impulse response $\mathcal{S}(t, t_0, 0, \delta_{t_i})$; and the summation becomes an integral:

$$y(t) = \int_{-\infty}^{\infty} \mathcal{S}(t, t_0, 0, \delta_{t_i}) u(t_i)\, dt_i$$

which is just

$$y(t) = \int_{-\infty}^{\infty} g(t, \xi) u(\xi)\, d\xi$$

Our requirement of smoothness,† which meant that $\mathcal{S}$ was continuous in u and let us pass limits in and out of $\mathcal{S}$, means that for smooth linear systems, *infinite superposition* holds. Infinite series and integrals, being regarded as limits of finite linear combinations

$$a_1 u_1 + \cdots + a_n u_n$$

can then be passed in and out of the operator $\mathcal{S}(t, t_0, 0, \cdot)$. Because of the usefulness of representing an input u by a Fourier series or Fourier or Laplace transform integral (which we develop in Section 6–3), this property of smoothness is often tacitly assumed by workers in linear systems. The following example, due to Shefi and Kailath, illustrates the need for caution; there are such things as simple linear systems which are not smooth.

Example 1

Consider the system with input space Ω of all piecewise continuous functions having only finitely many finite discontinuities (jumps). Let the system output at time t be the algebraic sum of all the jumps in the input up to time t:

$$y(t) = \mathcal{S}(t, -\infty, 0, u) = \sum_{-\infty < t_i < t} [u(t_i^+) - u(t_i^-)] \tag{5}$$

where the $\{t_i\}$ are the points of discontinuity and $u(t_i^+)$ and $u(t_i^-)$ denote limits of $u(t)$ as t approaches t_i from the right and from the left, respectively.

For example, if u_1 is a unit step at t_0; $u_2 = P_\varepsilon^{t_1}$; and $u_3(t) = \sin(t)$; then the corresponding outputs are

$$y_1(t) = \begin{cases} 0 \text{ for } t < t_0 \\ 1 \text{ for } t_0 < t \end{cases}$$

$$y_2(t) = \begin{cases} 0 \text{ for } t < t_1 \\ \dfrac{1}{\varepsilon} \text{ for } t_1 < t < t_1 + \varepsilon \\ 0 \text{ for } t_1 + \varepsilon < t \end{cases}$$

and $$y_3(t) = 0 \text{ for all } t.$$

It is easy to check that $\mathcal{S}$ is linear in u, since adding two inputs causes their jumps to be added, and multiplying an input by a constant multiplies

---

† Recall from Section 5–2 that smoothness was guaranteed by the local transition function satisfying a Lipschitz condition.

all the jumps by that constant. But $\mathcal{S}$ is not smooth. To see this, consider the inputs

$$u_n(t) = \begin{cases} 0 & t \leq 0 \\ nt & 0 < t \leq \dfrac{1}{n} \\ 1 & \dfrac{1}{n} \leq t \end{cases}$$

shown in Figure 5–7.

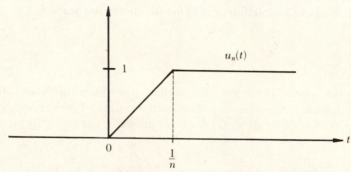

Figure 5–7   A family of inputs with no jumps, whose limit has a jump.

The response to each u_n is zero, since u_n is continuous everywhere. But the response to $\lim\limits_{n \to \infty} u_n = 1_0$, the unit step at $t = 0$, is not zero:

$$1_0(t) = \mathcal{S}(t, -\infty, 0, \lim_{n \to \infty} u_n) \neq \lim_{n \to \infty} \mathcal{S}(t, -\infty, 0, u_n) = 0 \qquad \Diamond$$

For a general time-varying linear system it is difficult and often impossible to determine exactly the state transition matrix $\Phi(t, t_0)$ and the impulse response matrix $g(t, \tau) = [g_{ij}(t, \tau)]$ from (3). In these cases it is often easier to determine the input-output characteristics of the system directly, applying at the jth input terminal a "unit impulse" at time τ and determining the response $g_{ij}(t, \tau)$ at the ith output terminal.

Example 2

Consider the two-input, two-output, time-varying system shown in Figure 5–8. It is clear from the diagram that $g_{21}(t, \tau) = 0$ since, with all initial conditions zero and holding $u_2 = 0$, an impulse δ_τ applied as u_1 produces no effect at output y_2. We will outline how to find $g_{12}(t, \tau)$ and leave the simpler problems of finding g_{11} and g_{22} to the reader.

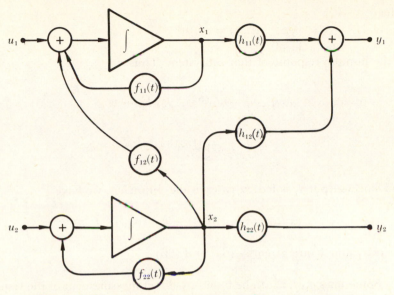

Figure 5-8 A two-input, two-output, time-varying system.

Since for g_{12} we hold $u_1 = 0$ and determine y_1 when $u_2 = \delta_\tau$, we may redraw the relevant portions of Figure 5-8 for greater clarity as Figure 5-9.

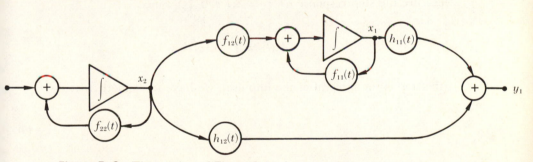

Figure 5-9 The portion of Figure 5-8 relevant to finding $g_{12}(t, \tau)$.

From Exercise 2(c), with $u_2 = \delta_\tau$, we know the signal at x_2 to be

$$x_2(t) = e^{\int_\tau^t f_{22}(\sigma)\, d\sigma}, \quad t > \tau \tag{6a}$$

Considering $f_{12}(t)x_2(t)$ as the "input" to the subsystem whose output is x_1, we may use Exercise 2(c) again to get

$$x_1(t) = \int_\tau^t h(t, \xi) f_{12}(\xi) x_2(\xi)\, d\xi$$

where

$$h(t, \xi) = e^{\int_{\xi}^{t} f_{11}(\alpha)\, d\alpha}, \quad t > \xi$$

is the impulse response of that subsystem. Thus

$$x_1(t) = \int_{\tau}^{t} e^{\int_{\xi}^{t} f_{11}(\alpha)\, d\alpha} f_{12}(\xi) e^{\int_{\tau}^{\xi} f_{22}(\sigma)\, d\sigma}\, d\xi \tag{6b}$$

$$= \int_{\tau}^{t} f_{12}(\xi) e^{[\int_{\xi}^{t} f_{11}(\alpha)\, d\alpha + \int_{\tau}^{\xi} f_{22}(\sigma)\, d\sigma]}\, d\xi$$

and since output y_1 is fed by two parallel branches, we have

$$g_{12}(t, \tau) = h_{11}(t)x_1(t) + h_{12}(t)x_2(t), \quad t > \tau \tag{7}$$

where x_1 and x_2 are given by (6a) and (6b). ◊

Sometimes $g_{ij}(t, \tau)$ can be found by direct measurement of the response at the ith output due to a "sharp," "narrow" pulse at the jth input. Just how sharp and narrow a pulse must be in order to look like an impulse is a difficult question to answer in general;† see Exercise 4 for some standard pulse approximations to impulses. Often, instead of trying to generate an "impulsive" input, we excite the system with a unit step 1_τ instead, and measure the **step response** $r(t, \tau) = \mathcal{S}(t, \tau, 0, 1_\tau)$. Since

$$1_\tau(t) = \int_{-\infty}^{t} \delta_\tau(\xi)\, d\xi$$

(the step is the integral of the impulse), we have for smooth systems

$$r(t, \tau) = \int_{\tau}^{t} g(t, \xi)\, d\xi \tag{8}$$

whence [Exercise 5]

$$g(t, \tau) = -\frac{\partial}{\partial t} r(t, \tau) \tag{9}$$

and the impulse response can be gotten by a differentiation of the step response.

We shall call two smooth linear systems **zero-state equivalent** if they have the same impulse response matrix. Thus, if the two zero-state equivalent systems have the same input $u \in \Omega$, they will have identical outputs y. The

---

† For a discussion of this problem, see H. D'Angelo, *Linear Time-Varying Systems* (Allyn & Bacon, 1969), Section 3.2.3.

reader should have no trouble† constructing an example of two zero-state equivalent systems whose state spaces have different dimensions.

To emphasize the importance of smoothness in this type of equivalence, we observe that the step response of the "rough" system of Example 1 is a unit step [verify!]. This is the same step response that we get from the smooth (but boring) system of Figure 5–10.

$$u \bullet \!\!\!\longrightarrow\!\!\!\bigcirc\!\!\!\!\! 1 \longrightarrow\!\!\bullet y$$ **Figure 5-10**

However, we cannot differentiate both step responses, equation (9), to conclude that both systems have identical impulse responses‡ and hence are zero-state equivalent. This is made even clearer by observing that an identical input of $u(t) = \sin(\omega t)$ produces an output of $y(t) = 0$ in Example 1, but an output of $y(t) = \sin(\omega t)$ in Figure 5–10.

In the sequel, it will be assumed, unless specifically stated to the contrary, that all linear continuous-time systems are also smooth.

Since the impulse response only characterizes the zero-state response of smooth linear systems, the natural or unforced behaviors of zero-state equivalent systems may be drastically different [Exercises 6 and 8].

The complete behavior of the system with matrices $F(\cdot)$, $G(\cdot)$, and $H(\cdot)$ is described by the state transition matrix $\Phi(t, t_0)$. The impulse response matrix

$$g(t, \tau) = \begin{cases} H(t)\Phi(t, \tau)G(\tau) & t > \tau \\ 0 & t < \tau \end{cases}$$

has the information in Φ masked by the operations of G and H. To determine the complete response,

$$y(t) = H(t)\Phi(t, t_0)x(t_0) + \int_{t_0}^{t} H(t)\Phi(t, \tau)G(\tau)u(\tau)\, d\tau$$

we need to know the initial condition $x(t_0)$ and the operator $H(t)\Phi(t, t_0)$. We get absolutely no information about $x(t_0)$ from $g(t, \tau)$, and the information about $H(t)\Phi(t, \tau)$ is scrambled by $G(\tau)$ before we see it in $g(t, \tau)$.

EXERCISES FOR SECTION 5-4

✓ **1.**   Suppose that the input-output behavior of a time-varying linear system is given by the differential equation

$$y^{(n)}(t) + a_1(t)y^{(n-1)}(t) + \cdots + a_{n-1}(t)y^{(1)}(t) + a_n(t)y(t) = u(t)$$

---

†Or else peek at Exercise 6 or Example 5–5–1.
‡In fact, what would "impulse response" mean for the system of Example 1?

(a)   Choosing state variables in the so-called "phase canonical form":

$$x_1 = y, \ x_2 = \dot{y}, \ldots, \ x_n = y^{(n-1)}$$

and calling $\Phi(t, t_0)$ the state transition matrix of the realization

$$\dot{x}(t) = F(t)x(t) + G(t)u(t)$$
$$y(t) = H(t)x(t)$$

show that the impulse response matrix $g(t, \tau)$ is a scalar (1×1), for this single-input single-output system. Show that $g(t, \tau)$ satisfies the equation:

$$\frac{\partial^n}{\partial t^n} g(t, \tau) + a_1(t) \frac{\partial^{n-1}}{\partial t^{n-1}} g(t, \tau) + \cdots + a_n(t)g(t, \tau) = \delta_\tau(t)$$

for $t_0 \le \tau \le t$, with the initial conditions:

$$g(\tau^-, \tau) = 0, \frac{\partial}{\partial t} g(t, \tau)|_{t=\tau^-} = 0, \ldots, \frac{\partial^{n-1}}{\partial t^{n-1}} g(t, \tau)|_{t=\tau^-} = 0$$

[Impulse response requires system to be relaxed at τ^-, just before impulse is applied.]
(b)   Show that

$$g(\tau^+, \tau) = \frac{\partial}{\partial t} g(t, \tau)|_{t=\tau^+} = 0, \ldots, \frac{\partial^{n-2}}{\partial t^{n-2}} g(t, \tau)|_{t=\tau^+} = 0$$

but that

$$\frac{\partial^{n-1}}{\partial t^{n-1}} g(t, \tau)|_{t=\tau^+} = 1$$

$$\left[\textit{Hint:} \text{ Integrate } \int_{\tau^-}^{\tau^+} \frac{\partial^n}{\partial \xi^n} g(\xi, \tau) \, d\xi = \frac{\partial^{n-1}}{\partial \xi^{n-1}} g(\xi, \tau)|_{\tau^-}^{\tau^+} \right.$$

$$\left. = \int_{\tau^-}^{\tau^+} \left\{ \delta_\tau(\xi) - a_1(\xi) \frac{\partial^{n-1}}{\partial \xi^{n-1}} g(\xi, \tau) - \cdots - a_n(\xi)g(\xi, \tau) \right\} d\xi \right]$$

How do you interpret this "step" in $\dfrac{\partial^{n-1}}{\partial t^{n-1}} g(t, \tau)$ at $t = \tau$?

Illustrate this by determining the values $y(0^+)$ and $\dot{y}(0^+)$ from the differential equation

$$\ddot{y}(t) + 2\dot{y}(t) + 5y(t) = \delta(t)$$

with

$$y(0^-) = 0, \dot{y}(0^-) = 0$$

[*Answer:* $y(0^+) = y(0^-)$ since y must be continuous at $t = 0$, for a step in y at zero produces an impulse in $\dot{y}$ and a doublet in $\ddot{y}$, and there is no doublet on the right-hand side of the differential equation. From the "hint," integrating $\dot{y}$ from 0^- to 0^+ yields $\dot{y}(0^+) = 1 + \dot{y}(0^-) = 1$.]
(c)   Find $y(0^+)$ and $\dot{y}(0^+)$ given the differential equation $\ddot{y}(t) + y(t) = \delta^{(1)}(t) + 2\delta(t)$, where $\delta^{(1)}(t) = \dfrac{d}{dt} \delta(t)$, the **unit doublet**, and the "initial data" $y(0^-) = 1, \dot{y}(0^-) = 2$. [*Answer:* $\dot{y}(0^+) = 2 + \dot{y}(0^-) = 4$ and $y(0^+) = 1 + y(0^-) = 2$.]

2.   For $t > \tau$, observe that $g(t, \tau)$ is a solution of the homogeneous differential equation

$$\dot{x}(t) = F(t)x(t) \ t > \tau$$
$$x(\tau^-) = 0, x(\tau^+) = c$$
$$y(t) = H(t)x(t)$$

since $u(t) = \delta_\tau(t) = 0$ for $t > \tau$.

The effect of the impulse $\delta_\tau(t)$ can be thought of as an instantaneous resetting of the initial conditions from the relaxed state $x(\tau^-) = 0$ to the non-zero state $x(\tau^+) = c$.

(a)   Show that $g(t, \tau) = H(t)\Phi(t, \tau)c$, for $t > \tau$, where c is the "initial state" vector $c = x(\tau^+)$ which must be set.

(b)   Equate $\mathcal{S}(t, \tau, 0, \delta_\tau(t)) = \mathcal{S}(t, \tau, c, 0)$ using equations (2), and show that $c = G(\tau)$ is an initial state setting such that the natural response of the system $\dot{x} = Fx$, $x(\tau^+) = c$ will be the impulse response.

(c)   Find the impulse response of the scalar system

$$\dot{y}(t) + \alpha(t)\,y(t) = u(t)$$

with $y(t) = 0$ for $t < \tau$.

[*Hint:* The homogeneous equation $\dot{y} + \alpha(t)\,y = 0$ for $t > \tau$ has the solution

$$g(t, \tau) = k(\tau)e^{-\int_\tau^t \alpha(\xi)d\xi}, \ t > \tau$$

where $k(\tau) = g(\tau^+, \tau)$. Evaluating $g(\tau^+, \tau)$ from the expression

$$g(\tau^+, \tau) - g(\tau^-, \tau) = \int_{\tau^-}^{\tau^+} \{\delta\tau(t) - \alpha(t)g(t, \tau)\}\,dt$$

$$= 1 - 0$$

since $g(\cdot, \tau)$ has no impulse at τ, get $g(\tau^+, \tau) = 1$ as the initial condition set up by the impulse.]

Verify that

$$g(t, \tau) = 1e^{-\int_\tau^t \alpha(\xi)d\xi}, \ t > \tau$$

is also gotten from $H(t)\Phi(t, \tau)G(\tau)$ by letting the state $x(t) = y(t)$.

3.   Consider the infinite-dimensional state space system of Figure 5–11, where the output is delayed by T seconds, multiplied by k, and subtracted from the input.

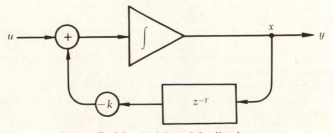

Figure 5–11   A delayed-feedback system.

(a) Write the dynamic equation describing this system, and observe that it is *not* a differential equation but, rather, a differential-difference equation. Does our analysis of going from "local" to "global" behavior by the methods of this chapter apply to this kind of system?

(b) Find an equivalent integral-difference equation for the system.

(c) Laplace transform your dynamic equation of part (a) and "solve" for $x(t)$ in terms of inverse transforms.

(d) Find the impulse response $g(t, \tau)$ of the system directly in the time domain by letting $u(t) = \delta_\tau(t)$ and chasing it around the feedback loop and integrating. Sketch $g(t, 0)$ for $k = T = 1$.

EE 4. This exercise illustrates that for short enough pulses, the actual shape of the pulse is unimportant as far as its use in approximating the impulse response is concerned.

(a) Calculate the zero state response for the system of Figure 5–12, corresponding to inputs of the forms shown in Figure 5–13(a–c). Do the calculations both in the time-domain (differential equations) and by Laplace transforms.

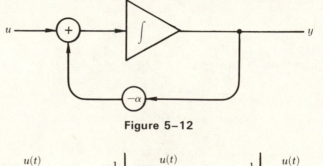

Figure 5–12

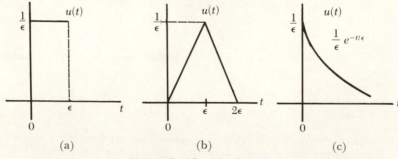

Figure 5–13 (a), (b), (c).

(b) Find the limits of the solutions as $\varepsilon \to 0$. What do you observe? What happens if $|\varepsilon| \approx |\alpha|$, the system "time constant"?

(c) Use linearity and your results from part (a) to find the responses to the forms shown in Figure 5–13(d–f).

(d) Observe that input (d) is the derivative of (b). What do you notice about the corresponding responses?

[*Note:* As $\varepsilon \to 0$, input (d) approaches a *unit-doublet.*]

✔ **5.** **(a)** Using smoothness, justify equation (8), expressing the step response as an integral of the impulse response.

(b) Differentiate equation (8) with respect to τ to get equation (9). What do you get if you differentiate equation (8) with respect to t?

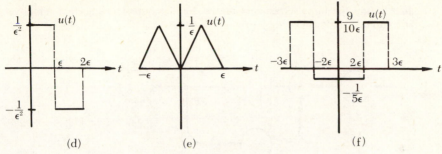

(d)   (e)   (f)

Figure 5–13   **(d), (e), (f).**

(c)   Show that the zero-state response is given by

$$y(t) = \int_{-\infty}^{\infty} \frac{du(\xi)}{d\xi} r(t, \xi) \, d\xi$$

EE 6.   Let u and y be input and output voltages of the two systems shown in Figure 5–14. Show that Σ and Σ' are zero-state equivalent, even though Σ has a three-dimensional state space and Σ' is two-dimensional.

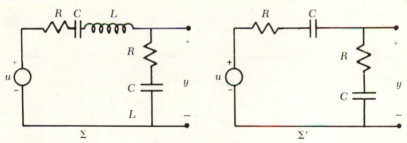

Figure 5–14

7.   How would you determine the "frequency response" of the system of Example 1?

✔ **8.**   **(a)**   Find the impulse response matrix for the system of Figure 5–15. Give the scalar differential equation relating output y to input u.

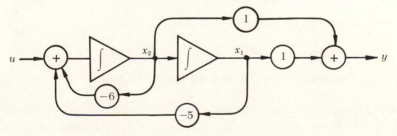

Figure 5–15   An interesting system.

(b)   Find the impulse response matrix for the system in Figure 5–16. Give the differential equation relating y to u.

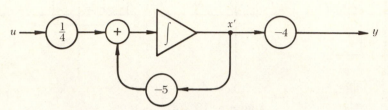

Figure 5–16   A system relevant to Figure 5–15.

(c)   Are these systems zero-state equivalent?

(d)   Compute the step response of each system directly, and use your results to verify equations (8) and (9).

(e)   Find the zero-state outputs due to the inputs shown in Figures 5–17(a) and (b). Use superposition and Exercise 5(c).

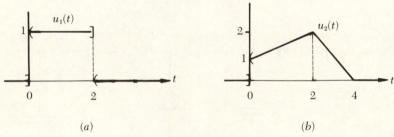

(a) (b)

Figure 5–17

9.   (a)   Find all four entries of the matrix $g(t, \tau)$ for the system of Example 2.

(b)   As an interesting special case, let all gains $f_{11}, f_{12}, f_{22}, h_{11}, h_{12}, h_{22}$ be *constants* and evaluate equations (6a), (6b), and (7) to get $g_{12}(t, \tau)$. Does $g_{12}(t, \tau) = g_{12}(t - \tau, 0)$?

(c)   For the constant coefficient case of part (b), find the zero-state response vector, $\mathcal{S}(t, t_0, 0, u)$, for the input vector $u(t) = e^{st} \begin{bmatrix} 0 \\ 1 \end{bmatrix}$, where s is a constant.

(d)   Using the result of part (c), does the limit $\lim\limits_{t_0 \to -\infty} \mathcal{S}(t, t_0, 0, u)|_{u(t) = e^{st}\begin{bmatrix} 0 \\ 1 \end{bmatrix}}$ make sense? Does it make sense for all exponents s?

5–5   Equivalent Systems and Realizations

In Section 4–2 we discussed the notions of distinguishable and indistinguishable events (τ, q'), where we considered trying to tell from input-output behavior whether or not a system was in state q or state q' at time τ. In this section we will formalize the intuitive notion of "two systems being equivalent so far as input-output behavior is concerned."

Suppose that we have two systems Σ and Σ' which have the same set of admissible input functions Ω.

DEFINITION 1

We say that state q of Σ and state q' of Σ' are **equivalent at time** τ, if when they are driven by any input $u \in \Omega$, Σ started in state q and Σ' started in q' at time τ yield identical outputs for $t > \tau$; i.e.,

$$\mathcal{S}_\Sigma(t, \tau, q, u) = \mathcal{S}_{\Sigma'}(t, \tau, q', u)$$

Thus, if $\Sigma = \Sigma'$, then q and q' are equivalent at time τ precisely when the events (τ, q) and (τ, q') are indistinguishable in the future. ○

We shall call two systems which have the same set of input-output pairs equivalent systems:

DEFINITION 2

Two systems Σ and Σ' are **equivalent** if, for any time $t_0 \in T$, for every state q of Σ there is a state q' of Σ' which is equivalent to q at t_0 and, conversely, for all $t_0 \in T$, for every state q' of Σ', there exists a state q of Σ such that q is equivalent to q' at t_0. ○

Example 1

The systems Σ and Σ' of Figure 5–18(a) and (b) are equivalent (even though Σ may smoke a bit when Σ' doesn't) under the correspondences non-unique

$$x'(t_0) = x_1(t_0) \text{ and } \begin{bmatrix} x_1(t_0) \\ x_2(t_0) \end{bmatrix} = \begin{bmatrix} x'(t_0) \\ 0 \end{bmatrix}$$

However, the system Σ'' of Figure 5–19 is *not* equivalent to Σ' (nor to Σ) since their responses are given by

$$y'(t) = 5e^{-3(t-t_0)}x'(t_0) + 5 \int_{t_0}^{t} e^{-3(t-\zeta)}u(\zeta)\,d\zeta$$

and

$$y''(t) = 5e^{2(t-t_0)}x''(t_0) + 5 \int_{t_0}^{t} e^{2(t-\zeta)}u(\zeta)\,d\zeta$$

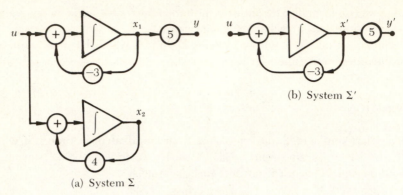

(a) System Σ

(b) System Σ'

Figure 5–18

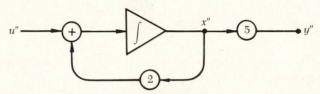

Figure 5–19   System Σ''

[verify!]. If, for any t_0, we take state $x'(t_0) = 0$, there is no value of $x''(t_0)$ which can yield output $y''(t)$ equal to output $y'(t)$ when both Σ'' and Σ' are driven by the step input $u(t) = 1_{t_0}(t)$:

$$5e^{-2(t-t_0)}x''(t_0) + \frac{5}{2}(1 - e^{-2(t-t_0)}) \neq \frac{5}{3}(1 - e^{-3(t-t_0)}) \text{ for } t > t_0 \qquad \Diamond$$

In the last section we called two smooth linear systems having the same impulse response zero-state equivalent. We formalize this notion in a more general setting.

DEFINITION 3

Two systems Σ and Σ' are **zero-state equivalent** if for every zero state 0 of Σ there is a zero state $0'$ of Σ' such that

$$\mathcal{S}_\Sigma(t, t_0, 0, u) = \mathcal{S}_{\Sigma'}(t, t_0, 0', u)$$

for every input function u. ○

Thus, two systems having identical zero-state responses are zero-state equivalent. Of course, if they are smooth and linear, so that the zero-state response is characterized by the impulse response, zero-state equivalent systems have identical impulse responses.

The reader should note that in Definitions 2 and 3, no mention is made of linearity, nor are there any assertions made about the dimension of the state spaces of equivalent systems.†

It is easy to see that if two *linear* systems are equivalent, then they are also zero-state equivalent [i.e., $\mathcal{S}_\Sigma(t, t_0, 0, u) = \mathcal{S}_{\Sigma'}(t, t_0, 0, u)$], but that the converse does not hold (Exercises 1 and 2).

Suppose now that we consider the linear differential system

$$\dot{x}(t) = F(t)x(t) + G(t)u(t)$$
$$y(t) = H(t)x(t) \tag{1}$$

hereafter denoted by the triple $(F(\cdot), G(\cdot), H(\cdot))$. We know that the output y is given by

$$y(t) = H(t)\Phi(t, t_0)x(t_0) + \int_{t_0}^{t} g(t, \tau)u(\tau)\, d\tau$$

where

$$g(t, \tau) = H(t)\Phi(t, \tau)G(\tau) \quad \text{for } t > \tau$$

Let us see what happens when we "change variables" by the state space transformation

$$x(t) = P(t)x'(t)$$

or

$$x'(t) = P^{-1}(t)x(t)$$

where $P(\cdot)$ is an invertible, differentiable matrix. Since

$$\frac{d}{dt}x(t) = \frac{d}{dt}(P(t)x'(t)) = \dot{P}(t)x'(t) + P(t)\dot{x}'(t)$$

equations (1) become

$$\dot{x}'(t) = [P^{-1}(t)F(t)P(t) - P^{-1}(t)\dot{P}(t)]x'(t) + [P^{-1}(t)G(t)]u(t)$$
$$y(t) = [H(t)P(t)]x'(t)$$

The invertible matrix $P(\cdot)$ which associates with each state x' the state $x = Px'$ is just a linear transformation on the state space. Because of this algebraic relabeling of the states, we have the following result.

---

†Example 1 showed a two-dimensional system Σ equivalent to the one-dimensional system Σ'.

DEFINITION 4

Two linear systems $(F(\cdot), G(\cdot)H(\cdot))$ and $(F'(\cdot), G'(\cdot), H'(\cdot))$ are said to be **algebraically equivalent** if there exists a nonsingular differentiable matrix $P(\cdot)$ such that $F'(\cdot)$, $G'(\cdot)$, and $H'(\cdot)$ are related to $F(\cdot)$, $G(\cdot)$, and $H(\cdot)$ by the equations (2), (3), and (4):

$$F'(t) = P^{-1}(t)F(t)P(t) - P^{-1}(t)\dot{P}(t) \tag{2}$$

$$G'(t) = P^{-1}(t)G(t) \tag{3}$$

$$H'(t) = H(t)P(t) \tag{4}$$

○

Example 2

The systems shown in Figure 5–20(a) and (b) are algebraically equivalent under the transformation

$$P(t) = \begin{bmatrix} (2e^t - e^{2t}) & (e^t - e^{2t}) \\ -2(e^t - e^{2t}) & (-e^t + 2e^{2t}) \end{bmatrix}$$

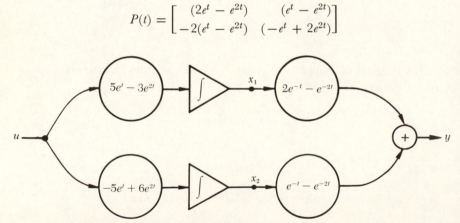

Figure 5–20(a)   A time-varying system $(F(\cdot), G(\cdot), H(\cdot))$.

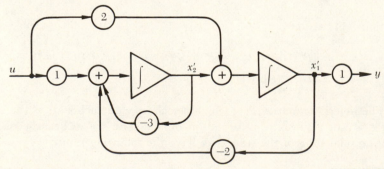

Figure 5–20(b)   An algebraically equivalent (constant!) system (F', G', H').

as the reader is urged to verify by writing the matrices (F, G, H) and (F', G', H') from the wiring-diagrams of Figure 5–20 and showing that equations (2) through (4) are satisfied. ◊

Example 2 illustrates that a time-varying system may be algebraically equivalent to a constant system. It also illustrates that by the proper choice of the state transformation $P(\cdot)$, it may be possible to end up with a simpler system to solve: we would certainly rather solve $\dot{x} = Fx + Gu$ in Example 2 *after* transforming it to the algebraically equivalent $\dot{x}' = F'x' + G'u$. [See Exercise 3.]

How are the solutions of two algebraically equivalent systems related? Calling $\Phi'(t, t_0)$ the state transition matrix of the homogeneous system

$$\dot{x}'(t) = F'(t)x'(t)$$

we have

$$x'(t) = \Phi'(t, t_0)x'(t_0)$$

From $x'(t) = P^{-1}(t)x(t)$, this becomes

$$P^{-1}(t)x(t) = \Phi'(t, t_0)P^{-1}(t_0)x(t_0)$$

whence

$$x(t) = (P(t)\Phi'(t, t_0)P^{-1}(t_0))x(t_0)$$

But $x(t) = \Phi(t, t_0)x(t_0)$ so, by uniqueness,

$$\Phi(t, t_0) = P(t)\Phi'(t, t_0)P^{-1}(t_0)$$

and equivalently,

$$\Phi'(t, t_0) = P^{-1}(t)\Phi(t, t_0)P(t_0) \tag{5}$$

What about the input-output description of (F', G', H')? If we let $g'(t, \tau)$ denote the impulse response matrix of (F', G', H'), then by equation (4) of Section 5–4, we have

$$g'(t, \tau) = H'(t)\Phi'(t, \tau)G'(\tau)$$

so, from equations (2) to (4),

$$
\begin{aligned}
g'(t, \tau) &= (H(t)P(t))(P^{-1}(t)\Phi(t, \tau)P(\tau))(P^{-1}(\tau)G(\tau)) \\
&= H(t)\Phi(t, \tau)G(\tau) \\
&= g(t, \tau)
\end{aligned}
\tag{6}
$$

Thus, algebraically equivalent systems have identical impulse responses and hence are zero-state equivalent also. The converse is clearly not true, since algebraically equivalent systems necessarily have state spaces of the same dimension (P is invertible), but we have seen zero-state equivalent systems with different dimensions in Example 1.

A similarly simple argument [Exercise 4] shows that algebraically equivalent systems are equivalent, and that, again, the converse statement does not hold.

We call the algebraic state transformation $P(\cdot)$ relating two algebraically equivalent systems an **equivalence transformation.** This generalizes the special case where $P(t) = P_0$, constant, and (2) and (5) reduce to

$$F'(t) = P_0^{-1}F(t)P_0, \qquad P_0 \text{ constant}$$

and

$$\Phi'(t, t_0) = P_0^{-1}\Phi(t, t_0)P_0, \quad P_0 \text{ constant}$$

and we call P_0 a **similarity transformation.** In this case, we say that F' is **similar** to F and Φ' is **similar** to Φ. We will see a lot more of similarity transformations in the next chapter, when we study time-invariant systems.

We close this section with the notion of a "realization" of a given impulse response matrix, a topic to which we shall return in Chapter 6, again briefly in Section 7–3, and then in more detail in Chapter 8. We have seen (equation (4) of Section 5–4), that it is easy to obtain the impulse response matrix $g(t, \tau)$ from a given dynamic system description $(F(\cdot), G(\cdot), H(\cdot))$ as $g(t, \tau) = H(t)\Phi(t, \tau)G(\tau)$. Suppose we now ask: "Given an impulse response matrix $g(t, \tau)$ of a system, is it possible to find a choice of state variable x and a system $(F(\cdot), G(\cdot), H(\cdot))$ whose impulse response will be $g(t, \tau)$?" If the answer to this question is "yes," we say that the system $(F(\cdot), G(\cdot), H(\cdot))$ is a *realization* of $g(t, \tau)$.

DEFINITION 5

A **realization** of a given matrix $g(t, \tau)$ is just a linear system $(F(\cdot), G(\cdot), H(\cdot))$ whose impulse response is $g(t, \tau)$. ○

Clearly, if there is any realization $(F(\cdot), G(\cdot), H(\cdot))$ of $g(t, \tau)$, then there are infinitely many others—all the algebraically equivalent systems $(F'(\cdot), G'(\cdot), H'(\cdot))$, for example. If $g(t, \tau)$ is the impulse response of a given system Σ, then any realization $(F(\cdot), G(\cdot), H(\cdot))$ of $g(t, \tau)$ need only be zero-state equivalent to Σ; their natural and complete responses may be unrelated.

We have the following result:

THEOREM 1

An impulse response matrix $g(t, \tau)$ has a realization $(F(\cdot), G(\cdot), H(\cdot))$ if and only if there exist matrices $A(\cdot)$ and $B(\cdot)$ such that g can be factored as

$$g(t, \tau) = A(t)B(\tau), t \geq \tau$$

Proof

The "only if" part is easy. We know that

$$g(t, \tau) = H(t)\Phi(t, \tau)G(\tau)$$

and

$$\Phi(t, \tau) = \Phi(t, 0)\Phi(0, \tau)$$

so

$$g(t, \tau) = (H(t)\Phi(t, 0))(\Phi(0, \tau)G(\tau))$$

and the assertion follows by letting

$$A(t) = H(t)\Phi(t, 0)$$

and

$$B(\tau) = \Phi(0, \tau)G(\tau)$$

For the "if" part, suppose that we can write

$$g(t, \tau) = A(t)B(\tau)$$

for some matrices $A(\cdot)$ and $B(\cdot)$. Consider the linear system $(0, B(\cdot), A(\cdot))$ described by

$$\dot{x}(t) = B(t)u(t)$$
$$y(t) = A(t)x(t)$$

The state transition matrix for this system is I, the identity, and its impulse response matrix is just $A(t)B(\tau) = g(t, \tau)$ [see Exercise 6]. Thus, taking

$$(F(\cdot), G(\cdot), H(\cdot)) = (0, B(\cdot), A(\cdot))$$

we have the asserted realization.   □

The algebraically equivalent systems presented in Example 2 illustrate Theorem 1 nicely:

Example 3

For the system of Figure 5–20(a), we have from the wiring-diagram that

$$(F(t), G(t), H(t)) = (0, B(t), A(t))$$

where

$$B(t) = \begin{bmatrix} (5e^t - 3e^{2t}) \\ (-5e^t + 6e^{2t}) \end{bmatrix}$$

and

$$A(t) = [(2e^{-t} - e^{-2t})\,(e^{-t} - e^{-2t})]$$

Thus, $\Phi(t, t_0) = I$ and

$$g(t, \tau) = A(t)B(\tau) = 5e^{-1(t-\tau)} - 3e^{-2(t-\tau)}, \, t \geq \tau,$$

as the reader should check directly [Exercise 5]. ◊

In the next three chapters, we shall study in more detail the interesting problem of finding useful realizations of a given input-output description.

EXERCISES FOR SECTION 5–5

✔ **1.** Show that if Σ and Σ' are equivalent linear systems, they are necessarily zero-state equivalent. [*Hint:* Take $q = 0$, and let q' be its equivalent state in Σ'. Since

$$\mathcal{S}_\Sigma(t, t_0, 0, u) = \mathcal{S}_{\Sigma'}(t, t_0, q', u)$$

for all u, let $u = 0$ to deduce that q' is a zero state of Σ', from which zero-state equivalence follows].

EE 2. Consider the two electrical systems shown in Figure 5–21(a) and (b).

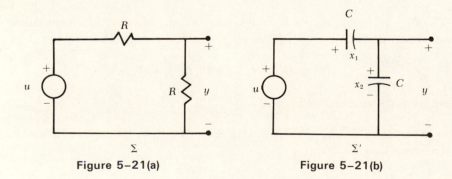

Figure 5–21(a)   **Figure 5–21(b)**

(a)   Show that for $t \geq 0$, we have $y_\Sigma(t) = \dfrac{1}{2} u(t)$ and

$$y_{\Sigma'}(t) = \frac{1}{2}[x_2(0) - x_1(0)] + \frac{1}{2}u(t)$$

where x_1 and x_2 are the voltages across the capacitors.

(b)   Show that Σ and Σ' are not equivalent systems but that they are zero-state equivalent.

(c)   Construct an example of your own illustrating this point.

✓ **3. (a)**   Show that the transformation $P(\cdot)$ in Example 2 was such that

$$0 = F(t) = P(t)F'(t)P^{-1}(t) - \dot{P}(t)P^{-1}(t)$$

(b)   Show that this is equivalent to

$$0 = F(t) = Q^{-1}(t)F'(t)Q(t) - Q^{-1}(t)\dot{Q}(t)$$

where $Q(t) = P^{-1}(t)$, and hence that $Q(t)$ is a fundamental matrix of the system $\dot{x}' = F'x'$.

(c)   Use (b) to determine the state transition matrix $\Phi'(t, t_0)$ for the system (F', G', H') in Figure 5–20(b).

✓ **4.**   Let (F, G, H) and (F', G', H') be algebraically equivalent systems. Prove that they must be equivalent. Give an example of two equivalent systems which are not algebraically equivalent.

✓ **5.**   Find the impulse response of the system (F, G, H) of Figure 5–20(a) directly by letting $[x_1(t_0), x_2(t_0)]^T = [0, 0]^T$ and $u = \delta_{t_0}$. Give the output of (F', G', H') of Figure 5–20(b), due to the input $u(t) = \begin{cases} 2, 0 < t \leq 1 \\ 0 \text{ otherwise} \end{cases}$ and the initial condition $[x_1'(0), x_2'(0)]^T = [1, -1]^T$.

6.   Prove that the state transition matrix of the system $(0, B(\cdot), A(\cdot))$ is I and that its impulse response is $g(t, \tau) = A(t)B(\tau)$.

7. (a)   Can you find a realization of the impulse response $g(t, \tau) = \delta(t - \tau - 3)$? Can you find a finite-dimensional realization?

(b)   Is $g(t, \tau) = e^{t\tau}$ the impulse response of a linear system?

(c)   Find two realization for $g(t, 0) = 2e^{-3t} - e^{-2t}$.

8.   Give the response functions

$$y = \mathcal{S}(t, t_0, x_0, u)$$

for the systems shown in Figure 5–22(a) and (b), and decide whether or not they are equivalent to the system Σ' of Figure 5–18(b). What about zero-state equivalence?

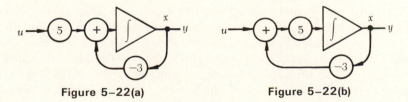

Figure 5–22(a)   **Figure 5–22(b)**

CHAPTER 6

Constant Linear Systems

In this chapter, we present the core material for the study of a constant linear system represented by $\dot{x}(t) = Fx(t) + Gu(t)$ and $y(t) = Hx(t)$, an almost self-contained exposition of the classical transform approach to linear systems together with an exposition of the basic state-variable results. This will provide a firm basis for our study, in a modern state-variable setting, of the basic notions of controllability, observability, and stability in Chapter 7.

First, in Section 6–1, we note that the matrix exponential provides the complete solution of the above system equations. Although the techniques here are justified by the general theory of Chapter 5, they will serve as an adequate reference for the reader who is simply content to check the formulas by differentiation. In Section 6–2 we introduce matrix diagonalization and eigenvector expansions as useful techniques for simplifying the computation of the matrix exponential. In particular, we learn how to separate the various "modes" of a linear system, so that we may study their evolutions separately. With this as background, we may, in Section 6–3, relate our growing understanding of the solution of system equations in the time domain to the classical methods, based on the Laplace transform, which solve these equations in the frequency domain. Those readers who are not familiar with block diagram manipulations and signal flow graphs will find a detailed introduction to these useful concepts in Section 6–4. In any case, in Section 6–5 we give special

378

attention to the transfer function of a system whose input and output spaces are one-dimensional, noting how easily these results generalize to the multi-dimensional case. Finally, in Section 6–6 we return to the problem of diagonalization introduced in Section 6–2, and note that even when that process cannot succeed completely there is still available a very useful canonical form which decouples the modes of oscillation of a linear system as much as possible—namely, the Jordan canonical form. Both the theory of normal forms and the theory of transfer functions, developed in the present chapter, provide basic tools in the analysis of linear systems. It is to basic applications of these tools that we will turn in Chapter 7.

Throughout this chapter we will let the typical finite-dimensional, time-invariant linear system be represented by

$$\dot{x}(t) = Fx(t) + Gu(t)$$

$$y(t) = Hx(t)$$

hereafter referred to as (F, G, H), where F, G, and H are respectively $n \times n$, $n \times m$, and $q \times n$ matrices whose entries are real (or complex) constants.

6–1   The Matrix Exponential and the Complete Solution

We may solve the homogeneous equation

$$\dot{x}(t) = Fx(t)$$

$$x(t_0) = x_0$$

by the method of successive approximations on the equivalent integral equation

$$x(t) = x(t_0) + \int_{t_0}^{t} Fx(\tau)\, d\tau$$

as we justified in the more general discussion of Section 5–2. If we start with the initial approximation

$$x_0(t) = x(t_0) = x_0 \text{ for all } t$$

we have

$$x_1(t) = x(t_0) + \int_{t_0}^{t} Fx_0(\tau)\, d\tau$$

$$= x(t_0) + F(t - t_0)x(t_0)$$
$$= [I + F(t - t_0)]x(t_0)$$

$$x_2(t) = x(t_0) + \int_{t_0}^t Fx_1(\tau)\, d\tau$$

$$= x(t_0) + \int_{t_0}^t F[I + F(\tau - t_0)]x(t_0)\, d\tau$$

$$= x(t_0) + \int_{t_0}^t F[I + F(\tau - t_0)]x(t_0)\, d\tau$$

$$= x(t_0) + F(t - t_0)x(t_0) + F^2 \frac{(t - t_0)^2}{2} x(t_0)$$

$$= \left[1 + F(t - t_0) + F^2 \frac{(t - t_0)^2}{2} \right] x(t_0)$$

from which an easy induction yields

$$x_n(t) = x(t_0) + \int_{t_0}^t Fx_{n-1}(\tau)\, d\tau$$

$$= \left[I + F(t - t_0) + F^2 \frac{(t - t_0)^2}{2!} + \cdots + F^n \frac{(t - t_0)^n}{n!} \right] x(t_0)$$

Thus

$$x(t) = \lim_{n \to \infty} x_n(t) = \left(\sum_{k=0}^{\infty} F^k \frac{(t - t_0)^k}{k!} \right) x(t_0) \tag{1}$$

We agreed in Section 5–3 to call the infinite matrix series within parentheses the **matrix exponential** $e^{F(t-t_0)}$, and since the solution $x(t)$ may be written uniquely as $x(t) = \Phi(t, t_0)x(t_0)$, we have just shown again that for the constant system (F, G, H)

$$\Phi(t, t_0) = e^{F(t-t_0)} = \sum_{k=0}^{\infty} F^k \frac{(t - t_0)^k}{k!} \tag{2}$$

a result we could have written from Example 9 or Theorem 6 of that section. [See Exercise 1.]

Example 1

If $F = \begin{bmatrix} d_1 & 0 \\ 0 & d_2 \end{bmatrix}$ is a diagonal matrix, then

$$F^2 = FF = \begin{bmatrix} d_1^2 & 0 \\ 0 & d_2^2 \end{bmatrix}$$

$$F^3 = FF^2 = \begin{bmatrix} d_1^3 & 0 \\ 0 & d_2^3 \end{bmatrix}$$

$$F^n = FF^{n-1} = \begin{bmatrix} d_1^n & 0 \\ 0 & d_2^n \end{bmatrix}$$

so

$$e^{Ft} = I + Ft + F^2 \frac{t^2}{2!} + F^3 \frac{t^3}{3!} + \cdots$$

$$= \begin{bmatrix} 1 & 0 \\ 0 & 1 \end{bmatrix} + \begin{bmatrix} d_1 t & 0 \\ 0 & d_2 t \end{bmatrix} + \begin{bmatrix} d_1^2 \frac{t^2}{2} & 0 \\ 0 & d_2^2 \frac{t^2}{2} \end{bmatrix} + \begin{bmatrix} d_1^3 \frac{t^3}{3!} & 0 \\ 0 & d_2^3 \frac{t^3}{3!} \end{bmatrix} + \cdots$$

$$= \begin{bmatrix} 1 + d_1 t + d_1^2 \frac{t^2}{2} + d_1^3 \frac{t^3}{3!} + \cdots & 0 \\ 0 & 1 + d_2 t + d_2^2 \frac{t^2}{2} + d_2^3 \frac{t^3}{3!} + \cdots \end{bmatrix}$$

$$= \begin{bmatrix} e^{d_1 t} & 0 \\ 0 & e^{d_2 t} \end{bmatrix}$$

More generally, for any diagonal matrix, the series is summable in closed form and

$$e^{\begin{bmatrix} \lambda_1 & 0 & 0 & \cdots & 0 \\ 0 & \lambda_2 & 0 & \cdots & 0 \\ \vdots & & & & \vdots \\ 0 & 0 & \cdots & & \lambda_n \end{bmatrix} t} = \begin{bmatrix} e^{\lambda_1 t} & 0 & \cdots & 0 \\ 0 & e^{\lambda_2 t} & \cdots & 0 \\ \vdots & \vdots & & \vdots \\ 0 & 0 & \cdots & e^{\lambda_n t} \end{bmatrix}$$ (3)

$\diamond$

Of course, in general we won't be so lucky and we will have to settle for approximating e^{Ft} by using a sufficient number of terms of the infinite series (2). [See Exercises 2, 3, and 4.] In Section 6–2, after we have discussed eigenvectors, and in Section 6–3, after we have discussed Laplace transforms, we shall present other methods of computing e^{Ft}.

Let us now develop some of the properties of the matrix exponential.

THEOREM 1

For the constant square matrix, F, the matrix exponential satisfies the following conditions:

(i) $\dfrac{d}{dt}\, e^{Ft} = Fe^{Ft} = e^{Ft}F$

(ii) $(e^{Ft})^{-1} = e^{-Ft}$, for all t

(iii) $e^{F(t_1+t_2)} = e^{Ft_1}e^{Ft_2}$, for all t_1, t_2

(iv) $(e^{Ft})^* = e^{F^*t}$, where $*$ denotes the adjoint as in Section 4–4

(v) $\det e^{Ft} = e^{t(\mathrm{trace}\,F)}$, where trace $F = \displaystyle\sum_{i=1}^{n} [F]_{ii}$

(vi) $e^{(P^{-1}FP)t} = P^{-1}e^{Ft}P$ for any invertible constant matrix P

Proof

(i) Since $\dfrac{\partial}{\partial t}\, \Phi(t, t_0) = F\Phi(t, t_0)$ and

$$e^{Ft} = \Phi(t, 0) = \sum_{k=0}^{\infty} F^k \frac{t^k}{k!}$$

we have

$$\frac{d}{dt}\, e^{Ft} = Fe^{Ft}$$

Furthermore, since

$$Fe^{Ft} = e^{Ft}F$$

as the reader should verify using the series (2), we also have that

$$\frac{d}{dt}\, e^{Ft} = e^{Ft}F$$

(ii) Since $\Phi(0, t_0) = e^{-Ft_0}$ for all t_0, then e^{Ft} is invertible and

$$(e^{Ft})^{-1} = \Phi^{-1}(t, 0) = \Phi(0, t) = e^{-Ft}$$

(iii) For constant systems $\Phi(t, 0)\Phi(s, 0) = \Phi(t + s, 0)$ [Example 5–3–9], so

$$e^{Ft_1}e^{Ft_2} = e^{F(t_1+t_2)} \text{ for all } t_1,\, t_2$$

(iv) Since $(F^n)^* = (F^*)^n$,

$$(e^{Ft})^* = \left(I + Ft + F^2\frac{t^2}{2!} + \cdots + F^n\frac{t^n}{n!} + \cdots\right)^*$$

$$= I^* + F^*t + (F^*)^2\frac{t_2}{2!} + \cdots + (F^*)^n\frac{t^n}{n!} + \cdots$$

so

$$(e^{Ft})^* = e^{F^*t}$$

(v) Since $\det \Phi(t_1, t_0) = e^{\int_{t_0}^{t_1} \text{trace } F(\tau)\,d\tau}$ [Theorem 5 of Section 5–3], and since F is constant, $\det e^{Ft} = e^{(\text{trace } F)(t)}$.

(vi) For any invertible matrix P, we have

$$(P^{-1}FP)^2 = (P^{-1}FP)(P^{-1}FP) = P^{-1}F^2P$$
$$(P^{-1}FP)^3 = (P^{-1}FP)(P^{-1}FP)^2 = (P^{-1}FP)(P^{-1}F^2P) = P^{-1}F^3P$$
$$\vdots$$
$$(P^{-1}FP)^n = P^{-1}F^nP, \quad n = 1, 2, \ldots$$

so

$$e^{(P^{-1}FP)t} = \sum_{n=0}^{\infty} (P^{-1}FP)^n\frac{t^n}{n!} = \sum_{n=0}^{\infty} P^{-1}F^nP\frac{t^n}{n!}$$

$$= P^{-1}\left(\sum_{n=0}^{\infty} F^n\frac{t^n}{n!}\right)P$$

whence

$$e^{(P^{-1}FP)t} = P^{-1}e^{Ft}P \qquad \qquad \square$$

Example 2

Consider the matrices $F = \begin{bmatrix} 0 & 1 \\ 0 & -2 \end{bmatrix}$ and $P = \begin{bmatrix} 1 & 1 \\ 0 & -2 \end{bmatrix}$.

Since P is invertible and $P^{-1} = \begin{bmatrix} 1 & 1/2 \\ 0 & -1/2 \end{bmatrix}$ and since

$P^{-1}FP = \begin{bmatrix} 0 & 0 \\ 0 & -2 \end{bmatrix}$ is a *diagonal* matrix, we have from (3) that

$$e^{(P^{-1}FP)t} = e^{\begin{bmatrix} 0 & 0 \\ 0 & -2 \end{bmatrix}t} = \begin{bmatrix} e^{0t} & 0 \\ 0 & e^{-2t} \end{bmatrix}$$

Thus, by Theorem 1(vi),

$$e^{Ft} = e^{P(P^{-1}FP)P^{-1}t} = Pe^{(P^{-1}FP)t}P^{-1}$$

$$= \begin{bmatrix} 1 & 1 \\ 0 & -2 \end{bmatrix} \begin{bmatrix} 1 & 0 \\ 0 & e^{-2t} \end{bmatrix} \begin{bmatrix} 1 & 1/2 \\ 0 & -1/2 \end{bmatrix}$$

$$= \begin{bmatrix} 1 & (1/2)(1 - e^{-2t}) \\ 0 & e^{-2t} \end{bmatrix}$$

as the reader should have found directly in Exercise 2.   ◊

The reader, not yet having bothered to carry out the manipulation of the power series, may have thought that having proved $e^{A(s+t)} = e^{As} \cdot e^{At}$ in Theorem 1 (iii) it would be a trivial extension to prove that $e^{(A+B)t} = e^{At} \cdot e^{Bt}$ is true for any pair of linear transformations A and B of the same size. However, this conclusion is *false*, as can be seen from comparing the following expansions:

$$e^{(A+B)t} = I + (A + B)t + \frac{1}{2}(A + B)^2 t^2 + \cdots$$

$$= I + (A + B)t + \frac{1}{2}(A^2 + AB + BA + B^2)t^2 + \cdots$$

$$e^{At} \cdot e^{Bt} = \left(I + At + \frac{A^2 t^2}{2} + \cdots\right) \cdot \left(I + Bt + \frac{B^2 t^2}{2} + \cdots\right)$$

$$= I + (A + B)t + \frac{1}{2}(A^2 + 2AB + B^2)t^2 + \cdots$$

We thus see that $e^{(A+B)t} = e^{At} \cdot e^{Bt}$ can hold if and only if, equating the coefficients of t^2,

$$A^2 + AB + BA + B^2 = A^2 + AB + AB + B^2$$

holds—but this can hold if and only if A and B commute, i.e., $AB = BA$. It is the fact that As and At commute that allows us to prove so readily that $e^{A(s+t)} = e^{As} \cdot e^{At}$, because, not having to worry about the order in which our matrix factors occur, we can manipulate our matrices just as if they were scalars, to get the result that holds in the scalar case.

Since the state transition matrix is given by the matrix exponential

$$\Phi(t, t_0) = e^{F(t-t_0)}$$

when F is constant, we can write the solution of the time-invariant linear system (F, G, H) as

$$x(t) = e^{F(t-t_0)}x(t_0) + \int_{t_0}^{t} e^{F(t-\tau)}Gu(\tau)\,d\tau \tag{4}$$

$$y(t) = He^{F(t-t_0)}x(t_0) + \int_{t_0}^{t} g(t, \tau)u(\tau)\,d\tau \tag{5}$$

where

$$g(t, \tau) = He^{F(t-\tau)}G \tag{6}$$

is the impulse response matrix. We observe from (6) that

$$g(t, \tau) = g(t - \tau, 0)$$

so that the zero-state response of a constant system at time t to an impulse applied at time $\tau < t$ is a function of the "age" or "memory" variable $(t - \tau)$ only.

It is customary to abuse the notation when dealing with constant systems and to write things like

$$\Phi(t - \tau) = \Phi(t - \tau, 0)$$
$$g(t - \tau) = g(t - \tau, 0)$$

since it is only the difference of t and τ that matters. With this notation, equations (5) and (6) become

$$y(t) = He^{F(t-t_0)}x(t_0) + \int_{t_0}^{t} g(t - \tau)u(\tau)\,d\tau$$

and

$$g(\tau) = \begin{cases} He^{F\tau}G & \text{for } \tau > 0 \\ 0 & \text{for } \tau < 0 \end{cases}$$

If we take $t_0 = 0$ as we usually do with time-invariant systems,† we have the following result:

---

†Recall from Section 1–6 on time-invariance that

$$x(t) = \phi(t, t_0, \tilde{x}, u) = \phi(t - t_0, \tilde{x}, z^{t_0}u)$$

and

$$y(t) = \eta(t, x(t)) = \eta(x(t)) \text{ for } t \geq t_0$$

where $(z^{t_0}u)(t) = u(t + t_0)$.

THEOREM 2

The complete solution of the constant system (F, G, H) is given by

$$x(t) = e^{Ft}x(0) + \int_0^t e^{F(t-\tau)}Gu(\tau)\,d\tau \tag{7}$$

$$y(t) = He^{Ft}x(0) + \int_0^t g(t-\tau)u(\tau)\,d\tau \tag{8}$$

where

$$g(t) = \begin{cases} He^{Ft}G & \text{for } t > 0 \\ 0 & \text{for } t < 0 \end{cases} \tag{9}$$
□

The superposition integral

$$y(t) = \int_{t_0}^t g(t, \tau)u(\tau)\,d\tau$$

which we saw in Section 5–4 described the input-output behavior of a linear system relaxed at t_0, has now become the famous **convolution integral**†

$$y(t) = \int_{t_0}^t g(t-\tau)u(\tau)\,d\tau \tag{10}$$

and we say that $y(\cdot)$ is the **convolution** of $g(\cdot)$ and $u(\cdot)$. Because of the simple algebraic properties of the convolution operation on a pair of functions g and u [see Exercise 7], we often write equation (10) as

$$y = g \star u \tag{11}$$

with the binary operation $\star$ defined by the integral (10). [See Exercise 8.]
Often we write

$$y(t) = \int_{-\infty}^\infty g(t-\tau)u(\tau)\,d\tau$$

for the convolution of g and u, where we assume the system to have been relaxed at $t = -\infty$ and to be *causal*‡ so that

$$g(t - \tau) = 0 \text{ for } t < \tau$$

and assume u to have been zero before the time of application $t_0 = 0$.

---

† Also called the *faltung* or *Duhamel* integral. See R. M. Bracewell, *The Fourier Transform and Its Applications* (McGraw-Hill, 1965) for a leisurely, detailed discussion of convolution.
‡ See Exercise 3 of Section 1–5.

In Section 6–3 we will discuss the Laplace transform, an integral transformation specifically designed to transform the complicated operation of convolution of g and u into the simpler operation of ordinary multiplication of the transforms of g and u, just as in Section 4–6 we saw that the Z-transform would transform a convolution sum into a multiplication of functions of z. Since in most "real world" problems we are not given nice little formulas for the functions $g(\cdot)$ and $u(\cdot)$ we will usually not be able to work out the integral (10) in closed form to give a formula for $y(\cdot)$. Often we will know $g(\cdot)$ and $u(\cdot)$ only graphically, or from computer print-outs as tables $\{g(t_0), g(t_0 + h), g(t_0 + 2h), \ldots, g(t_0 + nh), \ldots\}$ of numbers. In this case we use an approximate integration rule to generate a table of values $\{ y(t_0), \ldots, y(t_0 + nh), \ldots\}$ for y [see Exercise 9].

ALGEBRAICALLY EQUIVALENT CONSTANT SYSTEMS

Following our discussion in Section 5–5, we call the constant linear systems represented by (F, G, H) and $(\bar{F}, \bar{G}, \bar{H})$ algebraically equivalent if there exists a *constant* invertible matrix P such that $x = P\bar{x}$ relates their respective states x and $\bar{x}$ and

$$\bar{F} = P^{-1}FP$$
$$\bar{G} = P^{-1}G$$
$$\bar{H} = HP$$

Throughout the rest of this chapter we will encounter the interesting problem of trying to find such a matrix P to transform a given (F, G, H) into some more desirable, equivalent set of matrices $(\bar{F}, \bar{G}, \bar{H})$.

EXERCISES FOR SECTION 6–1

1.   Here is an alternate derivation of $\Phi(t, t_0) = e^{F(t-t_0)}$: Assume the solution of $\dot{\Phi}(t, 0) = F\Phi(t, 0)$ with

$$\Phi(0, 0) = I$$

has the form of a power series

$$\Phi(t, 0) = A_0 + A_1 t + A_2 \frac{t^2}{2} + \cdots + A_m \frac{t^m}{m!} + \cdots$$

where the A_i are $n \times n$ matrices to be determined. Substitute the power series into the differential equation for $\Phi(t, 0)$ and equate like powers of t to deduce $A_0 = I$, $A_1 = F, \ldots, A_m = F^m$, whence

$$\Phi(t, 0) = e^{Ft},$$

and consequently that

$$\Phi(t, t_0) = e^{F(t-t_0)}$$

✔ **2.**   Find e^{Ft} from its defining series for the following matrices:

(a) $F = I$

(b) $F = 0$

(c) $F = \begin{bmatrix} 0 & 3 \\ 0 & 0 \end{bmatrix}$ (a *nilpotent* matrix)

(d) $F = \begin{bmatrix} 0 & 1 \\ -5 & -6 \end{bmatrix}$ (find the first 4 terms)

(e) $F = \begin{bmatrix} 0 & 1 \\ 0 & -2 \end{bmatrix}$

(f) $F = \begin{bmatrix} 0 & 1 \\ -1 & 0 \end{bmatrix}$ (compare your result with Example 5–2–4)

(g) $F = I + B$, where B is any square matrix.

✔ **3.**   Show that if $J_n = \begin{bmatrix} \lambda & 1 & 0 & \cdots & 0 & 0 \\ 0 & \lambda & 1 & & 0 & \vdots \\ \vdots & \vdots & \lambda & & & \\ & & & & & 0 \\ \vdots & & & & \lambda & 1 \\ 0 & 0 & 0 & & 0 & \lambda \end{bmatrix}$

a so-called $n \times n$ "Jordan block," then

$$e^{J_n t} = \begin{bmatrix} e^{\lambda t} & te^{\lambda t} & \dfrac{t^2}{2}e^{\lambda t} & \cdots & \dfrac{t^{n-1}}{(n-1)!}e^{\lambda t} \\ 0 & e^{\lambda t} & te^{\lambda t} & \cdots & \dfrac{t^{n-2}}{(n-2)!}e^{\lambda t} \\ \vdots & \vdots & & & \vdots \\ 0 & 0 & & & te^{\lambda t} \\ 0 & 0 & & \cdots & e^{\lambda t} \end{bmatrix}$$

✔ **4.**   Let $A = \begin{bmatrix} 0 & 3 \\ 0 & 0 \end{bmatrix}$ and $B = \begin{bmatrix} 0 & 1 \\ 0 & -2 \end{bmatrix}$ as studied in Exercise 2. Let $F = A + B$ and compute e^{Ft}. Does $e^{(A+B)t} = e^{At} \cdot e^{Bt}$? Does $e^{BA} = e^{AB}$?

5.   Prove part (vi) of Theorem 1, using the results of Section 5–5 on algebraically equivalent systems.

✔ **6.**   This exercise shows that for any constant $n \times n$ matrix F, e^{Ft} can be expressed in terms of only a *finite* number of powers of F.

(a)   Let $V = M_{n \times n}(\mathbf{R})$, the set of all real $n \times n$ matrices. Show that V is a vector space and that the n^2 matrices $\begin{bmatrix} 1 & 0 & \cdots & 0 \\ \vdots & & & \\ 0 & & & 0 \end{bmatrix}, \begin{bmatrix} 0 & 1 & \cdots & 0 \\ \vdots & & & \\ 0 & & & 0 \end{bmatrix}, \ldots,$

$\begin{bmatrix} 0 & \cdots & 0 & 1 \\ \vdots & & & \\ 0 & & & 0 \end{bmatrix}, \ldots, \begin{bmatrix} 0 & \cdots & 0 \\ \vdots & & \\ 0 & & 1 \end{bmatrix}$ are a basis for V.

(b)   Since by part (a) dim $V = n^2$, the $n^2 + 1$ vectors $I = F^0, F, F^2, F^3, \ldots,$ F^{n^2} cannot be linearly independent. Show that this implies that there is a power $r \leq n^2$ such that $F^r = c_0 I + c_1 F + \cdots + c_{r-1} F^{r-1}$ for some constants $c_0, \ldots, c_{r-1}$.

(c)   Show that for any $k \geq r$

$$F^k = d_0 I + d_1 F + \cdots + d_{r-1} F^{r-1} \tag{12}$$

for appropriate constants $d_0, \ldots, d_{r-1}$ and hence that

$$e^{Ft} = \alpha_0(t)I + \alpha_1(t)F + \cdots + \alpha_{r-1}(t)F^{r-1} \tag{13}$$

for some appropriate functions $\alpha_0(\cdot), \ldots, \alpha_{r-1}(\cdot)$.

(d) Use the Cayley-Hamilton Theorem (proved in Section 4–6 and also in Exercise 1 of Section 6–5) as we did in the discrete-time system of Example 4–6–7, on parts (b) and (c) to get a sharper bound on the power r than n^2.

(e) Determine r and find functions $\alpha_0(t), \ldots, \alpha_{r-1}(t)$ to express e^{Ft} in the form of equation (13) for

$$F = \begin{bmatrix} \lambda_1 & 0 \\ 0 & \lambda_2 \end{bmatrix}$$

Repeat for the F's of Exercise 2(c) and (d).

✔ **7.** Define the "**convolution product**" of two scalar functions $g(\cdot)$ and $u(\cdot)$ by the formula (10):

$$y(t) = \int_{t_0}^{t} g(t - \tau)u(\tau)\, d\tau \text{ for } t_0 \le \tau \le t$$

and express this "product" by the symbol $\star$ so that $y = g \star u$.

(a) Prove that $g \star u = u \star g$ for all u and g.
(b) Prove that $g \star (u + w) = (g \star u) + (g \star w)$ for all u, w, and g.
(c) Show that $g \star \delta = g$, for all g where δ is the unit impulse.
(d) Prove that $g \star (u \star w) = (g \star u) \star w$, for all g, u, and w.
Which of these properties still hold when g and u are matrices?

8. Consider a relaxed system having impulse response g and input u, as shown in Figure 6–1.

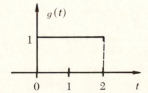

Figure 6–1

(a) Compute $y(t) = \int_{0}^{t} g(t - \tau)u(\tau)\, d\tau$ and sketch $y(t)$ versus t.

(b) Repeat for the inputs u_1 and u_2 shown in Figure 6–2.

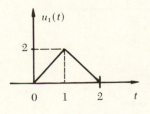
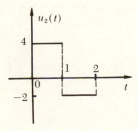

Figure 6–2

9.   Use the trapezoidal rule of integration on the convolution integral $y(t) = \int_0^t g(t - \tau)u(\tau)\,d\tau$, letting $t_n = 0 + nh$, where h is the time increment, and derive a discrete approximation for $y(0), y(h), \ldots, y(nh), \ldots$. Check your work by sampling the $g(\cdot)$ and $u(\cdot)$'s of Exercise 8.

10.   Given a matrix F, the result of performing elementary row operations on F can be expressed as pre-multiplication of F by an invertible matrix L. Similarly, elementary column operations on F are describable by post-multiplication by an invertible matrix R. Let F' be the result of elementary row and/or column operations on F, so that $F' = LFR$ for some invertible matrices L and R. What can you say about $e^{F't}$?

11.   We have defined the matrix exponential e^{At} for a square constant matrix A and know how to differentiate this function:

$$\frac{d}{dt}(e^{At}) = Ae^{At}$$

(a)   Can you define the following *integral* of the matrix exponential?

$$\int_{t_0}^t e^{A\tau}\,d\tau = B(t)$$

(b)   If $A = 0$, give $e^{At}, \dfrac{d}{dt}e^{At}$, and $\displaystyle\int_{t_0}^t e^{A\tau}\,d\tau$.

(c)   If $A = \begin{bmatrix} 1 & 0 \\ 0 & 0 \end{bmatrix}$, give $e^{At}, \dfrac{d}{dt}e^{At}$, and $\displaystyle\int_{t_0}^t e^{A\tau}\,d\tau$.

(d)   Repeat part (c) for $A = \begin{bmatrix} 0 & 1 \\ -6 & -5 \end{bmatrix}$.

✔ **12.**   This exercise outlines the approximation of the continuous-time dynamic
COMP system (F, G, H) by a discrete-time system.
     (a)   Suppose that the input $u(t)$ is smooth enough for it to be approximated by the piecewise constant, rectangular waveform $u_h(t)$ as shown in Figure 6–3. Using the notation $x(n) \overset{\Delta}{=} x(nh)$, show that at the sampling instants, the system behavior

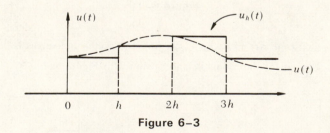

Figure 6–3

is approximated by the discrete system

$$x(n + 1) = Ax(n) + Bu(n)$$
$$y(n) = Cx(n)$$

where

$$A = e^{Fh}, \quad B = \left(\int_0^h e^{F\tau} \, d\tau \right) G, \quad C = H$$

[*Hint:* Use equations (4) and (5) with $t_0 = nh$ and $t = (n + 1)h$; then change variable of integration by $\xi = (n + 1)h - \tau$.]

Notice that A is invertible; hence, unlike most discrete-time systems, this system (A, B, C) will have an invertible state transition matrix $\Phi(n, m)$ [see Exercise 9 of Section 2–1].

(b)   Suppose that the sampling interval h is also small enough that $e^{Ft} \doteq I + Ft$ for $t \in [0, h]$. What do A, B, and C look like now?

(c)   If the sampling interval h is chosen to approximate u well, it might not be small enough for the approximation to e^{Ft} used in part (b). What approximation can you use to e^{Ft} over $[0, h]$ even if h is pretty big?

(d)   Go back and consider using the rectangular rule of integration

$$\int_a^b f(\tau) \, d\tau \doteq f(a)(b - a)$$

on the integral in equation (4), letting $t_0 = nh$ and $t = (n + 1)h$. What are A, B, and C now for the resulting discrete-time approximations?

(e)   Do part (d) with trapezoidal rule integration.

(f)   Let

$$F = \begin{bmatrix} -2 & 0 \\ 0 & -3 \end{bmatrix}, \ G = \begin{bmatrix} 1 \\ 1 \end{bmatrix}, \ H = [-1 \quad 2], \ x(0) = \begin{bmatrix} 1 \\ 0 \end{bmatrix}, \ u(t) = \sin wt$$

Run this method on a digital computer for $w = \frac{1}{5}$, 1, and 5, choosing h accordingly. Compare with the exact result.

13.   Let

$$F = \begin{bmatrix} -1 & 1 \\ -2 & -4 \end{bmatrix}, \ G = \begin{bmatrix} 0 \\ 1 \end{bmatrix}, \ H = [0 \quad 1]$$

Try to find a matrix P such that

$$\bar{F} = P^{-1}FP, \quad \bar{G} = P^{-1}G, \quad \bar{H} = HP$$

for the following matrices:

(a) $\bar{F} = \begin{bmatrix} -2 & 0 \\ 0 & -3 \end{bmatrix}, \ \bar{G} = \begin{bmatrix} 1 \\ 1 \end{bmatrix}, \ \bar{H} = [-1 \quad 2]$

(b) $\bar{F} = \begin{bmatrix} 0 & 1 \\ -6 & -5 \end{bmatrix}, \ \bar{G} = \begin{bmatrix} 0 \\ 1 \end{bmatrix}, \ \bar{H} = [1 \quad 1]$

14.   Give a formula for

$$\frac{\partial}{\partial t_0} e^{F(t-t_0)}$$

6-2 Diagonalization and Eigenvector Expansions

We saw in the last section that the complete solution [Theorem 2] of the system represented by matrices (F, G, H) was expressible in terms of the matrix exponential e^{Ft} defined as an infinite matrix-power series. We saw in Example 6-1-1 that if F is diagonal or, as in Example 6-1-2, if F is algebraically similar to a diagonal matrix, then e^{Ft} is especially easy to compute and may be expressed in closed form.

In this section we shall study conditions under which we may obtain a constant similarity transformation P such that by the change of variables

$$x = P\bar{x} \tag{1}$$

we change the representation

$$(F, G, H): \dot{x} = Fx + Gu$$
$$y = Hx$$

into the equivalent representation

$$(\bar{F}, \bar{G}, \bar{H}): \dot{\bar{x}} = \bar{F}\bar{x} + \bar{G}u$$
$$y = \bar{H}\bar{x}$$

where $\bar{F}$ is *diagonal*. If such an $\bar{F}$ exists, we say that F is **diagonalizable.** In Section 6-6, we shall study the Jordan canonical form of F, which is just $\bar{F}$ if F is diagonalizable and is "as close to diagonal as possible" if not.

Under any constant similarity transformation P of equation (1), we have, of course, that

$$\dot{\bar{x}} = P^{-1}FP\bar{x} + P^{-1}Gu$$
$$y = HP\bar{x}$$

so that

$$\bar{F} = P^{-1}FP$$
$$\bar{G} = P^{-1}G$$
$$\bar{H} = HP$$

and $$\bar{x}(0) = P^{-1}x(0)$$

just as provided by Definition 5-5-4 with $\dot{P} = 0$.

Now we want a P such that $\bar{F} = P^{-1}FP$ is diagonal, say,

$$P^{-1}FP = \bar{F} = \begin{bmatrix} \lambda_1 & 0 & \cdots & 0 \\ 0 & \lambda_2 & \cdots & 0 \\ \vdots & & & \\ 0 & \cdots & & \lambda_n \end{bmatrix}$$

or equivalently, that

$$FP = P\bar{F}$$

Partitioning P into its columns $P_1, P_2, \ldots, P_n$, this last equation becomes

$$F \left[P_1 \mid P_2 \mid \cdots \mid P_n \right] = \left[P_1 \mid P_2 \mid \cdots \mid P_n \right] \begin{bmatrix} \lambda_1 & 0 & \cdots & 0 \\ 0 & \lambda_2 & \cdots & 0 \\ \vdots & \vdots & & \vdots \\ 0 & 0 & \cdots & \lambda_n \end{bmatrix} \tag{2}$$

The jth column of the product of two matrices AB is the first matrix times the jth column of the second: $(AB)_j = (A)B_j$, so equating the jth columns of each side of (2) we have

$$FP_j = P \begin{bmatrix} 0 \\ 0 \\ \vdots \\ \lambda_j \\ \vdots \\ 0 \end{bmatrix} \leftarrow j\text{th row} = \lambda_j P_j$$

In other words, the jth column of the matrix P we seek must have the property that when operated on by F, the result is just itself, stretched or shrunk by the scalar λ_j. The columns of P are very special vectors as far as F is concerned—they are the vectors that F leaves pretty much alone, changing only their magnitudes but not their directions. We call the $\{P_j\}$ **characteristic vectors** or **eigenvectors** of F, and we call the scalars $\{\lambda_j\}$ **characteristic values** or **eigenvalues** of F. More formally, we have the following definition:

DEFINITION 1

Let $A : X \to X$ be a linear transformation of a space into itself. We say that a nonzero vector v in X is an **eigenvector** of X if there exists a scalar λ such that

$$Av = \lambda v \tag{3}$$

Any λ for which equation (3) has a nonzero solution v is called an **eigenvalue** of A, and we call any such v an **eigenvector corresponding to** the eigenvalue λ. Sometimes we say that v is an eigenvector with the eigenvalue λ.   ○

Note our insistence that the v which solves equation (3) be nonzero—for if we allow v to be zero, then (3) is satisfied for *every* value of λ, and we wish to avoid this nuisance.

If λ is an eigenvalue of A, then what we are saying is that

$$(\lambda I - A)v = \lambda v - Av = 0$$

has a nonzero solution, where I is the identity transformation on X.

Thus, if A is an $n \times n$ matrix, then λ is an eigenvalue of A if, and only if, $\det(\lambda I - A) = 0$. This determinant, when written out,

$$\det(\lambda I - A) = \det \begin{bmatrix} \lambda - a_{11} & -a_{12} & \cdots & -a_{1n} \\ -a_{21} & \lambda - a_{22} & \cdots & -a_{2n} \\ \vdots & \vdots & \cdots & \vdots \\ -a_{n1} & -a_{n2} & \cdots & \lambda - a_{nn} \end{bmatrix} \triangleq \chi_A(\lambda)$$

is seen to be a polynomial of the form

$$\chi_A(\lambda) \triangleq \lambda^n + \alpha_1 \lambda^{n-1} + \alpha_2 \lambda^{n-2} + \cdots + \alpha_n \tag{4}$$

of degree n with high-order coefficient 1. We call $\chi_A(\lambda)$ the **characteristic polynomial** of the matrix A and see that λ is an eigenvalue of A iff it is a zero, or root, of χ_A. Thus, an n-dimensional matrix has **at most** n distinct eigenvalues.

Example 1

For $F = \begin{bmatrix} 0 & 1 \\ 0 & -2 \end{bmatrix}$ we have $(\lambda I - F) = \begin{bmatrix} \lambda & -1 \\ 0 & \lambda + 2 \end{bmatrix}$ and $\chi_F(\lambda) = \det \begin{bmatrix} \lambda & -1 \\ 0 & \lambda + 2 \end{bmatrix} = \lambda(\lambda + 2) = \lambda^2 + 2\lambda$. The eigenvalues are found by setting

$$\chi_F(\lambda) = 0 = \lambda(\lambda + 2)$$

for which the roots are $\lambda_1 = 0$ and $\lambda_2 = -2$; so the matrix F has two distinct eigenvalues.

The eigenvectors P_1 and P_2 corresponding to λ_1 and λ_2 respectively are found by solving the matrix equations

$$(\lambda_1 I - F)P_1 = 0 \qquad \text{and} \qquad (\lambda_2 I - F)P_2 = 0$$

If we call $P_1 = \begin{bmatrix} x \\ y \end{bmatrix}$, the first of these becomes

$$\begin{bmatrix} 0 & -1 \\ 0 & 2 \end{bmatrix}\begin{bmatrix} x \\ y \end{bmatrix} = \begin{bmatrix} 0 \\ 0 \end{bmatrix}$$

whence

$$0x - 1y = 0$$
$$0x + 2y = 0$$

which must be solved for x and y. There are clearly infinitely many solutions of the form $y = 0$, x arbitrary:

$$\begin{bmatrix} x \\ y \end{bmatrix} = \begin{bmatrix} a \\ 0 \end{bmatrix} \text{ for any } a \neq 0 \text{ (cf. Exercise 4)}$$

For convenience we take $a = 1$ and $P_1 = \begin{bmatrix} 1 \\ 0 \end{bmatrix}$ as the eigenvector of F corresponding to $\lambda_1 = 0$. In a similar manner (verify!), we find that $P_2 = \begin{bmatrix} 1 \\ -2 \end{bmatrix}$ is an eigenvector corresponding to $\lambda_2 = -2$, so that $P = \begin{bmatrix} 1 & 1 \\ 0 & -2 \end{bmatrix}$. Checking this result, we compute $P^{-1} = \begin{bmatrix} 1 & \frac{1}{2} \\ 0 & -\frac{1}{2} \end{bmatrix}$, so that

$$P^{-1}FP = \begin{bmatrix} 1 & \frac{1}{2} \\ 0 & -\frac{1}{2} \end{bmatrix}\begin{bmatrix} 0 & 1 \\ 0 & -2 \end{bmatrix}\begin{bmatrix} 1 & 1 \\ 0 & -2 \end{bmatrix}$$

$$= \begin{bmatrix} 0 & 0 \\ 0 & -2 \end{bmatrix} = \begin{bmatrix} \lambda_1 & 0 \\ 0 & \lambda_2 \end{bmatrix}$$

as in Example 6–1–2.   ◊

So far, we have shown that if an $n \times n$ matrix A is diagonalizable by a similarity transformation P, then necessarily, the columns of P are eigenvectors of A. Since P must be invertible, its columns $P_1, P_2, \ldots, P_n$ are linearly independent, and we see that A is diagonalizable only if it has n linearly independent eigenvectors.

Conversely, suppose that a matrix A has n independent eigenvectors v_1, $v_2, \ldots, v_n$ corresponding to eigenvalues $\lambda_1, \lambda_2, \ldots, \lambda_n$ which may or may not be distinct. The set of eigenvectors $\mathcal{B} = \{v_1, v_2, \ldots, v_n\}$ may be used as a basis for the underlying n-dimensional space X upon which A operates. With respect to this basis, the basis vectors of course have the representations†

$$[v_1]_\mathcal{B} = \begin{bmatrix} 1 \\ 0 \\ \vdots \\ 0 \end{bmatrix}, \quad [v_2]_\mathcal{B} = \begin{bmatrix} 0 \\ 1 \\ \vdots \\ 0 \end{bmatrix}, \quad \ldots, \quad [v_n]_\mathcal{B} = \begin{bmatrix} 0 \\ 0 \\ \vdots \\ 1 \end{bmatrix}$$

What is the representation of our given linear transformation A with respect to this basis? Well, we know that the jth column of the matrix $^\mathcal{B}A^\mathcal{B}$ for A must be the coordinates, $[Av_j]_\mathcal{B}$, of the image of the jth basis vector when acted upon by A. But

$Av_1 = \lambda_1 v_1$ and so is represented by the column vector $\begin{bmatrix} \lambda_1 \\ 0 \\ \vdots \\ 0 \end{bmatrix}$

$Av_2 = \lambda_2 v_2$ and so is represented by the column vector $\begin{bmatrix} 0 \\ \lambda_2 \\ \vdots \\ 0 \end{bmatrix}$

$\ldots\ldots$

$Av_n = \lambda_n v_n$ and so is represented by the column vector $\begin{bmatrix} 0 \\ 0 \\ \vdots \\ \lambda_n \end{bmatrix}$

and so the matrix of A takes the diagonal form $\begin{bmatrix} \lambda_1 & 0 & \cdots & 0 \\ 0 & \lambda_2 & \cdots & 0 \\ \vdots & \vdots & \ddots & \vdots \\ 0 & 0 & \cdots & \lambda_n \end{bmatrix}$ in which

the eigenvalues appear along the diagonal. (See Exercise 14.)

We emphasize that the preceding discussion makes no assumption about distinctness of the eigenvalues, but merely about the existence of a basis of eigenvectors. We have proven the general result:

THEOREM 1

An $n \times n$ matrix $A: X \to X$ is similar to a diagonal matrix $\Lambda = \mathrm{diag}(\lambda_1, \lambda_2, \ldots, \lambda_n)$ if and only if $\lambda_1, \lambda_2, \ldots, \lambda_n$ are the eigenvalues of A and X has a basis $v_1, v_2, \ldots, v_n$ of eigenvectors of A with $Av_j = \lambda_j v_j$ for $1 \le j \le n$. $\qquad\qquad\square$

---

†The reader who is rusty with matrix representations of linear operators should review Section 3-3.

Example 2

The identity matrix I has characteristic polynomial

$$\chi_I = \det(\lambda I - I) = (\lambda - 1)^n$$

and so has only one eigenvalue, 1, but with multiplicity n. But *any* basis for X is a basis of eigenvectors, since every nonzero vector is an eigenvector:

$$Iv = 1v \text{ for all } v$$

and I certainly has a diagonal matrix representation

$$I = \begin{bmatrix} 1 & 0 & \dots & 0 \\ 0 & 1 & \dots & 0 \\ \vdots & \vdots & \dots & \vdots \\ 0 & 0 & \dots & 1 \end{bmatrix}$$

$\Diamond$

Now the characteristic polynomial (4) of a matrix is an nth degree polynomial whose coefficients are elements of the field of scalars K. Not all polynomials in general fields can be factored†, but if $K = \mathbf{C}$, the complex number field, we can always write

$$\chi_A(\lambda) = (\lambda - \lambda_1)(\lambda - \lambda_2) \dots (\lambda - \lambda_n)$$

where the roots $\lambda_1, \dots, \lambda_n$ may or may not be distinct and the term $(\lambda - \lambda_i)$ is written out as many times as the multiplicity of the root λ_i.

In most mathematical treatments it proves to be convenient to work in the complex vector space, even if the system we are working with is real—this is because the behavior of the complex eigenvalues (which for a real matrix must always occur in conjugate pairs—why?) can provide valuable insight into real dynamics, and in particular will tell us much about the stability properties of the real system, as we shall see in Section 7–4. [See Exercise 15.]

We will now show that the distinct eigenvalues correspond to linearly independent eigenvectors.

LEMMA 2

Let $v_1, \dots, v_k$ be k eigenvectors of A corresponding to k *distinct* eigenvalues

---

†For example, if K is the field of rational numbers, $\lambda^2 - 2$ cannot be factored as $(\lambda - a)(\lambda + a)$ since there is no $a \in K$ such that $a^2 = 2$.

$\lambda_1, \ldots, \lambda_k$:

$$(\lambda_j I - A)v_j = 0$$

Then the vectors $v_1, \ldots, v_k$ are linearly independent.

Proof

Consider a linear combination of the v_j's which sums to zero:

$$\sum_j c_j v_j = 0$$

Our task is to show that each c_j is 0.

First, consider the action of $(\lambda_j I - A)$ upon v_k:

$$Av_k = \lambda_k v_k$$

since v_k is an eigenvector. Thus,

$$(\lambda_j I - A)v_k = (\lambda_j - \lambda_k)v_k$$

and this is 0 if $j = k$, and nonzero if $j \neq k$ (using the fact that the eigenvalues $\lambda_1, \ldots, \lambda_k$ are distinct).

Now fix any integer i, $1 \leq i \leq k$, and operate on the equation

$$\sum_j c_j v_j = 0$$

with the operator† $\displaystyle\prod_{\ell \neq i} (\lambda_\ell I - A)$ to obtain

$$\sum_j c_j \prod_{\ell \neq i} (\lambda_\ell I - A) \cdot v_j = 0$$

and so

$$\sum_j c_j \prod_{\ell \neq i} (\lambda_\ell - \lambda_j)v_j = 0$$

Thus, every term vanishes except the ith, and we have

$$c_i \cdot \prod_{\ell \neq i} (\lambda_\ell - \lambda_i) \cdot v_i = 0$$

---

†The symbol Π denotes the "continued product" as in Exercise 5-1-1.

Since the λ_j's are distinct, and $v_i \neq 0$, this implies $c_i = 0$. Since this holds for each i, we conclude that the v_j's are linearly independent.   □

What happens if some of the eigenvalues are not distinct? Let's quickly see that when an eigenvalue is repeated, we may or may not be able to find a basis of eigenvectors:

Example 3

Consider the matrices $\begin{bmatrix} 1 & 0 \\ 0 & 1 \end{bmatrix}$ and $\begin{bmatrix} 1 & 1 \\ 0 & 1 \end{bmatrix}$. Both have 1 as a double eigen-

value, but whereas *any* basis is a basis of eigenvectors for $\begin{bmatrix} 1 & 0 \\ 0 & 1 \end{bmatrix}$, the vector

$\begin{bmatrix} a \\ b \end{bmatrix}$ can only be an eigenvector of $\begin{bmatrix} 1 & 1 \\ 0 & 1 \end{bmatrix}$ if $\begin{bmatrix} 1 & 1 \\ 0 & 1 \end{bmatrix}\begin{bmatrix} a \\ b \end{bmatrix} = 1 \begin{bmatrix} a \\ b \end{bmatrix}$. But

$$\begin{bmatrix} 1 & 1 \\ 0 & 1 \end{bmatrix}\begin{bmatrix} a \\ b \end{bmatrix} = \begin{bmatrix} a + b \\ b \end{bmatrix} = \begin{bmatrix} a \\ b \end{bmatrix}$$

implies $b = 0$. Thus, the only eigenvectors of $\begin{bmatrix} 1 & 1 \\ 0 & 1 \end{bmatrix}$ lie in the 1-dimensional

space spanned by $\begin{bmatrix} 1 \\ 0 \end{bmatrix}$.   ◊

We thus see that there are three genuinely different possibilities when we consider a linear transformation of an n-dimensional vector space:

(i) A has n distinct eigenvalues—this *implies* that X has a basis of eigenvectors of A.

(ii) A has at least one multiple eigenvalue—but X *does* have a basis of eigenvectors of A.

(iii) A has at least one multiple eigenvalue—but X does *not* have a basis of eigenvectors of A.

We have seen that a linear operator can be represented, with respect to some suitable basis, by a diagonal matrix whose diagonal entries are the eigenvalues, if and only if it falls into one of the first two categories above. When we discuss the Jordan canonical form in Section 6–6 we will see how close we can come to diagonal form when we have a transformation possessing multiple eigenvalues and lacking a full set of n independent eigenvectors.

Before continuing, let us give a simple two-dimensional example which graphically emphasizes the simplicity of the diagonal form representation when it is available.

Example 4

Let (F, G, H) be a representation of a system where

$$F = \begin{bmatrix} 13 & -30 \\ 4 & -9 \end{bmatrix}, \quad G = \begin{bmatrix} g_1 \\ g_2 \end{bmatrix}, \quad H = [h_1 \quad h_2]$$

The analog computer diagram is shown in Figure 6–4(a).

Since $\chi_F(\lambda) = (\lambda - 13)(\lambda + 9) + 120 = \lambda^2 - 4\lambda + 3 = (\lambda - 3)(\lambda - 1)$, F has the distinct eigenvalues 3 and 1 and can be diagonalized to yield the equivalent representation $(\bar{F}, \bar{G}, \bar{H})$, where $\bar{F} = \begin{bmatrix} 1 & 0 \\ 0 & 3 \end{bmatrix}$, a diagonal matrix. Calling $\bar{G} = \begin{bmatrix} \bar{g}_1 \\ \bar{g}_2 \end{bmatrix}$ and $\bar{H} = [\bar{h}_1 \quad \bar{h}_2]$, the wiring-diagram for $(\bar{F}, \bar{G}, \bar{H})$ is given in Figure 6–4(b).

We see that the diagonal representation has no cross-coupling, and that its wiring-diagram therefore requires fewer multipliers and simpler adders. ◊

In general, if an n-dimensional system (F, G, H) is diagonalizable to $(\bar{F}, \bar{G}, \bar{H})$, the wiring-diagram for $(\bar{F}, \bar{G}, \bar{H})$ will consist of n uncoupled, first-order systems in parallel, each branch driven by the same input vector u, operating independently of the others and contributing its share to the output vector y. Since the diagonalization is accomplished by the change of variable $x = P\bar{x}$, where the columns of P are just the eigenvectors of F, the new state variables $\bar{x}_1(t), \ldots, \bar{x}_n(t)$ are just the coordinates of the original state vector

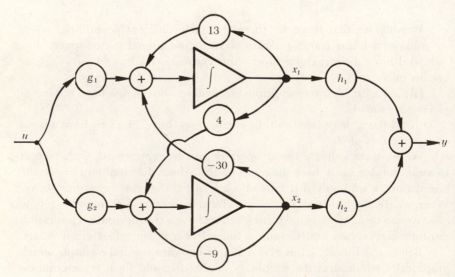

Figure 6–4(a)   Coupled representation (F, G, H).

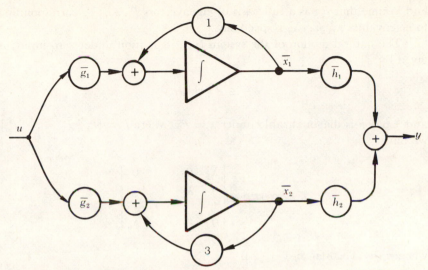

Figure 6–4(b)   An equivalent uncoupled representation $(\overline{F}, \overline{G}, \overline{H})$.

$x(t)$ with respect to the eigenvector basis:

$$x(t) = \overline{x}_1(t)P_1 + \cdots + \overline{x}_n(t)P_n \tag{5}$$

Thus, the outputs of the integrators in the uncoupled representation $(\overline{F}, \overline{G}, \overline{H})$ decompose the state trajectory $x(t)$ into its components along the eigenvectors of F. We call these fundamental motions along the eigenvectors of F the fundamental **modes**† of the system (F, G, H), and we call the expansion in (5) the **eigenvector expansion** of $x(t)$.

DEFINITION 2

Let (F, G, H) be a constant linear system. Let v be any eigenvector of F, say with $Fv = \lambda v$. Then the zero-input motion

$$x(t) = e^{\lambda t} k v$$

of the system, determined by initial state $x(0) = kv$ for some scalar k, is called a **mode** (normal mode, natural mode) of the system. It is customary also to speak of λ itself as being a mode, and of $e^{\lambda t}$ as being a mode.   ○

To deepen our understanding of the role of eigenvectors in the dynamics of a system, let us consider the homogeneous system $\dot{x}(t) = Fx(t)$

---

†Sometimes they are called "normal modes" or "natural modes," as in the study of oscillators in classical dynamics.

and assume that F has a full basis of eigenvectors $P_1, \ldots, P_n$ corresponding to eigenvalues $\lambda_1, \ldots, \lambda_n$ respectively.

The natural motion of the system (i.e., its motion under zero input) is given by

$$x(t) = e^{F(t-t_0)}x(t_0)$$

and, since F is diagonalizable under $x = P\bar{x}$, where $P = [P_1 \mid P_2 \mid \ldots \mid P_n]$, we have

$$\bar{F} = P^{-1}FP = \begin{bmatrix} \lambda_1 & 0 & & 0 \\ 0 & \lambda_2 & & \vdots \\ \vdots & & \ddots & \\ 0 & 0 & \cdots & \lambda_n \end{bmatrix}$$

whence (by Theorem 6-1-1 (vi))

$$e^{Ft} = Pe^{\bar{F}t}P^{-1}$$

$$= P\begin{bmatrix} e^{\lambda_1 t} & 0 & \cdots & 0 \\ 0 & e^{\lambda_2 t} & & 0 \\ \vdots & & \ddots & \vdots \\ 0 & 0 & \cdots & e^{\lambda_n t} \end{bmatrix}P^{-1}$$

so that

$$x(t) = P\begin{bmatrix} e^{\lambda_1 t} & 0 & \cdots & 0 \\ 0 & e^{\lambda_2 t} & & 0 \\ \vdots & \vdots & & \vdots \\ 0 & 0 & & e^{\lambda_n t} \end{bmatrix}P^{-1}x(0)$$

for $t \geq 0$.

Since $\bar{x}(0) = P^{-1}x(0)$ and $x = P\bar{x}$, the above equation is just

$$x(t) = P\bar{x}(t) \tag{6}$$

where

$$\bar{x}(t) = \begin{bmatrix} e^{\lambda_1 t} & \cdots & 0 \\ \vdots & \ddots & \vdots \\ 0 & \cdots & e^{\lambda_n t} \end{bmatrix}\bar{x}(0)$$

or

$$\bar{x}(t) = \begin{bmatrix} e^{\lambda_1 t} & \cdots & 0 \\ \vdots & \ddots & \vdots \\ 0 & \cdots & e^{\lambda_n t} \end{bmatrix}P^{-1}x(0) \tag{7}$$

From this last equation we may explicitly solve for $\bar{x}_1(t), \ldots, \bar{x}_n(t)$, the components of $x(t)$ along the eigenvectors $P_1, \ldots, P_n$ of F, as expressed by equations (5) or (6).

Example 5

If $F = \begin{bmatrix} 0 & 1 \\ -5 & -6 \end{bmatrix}$, we find eigenvalues $\lambda_1 = -5$ and $\lambda_2 = -1$ and corre-

sponding eigenvectors $P_1 = \begin{bmatrix} 1 \\ -5 \end{bmatrix}$ and $P_2 = \begin{bmatrix} 1 \\ -1 \end{bmatrix}$, so that $P = \begin{bmatrix} 1 & 1 \\ -5 & -1 \end{bmatrix}$

and $P^{-1} = \dfrac{1}{4}\begin{bmatrix} -1 & -1 \\ 5 & 1 \end{bmatrix}$. Then, by equation (7),

$$\bar{x}(t) = \begin{bmatrix} \bar{x}_1(t) \\ \bar{x}_2(t) \end{bmatrix} = \begin{bmatrix} e^{-5t} & 0 \\ 0 & e^{-1t} \end{bmatrix} \cdot \left(\frac{1}{4}\right) \begin{bmatrix} -1 & -1 \\ 5 & 1 \end{bmatrix} \begin{bmatrix} x_1(0) \\ x_2(0) \end{bmatrix}$$

$$= \frac{1}{4}\begin{bmatrix} -(x_1(0) + x_2(0))e^{-5t} \\ (5x_1(0) + x_2(0))e^{-1t} \end{bmatrix}$$

Substituting the above expression for $\bar{x}$ into equation (6), we get

$$x(t) = -\frac{1}{4}(x_1(0) + x_2(0))e^{-5t}\begin{bmatrix} 1 \\ -5 \end{bmatrix} + \frac{1}{4}(5x_1(0) + x_2(0))e^{-t}\begin{bmatrix} 1 \\ -1 \end{bmatrix}$$

which shows one mode at the rate of e^{-5t} along the direction of the eigenvector $\begin{bmatrix} 1 \\ -5 \end{bmatrix}$ corresponding to -5 and the other mode with time variation e^{-1t} along the eigenvector $\begin{bmatrix} 1 \\ -1 \end{bmatrix}$. Since e^{-5t} dies out five times faster than e^{-1t}, we call $P_1 = \begin{bmatrix} 1 \\ -5 \end{bmatrix}$ the "fast" eigenvector and $P_2 = \begin{bmatrix} 1 \\ -1 \end{bmatrix}$ the "slow" eigen-vector. We see that as t increases, $x(t)$ asymptotically approaches zero along the direction of its slowest eigenvector.

This is illustrated nicely in Figure 6–5 by the phase plane† plot of the trajectory corresponding to the initial condition $x(0) = \begin{bmatrix} 5 \\ 15 \end{bmatrix}$.

The reader is urged to pick some other initial states starting in other portions of the phase plane and to sketch the corresponding trajectories as $t \to \infty$. What happens if you choose an $x(0)$ lying right on one of the eigenvectors? ◊

In general, if we express the initial condition vector $x(0)$ in terms of the eigenvector basis, say

$$x(0) = a_1 P_1 + \cdots + a_n P_n$$

---

†The phase plane is just a plot of $x_2(t)$ versus $x_1(t)$ with the parameter t eliminated. We discussed this in Section 2–5 in the subsection "State Trajectories in Phase Space."

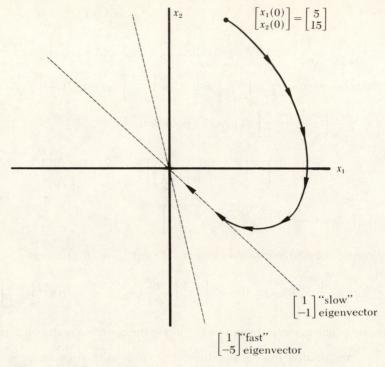

Figure 6–5 The eigenvector expansion in the phase plane, i.e., the plane in which trajectories are drawn in (position, velocity) coordinates.

and rewrite it as

$$x(0) = [P_1 \mid \ldots \mid P_n] \begin{bmatrix} a_1 \\ \vdots \\ a_n \end{bmatrix}$$

$$= P \begin{bmatrix} a_1 \\ \vdots \\ a_n \end{bmatrix}$$

Then

$$\bar{x}(0) = \begin{bmatrix} a_1 \\ \vdots \\ a_n \end{bmatrix}$$

and (7) becomes

$$\bar{x}(t) = \begin{bmatrix} e^{\lambda_1 t} & & 0 \\ & \cdot & \cdot \\ 0 & & \cdot & e^{\lambda_n t} \end{bmatrix} \begin{bmatrix} a_1 \\ \vdots \\ a_n \end{bmatrix}$$

or

$$\bar{x}(t) = \begin{bmatrix} a_1 e^{\lambda_1 t} \\ \vdots \\ a_n e^{\lambda_n t} \end{bmatrix}$$

Thus

$$x(t) = P \begin{bmatrix} a_1 e^{\lambda_1 t} \\ \vdots \\ a_n e^{\lambda_n t} \end{bmatrix}$$

or

$$x(t) = a_1 e^{\lambda_1 t} P_1 + \cdots + a_n e^{\lambda_n t} P_n \tag{8}$$

revealing the explicit dependence of the trajectory on the initial state coordinates and the modes. To answer the question posed in the last example (and previously answered by Definition 2!), if $x(0)$ lies on an eigenvector, say P_j, then $a_i = 0$ for $i \neq j$ and $x(t) = a_j e^{\lambda_j t} P_j$—the trajectory remains on P_j for all time.

Since we are working with complex scalars and the eigenvalues $\lambda_1, \ldots, \lambda_n$ are the roots of the polynomial $\chi_F(\lambda)$, they will in general be complex numbers. If we write

$$\lambda_k = \sigma_k + i\omega_k$$

with $i = \sqrt{-1}$ and with σ_k and ω_k as the real and imaginary parts, respectively, of λ_k, equation (8) may be written

$$x(t) = \sum_{k=1}^{n} a_k e^{\lambda_k t} P_k = \sum_{k=1}^{n} a_k e^{\sigma_k t} e^{i\omega_k t} P_k \tag{9}$$

Thus, if all the eigenvalues have negative real parts, $\sigma_k < 0$, $k = 1, \ldots, n$, then $x(t)$ dies out as t increases, or

$$\lim_{t \to \infty} x(t) = 0$$

and we call the origin 0 a "stable node," since no matter where we start the system (any $\{a_1, \ldots, a_n\}$) it heads for the origin. We shall see more of this notion of stability in Section 7–4.

Since $e^{i\omega t} = \cos \omega t + i \sin \omega t$, we see that each complex eigenvalue λ_k has an "oscillatory" part $e^{i\omega_k t}$ and a "damping" part $e^{\sigma_k t}$.

In fact, since for a *real* matrix F we know by equation (4) that $\chi_F(\lambda)$ has real coefficients, then any complex roots of $\chi_F(\lambda)$ must occur in conjugate pairs, say λ_k and $\overline{\lambda_k}$, where the bar denotes conjugation:

$$\overline{\lambda_k} = \sigma_k - i\omega_k$$

If P_k corresponds to λ_k so that

$$FP_k = \lambda_k P_k$$

then taking conjugates of both sides, since $\overline{F} = F$, we have

$$\overline{FP_k} = \overline{\lambda_k}\overline{P_k} \tag{10}$$

and we see that $\overline{P_k}$ is an eigenvector corresponding to $\overline{\lambda_k}$.

It is easy to show [Exercise 15] that, if the initial state $x(0)$ lies in the plane spanned by the real part of P_k and the imaginary part of P_k, corresponding to the conjugate pair $\lambda_k, \overline{\lambda_k}$, then the natural trajectory $x(t)$ remains in that plane forever. Thus, each conjugate pair of eigenvalues $\lambda_k, \overline{\lambda_k}$ determines a two-dimensional subspace V_k of X such that

$$V_k = \text{span } \{\mathcal{R}e(P_k), \mathcal{I}m(P_k)\} \text{ and } x(0) \in V_k \Rightarrow x(t) \in V_k \text{ for all } t > 0.$$

This ends our discussion of diagonalization and eigenvector expansions in the time domain. In the next section we will give an interpretation of eigenvectors in the world of Laplace transforms—the "frequency domain." Before we do, however, let us make a philosophical point about diagonalization.

If we pick n complex numbers at random, then there is "probability 0" that any pair of numbers will be the same. By the same token, if we pick those n numbers at random to be the coefficients of a characteristic polynomial of a matrix, it is again an event of "probability 0" that two of the roots should be equal. Thus, if we were to come upon a linear system "in nature" and assume that its various parameters were determined in a completely random way, then we would be safe in betting that it would have distinct eigenvalues and that it would therefore be reducible to diagonal form. Thus the reader, who a moment ago might have been worrying that it would be a rather lucky thing to be able to diagonalize a matrix, may now begin to wonder whether there is any point in bothering with nondiagonalizable matrices at all! Of course, the point is that in actually building systems, or in coming upon the more symmetric systems of nature, we expect certain substructures to be repeated and interleaved in the overall structure, and we should thus expect certain parameters of the system to occur again and again—in particular, we might expect the eigenvalues to be repeated, and thus enter the area in which nondiagonalizability is a very real possibility. In fact, it may well be crucial to a system design that two parameters be coupled, and that there be no way of decoupling them. Similarly, in biological systems which have resulted from a long period of evolution, we expect correlations between the various subsystems to be very strong indeed—and again, a model which is based on complete decoupling does not become as probable as our initial numerical argument would have suggested. One further point: If you actually build a coupled system—one which cannot realistically be put into diagonal form—it will still be the fact that, because you cannot make it with infinite accuracy, measurements upon it will yield an estimate for the F matrix which will have slightly different eigenvalues. Thus, it is a good strategy to seek an appropriate canonical form as presented in Section 6–6, rather than try to push it into a diagonal form which exploits an accuracy in measurement that is not available. The next example is designed to make this point.

Example 6

Consider the system of Figure 6–6. If built and measured with complete accuracy, it would have the matrix $F = \begin{bmatrix} 1 & 0 \\ 1 & 1 \end{bmatrix}$. However, actual measurements on a real-life version might yield the estimate

$$\begin{bmatrix} 0.99 & 0.01 \\ 1.01 & 1.02 \end{bmatrix}$$

which can only be put in diagonal form by using eigenvectors whose components are very different in magnitude—and which change wildly with small changes in the matrix coefficients. The reader is invited to work out the numerical details.   ◊

Thus, a wise model-builder will not force a system into diagonal form if the resultant parameters are highly sensitive to small perturbations

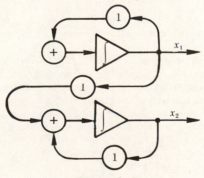

Figure 6–6

in the measured system parameters. We all know that we must not divide by zero. The above example is meant to indicate that it is also unwise to invert matrices whose determinant is close to zero. We shall return to this point briefly in our discussion of structural stability in Section 7–4.

EXERCISES FOR SECTION 6–2

1.   Try to establish a connection between eigenvectors of a linear operator $A : X \to X$ and fixed points of the operator. When will A be a contraction in the sense of Theorem 2–4–1?

2.   Show that F is invertible iff $\lambda = 0$ is not an eigenvalue of F.

✔ **3.**   Show that if v is an eigenvector of F corresponding to the eigenvalue λ, then v is an eigenvector of F^m for any integer $m \geq 1$. To what eigenvalue of F^m does v correspond? What can you say about negative powers of F? Is v an eigenvector of F^{-1}? Show that v is an eigenvector of e^{Ft} corresponding to eigenvalue $e^{\lambda t}$.

✔ **4.**   Show that every scalar multiple of an eigenvector is an eigenvector. Is the sum of two eigenvectors of F an eigenvector of F?

5.   Suppose that v is an eigenvector of F^2; can you find an eigenvector of F?

6.   Consider the characteristic polynomial (equation (4)), $\chi_A(\lambda) = \lambda^n + \alpha_1 \lambda^{n-1} + \cdots + \alpha_n$, of A. Show that the "constant term" is given by $\alpha_n = (-1)^n \det A$. Compare with Exercise 2.

Show that $\det A = (\lambda_1)(\lambda_2) \ldots (\lambda_n)$, where $\{\lambda_i\}$ are the eigenvalues of A.

7.   **(a)**   Find the eigenvectors P_1 and P_2 and the diagonalizing matrix P wherever possible for each of the matrices of Exercise 2 of Section 6–1. In the seventh matrix, $F = I + B$, let $B = \begin{bmatrix} 0 & 1 \\ 0 & 0 \end{bmatrix}$.

(b)   Find e^{Ft} for each of the above matrices using a similarity transformation P as in Example 1. Check your answers with your previous results using the series.

8.   **(a)**   Show that $\begin{bmatrix} 1 \\ 2 \\ 1 \end{bmatrix}$, $\begin{bmatrix} 1 \\ 0 \\ -1 \end{bmatrix}$, and $\begin{bmatrix} 0 \\ 1 \\ 3 \end{bmatrix}$ are eigenvectors of the matrix

$$F = \begin{bmatrix} 3 & -3 & 1 \\ 3 & -5 & 3 \\ 4 & -6 & 6 \end{bmatrix}.$$

(b)   Find e^{Ft}.

(c)   Find $\chi_F(\lambda)$.

✔ **9.**   A system has its input u and output y related by the equation $\dddot{y} + 6\ddot{y} + 11\dot{y} + 6y = u$.

(a)   Obtain one representation (F, G, H) of this system by defining the state vector as $x_1 = y$, $x_2 = \dot{y}$, $x_3 = \ddot{y}$. Draw the wiring diagram. Find χ_F.

(b)   Obtain another, equivalent representation (F', G', H') by diagonalizing the F of part (a). Draw the wiring diagram. Find $\chi_{F'}$.

(c)   Find e^{Ft}, the impulse response $g(t)$, and the solution for the output in terms of the input using Theorem 2 of Section 6–1. Check by solving the differential equation directly.

✔ **10.**   **(a)**   Find χ_F for $F = \begin{bmatrix} 0 & 1 & 0 \\ 0 & 0 & 1 \\ 3 & 4 & 5 \end{bmatrix}$.

[*Answer:* $\chi_F(\lambda) = \lambda^3 - 5\lambda^2 - 4\lambda - 3$. Compare with bottom row of F; note order.]

(b)   Let $F = \begin{bmatrix} 0 & 1 & 0 & \cdots & 0 & 0 \\ 0 & 0 & 1 & \cdots & 0 & 0 \\ \vdots & \vdots & \vdots & & \vdots & \vdots \\ 0 & 0 & 0 & & 0 & 1 \\ -\alpha_n & -\alpha_{n-1} & -\alpha_{n-2} & \cdots & -\alpha_2 & -\alpha_1 \end{bmatrix}$

with ones on the diagonal above the main diagonal, α's in the bottom row, and zeros elsewhere. Show that $\chi_F(\lambda) = \lambda^n + \alpha_1\lambda^{n-1} + \cdots + \alpha_{n-1}\lambda + \alpha_n$. Observe that the coefficients of the characteristic polynomial appear in reverse order, with minus signs, as the bottom row of F. Thus, we may consider F as a realization of a given characteristic polynomial. We call F a **companion matrix** of the polynomial

$$\lambda^n + \alpha_1\lambda^{n-1} + \cdots + \alpha_n$$

(c)   Let s be an eigenvalue of the F in part (b). Find an eigenvector of F corresponding to s. Repeat (b) and (c) for the matrix F^T.

(d)   Give a companion matrix for $\lambda^3 + 6\lambda^2 + 11\lambda + 6$. Compare with Exercise 9.

✔ **11.**   Let V be the vector space of all real-valued functions having derivatives of all orders:

$$V = \{f \mid f : \mathbf{R} \to \mathbf{R}, f^{(n)}(t) \text{ exists for all } n\}$$

Let $D : V \to V$ by the rule $D(f) = \dot{f}$, ordinary differentiation.

(a)   Show that the linear operator D has as an eigenvector the exponential function g such that $g(t) = e^{st}$ and that the number s is the corresponding eigenvalue.

(b)   Show that the scalar, linear differential equation with constant coefficients $\dddot{y} + 6\ddot{y} + 11\dot{y} + 6y = u$ can be written in terms of our differential operator D as

$$D^3y + 6D^2y + 11Dy + 6y = u$$

where $D^2 = D \circ D$, or D followed by D (composition).

(c)   Let $A = D^3 + 6D^2 + 11D + 6I$ be a "polynomial" operator. Show that $A : V \to V$ is linear and that the g of part (a) is an eigenvector of A and corresponds to the eigenvalue $\lambda = s^3 + 6s^2 + 11s + 6$. Generalize.

12.   Find a matrix F which has $P_1 = \begin{bmatrix} 1 \\ -2 \end{bmatrix}$ and $P_2 = \begin{bmatrix} -1 \\ 3 \end{bmatrix}$ as its eigenvectors.

What are the eigenvalues of F? $\left(Answer: F = \begin{bmatrix} 0 & 1 \\ -6 & -5 \end{bmatrix}. \right)$ Can this always be done?

13.   **(a)**   Let $F = \begin{bmatrix} 0 & 1 \\ 0 & -2 \end{bmatrix}$. Find e^{-Ft} by diagonalizing $-F$ as we did with F in Example 1. Verify Theorem 6-1-1 (ii) by inverting the matrix e^{Ft} found in Example 6-1-2.

(b)   Replace t by $-t$ in the expression for e^{Ft} found in Example 6-1-2 and compare with part (a).

(c)   Find e^{F^2t} and e^{F^3t}.

✔ **14.**   Let $A : X \to X$ be linear, let $\mathfrak{B}$ and $\mathfrak{B}'$ be any two bases for X, and let I_X be the identity map on X.

(a)   Write $A = I_X \circ A \circ I_X$, representing the change of bases pictorially by Figure 6-7, and use the fact [Exercise 9 of Section 3-3] that the matrix of a composition of two operators is the product of the two matrices of the respective operators to get

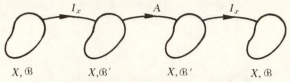

Figure 6-7   Changing bases.

to get

$$^{\mathcal{B}}A^{\mathcal{B}} = {}^{\mathcal{B}}I_X^{\mathcal{B}'} \cdot {}^{\mathcal{B}'}A^{\mathcal{B}'} \cdot {}^{\mathcal{B}'}I_X^{\mathcal{B}} \quad \text{(the "change of basis" formula)}$$

How are ${}^{\mathcal{B}}I_X^{\mathcal{B}'}$ and ${}^{\mathcal{B}'}I_X^{\mathcal{B}}$ related?

(b)   Consider $\mathcal{B}'$ to be the original basis and let it be the usual one, so that ${}^{\mathcal{B}'}A^{\mathcal{B}'} = A$, and let $\mathcal{B}$ be a basis of eigenvectors of A so that ${}^{\mathcal{B}}A^{\mathcal{B}}$ is diagonal. Find the matrices ${}^{\mathcal{B}}I_X^{\mathcal{B}'}$ and ${}^{\mathcal{B}'}I_X^{\mathcal{B}}$.

(c)   Is it true that "A is diagonalizable iff there exists an invertible matrix P such that $P^{-1}AP$ is diagonal"? If so, what is P?

15.   Let F be a real $m \times n$ matrix; let λ_k and $\overline{\lambda}_k$ be complex conjugate eigenvalues and let P_k and $\overline{P}_k$ be their corresponding eigenvectors. Set $v_R = \Re e(P_k)$ and $v_I = \Im m(P_k)$ the real and imaginary parts, respectively, of the vector P_k.

(a)   Show that if $x(0) = v_R$, then the solution of the homogeneous system $\dot{x} = Fx$ is given by $x(t) = \Re e(e^{\sigma_k t}e^{i\omega_k t}P_k)$.

(b)   Show that if $x(0) = v_I$, then the homogeneous solution is

$$x(t) = \Im m(e^{\sigma_k t}e^{i\omega_k t}P_k).$$

(c)   Use parts (a) and (b) and linearity to show that if $x(0)$ is in the plane spanned by v_R and v_I, then the unforced trajectory will always remain in that plane.

(d)   Illustrate parts (a) and (b) by sketching the phase plane portraits of the matrix $F = \begin{bmatrix} -1 & 3 \\ -3 & -1 \end{bmatrix}$, finding the eigenvectors P_1 and P_2, and using the initial conditions $x(0) = \begin{bmatrix} 1 \\ 0 \end{bmatrix}$ and $x(0) = \begin{bmatrix} 0 \\ 1 \end{bmatrix}$ respectively. [*Answer:*

$$P_1 = P_2 = \begin{bmatrix} 1 \\ i \end{bmatrix} = \begin{bmatrix} 1 \\ 0 \end{bmatrix} + i\begin{bmatrix} 0 \\ 1 \end{bmatrix}; \lambda_1 = \lambda_2 = -1 + i3;$$

both trajectories are spirals.]

✔ **16.**   Here we generalize the example $\begin{bmatrix} 1 & 1 \\ 0 & 1 \end{bmatrix}$ to the useful $k \times k$ matrix

$$J_k(\lambda) = \begin{bmatrix} \lambda & 1 & 0 & \dots & 0 & 0 \\ 0 & \lambda & 1 & \dots & 0 & 0 \\ 0 & 0 & \lambda & \dots & 0 & 0 \\ \dots & \dots & \dots & \dots & \dots & \dots \\ 0 & 0 & 0 & \dots & \lambda & 1 \\ 0 & 0 & 0 & \dots & 0 & \lambda \end{bmatrix}$$

in which all the entries in the main diagonal equal λ, in which the diagonal above the main diagonal has all entries equal to 1, and in which all other entries are 0. We will see that this matrix has repeated eigenvalues and does not have a basis of eigenvectors.

(a)   Compute the characteristic polynomial of $J_k(\lambda)$ as a function of the variable s. [*Answer:* $\chi_{J_k(\lambda)}(s) = (s - \lambda)^k$. That's why we call $J_k(\lambda)$ the **Jordan block of order k with eigenvalue λ.**]

(b)   Let us now look for the eigenvectors of our Jordan block—in other words,

we look for column vectors $\begin{bmatrix} a_1 \\ \vdots \\ a_k \end{bmatrix}$ for which it is true that $J_k(\lambda) \cdot a = \lambda a$. By actually

multiplying out the terms in the above equations, we obtain the vector equality

$$\begin{bmatrix} \lambda a_1 & + a_2 \\ \lambda a_2 & + a_3 \\ \vdots & \\ \lambda a_{k-1} & + a_k \\ & a_k \end{bmatrix} = \begin{bmatrix} \lambda a_1 \\ \lambda a_2 \\ \vdots \\ \lambda a_{k-1} \\ \lambda a_k \end{bmatrix}$$

Deduce that the eigenvectors of $J_k(\lambda)$ must all be of the form $\begin{bmatrix} a_1 \\ 0 \\ \vdots \\ 0 \end{bmatrix}$; in other words,

the eigenvectors of $J_k(\lambda)$ span only a one-dimensional subspace of the underlying vector space—a far cry from a basis, which would have to span the full k dimensions.

✔ **17.**   Suppose that F and $\bar{F}$ are any similar square matrices: $\bar{F} = Q^{-1}FQ$ for some invertible matrix Q. Prove that $\chi_F(\lambda) = \chi_{\bar{F}}(\lambda)$.

18.   (a)   For $F = \begin{bmatrix} 0 & 1 \\ -5 & -6 \end{bmatrix}$ as in Example 5, use Exercise 10 to write $\chi_F(\lambda)$.

Show that F satisfies its own characteristic equation (Cayley-Hamilton).

(b)   Find the matrix $(sI - F)$ and compute its inverse $(sI - F)^{-1}$ using determinants (Cramer's rule):

$$(sI - F)^{-1} = \frac{\text{adjugate } (sI - F)}{\det(sI - F)}$$

Note that the entries of adjugate $(sI - F)$ are polynomials of degree $\leq n - 1$.

(c)   Repeat part (b) for the F of Exercise 10(a).

✔ **19.**   This exercise gives an informal proof that complex exponentials are eigen-functions of constant linear systems.

Suppose that such a system is driven by input $u_s(t) = e^{st}$ for a long time (from $t_0 = -\infty$) and that it has a steady state response $\mathcal{S}(t, -\infty, 0, u_s)$ which we denote by $r_s(t)$. (The subscript s reminds us that we expect the response to depend on the complex exponent s). Symbolically, we say that input e^{st} causes (steady state) response $r_s(t)$, as illustrated in Figure 6-8.

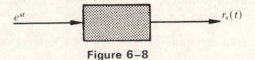

$e^{st} \longrightarrow \boxed{} \longrightarrow r_s(t)$

Figure 6-8

(a)   Show that time invariance implies that $e^{s(t-\tau)}$ causes response $r_s(t - \tau)$.

(b)   Write $e^{s(t-\tau)}$ as $(e^{-\tau s}) e^{st}$ and use linearity (homogeneity) to show that this input causes response $(e^{-\tau s}) r_s(t)$.

(c)   Equate the two response expressions to get $r_s(t) = e^{s\tau} r_s(t - \tau)$ and, since this holds for all t and τ, let $\tau = t$ to get $r_s(t) = r_s(0)e^{st}$. Since $r_s(0)$ does not depend upon t (but only on s), this shows that inputs of the form e^{st} produce outputs (steady state) of the form λe^{st}.

(d)   We often call $r_s(0) = \mathfrak{L}(s)$ the **transfer function** of the system. Show that it can be defined as a ratio of output to input, where the input is e^{st}.

(e)   If $s_1, s_2, \ldots, s_k$ are distinct complex numbers, show that the functions $e^{s_1(t)}, e^{s_2(t)}, \ldots, e^{s_k(t)}$ are linearly independent.

(f)   What is the steady state response to the input

$$u(t) = c_1 e^{s_1(t)} + c_2 e^{s_2(t)} + \cdots + c_k e^{s_k(t)}?$$

Do you see why we like the additivity part of linearity as well as homogeneity?

6–3   Transform Methods and the Frequency Domain

Since most of so-called *classical* control theory uses Laplace transform theory, it might be worth noting here that our state-variable theory lets us quickly derive the classical theory and, at the same time, gain insight into the cases which are more appropriately handled by state-variable methods.

THE CONNECTION WITH LAPLACE TRANSFORMS

First, let us get a better feeling for the Laplace transform by seeing how it may be obtained in a natural way from the Fourier transform. The ensuing discussion assumes only a mild familiarity with the Fourier transform, and readers completely ignorant of Fourier methods may turn directly to Definition 1 and take the formulas of equation (2) as the definition of the Laplace transform.†

We recall that the Fourier transform tells us how to build up functions from the basic trigonometric functions $\sin \omega t$ and $\cos \omega t$. Using the compact expression $e^{i\omega t}$ for $\cos \omega t + i \sin \omega t$, we may write the desired linear combination in the form

$$\frac{1}{2\pi} \int_{-\infty}^{\infty} F(\omega)e^{i\omega t} \, d\omega$$

Fourier's theorem tells us that $f(t)$ can be expressed in this form if, as t goes to infinity, f goes to zero quickly enough for $|f|^2$ to be integrable. We write $f \in L^2$ as an abbreviation for $\int_{-\infty}^{\infty} |f(t)|^2 \, dt < \infty$ and have the following (where a few minor side-conditions have been omitted):

---

†For simple, very readable discussions of Fourier analysis, see R. M. Bracewell, *The Fourier Transform,* McGraw-Hill, 1965, or H. H. Skilling, *Electrical Engineering Circuits,* Second Edition, Chapters 14 and 15 (Wiley, 1965).

FOURIER'S THEOREM

If f is in L^2, it can be expressed by the formula:

$$f(t) = \frac{1}{2\pi} \int_{-\infty}^{\infty} F(\omega)e^{i\omega t}\, d\omega$$

where "Fourier Transform Pair" (1)

$$F(\omega) = \int_{-\infty}^{\infty} f(t)e^{-i\omega t}\, dt$$

$\square$

Now, in system theory we shall often be interested in functions which are *not* square integrable but which *do* vanish for $t < 0$. Such a function f can often be made square integrable by "exponential damping"; i.e., we multiply $f(t)$ by $e^{-\sigma t}$ to obtain a function $f(t)e^{-\sigma t}$ which *is* square integrable for a sufficiently large positive σ.

Example 1

(a) Let $1(t) = \begin{cases} 1 & \text{if } t > 0 \\ 0 & \text{if } t \le 0 \end{cases}$ be the Heaviside step function or "unit step."

Then $1(t) \notin L^2$ but $1(t)e^{-\sigma t} \in L^2$ for any $\sigma > 0$:

$$\int_{-\infty}^{\infty} (1(t)e^{-\sigma t})^2\, dt = \int_{0}^{\infty} e^{-2\sigma t}\, dt = \frac{1}{2\sigma}$$

(b) If $f(t) = \begin{cases} e^{-\sigma_0 t} & \text{for } t > 0 \\ 0 & \text{for } t \le 0 \end{cases}$ then $f(t)e^{-\sigma t} \in L^2$ as long as $\sigma > \sigma_0$.

(c) If f is constant (and so does not vanish on $(-\infty, 0)$) then $fe^{-\sigma t}$ is *not* in L^2 for any σ. $\Diamond$

Let us pick some σ_0 such that $f(t)e^{-\sigma t}$ is square integrable for $\sigma > \sigma_0$ and let F_σ be the Fourier transform of $f(t)e^{-\sigma t}$. Recalling that f vanishes on $(-\infty, 0)$, our Fourier transform pair becomes

$$F_\sigma(\omega) = \int_{0^-}^{\infty} [f(t)e^{-\sigma t}]e^{-i\omega t}\, dt$$

for $\sigma > \sigma_0$

$$f(t)e^{-\sigma t} = \frac{1}{2\pi i} \int_{-\infty}^{\infty} F_\sigma(\omega)e^{i\omega t}\, d(i\omega)$$

where we use 0^- as the lower limit of integration to make sure that the integration will include any impulses at $t = 0$ which may be present in $f(t)$ or its derivatives. [See Exercise 9.]

Now we introduce a complex variable $s = \sigma + i\omega$ and define the function† $F(s)$ by the equation

$$F(\sigma + i\omega) = F_\sigma(\omega)$$

Then $F(s)$ is defined for $\Re e(s) > \sigma_0$ by the equations

$$\left. \begin{aligned} F(s) &= \int_{0^-}^{\infty} f(t)e^{-st}\,dt \\ f(t) &= \frac{1}{2\pi i} \int_{\sigma-i\infty}^{\sigma+i\infty} F(s)e^{st}\,ds \end{aligned} \right\} \quad \sigma = \Re e(s) > \sigma_0 \quad \text{(2)}$$

where we have used the equality $e^{st} = e^{\sigma t} \cdot e^{i\omega t}$ and noted that as ω runs from $-\infty$ to $+\infty$, so $s = \sigma + i\omega$ runs from $\sigma - i\infty$ to $\sigma + i\infty$ along the line parallel to the imaginary axis, with abscissa σ.

DEFINITION 1

We call the $F(s)$ of equation (2) the **one-sided** (since we only use $(0, \infty)$ and not $(-\infty, 0)$ to define it) **Laplace transform** of $f(t)$ and sometimes denote it as $\mathcal{L}\{f\}$. Similarly, we call $f(t)$ the **inverse Laplace transform** of $F(s)$ and sometimes denote this by $f(t) = \mathcal{L}^{-1}\{F(s)\}$.   ○

The reader should note that the pair of formulas in (2) remain valid if we consider $f(t)$ and $F(s)$ to take values lying in the same Banach space. In $\mathbf{R}^n$ and $\mathbf{C}^n$ this reduces to applying the formulas component by component.

Having seen that the Laplace transform is a simple extension of the range of validity of the Fourier transform and allows us to express "any" function as a (nondenumerably infinite) linear combination $\dfrac{1}{2\pi} \int_{\sigma-i\infty}^{\sigma+i\infty} F(s)e^{st}\,ds$ of "complex sinusoids" e^{st}, let us see what happens to a linear system when its input is the sinusoid e^{st} with "complex frequency" s.

---

† This is a traditional abuse of notation; ordinarily $F(s)$ would be understood to be the value of the function $F(\cdot)$ evaluated at the point s and $f(t)$ would be the value of $f(\cdot)$ at t. To stress that the independent variable of $f(\cdot)$ is t and the independent variable of $F(\cdot)$ is the complex variable s, we agree to say "the function (whose formula is) $F(s)$ is the Laplace transform of the function (whose formula is) $f(t)$" with the parenthetical expressions understood. In a similar manner we speak about the "exponential function e^{st}" instead of "the function $e^{s(\cdot)}$ whose value at any time t is given by e^{st}."

Example 2

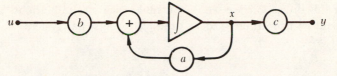

Figure 6–9   A one-dimensional system.

Consider, for simplicity, the one-dimensional system of Figure 6–9, described by

$$\left.\begin{array}{l} \dot{x} = ax + bu \\ y = cx \end{array}\right\} \tag{3}$$

If we start it *in the zero state* at time t_0, its output at time t will be

$$y(t) = \int_{t_0}^{t} ce^{a(t-\tau)}bu(\tau)\, d\tau$$

$$= [cbe^{at}] \int_{t_0}^{t} e^{-a\tau}u(\tau)\, d\tau$$

where we have used the commutativity of the scalar case (which is not available in general for matrices) to pull a few terms outside. Then if $u(\tau) = e^{s\tau}$, the last integral becomes

$$\int_{t_0}^{t} e^{(s-a)\tau}\, d\tau = \begin{cases} \dfrac{1}{s-a}[e^{(s-a)t} - e^{(s-a)t_0}] & \text{as long as } s \neq a \\[2mm] t - t_0 & \text{if } s = a \end{cases}$$

yielding

$$y(t) = \begin{cases} \dfrac{cb}{s-a}e^{st} - \dfrac{cbe^{at}}{s-a}e^{(s-a)t_0} & \text{for } s \neq a \\[2mm] cbe^{at}(t - t_0) & \text{if } s = a \end{cases}$$

What happens as $t_0 \to -\infty$, i.e., as we study the effect of having applied e^{st} for a longer and longer interval in the past? If $s = a$, the output "blows up" since $(t - t_0) \to \infty$. This "blow-up" is not unexpected, in view of the appearance of a term $\dfrac{1}{s-a}$ in the general formula. Because of this $\dfrac{1}{s-a}$ term, we call $s = a$ a "pole" of our system (3), and also call a a "resonant frequency" of (3).

For $s \neq a$, we see that only the second term depends on t_0, and that it goes to zero if $\mathfrak{Re}(s - a) > 0$ and goes to infinity if $\mathfrak{Re}(s - a) < 0$. Summarizing this discussion, we see that it makes sense to talk about "applying the sinusoid e^{st} from time $-\infty$ to time t" to the scalar system

$$\dot{x} = ax + bu$$
$$y = cx$$

if and only if $\mathfrak{Re}(s) > \mathfrak{Re}(a)$. In that case, the forced response is given by

$$y(t) = \left[\frac{cb}{s - a} \right] e^{st}$$

i.e., it is just the input multiplied by a complex constant which depends only on s and not on t. We denote this complex multiplier by $\mathfrak{H}(s)$:

$$\mathfrak{H}(s) = \frac{cb}{s - a} \qquad \qquad \Diamond$$

Now, with this as background, we can turn to the input-output analysis of linear systems, in which we specify the input which has been applied to the system since time $-\infty$ and ask what the resulting output function will be. The point of the above discussion is that such an analysis only makes sense if the input vanishes rapidly enough as we go back to $-\infty$ [in the sense that $\mathfrak{Re}(s) > \mathfrak{Re}(a)$, so that $e^{(s-a)t_0} \to 0$ as $t_0 \to -\infty$] to avoid remote inputs having catastrophically large effects on the current state.

To formalize this idea we make the following definition:

DEFINITION 2

The **steady-state response** of a system is given by

$$y_{ss}(t) = \mathcal{S}(t, -\infty, x_0, u)$$
$$= \lim_{t_0 \to -\infty} \mathcal{S}(t, t_0, x_0, u)$$

provided that the limit exists and is independent of the initial state x_0. $\bigcirc$

A system may fail to have a steady-state response:

Example 3

The linear system

$$\begin{bmatrix} \dot{x}_1 \\ \dot{x}_2 \end{bmatrix} = \begin{bmatrix} -2 & 0 \\ 0 & 0 \end{bmatrix} \begin{bmatrix} x_1 \\ x_2 \end{bmatrix} + \begin{bmatrix} 1 \\ 0 \end{bmatrix} u; \quad x_0 = \begin{bmatrix} x_{10} \\ x_{20} \end{bmatrix}$$

$$y = \begin{bmatrix} 1 & 1 \end{bmatrix} \begin{bmatrix} x_1 \\ x_2 \end{bmatrix}$$

has a natural response given by

$$y(t) = \mathcal{S}(t, t_0, x_0, 0) = x_{10} e^{-2(t-t_{})} + x_{20} \qquad \text{[verify]}$$

Here the limit exists but depends upon x_0:

$$\lim_{t_0 \to -\infty} \mathcal{S}(t, t_0, x_0, 0) = x_{20}$$

See Exercises 1 and 2. $\diamondsuit$

Since the steady-state response (when it exists) does not depend upon the initial state x_0, it is usually computed with x_0 set to 0:

$$y_{ss}(t) = \lim_{t_0 \to -\infty} \mathcal{S}(t, t_0, 0, u) \qquad \qquad \textbf{(4)}$$

For smooth linear systems, the zero-state response is given by the super-position integral

$$\mathcal{S}(t, t_0, 0, u) = \int_{t_0}^{t} g(t, \tau) u(\tau) \, d\tau \qquad \qquad \textbf{(5)}$$

where $g(t, \tau)$ is the zero-state response at time t due to an impulse applied at time τ.

Consider then a smooth linear system and an input u which is small enough in the past for (5) to have a limit as $t_0 \to -\infty$. Let us then denote by Lu the output function whose value at time t is given by the formula

$$(Lu)(t) = \int_{-\infty}^{t} g(t, \tau) u(\tau) \, d\tau = \lim_{t_0 \to -\infty} \int_{t_0}^{t} g(t, \tau) u(\tau) \, d\tau \qquad \textbf{(6a)}$$

Then, for a linear combination of two inputs for which Lu_1 and Lu_2 are well defined, we have

$$L(k_1 u_1 + k_2 u_2) = k_1 Lu_1 + k_2 Lu_2$$

since

$$\int_{-\infty}^{t} g(t, \tau)[k_1 u_1(\tau) + k_2 u_2(\tau)] \, d\tau = k_1 \int_{-\infty}^{t} g(t, \tau) u_1(\tau) \, d\tau + k_2 \int_{-\infty}^{t} g(t, \tau) u_2(\tau) \, d\tau$$

by the linearity of g (which is in general a matrix, rather than a scalar, function) and of integration.

In other words, the input-output behavior of a linear system started in the 0 state (at t_0, which may be $-\infty$) is in fact a linear function.

Now recall our delay operator z^{-t_0}:

$$[z^{-t_0} u](t) = u(t - t_0)$$

If our system is time-invariant, we have $g(t, \tau) = g(t - \tau, 0)$ and we see that

$$(Lu)(t) = \int_{-\infty}^{t} g(t - \tau) u(\tau) \, d\tau \qquad \textbf{(6b)}$$

the **convolution** of g and u, and that

$$Lz^{-t_0} = z^{-t_0} L \qquad \textbf{(6c)}$$

for all t_0. For those readers who did not already prove this in Exercise 9 of Section 3–1, the last equation holds because

$$(Lz^{-t_0})(t) = \int_{-\infty}^{t} g(t, \tau)(z^{-t_0} u)(\tau) \, d\tau$$

$$= \int_{-\infty}^{t} g(t - \tau, 0) u(\tau - t_0) \, d\tau$$

$$= \int_{-\infty}^{t-t_0} g(t - t_0, \tau) u(\tau) \, d\tau$$

and

$$(z^{-t_0} Lu)(t) = z^{-t_0} \left(\int_{-\infty}^{t} g(t, \tau) u(\tau) \, d\tau \right)$$

$$= \int_{-\infty}^{t-t_0} g(t - t_0, \tau) u(\tau) \, d\tau$$

[See Exercise 3.]

At this point we should clear up a technical point which we shall encounter in the details of the next theorem. In general, the inputs to our system are vectors, and so a scalar sinusoidal input e^{st} does not make sense.

However, if we take a *fixed* vector v in U, we can define an input function $u: T \to U$ by $t \mapsto e^{st}v$.

Now let us verify that these two properties of the steady-state response:

$$\text{Linearity: } L(k_1 u_1 + k_2 u_2) = k_1 L u_1 + k_2 L u_2$$
$$\text{Stationarity: } z^{-t_0} L = L z^{-t_0} \text{ for all } t_0,$$

guarantee that exponential inputs produce exponential outputs:

THEOREM 1

For any sinusoid $e^{s(\cdot)}: t \mapsto e^{st}$ for which $L(e^{s(\cdot)}v)$ is defined for each fixed v, there exists a complex *matrix* $\mathfrak{H}(s)$, depending on s but not on t, such that

$$\mathcal{S}(t, -\infty, 0, e^{s(\cdot)}v) = L[e^{s(\cdot)}v](t) = \mathfrak{H}(s)e^{st}v \tag{7a}$$

Proof

By stationarity

$$z^t L[e^{s(\cdot)}v](t_0) = L z^t[e^{s(\cdot)}v](t_0)$$

that is,

$$L[e^{s(\cdot)}v](t + t_0) = L[e^{s(\cdot + t)}v](t_0) = L[e^{st}e^{s(\cdot)}v](t_0)$$

Using the linearity of L, we can pull out the e^{st} on the right-hand side to get

$$L[e^{s(\cdot)}v](t + t_0) = e^{st}L[e^{s(\cdot)}v](t_0)$$

Setting $t_0 = 0$, we deduce that

$$L[e^{s(\cdot)}v](t) = \{L[e^{s(\cdot)}v](0)\} \cdot e^{st}$$

Using linearity to "pass out" v, this last expression can be written as

$$L[e^{s(\cdot)}v](t) = \mathfrak{H}(s)e^{st}v$$

where

$$\mathfrak{H}(s)v \overset{\Delta}{=} L[e^{s(\cdot)}v](0) \tag{7b}$$

is an expression which depends on s but not on t. Thus, $\mathfrak{H}(s)v$ is simply the output of our system at time 0 if it has received input $e^{st}v$ from time $-\infty$ up to 0. [See Exercise 6.]   □

We emphasize that Theorem 1 only holds if $\Re e(s)$ is sufficiently large for $L[e^{s(\cdot)}v]$ to be defined.

When it exists, the matrix $\mathfrak{H}(s)$ is called the **system function** or **transfer function matrix** of the system, and we interpret equation (7a) as saying that the steady-state response of a constant linear system to an exponential input, $u(t) = e^{st}v$, is just $\mathfrak{H}(s)$ times the input:

$$(\text{Output}) = \mathfrak{H}(s) \cdot (\text{Input})|_{\text{Input}=e^{st}v} \tag{8}$$

If the input and output are *scalars*, so that v is just a number, we can solve (7a) and (8) to characterize $\mathfrak{H}(s)$ as a ratio:

$$\mathfrak{H}(s) = \frac{(\text{output})}{(\text{input})}\bigg|_{\text{exponential input } e^{st}}$$

or

$$\mathfrak{H}(s) = \frac{y_{ss}(t)}{u(t)}\bigg|_{u(t)=e^{st}}$$

This last equation shows that for scalar input systems, the system function can be obtained by measurement. To get $\mathfrak{H}(s_0)$ we apply an input $u(t) = e^{s_0 t}$ for a long time, till steady state is reached at t_1; then we measure the output y_{ss} at time t_1 to get $\mathfrak{H}(s_0) = \dfrac{y_{ss}(t_1)}{u(t_1)}$.

For multiple-input, multiple-output systems we can, of course, extend this notion and express the i, jth entry of the matrix $\mathfrak{H}(s)$ as the ratio of the ith output to the jth input, holding all other inputs equal to zero:

$$\mathfrak{H}_{ij}(s) = [\mathfrak{H}(s)]_{ij} = \frac{y_{i_{ss}}(t)}{u_j(t)}\bigg|_{\substack{u_j(t)=e^{st} \\ u_k(t)=0, k \neq j}} \tag{9}$$

Example 4

Consider the two-input, two-output system of Figure 6–10. This constant system is a special case of the time varying system of Figure 5–8 which we studied in Example 5–4–2.

It is clear from the diagram that $\mathfrak{H}_{21}(s) = 0$ since, with all initial conditions zero and holding $u_2 = 0$, a signal e^{st} applied as $u_1(t)$ produces no effect at output y_2.

We will outline how to find $\mathfrak{H}_{12}(s)$ and leave the simpler problems of finding $\mathfrak{H}_{11}(s)$ and $\mathfrak{H}_{22}(s)$ to the reader. Just as in Example 5–4–2, holding $u_1 = 0$ and determining $y_{1_{ss}}(t)$ when $u_2(t) = e^{st}$, we may redraw the system of Figure 6–10 to show only the relevant portions. We get the system shown in Figure 5–9 but with all the gains $f_{22}, f_{12}, f_{11}, h_{12}$, and h_{11} now constants.

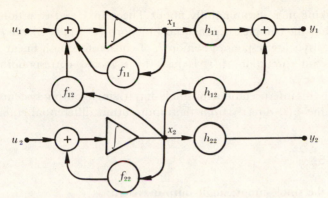

Figure 6–10   A 2-input, 2-output constant system.

From Example 2 with $u_2(t) = e^{st}$ we know the steady state signal at x_2 to be

$$x_{2_{ss}}(t) = \frac{1}{s - f_{22}} e^{st} \quad \text{(exponential)}$$

Considering $x_{2_{ss}}$ to be the exponential "input" to each of the two simple subsystems we have connected in parallel between x_2 and y_1, we use Example 2 again to get

$$x_{1_{ss}}(t) = \frac{f_{12}}{s - f_{11}} x_{2_{ss}}(t)$$

$$= \frac{f_{12}}{(s - f_{11})(s - f_{22})} e^{st} \tag{10}$$

whence

$$y_{1_{ss}}(t) = h_{11} x_{1_{ss}}(t) + h_{12} x_{2_{ss}}(t)$$

$$= \frac{h_{11} f_{12}}{(s - f_{11})(s - f_{22})} e^{st} + \frac{h_{12}}{(s - f_{22})} e^{st}$$

$$= \frac{h_{12}(s - f_{11}) + h_{11} f_{12}}{(s - f_{11})(s - f_{22})} e^{st} \tag{11}$$

Thus

$$\mathcal{Y}_{12}(s) = \left. \frac{y_{1_{ss}}(t)}{u_2(t)} \right|_{u_2(t) = e^{st}, \, u_1(t) = 0}$$

$$= h_{12} \frac{s - \left(f_{11} - \dfrac{h_{11}}{h_{12}} f_{12} \right)}{(s - f_{11})(s - f_{22})}$$

which is of the form $k \dfrac{(s - a)}{(s - b)(s - c)}$, a ratio of polynomials in s. [See Exercise 8.]   ◊

Speaking most meaningfully about Theorem 1 and equation (7a), we can say that the complex exponential functions $\{e^{s(\cdot)}v \mid s \in \mathbf{C}\}$ are *eigenvectors* for the steady state response function L of constant smooth linear systems.† With this interpretation, $\mathfrak{H}(s)$ is just the *eigenvalue* corresponding to the eigenvector $e^{s(\cdot)}v$ of L.

Using this interpretation, for single-input, single-output systems it is easy to determine $\mathfrak{H}(s)$ directly from the input-output differential equation.

Example 5

Consider the single-input, single-output system:

$$\begin{bmatrix} \dot{x}_1 \\ \dot{x}_2 \end{bmatrix} = \begin{bmatrix} 0 & 1 \\ -6 & -5 \end{bmatrix} \begin{bmatrix} x_1 \\ x_2 \end{bmatrix} + \begin{bmatrix} 0 \\ 1 \end{bmatrix} u$$

$$y = \begin{bmatrix} 1 & 0 \end{bmatrix} \begin{bmatrix} x_1 \\ x_2 \end{bmatrix}$$

We have $y = x_1$ so that $\dot{y} = \dot{x}_1 = x_2$ and

$$\ddot{y} = \dot{x}_2 = -6x_1 - 5x_2 + u$$
$$= -6y - 5\dot{y} + u$$

Thus, y satisfies the differential equation

$$\ddot{y} + 5\dot{y} + 6y = u$$

Now we want to set $u(t) = e^{st}$ and look for a solution $y(t) = \mathfrak{H}(s)e^{st}$. Then we clearly have $\dot{y}(t) = s\mathfrak{H}(s)e^{st}$ and $\ddot{y}(t) = s^2\mathfrak{H}(s)e^{st}$, so that

$$s^2\mathfrak{H}(s)e^{st} + 5s\mathfrak{H}(s)e^{st} + 6\mathfrak{H}(s)e^{st} = e^{st}$$

Since $e^{st} \neq 0$, we may divide both sides by it to get

$$(s^2 + 5s + 6)\mathfrak{H}(s) = 1$$

whence

$$\mathfrak{H}(s) = \frac{1}{s^2 + 5s + 6} = \frac{1}{(s+2)(s+3)} \text{ for } s \neq -2, -3$$

[See Exercise 12.]

---

†A result we deduced elegantly though somewhat less rigorously in Exercise 19 of Section 6–2.

If we now look again at the definition of the Laplace transform (2), we see that if the input $u(t)$ has transform $U(s)$, then the inversion integral

$$u(t) = \frac{1}{2\pi i} \int_{\sigma-i\infty}^{\sigma+i\infty} U(s)e^{st} \, ds \tag{12}$$

written out as the limit of a sum

$$u(t) = \lim_{\substack{\Delta s_j \to 0 \\ N \to \infty}} \frac{1}{2\pi i} \sum_{j=-N}^{N} U(s_j)e^{s_j t}\Delta s_j$$

is just the eigenvector expansion of $u(\cdot)$. Recalling that the integral (12) can be thought of as an *inner product* [Example 4–4–2], the value $U(s_j)$ is just the "projection" of the transform $U(\cdot)$ on the vector $e^{(\cdot)t}$. Thus, $U(s_j)$ tells us "how much" of the exponential function $e^{s_j t}$ is "in" the input $u(t)$.

It is precisely because the complex exponential functions e^{st} are eigenvectors of constant smooth linear systems that the Laplace transform is defined the way it is and is the reason why Fourier methods for decomposing functions into exponential components are so useful in the analysis of such systems.

To further emphasize the importance of each of the adjectives *constant*, *linear*, and *smooth*, consider such a system. Again denoting its steady-state, zero-state response by the input-output operator L as in equation (6a),

$$(Lu)(t) = \mathcal{S}(t, -\infty, 0, u(\cdot)) \tag{13}$$

and using the integral representation (12) for the function u, we have

$$(Lu)(t) = \mathcal{S}\left(t, -\infty, 0, \frac{1}{2\pi i} \int_{\sigma-i\infty}^{\sigma+i\infty} U(s)e^{s(\cdot)} \, ds\right)$$

Since $\mathcal{S}(t, -\infty, 0, \cdot)$ is *smooth* [as in Section 5–4], we may pull the limiting process of integration outside:

$$(Lu)(t) = \int_{\sigma-i\infty}^{\sigma+i\infty} \mathcal{S}\left(t, -\infty, 0, \frac{1}{2\pi i} U(s)e^{s(\cdot)}\right) ds$$

By *linearity*, the constants $\frac{1}{2\pi i}$ and $U(s)$ can now be pulled outside $\mathcal{S}$:

$$(Lu)(t) = \int_{\sigma-i\infty}^{\sigma+i\infty} \frac{1}{2\pi i} \mathcal{S}(t, -\infty, 0, e^{s(\cdot)})U(s) \, ds$$

and by *time-invariance* and *linearity*, Theorem 1 can be used to give

$$\mathscr{S}(t, -\infty, 0, e^{s(\cdot)}) = \mathscr{G}(s)e^{s(t)}$$

(which is just equation (8)) so that our expression for L becomes

$$(Lu)(t) = \frac{1}{2\pi i} \int_{\sigma - i\infty}^{\sigma + i\infty} \mathscr{G}(s)U(s)e^{s(t)} \, ds \tag{14}$$

This is just the inverse Laplace transform of the product $\mathscr{G}(s)U(s)$, so

$$(Lu)(t) = \mathscr{L}^{-1}\{\mathscr{G}(s)U(s)\}$$

or

$$\mathscr{L}\{(Lu)(t)\} = \mathscr{G}(s)U(s) \tag{15}$$

But for *constant, smooth* systems we know that we can also express the operator L of equation (13) by the convolution integral

$$(Lu)(t) = \int_{-\infty}^{t} g(t - \tau)u(\tau) \, d\tau$$

Taking Laplace transforms of both sides,

$$\mathscr{L}\{(Lu)t\} = \mathscr{L}\left\{ \int_{-\infty}^{t} g(t - \tau)u(\tau) \, d\tau \right\} \tag{16}$$

and equating the two expressions (15) and (16), we have proven the important theorem:

THEOREM 2

For a smooth, constant linear system with impulse response $g(t)$ to a unit impulse at $t = 0$, and with system function $\mathscr{G}(s)$ defined by (7b), we have

$$\mathscr{L}\left\{ \int_{-\infty}^{t} g(t - \tau)u(\tau) \, d\tau \right\} = \mathscr{G}(s)U(s) \tag{17}$$

where $U(s) = \mathscr{L}\{u(t)\}$ and $\mathscr{L}$ denotes the Laplace transform.   □

Without smoothness, this analysis breaks down:

Example 6

In Example 5–4–1 we presented a linear, time-invariant system, the "jump-summer," which was not smooth. What is the steady-state response

$\mathcal{S}(t, -\infty, 0, e^{s(\cdot)})$ of that system to input e^{st}? Since e^{st} has no discontinuities, the output due to it is zero and $\mathcal{S}(s) \equiv 0$.

   With $\mathcal{S}(s)$ identically zero for this system, equation (14), if it were valid, would imply that $Lu = 0$ for *all* inputs u. We have already seen in Example 5–4–1 that when u is the unit step, $Lu = u \neq 0$, so equations (14) to (17) are not valid for this system. $\Diamond$

THE SYSTEM FUNCTION AND IMPULSE RESPONSE

Let us now develop some further properties of the system function $\mathcal{S}(s)$. In our one-dimensional system of Example 2, we saw that $\mathcal{S}(s)$ took the value $\dfrac{bc}{s-a}$. Let us see what happens for the constant linear system

$$(F, G, H): \quad \dot{x} = Fx + Gu$$
$$y = Hx$$

whose impulse response matrix is given by

$$g(t - \tau) = He^{F(t-\tau)}G, \quad t > \tau$$

Again we fix a vector v and pick $\mathcal{R}e(s)$ sufficiently large that, from equation (7a),

$$\mathcal{S}(s)e^{st}v = \int_{-\infty}^{t} He^{F(t-\tau)}Ge^{s\tau}v \, d\tau$$

is well-defined. We may rewrite the right-hand side as

$$H\left[\int_{-\infty}^{t} e^{F(t-\tau)}e^{s(\tau-t)} \, d\tau\right] Ge^{st}v = H\left[\int_{-\infty}^{0} e^{(sI-F)\xi} \, d\xi\right] Ge^{st}v$$

and, evaluating the integral [Exercise 5] as

$$H(sI - F)^{-1}Ge^{st}v$$

we thus get the formula

$$\mathcal{S}(s) = H(sI - F)^{-1}G \tag{18}$$

for the system function of (F, G, H).

   We see that $\mathcal{S}(s)$ is well-defined unless s is such that $(sI - F)$ is not invertible; i.e., such that $\det(sI - F) = 0$. In line with our earlier discussion, we call such an s a "pole" of $(sI - F)^{-1}$, and in Definition 6–2–1 we saw that such an s (namely, a solution of $\det(sI - F) = 0$) is called an *eigenvalue* of F.

It is worth relating $\mathfrak{H}(s)$ and our "impulse response" function $g(t - \tau)$. From

$$g(t) = He^{Ft}G \quad \text{for } t > 0$$

and

$$\mathfrak{H}(s) = \int_{-\infty}^{t} He^{F(t-\tau)}Ge^{s(\tau-t)} \, d\tau$$

by changing the variable of integration, we see that

$$\mathfrak{H}(s) = \int_{0}^{\infty} He^{F\tau}Ge^{-s\tau} \, d\tau = \mathcal{L}\{g(t)\} \tag{19}$$

In other words, we have shown the following result:

THEOREM 3

For the constant linear system (F, G, H), the system function $\mathfrak{H}(s) = H(sI - F)^{-1}G$ is simply the Laplace transform of the impulse response $g(t) = He^{Ft}G$. □

If we use the relationship (19) that

$$\mathfrak{H}(s) = \mathcal{L}\{g(t)\}$$

in Theorem 2, we have also proven the corollary:

COROLLARY 4

For *causal* functions g and u, i.e., those which vanish for negative arguments,

$$\mathcal{L}\left\{\int_{-\infty}^{t} g(t - \tau)u(\tau) \, d\tau\right\} = \mathcal{L}\{g(t)\} \cdot \mathcal{L}\{u(t)\} \qquad \Box$$

This is the well-known result that the product of the Laplace transforms of two functions is the transform of the convolution of the two functions. [See Exercises 10 and 11.]

We now have two distinct ways of determining the system function $\mathfrak{H}(s)$: (i) as the steady-state response to an exponential input e^{st}, and (ii) as the Laplace transform of the impulse response. When $\mathfrak{H}(s)$ is a matrix, equation

(9) may now be supplemented, using Laplace transforms, by

$$\mathscr{Y}_{ij}(s) = [\mathscr{Y}(s)]_{ij} = \frac{\mathcal{L}\{y_i(t)\}}{\mathcal{L}\{u_j(t)\}}\bigg|_{u_k=0,\, k\neq j} \tag{20}$$

To amplify this discussion, let us give an alternative derivation of equation (19) and then apply it to an example.

Let $x(t)$ be a state function and $X(s)$ its Laplace transform. Then, if we integrate by parts the Laplace transform of the derivative $\dot{x}(t)$, we obtain (cf. Exercise 9)

$$\int_{0^-}^{\infty} \dot{x}(t)e^{-st}\, dt = [x(t)e^{-st}]_{0^-}^{\infty} - \int_{0^-}^{\infty} x(t)(-se^{-st})\, dt$$

so that

$$\mathcal{L}\{\dot{x}(t)\} = sX(s) - x(0^-) \tag{21}$$

(assuming that $\mathcal{R}e(s)$ is sufficiently large that $x(t)e^{-st} \to 0$ as $t \to \infty$).

Note further that the Laplace transform commutes with the application of constant matrices, e.g.,

$$\mathcal{L}\{Hy(t)\} = HY(s)$$

Thus we see that the homogeneous system equation

$$\dot{x}(t) = Fx(t) \tag{22a}$$

takes the following form in the transform domain:

$$sX(s) - x(0) = FX(s) \tag{22b}$$

whence

$$(sI - F)X(s) = x(0)$$

where *by the initial condition* $x(0)$ *we always mean* $x(0^-)$, as explained before Definition 1 and in Exercise 9.

Multiplying both sides by $(sI - F)^{-1}$ yields

$$X(s) = (sI - F)^{-1}x(0)$$

and inverse Laplace transforming both sides gives

$$x(t) = \mathcal{L}^{-1}\{X(s)\} = \mathcal{L}^{-1}\{(sI - F)^{-1}\}x(0)$$

But the solution of (22a) is

$$x(t) = e^{Ft}x(0)$$

and hence, by uniqueness,

$$e^{Ft} = \mathcal{L}^{-1}\{(sI - F)^{-1}\} \tag{23a}$$

and

$$\mathcal{L}\{e^{Ft}\} = (sI - F)^{-1} \tag{23b}$$

Using this result, the complete solution to the inhomogeneous system (F, G, H) has the equations

$$X(s) = (sI - F)^{-1}x(0) + (sI - F)^{-1}GU(s) \tag{24a}$$

and

$$Y(s) = H(sI - F)^{-1}x(0) + H(sI - F)^{-1}GU(s) \tag{24b}$$

which are the transformed versions of

$$x(t) = e^{Ft}x(0) + \int_0^t e^{F(t-\tau)}Gu(\tau)\,d\tau$$

and

$$y(t) = He^{Ft}x(0) + \int_0^t g(t - \tau)u(\tau)\,d\tau$$

Thus, we see once again that $\mathcal{L}\{g(t)\} = H(sI - F)^{-1}G = \mathcal{G}(s)$.

Example 7

Let us use Laplace transform methods to study the system

$$\begin{bmatrix} \dot{x}_1(t) \\ \dot{x}_2(t) \end{bmatrix} = \begin{bmatrix} 0 & 1 \\ -6 & -5 \end{bmatrix}\begin{bmatrix} x_1(t) \\ x_2(t) \end{bmatrix} + \begin{bmatrix} 0 \\ 1 \end{bmatrix}u(t)$$

$$y(t) = \begin{bmatrix} 1 & 1 \end{bmatrix}\begin{bmatrix} x_1(t) \\ x_2(t) \end{bmatrix}$$

Then

$$(sI - F) = \begin{bmatrix} s & -1 \\ 6 & s + 5 \end{bmatrix}$$

and a routine computation yields

$$(sI - F)^{-1} = \begin{bmatrix} \dfrac{s+5}{(s+2)(s+3)} & \dfrac{1}{(s+2)(s+3)} \\[3mm] \dfrac{-6}{(s+2)(s+3)} & \dfrac{s}{(s+2)(s+3)} \end{bmatrix} \quad \text{for } s \neq -2, -3$$

Now each element of $(sI - F)^{-1}$ is a ratio of two polynomials in s and may be inverse Laplace transformed from tables such as our Appendix A-1 or from partial fraction expansions. We illustrate the latter method for $[e^{Ft}]_{11}$: Set

$$\frac{s+5}{(s+2)(s+3)} = \frac{a}{(s+2)} + \frac{b}{(s+3)}$$

and find a and b. Putting the right side over lowest common denominator,

$$\frac{s+5}{(s+2)(s+3)} = \frac{(a+b)s + (3a+2b)}{(s+2)(s+3)}$$

and equating numerators

$$a + b = 1$$
$$3a + 2b = 5$$

we have $a = 3$ and $b = -2$. Thus

$$\frac{s+5}{(s+2)(s+3)} = \frac{3}{s+2} + \frac{-2}{s+3}$$

Since the Laplace transform of $e^{\alpha t}$ is $\displaystyle\int_0^\infty e^{\alpha t} e^{-st}\, dt = \frac{1}{s-\alpha}$ for $\Re e(s) > \Re e(\alpha)$,

we know that $\dfrac{1}{s+2}$ and $\dfrac{1}{s+3}$ are the transforms of e^{-2t} and e^{-3t} respectively and

$$\mathcal{L}^{-1}\left\{\frac{s+5}{(s+2)(s+3)}\right\} = 3e^{-2t} - 2e^{-3t}, \ t \geq 0$$

Similarly,

$$\frac{1}{(s+2)(s+3)} = \frac{1}{s+2} + \frac{-1}{s+3}$$

and

$$\frac{s}{(s+2)(s+3)} = \frac{-2}{s+2} + \frac{3}{s+3}$$

Thus

$$e^{Ft} = \mathcal{L}^{-1}\{(sI - F)^{-1}\}$$

yields

$$e^{Ft} = \begin{bmatrix} (3e^{-2t} - 2e^{-3t}) & (e^{-2t} - e^{-3t}) \\ (-6e^{-2t} + 6e^{-3t}) & (-2e^{-2t} + 3e^{-3t}) \end{bmatrix}, \ t \geq 0$$

in terms of which any aspect of the system behavior may now be determined. For example,

$$\mathfrak{H}(s) = \begin{bmatrix} 1 & 1 \end{bmatrix}(sI - F)^{-1}\begin{bmatrix} 0 \\ 1 \end{bmatrix} = \begin{bmatrix} 1 & 1 \end{bmatrix}\begin{bmatrix} \dfrac{1}{(s + 2)(s + 3)} \\ \dfrac{s}{(s + 2)(s + 3)} \end{bmatrix}$$

$$= \frac{1}{(s + 2)(s + 3)} + \frac{s}{(s + 2)(s + 3)}$$

so

$$\mathfrak{H}(s) = \frac{(s + 1)}{(s + 2)(s + 3)} \quad s \neq -2, -3$$

and

$$g(t) = (-e^{-2t} + 2e^{-3t})1(t)$$

where $1(t)$ is the unit-step function.   ◊

The reader should compare this last example with his solution of Exercise 7 and with his determination of $\mathfrak{H}(s)$ in Exercise 12. [See also Exercise 10.]

Since we have shown in equation (19) and Theorem 3 that the system function $\mathfrak{H}(s)$ is the Laplace transform of the impulse response for a constant linear system (F, G, H), it follows that any two zero-state equivalent systems have the same system function. [See Exercise 16.] Moreover, if $(\widehat{F}, \widehat{G}, \widehat{H})$ is another system related to (F, G, H) by an invertible state transformation $x = P\widehat{x}$, so that

$$\widehat{F} = P^{-1}FP, \quad \widehat{G} = P^{-1}G, \quad \widehat{H} = HP$$

then

$$(sI - \widehat{F}) = (sI - P^{-1}FP) = P^{-1}(sI - F)P$$

and the characteristic function of $\widehat{F}$ is

$$\begin{aligned} \chi_{\widehat{F}}(s) &= \det(sI - \widehat{F}) = \det(P^{-1})\det(sI - F)\det(P) \\ &= \det(sI - F)\det(P^{-1})\det(P) \\ &= \det(sI - F)\det(P^{-1}P) \\ &= \det(sI - F) = \chi_F(s) \end{aligned}$$

Also,

$$(sI - \widehat{F})^{-1} = (P^{-1}(sI - F)P)^{-1} = P^{-1}(sI - F)^{-1}(P^{-1})^{-1}$$
$$= P^{-1}(sI - F)^{-1}P$$

so

$$\widehat{\mathfrak{H}}(s) = \widehat{H}(sI - \widehat{F})^{-1}\widehat{G} = (HP)P^{-1}(sI - F)^{-1}P(P^{-1}G)$$
$$= H(sI - F)^{-1}G = \mathfrak{H}(s)$$

and we see that *algebraically equivalent systems have identical characteristic equations and identical system functions.*

We shall use this fact later, in Sections 6–5 and 6–6, when we try to synthesize a system with a given system function $\mathfrak{H}(s)$ by finding matrices (F, G, H) such that $H(sI - F)^{-1}G = \mathfrak{H}(s)$. Once we have found one such realization (F, G, H), we often seek other equivalent realizations whose matrices $(\widehat{F}, \widehat{G}, \widehat{H})$ have some nicer, "canonical" form.

STATE RESPONSE AND MODE SUPPRESSION

So far we have utilized the notions of system function $\mathfrak{H}(s)$ and impulse response $g(t)$ to characterize the input-output behavior of a system. Since the output y is just a simple read-out of the state x through the matrix H, and it is in fact the state which contains all the dynamic information about the system, we see that $\mathfrak{H}(s)$ and $g(t)$ offer at best only an incomplete characterization of the system.

To use the appealing "cause and effect" ideas inherent in state-variable analysis, we often employ the notions of "state system function" $\mathfrak{H}_X(s)$ and "state impulse response" $g_X(t)$, where we pretend that the "output" is the state system function. This is equivalent, of course, to finding the system function of the system (F, G, I) and then multiplying the "output" of (F, G, I) by the H matrix to get the actual output y.

Thus, for the system (F, G, H)

$$\mathfrak{H}_X(s) = (sI - F)^{-1}G \tag{25}$$
$$g_X(t) = \begin{cases} e^{Ft}G & t > 0 \\ 0 & t < 0 \end{cases}$$

and the steady-state response is given by

$$X_{ss}(s) = \mathfrak{H}_X(s)U(s)$$
$$x_{ss}(t) = \int_{-\infty}^{t} g_X(t - \tau)u(\tau)\, d\tau$$

Example 8

Find the state $x(t)$ for the system (F, G, H) of Example 7 due to an input $u(t) = e^{-t}1(t)$. From the last example we have

$$\mathscr{L}_X(s) = \begin{bmatrix} \dfrac{1}{(s+2)(s+3)} \\[3mm] \dfrac{s}{(s+2)(s+3)} \end{bmatrix}$$

and

$$g_X(t) = \begin{bmatrix} (e^{-2t} - e^{-3t})1(t) \\ (-2e^{-2t} + 3e^{-3t})1(t) \end{bmatrix}$$

Since $U(s) = \mathscr{L}\{e^{-t}1(t)\} = \dfrac{1}{s+1}$ (for $\mathscr{R}e(s) > -1$), it follows that

$$X_{ss}(s) = \begin{bmatrix} \dfrac{1}{(s+1)(s+2)(s+3)} \\[3mm] \dfrac{s}{(s+1)(s+2)(s+3)} \end{bmatrix}$$

and (verify!)

$$x_{ss}(t) = \begin{bmatrix} \left(\dfrac{1}{2}e^{-t} - e^{-2t} + \dfrac{1}{2}e^{-3t}\right)1(t) \\[3mm] \left(-\dfrac{1}{2}e^{-t} + 2e^{-2t} - \dfrac{3}{2}e^{-3t}\right)1(t) \end{bmatrix}$$

The complete state response is gotten from equations (24) and (25):

$$\begin{aligned} X(s) &= (sI - F)^{-1}x(0) + \mathscr{L}_X(s)U(s) \\ &= (sI - F)^{-1}x(0) + X_{ss}(s) \end{aligned}$$

For instance, suppose we are given the initial state

$$x(0) = \begin{bmatrix} \dfrac{1}{2} \\[3mm] -\dfrac{1}{2} \end{bmatrix} = \left(\dfrac{1}{2}\right)\begin{bmatrix} 1 \\ -1 \end{bmatrix}$$

Then

$$(sI - F)^{-1}x(0) = \left(\dfrac{1}{2}\right)\begin{bmatrix} \dfrac{(s+4)}{(s+2)(s+3)} \\[3mm] \dfrac{-(s+6)}{(s+2)(s+3)} \end{bmatrix}$$

and

$$X(s) = \left(\frac{1}{2}\right)\begin{bmatrix} \dfrac{(s+4)}{(s+2)(s+3)} \\[2mm] \dfrac{-(s+6)}{(s+2)(s+3)} \end{bmatrix} + \begin{bmatrix} \dfrac{1}{(s+1)(s+2)(s+3)} \\[2mm] \dfrac{s}{(s+1)(s+2)(s+3)} \end{bmatrix}$$

$$= \begin{bmatrix} \left(\dfrac{1}{2}\right)\dfrac{s^2+5s+6}{(s+1)(s+2)(s+3)} \\[3mm] \left(-\dfrac{1}{2}\right)\dfrac{s^2+5s+6}{(s+1)(s+2)(s+3)} \end{bmatrix}$$

$$X(s) = \left(\frac{1}{2}\right)\frac{s^2+5s+6}{(s+1)(s+2)(s+3)}\begin{bmatrix} 1 \\ -1 \end{bmatrix}$$

But $s^2 + 5s + 6 = (s+2)(s+3)$, so a *cancellation* is possible and

$$X(s) = \frac{1}{2}\frac{1}{s+1}\begin{bmatrix} 1 \\ -1 \end{bmatrix}$$

so

$$x(t) = \frac{1}{2}e^{-t}1(t)\begin{bmatrix} 1 \\ -1 \end{bmatrix}$$

which we recognize to be

$$u(t)x(0) = e^{(-1)t}1(t)\underset{x}{\&}(-1)$$

This is curious; the state vector always remains along the given initial condition vector $x(0)$ and has its time variation exactly like the exponential input. We did not have to wait a "long time" before the state began to look like the forcing function.   ◊

In the last example we have sneaked up upon the important notion of "mode suppression." Why were we able to choose an initial state $x(0)$ so that the state immediately followed the input with no transients? Can we always do this?

To this end, let $u(t) = e^{s_1 t}1(t)v$, where v is a fixed vector and s_1 is a complex number. From equations (24) and (25) with $U(s) = \dfrac{1}{s - s_1}v$,

$$X(s) = (sI - F)^{-1}x(0) + \underset{x}{\&}(s)U(s)$$

$$= (sI - F)^{-1}x(0) + (sI - F)^{-1}\frac{1}{s - s_1}Gv$$

But from Exercise 18, we can use the matrix identity

$$(sI - F)^{-1} \frac{1}{s - s_1} = (s_1 I - F)^{-1} \frac{1}{s - s_1} - (sI - F)^{-1}(s_1 I - F)^{-1}$$

and get

$$X(s) = (sI - F)^{-1}[x(0) - (s_1 I - F)^{-1} Gv] + (s_1 I - F)^{-1} \frac{1}{s - s_1} Gv$$

$$= (sI - F)^{-1}[x(0) - \mathcal{Z}_X(s_1)v] + \mathcal{Z}_X(s_1) \frac{1}{s - s_1} v$$

Thus, for $U(s) = \dfrac{1}{s - s_1} v$ we have

$$X(s) = (sI - F)^{-1}[x(0) - \mathcal{Z}_X(s_1)v] + \mathcal{Z}_X(s_1)U(s) \qquad (26)$$

Clearly, if we choose

$$x(0) = \mathcal{Z}_X(s_1)v \qquad (27)$$

then

$$X(s) = \mathcal{Z}_X(s_1) \frac{1}{s - s_1} v$$

and thus the terms involving F (through $(sI - F)^{-1}$) drop out and $X(s)$ contains none of the natural frequencies of (F, G, H) but depends only on the input exponential. It is interesting to note that this initial state $x(0)$ is precisely the state to which $e^{s_1 t}v$ would have driven the system starting from the zero state at $t_0 = -\infty$ (verify!). Thus, when $u(t) = e^{s_1 t}1(t)v$ and $x(0) = \mathcal{Z}_X(s_1)v$, we have

$$x(t) = \mathcal{Z}_X(s_1)e^{s_1 t}1(t)v \qquad (28)$$

In Section 6–5 when we study single-input, single-output systems where the system function (transfer function) is a scalar, we will use our results on mode suppression to give a characterization of a "zero of transmission" of such a system.

Suppose that we do not want to suppress all of the modes of the system (F, G, H), but only some of them. It should be clear from equation (26) and our experience with eigenvectors that we can choose an initial state $x(0)$ to accomplish this. Instead of constraining $x(0) - \mathcal{Z}_X(s_1)v$ to be 0 (i.e., to be in the subspace spanned by *none* of the modes of the system), we constrain it to lie in a subspace containing none of the modes we want suppressed (i.e., containing none of the corresponding eigenvectors of F—see Definition 6–2–2).

Let us illustrate with the system of Examples 7 and 8:

Example 9

Since $F = \begin{bmatrix} 0 & 1 \\ -6 & -5 \end{bmatrix}$, its eigenvalues are $\lambda_1 = -2$ and $\lambda_2 = -3$; corresponding to these are the eigenvectors

$$P_1 = \begin{bmatrix} 1 \\ -2 \end{bmatrix} \quad \text{and} \quad P_2 = \begin{bmatrix} -1 \\ 3 \end{bmatrix}$$

Suppose that we want to suppress the mode $\lambda_2 = -3$ corresponding to P_2. We can pick as our candidate for $x(0) - \mathfrak{F}_X(s_1)v$ a vector $w \in X$ such that w has no P_2 in it. We do this by letting w be in the subspace of X spanned by the other eigenvectors. In this two-dimensional case, P_1 is the only other eigenvector, so $w = cP_1$ for some constant c. We know that P_1 is also an eigenvector of e^{Ft} and of $(sI - F)^{-1}$, so that $(sI - F)^{-1}w$ will also have no P_2 in it.

Using the numbers of our example, we now take

$$w = P_1 = \begin{bmatrix} 1 \\ -2 \end{bmatrix}$$

and set

$$\begin{bmatrix} 1 \\ -2 \end{bmatrix} = x(0) - \begin{bmatrix} \dfrac{1}{(s_1 + 2)(s_1 + 3)} \\[2mm] \dfrac{s_1}{(s_1 + 2)(s_1 + 3)} \end{bmatrix}$$

to yield the initial condition

$$x(0) = \begin{bmatrix} \dfrac{s_1^2 + 5s_1 + 7}{(s_1 + 2)(s_1 + 3)} \\[3mm] \dfrac{-2s_1^2 - 9s_1 - 12}{(s_1 + 2)(s_1 + 3)} \end{bmatrix}$$

Checking,

$$(sI - F)^{-1}w = \frac{1}{(s + 2)(s + 3)} \begin{bmatrix} s + 5 & 1 \\ -6 & s \end{bmatrix} \begin{bmatrix} 1 \\ -2 \end{bmatrix}$$

$$= \frac{1}{(s + 2)(s + 3)} \begin{bmatrix} (s + 3) \\ -2(s + 3) \end{bmatrix} = \frac{1}{(s + 2)(s + 3)} (s + 3) \begin{bmatrix} 1 \\ -2 \end{bmatrix}$$

$$= \frac{1}{s + 2} \begin{bmatrix} 1 \\ -2 \end{bmatrix}$$

which contains no $(s + 3)$ term. ◊

In the next section we shall focus our attention on some graph-theoretic methods of system analysis and their application to the problem of determination and realization of the system function $\mathfrak{H}(s)$.

EXERCISES FOR SECTION 6–3

1. Since the notion of steady-state response involves having the input u applied for an infinitely long time, it implies that the natural behavior of the system must settle down. We make this notion precise by defining the **ground state** x_g of a system as

$$x_g = \lim_{t \to \infty} \phi(t, t_0, x_0, 0)$$

when the limit exists and is independent of x_0, the state at t_0.

    (a) Compute this limit for the system of Example 2. Is there a ground state? Repeat for the system of Example 3.

    (b) Show that if a system has a ground state, then it has a steady-state response and

$$y_{ss}(t) = \lim_{t_0 \to -\infty} \mathcal{S}(t, t_0, x_g, u)$$

2. Let $u_0(t) = e^{s_0 t}$ for the system of Example 3, and give the complete response. Consider only the zero-state (forced) response. Does $\mathcal{S}(t, -\infty, 0, u_0)$ look like the input?

3. Here we outline a proof of equation (6c) which uses only the response function $\mathcal{S}(t, t_0, x_0, u)$ of a system and does not require that the response be given by an integral as in equation (5). Thus, if we define L by $(Lu)(t) = \mathcal{S}(t, -\infty, 0, u)$ as long as this steady-state response makes sense, prove, for discrete-time or continuous-time, *time-invariant* systems, that $z^{-\tau}L = Lz^{-\tau}$ for all delays $\tau \in T$.

Outline:

$$
\begin{aligned}
(Lz^{-\tau}u)(t) &= \mathcal{S}(t, -\infty, 0, z^{-\tau}u) \\
&= \mathcal{S}(t - \tau, -\infty - \tau, 0, z^{\tau}(z^{-\tau}u)) \\
&= \mathcal{S}(t - \tau, -\infty, 0, u) = (Lu)(t - \tau) \\
&= (z^{-\tau}(Lu))(t)
\end{aligned}
$$

4. Go back to the proof of Theorem 1 and see that we did not really need all of "linearity" to prove it. Homogeneity of the operator L was all we needed; additivity of L did not matter.

✔ **5.** For a constant invertible matrix A, show that

$$\int_a^b e^{A\tau}\, d\tau = A^{-1} \cdot e^{A\tau}\big|_a^b = A^{-1}(e^{Ab} - e^{Aa})$$

$$\left[\textit{Hint:} \qquad \int_a^t e^{A\tau}\, d\tau = B(t) \text{ if } \frac{d}{dt} B(t) = e^{At}; \text{ use } \frac{d}{dt} e^{At} = A e^{At}. \right]$$

6. Carry out all the steps in the proof of Theorem 1 using the integral formula

(6a) for the operator L. In particular, check that

$$\mathfrak{L}(s)v = (Le^{s(\cdot)}v)(0) = \int_{-\infty}^{0} g(0 - \tau)e^{s\tau}v \, d\tau$$

is the output at time 0 due to input $e^{st}v$ from $-\infty$ up to 0.

✔ **7.**   Consider the system

$$\begin{bmatrix} \dot{x}_1(t) \\ \dot{x}_2(t) \end{bmatrix} = \begin{bmatrix} 0 & 1 \\ -6 & -5 \end{bmatrix} \begin{bmatrix} x_1(t) \\ x_2(t) \end{bmatrix} + \begin{bmatrix} 0 \\ 1 \end{bmatrix} u(t)$$

$$y(t) = \begin{bmatrix} 1 & 1 \end{bmatrix} \begin{bmatrix} x_1(t) \\ x_2(t) \end{bmatrix}$$

(a)   Solve the matrix differential equation and give the complete response.
(b)   Does this system have a ground state [Exercise 1]? Does it have a steady-state response?
(c)   Let $u_0(t) = e^{s_0 t}$ and find the steady-state response using equation (7a) and part (a). Does $\mathcal{S}(t, -\infty, 0, u_0)$ look like the input? What is $\mathcal{S}(t, -\infty, 0, u_0)$ when $s_0 = -1$? What is the system function $\mathfrak{L}(s)$? For which values of s does it exist? What is $\mathfrak{L}(-1)$?

8. (a)   Find all four entries of the matrix $\mathfrak{L}(s)$ for the system of Example 4.
(b)   Check the results of equations (10) and (11) by using the results of Example 5–4–2, which give the impulse response $g_{12}(t, \tau)$. Then compute

$$y_{1_{ss}}(t) = \int_{-\infty}^{t} g_{12}(t, \tau)u_2(\tau) \, d\tau$$

with $u_2(\tau) = e^{s\tau}$. [*Note:* If you do not wish to do this for general coefficients, use the values $f_{11} = -1$, $f_{12} = 2$, $f_{22} = -3$, $h_{11} = 4$, $h_{12} = 5$, $h_{22} = 6$ to obtain
$$\mathfrak{L}_{12}(s) = \frac{5s + 13}{(s + 1)(s + 3)}.]$$

✔ **9.**   This exercise illustrates why we use 0^- as the lower limit of integration in the Laplace transform (2).

(a)   Show that the unit step $\mathbf{1}(t)$ has Laplace transform $\dfrac{1}{s}$ for $\mathfrak{Re}(s) > 0$ and that the unit impulse has transform 1 for all s.
(b)   Let $f(t) = \mathbf{1}(t)$ be the unit step, a function which has a Dirac delta function at $t = 0$ in its derivative:

$$\dot{f}(t) = \delta(t) \text{ (the unit impulse)}$$

Here $f(0^-) = 0$ and $f(0^+) = 1$. Use the formula for the transform of $\dot{f}(t)$, equation (21), and the results of part (a) to see what goes wrong if we use 0^+ instead of 0^-. [For a thorough discussion of the Laplace transform with 0^- as the lower limit, see Appendix B of L. A. Zadeh and C. A. Desoer, *Linear System Theory* (McGraw-Hill, 1963).]

✔ **10.**   Let $g(t) = e^{-2t}\mathbf{1}(t)$ and $u(t) = e^{-3t}\mathbf{1}(t)$.
(a)   Convolve $g(t)$ with $u(t)$ to get $y(t)$.

(b)   Laplace transform $g(t)$, $u(t)$, and the resulting time function $y(t)$ of part (a) and verify that Corollary 4 is valid: $Y(s) = G(s)U(s)$.

(c)   Use Corollary 4 to determine

$$\mathcal{L}^{-1}\left\{\frac{1}{(s+2)(s+3)}\right\} \text{ and } \mathcal{L}^{-1}\left\{\frac{s}{(s+2)(s+3)}\right\}$$

by convolution, thereby checking the partial fraction expansion of Example 7.

11.   **(a)**   Prove Corollary 4 directly by taking the Laplace transform of the convolution integral. Notice that you must interchange the order of some of the integrations—smoothness again!

(b)   Use the result of part (a) and Theorem 2 to get another proof of Theorem 3.

(c)   To what extent is the operation $\mathcal{L}$ of Laplace transformation an **isomorphism** (linear, one-to-one, and onto) between the vector spaces of time-domain functions $\{f(t)\}$ and frequency-domain functions $\{F(s)\}$? In what other sense is $\mathcal{L}$ also a **homomorphism** (preserver of binary operations) between these spaces? [*Hint:* Consider the binary operations relating g and u to G and U in Corollary 4.]

✔ **12.**   For the system (F, G, H) studied in Exercise 7 and in Example 7,

(a)   Draw the analog computer diagram.

(b)   Eliminate x_1, x_2 to get a scalar differential equation relating output y to input u. [*Answer:* $\ddot{y} + 5\dot{y} + 6y = \dot{u} + u$]

(c)   Find the system function $\mathcal{S}(s)$ from the input-output equation of part (b), using the fact that when $u(t) = e^{st}$, then $y(t) = \mathcal{S}(s)e^{st}$ as in Example 5.

(d)   Compare this method with that of Example 7 and Exercise 7.

(e)   Suppose that you were given

$$\mathcal{S}(s) = \frac{s + a}{(s + b)(s + c)}$$

Can you give a differential equation relating y and u for a system which has this $\mathcal{S}(s)$? [*Answer:* $\ddot{y} + (b + c)\dot{y} + bcy = \dot{u} + au$]

✔ **13.**   Several times in this book (such as in Exercise 12(b)) we have had to find the **input-output differential equation from** a given **state variable** formulation $\dot{x} = Fx + Gu, y = Hx$. In this exercise we illustrate a systematic method for accomplishing this. Let $\dot{x} = Fx + Gu$ and $y = Hx$ with

$$F = \begin{bmatrix} 0 & 6 \\ 1 & -1 \end{bmatrix}, G = \begin{bmatrix} 0 \\ 1 \end{bmatrix}, H = \begin{bmatrix} 1 & 1 \end{bmatrix}$$

(a)   Use the Laplace transform to show that

(i)
$$\begin{bmatrix} s & -6 & 0 \\ -1 & s+1 & 0 \\ -1 & -1 & 1 \end{bmatrix}\begin{bmatrix} X_1(s) \\ X_2(s) \\ Y(s) \end{bmatrix} = \begin{bmatrix} 0 \\ 1 \\ 0 \end{bmatrix} U(s)$$

(b)   Using elementary row operations on the above matrix, we can derive several equivalent representations. For example, multiplying the second row by s and adding the result to the top row yields the equivalent system

(ii)
$$\begin{bmatrix} 0 & s^2+s-6 & 0 \\ -1 & s+1 & 0 \\ -1 & -1 & 1 \end{bmatrix}\begin{bmatrix} X_1(s) \\ X_2(s) \\ Y(s) \end{bmatrix} = \begin{bmatrix} s \\ 1 \\ 0 \end{bmatrix} U(s)$$

having the same solution as (i). Carry out the corresponding steps on the time-domain differential equations which would yield this Laplace transform system (ii).

(c)   Continue making elementary row operations (which leave the solutions unchanged) to get the triangularized system

$$
\begin{bmatrix}
1 & 1 & -1 \\
0 & (s+2)(s^2+s-6) & -(s^2+s-6) \\
0 & 0 & (s^2+s-6)
\end{bmatrix}
\begin{bmatrix}
X_1(s) \\
X_2(s) \\
Y(s)
\end{bmatrix}
=
\begin{bmatrix}
0 \\
(s^2+s-6) \\
s+6
\end{bmatrix}
U(s)
$$

whose bottom row has no X_1 or X_2 terms in it. From this get $\ddot{y} + \dot{y} - 6y = \dot{u} + 6u$, the scalar differential equation relating input to output. [*Ans:* Starting from (b), multiply third row by (-1) and interchange first and third rows; in result, add first to second; in result, multiply second by (s^2+s-6) and third by $(s+2)$; then subtract second from third.]

(d)   Find a scalar difference equation relating input and output of the discrete-time system $x(k+1) = Fx(k) + Gu(k)$, $y(k) = Hx(k)$ for the F, G, and H of this problem.

✔ **14.**   Example 4 illustrated how the transfer function of a pair of cascaded or parallel subsystems can be computed. To explore this further, let Σ_1 and Σ_2 be two systems having inputs u_1 and u_2, outputs y_1 and y_2, and transfer functions $\mathscr{G}_1(s)$ and $\mathscr{G}_2(s)$ respectively.

(a)   Consider the **cascade or series connection** in Figure 6–11; let $u_1 = e^{st}v_1$

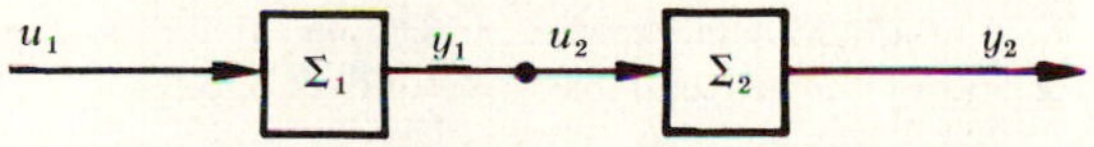

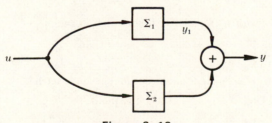

Figure 6–11

and compute the steady state y_2 to determine that the overall transfer function $\mathscr{G}(s)$ is given by the product

$$\mathscr{G}(s) = \mathscr{G}_2(s)\mathscr{G}_1(s) \quad \text{(note order)}$$

(b)   Show that the overall transfer function $\mathscr{G}(s)$ of the **parallel connection** of Figure 6–12 is given by the sum $\mathscr{G}(s) = \mathscr{G}_1(s) + \mathscr{G}_2(s)$.

Figure 6–12

(c)   Show that the transfer function matrix $\mathscr{G}(s)$ for the system comprised of Σ_1 and Σ_2 in the **feedback connection** of Figure 6–13 is given by

$$\mathscr{G}(s) = (I - \mathscr{G}_1(s)\mathscr{G}_2(s))^{-1}\mathscr{G}_1(s) \quad \text{(note order)}$$

What assumptions did you need in order to derive this expression?

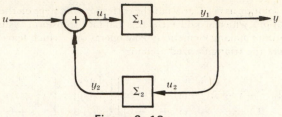

Figure 6–13

(d)   Illustrate your results of parts (a) and (c) by computing the overall transfer functions for the case when Σ_1 and Σ_2 have scalar transfer functions $\mathfrak{L}_1(s) = \dfrac{s-2}{s+2}$ and $\mathfrak{L}_2(s) = \dfrac{1}{s-2}$. Notice that Σ_2 has a "bad" pole in the right half plane at $s = +2$ causing a growing impulse response $g_2(t) = e^{+2t}1(t)$, while Σ_1 is "good." Carefully compare the cascade and feedback approaches to removing the bad behavior from the zero-state response by *cancellation*. Is the bad behavior still present in the complete response?

(e)   Write out the time domain expressions for the overall impulse response for each of the configurations of parts (a), (b), and (c).

✔ **15.**   This exercise further illustrates the pitfalls in forgetting the hypotheses under which we derived Theorem 3 and equations (15), (19), and (20).

Consider the system shown in Figure 6–14, consisting of a periodically operated switch. Let the switch start out closed at $t = 0$, open at $t = 1$, close at $t = 2$, and so forth, so the period of the switching is 2 seconds.

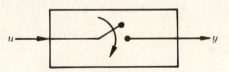

Figure 6–14

(a)   Let $u_1(t) = 1(t)$ (unit step) and find the output $y_1(t)$. Sketch $u_1(t)$ and $y_1(t)$, and compute their Laplace transforms $U_1(s)$ and $Y_1(s)$. Give $\mathfrak{L}_1(s) = \dfrac{Y_1(s)}{U_1(s)}$. Is $\mathfrak{L}_1(s)$ the transfer function? [*Answer:* $Y_1(s) = \dfrac{1}{s}\left(\dfrac{1}{1 + e^{-s}}\right)$, $U_1(s) = \dfrac{1}{s}$, $\mathfrak{L}_1(s) = \dfrac{1}{1 + e^{-s}}$.]

(b)   Let $u_2(t) = \sin \pi t$. Find the output $y_2(t)$ and sketch both waveforms. Find $U_2(s)$ and $Y_2(s)$. Does $Y_2(s) = \mathfrak{L}_1(s)U_2(s)$? Explain.

[*Answer:* $Y_2(s) = \dfrac{\pi}{\pi^2 + s^2}\left(\dfrac{1}{1 + e^{-s}}\right)$, $U_2(s) = \dfrac{\pi}{\pi^2 + s^2}$, $Y_2(s) \neq \mathfrak{L}_1(s)U_2(s)$.]

16.   We showed that algebraically equivalent systems have the same transfer functions. What about two *zero-input equivalent* systems; are their transfer functions equal? How about their characteristic equations?

What can you say about two *equivalent* systems and their characteristic equations and system functions?

✔ **17.   (a)**   Convince yourself that the diagram manipulations in Figure 6–15 are

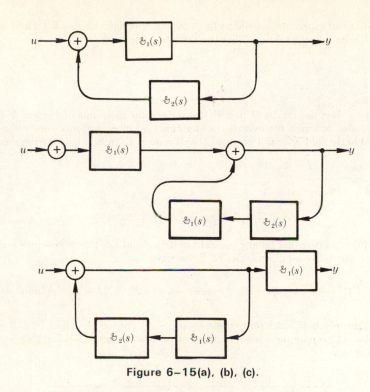

Figure 6–15(a), (b), (c).

equivalent to each other as far as input-output behavior is concerned. [Note order of operators $\mathcal{G}_1$ and $\mathcal{G}_2$.]

   (b)   Both Figures 6–15(b) and (c) contain a feedback portion, which is of the form shown in Figure 6–15(d). Show that this configuration has a transfer function (from A to B) of $(I - \mathcal{G}_3(s))^{-1}$ and hence that Figure 6–15(d) is equivalent (from A to B) to Figure 6–15(e).

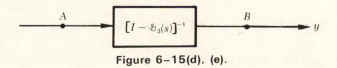

Figure 6–15(d), (e).

   (c)   Try to make sense out of the operator equation

$$(I - \mathcal{G}_3(s))^{-1} = I + \mathcal{G}_3(s) + [\mathcal{G}_3(s)]^2 + [\mathcal{G}_3(s)]^3 + \cdots$$

[*Hint:* Put a signal in at A and chase it around the feedback loop in Figure 6–15(d).]

Prove that this expression is valid by multiplying both sides by $I - \mathscr{B}_3(s)$ and getting $I = I$. Under what conditions does the infinite (power) series

$$\sum_{n=0}^{\infty} [\mathscr{B}(s)]^n$$

converge?

(d) Use the results of part (b) to reduce the diagrams of Figures 6–15 (b) and (c), and compute the overall transfer function of each. Since they must both be equal to that of Figure 6–15(a), which we found in Exercise 14, derive the theorem:

$$[I - \mathscr{B}_1(s)\mathscr{B}_2(s)]^{-1}\mathscr{B}_1(s) = \mathscr{B}_1(s)[I - \mathscr{B}_2(s)\mathscr{B}_1(s)]^{-1}$$

(e) Prove that

$$\det[I + \mathscr{B}_1(s)\mathscr{B}_2(s)] = \det[I + \mathscr{B}_2(s)\mathscr{B}_1(s)]$$

✔ **18.** **(a)** Prove that for any constant matrix F and any pair of complex numbers s and s_1 which are *not* eigenvalues of F, we have

$$(sI - F)^{-1}\frac{1}{s - s_1} = (s_1 I - F)^{-1}\frac{1}{s - s_1} - (sI - F)^{-1}(s_1 I - F)^{-1} \quad \text{(note order)}$$

[*Hint:* Multiply both sides on the left by $(sI - F)$ and on the right by $(s_1 I - F)$.]

(b) Compare the matrix identity of part (a) with the following identity for the *scalar* case:

$$\frac{1}{(s - A)}\frac{1}{s - s_1} = \frac{1}{(s_1 - A)}\frac{1}{s - s_1} - \frac{1}{(s - A)}\frac{1}{(s_1 - A)}$$

[*Hint:* Make a partial fraction expansion of the left-hand side. This shows that our matrix identity is a generalization of partial fractions.]

(c) Suppose that A and B are two matrices. Inspired by your scalar experience in part (b), seek an identity yielding another expression for

$$(sI - A)^{-1}(sI - B)^{-1} = ??$$

(d) Repeat part (c) for the product

$$(sI - C)(sI - D)^{-1} = ??$$

6–4 Block Diagrams and Signal Flow Graphs

Since the Laplace transform permits us to move freely back and forth from the time domain to the frequency domain, let us now see how our analog computer time-operation diagrams transform into diagrams involving complex-frequency operations.

Consider first an integrator in the time domain, shown in Figure 6–16. What is its transfer function? Recalling equation (22b) of Section 6–3, we have

$$\mathcal{L}\{\dot{x}(t)\} = sX(s) - x(0)$$

Figure 6-16 An integrator
in the time
domain.

and setting $x(0) = 0$ for zero state response, dividing output by input transforms gives

$$\mathfrak{H}(s) = \frac{X(s)}{sX(s)} = \frac{1}{s}$$

Thus, we say integration in the time domain corresponds to a multiplication by $\frac{1}{s}$ in the frequency domain, as shown in Figure 6-17.

Figure 6-17 Block diagram showing the transfer function of an integrator.

Quite often we get sloppy and mix up time domain quantities with *s*-domain quantities; we will use hybrid diagrams such as Figures 6-18(a) and (b) as long as no confusion arises.

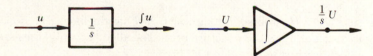

Figure 6-18(a) A mixed-up diagram. **(b)** Another mixed-up diagram.

We may now go back through this book and, wherever we find an analog computer diagram, replace each integration with a $\frac{1}{s}$ multiplication or *gain†* to get a frequency domain block diagram.

Example 1

Consider the one-dimensional system (a, b, c) studied in Example 6-3-2, with the time-domain diagram shown in Figure 6-9. The corresponding *s*-domain diagram is thus shown in Figure 6-19.

---

†Use of the term "gain" to stand for the quantity a signal gets multiplied by is commonplace in system theory; it grew out of years of working with block diagrams of electrical circuits, where most operational blocks represented amplification or attenuation (i.e., gains greater than 1 or less than 1, respectively).

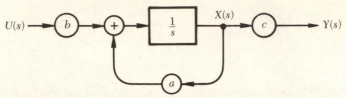

Figure 6–19   Block diagram corresponding to Figure 6–9.

If we wish to show even more economically the overall transfer function from input to output, we may use the result of Example 6–3–2 to get the block diagrams of Figures 6–20(a) and (b).   ◊

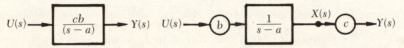

Figure 6–20(a)   Transfer function block diagram **(b)** Equivalent block diagram.

In Exercise 14 of Section 6–3 we derived expressions for the overall transfer functions of systems composed of series, parallel, or feedback connections of subsystems. Let us now use those results, along with some obvious manipulations of block diagrams, to illustrate another way to compute a system's transfer function.

Example 2

The system (F, G, H) studied in Example 6–3–7 has the block diagram shown in Figure 6–21, and has the structure of the so-called "controllable canonical form" which we first encountered in Figure 2–43.

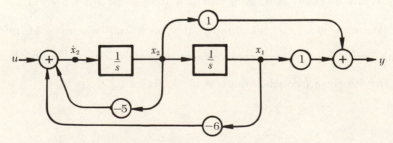

Figure 6–21   Block diagram of the (F, G, H) of Example 6–3–7 is of the controllable canonical form.

In Figures 6–22(a) through (d) we show a sequence of simple modifications of the block diagram, each of which clearly does not alter the influence of the input u upon the output y.

Now, the last block diagram clearly reveals a feedback portion in series with a pair of parallel branches.

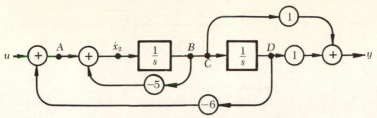

Figure 6-22(a)   Inserting another summer.

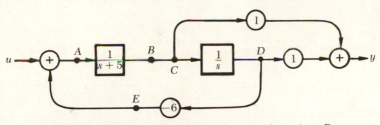

Figure 6-22(b)   Using transfer function from A to B.

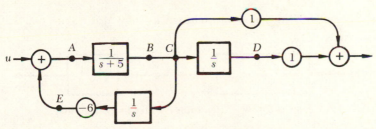

Figure 6-22(c)   Signal from C produces same effect at E as in preceding diagram.

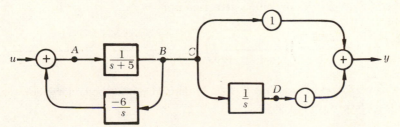

Figure 6-22(d)   Redrawn to exhibit a feedback subsystem cascaded with a pair of parallel branches.

Using Exercise 14(b) of Section 6-3, the transfer function of the parallel pair is just the sum of the two branch gains, $1 + (1)\dfrac{1}{s} = \dfrac{s+1}{s}$, so Figure 6-23 results.

For the feedback portion, Exercise 6-3-14(c) tells us that

$$\mathscr{Y}_B(s) = (I - \mathscr{Y}_1(s)\mathscr{Y}_2(s))^{-1}\mathscr{Y}_1(s), \text{ where } \mathscr{Y}_1(s) = \frac{1}{s+5}, \ \mathscr{Y}_2(s) = \frac{-6}{s},$$

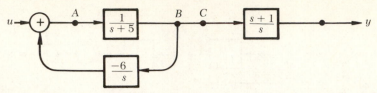

Figure 6–23   The result of combining the parallel branches of Figure 6–22(d).

and $\mathfrak{H}_B(s)$ is the transfer function from the input to point B:

$$\mathfrak{H}_B(s) = \left[1 + \frac{6}{s(s+5)}\right]^{-1} \frac{1}{s+5} = \left[\frac{s^2 + 5s + 6}{s(s+5)}\right]^{-1} \frac{1}{s+5}$$

$$= \frac{s(s+5)}{s^2 + 5s + 6} \frac{1}{(s+5)}$$

$$= \frac{s}{s^2 + 5s + 6}$$

Thus, Figure 6–23 may now be redrawn as shown in Figure 6–24.

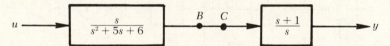

Figure 6–24   After using the feedback formula.

At last we have two subsystems in cascade, so we multiply their transfer functions to get the overall system function $\mathfrak{H}(s)$:

$$\mathfrak{H}(s) = \left(\frac{s+1}{s}\right)\left(\frac{s}{s^2 + 5s + 6}\right) = \frac{s+1}{s^2 + 5s + 6} = \frac{(s+1)}{(s+3)(s+2)}$$

just as we found in Example 6–3–7. The input-output behavior is summarily represented by Figure 6–25.

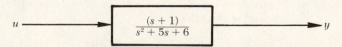

Figure 6–25   Overall block diagram of Figure 6–21.

◊

The determination of the overall operational characteristics (in this case the transfer function) of a system via step-by-step manipulation of its block diagram showing how signals flow from input to output through the various subsystems is a topological or graphical technique. Block diagrams and analog

computer diagrams are both special cases of the theory of **signal flow graphs**†
presented by Mason in 1953. In signal flow graphs we simplify our system
diagrams by eliminating the need for drawing blocks and summing elements.
We simply replace each block by a path with an arrow indicating the direction
in which the signal flows, and just above the arrow we write the operation
done by the block.

Example 3

The signal flow graph corresponding to the transfer function block diagram
of Figure 6–25 is shown in Figure 6–26.   ◊

Figure 6–26   Signal flow
graph of Figure
6–25.

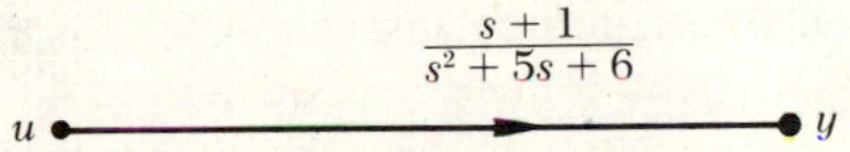

$$u \bullet \xrightarrow{\quad\quad\dfrac{s+1}{s^2+5s+6}\quad\quad} \bullet y$$

Notice that a **path** starts and ends at a **node,** and that the arrows represent
unidirectional flow of a signal from the originating node through the indicated operation (called the path **gain** or **transmittance**) to the terminal node.
At each node there may be several incoming branches and several outgoing
branches. The *signal at a node* is defined to be *the sum of the signals due to all
incoming branches at that node.* This convention lets us do away with summing
devices.

Example 4

The signal flow graph of the block diagram of Figure 6–21 is shown in Figure
6–27. Notice that the signal $\dot{x}_2$ now appears at the node shown and is given

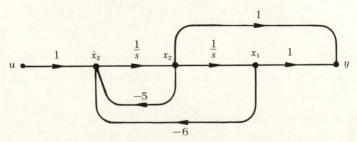

Figure 6–27   Signal flow graph of Figure 6–21.

---

† See Zadeh and Desoer's *Linear System Theory* for references and derivations. Signal flow graphs
are themselves subsumed under the more general theories of *oriented graphs* and *graph theory.*
See D. E. Johnson and J. R. Johnson, *Graph Theory* (Ronald Press, 1972) for an elementary
survey of these areas.

by

$$\dot{x}_2 = -6x_1 - 5x_2 + 1u$$

which is the sum of signals at nodes x_1, x_2, and u operated on by the respective path transmittances -6, -5, and $+1$.   ◊

Since most systems represented by block diagrams or signal flow graphs are made up of certain commonly recurring subsystems, it pays to develop facility with a small number of easily recognizable subsystem structures. [See Exercise 2.]

As a matter of fact, it is possible, as we shall now see, to compute the overall gain or transmittance of a signal flow graph using *Mason's rule,*† and with a little practice at picking out forward paths and loops, transfer functions can often be written by inspection.

A **path** is a connected sequence of branches which can be traversed without traversing any branch in the wrong direction. A **forward path** between two nodes is a path along which no node is met twice. A **loop** is a path starting and ending on the same node and not meeting any other node more than once. The **path gain** is the total transmittance of the path, and the **loop gain** is the total transmittance of the loop.

Mason's rule for overall gain states that

$$\mathcal{G} = \frac{1}{\Delta} \sum_{k=1}^{q} \mathcal{G}_k \Delta_k$$

where

Δ = the graph determinant

= 1 − (sum of all different loop gains)

   + (sum of gain products of all possible combinations of
      two non-touching loops)

   − (sum of gain products of all possible combinations of
      three non-touching loops)

   + ⋯;

$\mathcal{G}_k$ = gain of the kth forward path from input node to output node;

Δ_k = value of Δ for that part of the graph which does not touch

      the kth forward path;

q = number of forward paths.

Two loops are said to be **non-touching** if they do not have any node in common.

---

†Or a similar rule due to Coates; see D. E. Johnson and J. R. Johnson, *Graph Theory* (Ronald Press, 1972). Mason's rule is derived in Zadeh and Desoer's *Linear System Theory* (McGraw Hill, 1963), p. 462.

Example 5

The system of Figure 6–28

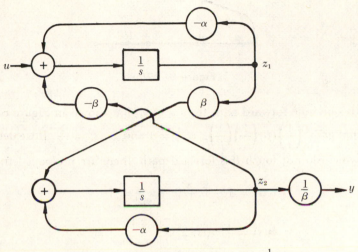

Figure 6–28   A real realization of $\dfrac{1}{(s + \alpha)^2 + \beta^2}$.

is a realization of the transfer function $\mathscr{G}(s) = \dfrac{1}{(s + \alpha)^2 + \beta^2}$ which employs only *real* coefficients, α and β.

This system has three loops, as shown in Figure 6–29(a), with loop gains of $-\dfrac{\alpha}{s}$, $-\dfrac{\beta^2}{s^2}$, and $-\dfrac{\alpha}{s}$, respectively. There are only two loops (namely

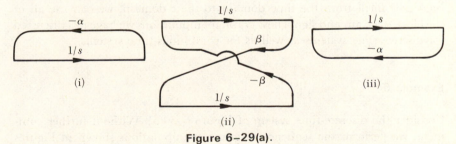

Figure 6–29(a).

(i) and (iii)) that do not touch each other, so

$$\Delta = 1 - \left(-\frac{\alpha}{s} - \frac{\beta^2}{s^2} - \frac{\alpha}{s} \right) + \left[\left(-\frac{\alpha}{s} \right)\left(-\frac{\alpha}{s} \right) \right] - 0 + \cdots$$

$$= \frac{s^2 + 2\alpha s + \beta^2 + \alpha^2}{s^2}$$

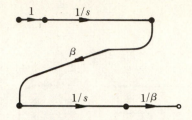

Figure 6–29(b).

There is only one forward path from u to y, that shown in Figure 6–29(b), with total gain $\left(\dfrac{1}{s}\right)(\beta)\left(\dfrac{1}{s}\right)\left(\dfrac{1}{\beta}\right)$, so $q = 1$ and $\mathcal{Y}_1(s) = \dfrac{1}{s^2}$. If we delete the loops which do not touch this forward path, there are no loops left, so

$$\Delta_1 = 1 - 0 + 0 - \cdots = 1$$

Thus,

$$\mathcal{Y}(s) = \frac{1}{\Delta}\,\mathcal{Y}_1(s)\,\Delta_1$$

$$= \frac{s^2}{s^2 + 2\alpha s + \beta^2 + \alpha^2} \cdot \frac{1}{s^2} \cdot 1$$

$$= \frac{1}{(s + \alpha)^2 + \beta^2}$$

as claimed.   ◇

If we recall from Section 4–6 that the Z-transform permitted us to move back and forth from the time domain to the z-domain, we can use all of the block diagram and signal flow graph manipulations we have just discussed for discrete-time systems as well as for continuous-time systems.

Example 6

Consider the discrete-time system of Figure 6–30(a). Without further comment, we perform the sequence of graph manipulations shown in Figures 6–30(b), (c), (d), (e), and (f).
The overall discrete-time transfer function is found to be

$$\mathcal{Y}(z) = \frac{z(ac + bd) + bc - \alpha(ac + bd)}{(z - \alpha)^2}$$

We point out that this system represents a Jordan block or chain, and was studied in Example 7 of Section 4–3. The reader should review that example

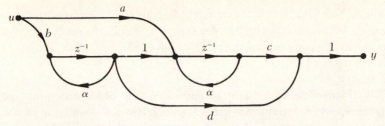

Figure 6–30(a)   A discrete-time signal flow graph.

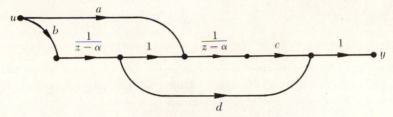

Figure 6–30(b).

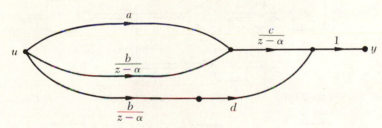

Figure 6–30(c).

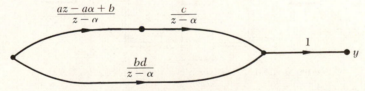

Figure 6–30(d).

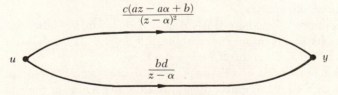

Figure 6–30(e).

$$\frac{z(ac + bd) + bc - \alpha(ac + bd)}{(z - \alpha)^2}$$

$u \bullet \longrightarrow \bullet y$

Figure 6–30(f)   The overall transfer function $\mathscr{G}(z)$.

and notice here that if $b = 0$, corresponding to loss of controllability, then there is a cancellation in the transfer function. Similarly we see that loss of observability, corresponding to $c = 0$, also shows up as a cancellation in the transfer function.† ◊

It will often be convenient to employ *matrix* block diagrams or flow graphs to represent dynamic systems (F, G, H). Instead of drawing all n scalar integrators, all n^2 elements of F, and so on, for the components of the equations

$$\begin{cases} \dot{x} = Fx + Gu \\ y = Hx \end{cases}$$

we represent the matrix operations on the vectors x, y, u as illustrated in Figures 6–31(a) and (b). [Sometimes, as here, we use "fat" arrows to remind us that we are dealing with vectors instead of scalars.]

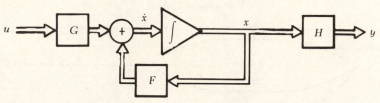

Figure 6–31(a)   Matrix analog computer diagram of (F, G, H).

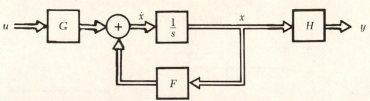

Figure 6–31(b)   Matrix block diagram of (F, G, H).

In these matrix diagrams the "integrator" or "$\frac{1}{s}$ block," in fact represents n scalar integrators.

Sometimes, when there is room on the diagram, we label each matrix block with its actual value. For example, in Figure 6–32 we show the equivalent matrix block diagram of Figure 6–21.

In the remainder of this book we will use signal flow graphs, block diagrams, or analog computer diagrams, in either scalar or matrix versions, as the spirit moves us. The reader should have no trouble moving back and forth from one representation to another, and should not be disturbed if we use regular lines rather than "fat lines" in matrix block diagrams.

---

†In Section 7–3 we prove that observability and controllability of a single-input, single output system together imply that there are no cancellations in the transfer function.

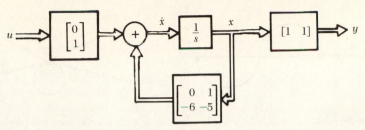

Figure 6-32   Matrix block diagram for Figure 6-21.

EXERCISES FOR SECTION 6-4

✔ **1. (a)**   Replace integrators by $\frac{1}{s}$ gains in the analog computer system of Exercise 3 of Section 2–5 and draw the block diagram which corresponds to Figure 2–43.

(b)   Manipulate the diagram as we did in the steps of Example 2 to compute the overall transfer function.

$$\left[Answer: \ \mathfrak{Y}(s) = \frac{b_1 s^2 + b_2 s + b_3}{s^3 + a_1 s^2 + a_2 s + a_3} \right]$$

This means that we can think of this system as a realization of this transfer function. [It is called the *controllable canonical realization* of $\mathfrak{Y}(s)$.]

(c)   From the transfer function, give the differential equation relating input to output, and compare with the ways you found it in Exercise 2–5–3 and Exercise 6–3–13.

(d)   From the diagram of part (a), write the dynamic state equations; and from the matrices (F, G, H) compute $(sI - F)$, adjugate$(sI - F)$, $(sI - F)^{-1}$, $(sI - F)^{-1}G$, and $H(sI - F)^{-1}G = \mathfrak{Y}(s)$.

(e)   Describe how you would find a realization (i.e., choose state variables) and an analog computer representation of an nth order, proper rational transfer function

$$\mathfrak{Y}(s) = \frac{b_1 s^{n-1} + b_2 s^{n-2} + \cdots + b_{n-1} s + b_n}{s^n + a_1 s^{n-1} + a_2 s^{n-2} + \cdots + a_{n-1} s + a_n}$$

Notice that this choice (F, G, H) will have F as the *companion matrix* (Exercise 6–2–10(b)) of the denominator of $\mathfrak{Y}(s)$; G will be the unit vector $e_n = [0 \ 0 \ \cdots \ 0 \ 1]^T$; and H is just the string of numerator coefficients. [*Answer:* See Figure 7–6 in Section 7–1.]

✔ **2.**   Here we study some common flow graph structures shown in Figure 6–33.

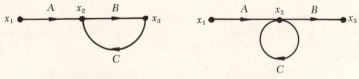

Figure 6-33(a)   A loop. **(b)** A self-loop at a node.

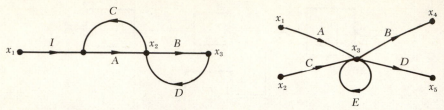

Figure 6–33(c) Cascaded loop. **(d)** Self-loop with several incoming branches.

   (a)   Draw equivalent block diagrams using summers and boxes labeled with appropriate transmittances $A, B, C, \ldots$. [*Hint:* Figure 6–33(b) is equivalent to the block diagram of Figure 6–34(a).]
   (b)   Do simple graph manipulations to get equivalent overall graphs for each system of part (a). [*Hint:* Figure 6–33(d) is equivalent to Figure 6–34(b), where I is the identity.]

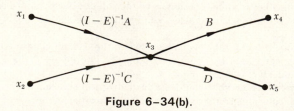

Figure 6–34(a).

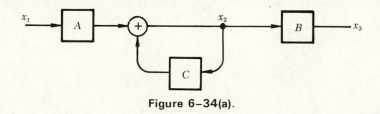

Figure 6–34(b).

   (c)   Figure 6–33(c) lends itself nicely to the sequence of manipulations shown in Figure 6–35. [Note order of transmittances!] So $x_3 = \{B[I - (AC + DB)]^{-1}A\}x_1$. Check each step.
   (d)   Replace the second step of part (c) with the signal flow graph of Figure 6–36, and get another expression for the overall transmittance:

$$x_3 = \{(I - BD)^{-1}B(I - AC)^{-1}A\}x_1$$

What operator identity did you just "derive"? Check by letting $C = D = 0$. What happens if $A = B = I$? Do you believe it? What was wrong with this new second step?
3.   **(a)**   Use Mason's rule to find $\mathfrak{d}(s)$ for the graph of Figure 6–27. [*Hint:* There are two forward paths and two loops; each loop touches each forward path.]
   (b)   Use Mason's rule on the three-dimensional system you drew in Exercise 1(a) and find $\mathfrak{d}(s)$.
   (c)   Use Mason's rule to find $\mathfrak{d}(s)$ for an n-dimensional flow graph such as you used in Exercise 1(e). [*Hint:* There are n forward paths and n loops; each loop

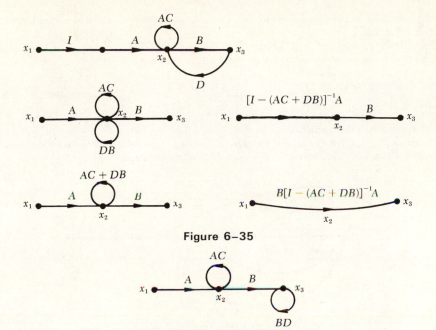

Figure 6–35

Figure 6–36

touches each forward path, so $\Delta_k = 1$; $\Delta = 1 + \dfrac{a_1}{s} + \dfrac{a_2}{s^2} + \cdots + \dfrac{a_n}{s^n}$; $\mathcal{Y}_k(s) = \dfrac{b_k}{s^k}.$]

4.   Find the overall transfer function $\mathcal{Y}(s)$ for the system of Figure 6–37.

   (a)   Use Mason's rule. [*Hint:* There are three forward paths and three loops; all loops touch each other; all loops touch each forward path, so $\Delta_1 = \Delta_2 = \Delta_3 = 1.$]

   (b)   Use simple graph manipulations and the results from Exercise 2 to find $\mathcal{Y}(s)$ and check your work in (a).

   (c)   From the flow graph, write the matrix dynamic equations (F, G, H) for this system and compare with the system of Exercise 1(a), which had the same transfer function.

   (d)   Draw a graph, labelling state variables, and give the matrices (F, G, H) for this approach to realizing the *n*th-order proper rational transfer function you realized with the controllable canonical realization in Exercise 1(e). This latest realization is called the *observable canonical realization*. Compare the graphs of these

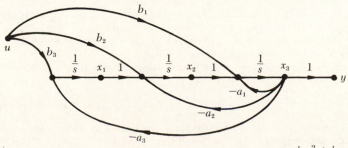

Figure 6–37   The observable canonical realization of $\mathcal{Y}(s) = \dfrac{b_1 s^2 + b_2 s + b_3}{s^3 + a_1 s^2 + a_2 s + a_3}.$

two canonical realizations. Except for labelling and direction of arrows, don't they look a lot alike?

5. Figures 6–38 and 6–39 present two flow graphs, with state variables labelled, which give two more realizations of the transfer function

$$\mathfrak{H}(s) = \frac{b_1 s^2 + b_2 s + b_3}{s^3 + a_1 s^2 + a_2 s + a_3}$$

(You saw two other realizations in Exercises 1 and 4.) Show that you can specify the gains $\beta_1, \beta_2, \beta_3$ in Figure 6–38 and $\alpha_1, \alpha_2, \alpha_3$ in Figure 6–39 to make both systems realize $\mathfrak{H}(s)$. Give the matrices (F, G, H) for each realization.

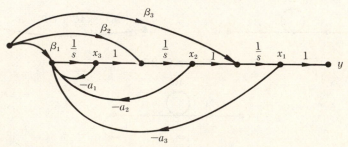

Figure 6–38 A realization of $\mathfrak{H}(s)$.

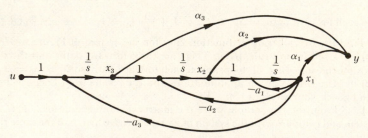

Figure 6–39 Another realization of $\mathfrak{H}(s)$.

✔ **6.** In the matrix block diagram for (F, G, H) shown in Figure 6–31(b), suppose that you let $u = Kx$, where K is a matrix of the right size.

(a) Draw the matrix block diagram for this "closed loop" system. [We say that you have just introduced linear state feedback.]

(b) Write the matrix differential equations describing the closed loop system. How does K affect the eigenvalues of the closed loop system?

(c) Manipulate the block diagram of part (a) to get a feedback branch (involving K and G) directly in parallel with F.

Combine these branches to get a new diagram like Figure 6–31(b) but now described by the matrices $(F + GK, G, H)$.

(d) Let $F = \begin{bmatrix} -2 & 0 \\ 0 & -3 \end{bmatrix}$, $G = \begin{bmatrix} 1 \\ 1 \end{bmatrix}$, $H = [-1 \quad 2]$ and find the modes of the open loop system (F, G, H). Introduce state feedback $u = Kx$ as above, and find a K such that the eigenvalues of the closed loop system $(F + GK, G, H)$ are -1 and -10. [*Hint:* Just call $K = [k_1 \quad k_2]$. Set $\chi_{F+GK}(\lambda)$ equal to what you want and solve for k_1 and k_2.] [*Answer:* $K = [8 \quad -14]$.]

6-5   Scalar Transfer Functions

In this section we investigate the properties of single-input, single-output systems (F, G, H), where G is an $(n \times 1)$ column matrix, H is a $(1 \times n)$ row matrix, and F is $(n \times n)$.

For such systems the state transfer function $\mathfrak{S}_X(s) = (sI - F)^{-1}G$ is a column matrix, and the transfer function

$$\mathfrak{S}(s) = H\mathfrak{S}_X(s) = H(sI - F)^{-1}G$$

is a scalar function of s.

We shall be interested in seeing how various properties of the matrices F, G, and H are reflected in the transfer function and, alternatively, how to synthesize a given $\mathfrak{S}(s)$ by finding matrices F, G, and H such that the system (F, G, H) has $\mathfrak{S}(s)$ as its transfer function.

POLES AND ZEROS

For a given system (F, G, H), the transfer function $\mathfrak{S}(s)$ may be computed from the given matrices F, G, and H by

$$\mathfrak{S}(s) = H(sI - F)^{-1}G$$

By Cramer's rule, $(sI - F)^{-1}$ can be expressed in terms of its adjugate† matrix, adj$(sI - F)$, by the formula

$$(sI - F)^{-1} = \frac{1}{\det(sI - F)} \text{adj}(sI - F) \tag{1}$$

where the i, j element of adj$(sI - F)$ is the cofactor given by

$$[\text{adj}(sI - F)]_{ij} = (-1)^{i+j} \det(M_{ji}) \tag{2}$$

and M_{ji} is the $(n - 1)$ by $(n - 1)$ submatrix obtained by deleting the jth row and the ith column of $(sI - F)$. Since $(sI - F)$ is an $n \times n$ matrix with the only s's occurring in the diagonal entries $(s - f_{ii})$, M_{ji} will have one less s term in it than $(sI - F)$, and thus $\det(M_{ji})$ will be a polynomial of degree at most $n - 1$. Since $\det(sI - F) = \chi_F(s)$ (the characteristic equation of F)

---

† Sometimes called the conjoint of F or the *classical adjoint* so as not to confuse it with the adjoint of an operator, which we studied in Section 4–4. See Serge Lang, *Linear Algebra,* 2nd edition (Addison Wesley, 1971), p. 182 and p. 198.

is an nth degree polynomial in s, and since (2) shows that each entry of $\text{adj}(sI - F)$ is a polynomial of degree $\leq (n - 1)$, we see from equation (1) that the entries of $(sI - F)^{-1}$ are *proper rational functions;* i.e., the ratios of two polynomials in s with denominators of higher degree than numerators. At this point, the reader should go back to Example 7 of Section 6–3 and use equations (1) and (2) to compute the matrix $(sI - F)^{-1}$ found in that example. In Exercise 1 we present a recursive method of computing $\text{adj}(sI - F)$ and $(sI - F)^{-1}$ which avoids the evaluation of all those $(n - 1)$ by $(n - 1)$ determinants called for in equation (2).

For scalar systems, H is a row and G is a column, so $(sI - F)^{-1}G$ is just a linear combination of the columns of $(sI - F)^{-1}$; hence, it is a column of rational functions as in Example 6–3–7. Multiplying this column of rational functions by the row matrix H yields the scalar proper rational transfer function $\mathfrak{H}(s)$ of the form

$$\mathfrak{H}(s) = \frac{b_1 s^{n-1} + b_2 s^{n-2} + \cdots + b_{n-1}s + b_n}{s^n + a_1 s^{n-1} + a_2 s^{n-2} + \cdots + a_{n-1}s + a_n} \tag{3}$$

where $\chi_F(s) = s^n + a_1 s^{n-1} + \cdots + a_{n-1}s + a_n$.

If $\mathfrak{H}(\lambda) = 0$ for some complex number λ, we say that λ is a **zero of** $\mathfrak{H}(s)$. We call λ a **pole of** $\mathfrak{H}(s)$ if $|\mathfrak{H}(\lambda)| = \infty$. Factoring both numerator and denominator polynomials in equation (3),

$$\mathfrak{H}(s) = b \frac{(s - s_1)(s - s_2) \cdots (s - s_m)}{(s - p_1)(s - p_2) \cdots (s - p_n)} \tag{4}$$

(where b is the non-zero coefficient of the highest power s^m actually present in the numerator), displays the zeros and poles as the roots $\{s_1, \ldots s_m\}$ and $\{p_1, \ldots p_n\}$ respectively.

For example, we saw in Example 6–3–7 that

$$\mathfrak{H}(s) = \frac{s + 1}{s^2 + 5s + 6} = \frac{s + 1}{(s + 2)(s + 3)}$$

so that system had a zero at $s = -1$ and poles at $s = -2$ and -3.

Since knowing the poles and zeros determines $\mathfrak{H}(s)$ to within the "scale factor" b in equation (4), we often display them graphically in the complex plane as a "pole-zero" diagram. To illustrate, we show in Figure 6–40 the pole-zero diagram for the $\mathfrak{H}(s)$ of Example 6–3–7.

Clearly, every pole is a root of $\chi_F(s)$, and hence is an eigenvalue of F. The converse is not true, however, since some eigenvalue of F, say λ_j, might turn out also to be a root of the numerator so that $\lambda_j = p_j = s_j$; and because of this **pole zero cancellation,**† $|\mathfrak{H}(\lambda_j)| \neq \infty$. The reader should notice in

---

† As we shall see in Section 7–3, such cancellations in the transfer function have great significance for the system theorist, since they reveal a lack of controllability or observability and may hide an unstable mode.

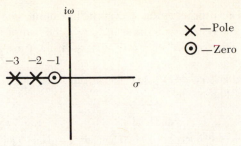

Figure 6–40   Pole zero diagram for Example 6–3–7.

Example 6–3–7 that we could have picked the H matrix to cause such a cancellation; changing H from $[1 \quad 1]$ to $[2 \quad 1]$ does not affect $(sI - F)^{-1}G$ at all but yields $\mathscr{Y}(s) = \dfrac{s + 2}{(s + 2)(s + 3)} = \dfrac{1}{s + 3}$, in which case the eigenvalue -2 is not a pole.

With this interpretation, and recalling Definition 6–2–2, the poles of the transfer function are modes or natural frequencies of the system. What, then, is a zero? By the defining property of the transfer function given in Theorem 6–3–1, when $u(t) = e^{s_0 t}$ is applied for $t > t_0 = -\infty$, the steady state response is

$$\mathscr{S}(t, -\infty, 0, u) = \mathscr{Y}(s_0)e^{s_0 t}$$

It is clear that this response will be zero if s_0 is a zero of $\mathscr{Y}(s)$. To illustrate, consider Example 6–3–7 again. From

$$y(t) = \mathscr{S}(t, t_0, x_0, u) = He^{F(t-t_0)}x_0 + \int_{t_0}^{t} g(t - \tau)u(\tau)\, d\tau$$

since F has eigenvalues -2 and -3,

$$\lim_{t_0 \to \infty} e^{F(t-t_0)} = 0 \quad \text{(verify)}$$

so the steady state response exists and is given by

$$\mathscr{S}(t, -\infty, 0, u) = \int_{-\infty}^{t} g(t - \tau)u(\tau)\, d\tau$$

For $u(t) = e^{s_0 t}$, the exponential, we have

$$\mathscr{S}(t, -\infty, 0, u) = \mathscr{Y}(s_0)e^{s_0 t} = \frac{s_0 + 1}{(s_0 + 2)(s_0 + 3)}\, e^{s_0 t}$$

and for $u(t) = e^{-1t}$, with $t \geq -\infty$,

$$\mathscr{S}(t, -\infty, 0, u) = \mathscr{Y}(-1)e^{-1t} = 0$$

Thus, if $u(t)$ is any input having a certain harmonic content given by the transform $U(s)$, this system will "filter out" of u the component with $s = -1$. What happens if we apply e^{-1t} for only a finite time rather than from $t_0 = -\infty$?

Suppose that $u(t) = e^{s_0 t}1(t)$, so $e^{s_0 t}$ is applied only for $t > 0$, letting $t_0 = 0$ for convenience. Then

$$y(t) = \mathcal{S}(t, 0, x_0, u) = He^{Ft}x_0 + \int_0^t g(t - \tau)u(\tau)\, d\tau$$

and, in terms of transforms,

$$Y(s) = H(sI - F)^{-1}x_0 + \mathcal{G}(s)U(s)$$
$$= H(sI - F)^{-1}x_0 + H(sI - F)^{-1}GU(s)$$

where

$$U(s) = \mathcal{L}\{e^{s_0 t}1(t)\} = \frac{1}{s - s_0}$$

But, by using H times the expression in equation (26) of Section 6–3, which we encountered in our study of mode suppression, we can rewrite

$$Y(s) = H\left\{(sI - F)^{-1}x_0 + (sI - F)^{-1}\frac{1}{s - s_0}G\right\}$$

as

$$Y(s) = H(sI - F)^{-1}[x_0 - \mathcal{G}_x(s_0)] + H\mathcal{G}_x(s_0)\frac{1}{s - s_0}$$

$$= H(sI - F)^{-1}[x_0 - (s_0 I - F)^{-1}G] + \mathcal{G}(s_0)\frac{1}{s - s_0}$$

By choosing $x_0 = (s_0 I - F)^{-1}G$, we can suppress all modes of F and get

$$y(t) = \mathcal{G}(s_0)e^{s_0 t}1(t)$$

so that the output looks like $e^{s_0 t}1(t)$ even for small t and not only in the steady state. Moreover, the transfer function $\mathcal{G}(s)$ evaluated at $s = s_0$ tells us "how much" $e^{s_0 t}1(t)$ is in the output. This is hardly surprising when we observe that we have taken x_0 to be the state to which $e^{s_0 t}$ would have driven the system starting from the zero state at time $t_0 = -\infty$.

Clearly, if we choose s_0 to be a zero of $\mathcal{G}(s)$, so that $\mathcal{G}(s_0) = 0$, then the response to the input $e^{s_0 t}1(t)$ can be made to vanish identically for all $t \geq 0$ by selecting x_0 in the manner described.

REALIZATIONS OF TRANSFER FUNCTIONS

In Examples 6-3-7 and 6-4-2 we saw a system described by the dynamic matrix representation (F, G, H), where

$$F = \begin{bmatrix} 0 & 1 \\ -6 & -5 \end{bmatrix}, \; G = \begin{bmatrix} 0 \\ 1 \end{bmatrix}, \text{ and } H = [1 \quad 1]$$

The system is represented by the block diagram of Figure 6-21. We found that the transfer function of this system was

$$\mathfrak{H}(s) = \frac{s+1}{s^2 + 5s + 6}$$

and, in general, that given any system represented either by (F, G, H) or by a block diagram, we can compute the transfer function by $\mathfrak{H}(s) = H(sI - F)^{-1}G$, by graphical manipulations, or by Mason's rule.

Suppose instead that we are given a scalar transfer function $\mathfrak{H}(s)$. Can we find a system (F, G, H) which has the given $\mathfrak{H}(s)$ as its transfer function? If the given $\mathfrak{H}(s)$ is a proper rational function (i.e., the ratio of two polynomials with numerator of degree no greater than the denominator), the reader who has worked Exercises 1, 4, or 5 of Section 6-4 already knows that the answer is yes. In those exercises four distinct realizations were presented for the same transfer function $\mathfrak{H}(s)$; the first two can be written by inspection directly from the coefficients of the numerator and denominator polynomials comprising $\mathfrak{H}(s)$.

For example, if we were given

$$\mathfrak{H}(s) = \frac{s+1}{s^2 + 5s + 6}$$

the *controllable canonical realization* of Exercise 6-4-1 (try it) results in the block diagram of Figure 6-21 and the matrices (F, G, H) of Example 6-3-7. In later sections we shall make frequent use of this canonical form, as well as the *observable canonical form* of Exercise 6-4-4.

For the present, however, let us consider a few simple manipulations with the polynomials of $\mathfrak{H}(s)$ which result in other useful realizations.

Suppose that we factor the denominator polynomial (this is always possible if we allow complex roots):

$$s^2 + 5s + 6 = (s+3)(s+2)$$

and write

$$\mathfrak{H}(s) = \frac{(s+1)}{(s+3)(s+2)} = \left(\frac{1}{s+3}\right)\left(\frac{s+1}{s+2}\right)$$

a product which can be thought of as corresponding to two subsystems in cascade as shown in Figure 6–41(a). Now, each of these subblocks looks easy;

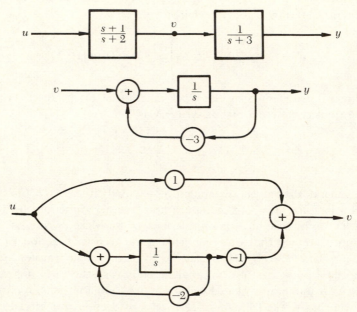

Figure 6–41 **(a), (b), (c).**

in fact, $\dfrac{1}{s+3}$ corresponds to Figure 6–41(b), and $\dfrac{s+1}{s+2} = \dfrac{(s+2)-1}{(s+2)}$ $= 1 - \dfrac{1}{s+2}$ corresponds to Figure 6–41(c), so that we can draw the realization of Figure 6–42.

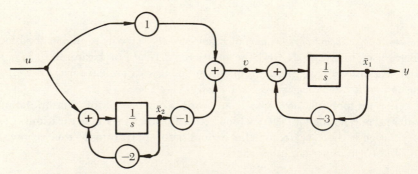

Figure 6–42 A cascade realization of the $\mathscr{G}(s)$ of Figure 6–41(a).

Now, to find the matrices $(\bar{F}, \bar{G}, \bar{H})$ of this realization we must pick the state variables. Since the $\dfrac{1}{s}$ blocks are integrators, we can safely take their outputs $\bar{x}_1$, $\bar{x}_2$ as state variables, and from the diagram we write

$$\dot{\bar{x}}_2 = u - 2\bar{x}_2, \quad \dot{\bar{x}}_1 = u - 3\bar{x}_1 - \bar{x}_2, \quad y = \bar{x}_1$$

whence

$$\bar{F} = \begin{bmatrix} -3 & -1 \\ 0 & -2 \end{bmatrix}, \quad \bar{G} = \begin{bmatrix} 1 \\ 1 \end{bmatrix}, \quad \bar{H} = [1 \quad 0]$$

This example illustrates that it is possible to realize a given $\mathcal{B}(s)$ as a series connection of simple, first-order, basic blocks of the form $\dfrac{as + b}{s + c}$ and that corresponding matrices ($\bar{F}, \bar{G}, \bar{H}$) will have $\bar{F}$ *triangular,* with the denominator roots of $\mathcal{B}(s)$ on the diagonal. We could call this realization a "triangular canonical form." [See Exercise 2.]

Now let's try some other tricks on $\mathcal{B}(s)$ and see what happens. Instead of decomposing $\mathcal{B}(s)$ as a product, suppose we use partial fractions to write it as a *sum:*

$$\mathcal{B}(s) = \frac{s + 1}{(s + 3)(s + 2)} = \frac{2}{s + 3} + \frac{-1}{s + 2}$$

as the reader may readily verify by adding the two fractions on the right-hand side. This suggests the connection of two subsystems in parallel, as shown in Figure 6–43.

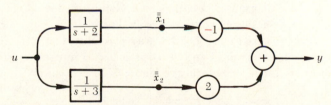

Figure 6–43   A parallel realization of the $\mathcal{B}(s)$ of Figure 6–41(a).

If the first order sub-blocks are broken down, Figure 6–44 results, and the outputs of the blocks are seen to be the outputs of integrators and hence respectable state variables.

From the diagram of Figure 6–44 we may write by inspection the matrices ($\bar{\bar{F}}, \bar{\bar{G}}, \bar{\bar{H}}$):

$$\bar{\bar{F}} = \begin{bmatrix} -2 & 0 \\ 0 & -3 \end{bmatrix}, \quad \bar{\bar{G}} = \begin{bmatrix} 1 \\ 1 \end{bmatrix}, \quad \bar{\bar{H}} = [-1 \quad 2]$$

and, lo and behold, $\bar{\bar{F}}$ is diagonal [Exercise 3]. The partial fraction expansion of $\mathcal{B}(s)$ exposes the modes e^{-3t} and e^{-2t} in uncoupled form. We thus see that

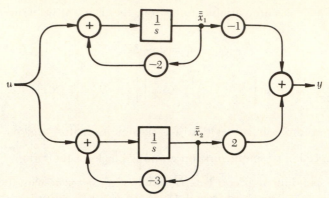

Figure 6–44   State selection for parallel realization.

if $\mathfrak{F}(s)$ has distinct poles,† we can obtain a diagonal representation simply by drawing the block diagram with one state variable for each distinct summand in the partial fraction expansion of $\mathfrak{F}(s)$.

What happens if the denominator of a $\mathfrak{F}(s)$ with real coefficients has complex roots? In this case, the roots occur in conjugate pairs, and the partial fraction expansion results in complex coefficients. For example, if

$$\mathfrak{F}(s) = \frac{1}{(s + \alpha)^2 + \beta^2} = \frac{\dfrac{i}{2\beta}}{s + \alpha + i\beta} + \frac{-\dfrac{i}{2\beta}}{s + \alpha - i\beta} \quad \text{(verify)}$$

we would get the diagram of Figure 6–45. Complex roots, state variables and complex representation matrices result. Fortunately, however, we saw in Example 6–4–5 and Figure 6–28 a way to realize this $\mathfrak{F}(s)$ and retain *real* coefficients (if we choose) and still use just two integrators. Instead of complex matrices, the representation of Figure 6–28 with states z_1, z_2 as shown

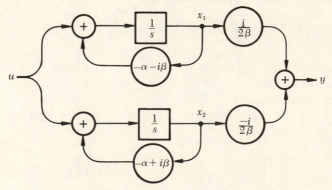

Figure 6–45   A complex diagonal realization.

---

†This corresponds to the original F having distinct eigenvalues and no cancellation.

has

$$F = \begin{bmatrix} -\alpha & -\beta \\ \beta & -\alpha \end{bmatrix}, G = \begin{bmatrix} 1 \\ 0 \end{bmatrix}, H = \begin{bmatrix} 0 & \dfrac{1}{\beta} \end{bmatrix}$$

where F is real but not diagonal.

It will be instructive to try to relate our finding different realizations of $\mathcal{G}(s)$ to changing the basis of the state space.

Example 1

Consider the controllable canonical representation (F, G, H) of Figure 6–21 and the diagonal representation $(\bar{\bar{F}}, \bar{\bar{G}}, \bar{\bar{H}})$ of Figure 6–44, obtained by partial fractions, both of which correspond to $\mathcal{G}(s) = \dfrac{(s + 1)}{(s + 2)(s + 3)}$.

How are the state variables x_1, x_2 related to $\bar{\bar{x}}_1$, $\bar{\bar{x}}_2$? From (F, G, H) we can write $y = x_1 + x_2$, $\dot{x}_1 = x_2$, and $\dot{x}_2 = -6x_1 - 5x_2 + u$. We can solve [Exercise 5] these equations to express x_1 and x_2 in terms of y and u:

$$\begin{bmatrix} x_1 \\ x_2 \end{bmatrix} = \begin{bmatrix} -2 & -\dfrac{1}{2} \\ 3 & \dfrac{1}{2} \end{bmatrix} \begin{bmatrix} y \\ \dot{y} - u \end{bmatrix}$$

From $(\bar{\bar{F}}, \bar{\bar{G}}, \bar{\bar{H}})$ we can express y (and hence $\dot{y}$) in terms of $\bar{\bar{x}}_1$ and $\bar{\bar{x}}_2$. Since $y = -\bar{\bar{x}}_1 + 2\bar{\bar{x}}_2$, $\dot{\bar{\bar{x}}}_1 = -2\bar{\bar{x}}_1 + u$, and $\dot{\bar{\bar{x}}}_2 = -3\bar{\bar{x}}_2 + u$, we have

$$\begin{aligned} \dot{y} &= -\dot{\bar{\bar{x}}}_1 + 2\dot{\bar{\bar{x}}}_2 \\ &= -(-2\bar{\bar{x}}_1 + u) + 2(-3\bar{\bar{x}}_2 + u) \\ &= 2\bar{\bar{x}}_1 - 6\bar{\bar{x}}_2 + u \end{aligned}$$

so

$$\begin{bmatrix} y \\ \dot{y} - u \end{bmatrix} = \begin{bmatrix} -1 & 2 \\ 2 & -6 \end{bmatrix} \begin{bmatrix} \bar{\bar{x}}_1 \\ \bar{\bar{x}}_2 \end{bmatrix}$$

Thus

$$\begin{bmatrix} x_1 \\ x_2 \end{bmatrix} = \begin{bmatrix} -2 & -\dfrac{1}{2} \\ 3 & \dfrac{1}{2} \end{bmatrix} \begin{bmatrix} -1 & 2 \\ 2 & -6 \end{bmatrix} \begin{bmatrix} \bar{\bar{x}}_1 \\ \bar{\bar{x}}_2 \end{bmatrix}$$

$$\begin{bmatrix} x_1 \\ x_2 \end{bmatrix} = \begin{bmatrix} 1 & -1 \\ -2 & 3 \end{bmatrix} \begin{bmatrix} \bar{\bar{x}}_1 \\ \bar{\bar{x}}_2 \end{bmatrix}$$

Setting $P = \begin{bmatrix} 1 & -1 \\ -2 & 3 \end{bmatrix}$ with inverse $P^{-1} = \begin{bmatrix} 3 & 1 \\ 2 & 1 \end{bmatrix}$, we can immediately check that under $x = P\bar{\bar{x}}$,

$$\bar{\bar{F}} = P^{-1}FP = \begin{bmatrix} 3 & 1 \\ 2 & 1 \end{bmatrix} \begin{bmatrix} 0 & 1 \\ -6 & -5 \end{bmatrix} \begin{bmatrix} 1 & -1 \\ -2 & 3 \end{bmatrix} = \begin{bmatrix} -2 & 0 \\ 0 & -3 \end{bmatrix}$$

and

$$\bar{\bar{G}} = P^{-1}G = \begin{bmatrix} 3 & 1 \\ 2 & 1 \end{bmatrix} \begin{bmatrix} 0 \\ 1 \end{bmatrix} = \begin{bmatrix} 1 \\ 1 \end{bmatrix}$$

and

$$\bar{\bar{H}} = HP = \begin{bmatrix} 1 & 1 \end{bmatrix} \begin{bmatrix} 1 & -1 \\ -2 & 3 \end{bmatrix} = \begin{bmatrix} -1 & 2 \end{bmatrix}$$

which are the same results we get by computing the basis of eigenvectors of F.   ◊

In the last example, it turned out that the two realizations (F, G, H) and $(\bar{F}, \bar{G}, \bar{H})$ were algebraically equivalent, as we established by finding the similarity transformation P relating them. Of course, this will not happen in general.

Example 2

Consider the system (F, G, H) with $F = \begin{bmatrix} 3 & 1 \\ 0 & 3 \end{bmatrix}, G = \begin{bmatrix} 1 \\ 0 \end{bmatrix}, H = \begin{bmatrix} 1 & 1 \end{bmatrix}$ which has transfer function $\mathcal{Z}(s)$ given by

$$\mathcal{Z}(s) = H(sI - F)^{-1}G = \begin{bmatrix} 1 & 1 \end{bmatrix} \begin{bmatrix} \dfrac{1}{s-3} & \dfrac{1}{(s-3)^2} \\ 0 & \dfrac{1}{(s-3)} \end{bmatrix} \begin{bmatrix} 1 \\ 0 \end{bmatrix}$$

$$= \frac{1}{s-3}$$

This transfer function also has the realizations $(\bar{F}, \bar{G}, \bar{H})$ and $(\bar{\bar{F}}, \bar{\bar{G}}, \bar{\bar{H}})$ given by

$$\bar{F} = \begin{bmatrix} 3 & 0 \\ 0 & 3 \end{bmatrix}, \bar{G} = \begin{bmatrix} 1 \\ 0 \end{bmatrix}, \bar{H} = \begin{bmatrix} 1 & 1 \end{bmatrix}$$

and

$$\bar{\bar{F}} = [3], \bar{\bar{G}} = [1], \bar{\bar{H}} = [1]$$

It is obvious that $(\bar{\bar{F}}, \bar{\bar{G}}, \bar{\bar{H}})$ is not algebraically equivalent to the two-dimensional (F, G, H), but what about $(\bar{F}, \bar{G}, \bar{H})$? If the states x of (F, G, H) and $\bar{x}$ of $(\bar{F}, \bar{G}, \bar{H})$ are to be related by an invertible matrix P, then $\bar{F} = P^{-1}FP$ whence $F = P\bar{F}P^{-1}$. Since $\bar{F} = 3I$, it commutes with P and we have the contradiction $F = \bar{F}$. Thus F and $\bar{F}$ are not similar.   ◊

In general, when we find different realizations of a transfer function, they will only be zero-state equivalent. We will be especially interested in those realizations which are also algebraically equivalent since, in this case, any operations or items we care about in one realization can be coded into equivalent ones in the other realization.

EXERCISES FOR SECTION 6–5

✔ **1.**   Since the entries of $\mathrm{adj}(sI - F)$ are polynomials of degree $\leq n - 1$, we can write

$$\mathrm{adj}(sI - F) = s^{n-1}B_0 + s^{n-2}B_1 + \cdots + sB_{n-2} + B_{n-1} \tag{5}$$

where $B_0, B_1, \ldots, B_{n-1}$ are constant matrices.

(a)   Why are the matrices $B_0, \ldots, B_{n-1}$ unique? What are they for the F of Example 6–3–7?

(b)   Since $(sI - F)^{-1} = \dfrac{1}{\chi_F(s)} \mathrm{adj}(sI - F)$ where $\chi_F(s) = s^n + a_1 s^{n-1} + a_2 s^{n-2} + \cdots + a_n$, set

$$\chi_F(s)I = (sI - F)\,\mathrm{adj}(sI - F) \tag{6}$$

Substitute equation (5) for $\mathrm{adj}(sI - F)$ in equation (6) and multiply out on the right side; collect terms in descending powers of s. Equate coefficients of s term by term on both sides of your last equation, and solve for the $B_0, \ldots, B_{n-1}$.

$$[Answer: B_0 = I$$
$$B_1 = FB_0 + a_1 I$$
$$B_2 = FB_1 + a_2 I$$
$$\cdots$$
$$B_{n-1} = FB_{n-2} + a_{n-1}I$$
$$0 = FB_{n-1} + a_n I]$$

(c)   Use "back substitution" in $0 = FB_{n-1} + a_n I$ to successively eliminate $B_{n-1}, B_{n-2}, \ldots, B_0$ and derive

$$F^n + a_1 F^{n-1} + a_2 F^{n-2} + \cdots + a_{n-1}F + a_n I = 0$$

the famous **Cayley-Hamilton** theorem.

2.   **(a)**   Compute the transfer function of the block diagram of Figure 6–42 by graph reduction or Mason's rule.

 (b) Find $\mathfrak{L}(s)$ from the matrices $\bar{F}, \bar{G}, \bar{H}$. Is $(sI - \bar{F})^{-1}$ easy with $\bar{F}$ triangular?

 (c) Can you find an invertible matrix P such that $x = P\bar{x}$ makes (F, G, H) algebraically equivalent to $(\bar{F}, \bar{G}, \bar{H})$?

✔ **3.** **(a)** Compute $\mathfrak{L}(s)$ from the matrices $(\bar{\bar{F}}, \bar{\bar{G}}, \bar{\bar{H}})$ for the diagonal realization of Figure 6–44. Notice how simple $(sI - \bar{\bar{F}})^{-1}$ is with $\bar{\bar{F}}$ diagonal.

 (b) Can you find a similarity transformation P relating (F, G, H) and $(\bar{\bar{F}}, \bar{\bar{G}}, \bar{\bar{H}})$? [*Caution:* Check $\bar{\bar{G}}$ and $\bar{\bar{H}}$ too; not just $\bar{\bar{F}} = P^{-1}FP$]

 (c) Could we just as easily have moved the gains $\boxed{-1}$ and $\boxed{2}$ to before the blocks of Figure 6–43 and gotten the realization

$$\bar{\bar{F}} = \begin{bmatrix} -2 & 0 \\ 0 & -3 \end{bmatrix}, \ \bar{\bar{G}} = \begin{bmatrix} -1 \\ 2 \end{bmatrix}, \ \bar{\bar{H}} = [1 \ \ 1]?$$

Are there any troubles with $\bar{\bar{H}}(sI - \bar{\bar{F}})^{-1}\bar{\bar{G}}$?

✔ **4.**

 (a) What is the partial fraction expansion of $\mathfrak{L}(s) = \dfrac{1}{(s + a)^2}$? Give a series realization (F, G, H). Notice that the resulting F is a Jordan block.

$$\left[Answer: F = \begin{bmatrix} a & 1 \\ 0 & a \end{bmatrix}, \ G = \begin{bmatrix} 0 \\ 1 \end{bmatrix}, \ H = [1 \ \ 0] \right]$$

 (b) Give the controllable canonical realization diagram and its matrices.

5. **(a)** For Figure 6–21, use $y = x_1 + x_2$ and $\dot{y} = \dot{x}_1 + \dot{x}_2 = -6x_1 - 4x_2 + u$ to get

$$\begin{bmatrix} 1 & 1 \\ -6 & -4 \end{bmatrix}\begin{bmatrix} x_1 \\ x_2 \end{bmatrix} = \begin{bmatrix} y \\ \dot{y} - u \end{bmatrix}$$

Solve for x_1 and x_2 in terms of y, $\dot{y}$, and u.

$$\left[Answer: \begin{cases} x_1 = -\dfrac{1}{2}\dot{y} - 2y + \dfrac{1}{2}u \\[2mm] x_2 = \dfrac{1}{2}\dot{y} + 3y - \dfrac{1}{2}u \end{cases} \right]$$

 (b) How are the state variables $\bar{x}_1$ and $\bar{x}_2$ related to input and output for Figure 6–44?

$$\left[Answer: \begin{cases} \bar{x}_1 = -\dot{y} - 3y + u \\[2mm] \bar{x}_2 = -\dfrac{1}{2}\dot{y} - y + \dfrac{1}{2}u \end{cases} \right]$$

6. Which is easier in general: finding P by computing eigenvectors of F or by a partial fraction expansion of $\mathfrak{L}(s)$ gotten from $H(sI - F)^{-1}G = \mathfrak{L}(s)$? Do you actually need to know P, or is just the diagonal realization $(\bar{F}, \bar{G}, \bar{H})$ enough?

✔ **7.** Study repeated poles and how they show up in the time domain.

 (a) Write

$$\frac{1}{(s + a)^2} = \frac{1}{(s + a)(s + [a + \varepsilon])}\bigg|_{\varepsilon = 0}$$

Make a partial fraction expansion

$$\frac{1}{(s + a)(s + (a + \varepsilon))} = \frac{b_\varepsilon}{s + a} + \frac{c_\varepsilon}{s + a + \varepsilon} = \mathcal{Y}_\varepsilon(s)$$

Inverse Laplace transform to get

$$(b_\varepsilon + c_\varepsilon e^{-\varepsilon t})e^{-at}1(t)$$

and then take the limit as $\varepsilon \to 0$ (use L'Hospital's rule) to get $\dfrac{1}{(s + a)^2}$ corresponding to $te^{-at}1(t)$.

    (b)   Draw a block diagram for $\mathcal{Y}_\varepsilon(s)$ and examine the gains as $\varepsilon \to 0$.

    (c)   Since $\dfrac{1}{(s + a)^2} = \dfrac{1}{s + a} \cdot \dfrac{1}{s + a}$, and since multiplication in the s-do-
main corresponds to convolution in the time domain, convolve $e^{-at}1(t)$ with itself
and check your result of part (a). Show that

$$\frac{1}{(s + a)^{k+1}} = \mathcal{L}\left\{\frac{1}{k!} t^k e^{-at}1(t)\right\}$$

    (d)   Prove that the time functions $\left\{e^{-at}, \ te^{-at}, \ \dfrac{1}{2!} t^2 e^{-at}, \ldots, \dfrac{t^k}{k!}e^{-at}\right\}$ are
linearly independent over **R** or **C**. What about the complex functions
$\left\{\dfrac{1}{s + a}, \dfrac{1}{(s + a)^2}, \ldots, \dfrac{1}{(s + a)^k}\right\}$ over **C**?

✔ **8.**   Here we consider how to realize any proper rational *matrix* system function
by realizing separately each of its scalar entries.

    (a)   Let

$$\mathcal{Y}(s) = \begin{bmatrix} \dfrac{1}{s} & \dfrac{3s + 8}{s^2 + 4s} \\[2mm] \dfrac{1}{s} & \dfrac{2s + 4}{s^2 + 4s} \end{bmatrix}$$

and find a realization (F, G, H). Draw a block diagram.

    (b)   How would you realize

$$\mathcal{Y}(s) = \begin{bmatrix} \mathcal{Y}_{11}(s) & \mathcal{Y}_{12}(s) \\ \mathcal{Y}_{21}(s) & \mathcal{Y}_{22}(s) \end{bmatrix}$$

if you already had a realization (F_{ij}, G_{ij}, H_{ij}) for each $\mathcal{Y}_{ij}(s)$? Hint: Consider the
block partitioned matrix

$$F = \begin{bmatrix} F_{11} & 0 & 0 & 0 \\ 0 & F_{12} & 0 & 0 \\ 0 & 0 & F_{21} & 0 \\ 0 & 0 & 0 & F_{22} \end{bmatrix}$$

What do you think about minimality of this method of realization?

✔ **9.**   Suppose that you don't like where the poles of a system are. Here we study the effect of *state feedback* on pole locations.
   (a)   For the system (F, G, H) of Examples 6–3–7 and 6–4–2 we found

$$\mathcal{E}(s) = \frac{s + 1}{(s + 3)(s + 2)}$$

Let the input u be Kx (state feedback) for an appropriate matrix K. Draw the signal flow graph for this closed loop system. Write the new matrix equations describing the closed loop system. What are the eigenvalues of this system? Solve for K so that the closed loop transfer function becomes

$$\frac{s + 1}{(s + 3)(s + 5)}$$

Can this always be done? [*Answer:* $K = [-9 \quad -3]$.]
   (b)   Let

$$F = \begin{bmatrix} -2 & 0 & 0 \\ 0 & -3 & 0 \\ 0 & 0 & +1 \end{bmatrix}, G = \begin{bmatrix} 1 \\ 1 \\ 1 \end{bmatrix}, H = [1 \quad 0 \quad 3]$$

Draw the signal flow graph and find $\mathcal{E}(s)$. Notice that $\mathcal{E}(s)$ has one "bad" pole at $+1$, corresponding to a growing exponential. Let $u = Kx$. Draw the closed loop diagram and give the new matrix equations. Solve for K to place poles at $\{-2, -3, -1\}$. [*Answer:* $K = [0 \quad 0 \quad -2]$.]

10.   Consider the system (F, G, H) with $F = \begin{bmatrix} 0 & 6 \\ 1 & -1 \end{bmatrix}$, $G = \begin{bmatrix} 0 \\ 1 \end{bmatrix}$, $H = [1 \quad 1]$. It is easy to check that F has $v_1 = \begin{bmatrix} 3 \\ 1 \end{bmatrix}$ and $v_2 = \begin{bmatrix} 2 \\ -1 \end{bmatrix}$ as eigenvectors.

   (a)   Determine $\Phi(t, 0)$.

   (b)   Express the solution to the unforced system $\dot{x} = Fx$, with $x(0) = \begin{bmatrix} x_1(0) \\ x_2(0) \end{bmatrix}$, as a linear combination of v_1 and v_2.

   (c)   Give a value of initial state $x(0) = x_0$ such that the state response $x(t) = \phi(t, 0, x_0, 0) = f(t)v_2$. Determine $f(t)$. [You have just suppressed e^{+2t}.] What is $\mathcal{E}(t, 0, (-3)x_0, 0)$?
   (d)   Find an initial state $x(0) = \bar{x}_0$ such that for input $u(t) = e^{dt}1(t)$, the output will be $y(t) = \mathcal{E}(t, 0, \bar{x}_0, u(\cdot)) = 7e^{dt}1(t)$. Can you do this for all d? What happens if $d = -1$? Explain the significance of this. What happens if $d = -2$ or -3? Explain.
✔   **(e)**   Finally, compute an initial state $x(0) = \bar{\bar{x}}_0$ such that $y(t) = \mathcal{E}(t, 0, \bar{\bar{x}}_0, 0) = g(t, 0)$, the impulse response.

6–6   The Jordan Canonical Form

When attempting a partial fraction expansion of a given proper rational function $\mathcal{E}(s)$, what happens if not all n roots of the denominator polynomial are distinct? In Exercise 4 of the last section we saw that an expression $\frac{1}{(s + a)^2}$ cannot be further broken down as a sum $\frac{b}{s + a} + \frac{c}{s + a}$, but is most simply realized as two basic $\frac{1}{s + a}$ blocks in series, as in Figure 6–46.

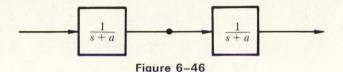

Figure 6–46

Thus, if a root λ occurs with multiplicity k, we shall expect to find terms like $\frac{1}{(s - \lambda)^k}$ in the partial fraction expansion and to realize these with a cascade of k basic $\frac{1}{s - \lambda}$ blocks.

Example 1

To find a realization by partial fractions of

$$\mathcal{Y}(s) = \frac{s + 1}{(s + 3)^2(s + 2)}$$

we try an expansion of the form

$$\frac{s + 1}{(s + 3)^2(s + 2)} = \frac{a}{(s + 3)^2} + \frac{b}{s + 3} + \frac{c}{s + 2}$$

and solve [Exercise 1] for $a = 2$, $b = 1$, $c = -1$. Thus, we can realize $\mathcal{Y}(s)$ by the three parallel branches of Figure 6–47. If we break down (realize) the $\frac{1}{(s + 3)^2}$ block as a cascade, we get the diagram of Figure 6–48.

In terms of the actual number of dynamic elements (the integrators or $\frac{1}{s}$ blocks), Figure 6–48 is disturbing because it requires four integrators to realize this third-order $\mathcal{Y}(s)$; we have a *non-minimal* realization [Exercise 2]. The alert reader may have already figured out how we can get along without one of the $\frac{1}{s + 3}$ blocks. Since the signal at x_4 is exactly that at x_2, we can eliminate the redundant state variable x_4 and have the three-dimensional realization shown in Figure 6–49.

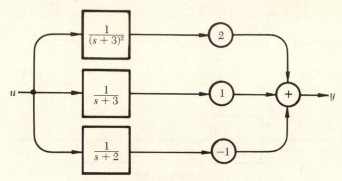

Figure 6–47 Partial fraction realization.

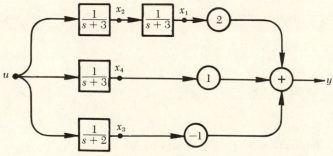

Figure 6–48 Broken down into basic blocks.

With the outputs of the basic blocks x_1, x_2, x_3 as state variables, we can write by inspection for our minimal realization (F_N, G_N, H_N) the dynamic equations

$$\begin{bmatrix} \dot{x}_1 \\ \dot{x}_2 \\ \dot{x}_3 \end{bmatrix} = \begin{bmatrix} -3 & 1 & 0 \\ 0 & -3 & 0 \\ 0 & 0 & -2 \end{bmatrix} \begin{bmatrix} x_1 \\ x_2 \\ x_3 \end{bmatrix} + \begin{bmatrix} 0 \\ 1 \\ 1 \end{bmatrix} u$$

$$y = \begin{bmatrix} 2 & 1 & -1 \end{bmatrix} \begin{bmatrix} x_1 \\ x_2 \\ x_3 \end{bmatrix}$$

where we inserted dashed partitioning lines to identify the portion associated with the multiple root (-3). ◊

Notice in the last example that because of the repeated root we did not get a diagonal matrix representation. The presence of the 1 in the 1, 2 place of F_N indicates the coupling between x_1 and x_2, and because of this coupling [See Exercise 7 of Section 6–5], the normal modes will have time variations

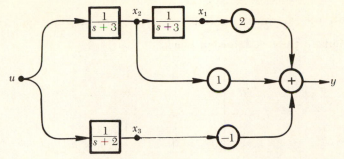

Figure 6-49   A minimal realization (F_N, G_N, H_N) using 3 basic blocks.

of the form e^{-3t}, te^{-3t}, and e^{-2t}. We used the subscript N for *normal* since (F_N, G_N, H_N) displayed the modes and was as uncoupled as we could get. The matrix F_N gotten by partial fractions is of the form

$$F_N = \left[\begin{array}{c|c} J_2(-3) & 0 \\ \hline 0 & J_1(-2) \end{array}\right]$$

where

$$J_2(-3) = \begin{bmatrix} -3 & 1 \\ 0 & -3 \end{bmatrix} \text{ and } J_1(-2) = [-2]$$

are *Jordan blocks*† of order 2 and 1 corresponding to roots -3 and -2 of multiplicities 2 and 1 respectively. The matrix G_N is of the form

$$G_N = \left[\begin{array}{c} G_1 \\ \hline G_2 \end{array}\right]$$

where $G_1 = \begin{bmatrix} 0 \\ 1 \end{bmatrix}$ and $G_2 = [1]$ are columns of zeros with final entry 1 of lengths 2 and 1 respectively (the multiplicities of $\lambda_1 = -3$ and $\lambda_2 = -2$). The read-out matrix H_N is of the form

$$H_N = [H_1 \mid H_2]$$

where $H_1 = [2 \ \ 1]$ and $H_2 = [-1]$ are rows of lengths 2 and 1 and whose entries are the coefficients‡ of the partial fraction expansions

$$\frac{2}{(s+3)^2} + \frac{1}{s+3}, \quad \frac{-1}{s+2}$$

---

† See Section 6-2, Exercise 16 for a discussion of Jordan blocks.
‡ These coefficients are called the *residues in the poles;* see Exercise 1.

With this approach to state selection, each multiple pole $\dfrac{1}{(s-\lambda)^k}$ corresponds to the $k \times k$ **Jordan block:**

$$J_k(\lambda) = \begin{bmatrix} \lambda & 1 & 0 & \cdots & 0 \\ 0 & \lambda & 1 & & 0 \\ 0 & 0 & \lambda & & \vdots \\ \vdots & \vdots & \vdots & & 1 \\ 0 & 0 & 0 & \cdots & \lambda \end{bmatrix}, k \times k \tag{1}$$

Thus, if $\mathfrak{H}(s)$ has the complete partial fraction expansion

$$
\begin{aligned}
\mathfrak{H}(s) = {} & \frac{h_{11}}{(s-\lambda_1)^{k_1}} + \frac{h_{12}}{(s-\lambda_1)^{k_1-1}} + \cdots + \frac{h_{1,k_1}}{(s-\lambda_1)} \\
& + \frac{h_{21}}{(s-\lambda_2)^{k_2}} + \frac{h_{22}}{(s-\lambda_2)^{k_2-1}} + \cdots + \frac{h_{2,k_2}}{(s-\lambda_2)} + \cdots \\
& + \frac{h_{r1}}{(s-\lambda_r)^{k_r}} + \frac{h_{r2}}{(s-\lambda_r)^{k_r-1}} + \cdots + \frac{h_{r,k_r}}{(s-\lambda_r)}
\end{aligned}
\tag{2}
$$

corresponding to poles $\lambda_1, \lambda_2, \ldots, \lambda_r$ of multiplicities $k_1, k_2, \ldots, k_r$ respectively, then we obtain the realization shown in Figure 6–50.

Obviously, $k_1 + k_2 + \cdots + k_r = n$ and we get the matrix representation (F_N, G_N, H_N), where

$$F_N = \begin{bmatrix} J_{k_1}(\lambda_1) & 0 & \cdots & 0 \\ 0 & J_{k_2}(\lambda_2) & & 0 \\ \vdots & & \ddots & \vdots \\ 0 & 0 & \cdots & J_{k_r}(\lambda_r) \end{bmatrix} \begin{matrix} {\scriptstyle\}}\, k_1 \\ {\scriptstyle\}}\, k_2 \\ \\ {\scriptstyle\}}\, k_r \end{matrix} \tag{3a}$$

$$G_N = \begin{bmatrix} 0 \\ 0 \\ \vdots \\ 1 \\ \hline 0 \\ 0 \\ \vdots \\ 1 \\ \hline \vdots \\ \hline 0 \\ 0 \\ \vdots \\ 1 \end{bmatrix} \begin{matrix} {\scriptstyle\}}\, k_1 \\ \\ \\ {\scriptstyle\}}\, k_2 \\ \\ \\ \\ {\scriptstyle\}}\, k_r \end{matrix} \tag{3b}$$

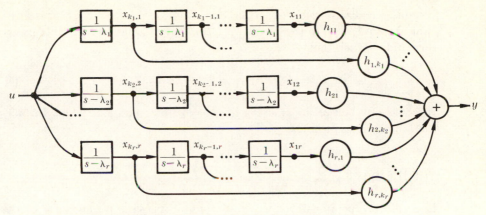

Figure 6–50   Jordan canonical realization of $\mathfrak{H}(s)$.

and

$$H_N = [\underbrace{h_{11}h_{12}\ldots h_{1k_1}}_{k_1} \mid \cdots \mid \underbrace{h_{r1}\ldots h_{r,k_r}}_{k_r}] \tag{3c}$$

corresponding to the state variable

$$x_N = \begin{bmatrix} x_{11} \\ \vdots \\ x_{k_1,1} \\ \hline x_{12} \\ \vdots \\ x_{k_2,2} \\ \hline \vdots \\ \hline x_{1r} \\ \vdots \\ x_{k_r,r} \end{bmatrix} \begin{matrix} \Big\} k_1 \\ \\ \Big\} k_2 \\ \\ \\ \Big\} k_r \end{matrix} \tag{3d}$$

We thus see that any matrix F can be reduced to the *block diagonal* form in which each block is a Jordan block. We call F_N the **Jordan canonical form** of F, and we call (F_N, G_N, H_N) the **Jordan canonical realization**† of the transfer function $\mathfrak{H}(s)$. Of course, if each multiplicity $k_j = 1$, then F_N is just the diagonal form of F.

Let us pause for a moment to relate this discussion to similarity transformations and change of bases of the state space. Suppose that we are given three matrices (F, G, H). From them we can compute $\mathfrak{H}(s) = H(sI - F)^{-1}G$ and then, making a partial fraction expansion of $\mathfrak{H}(s)$, arrive at the Jordan

---

†It should be clear that the gains $h_{11}, \ldots, h_{r,k_r}$ could have been associated with the read-in matrix G_N instead of the output matrix H_N, in which case $G_N \mapsto H_N^T$ and $H_N \mapsto G_N^T$.

canonical realization (F_N, G_N, H_N). Does it follow that (F_N, G_N, H_N) is algebraically equivalent to the given (F, G, H)? Can we find an invertible transformation P such that

$$F_N = P^{-1}FP, \; G_N = P^{-1}G, \; H_N = HP? \tag{4}$$

It is a happy fact of life that the answer to these questions is yes. In fact, the following result holds:

THEOREM 1

Every linear operator over the complex numbers has a Jordan form representation (3a) with respect to a proper choice of basis.   □

    Because of multiple roots, this basis will not just consist of eigenvectors of F, but will in general also contain "generalized eigenvectors" of F. We defer discussion of the relatively complicated† procedure for constructing the necessary basis vectors of Theorem 1 until Section 8–3.

Example 2

The matrix

$$F = \begin{bmatrix} 1 & 0 & 0 \\ 1 & 2 & 0 \\ 1 & 1 & 2 \end{bmatrix}$$

has characteristic equation $\chi_F(s) = (s - 2)^2(s - 1)$ and a Jordan basis

$$P_1 = \begin{bmatrix} 1 \\ -1 \\ 0 \end{bmatrix}, \; P_2 = \begin{bmatrix} 0 \\ 0 \\ 1 \end{bmatrix}, \; P_3 = \begin{bmatrix} 0 \\ 1 \\ 0 \end{bmatrix}$$

as we now verify. To get the representation of F with respect to the basis $\mathcal{B} = \{P_1, P_2, P_3\}$, we compute

$$FP_1 = \begin{bmatrix} 1 \\ -1 \\ 0 \end{bmatrix}, \; FP_2 = \begin{bmatrix} 0 \\ 0 \\ 2 \end{bmatrix}, \; FP_3 = \begin{bmatrix} 0 \\ 2 \\ 1 \end{bmatrix}$$

---

† As an example of what passes for a "conceptually simple algorithm" for doing this, see Rubin, "A Simple Method for Finding the Jordan Form of a Matrix," *IEEE Transactions on Automatic Control,* Feb. 1972, p. 145–146.

and notice that $FP_1 = 1P_1$ and $FP_2 = 2P_2$, so P_1 and P_2 are eigenvectors of F, but $FP_3 \neq \lambda P_3$ for any λ; hence P_3 is not an eigenvector. However, $FP_3 = 1P_2 + 2P_3$ so

$$\mathcal{B}F\mathcal{B} = \begin{bmatrix} 1 & 0 & 0 \\ 0 & 2 & 1 \\ 0 & 0 & 2 \end{bmatrix}$$

which is of the Jordan canonical form

$$F_N = \left[\begin{array}{c|c} J_1(1) & 0 \\ \hline 0 & J_2(2) \end{array} \right]$$

As a check, see Exercise 3.   ◊

MANIPULATIONS WITH THE JORDAN FORM

When we use partial fractions on the $\mathfrak{L}(s)$ (or equally well, on the $\mathfrak{Z}(z)$ of the discrete-time system described by (F, G, H)) to come up with the equivalent Jordan canonical form representation (F_N, G_N, H_N), we can always relabel the state variables in Figure 6–50 and change the order of the sub-blocks of F_N around.† This amounts to renumbering the basis $P_1, P_2, \ldots, P_n$, and it can in fact be shown that the Jordan canonical form is unique except for this rearrangement of the Jordan blocks. The reader who is rusty on the manipulations of block-partitioned matrices would do well to brush up by working Exercise 13 of Section 4–2 at this time.

To free us from some of the pesky subscripts which clutter equations $(3a, b, c, d)$, let us temporarily denote the Jordan matrix by J and reorder the Jordan blocks of $(3a)$ in such a way that all the 1×1 blocks occur first—and thus can be gathered together into a single pure diagonal block J_0. We then have that

$$J = \begin{bmatrix} J_0 & 0 & 0 & \cdots & 0 \\ 0 & J_1 & 0 & \cdots & 0 \\ 0 & 0 & J_2 & \cdots & 0 \\ \vdots & \vdots & \vdots & \ddots & \vdots \\ 0 & 0 & 0 & \cdots & J_q \end{bmatrix} \tag{5a}$$

where

$$J_0 = \begin{bmatrix} \lambda_1' & 0 & \cdots & 0 \\ 0 & \lambda_2' & \cdots & 0 \\ \vdots & \vdots & \ddots & \vdots \\ 0 & 0 & \cdots & \lambda_m' \end{bmatrix} \tag{5b}$$

---

†Try it in Example 1, Figure 6–49, letting $x_1 \mapsto x_2$, $x_2 \mapsto x_3$, $x_3 \mapsto x_1$.

is a pure diagonal matrix, corresponding to that part of the Jordan basis $P_1, P_2, \ldots, P_m$ consisting of eigenvectors, while all the other blocks correspond to "super-eigenvectors" (see Section 8–3):

$$J_i = \begin{bmatrix} \lambda_i & 1 & \ldots & 0 & 0 \\ 0 & \lambda_i & \ldots & 0 & 0 \\ \vdots & \vdots & \ddots & \vdots & \vdots \\ 0 & 0 & \ldots & \lambda_i & 1 \\ 0 & 0 & \ldots & 0 & \lambda_i \end{bmatrix} \quad \text{is } k_i \times k_i \text{ for some } k_i \geq 2 \qquad \textbf{(5c)}$$

Of course, in some cases we have $J = J_0$, while in other cases m may be 0, and so J_0 will not appear in the above expression for J.

With this relabeling we now have $J = P^{-1}FP$, $G_N = P^{-1}G$, $H_N = HP$, $(sI - F)^{-1} = P(sI - J)^{-1}P^{-1}$ and

$$e^{Ft} = Pe^{Jt}P^{-1}$$

Because of the block diagonal form of J,

$$(sI - J) = \begin{bmatrix} (sI - J_0) & 0 & \ldots & 0 \\ 0 & (sI - J_1) & \ldots & 0 \\ \vdots & \vdots & \ddots & \vdots \\ 0 & 0 & \ldots & (sI - J_q) \end{bmatrix}$$

where the different Is may be of different dimensions, and the inverse of this matrix [Exercise 4–2–13] is

$$(sI - J)^{-1} = \begin{bmatrix} (sI - J_0)^{-1} & 0 & \ldots & 0 \\ 0 & (sI - J_1)^{-1} & \ldots & 0 \\ \vdots & \vdots & \ddots & \vdots \\ 0 & 0 & \ldots & (sI - J_q)^{-1} \end{bmatrix} \qquad \textbf{(5d)}$$

Note that $(sI - J)^{-1}$ is defined only if s is not an eigenvalue of J, and that this is precisely the condition for all the blocks on the right-hand side diagonal of (5d) to be well-defined.

Now let us form J^k. Since J is block-diagonal, we know, from the multiplication of partitioned matrices, that we simply have

$$J^k = \begin{bmatrix} J_0^k & 0 & 0 & \ldots & 0 \\ 0 & J_1^k & 0 & \ldots & 0 \\ 0 & 0 & J_2^k & \ldots & 0 \\ \vdots & \vdots & \vdots & \ddots & \vdots \\ 0 & 0 & 0 & \ldots & J_q^k \end{bmatrix}$$

i.e., we just raise each block of the diagonal to the appropriate power.

If we now express the exponential e^{Jt} as an infinite power series, we immediately deduce that

$$e^{Jt} = \sum_{n=0}^{\infty} J^n \frac{t^n}{n!} = \begin{bmatrix} \sum_{n=0}^{\infty} J_0^n \frac{t^n}{n!} & \cdots & 0 \\ \vdots & \ddots & \vdots \\ 0 & \cdots & \sum_{n=0}^{\infty} J_q^n \frac{t^n}{n!} \end{bmatrix} = \begin{bmatrix} e^{J_0 t} & \cdots & 0 \\ \vdots & \ddots & \vdots \\ 0 & \cdots & e^{J_q t} \end{bmatrix}$$

Thus, the exponential of a block diagonal matrix is obtained by just forming the exponentials of the blocks—and so what we have to do now is compute the exponentials of the blocks in our Jordan canonical form. Now, the argument we have just given for reducing the exponential of a block diagonal matrix to the exponentials of the individual blocks certainly applies when each of the blocks is 1×1, and so we may apply the above argument to J_0 to deduce that

$$e^{J_0 t} = \begin{bmatrix} e^{\lambda_1 t} & 0 & \cdots & 0 \\ 0 & e^{\lambda_2 t} & \cdots & 0 \\ \vdots & \vdots & \ddots & \vdots \\ 0 & 0 & \cdots & e^{\lambda_m t} \end{bmatrix}$$

All that remains, then, is to show how to compute the exponential of a Jordan block. The smug reader who already did this in Exercise 6-1-3 may now relax as we carry out the gory details. To do this, we recall that

$$e^{At} \cdot e^{Bt} = e^{(A+B)t}$$

holds so long as A and B *commute,* and apply the fact that the identity matrix commutes with everything:

$$J_i = \begin{bmatrix} \lambda_i & 1 & \cdots & 0 \\ 0 & \lambda_i & \cdots & 0 \\ \vdots & \vdots & \ddots & 1 \\ 0 & 0 & \cdots & \lambda_i \end{bmatrix} = \lambda_i I + Z_i \qquad \text{(5e)}$$

where I is the identity matrix of the same size as J_i, and Z_i is the matrix, again of the same size as J_i, which is zero everywhere save for the fact that it has ones on each entry immediately above the main diagonal. Therefore, our above observation allows us to write

$$e^{J_i t} = e^{\lambda_i I t} \cdot e^{Z_i t}$$

Now, $\lambda_i I$ is a pure diagonal matrix,

$$e^{\lambda_i I t} = \begin{bmatrix} e^{\lambda_i t} & \cdots & 0 \\ \vdots & \ddots & \vdots \\ 0 & \cdots & e^{\lambda_i t} \end{bmatrix}$$

and our task simply reduces to computing $e^{Z_i t}$.

It is easy to see that the kth power of Z_i is the matrix that is zero everywhere save for having ones all along the kth diagonal above the main diagonal. Thus if Z_i were a 4×4 matrix, we would have

$$Z_i^0 = \begin{bmatrix} 1 & 0 & 0 & 0 \\ 0 & 1 & 0 & 0 \\ 0 & 0 & 1 & 0 \\ 0 & 0 & 0 & 1 \end{bmatrix}, Z_i^1 = \begin{bmatrix} 0 & 1 & 0 & 0 \\ 0 & 0 & 1 & 0 \\ 0 & 0 & 0 & 1 \\ 0 & 0 & 0 & 0 \end{bmatrix},$$

$$Z_i^2 = \begin{bmatrix} 0 & 0 & 1 & 0 \\ 0 & 0 & 0 & 1 \\ 0 & 0 & 0 & 0 \\ 0 & 0 & 0 & 0 \end{bmatrix}, Z_i^3 = \begin{bmatrix} 0 & 0 & 0 & 1 \\ 0 & 0 & 0 & 0 \\ 0 & 0 & 0 & 0 \\ 0 & 0 & 0 & 0 \end{bmatrix}$$

whereas

$$Z_i^4 = \begin{bmatrix} 0 & 0 & 0 & 0 \\ 0 & 0 & 0 & 0 \\ 0 & 0 & 0 & 0 \\ 0 & 0 & 0 & 0 \end{bmatrix}$$

and [See Exercise 5]

$$Z_i^n = \begin{bmatrix} 0 & 0 & 0 & 0 \\ 0 & 0 & 0 & 0 \\ 0 & 0 & 0 & 0 \\ 0 & 0 & 0 & 0 \end{bmatrix} \quad \text{for } n \geq 4$$

In general, if Z_i is $k \times k$, we have that Z_i^n is zero if and only if $n \geq k$, and so we have

$$e^{Z_i t} = \sum_{n=0}^{\infty} Z_i^n \frac{t^n}{n!} = \sum_{n=0}^{k-1} Z_i^n \frac{t^n}{n!} = \begin{bmatrix} 1 & t & \dfrac{t^2}{2!} & \cdots & \dfrac{t^{k-3}}{(k-3)!} & \dfrac{t^{k-2}}{(k-2)!} & \dfrac{t^{k-1}}{(k-1)!} \\ 0 & 1. & t. & \cdots & \dfrac{t^{k-4}}{(k-4)!} & \dfrac{t^{k-3}}{(k-3)!} & \dfrac{t^{k-2}}{(k-2)!} \\ \vdots & \vdots & \vdots & & \vdots & \vdots & \vdots \\ 0 & 0 & 0 & \cdots & 1 & t & \dfrac{t^2}{2!} \\ 0 & 0 & 0 & \cdots & 0 & 1 & t \\ 0 & 0 & 0 & \cdots & 0 & 0 & 1 \end{bmatrix}$$

Returning now to our exponential, we see that multiplying a matrix by a scalar multiple of the identity corresponds to multiplying every entry of the given matrix by the given scalar, and so we finally deduce that

$$e^{J_i t} = e^{\lambda_i I t} \cdot e^{Z_i t} = \begin{bmatrix} e^{\lambda_i t} & e^{\lambda_i t} t & \cdots & e^{\lambda_i t} \dfrac{t^{k_i - 2}}{(k_i - 2)!} & e^{\lambda_i t} \dfrac{t^{k_i - 1}}{(k_i - 1)!} \\ 0 & e^{\lambda_i t} & \cdots & e^{\lambda_i t} \dfrac{t^{k_i - 3}}{(k_i - 3)!} & e^{\lambda_i t} \dfrac{t^{k_i - 2}}{(k_i - 2)!} \\ \vdots & \vdots & \cdots & \vdots & \vdots \\ 0 & 0 & \cdots & e^{\lambda_i t} & e^{\lambda_i t} t \\ 0 & 0 & \cdots & 0 & e^{\lambda_i t} \end{bmatrix} \tag{6a}$$

Recalling

$$e^{J_0 t} = \begin{bmatrix} e^{\lambda_1' t} & 0 & \cdots & 0 \\ 0 & e^{\lambda_2' t} & \cdots & 0 \\ \vdots & \vdots & \ddots & \vdots \\ 0 & 0 & \cdots & e^{\lambda_m' t} \end{bmatrix} \tag{6b}$$

we now see that we have an explicit formula for

$$e^{J t} = \begin{bmatrix} e^{J_0 t} & 0 & \cdots & 0 \\ 0 & e^{J_1 t} & \cdots & 0 \\ \vdots & \vdots & \ddots & \vdots \\ 0 & 0 & \cdots & e^{J_q t} \end{bmatrix} \tag{6c}$$

Let us now apply this to the zero input response of our linear system $x(t) = Jx(t)$, where we shall write out the state vector as

$$x(t) = \begin{bmatrix} \begin{matrix} x_1(t) \\ \vdots \\ x_m(t) \end{matrix} \\ \hline \begin{matrix} x_{11}(t) \\ \vdots \\ x_{1k_1}(t) \end{matrix} \\ \hline \vdots \\ \hline \begin{matrix} x_{q1}(t) \\ \vdots \\ x_{qk_q}(t) \end{matrix} \end{bmatrix} \begin{matrix} \Big\} \text{ corresponds to } J_0 \\ \\ \Big\} \text{ corresponds to } J_1 \\ \\ \\ \Big\} \text{ corresponds to } J_q \end{matrix} \qquad \text{(6d)}$$

to conform with the basis which puts our matrix F into Jordan canonical form, J. If we apply the formula

$$x(t) = e^{Jt}x(0) \qquad \text{(6e)}$$

we see that

$$x_i(t) = e^{\lambda_i t}x_i(0) \text{ for } 1 \leq i \leq m \qquad \text{(6f)}$$

corresponding to the completely decoupled component x_i of the state vector shown in Figure 6–51, whereas we have the formulas

$$x_{k_i,i}(t) = e^{\lambda_i t}x_{k_i,i}(0)$$
$$x_{k_i-1,i}(t) = e^{\lambda_i t}x_{k_i-1,i}(0) + te^{\lambda_i t}x_{k_i,i}(0)$$
$$x_{1,i}(t) = e^{\lambda_i t}x_{1,i}(0) + te^{\lambda_i t}x_{2,i}(0) + \frac{t^2}{2!}e^{\lambda_i t}x_{3,i}(0) + \cdots \qquad \text{(6g)}$$
$$+ \frac{t^{k_i-1}}{(k_i-1)!}e^{\lambda_i t}x_{k_i,i}(0)$$

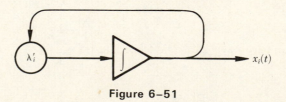

Figure 6–51

corresponding to the ith block of modes of the state, which are coupled in the form shown in Figure 6–52. This form should be recognizable as one

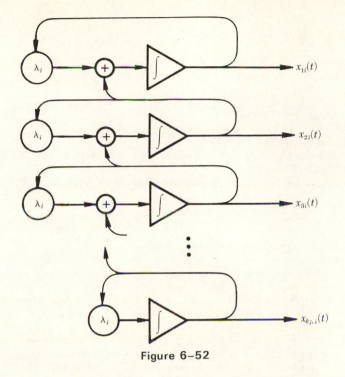

Figure 6-52

of the horizontal chains of blocks shown in Figure 6–50, except that we have stacked the component $\dfrac{1}{s - \lambda_i}$ blocks one on top of the other in Figure 6–52 instead of stringing them out horizontally.

We note that each of the first m modes is completely decoupled, and that each of the consecutive blocks is decoupled from all the other blocks. We further notice that although the matrix J, and the consequent state diagram we have drawn above, only couples a component to the component ' immediately below it in its own block, we see that over any period of time the interaction propagates. Thus, once we look at the behavior of the system over any time interval, we see that the output of any component is influenced by the initial state not only of itself and the component immediately below it, but also of *all* the components below it within the same block. The reader should note that if we integrate 1 j times, we get $\dfrac{t^j}{j!}$; thus, it is not surprising that this coefficient appears in the above formulas after the passage through j integrators.

We shall study stability in quite general terms in Section 7–4, but we should pause to point out here that the system (F, G, H) will have its natural response die out as $t \to \infty$ precisely when the eigenvalues of F have negative real parts [see Exercise 9]. This is clearly evidenced in the modes of e^{Jt}, which comprise, through $Pe^{Jt}P^{-1}$, the modes of e^{Ft}.

EXERCISES FOR SECTION 6–6

✔ **1.** The coefficients of the partial fraction expansion where $\mathscr{B}(s)$ has multiple poles, such as in equation (2) (called *residues*), can be easily calculated from

$$h_{i,j} = \frac{1}{(j-1)!} \left[\frac{d^{j-1}}{ds^{j-1}} (s - \lambda_i)^{k_i} \mathscr{B}(s) \right] \Bigg|_{s=\lambda_i} \quad \text{for } j = 1, 2, \ldots, k_i$$

where k_i is the multiplicity of pole λ_i.

(a) Use this formula to find the constants a, b, and c in the expansions of Example 1 and of Example 6–3–7.

(b) Find the Jordan canonical realization (F_N, G_N, H_N) of

$$\mathscr{B}(s) = \frac{(s+1)}{(s+3)^3(s+2)^2(s-1)}$$

Draw the block diagram and plot poles and zeros.

✔ **2. (a)** Give five different realizations of the $\mathscr{B}(s)$ of Example 1. Notice that all five use only three integrators. Can you prove that there is no two-integrator realization of this $\mathscr{B}(s)$? [*Hint:* Think about Mason's rule.]

(b) Write the matrices (F, G, H) for the non-minimal realization of Figure 6–48 and compare them with the matrices (F_N, G_N, H_N) of the minimal realization of Figure 6–49.

(c) Compute the transfer function of Figure 6–48, keeping your eyes open to see where a cancellation must occur if that fourth-order system is to end up with a third-order transfer function. [*Hint:* Look at $(sI - F)^{-1}G$ and $H(sI - F)^{-1}$.]

(d) How would you find some other third-order realization different from the "canonical" ones you used in part (a)? Is there any limit to the number of minimal realizations? [*Hint:* Pick yourself any non-singular matrix P.]

(e) Find a similarity transformation P which relates the controllable canonical representation (F_c, G_c, H_c) to the Jordan canonical (F_N, G_N, H_N) found in Example 1.

✔ **3.** Check the results of Example 2 in two ways:

(a) Compute $P^{-1}FP$ with $P = [P_1 \mid P_2 \mid P_3]$ whose columns are the given Jordan basis. Verify that $FP = [FP_1 \mid FP_2 \mid FP_3]$. Make sure you see that finding the coordinates of FP_1, FP_2, FP_3 with respect to the basis $\mathscr{B} = \{P_1, P_2, P_3\}$ to get $^\mathscr{B}F^\mathscr{B}$, the matrix of F with respect to $\mathscr{B}$, is equivalent to solving the three sets of equations

$$FP_1 = c_{11}P_1 + c_{21}P_2 + c_{31}P_3$$
$$FP_2 = c_{12}P_1 + c_{22}P_2 + c_{32}P_3$$
$$FP_3 = c_{13}P_1 + c_{23}P_2 + c_{33}P_3$$

for the constants $\{c_{ij}\}$. Write this out in matrix form

$$FP = [FP_1 \mid FP_2 \mid FP_3] = [P_1 \mid P_2 \mid P_3] \begin{bmatrix} c_{11} & c_{12} & c_{13} \\ c_{21} & c_{22} & c_{23} \\ c_{31} & c_{32} & c_{33} \end{bmatrix} = (P)(^\mathscr{B}F^\mathscr{B})$$

whence

$$^\mathscr{B}F^\mathscr{B} = P^{-1}FP$$

(b) Make a partial fraction expansion of $\mathscr{B}(s) = \dfrac{1}{\chi_F(s)}$, draw the block diagram, label the states, and write the dynamic equations to find F_N as in Example 1.

4. Let $\mathcal{B} = \{P_1, P_2, \ldots, P_n\}$ be a Jordan basis for F and $\mathcal{B}' = \{q_1, q_2, \ldots, q_n\}$ be a Jordan basis for F^T. How is $\mathcal{B}'$ related to $\mathcal{B}$? What happens if $\mathcal{B}'$ is a basis for F^*, the *adjoint* of F? Illustrate with the F of Example 2. [*Answer:* If $P = [P_1 \mid P_2 \mid \cdots \mid P_n]$ and $Q = [q_1 \mid q_2 \mid \cdots \mid q_n]$, then $Q = (P^{-1})^*$]

✔ **5.** Let A be a matrix such that $A^k = 0$ but $A^{k-1} \neq 0$. We say that A is **nilpotent of index** k.

    (a) Show that every strictly triangular matrix (such as Z_i in equation (5e)) with $A_{ij} = 0$ for $j \leq i$ is nilpotent.

    (b) If A is nilpotent of index k, show that

$$(sI - A)^{-1} = \frac{1}{s}\left(I + \frac{A}{s} + \frac{A^2}{s^2} + \cdots + \frac{A^{k-1}}{s^{k-1}}\right)$$

and give the finite matrix polynomial for the exponential e^{At}.

    (c) Notice that, with $A \neq 0$ and $A^{k-1} \neq 0$, we have the product of two non-zero matrices turning out to be zero. In general, if $AB = 0$ with neither A nor B zero, show that both A and B are singular.

✔ **6.** Here we make a partial fraction expansion directly on the matrix $(sI - F)^{-1} = \dfrac{\text{adj}(sI - F)}{\chi_F(s)}$. For simplicity, suppose that $\chi_F(s)$ has n distinct roots: $\chi_F(s) = (s - \lambda_1)(s - \lambda_2) \ldots (s - \lambda_n)$.

    (a) Write

$$(sI - F)^{-1} = \frac{E_1}{s - \lambda_1} + \frac{E_2}{s - \lambda_2} + \cdots + \frac{E_n}{s - \lambda_n}$$

where $E_1, \ldots, E_n$ are $n \times n$ constant *matrices* to be determined (called **residue matrices** or **constituent idempotents**). Show that

$$E_i = \lim_{s \to \lambda_i}\left[(s - \lambda_i)\frac{\text{adj}(sI - F)}{\chi_F(s)}\right]$$

$$= \frac{\text{adj}(\lambda_i I - F)}{(s - \lambda_1) \ldots (s - \lambda_{i-1})(s - \lambda_{i+1}) \ldots (s - \lambda_n)}$$

    (b) From $(sI - F)(sI - F)^{-1} = I$, show that

$$I = \sum_{i=1}^{n}\frac{(sI - F)E_i}{(s - \lambda_i)}$$

Multiply both sides by $(s - \lambda_j)$ and let $s \to \lambda_j$ in the result to get

$$0 = (\lambda_j I - F)E_j$$

whence $FE_j = \lambda_j E_j$. (This shows that the non-zero columns of E_j are eigenvectors of F corresponding to λ_j.) Is this a good way to find eigenvectors?

    (c) Show that

$$e^{Ft} = \sum_{i=1}^{n} E_i e^{\lambda_i t}$$

and from this that

$$\sum_{i=1}^{n} E_i = I$$

(d)   Verify all these expressions for

$$F = \begin{bmatrix} 0 & 1 \\ -6 & -5 \end{bmatrix}$$

and compare with Example 6–3–7 and Figure 6–44. What do the E matrices have to do with the B matrices of Exercise 1 of Section 6–5?

(e)   Redo Example 6–2–5, finding eigenvectors by part (b). [*Note:* See Zadeh and Desoer [1963], page 307, and Rubio [1971], Section 4.5, for extensive discussions of these residue matrices.]

COMP 7.   Let

$$F = \begin{bmatrix} 38 & 7 & 66 \\ 30 & 7 & 54 \\ -23\tfrac{1}{2} & -4\tfrac{1}{2} & -41 \end{bmatrix}$$

(a)   Compute the eigenvalues of F. [In fact, there are two distinct eigenvalues, both small integers.]

(b)   Let the eigenvalues be $\lambda_1, \lambda_1,$ and λ_2. Find solutions for the three equations

$$(\lambda_1 I - F)P_1 = 0$$
$$(\lambda_1 I - F)^2 P_2 = 0$$
$$(\lambda_2 I - F)P_3 = 0$$

(c)   By direct computation show that the resultant change of basis does yield

$$P^{-1}FP = \begin{bmatrix} \lambda_1 & 1 & 0 \\ 0 & \lambda_1 & 0 \\ 0 & 0 & \lambda_2 \end{bmatrix}$$

Call this Jordan matrix J.

(d)   Give a formula for J^n for each n and then compute e^{Jt}. Then compute e^{Ft}.

(e)   Compute e^{Ft} by Laplace transforms and compare results.

✔ **8.**   Since it is so easy to take powers of block diagonal matrices, the Jordan form is very convenient for evaluating functions of a square matrix A.

(a)   Let

$$f(x) = \sum_{i=0}^{\infty} c_i x^i$$

where the power series converges in some region $|x| < r$ which contains all of the eigenvalues $\lambda_1, \ldots, \lambda_n$ of F. [Polynomials are easy special cases where $f(x) =$

$c_0 + c_1 x + \cdots + c_k x^k$ and $r = \infty$.] Define $f(A) = c_0 I + c_1 A + c_2 A^2 + \cdots$ and show that this series makes sense (converges).

   (b)   Let J be the Jordan matrix (equation (5a)) of A and let the columns of P be the Jordan basis. Show that

$$f(A) = Pf(J)P^{-1}$$

and that

$$f(J) = \begin{bmatrix} f(J_0) & 0 & \cdots & 0 \\ 0 & f(J_1) & \cdots & 0 \\ \vdots & \vdots & \ddots & \vdots \\ 0 & 0 & \cdots & f(J_q) \end{bmatrix}$$

   (c)   Make up series to define the matrix functions $\cos A$, A^{-1}, $(I - A)^{-1}$, and $\left(I - \dfrac{1}{s} A\right)^{-1}$; give conditions on A necessary for convergence.

   (d)   If v is an eigenvector of A, show that it is also an eigenvector of $f(A)$.
9.   Show that if $\mathcal{R}e(\lambda) < 0$, then

$$\lim_{t \to \infty} t^k e^{\lambda t} = 0 \text{ for any integer } k$$

Check your work by using the **final value theorem** of Laplace transforms given in Appendix A–1.

CHAPTER 7

Controllability, Observability, and Stability

The central task of control theory, obviously, is to analyze how to control a system to behave in some desired way. We may be interested in such questions as "What is the quickest way to get a system to a desired state with the least expenditure of energy?" It is to questions like these of optimal control that we shall turn in Section 7–6 and Appendix A–3. But, first, we must look at basic problems which underlie any question of optimal control; namely, can we control the system at all, let alone optimally? To analyze this, we shall have to see what the concepts of reachability and controllability look like for continuous-time linear systems. Again, a crucial factor in the control of a system must be the knowledge of what the system is doing now. If we do not know what state the system is in, it is hard to suggest appropriate control signals to apply to it. Thus, the theory of observability must also be developed for continuous-time linear systems. An even more disturbing question may arise—namely, do we know what system it is that we are controlling? It is this realization problem—of formulating an accurate model of the system being observed—that will provide the focal theme of Chapter 8. For now, let us concentrate upon the topics of controllability and observability, and the related problem of stability: can we be sure that the system will not "blow up"?

In Chapter 4 we introduced the important concepts of reachability, controllability, observability, and system identification (realizability), in their most general formulations. We then quickly brought these abstract definitions down to earth by applying them to simple, concrete systems. For the special class of finite-dimensional, constant, linear, discrete-time systems described by the matrices (F, G, H) we used a little linear algebra to derive conditions on F, G, and H under which controllability, reachability, and observability were achieved. Using the notions of inner products and adjoints of linear operators, we formulated the concept of duality and showed that controllability and observability were duals of each other.

In this chapter we shall focus our attention on continuous-time, finite-dimensional linear systems $(F(\cdot), G(\cdot), H(\cdot))$ and see what the controllability and observability conditions look like for these systems. We will, of course, be greatly interested in the special case where $F(\cdot)$, $G(\cdot)$, and $H(\cdot)$ are constant matrices.

In Section 7-1, we give a general discussion of feedback, and analyze the way in which the behavior of a system may be modified by adding a feedback loop to it. In Section 7-2 we explore the "dual" problem of estimating the state of a system by continued observation of the inputs applied to it and the corresponding outputs. Then, in Section 7-3 we provide the continuous-time theory which corresponds to the study of observability and controllability given in Chapter 4. It is our hope that the reader who has gained a firm feel for the concepts in the discrete-time context will find the current theory more accessible as a result. In Section 7-4 we develop a number of elementary results from stability theory, including a complete classification of the stabilities and instabilities of real two-dimensional constant linear systems.

Next, in Section 7-5 we apply the preceding theories to analyzing the way in which we may estimate the state and control the stability of a multiple-input, multiple-output linear system. Finally, in Section 7-6, we use our previous experience with Banach space derivatives from Section 3-4 to formulate and introduce the major approaches to optimal control theory, leaving most of the specialized mathematical details to Appendix A-3.

7-1   Feedback, Control, and Canonical Forms

The notion of monitoring a system's performance, comparing it with some desired "reference" performance, and using any discrepancy to generate a correction input, or "control," is fundamental and pervades all aspects of modern life.

Perhaps the most familiar feedback control system is the thermostatically controlled heater. A reference dial is set, say to 70°C, and a thermometer is monitored. When the temperature is below the desired 70°, the heater is turned on. It is kept on until the thermometer indicates a temperature higher than 70°, and at this point the heater is turned off. In this system the thermostat is the **controller** which, using a transducer,† converts performance data (temperature) into the **control signal** (on-off switch) for the **dynamic plant** (heater). We might design the controller to optimize some criterion of merit such as, say, our comfort, and have it not allow the temperature to drop below 69° nor to rise above 71°. Alternatively, we might design it to optimize heating efficiency and fix it so that the heater won't be "on" so much of the time.

We call such a control system a **closed loop** or **feedback control** system, since the output (performance) is processed (transduced?) and information from it is fed back into the input to control the performance.

Not every control system uses feedback. Consider an electric toaster: you set the dial to *L*ight, *M*edium or *D*ark and push the bread down; a timer leaves the heat on for T_L, T_M, or T_D seconds and then pops it up. Here the control is the dial setting, but there is no performance feedback. To utilize true feedback, the toaster would have to be modified to observe its output (the toast) and determine whether it is dark enough yet to pop it up. Quite descriptively, the use of feedback to regulate a system's performance is called **regulation,** and we call the determination of the feedback necessary to control a system a solution of **the regulator problem.**

There is another important use of feedback, which is to improve the characteristics of a system by making them less sensitive to changes in the values of certain parameters or elements. We illustrate this application with an example.

Example 1

In Exercise 17 of Section 6–3 we studied the basic feedback configuration shown in Figure 7–1.

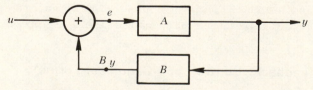

Figure 7–1   The basic feedback configuration.

Since
$$e = u + By$$

and
$$y = Ae$$

---

†Any device which converts information from one form of carrier to another: a loudspeaker transforms electrical signals into sounds.

we have $$y = Au + ABy$$

or $$(I - AB)y = Au$$

whence $$y = (I - AB)^{-1}Au$$

provided the operation $(I - AB)$ is invertible.

If the operations A and B are *scalars* (such as, perhaps, the transfer functions $\mathcal{B}_1(s)$ and $\mathcal{B}_2(s)$ respectively), then

$$y = \frac{A}{1 - AB}\,u$$

Suppose, for a given B, that A can be made very large—so large that $\frac{A}{1 - AB}\,u \doteq \frac{1}{B}\,u$, independent of A.

This shows that even if A cannot be determined with precision, as long as we know that it is large, we can make the system insensitive to A, and have the characteristics depend only on B. See Exercise 1 for some useful examples.   ◊

STATE FEEDBACK

In Chapter 4 of this book we began our study of the related concepts of controllability and observability, and through a sequence of examples tied them in with the notion of feedback. It all started in Example 2–5–7, with the simple dynamic system of a cart of unit mass on frictionless wheels. There, we wrote the differential equation $\ddot{y} = u$ relating its output (position) y to input (force) u and chose state variables $x_1 = y$, $x_2 = \dot{y}$ to get an analog computer realization represented by matrices (F, G, H). All of these relationships are summarized here for convenience in Figure 7–2.

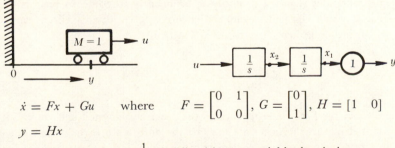

$$\dot{x} = Fx + Gu \quad \text{where} \quad F = \begin{bmatrix} 0 & 1 \\ 0 & 0 \end{bmatrix},\ G = \begin{bmatrix} 0 \\ 1 \end{bmatrix},\ H = \begin{bmatrix} 1 & 0 \end{bmatrix}$$

$$y = Hx$$

Figure 7–2   A "$\frac{1}{s^2}$ plant", with state-variable description.

By studying the phase plane portraits of this system's state response to constant inputs $u = +1$ and $u = -1$, obtained earlier in Example 2–5–7, we had prepared ourselves for Example 4-2-4. There we introduced **state feedback** to generate the **"bang-bang"** control signal

$$u(x) = \mathcal{K}\left(\begin{bmatrix} x_1 \\ x_2 \end{bmatrix}\right) = -\text{sgn}(x_1)$$

This control signal made the system behave like a mechanical oscillator, as we saw in Figure 4-4. We can visualize this use of state feedback control with the diagram of Figure 7-3, where $\mathcal{K}$ denotes the operation of the controller. Notice that this operator $\mathcal{K}$ is non-linear† and represents the non-linear processing of state information by the "bang-bang" signum operation.

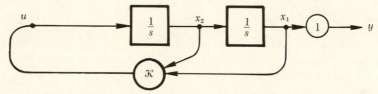

Figure 7-3 Introducing the controller $\mathcal{K}: u = \mathcal{K}(x)$.

In Exercise 10 of Section 4-2 we introduced the more complicated non-linear controller operation,

$$\mathcal{K}\left(\begin{bmatrix} x_1 \\ x_2 \end{bmatrix}\right) = -\text{sgn}(x_2 + 3x_1)$$

and we saw that the input $u(x) = \mathcal{K}(x)$ drove the system from the initial state $x(0) = \begin{bmatrix} y(0) \\ \dot{y}(0) \end{bmatrix} = \begin{bmatrix} 1 \\ 1 \end{bmatrix}$ toward the origin $\begin{bmatrix} 0 \\ 0 \end{bmatrix}$ as the time increased.

The cart could have had any initial position and velocity, and this last type of controller would still cause it eventually to come to rest with zero velocity at $y = 0$. In fact, some readers may have already discovered that the still more complicated non-linear controller

$$\mathcal{K}\left(\begin{bmatrix} x_1 \\ x_2 \end{bmatrix}\right) = -\text{sgn}(x_2 - f(x_1))$$

(where $x_2 = f(x_1)$ is the equation of the "switching line" shown in Figure 7-4) will drive any initial state $x(0)$ to the zero state in a finite time with at most one change of sign in the control signal.

---

†Is $\mathcal{K}\left(\begin{bmatrix} x_1 \\ x_2 \end{bmatrix} + \begin{bmatrix} \hat{x}_1 \\ \hat{x}_2 \end{bmatrix}\right) = \mathcal{K}\left(\begin{bmatrix} x_1 \\ x_2 \end{bmatrix}\right) + \mathcal{K}\left(\begin{bmatrix} \hat{x}_1 \\ \hat{x}_2 \end{bmatrix}\right)$?

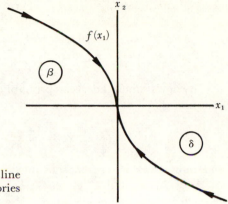

Figure 7–4   An "optimal" switching line
constructed from trajectories
β, δ of Figure 2–39.

The existence of this last controller established that our $\dfrac{1}{s^2}$ plant is state controllable.

Now let us turn our attention to *linear state feedback,* where the controller $\mathcal{K}$ performs only linear transformations on the state information to generate the feedback control signals. If $\mathcal{K}$ is linear it can be represented by a matrix, say K, so

$$u = \mathcal{K}(x) = Kx$$

For our $\dfrac{1}{s^2}$ system, since u is a scalar, K is a row, $K = [k_1 \quad k_2]$, and

$$u = [k_1 \quad k_2]\begin{bmatrix} x_1 \\ x_2 \end{bmatrix} = k_1 x_1 + k_2 x_2$$

is a linear combination of x_1 and x_2, we may now redraw Figure 7–3 to get Figure 7–5, showing the inner details of this linear controller. We have drawn the controller's inner parts spread out in Figure 7–5 so that we would recognize the feedback block diagram as having the structure of the *controllable*

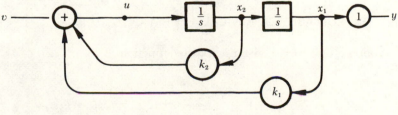

Figure 7–5   Linear state feedback. $\mathcal{K}\begin{pmatrix} x_1 \\ x_2 \end{pmatrix} = [k_1 \quad k_2]\begin{bmatrix} x_1 \\ x_2 \end{bmatrix}.$

canonical form,† say (F_k, G_k, H_k), where

$$F_k = \begin{bmatrix} 0 & 1 \\ k_1 & k_2 \end{bmatrix}, \ G_k = \begin{bmatrix} 0 \\ 1 \end{bmatrix}, \ H_k = \begin{bmatrix} 1 & 0 \end{bmatrix}$$

By now‡ some readers can probably write by inspection the characteristic equation $\chi_{F_k}(s)$ of this system:

$$\chi_{F_k}(s) = s^2 - k_2 s - k_1$$

and right away we see one possible use for linear state feedback. We can set the gains k_1 and k_2 to have any values we choose and thus get any characteristic equation we want.

Suppose that we wish the system to have eigenvalues at $\lambda_1 = -6$ and $\lambda_2 = -8$. Then we want

$$\chi_{F_k}(s) = (s - \lambda_1)(s - \lambda_2) = s^2 - (\lambda_1 + \lambda_2)s + (\lambda_1 \lambda_2)$$
$$= s^2 + 14s + 48$$

so we set $k_2 = (\lambda_1 + \lambda_2) = -14$ and $k_1 = -\lambda_1 \lambda_2 = -48$.

Thus, if we didn't like the modes of the original open loop system of Figure 7–2, $e^{0t}1(t)$ and $te^{0t}1(t)$ corresponding to the double pole of $\dfrac{1}{s^2}$ at $s = 0$, we can change them to $e^{-6t}1(t)$ and $e^{-8t}1(t)$ by closing the loop with state feedback $u = \begin{bmatrix} -48 & -14 \end{bmatrix} \begin{bmatrix} x_1 \\ x_2 \end{bmatrix}$. We shall refer to this procedure as **controlling the modes** and to this property as **modal controllability**.

Linear state feedback allowed us to change the open loop transfer function $\mathfrak{F}(s) = \dfrac{1}{s^2}$ to the closed loop transfer function $\mathfrak{F}_k(s) = \dfrac{1}{s^2 - k_2 s - k_1}$, so this process can also be called **pole-shifting** or **pole assignment** [see Exercises 2 and 3]. Motivated by this example, let us now delve more deeply into modal controllability by linear state feedback and its connection with the controllable canonical form realization of transfer functions.

THE CONTROLLABLE CANONICAL§ FORM (F_c, G_c, H_c)

Suppose that we are given a transfer function

$$\mathfrak{F}(s) = \frac{b_1 s^{n-1} + b_2 s^{n-2} + \cdots + b_{n-1} s + b_n}{s^n + a_1 s^{n-1} + a_2 s^{n-2} + \cdots + a_{n-1} s + a_n} \tag{1}$$

†Compare with Figure 2–43 and Example 6–4–1
‡Especially after Exercise 10 of 6–2 and Exercise 1 of Section 6–4.
§ Also known in the literature as the *phase canonical* form.

whose poles $\lambda_1, \lambda_2, \ldots, \lambda_n$ we wish to shift to $\widehat{\lambda}_1, \widehat{\lambda}_2, \ldots, \widehat{\lambda}_n$ by linear state feedback.

The controllable canonical form realization (F_c, H_c, G_c) of equation (1) can be written by inspection (though we give a formal proof below) from the coefficients of the numerator and denominator polynomials of $\mathfrak{H}(s)$:

$$
F_c = \begin{bmatrix}
0 & 1 & 0 & \cdots & 0 & 0 \\
0 & 0 & 1 & \cdots & 0 & 0 \\
0 & 0 & 0 & \cdots & 0 & 0 \\
\vdots & \vdots & \vdots & & \vdots & \vdots \\
0 & 0 & 0 & \cdots & 1 & 0 \\
0 & 0 & 0 & \cdots & 0 & 1 \\
-a_n & -a_{n-1} & -a_{n-2} & \cdots & -a_2 & -a_1
\end{bmatrix}
\tag{2}
$$

$$
G_c = \begin{bmatrix} 0 \\ 0 \\ 0 \\ \vdots \\ 0 \\ 1 \end{bmatrix}, \quad H_c = [b_n \quad b_{n-1} \quad b_{n-2} \quad \cdots \quad b_2 \quad b_1],
$$

and the block diagram of this realization is shown in Figure 7–6.

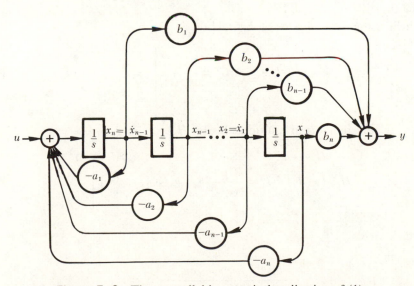

Figure 7–6 The controllable canonical realization of (1).

In Exercise 3(c) of Section 6–4, we outlined a proof, using Mason's rule, that the block diagram of Figure 7–6 described by the matrices (F_c, G_c, H_c) really does have as its transfer function the $\mathfrak{H}(s)$ of equation (1). It will be instructive now to prove this by actually computing the product $H_c(sI - F_c)^{-1}G_c$. We have, from equation (2), that

$$(sI - F_c) = \begin{bmatrix} s & -1 & 0 & \cdots & 0 & 0 \\ 0 & s & -1 & \cdots & 0 & 0 \\ 0 & 0 & s & \cdots & & \cdot \\ \vdots & \vdots & \vdots & & \vdots & \vdots \\ 0 & 0 & 0 & \cdots & s & -1 \\ a_n & a_{n-1} & a_{n-2} & \cdots & a_2 & s + a_1 \end{bmatrix} \tag{3}$$

Since $G_c = \begin{bmatrix} 0 \\ 0 \\ \vdots \\ 1 \end{bmatrix}$ is the standard unit vector e_n, then $(sI - F_c)^{-1}G_c$ will be the

last column of $(sI - F_c)^{-1}$. By Cramer's rule [equation (1) of Section 6–5], the last column of $(sI - F_c)^{-1}$ is $\dfrac{1}{\det(sI - F_c)}$ times the last column of adj$(sI - F_c)$; i.e., $\dfrac{1}{\chi_{F_c}(s)}$ times the column of cofactors of the last *row* of $(sI - F_c)$. These cofactors are easily computed from equation (3) to be $[1 \ s \ s^2 \ldots s^{n-1}]^T$ and, just as easily, $\det(sI - F_c)$ yields

$$\chi_{F_c}(s) = s^n + a_1 s^{n-1} + a_2 s^{n-2} + \cdots + a_{n-1}s + a_n$$

which is the denominator of $\mathscr{G}(s)$. Thus,

$$(sI - F_c)^{-1}G_c = \frac{1}{\chi_{F_c}(s)} \begin{bmatrix} 1 \\ s \\ \vdots \\ s^{n-1} \end{bmatrix} \tag{4}$$

It only remains to multiply this by H_c:

$$H_c(sI - F_c)^{-1}G_c = \frac{1}{\chi_{F_c}(s)}[b_n \quad b_{n-1} \quad \cdots \quad b_1] \begin{bmatrix} 1 \\ s \\ \vdots \\ s^{n-1} \end{bmatrix} \tag{5}$$

which yields

$$\frac{b_1 s^{n-1} + \cdots + b_n}{s^n + a_1 s^{n-1} + \cdots + a_n} = \mathscr{G}(s)$$

as claimed [see Exercise 5].

Let us now consider introducing linear state feedback, $u = Kx$. Again, since u is a scalar, K is a row matrix

$$K = [k_1 \quad k_2 \ldots k_n]$$

and

$$u = k_1 x_1 + k_2 x_2 + \cdots + k_n x_n$$

If we add this state feedback to the block diagram of Figure 7–6 as we did in the two-dimensional example of Figures 7–2, 7–3, and 7–5 (try it for $n = 3$), we see a curious thing. When we take state x_1, pass it through the gain k_1, and send it to the adder to construct the linear combination of the last equation, we have effectively placed a branch with gain k_1 in parallel with the existing branch with gain $-a_n$. Similarly, feeding back the ith variable x_i places the branch k_i precisely in parallel with the existing branch $-a_{n+1-i}$. Combining each parallel pair of branches, we get the closed loop system equivalent to the block diagram shown in Figure 7–7, which is still of the controllable canonical structure.

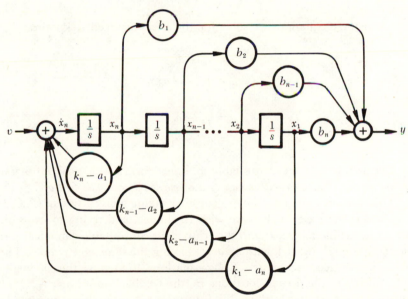

Figure 7–7   After introducing state feedback Kx in Figure 7–6.

In Figure 7–7, we also have decided to denote the input to the overall closed loop system by v (the reference) so that $u = v + Kx$.†

The effect of the state feedback Kx is now clear: each component k_i permits us to change the coefficient, a_{n+1-i}, of the s^{i-1} term in the denominator of equation (1).

---

†The signal $u = v + Kx$ still affects $\dot{x}_n$ through

$$\dot{x}_n = u - (a_n x_1 + a_{n-1} x_2 + \cdots + a_1 x_n).$$

To shift the roots from $\lambda_1, \ldots, \lambda_n$ to $\widehat{\lambda}_1, \ldots, \widehat{\lambda}_n$ we merely construct the polynomial

$$\widehat{\chi}(s) = (s - \widehat{\lambda}_1)(s - \widehat{\lambda}_2) \ldots (s - \widehat{\lambda}_n) \tag{6}$$

and multiply it out to get

$$\widehat{\chi}(s) = s^n + \widehat{a}_1 s^{n-1} + \widehat{a}_2 s^{n-2} + \cdots + \widehat{a}_n \tag{7}$$

We then set

$$
\begin{aligned}
-\widehat{a}_1 &= k_n - a_1 \\
-\widehat{a}_2 &= k_{n-1} - a_2 \\
&\ \vdots \\
-\widehat{a}_{n-1} &= k_2 - a_{n-1} \\
-\widehat{a}_n &= k_1 - a_n
\end{aligned}
\tag{8}
$$

From these equations we solve for the necessary feedback gains,

$$
\begin{aligned}
k_1 &= a_n - \widehat{a}_n \\
k_2 &= a_{n-1} - \widehat{a}_{n-1} \\
&\ \vdots \\
k_{n-1} &= a_2 - \widehat{a}_2 \\
k_n &= a_1 - \widehat{a}_1
\end{aligned}
\tag{9}
$$

The usefulness of the controllable canonical form should now be obvious; *it is the natural realization with which to accommodate linear state feedback.*

This leads us quite naturally to consider taking a given system (F, G, H) with state x whose eigenvalues we wish to change, and trying to transform it by an invertible matrix P into the equivalent controllable canonical form representation (F_c, G_c, H_c), with state x_c, having the same eigenvalues. Then we can find the feedback matrix K_c by equations (9) to move the eigenvalues of (F_c, G_c, H_c) to the desired locations by the feedback input $u = K_c x_c$. The invertibility of P lets us transform this information back to the original system and find the necessary feedback input $u = Kx$ in terms of the state x rather than x_c.

From $u = K_c x_c$, since $x = P x_c$, we have $u = K_c(P^{-1}x) = (K_c P^{-1})x$, which tells us to use the state feedback matrix

$$K = K_c P^{-1} \tag{10}$$

in our original representation to place its poles where we want them. We summarize this discussion with the following statement.

THEOREM 1

Any system (F, G, H) which is algebraically equivalent under $x = Px_c$ to the controllable canonical form (F_c, G_c, H_c) may have its eigenvalues arbitrarily assigned by using linear state feedback $u = Kx$.   □

We illustrate this procedure with an example.

Example 2

Consider the system (F, G, H) with $F = \begin{bmatrix} -2 & 0 \\ 0 & -3 \end{bmatrix}$, $G = \begin{bmatrix} 1 \\ 1 \end{bmatrix}$, $H = [-1 \quad 2]$ which has characteristic equation $\chi_F(s) = (s + 3)(s + 2) = s^2 + 5s + 6$ and transfer function $\mathfrak{H}(s) = \dfrac{s + 1}{s^2 + 5s + 6}$. Suppose that we want to shift the poles from $-3, -2$ to $-1, -10$.

Thanks to Example 6-5-1, we already happen to know that (F, G, H) is algebraically equivalent to $F_c = \begin{bmatrix} 0 & 1 \\ -6 & -5 \end{bmatrix}$, $G_c = \begin{bmatrix} 0 \\ 1 \end{bmatrix}$, and $H_c = [1 \quad 1]$, and we recognize (F_c, G_c, H_c) as the controllable canonical form. The invertible matrix P that we need for $x = Px_c$ is available as the inverse of the P of Example 6-5-1 (think about it), so

$$P = \begin{bmatrix} 3 & 1 \\ 2 & 1 \end{bmatrix}$$

Using equations (6) and (7), the desired polynomial is

$$\hat{\chi}(s) = (s + 1)(s + 10) = s^2 + 11s + 10$$

so we identify $\hat{a}_1 = 11$ and $\hat{a}_2 = 10$, and from the old $\chi(s)$ we read off $a_1 = 5$ and $a_2 = 6$. Equations (9) then yield

$$k_{c1} = a_2 - \hat{a}_2 = 6 - 10 = -4$$
$$k_{c2} = a_1 - \hat{a}_1 = 5 - 11 = -6$$

whence $K_c = [-4 \quad -6]$ and $u = K_c x_c$.

Translating this feedback back into x language, $u = Kx$, and

$$K = K_c P^{-1} = [-4 \quad -6] \begin{bmatrix} 1 & -1 \\ -2 & 3 \end{bmatrix} = [8 \quad -14]$$

As a check, the reader should now draw the block diagram of (F, G, H), introduce the state feedback loop $u = Kx$, simplify the diagram to get new matrices (F_k, G_k, H_k) for the closed loop system, and compute its transfer function.   ◊

Now, obviously, not every system (F, G, H) is going to be invertibly transformable to controllable canonical form [see Exercise 4]. Let us attempt to characterize those which are.

Given a system (F, G, H), we can compute $\chi_F(s)$ and $\mathfrak{H}(s) = H(sI - F)^{-1}G$ and write $\mathfrak{H}(s)$ in the full n-dimensional form of equation (1), i.e., without cancelling. Then, by inspection of $\mathfrak{H}(s)$, we can write the controllable canonical realization (F_c, G_c, H_c) using equations (2). With the matrices F, G, H and F_c, G_c, H_c now known, we seek an invertible transformation P such that $x = Px_c$, $P^{-1}FP = F_c$, $P^{-1}G = G_c$, and $HP = H_c$. Equivalently, if such a P exists, the matrix equations

$$G = PG_c$$

$$FP = PF_c$$

$$HP = H_c$$

must be satisfied by P.

Now, G_c is just $e_n = [0 \quad 0 \ldots 1]^T$, the nth unit vector of the standard basis for $\mathbf{C}^n$, so $G = PG_c = Pe_n$ is just the nth column of P. Partitioning P by columns,

$$P = [P_1 \mid P_2 \mid \ldots \mid P_n]$$

we have
$$P_n = G$$

Thus the last column of P is now known.

Recalling that the ith column of the product of two matrices is just the first matrix times the ith column of the second, from $FP = PF_c$ we have

$$FP_i = PF_{ci} \qquad \text{(11a)}$$

But the columns of F_c are exceedingly simple, and from equation (2) we can write

$$F_{ci} = e_{i-1} - a_{n+1-i}e_n, \quad i = 1, \ldots, n \qquad \text{(11b)}$$

where e_i denotes the column with all zeros except for a 1 in the ith place of the standard basis $e_1, \ldots, e_n$ for $\mathbf{C}^n$ and $e_0 \triangleq [0 \quad 0 \ldots 0]^T$. If A is any matrix, Ae_j is just the jth column of A, so substituting equation (11b) into equation (11a), we get

$$FP_i = P(e_{i-1} - a_{n+1-i}e_n)$$
$$= Pe_{i-1} - a_{n+1-i}Pe_n$$

whence

$$FP_i = P_{i-1} - a_{n+1-i}P_n \text{ for } i = 1, \ldots, n \qquad \textbf{(11c)}$$

Now, since we already know that $P_n = G$, if we start with $i = n$ in equation (11c) we get

$$FP_n = P_{n-1} - a_1 P_n$$
$$= P_{n-1} - a_1 G$$

which lets us solve for the $(n - 1)$st column of P in terms of F and G:

$$P_{n-1} = FP_n + a_1 P_n = FG + a_1 G$$

Stepping along to $i = n - 1$ in equation (11c),

$$FP_{n-1} = P_{n-2} - a_2 P_n$$

and with P_{n-1} and P_n now known, we have the $(n - 2)$nd column of P,

$$P_{n-2} = FP_{n-1} + a_2 P_n$$
$$= F^2 G + a_1 FG + a_2 G$$

Proceeding in this fashion, we recursively find all of the columns of P in terms of the given F and G:

$$P_{n-j} = \sum_{i=0}^{j} a_i F^{j-i} G, \quad j = 0, 1, \ldots, n - 1 \qquad \textbf{(12)}$$

where $a_0 \triangleq 1$. When $j = (n - 1)$, we get the first column of P as

$$P_1 = FP_2 + a_{n-1} P_n$$
$$= F^{n-1} G + a_1 F^{n-2} G + \cdots + a_{n-2} FG + a_{n-1} G$$

which can be written as the partitioned-matrix product

$$P_1 = [F^{n-1}G, F^{n-2}G, \ldots, FG, G] \begin{bmatrix} 1 \\ a_1 \\ \vdots \\ a_{n-2} \\ a_{n-1} \end{bmatrix}$$

Similarly,

$$P_2 = [F^{n-1}G, F^{n-2}G, \ldots, FG, G] \begin{bmatrix} 0 \\ 1 \\ a_1 \\ \vdots \\ a_{n-2} \end{bmatrix}$$

$$\vdots$$

$$P_n = [F^{n-1}G, F^{n-2}G, \ldots, FG, G] \begin{bmatrix} 0 \\ 0 \\ \vdots \\ 1 \end{bmatrix}$$

so we can lump all these expressions together into one big matrix equation

$$P = [P_1, P_2, \ldots, P_n]$$

$$= [F^{n-1}G, F^{n-2}G, \ldots, FG, G] \begin{bmatrix} 1 & 0 & \cdots & 0 & 0 \\ a_1 & 1 & & 0 & 0 \\ a_2 & a_1 & & \cdot & \cdot \\ \cdot & \cdot & & \cdot & \cdot \\ \cdot & \cdot & & \cdot & \cdot \\ a_{n-2} & a_{n-3} & \cdots & 1 & 0 \\ a_{n-1} & a_{n-2} & \cdots & a_1 & 1 \end{bmatrix} \tag{13}$$

Now P is of the form

$$P = \mathcal{C}\mathcal{Q} \tag{14}$$

where $$\mathcal{C} = [F^{n-1}G, F^{n-2}G, \ldots, FG, G]$$

is our old friend from Chapter 4, the **controllability matrix of F and G,** and

$$\mathcal{Q} = \begin{bmatrix} 1 & 0 & \cdots & 0 & 0 \\ a_1 & 1 & \cdots & 0 & 0 \\ a_2 & a_1 & \cdots & \cdot & \cdot \\ \cdot & \cdot & & \cdot & \cdot \\ \cdot & \cdot & & \cdot & \cdot \\ \cdot & \cdot & & 1 & 0 \\ a_{n-1} & a_{n-2} & \cdots & a_1 & 1 \end{bmatrix} \tag{15}$$

is an invertible matrix, made up of the known coefficients $a_1, \ldots, a_{n-1}$ of the given characteristic equation of F. [See Exercise 6.] Sometimes, we shall write $\mathcal{C}$ as $\mathcal{C}(F, G)$ when we wish to indicate that we are talking about the controllability matrix of the pair (F, G).

Since $\mathcal{A}$ is invertible, equation (14) tells us that P is invertible if and only if $\mathcal{C}$ is invertible, and we have proved the following result:

THEOREM 2

If (F, G, H) is algebraically equivalent under $x = Px_c$ to (F_c, G_c, H_c), then the controllability matrix of (F, G) is non-singular, i.e.,

$$\text{rank } [F^{n-1}G, \ldots, FG, G] = n. \qquad \square$$

Interestingly, rank $\mathcal{C}(F, G) = n$ will also prove to be the controllability condition for continuous-time systems, as we shall see in Section 7-3.

We now show that the non-singularity of $\mathcal{C}(F, G)$ is also sufficient.

THEOREM 3

If rank $[F^{n-1}G, \ldots, FG, G] = n$, then there exists an invertible matrix P such that $x = Px_c$ transforms the representation (F, G, H) to (F_c, G_c, H_c).

Proof

If we know that $\mathcal{C}$ has full rank we may *define* P to be given by equation (13) and know that P is invertible. Thus, its columns $\{P_1, \ldots, P_n\}$ are linearly independent and constitute a basis $\mathcal{B}$ for the state space X. We shall see what the given matrix F looks like when represented with respect to this new basis.† The nth column of $^{\mathcal{B}}F^{\mathcal{B}}$ is just $[F(P_n)]_{\mathcal{B}}$, the coordinates of $F(P_n)$ with respect to $P_1, \ldots, P_n$. But with P defined by (13), the relationships (12) and (11c) hold and

$$FP_n = P_{n-1} - a_1 P_n = 0P_1 + 0P_2 + \cdots + 1P_{n-1} + (-a_1)P_n$$

so

$$[FP_n]_{\mathcal{B}} = \begin{bmatrix} 0 \\ \vdots \\ 0 \\ 1 \\ -a_1 \end{bmatrix}$$

Similarly, the $(n-1)$ column of $^{\mathcal{B}}F^{\mathcal{B}}$ is

$$[FP_{n-1}]_{\mathcal{B}} = [P_{n-2} - a_2 P_n]_{\mathcal{B}} = \begin{bmatrix} 0 \\ \vdots \\ 0 \\ 1 \\ 0 \\ -a_2 \end{bmatrix} \leftarrow (n-2) \text{ row}$$

---

†We showed how to represent a matrix with respect to a new basis $\mathcal{B}$ in Section 3-3.

and, for $j = 0, 1, \ldots, n - 2$, the $(n - j)$ column is

$$[FP_{n-j}]_{\mathcal{B}} = [P_{n-j-1} - a_{j+1}P_n]_{\mathcal{B}} = \begin{bmatrix} 0 \\ \vdots \\ 1 \\ \vdots \\ 0 \\ \vdots \\ 0 \\ -a_{j+1} \end{bmatrix} \leftarrow (n - j - 1) \text{ row}$$

Only the first column requires a little thought.

Since
$$\begin{aligned} FP_1 &= F(F^{n-1}G + a_1 F^{n-2}G + \cdots + a_{n-1}G) \\ &= F^n G + a_1 F^{n-1}G + \cdots + a_{n-1}FG \\ &= (F^n + a_1 F^{n-1} + \cdots + a_{n-1}F)G \end{aligned}$$

we may use the Cayley-Hamilton theorem [Theorem 4-6-7],

$$F^n + a_1 F^{n-1} + \cdots + a_{n-1}F + a_n I = 0 \tag{16}$$

to write
$$F^n + a_1 F^{n-1} + \cdots + a_{n-1}F = -a_n I$$

whence
$$FP_1 = (-a_n I)G = -a_n P_n$$

and
$$[FP_1]_{\mathcal{B}} = \begin{bmatrix} 0 \\ 0 \\ \vdots \\ -a_n \end{bmatrix}$$

Thus
$$^{\mathcal{B}}F^{\mathcal{B}} = \begin{bmatrix} 0 & 1 & \cdots & 0 \\ 0 & 0 & \cdots & \cdot \\ \cdot & \cdot & & \cdot \\ \cdot & \cdot & & \cdot \\ \cdot & \cdot & \cdots & 0 \\ \cdot & \cdot & \cdots & 1 \\ -a_n & -a_{n-1} & \cdots & -a_1 \end{bmatrix} = F_c$$

and the controllable canonical form matrix F_c is just the representation of the given matrix F with respect to the basis $\mathcal{B} = \{P_1, \ldots, P_n\}$. Of course, $[G]_{\mathcal{B}} = [P_n]_{\mathcal{B}} = [0 \quad 0 \ldots 1]^T = G_c$ and the matrix $P = [P_1, \ldots, P_n]$ gives the desired similarity transformation† $x = Px_c$.   □

---

†The reader who worked Exercise 14(a) of Section 6-2 can let $\mathcal{B}' = \{e_1, \ldots, e_n\}$ (the usual basis), so $F = {}^{\mathcal{B}'}F^{\mathcal{B}'}$. Then with $F_c = {}^{\mathcal{B}}F^{\mathcal{B}} = ({}^{\mathcal{B}}I^{\mathcal{B}'})({}^{\mathcal{B}'}F^{\mathcal{B}'})({}^{\mathcal{B}'}I^{\mathcal{B}})$, we may identify $P = {}^{\mathcal{B}'}I^{\mathcal{B}}$ since $[I(P_i)]_{\mathcal{B}'} = [P_i]_{\mathcal{B}'} = P_i$.

Together, Theorems 2 and 3 completely characterize those systems which are invertibly transformable to controllable canonical form, so for convenience we combine them to state the following result:

THEOREM 4

The representation (F, G, H) is algebraically equivalent to the controllable canonical form (F_c, G_c, H_c) of equations (2) if and only if

$$\text{rank } [F^{n-1}G, F^{n-2}G, \ldots, FG, G] = n$$

If this condition is met, the form is obtained via the transformation $x = Px_c$, where P is given by equation (13). □

We may now use this characterization to sharpen our discussion of state feedback. For ease of visualization, we show in Figure 7–8 the matrix block diagram of the system represented by (F, G, H).

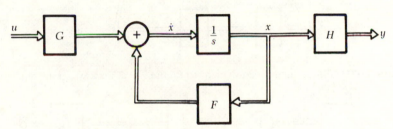

Figure 7–8   The open loop system (F, G, H).

In Figure 7–9 we introduce the linear controller represented by the matrix K which generates the feedback signal Kx. Notice that in Figure 7–9 we have decided to denote the input to the overall closed loop system by v (the reference) and we continue to let u denote the input to the block G.

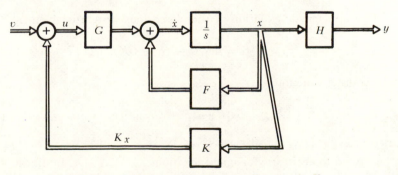

Figure 7–9   Adding the controller matrix K.

Thus $u = v + Kx$ and the dynamic equations describing the closed loop system are

$$\dot{x} = Fx + Gu$$
$$= Fx + G(v + Kx)$$

or $\qquad\qquad \dot{x} = (F + GK)x + Gv$

These are the dynamic equations of the representation $(F + GK, G, H)$, which we could have determined graphically by the manipulations shown in Figures 7–10 and 7–11.

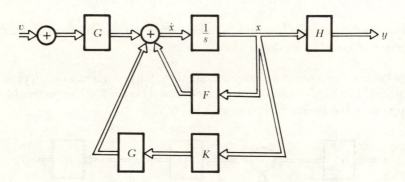

Figure 7–10   Block diagram equivalent to Figure 7–9 as far as $\dot{x}, v, y$ are concerned.

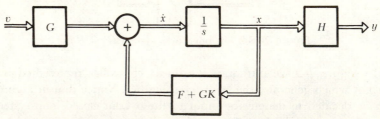

Figure 7–11   The equivalent system $(F + GK, G, H)$

The characteristic equation of the closed loop system is given by

$$\chi_{F+GK}(s) = \det[sI - (F + GK)]$$

and its transfer function by

$$\widehat{\mathfrak{Z}}(s) = H[sI - (F + GK)]^{-1}G \tag{17}$$

[see Exercise 8]. Thus, when we ask whether or not the poles of (F, G, H) can be shifted to arbitrarily assigned locations $\widehat{\lambda}_1, \widehat{\lambda}_2, \ldots, \widehat{\lambda}_n$ by linear state feedback, we are really asking whether or not there exists a matrix K such that the polynomial $\chi_{F+GK}(s) = \det[sI - (F + GK)]$ has $\widehat{\lambda}_1, \widehat{\lambda}_2, \ldots, \widehat{\lambda}_n$ as its roots. We present the answer to this question in the next theorem.

THEOREM 5

Let
$$\widehat{\chi}(s) = s^n + \widehat{a}_1 s^{n-1} + \cdots + \widehat{a}_n$$
$$= (s - \widehat{\lambda}_1) \ldots (s - \widehat{\lambda}_n)$$

be an arbitrary polynomial. There exists a matrix K such that

$$\chi_{F+GK}(s) = \det[sI - (F + GK)] = \widehat{\chi}(s)$$

if and only if rank $[F^{n-1}G, F^{n-2}G, \ldots, FG, G] = n$.

Proof

The "if" part is just a restatement of Theorems 1 and 4. We defer the proof of the "only if" part until Theorem 1 of Section 7–5, where we generalize this whole discussion to the multiple-input multiple-output case where G is $(n \times m)$. $\square$

As a matter of fact, we have an algorithm for constructing the K of Theorem 5.

COROLLARY 6

Calling $\chi_F(s) = s^n + a_1 s^{n-1} + \cdots + a_n$, with $\widehat{\chi}(s) = s^n + \widehat{a}_1 s^{n-1} + \cdots + \widehat{a}_n$ as the polynomial whose roots are the desired eigenvalues, we get K_c, the feedback of controllable canonical state variables, from equations (9) as the matrix

$$K_c = [(a_n - \widehat{a}_n) \, (a_{n-1} - \widehat{a}_{n-1}) \cdots (a_1 - \widehat{a}_1)]$$

and, by equations (10) and (13), the feedback matrix in the actual system is

$$K = K_c P^{-1}$$

$$= [(a_n - \widehat{a}_n) \ldots (a_1 - \widehat{a}_1)] \, [F^{n-1}G, \ldots, FG, G] \left(\begin{bmatrix} 1 & 0 & \cdots & 0 \\ a_1 & 1 & \cdots & 0 \\ a_2 & a_1 & \cdots & \cdot \\ \cdot & \cdot & & \cdot \\ \cdot & \cdot & & \cdot \\ a_{n-1} & a_{n-2} & \cdots & 1 \end{bmatrix} \right)^{-1} \square$$

To illustrate, we had in Example 2 a system (F, G, H) with $\chi_F(s) = s^2 + 5s + 6$ and a desired $\hat{\chi}(s) = s^2 + 11s + 10$. By Corollary 6,

$$K = [-4 \quad -6] \left(\begin{bmatrix} -2 & 1 \\ -3 & 1 \end{bmatrix} \begin{bmatrix} 1 & 0 \\ 5 & 1 \end{bmatrix} \right)^{-1} = [-4 \quad -6] \begin{bmatrix} 3 & 1 \\ 2 & 1 \end{bmatrix}^{-1}$$

$$= [-4 \quad -6] \begin{bmatrix} 1 & -1 \\ -2 & 3 \end{bmatrix} = [8 \quad -14]$$

which is the same result we found earlier.

In this section we have found that if $\mathcal{C} = [F^{n-1}G, \dots, FG, G]$ is nonsingular we can take the state x of (F, G, H), find a feedback signal Kx to add to the input v, and thereby shift the eigenvalues to wherever we want them. This approach tacitly assumes that the state x is available to feed back. What if it isn't? It's not hard to conceive of situations where one or more of the components, say x_2 and x_3, of a state vector $x = [x_1, x_2, x_3, \dots, x_n]^T$ may be inaccessible to us for physical reasons—x_2 might be the angular momentum of a particle inside a nuclear furnace or the velocity of a diaphragm inside an astronaut's heart.

In the next section we study ways of using feedback to provide estimates of inaccessible state variables from those quantities we can actually observe.

EXERCISES FOR SECTION 7–1

1. Consider the basic feedback configuration of Example 1. Suppose that A is an amplifier over whose gain we have very little control. For example, A may be built with very unreliable components and have its amplification wandering all over the place from, say, 10^7 to 10^{10}, quite a variation. Suppose that B can be designed with great precision.

(a) Let $B = \frac{1}{100}$, gotten by a "fractional pick-off," say. Show that the closed loop gain will be a very precise amplification of 100. Give a bound on the error or deviation from this desired gain of 100.

(b) Let $B(s) = \frac{1}{s}$, the transfer function of an integrator. Show that this gives us an analog computer differentiation capability.

(c) In Figure 7–12 we show a practical electronic feedback circuit used to construct the operational amplifiers used in analog computers. Assuming the gain A to be enormous, for v_2 to be finite we must have $e \approx 0$. Then $i_1 \approx i_2$, so

$$\frac{v_1 - e}{R_1} \approx i_2 = C \frac{d}{dt}(e - v_2)$$

Show that this implies $v_2(t) = v_2(t_0) + \int_{t_0}^{t} v_1(\tau) \, d\tau$ for appropriate R and C, and that we have an integrator. [We can set the initial condition voltage $v_2(t_0)$ by just putting an initial charge on the capacitor.] What happens if R_1 and C are interchanged?

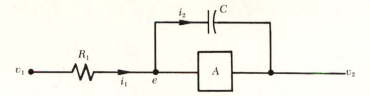

Figure 7-12   Electronic integration.

(d)   Think of our method of successive approximations, equation (20) of Section 5–3, as a feedback operation with B corresponding to the integral operator S. What is A? Draw a block diagram for the process.

(e)   Think of B as a small error, $B = E$, made in numerical calculations and fed back and used in some numerical operation A.

2.   (a)   Check the "work by inspection" which yielded the transfer function of Figure 7–5 by computing $(sI - F_k)$, $(sI - F_k)^{-1}$, $(sI - F_k)^{-1}G_k$, and $H_k(sI - F_k)^{-1}G_k$. Did you notice that the block diagrams of Figures 7–2 and 7–5 were in controllable canonical form? Find a feedback matrix K so that this system has poles at -6 and -8.

(b)   Carry out the analysis and modify Figures 7–2 and 7–5 and matrices (F_k, G_k, H_k) for the cart of mass M.

✔ 3.   Suppose that instead of feeding back state information, we only had partial state information available from which to generate control. In the example of Figure 7–2, let only the output y be available, and let a linear controller $\mathcal{K}$ generate $u = Ky$. Draw a closed loop diagram showing the controller as we did in Figures 7–3 and 7–5. Try to find a matrix K which yields closed loop poles at $\lambda_1 = -6$ and $\lambda_2 = -8$, as we did with full state feedback. Notice that you don't have complete freedom now in placing poles. Where can you place poles? In our cart example, **position** (y) **feedback** is not enough; we also need **rate** ($\dot{y}$) **feedback** for modal controllability.

✔ 4.   Let $F = \begin{bmatrix} 0 & -2 \\ 1 & -3 \end{bmatrix}$, $G = \begin{bmatrix} 1 \\ 1 \end{bmatrix}$, and $H = [0 \ \ 1]$.

(a)   Draw the block diagram for (F, G, H) and compute the transfer function $\mathcal{L}(s)$. Do not cancel.

(b)   From the uncancelled, second-order $\mathcal{L}(s)$ of part (a), give the controllable canonical realization (F_c, G_c, H_c) and draw the block diagram.

(c)   Show that the state variables of (F, G, H) and (F_c, G_c, H_c) are related by $x = Px_c$ where $P = \begin{bmatrix} 1 & 1 \\ 1 & 1 \end{bmatrix}$.

(d)   Notice that P is *not* invertible, but that $PF_c = FP$, $PG_c = G$, and $H_c = HP$.

✔ 5.   In Exercise 1 of Section 6–5 you found an expansion for $\text{adj}(sI - F)$

$$\text{adj}(sI - F) = s^{n-1}I + s^{n-2}B_1 + s^{n-3}B_2 + \cdots + sB_{n-2} + B_{n-1}$$

where

$$B_1 = F + a_1I, \quad B_2 = FB_1 + a_2I, \ldots$$

$$B_{n-i} = FB_{n-(i+1)} + a_{n-i}I, \ldots$$

$$B_{n-1} = FB_{n-2} + a_{n-1}I$$

and

$$0 = FB_{n-1} + a_nI.$$

(a)   Here is still another derivation of equations (4) and (5). Use the special column structure of F_c, equation (2), and the fact that multiplication on the right by $G_c = e_n$ picks out the nth column to work out $\text{adj}(sI - F_c)G_c$.

[*Hint:* $B_1 G_c = $ *n*th col of $(F_c + a_1 I) = e_{n-1}$, $B_2 G_c = F \, _{\text{col}}^{\text{nth}}(B_1) + a_2 \, _{\text{col}}^{\text{nth}}(I)$

$$= F e_{n-1} + a_2 e_n$$

$$= e_{n-2}, \text{ etc.}]$$

(b)   Show by eliminating the B_i that

$\text{adj}(sI - F) = s^{n-1}I + s^{n-2}(F + a_1 I) + s^{n-2}(F^2 + a_1 F + a_2 I) + \cdots$

$\qquad + s(F^{n-2} + a_1 F^{n-3} + \cdots + a_{n-2}I) + (F^{n-1} + a_1 F^{n-2} + \cdots + a_{n-1}I)$

Use this to express $(sI - F)^{-1}G$ (for a fixed s) as a linear combination of the matrices $F^k G$ for $0 \le k < n$.

(c)   From $(sI - F)^{-1} = \left[s\left(I - \frac{1}{s}F \right) \right]^{-1} = \frac{1}{s}\left[\left(I - \frac{1}{s}F \right)^{-1} \right] = \frac{1}{s} \sum_{j=0}^{\infty} \left(\frac{1}{s}F \right)^j$

[see Section 6–3, Exercise 17(c)], and Cramer's Rule (equation 6–5–1), derive the expression of part (b) in yet another way.

6.   Prove that $F^n G + a_n F^{n-1}G + a_{n-1}F^{n-2}G + \cdots + a_1 G = 0$ if and only if

$$\chi_F(\lambda) = \lambda^n + a_n \lambda^{n-1} + a_{n-1}\lambda^{n-2} + \cdots + a_1$$

provided $\mathcal{C}(F, G)$ is invertible.

7.   Try to carry out the reduction to controllable canonical form for

$$F = \begin{bmatrix} 2 & 0 & 0 \\ 0 & 1 & 0 \\ 0 & 0 & 3 \end{bmatrix}, \quad G_1 = \begin{bmatrix} 1 \\ 0 \\ 1 \end{bmatrix}$$

What goes wrong? Make the successful reduction when we replace G_1 by $G_2 = \begin{bmatrix} 1 \\ 1 \\ 1 \end{bmatrix}$.

✔ **8.**   Play around with equation (17), using

$$[sI - (F + GK)]^{-1} = [(sI - F)(I - (sI - F)^{-1}GK)]^{-1}$$
$$= [I - (sI - F)^{-1}GK]^{-1}(sI - F)^{-1}$$

and the matrix identity

$$(I - AB)^{-1} = I + A(I - BA)^{-1}B \text{ (verify)}$$

to relate the transfer function of the system of Figure 7–11 to that of (F, G, H).

✔ **9.**   In Exercise 4 of Section 6–4 we introduced the **observable canonical form,** which we now denote by (F_o, G_o, H_o) as a realization of a scalar transfer function.
Let (F, G, H) be given, and find necessary and sufficient conditions under which (F, G, H) is algebraically equivalent to (F_o, G_o, H_o), where in the notation of equation (1)

$$F_o = \begin{bmatrix} 0 & 0 & \cdots & 0 & -a_n \\ 1 & 0 & \cdots & 0 & -a_{n-1} \\ 0 & 1 & \cdots & \cdot & \cdot \\ \vdots & \vdots & \vdots & \vdots & \vdots \\ 0 & 0 & \cdots & 1 & -a_1 \end{bmatrix}, \quad G_o = \begin{bmatrix} b_n \\ b_{n-1} \\ \vdots \\ b_1 \end{bmatrix}$$

and $H_o = [0 \quad 0 \quad \cdots \quad 0 \quad 1]$.

[*Answer:* If $x = Px_o$, or $x_o = P^{-1}x$, then

$$P^{-1} = \begin{bmatrix} a_{n-1} & a_{n-2} & \cdots & a_1 & 1 \\ a_{n-2} & a_{n-3} & \cdots & 1 & 0 \\ \vdots & \vdots & & \vdots & \vdots \\ a_1 & 1 & \cdots & 0 & 0 \\ 1 & 0 & \cdots & 0 & 0 \end{bmatrix} \begin{bmatrix} H \\ HF \\ \vdots \\ HF^{n-2} \\ HF^{n-1} \end{bmatrix}$$

where $\mathcal{O} = [H^T, (HF)^T, \ldots, (HF^{n-1})^T]^T$ is the **observability matrix.**]

✔ **10.** **(a)** Compute the controllability matrix $\mathcal{C}_c$ of the matrices (F_c, G_c, H_c) of the controllable canonical form. Is (F_c, G_c, H_c) controllable?

(b) Compute the controllability matrix $\mathcal{C}_o$ of the observable canonical form (F_o, G_o, H_o) of Exercise 9 for the case $n = 3$. Can you tell whether (F_o, G_o, H_o) is controllable?

✔ **11.** **(a)** Show that $(F + GK, G, H)$ is controllable if and only if (F, G, H) is controllable.

(b) Show that controllability is preserved under any invertible state transformation P.

12. Redo Exercise 9 of Section 6–5, finding the K by Corollary 6. Check your previous solutions of Exercise 6–4–6(d) and the discrete-time case of Exercise 4–6–7(f) and (g).

✔ **13.** We are given the scalar transfer function

$$\mathcal{Y}(s) = \frac{-7s - 7}{s^2 + 6s + 5}$$

(a) Give the controllable canonical realization (F_c, G_c, H_c) of $\mathcal{Y}(s)$.

(b) Compute the controllability and observability matrices $\mathcal{C}_c$ and $\mathcal{O}_c$. Is the realization (F_c, G_c, H_c) controllable? Observable? [For the moment, define a realization as being observable if its observability matrix is nonsingular.]

(c) Find a feedback matrix K_c so that the closed loop system shown in Figure 7–13 has poles at -2 and -3.

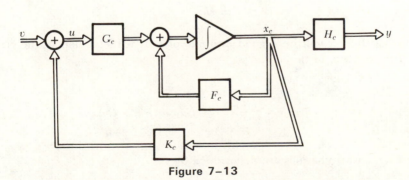

Figure 7–13

(d) Find the transfer function of the closed loop system of part (c).

(e) Compute the controllability and observability matrices $\mathcal{C}_{ck}$ and $\mathcal{O}_{ck}$ for the closed loop system of part (c). Compare with the results of part (b).

7–2   Observers and Controllers

Suppose that we have a dynamic system described by the equations

$$\dot{x} = Fx + Gu$$
$$y = Hx$$

and that the input u and output y are available for observation; but that for one reason or another† the state vector x is not directly available. Can we find out what x is from the known matrices F, G, H and the functions u and y? We know that $x(t)$ is given by

$$x(t) = e^{F(t-t_0)}x(t_0) + \int_{t_0}^{t} e^{F(t-\tau)}Gu(\tau)\, d\tau$$

which is the solution of the given differential equation, and that $x(t)$ clearly depends upon the initial condition $x(t_0)$ as well as the given F, G, and u; so finding $x(t)$ from the given data boils down to finding the initial state $x(t_0)$.

SIMULATION AND STATE ESTIMATION

Since we know the matrices F, G, and H, we can *simulate* the system, perhaps with an analog computer realization, and have a *model* obeying the same dynamic equations (F, G, H) as the actual system but whose state is accessible to us. Letting $\hat{x}$ denote the state vector of the model and $\hat{y}$ the output, if we drive it with the same input u as the system, then

$$\dot{\hat{x}} = F\hat{x} + Gu$$
$$\hat{y} = H\hat{x}$$

and hopefully $\hat{x}$ will be a good estimate of x. If we let $\tilde{x}$ denote the error in our estimate, so that

$$\tilde{x} = x - \hat{x}$$

then

$$\dot{\tilde{x}} = \dot{x} - \dot{\hat{x}} = (Fx + Gu) - (F\hat{x} + Gu)$$
$$= F(x - \hat{x}) = F\tilde{x}$$

and the error dynamics are determined by F.

---

†It may be helpful to think of (F, G, H) as a spaceship, to which we broadcast the control signal u and monitor the performance y from Earth. It would be very cumbersome to monitor every state variable x_i of the spacecraft and transmit all that information back down to the control center.

The solution of $\dot{\tilde{x}} = F\tilde{x}$ is

$$\tilde{x}(t) = e^{F(t-t_0)}\tilde{x}(t_0) = e^{F(t-t_0)}(x(t_0) - \hat{x}(t_0))$$

and this reveals that if we set the initial condition of the model exactly, $\hat{x}(t_0) = x(t_0)$, then $\tilde{x} \equiv 0$ and our model will estimate the state exactly with no error. If we make a mistake and $\hat{x}(t_0) \neq x(t_0)$, then the error in the model's estimate varies like $e^{F(t-t_0)}$. As long as the eigenvalues of F have negative real parts, the error dies out exponentially and the model does better and better as time goes on. If, however, any of the eigenvalues of F have positive real parts, then even a tiny initial error $\tilde{x}(t_0)$ will grow rapidly and the model will be unsatisfactory. This observation will be developed in our discussion of stability in Section 7-4.

The trouble with this approach, which is illustrated in Figure 7-14, is that it is an *open-loop* approach; the error is dictated by the given matrix F, over which we have no control. We are thus motivated to seek a *closed-loop* estimator whose error dynamics we can control.

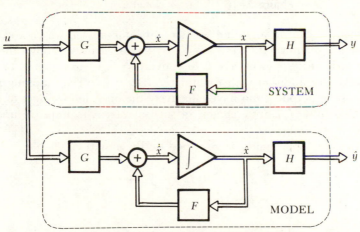

Figure 7-14   An open-loop model.

Since we have the actual output y available, we may compare it with our model output $\hat{y}$ and feed back the difference $(y - \hat{y})$ as a correction term. At first glance we might employ the diagram of Figure 7-15 for the feedback

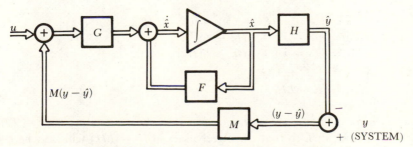

Figure 7-15   A tentative closed-loop model.

model, where the $(m \times q)$ matrix M is needed to convert the $(q \times 1)$ vector $(y - \hat{y})$ to an $(m \times 1)$ vector suitable for addition to input u.

Upon further thought, however, we see that we can make more effective use of the information in $(y - \hat{y})$ if we feed it right into the summer at the integrator input, as shown in Figure 7–16.

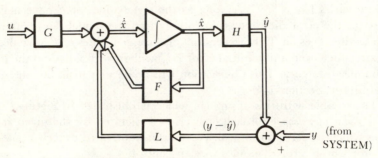

Figure 7–16 A better closed loop model.

Since we can do anything we like with our model, we choose not to risk losing any information in $(y - \hat{y})$ by having it pass through the block G. We need an $(n \times q)$ matrix L to convert $(y - \hat{y})$ to an $(n \times 1)$ vector compatible for addition with Fx and Gu. If $L = 0$, we are making no use of the system output y and have effectively the open loop system of Figure 7–14. Hopefully, L can be chosen in such a way as to improve upon the open-loop performance.

The closed loop model is governed by the equations

$$\dot{\hat{x}} = F\hat{x} + L(y - \hat{y}) + Gu$$
$$\hat{y} = H\hat{x}$$

which can be combined to yield

$$\dot{\hat{x}} = (F - LH)\hat{x} + (Gu + Ly).$$

Again letting $\tilde{x} = x - \hat{x}$, the error, we have

$$\dot{\tilde{x}} = \dot{x} - \dot{\hat{x}} = (Fx + Gu) - [(F - LH)\hat{x} + Gu + Ly]$$
$$= F(x - \hat{x}) + LH\hat{x} - Ly$$
$$= F(x - \hat{x}) + LH\hat{x} - L(Hx)$$
$$= (F - LH)(x - \hat{x})$$

or $\qquad \dot{\tilde{x}} = (F - LH)\tilde{x}$

and the error dynamics are now determined by the eigenvalues of $(F - LH)$. Can we choose an L so that the eigenvalues of $(F - LH)$ will be wherever we want them, say all with large negative real parts so the error will die out rapidly? Equivalently, can we find a matrix L such that the characteristic equation $\chi_{F-LH}(s) = \det[sI - (F - LH)]$ has any desired roots, say $\tilde{\lambda}_1, \ldots, \tilde{\lambda}_n$?

This question has a familiar ring to it. We saw in the last section, in Theorem 5, conditions under which we could find a matrix K to put the roots of $\det[sI - (F + GK)]$ anywhere we wished. To get the unknown matrix L into an analogous position with K, we take the transpose

$$(F - LH) = (F^T - H^T L^T)^T$$
$$= (F^T + H^T(-L^T))^T$$

which is of the form $(\tilde{F} + \tilde{G}\tilde{K})^T$ where

$$\tilde{F} = F^T, \; \tilde{G} = H^T, \; \tilde{K} = -L^T$$

Since the determinant of a matrix is equal to the determinant of its transpose,

$$\det[sI - (F - LH)] = \det[sI - (F - LH)]^T$$
$$= \det[(sI)^T - (F - LH)^T]$$
$$= \det[sI - (\tilde{F} + \tilde{G}\tilde{K})]$$

the eigenvalues of $(F - LH)$ are precisely those of $(\tilde{F} + \tilde{G}\tilde{K})$. Invoking Theorem 7-1-5, we can solve for $\tilde{K}$, and hence L, if and only if the matrix

$$\tilde{\mathcal{C}} = [\tilde{F}^{n-1}\tilde{G}, \ldots, \tilde{F}\tilde{G}, \tilde{G}]$$
$$= [(F^T)^{n-1}H^T, \ldots, F^T H^T, H^T]$$
$$= \begin{bmatrix} HF^{n-1} \\ HF^{n-2} \\ \vdots \\ HF \\ H \end{bmatrix}^T$$

is nonsingular. But this last matrix under the transpose operation is just our old friend from Chapter 4, the observability matrix $\mathcal{O}$ of (F, G, H), turned upside down:

$$\mathcal{O} = \begin{bmatrix} H \\ HF \\ \vdots \\ HF^{n-1} \end{bmatrix}$$

which we will sometimes also denote by $\mathcal{O}(F, H)$ to emphasize its dependence upon the matrices F and H.

We can wrap all this discussion up in the following theorem.

THEOREM 1

Let

$$\widetilde{\chi}(s) = s^n + \widetilde{a}_1 s^{n-1} + \cdots + \widetilde{a}_n$$
$$= (s - \widetilde{\lambda}_1) \ldots (s - \widetilde{\lambda}_n)$$

be an arbitrary polynomial. There exists a matrix L such that $\chi_{F-LH}(s) = \det[sI - (F - LH)] = \widetilde{\chi}(s)$ if and only if

$$\text{rank} \begin{bmatrix} H \\ HF \\ \vdots \\ HF^{n-1} \end{bmatrix} = n$$

☐

Thus, we can construct a system, such as the one in Figure 7–16, whose state $\widehat{x}(t)$ will approach the actual state $x(t)$ arbitrarily fast as t increases, no matter what the initial states $\widehat{x}(t_0)$ and $x(t_0)$ are, as long as the observability matrix $\mathcal{O}(F, H)$ is nonsingular. If we redraw Figure 7–16 to display the estimate $\widehat{x}(t)$ as the "output" and the signals u and y as "inputs," we have the diagram of the **asymptotic state estimator** shown in Figure 7–17.

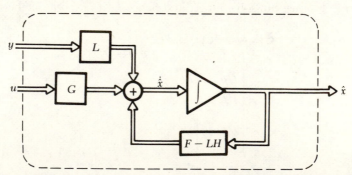

Figure 7–17 An asymptotic state estimator or observer.

Although we cannot observe the state $x(t)$ exactly without knowing the initial condition $x(t_0)$, we can observe it asymptotically [i.e., $\lim_{t \to \infty} \widehat{x}(t) = x(t)$]

if and only if the matrices (F, G, H) are *observable* in the sense of Chapter 4. Studying the asymptotic observability of the pair (F, H) has been transformed to studying the controllability of the pair $(\widetilde{F}, \widetilde{G})$, where $\widetilde{F} = F^T$ and $\widetilde{G} = H^T$—*duality* again. If $(\widetilde{F}, \widetilde{G})$ is controllable, we can use Corollary 7–1–6 to find the matrix $\widetilde{K}$ and hence the matrix $L = -\widetilde{K}^T$ which achieves the desired eigenvalues for the estimator.

The asymptotic state estimator of Figure 7–17 is also called an **observer,** since it provides information about the state of (F, G, H) from observations of the input and output. The selection of a "good" L accomplishes the design of the observer [see Exercise 5].

It should be noted that our observer is as complicated as the original system (F, G, H), and it may not always be economically feasible to construct an nth order model or even to simulate it on a computer. In Section 7–5, when we discuss multiple input-output systems, we will discuss state estimation with observers of lower dynamic order [see Exercise 8].

Let us now recall that one reason for trying to estimate the state was to be able to feed it back through K to control the modes of a given system whose state was not directly available. What happens if we feed back $\hat{x}$, our estimate, instead of the actual x? Will the resulting modes be greatly affected by the fact that $\hat{x}(t)$ only approaches $x(t)$ asymptotically?

In the next subsection we answer these questions for the single-input single-output case, and use state estimation to design a complete controller-observer system, called a **regulator.**

CONTROL BY STATE ESTIMATION

Suppose that we are given the dynamic equations

$$\dot{x} = Fx + Gu$$

$$y = Hx$$

of a single-input single-output system, and that the matrices $\mathcal{C}(F, G)$ and $\mathcal{O}(F, H)$ are nonsingular; i.e., (F, G, H) is both controllable and observable. Since (F, G) is controllable, if we don't like the characteristic equation

$$\chi_F(s) = s^n + a_1 s^{n-1} + \cdots + a_n$$

we can compute a feedback matrix K such that the closed loop system shown in Figure 7–18† has any desired coefficients for its characteristic equation

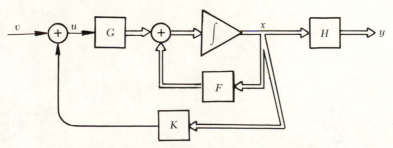

Figure 7–18   The feedback when x is accessible.

---

†Recall, we use thin lines to denote scalar quantities and fat lines to denote vector quantities.

$\chi_{F+GK}(s)$. This assumes that the state x is available with which to construct the control signal $u = v + Kx$.

If we cannot get at x, since (F, H) is observable we can design an asymptotic state estimator (observer) and feed $\hat{x}$ back instead of x, as shown in Figure 7–19.

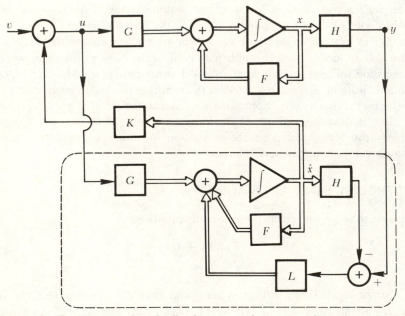

Figure 7–19 The complete feedback system when x must be estimated by $\hat{x}$.

The subsystem inside the dashed line is just the observer of Figure 7–17, which we drive with the signals u and y from the plant (F, G, H). The complete system of order $2n$ is described by the dynamic equations

$$\dot{x} = Fx + Gu$$
$$y = Hx$$
$$u = v + K\hat{x}$$
$$\dot{\hat{x}} = F\hat{x} + Gu + L(y - H\hat{x})$$

which can be combined to give

$$\dot{x} = Fx + GK\hat{x} + Gv$$
$$\dot{\hat{x}} = (F - LH + GK)\hat{x} + LHx + Gv$$
$$y = Hx$$

This last set of equations can be written in the block-partitioned matrix†
format as

$$
\begin{bmatrix} \dot{x} \\ \hline \dot{\hat{x}} \end{bmatrix} = \left[\begin{array}{c|c} F & GK \\ \hline LH & F - LH + GK \end{array} \right] \begin{bmatrix} x \\ \hline \hat{x} \end{bmatrix} + \begin{bmatrix} G \\ \hline G \end{bmatrix} v
$$

$$
y = [H \mid 0] \begin{bmatrix} x \\ \hline \hat{x} \end{bmatrix} \qquad \text{(verify)}
$$

which emphasizes that the $(2n)$-dimensional vector $\begin{bmatrix} x \\ \hline \hat{x} \end{bmatrix}$ is a valid choice

for the state of the composite system.

What is the characteristic equation of the composite system? If we call
the composite system (A, B, C) where

$$
A = \left[\begin{array}{c|c} F & GK \\ \hline LH & F - LH + GK \end{array} \right], \ B = \begin{bmatrix} G \\ \hline G \end{bmatrix}, \ C = [H \mid 0]
$$

this question is most simply answered by making a judicious change of

variables from $\begin{bmatrix} x \\ \hline \hat{x} \end{bmatrix}$ to $\begin{bmatrix} x \\ \hline \tilde{x} \end{bmatrix}$ where $\tilde{x} = x - \hat{x}$. This change of variables

is accomplished (verify) by the invertible transformation

$$
\begin{bmatrix} x \\ \hline \hat{x} \end{bmatrix} = P \begin{bmatrix} x \\ \hline \tilde{x} \end{bmatrix}
$$

where

$$
P = \left[\begin{array}{c|c} I & 0 \\ \hline I & -I \end{array} \right]
$$

and (verify)

$$
P^{-1} = \left[\begin{array}{c|c} I & 0 \\ \hline I & -I \end{array} \right] = P
$$

The algebraically equivalent system $(\bar{A}, \bar{B}, \bar{C})$, with

$$
\bar{A} = P^{-1}AP
$$
$$
\bar{B} = P^{-1}B
$$
$$
\bar{C} = CP
$$

will of course have the same characteristic equation and the same transfer
function as (A, B, C).

---

† In Exercise 13 of Section 4–2 we developed all of the necessary properties of block-partitioned
matrices.

Substituting for A and P

$$\bar{A} = \begin{bmatrix} I & 0 \\ I & -I \end{bmatrix} \begin{bmatrix} F & GK \\ LH & F - LH + GK \end{bmatrix} \begin{bmatrix} I & 0 \\ I & -I \end{bmatrix}$$

$$= \begin{bmatrix} I & 0 \\ I & -I \end{bmatrix} \begin{bmatrix} F + GK & -GK \\ F + GK & -F + LH - GK \end{bmatrix}$$

$$= \begin{bmatrix} F + GK & -GK \\ 0 & F - LH \end{bmatrix}$$

and, similarly,

$$\bar{B} = \begin{bmatrix} I & 0 \\ I & -I \end{bmatrix} \begin{bmatrix} G \\ G \end{bmatrix} = \begin{bmatrix} G \\ 0 \end{bmatrix}$$

and

$$\bar{C} = [H \mid 0] \begin{bmatrix} I & 0 \\ I & -I \end{bmatrix} = [H \mid 0]$$

Now, for any block-partitioned matrix $\begin{bmatrix} R & S \\ 0 & T \end{bmatrix}$ whose diagonal blocks R and T are invertible, multiplication by the matrix $\begin{bmatrix} R^{-1} & -R^{-1}ST^{-1} \\ 0 & T^{-1} \end{bmatrix}$ yields

$$\begin{bmatrix} R & S \\ 0 & T \end{bmatrix} \begin{bmatrix} R^{-1} & -R^{-1}ST^{-1} \\ 0 & T^{-1} \end{bmatrix} = \begin{bmatrix} I & 0 \\ 0 & I \end{bmatrix} = I_{(2n)}$$

so

$$\begin{bmatrix} R & S \\ 0 & T \end{bmatrix}^{-1} = \begin{bmatrix} R^{-1} & -R^{-1}ST^{-1} \\ 0 & T^{-1} \end{bmatrix}$$

Using this identity to compute $(sI - \bar{A})^{-1}$, we get

$$(sI - \bar{A})^{-1} = \begin{bmatrix} sI - (F + GK) & GK \\ 0 & sI - (F - LH) \end{bmatrix}^{-1}$$

$$= \begin{bmatrix} [sI - (F + GK)]^{-1} & -[sI - (F + GK)]^{-1}GK[sI - (F - LH)]^{-1} \\ 0 & [sI - (F - LH)]^{-1} \end{bmatrix}$$

If we let $\bar{\mathscr{E}}(s)$ denote the transfer function of $(\bar{A}, \bar{B}, \bar{C})$, and hence of (A, B, C), then

$$\bar{\mathscr{E}}(s) = \bar{C}(sI - \bar{A})^{-1}\bar{B}$$

$$= [H \mid 0] \begin{bmatrix} [sI - (F + GK)]^{-1} & -[sI - (F + GK)]^{-1}GK[sI - (F - LH)]^{-1} \\ 0 & [sI - (F - LH)]^{-1} \end{bmatrix} \begin{bmatrix} G \\ 0 \end{bmatrix}$$

$$= [H \mathbin{\vdots} 0] \left[\begin{array}{c} [sI - (F + GK)]^{-1}G \\ \hline 0 \end{array} \right]$$

$$= H[sI - (F + GK)]^{-1}G$$

which is precisely the same transfer function as if we had fed back x instead of $\hat{x}$!
Moreover, since

$$\det(sI - A) = \det(sI - \bar{A})$$

$$= \det \left[\begin{array}{c|c} sI - (F + GK) & GK \\ \hline 0 & sI - (F - LH) \end{array} \right]$$

because of the block triangular form, we have

$$\det(sI - A) = \det[sI - (F + GK)] \cdot \det[sI - (F - LH)]$$

so that the characteristic polynomial of the composite system is just the product of the characteristic polynomials of the matrices $(F + GK)$ and $(F - LH)$ which describe the exact state feedback system and the observer subsystem, respectively. This so-called **separation property** means that the dynamics of the control and of the observer are independent.

As long as the plant (F, G, H) is both controllable and observable, we can choose the control matrix K to shift the plant poles to wherever we want them and then, with complete freedom, choose the matrix L to make the observer have any desired dynamics, without disturbing the shifted plant poles.

It is customary to lump the matrix K in with the observer subsystem of Figure 7-19 and to call the resulting big feedback block the **controller** or **regulator,** as shown in Figure 7-20.

Once the matrices K and L have been chosen, the controller is just a

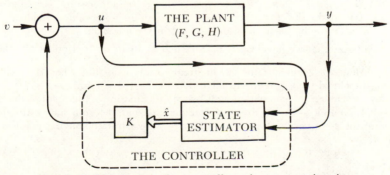

Figure 7-20 Design of a controller using state estimation.

dynamic system. Its transfer function can be computed (as an exercise?) and then be realized in any form the designer may choose, canonical or not.

In Section 7–5, when we discuss multivariable systems, we will present the design of a reduced order controller by estimating only those components $\{x_i\}$ of the state vector which are unavailable for direct measurement.

EXERCISES FOR SECTION 7–2

1.   Suppose that for some reason we have as our model a system (F, G, H), not all of whose innards are accessible. Suppose that we can't get access to $\hat{x}$ except by going through G as in Figure 7–15. Carry out the analysis to see what kind of state estimation we can achieve by proper choice of M.

2.   Find the invertible transformation $\bar{P}$ which turns a matrix upside down so that $\bar{\mathbb{C}} = (\bar{P}\mathbb{O})^T$ in the discussion preceding Theorem 1. Use this to derive an expression for L in terms F, H, $\chi_F(s)$, and the desired error polynomial $\hat{\chi}(s)$.

3.   Carry out the block diagram manipulations which reduce Figure 7–16 to Figure 7–17.

4.   We showed that we can set our model, Figure 7–16, with any initial condition $x(t_0)$ and, if $\mathbb{O}(F, H)$ is nonsingular, pick L such that our estimate $\hat{x}(t)$ converges as rapidly as we wish to the state $x(t)$. How can we use this to get an idea of what $x(t_0)$ was [i.e., an estimate of $x(t_0)$]?

5.   In designing an observer, why do we not always choose L so that the estimator eigenvalues have large negative real parts?

Calculate the transfer function matrix of the observer in Figure 7–17 and study its behavior as $\mathfrak{Re}(s) \to -\infty$.

✔ **6.   (a)**   Find the observability matrix $\mathbb{O}_o$ of the observable canonical form (F_o, G_o, H_o) of Section 7–1, Exercise 9. Is (F_o, G_o, H_o) observable? Controllable?

(b)   Find the observability matrix $\mathbb{O}_c$ for the controllable canonical form (F_c, G_c, H_c) for the case $n = 3$. Is (F_c, G_c, H_c) necessarily observable? Compare with Exercise 7–1–10.

7.   Is $(F + GK, G, H)$ *observable* if and only if (F, G, H) is observable? [See Exercises 7–1–11 and 7–1–13.]

✔ **8.**   Suppose that you have a second-order scalar system (F, G, H) and that you wish to estimate the state $x = [x_1, x_2]^T$ from input u and output y. Instead of designing a second-order observer as in Figure 7–17, suppose that you could estimate either x_1 *or* x_2. Since $y = Hx = h_1 x_1 + h_2 x_2$, you could get the other state variable free of charge. Try to do this for a second-order system in observable canonical form (F_o, G_o, H_o). Draw the block diagrams for aid in visualization.

✔ **9.**   Suppose that we have a "$\frac{1}{s^2}$ plant" such as a cart on frictionless wheels. We want to design a "compensator" as in Figure 7–21, such that when the switch is closed, the closed loop system will have poles at $(-1 + i2)$ and $(-1 - i2)$:

(a)   Design the compensator in two steps: First move the poles of the plant to $(-1 + i2)$ and $(-1 - i2)$ by state feedback; second, design a state estimator to generate the state vector. Use a second-order state estimator having poles at -2 and -3.

(b)   Draw the wiring diagram of the compensator and give its transfer function.

(c)   Compute the closed loop transfer function of the compensated system.

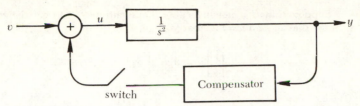

Figure 7-21 Compensation by feedback.

[*Hint:* Reference v has no influence on placing poles, so set $v = 0$. *Partial answers:*
$K = [-5 \quad -2]$, $L = \begin{bmatrix} 5 \\ 6 \end{bmatrix}$, overall transfer function is $\dfrac{1}{s^2 + 2s + 5}$.]

(d) Is the overall compensated system controllable? Observable? Check by writing matrices (A, B, C) for the overall system and computing the controllability and observability matrices $\mathcal{C}(A, B)$ and $\mathcal{O}(A, C)$ respectively for this four-dimensional system.

✔ **10.** In Exercise 14(d) of Section 6–3 we considered trying to replace an undesirable pole by a "nicer" one by using a cascade compensation network in the configuration of Figure 6–11. In this exercise we shall see how the controllability and observability matrices reveal the inherent shortcomings of this naive approach.

(a) In the connection of Figure 6–11, let Σ_1 have transfer function $\mathcal{Y}_1(s) = \dfrac{s - 2}{s + 2}$ and Σ_2 have transfer function $\mathcal{Y}_2(s) = \dfrac{1}{s - 2}$. Choose the output of the $\dfrac{1}{s - 2}$ block as the state variable x_1, call the output of the integrator in the $\dfrac{s - 2}{s + 2}$ block the state variable x_2, and write the dynamic state equations (F, G, H) for the cascaded second-order system, showing its structure as you did in Figure 6–42.

(b) Compute the controllability and observability matrices of your realization (F, G, H). What do you notice?

(c) Compute $(sI - F)^{-1}G$, $H(sI - F)^{-1}$, and $H(sI - F)^{-1}G = \mathcal{Y}(s)$. Where do cancellations occur?

(d) Now see what happens if you reverse the order of the cascaded blocks. Interchange Σ_1 and Σ_2 and take the output of the integrator in the $\dfrac{s - 2}{s + 2}$ block as x_1 and the output of the $\dfrac{1}{s - 2}$ block as x_2. Draw the block diagram and find the matrices $(\hat{F}, \hat{G}, \hat{H})$ of this realization. Repeat parts (b) and (c) for this realization and compare with what you found earlier.

(e) Transform each of your realizations (F, G, H) and $(\hat{F}, \hat{G}, \hat{H})$ to their algebraically equivalent diagonal representations (F_N, G_N, H_N) and $(\hat{F}_N, \hat{G}_N, \hat{H}_N)$, which reveal the modes $\lambda_1 = -2$, $\lambda_2 = +2$. What do lack of controllability and lack of observability look like in the block diagrams and matrices of these diagonal (Jordan) representations?

(f) In Exercise 6–3–14(d) you also tried pole cancellation by the feedback configuration shown in Figure 6–13. Carry out a detailed state variable analysis studying controllability, observability, and Jordan form manifestations of cancellations in the transfer function.

✔ **11.** This exercise shows how the observability matrix crops up in trying to compute the initial state $x(0)$ of a constant continuous-time system (F, G, H) from the zero-input (natural) response. Let $\hat{y}(t)$ denote the unforced response of the system, so that $\dot{x} = Fx$ and $\hat{y} = Hx$.

(a) Show that if H is invertible, then $x(0)$ is easily determined from a measurement of $\hat{y}(0)$.

(b)   Suppose that H is not invertible. Differentiate $\hat{y}(t)$ to get the equation $\dot{\hat{y}} = HFx$. Show that in *partitioned* format you get

$$
\left[\begin{array}{c} \hat{y} \\ \hline \dot{\hat{y}} \end{array}\right] = \left[\begin{array}{c} H \\ \hline HF \end{array}\right] x
$$

so x can be found from $\hat{y}$ and $\dot{\hat{y}}$ if the rank of $\left[\begin{array}{c} H \\ \hline HF \end{array}\right]$ is n.

(c)   Extend the approach begun in part (b) and show that

$$
\left[\begin{array}{c} y \\ \hline \dot{\hat{y}} \\ \hline \ddot{\hat{y}} \\ \hline \vdots \\ \hline \hat{y}^{(n-1)} \end{array}\right] = \left[\begin{array}{c} H \\ \hline HF \\ \hline HF^2 \\ \hline \vdots \\ \hline HF^{(n-1)} \end{array}\right] x = \Theta(F, H)x
$$

so that if the observability matrix has rank n, then x is known at every instant from $(n-1)$ differentiations of the unforced response.

(d)   Derive an expression relating the state $x(t)$, $\Theta(F, H)$, $(n-1)$ derivatives of the *complete* response $y(t)$, and $(n-2)$ derivatives of the input $u(t)$.

7–3   Controllability and Observability for Continuous Time

We now want to discuss controllability, reachability, and observability for the linear continuous-time *time-varying* system

$$
\dot{x}(t) = F(t)x(t) + G(t)u(t)
$$
$$
y(t) = H(t)x(t)
$$

where $F(\cdot)$, $G(\cdot)$, $H(\cdot)$ are $n \times n$, $n \times m$, and $q \times n$ matrix functions, respectively.

Recall that an "event" $(\tau, \bar{x})$—i.e., the system in state $\bar{x}$ at time τ—is **controllable** if there is some choice of the input which will transfer the system to some event $(t, 0)$ (i.e., to the zero state at some later time t). On the other hand, we say that $(\tau, \bar{x})$ is **reachable** if and only if for some time $t \leq \tau$ there is a choice of input function which will control the system from $(t, 0)$ to $(\tau, \bar{x})$.

Note that the specification of time τ in the event $(\tau, \bar{x})$ is crucial for *time-varying* systems, since the system dynamics may change radically with time.

Example 1

Let $a(t)$ be zero until time $t = 0$, and consider the system

$$
\left[\begin{array}{c} \dot{x}_1(t) \\ \dot{x}_2(t) \end{array}\right] = \left[\begin{array}{cc} 1 & 2 \\ a(t) & 1 \end{array}\right] \left[\begin{array}{c} x_1(t) \\ x_2(t) \end{array}\right] + \left[\begin{array}{c} 1 \\ 0 \end{array}\right] u(t)
$$

Before $t = 0$ we have $\dot{x}_2 = 0x_1 + 1x_2 + 0u$ (x_2 is unaffected by the input for $t < 0$), so the event $(\tau, \tilde{x}) = \left(-3, \begin{bmatrix} 0 \\ 1 \end{bmatrix}\right)$ is not reachable from $(t, 0) = \left(-5, \begin{bmatrix} 0 \\ 0 \end{bmatrix}\right)$. It is clear, for any state $\begin{bmatrix} b \\ c \end{bmatrix}$ with $c \neq 0$, that $\left(\tau, \begin{bmatrix} b \\ c \end{bmatrix}\right)$ cannot be reachable for any $\tau < 0$.   ◊

Of course, for a *time-invariant* system, if $(\tau, \tilde{x})$ is reachable, then so is $(\tau', \tilde{x})$ for any other time τ', and so we speak of the state $\tilde{x}$, rather than some event $(\tau, \tilde{x})$, as being reachable or controllable [Exercise 1].

Notice that these definitions only require that the input be able to drive the state from one value to another in a finite time. No constraint is made on the path followed in moving between the two states, nor on the type of input used. [See Exercise 7.] As before, we will call the system $(F(\cdot), G(\cdot), H(\cdot))$ **controllable (or reachable)**† *at time* τ if every event $(\tau, \bar{x})$ is controllable (or reachable) at time τ. Often we are interested in a particular interval of time, say $[t_0, t_1]$, so we say that a system is **controllable (or reachable) on** $[t_0, t_1]$ if every event $(t_0, \bar{x})$ can be transferred to $(t_1, 0)$ by some input $u_{[t_0, t_1]}$. Obviously, if a system is controllable on $[t_0, t_1]$, then it is controllable on any interval which contains $[t_0, t_1]$, but the same is not true for reachability (why?).

Recall our general formula for the change of state of the time-varying system $(F(\cdot), G(\cdot), H(\cdot))$:

$$x(t_1) = \Phi(t_1, t_0)x(t_0) + \int_{t_0}^{t_1} \Phi(t_1, \zeta)G(\zeta)u(\zeta)\, d\zeta \tag{1}$$

If we rewrite equation (1), which shows the transfer of $(t_0, x(t_0))$ to $(t_1, x(t_1))$ as

$$[x(t_1) - \Phi(t_1, t_0)x(t_0)] = \int_{t_0}^{t_1} \Phi(t_1, \zeta)G(\zeta)u(\zeta)\, d\zeta \tag{2a}$$

and multiply both sides by $\Phi^{-1}(t_1, t_0) = \Phi(t_0, t_1)$, we get

$$[\Phi(t_0, t_1)x(t_1) - x(t_0)] = \int_{t_0}^{t_1} \Phi(t_0, \zeta)G(\zeta)u(\zeta)\, d\zeta \tag{2b}$$

If we call

$$\hat{x} = [\Phi(t_0, t_1)x(t_1) - x(t_0)] \tag{3}$$

---

† In Section 4–2 we used the terms *completely controllable* and *completely reachable;* in the interest of brevity here we omit the adjective *completely*.

then reachability of $(t_1, \tilde{x})$ from $(t_0, 0)$ corresponds to

$$\hat{x} = +\Phi(t_0, t_1)\tilde{x} \tag{4}$$

while controllability of $(t_0, \bar{x})$ to $(t_1, 0)$ corresponds to

$$\hat{x} = -\bar{x} \tag{5}$$

Thus, both reachability and controllability are subsumed under the solution of

$$\hat{x} = \int_{t_0}^{t_1} \Phi(t_0, \zeta)G(\zeta)u(\zeta)\, d\zeta \tag{6}$$

for some admissible input segment $u_{[t_0, t_1)}$.

Because $\Phi(t_1, t_0)$ is invertible, $x(t_1)$ and $x(t_0)$ can be specified arbitrarily if and only if the same is true of $\hat{x}$ [Exercise 3], so we have the following result.

THEOREM 1

The continuous-time system $(F(\cdot), G(\cdot), H(\cdot))$ is both controllable and reachable on $[t_0, t_1]$ if and only if any arbitrary event $(t_0, x(t_0))$ can be transferred to the arbitrary event $(t_1, x(t_1))$.   □

Using equation (6), we also have the next result:

COROLLARY 2

The continuous-time system $(F(\cdot), G(\cdot), H(\cdot))$ is both controllable and reachable on $[t_0, t_1]$ if and only if the integral equation

$$\hat{x} = \int_{t_0}^{t_1} \Phi(t_0, \zeta)G(\zeta)u(\zeta)\, d\zeta$$

is solvable for $u_{[t_0, t_1)}$ for any specified state $\hat{x}$.   □

We have seen that a special feature of continuous-time linear systems is that even when the matrix F is not invertible, Φ is invertible and we may solve the state trajectory both forward and backward in time. In marked distinction to the discrete-time case (Corollary 4–3–9, where reachability was a stronger condition than controllability), *controllability and reachability of* (F, G, H) *are equivalent for continuous-time systems.*

Because of this, from now on we shall present the analyses in terms of controllability, expecting the reader to have no difficulty in translating our results into statements about reachability.

Solving the integral equation†

$$\hat{x} = \int_{t_0}^{t_1} \Phi(t_0, \zeta)G(\zeta)u(\zeta) \, d\zeta$$

of Corollary 2 for an admissible input function $u_{[t_0, t_1)}$, where $\hat{x}$ is a given state and Φ and G are known, is possible if and only if the integral operator $\mathcal{G}$ defined by

$$\mathcal{G}(u) = \int_{t_0}^{t_1} \Phi(t_0, \zeta)G(\zeta)u(\zeta) \, d\zeta \tag{7}$$

for each $u \in \Omega$ (the space of admissible inputs) has $\hat{x}$ in its range space. Thus, deciding controllability of $(F(\cdot), G(\cdot))$ reduces to deciding whether or not a certain operator $\mathcal{G}: \Omega \rightarrow X$ is *onto*.

In general this is a difficult problem, since it involves our searching for a vector function $u(\cdot)$ defined on $[t_0, t_1)$ which, when plugged into $\mathcal{G}(\cdot)$, yields the given $\hat{x}$. It is with some relief that we turn to the following lemma, which lets us replace the search for a *vector function* by the search for a *single vector*. In this lemma we will study a general matrix $A(\zeta)$ in the integrand rather than the particular case

$$A(\zeta) = \Phi(t_0, \zeta)G(\zeta)$$

that we are interested in for controllability (equation 6).

LEMMA 3

Let $A(\cdot)$ be an $(n \times m)$ smooth matrix function of time, and let $\hat{x}$ be a vector in X. Then there exists a vector function $u(\cdot)$ on $[t_0, t_1]$ such that

$$\hat{x} = \int_{t_0}^{t_1} A(\zeta)u(\zeta) \, d\zeta \tag{8}$$

if and only if there exists a single vector $z \in X$ such that

$$\hat{x} = \left[\int_{t_0}^{t_1} A(\zeta)A^*(\zeta) \, d\zeta \right] z \tag{9}$$

---

†A linear integral equation with kernel $\Phi(t_0, \cdot)G(\cdot)$.

(where $A^*(\zeta)$ is the adjoint of $A(\zeta)$); i.e., if and only if $\hat{x}$ is in $\mathfrak{R}(W)$, the range of W, for

$$W = \int_{t_0}^{t_1} A(\zeta)A^*(\zeta)\,d\zeta \tag{10}$$

Proof

"IF": Suppose that there exists a $z \in X$ such that

$$\hat{x} = \left[\int_{t_0}^{t_1} A(\zeta)A^*(\zeta)\,d\zeta\right]z$$

Then we may define a vector function of time in terms of z by setting

$$u(t) = A^*(t)z \tag{11}$$

for each $t \in [t_0, t_1]$. Note that since $A(t): U \to X$, we have $A^*(t): X \to U$ to assure us that $u(t)$ is indeed a vector in U and $u(\cdot) \in \Omega$. We then see that this $u(\cdot)$ does the job:

$$\int_{t_0}^{t_1} A(\zeta)u(\zeta)\,d\zeta = \int_{t_0}^{t_1} A(\zeta)A^*(\zeta)z\,d\zeta \qquad \text{(by definition of } u(\zeta))$$

$$= \left(\int_{t_0}^{t_1} A(\zeta)A^*(\zeta)\,d\zeta\right)z \qquad \begin{array}{l}\text{(using linearity}\\ \text{to "pull out" the}\\ \text{constant vector } z)\end{array}$$

$$= \hat{x} \qquad \text{(by definition of } z)$$

It is thus a very simple process to pass from a vector z which solves equation (9) to a vector function u which solves equation (8). Let us now turn to the more difficult part, in which we prove that if we can solve equation (8), then $\hat{x}$ must lie entirely within the range of W:

"ONLY IF": We note first that $W: X \to X$ and

$$W^* = \left(\int_{t_0}^{t_1} A(\zeta)A^*(\zeta)\,d\zeta\right)^* = \int_{t_0}^{t_1} (A(\zeta)A^*(\zeta))^*\,d\zeta$$

$$= \int_{t_0}^{t_1} A^{**}(\zeta)A^*(\zeta)\,d\zeta = \int_{t_0}^{t_1} A(\zeta)A^*(\zeta)\,d\zeta$$

$$= W$$

so that W is a **self-adjoint** or **Hermitian** operator. From our discussion in Section 4–4 (Theorem 4–4–11 and Exercise 4–4–21) we know that W has its range space and nullspace related by

$$\mathfrak{R}(W) = [\mathfrak{N}(W^*)]^{\perp} = [\mathfrak{N}(W)]^{\perp} \tag{12}$$

and that the state space X can be written as the *direct sum* of these subspaces:

$$\begin{aligned} X &= \mathfrak{R}(W) \oplus [\mathfrak{R}(W)]^{\perp} \\ &= \mathfrak{R}(W) \oplus \mathfrak{N}(W) \end{aligned} \tag{13}$$

Thus, every state vector $\hat{x}$ can be decomposed uniquely into the sum of two vectors

$$\hat{x} = x_1 + x_2$$

with $x_1 \in \mathfrak{R}(W)$ and $x_2 \in \mathfrak{N}(W)$, as indicated pictorially in Figure 7–22. In other words, we use the fact that W is Hermitian to express $\hat{x}$ in two parts, one of which is in the range of W, and the other of which is in the nullspace of W.

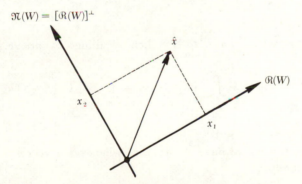

Figure 7–22   Illustrating the decomposition of a vector into components by projecting onto orthogonal subspaces.

First note that for any x_2 in $\mathfrak{N}(W)$ we have $Wx_2 = 0$, and so taking the inner product of x_2 and Wx_2,

$$0 = x_2^* W x_2 = \int_{t_0}^{t_1} x_2^* A(\zeta) A^*(\zeta) x_2 \, d\zeta = \int_{t_0}^{t_1} \|A^*(\zeta) x_2\|^2 \, d\zeta$$

Since the norm is never negative, and the integral of a positive function is positive, we must conclude that the integrand above must be identically zero;

in other words, we have

$$x_2 \in \mathfrak{N}(W) \Rightarrow A^*(\zeta)x_2 \equiv 0 \quad \text{on } [t_0, t_1] \tag{14}$$

Now let us suppose that our vector $\hat{x}$ actually satisfies an equality of the form

$$\hat{x} = \int_{t_0}^{t_1} A(\zeta)u(\zeta) \, d\zeta$$

We wish to prove that it lies entirely in $\mathfrak{R}(W)$—i.e., that its x_2 part is zero. By our proof of the "IF" part above we know that x_1, the projection of $\hat{x}$ on $\mathfrak{R}(W)$, must satisfy an equation of the form

$$x_1 = \int_{t_0}^{t_1} A(\zeta)u_1(\zeta) \, d\zeta$$

for some suitable function u_1, and so we deduce that

$$x_2 = \hat{x} - x_1 = \int_{t_0}^{t_1} A(\zeta)[u(\zeta) - u_1(\zeta)] \, d\zeta = \int_{t_0}^{t_1} A(\zeta)u_2(\zeta) \, d\zeta$$

where we have set $u_2 = u - u_1$. But then, recalling (14) above, we have

$$\|x_2\|^2 = x_2^* x_2 = x_2^* \int_{t_0}^{t_1} A(\zeta)u_2(\zeta) \, d\zeta = \int_{t_0}^{t_1} (A^*(\zeta)x_2)^* u_2(\zeta) \, d\zeta = 0$$

and so $x_2 = 0$, as required.

In other words, we have shown that whenever a vector $\hat{x}$ satisfies an equality of the form

$$\hat{x} = \int_{t_0}^{t_1} A(\zeta)u(\zeta) \, d\zeta$$

then its component orthogonal to the range of W is zero—and we thus conclude that $\hat{x}$ lies entirely within the range of W, as was to be proved.$\square$

Notice, in the proof of Lemma 3, that whenever the vector function $u(\cdot)$ exists we have an explicit formula for it: when

$$W = \int_{t_0}^{t_1} A(\zeta)A^*(\zeta) \, d\zeta \tag{15a}$$

is non-singular, we can solve equation (9)

$$\hat{x} = Wz \qquad \text{(15b)}$$

for the vector z

$$z = W^{-1}\hat{x} \qquad \text{(15c)}$$

and from equation (11), construct u by

$$u(t) = A^*(t)W^{-1}\hat{x} \qquad (t_0 \leq t \leq t_1) \qquad \text{(15d)}$$

so that

$$\int_{t_0}^{t_1} A(\zeta)u(\zeta)\,d\zeta = \hat{x} \qquad \text{(15e)}$$

Applying the above discussion and the lemma for the case in which $A(\zeta) = \Phi(t_0, \zeta)G(\zeta)$ and $\hat{x} = -\bar{x}$, we deduce the following two criteria:

THEOREM 4

The event $(t_0, \bar{x})$ is controllable if and only if, for some $t_1 > t_0$, $\bar{x}$ is in the range of

$$W(t_0, t_1) = \int_{t_0}^{t_1} \Phi(t_0, \zeta)G(\zeta)G^*(\zeta)\Phi^*(t_0, \zeta)\,d\zeta \qquad \text{(16)}$$

$\square$

COROLLARY 5

If for some $t_1 \geq t_0$ the matrix $W(t_0, t_1)$ is invertible, then $(F(\cdot), G(\cdot))$ is controllable and $(t_0, \bar{x})$ is transferred to $(t_1, 0)$ by the input function

$$u(t) = -G^*(t)\Phi^*(t_0, t)W^{-1}(t_0, t_1)\bar{x} \quad \text{for } t_0 \leq t \leq t_1 \qquad \text{(17)}$$

$\square$

In Exercise 5 the reader is asked to formulate similar criteria for reachability and to construct an input $u_{[t_0, t_1)}$ which will transfer (t_0, x^0) to (t_1, x^1). If we rummage around still further in the proof of Lemma 3, we find that equation (14) reveals quite a bit about the invertibility of W in terms of $A(\cdot)$. Since W is a linear map of X into X, it is invertible if and only if its nullspace is just the zero vector. From (14), a nonzero vector $x_2 \in \mathfrak{N}(W)$ implies that $A^*(\zeta)x_2 \equiv 0$ on $[t_0, t_1]$, whence the columns of $A^*(\cdot)$ are linearly dependent functions on $[t_0, t_1]$. Since the columns of $A^*(\cdot)$ are the rows of $A(\cdot)$, we have the useful corollary:

COROLLARY 6

The set of rows of an $n \times m$ matrix function $A(\cdot)$ is linearly independent on $[t_0, t_1]$ if and only if the $n \times n$ constant matrix

$$W = \int_{t_0}^{t_1} A(\zeta)A^*(\zeta)\, d\zeta$$

is non-singular.   □

Thus, if we have a set of n functions taking vector values in $\mathbf{C}^m$, say

$$
\begin{aligned}
r_1(t) &= [r_{11}(t) \quad r_{12}(t) \cdots r_{1m}(t)] \\
&\;\;\vdots \qquad\quad \vdots \qquad \vdots \qquad\quad \vdots \\
r_n(t) &= [r_{n1}(t) \quad r_{n2}(t) \qquad r_{nm}(t)]
\end{aligned}
\tag{18a}
$$

so that each $r_i(t) : \mathbf{R} \to \mathbf{C}^m$, then we can test them for linear independence by constructing the $n \times m$ matrix

$$
A(t) = \begin{bmatrix} r_1(t) \\ \vdots \\ r_n(t) \end{bmatrix}
\tag{18b}
$$

and studying the invertibility of the constant matrix W of Corollary 6 [see Exercises 8 and 9].

In particular, now, we can use this corollary together with Theorem 4 to get a simpler necessary and sufficient condition for controllability:

THEOREM 7

The system $(F(\cdot), G(\cdot), H(\cdot))$ is controllable on $[t_0, t_1]$ if and only if the rows of $\Phi(t_0, \cdot)G(\cdot)$ are linearly independent functions of time on $[t_0, t_1]$.

Proof

The system is controllable on $[t_0, t_1]$ iff for each $\bar{x} \in X$ the event $(t_0, \bar{x})$ is controllable to $(t_1, 0)$. From Theorem 4, this occurs iff

$$W(t_0, t_1) = \int_{t_0}^{t_1} \Phi(t_0, \varsigma)G(\varsigma)[\Phi(t_0, \varsigma)G(\varsigma)]^* \, d\varsigma$$

is non-singular and, by Corollary 6, this condition holds iff the rows of $\Phi(t_0, \cdot)G(\cdot)$ are linearly independent on $[t_0, t_1]$.   □

Example 2

Consider $(F(\cdot), G(\cdot))$ where

$$F(t) = \begin{bmatrix} 0 & e^{-5t} \\ 0 & 0 \end{bmatrix} \text{ and } G(t) = \begin{bmatrix} a(t) \\ b(t) \end{bmatrix}$$

The state transition matrix is (untypically) easy to find for this time-varying system

$$\Phi(t, \tau) = \begin{bmatrix} 1 & \frac{1}{5}(e^{-5\tau} - e^{-5t}) \\ 0 & 1 \end{bmatrix} \quad \text{(verify)}$$

Thus

$$\Phi(t_0, \varsigma)G(\varsigma) = \begin{bmatrix} a(\varsigma) + \frac{1}{5}b(\varsigma)(e^{-5\varsigma} - e^{-5t_0}) \\ b(\varsigma) \end{bmatrix}$$

so the rows are just the 1 by 1 functions:

$$r_1(\varsigma) = a(\varsigma) + \frac{1}{5}b(\varsigma)(e^{-5\varsigma} - e^{-5t_0})$$
$$r_2(\varsigma) = b(\varsigma)$$

Since a pair of nonzero vectors is dependent iff one is a constant multiple of the other, we can easily determine the conditions on $a(t)$ and $b(t)$ which cause $r_1(\cdot)$ and $r_2(\cdot)$ to be dependent [see Exercise 11].

For example, if $b(t) = 0$ so that $G(t) = \begin{bmatrix} a(t) \\ 0 \end{bmatrix}$, then we have $\Phi(t_0, \cdot)G(\cdot) = \begin{bmatrix} a(\cdot) \\ 0(\cdot) \end{bmatrix}$; the rows contain the zero function $0(\cdot)$ and hence are linearly dependent; and $(F(\cdot), G(\cdot))$ is uncontrollable for every t_0. This is just what we'd expect from inspection of $F(\cdot)$ and $G(\cdot)$, since the input u has no influence on x_2 when $b = 0$.   ◊

Since $\Phi(t_0, \varsigma) = \Phi(t_0, t_3)\Phi(t_3, \varsigma)$ for any time t_3, and since $\Phi(t_0, t_3)$ is invertible, the rows of $\Phi(t_0, \cdot)G(\cdot)$ are linearly independent on $[t_0, t_1]$ as and only as the rows of $\Phi(t_3, \cdot)G(\cdot)$ are independent. This lets us fix the first time variable of Φ at any convenient value to simplify calculations.

Example 3

Consider the constant system (F, G) where

$$F = \begin{bmatrix} 0 & 6 \\ 1 & -1 \end{bmatrix} \text{ and } G = \begin{bmatrix} b & 1 \\ 1 & a \end{bmatrix}$$

In Exercise 10 of Section 6–5, from the given eigenvectors v_1 and v_2 corresponding to $\lambda_1 = +2$ and $\lambda_2 = -3$, we get

$$\Phi(t, 0) = e^{Ft} = \frac{1}{5} \begin{bmatrix} (2e^{-3t} + 3e^{2t}) & (-6e^{-3t} + 6e^{2t}) \\ (-e^{-3t} + e^{2t}) & (3e^{-3t} + 2e^{2t}) \end{bmatrix}$$

Since

$$\Phi(t_0, \zeta) = e^{F(t_0 - \zeta)} = e^{Ft_0}e^{-F\zeta}$$

we have

$$\Phi(t_0, \zeta)G(\zeta) = e^{Ft_0}\, e^{-F\zeta}G(\zeta) \tag{19}$$

and since e^{Ft_0} is invertible, the rows of $\Phi(t_0, \cdot)G(\cdot)$ will be independent iff the rows of $e^{-F(\cdot)}G(\cdot)$ are. Multiplying out,

$$e^{F(-\zeta)}G = \frac{1}{5}\begin{bmatrix} (2e^{3\zeta}+3e^{-2\zeta}) & (-6e^{3\zeta}+6e^{-2\zeta}) \\ (-e^{3\zeta}+e^{-2\zeta}) & (3e^{3\zeta}+2e^{-2\zeta}) \end{bmatrix}\begin{bmatrix} b & 1 \\ 1 & a \end{bmatrix}$$

$$= \frac{1}{5}\begin{bmatrix} [e^{3\zeta}(-2)(3-b)+e^{-2\zeta}3(b+2)] & [e^{3\zeta}(-2)(-1+3a)+e^{-2\zeta}3(1+2a)] \\ [e^{3\zeta}(3-b)+e^{-2\zeta}(b+2)] & [e^{3\zeta}(-1+3a)+e^{-2\zeta}(1+2a)] \end{bmatrix}$$

Here each row of $\begin{bmatrix} r_1(\cdot) \\ r_2(\cdot) \end{bmatrix}$ is a (2×1) time function. For these nonzero functions to be dependent, one must be a scalar multiple of the other; so if we set $r_1(\zeta) = kr_2(\zeta)$, equating the first components yields the equation

$$e^{3\zeta}(3 - b)(-2 - k) + e^{-2\zeta}(b + 2)(3 - k) = 0, \quad \text{for all } \zeta$$

Similarly, equating the second components of r_1 and kr_2 yields the condition

$$e^{3\zeta}(-1 + 3a)(-2 + k) + e^{-2\zeta}(1 + 2a)(3 - k) = 0, \quad \text{for all } \zeta$$

But each of these last two equations is a linear combination of the two time functions $e^{3\zeta}$ and $e^{-2\zeta}$, which are known to be linearly independent on any interval; hence each coefficient of each equation must be zero:

$$\begin{cases} (3 - b)(-2 - k) = 0 \\ (b + 2)(3 - k) = 0 \\ (-1 + 3a)(-2 - k) = 0 \\ (1 + 2a)(3 - k) = 0 \end{cases}$$

Equivalently,

$$(-2 - k)[(3 - b) + (-1 + 3a)] = 0$$

$$(3 - k)[(b + 2) + (1 + 2a)] = 0$$

If the terms within the brackets are not zero, we get the contradictions $k = -2$ and $k = 3$; so the rows are not dependent for such a and b. If, however, a and b are chosen so that just one bracket vanishes, a value of k is found from the other equation. For instance, if $b = 3$ and $a = \frac{1}{3}$ then the second equation yields $k = 3$, and we see that for these values of a and b we have $r_1(\zeta) = 3r_2(\zeta)$. There is just one pair of values (a, b) which makes both brackets vanish [see Exercise 12(d)], and this yields a contradiction also. Thus, (F, G) is controllable for all values of (a, b) except the pairs $\left(3, \frac{1}{3}\right)$, $\left(-2, \frac{1}{2}\right)$.   ◊

As the last example illustrated, for constant systems (F, G, H) the rows of $\Phi(t_0, t)G(t)$ are just the rows of $e^{F(t_0 - \zeta)}G = e^{Ft_0}e^{-F\zeta}G$, and the rows of $e^{-F\zeta}G$ are linearly independent iff the rows of $e^{F\zeta}G$ are [Exercise 13]. Since $\mathcal{L}\{e^{Ft}G\} = (sI - F)^{-1}G$ and the Laplace transform is a one-to-one, onto linear operator, both $\mathcal{L}$ and $\mathcal{L}^{-1}$ preserve linear independence, and we have the following result:

COROLLARY 8

The constant system (F, G, H) is controllable at every t_0 iff the rows of $(sI - F)^{-1}G$ are linearly independent functions of the complex variable s, or, equivalently, iff the rows of $e^{-Ft}G$ are linearly independent functions of t on $[0, \infty)$.   □

The expression $(sI - F)^{-1}G$ of Corollary 8 is just the state transfer function, $\mathcal{X}_X(s)$, of Section 6-3, so it is not surprising to see it crop up in this section as we ask about driving the state vector from one value to another [see Exercise 14]. For scalar input systems $G = g$, an $n \times 1$ column, so $(sI - F)^{-1}g$ is a column. If we write, using Cramer's rule,

$$(sI - F)^{-1}g = \frac{1}{\det(sI - F)} \begin{bmatrix} r_1(s) \\ \vdots \\ r_n(s) \end{bmatrix}$$

we say that $(sI - F)^{-1}g$ **has no cancellations** if and only if the polynomials $r_1(s), \ldots, r_n(s), \det(sI - F)$ have no common factor; i.e., no entry $\dfrac{r_i(s)}{\det(sI - F)}$ has a pole-zero cancellation.

There is a fundamental relationship between the absence of cancellations in the state transfer function and controllability:

THEOREM 9

The single-input system (F, g) is controllable if and only if the state transfer function $\mathfrak{Y}_X(s) = (sI - F)^{-1}g$ has no cancellation.

Proof

We defer the proof of this theorem† until the discussion following Theorem 16. [See Exercise 14.]   □

Example 4

For $F = \begin{bmatrix} 0 & 6 \\ 1 & -1 \end{bmatrix}$ and $g = \begin{bmatrix} b \\ 1 \end{bmatrix}$,

$$(sI - F)^{-1} = \frac{1}{\chi_F(s)} \operatorname{adj}(sI - F) = \frac{1}{(s-2)(s+3)} \begin{bmatrix} s+1 & 6 \\ 1 & s \end{bmatrix}$$

so

$$(sI - F)^{-1}g = \frac{1}{(s-2)(s+3)} \begin{bmatrix} b\left(s + \dfrac{b+6}{b}\right) \\ s + b \end{bmatrix}$$

The polynomials $b\left(s + \dfrac{b+6}{b}\right)$, $s + b$, and $(s - 2)(s + 3)$ have no common factor unless b is -2 or $+3$, or $\dfrac{b+6}{b}$ is -2 or $+3$, or $\dfrac{b+6}{b} = b$. Testing these suspect cases, we see that $\dfrac{b+6}{b} = -2$ iff $b = -2$; $\dfrac{b+6}{b} = 3$ iff $b = 3$; and $\dfrac{b+6}{b} = b$ iff $(b - 3)(b + 2) = 0$. Thus, the system (F, g) is controllable for all values of b except $b = -2$ and $b = 3$.   ◊

---

†For an elementary but lengthy direct proof, see Butman and Sivan, "On Cancellations, Controllability and Observability," *IEEE Transactions on Automatic Control*, 1964, pp. 317–318.

CONTROLLABILITY OF CONSTANT SYSTEMS

We recall that for a *discrete-time* time-invariant linear system

$$x(t + 1) = Fx(t) + Gu(t), \quad F \text{ of dimension } n$$

we have reachability if and only if we have the equality

$$\text{rank}[G, FG, \dots, F^{n-1}G] = n$$

This criterion also holds for controllability if F is invertible, and we called the $n \times nm$ matrix

$$\mathcal{C} = \mathcal{C}(F, G) = [F^{n-1}G, F^{n-2}G, \dots, FG, G]$$

which has the same columns, just in different order, the **controllability matrix of F and G**.

It was perhaps surprising in our study of continuous-time constant systems in Section 7–1 to see this very same matrix pop up in a necessary and sufficient condition for shifting poles (**modal controllability**) by linear state feedback.

Still more surprisingly, we shall now see that the same criterion ensures the reachability and controllability of any *continuous-time* time-invariant linear system

$$\dot{x}(t) = Fx(t) + Gu(t)$$

This may seem surprising because in the discrete-time case the matrix F tells us how to update the state over one period of time, whereas in the continuous case, F tells us the instantaneous rate of change of the state, and it is the matrix e^{Ft} that tells us how to update the state over the interval of time t with zero input.

THEOREM 10

A continuous-time, time-invariant linear system

$$\dot{x}(t) = Fx(t) + Gu(t)$$

with state-space of dimension n is completely controllable if and only if

$$\text{rank}[G, FG, \dots, F^{n-1}G] = n$$

Proof

To say that our system is not controllable is to say that there is some state $\hat{x}$ which cannot be brought to the zero state. Let x_2 then be the nonzero component of $\hat{x}$ orthogonal to the subspace of controllable states. We then have that x_2 is in the nullspace of $W(0, t)$ for all $t \geq 0$, and so, by equation (14) in our proof of Lemma 3, we deduce that

$$G^*(\zeta)\Phi^*(0, \zeta)x_2 \equiv 0 \text{ for all } \zeta \geq 0$$

Using the fact that we are in the time-invariant case [see Exercise 13], this becomes

$$x_2^* e^{F\zeta} G \equiv 0$$

Then, equating like terms with the two sides considered as power series in ζ, we have that

$$x_2^* F^k G \equiv 0 \text{ for all } k \geq 0$$

But we know from our study of discrete-time systems, Lemma 4–3–7, that a nonzero vector can be orthogonal to the columns of the matrices $F^k G$ for all $k \geq 0$ if and only if it is orthogonal to all the columns of all the matrices $F^k G$ for all $0 \leq k < n$.† Since the system is controllable if and only if there are no nonzero vectors which satisfy all these orthogonality conditions, we conclude that our system is controllable if and only if

$$\text{rank}[G, FG, \ldots, F^{n-1}G] = n. \qquad \qquad \square$$

Example 5

For

$$F = \begin{bmatrix} 0 & 6 \\ 1 & -1 \end{bmatrix} \text{ and } G = \begin{bmatrix} b & 1 \\ 1 & a \end{bmatrix}$$

(the system we studied in Example 3), we construct the matrix

$$[G, FG] = \begin{bmatrix} b & 1 & \vdots & 6 & 6a \\ 1 & a & \vdots & (b-1) & (1-a) \end{bmatrix}$$

To determine the rank of this constant matrix we use so-called "elementary row operations." Multiplying the second row by b and subtracting the result

---

†This can also be deduced quite simply from the Cayley-Hamilton theorem.

from the first row, we get the row-equivalent matrix

$$\begin{bmatrix} 0 & (1-ab) & 6-b(b-1) & 6a-b(1-a) \\ 1 & a & b-1 & 1-a \end{bmatrix}$$

whose rank is the same as $\text{rank}[G, FG]$. If $ab \neq 1$, so that $(1-ab) \neq 0$, then the first two columns are independent, and hence $\text{rank}[G, FG] = 2$, and the system is controllable.

If $ab = 1$, the row-equivalent matrix is

$$\begin{bmatrix} 0 & 0 & (3-b)(2+b) & 6a-b+1 \\ 1 & a & (b-1) & (1-a) \end{bmatrix}$$

which, since $a = \dfrac{1}{b}$ for this case, is

$$\begin{bmatrix} 0 & 0 & (3-b)(2+b) & \dfrac{(3-b)(2+b)}{b} \\ 1 & \dfrac{1}{b} & (b-1) & \dfrac{(b-1)}{b} \end{bmatrix}$$

The second and fourth columns are just constant multiples of the first and third, respectively, and from the determinant of the first and third,

$$\det \begin{bmatrix} 0 & (3-b)(2+b) \\ 1 & (b-1) \end{bmatrix} = -(3-b)(2+b)$$

these two columns are independent for all b except for the values $b = 3$ and $b = -2$. For each of these "bad" values, all columns of the equivalent matrix are just multiples of $\begin{bmatrix} 0 \\ 1 \end{bmatrix}$ so the rank of $[G, FG]$ is one.

Summarizing, the given system (F, G) is controllable for all values of a and b except for the values $(b = 3, a = \frac{1}{3})$ and $(b = -2, a = -\frac{1}{2})$, just as we found in Example 3. [See Exercise 14.]   ◊

An interesting point about the proof of Theorem 10 is that the time t at which the state was controlled to 0 did *not* enter the criterion. We thus have the intriguing result:

COROLLARY 11

If a state $\bar{x}$ of the time-invariant system

$$\dot{x}(t) = Fx(t) + Gu(t)$$

is controllable, then the system can be brought from state $\bar{x}$ to state 0 in an arbitrarily short time.   □

Of course, in nature there are no such things as "arbitrarily-short-time" control signals but, as we saw in Section 5–4, we can consider the **unit-impulse,** $\delta(t)$, its derivative $\dot{\delta}(t)$ (the **unit-doublet**), and its higher derivatives, $\delta^{(2)}(t), \ldots, \delta^{(k)}(t)$ (the **unit-singularity functions**†) as idealized mathematical models of very short, abruptly changing functions. This motivates us to try a linear combination of the singularity functions $\{\delta, \dot{\delta}, \delta^{(2)}, \ldots, \delta^{(k)}, \ldots\}$ as an input, and to see what kind of control this type of input allows us over the state.

Example 6

Consider an input

$$u_\delta(t) = c_0\, \delta(t) + c_1\, \dot{\delta}(t) + \cdots + c_k\, \delta^{(k)}(t) \tag{20}$$

where

$$\delta^{(k)}(t) \triangleq \frac{d^k}{dt^k}\, \delta(t)$$

to the constant system (F, G, H). The c_i are constant m-vectors to be found. Recall from equation (5) that an initial state $\bar{x}$ can be driven to zero at time t_1 if we can find an input u such that

$$-\bar{x} = \int_0^{t_1} e^{F(0-\zeta)} Gu(\zeta)\, d\zeta$$

If we try $u_\delta(\cdot)$ as the input in this integral we get

$$-\bar{x} = \int_0^{t_1} e^{-F\zeta} G[c_0\, \delta(\zeta) + c_1\, \dot{\delta}(\zeta) + \cdots + c_k\, \delta^{(k)}(\zeta)]\, d\zeta$$

where a typical term in the integrand is

$$\int_0^{t_1} e^{-F\zeta} Gc_j\, \delta^{(j)}(\zeta)\, d\zeta$$

But the defining operational property of the unit-singularity function $\delta^{(j)}(t - a)$ is that it "samples" the jth derivative of the integrand at $t = a$:

$$\int_0^{t_1} f(\zeta)\, \delta^{(j)}(\zeta - a)\, d\zeta \triangleq (-1)^j \frac{d^j}{dt^j} f(t)\bigg|_{t=a} \tag{21}$$

---

†For a complete discussion of the unit-singularity functions and their operational properties, see Zadeh and Desoer, *Linear System Theory* (McGraw Hill, 1963).

so our jth term becomes [Exercise 15]

$$(-1)^j \frac{d^j}{dt^j} (e^{F(-t)}G) \Big|_{t=0} c_j$$

$$= F^j G c_j$$

Thus

$$-\bar{x} = [G c_0 + F G c_1 + \cdots + F^k G c_k]$$

or

$$-\bar{x} = \underbrace{\left[G \mid FG \mid \ldots \mid F^k G \right]}_{(n) \times (mk)} \underbrace{\begin{bmatrix} c_0 \\ \hline c_1 \\ \hline \vdots \\ \hline c_k \end{bmatrix}}_{(mk) \times 1} \tag{22}$$

in block-partitioned form. The last equation can be solved for the c_i for an arbitrary $\bar{x}$ iff the rank of $[G, FG, \ldots, F^k G]$ is n. We now see clearly that if (F, G) is controllable we may take $k = n - 1$, let $t_1 = 0^+$, and by the full rank of $[G, FG, \ldots F^{n-1}G]$ solve the last equation for $c_0, c_1, \ldots, c_{n-1}$ to drive the given initial state $x(0^-) = \bar{x}$ to $x(0^+) = 0$.   ◊

Of course, in time-varying systems $F(t)$ may vary in such a way that we can only control different state variables at widely separated times.

Let us now check that controllability is preserved by similarity transformations.

THEOREM 12

Let $\bar{F} = P^{-1}FP$ and $\bar{G} = P^{-1}G$. Then $(\bar{F}, \bar{G})$ is controllable iff (F, G) is controllable.

Proof

$(\bar{F}, \bar{G})$ is controllable iff

$$\text{rank}[\bar{G}, \bar{F}\bar{G}, \ldots, \bar{F}^{n-1}\bar{G}] = n$$

But

$$[\bar{G}, \bar{F}\bar{G}, \ldots, \bar{F}^{n-1}\bar{G}] = [P^{-1}G, P^{-1}FPP^{-1}G, \ldots, P^{-1}F^{n-1}PP^{-1}G]$$
$$= P^{-1}[G, FG, \ldots, F^{n-1}G]$$

Since P^{-1} is non-singular,

$$\mathrm{rank}[\bar{G}, \bar{F}\bar{G}, \ldots, \bar{F}^{n-1}\bar{G}] = \mathrm{rank}[G, FG, \ldots, F^{n-1}G]$$

and the theorem follows.   □

It is particularly illuminating to see how controllability is revealed when a system (F, G, H) is transformed to its equivalent Jordan canonical form (F_N, G_N, H_N). Consider, for simplicity, the case where F has n distinct eigenvalues $\lambda_1, \ldots, \lambda_n$ and let P be the transformation matrix whose jth column P_j is the eigenvector of F corresponding to λ_j. Then letting q denote the state of the system (F_N, G_N, H_N) and $\bar{G}_1, \ldots, \bar{G}_n$ the *rows* of G_N, we have

$$
\begin{bmatrix} \dot{q}_1 \\ \dot{q}_2 \\ \vdots \\ \dot{q}_n \end{bmatrix} = \begin{bmatrix} \lambda_1 & 0 & 0 & \cdots & 0 \\ 0 & \lambda_2 & 0 & & 0 \\ \vdots & \vdots & & & \vdots \\ 0 & 0 & & \cdots & \lambda_n \end{bmatrix} \begin{bmatrix} q_1 \\ q_2 \\ \vdots \\ q_n \end{bmatrix} + \begin{bmatrix} \bar{G}_1 \\ \hline \bar{G}_2 \\ \hline \vdots \\ \hline \bar{G}_n \end{bmatrix} \begin{bmatrix} u_1 \\ u_2 \\ \vdots \\ u_m \end{bmatrix} \tag{23a}
$$

where

$$\bar{G}_i = i\text{th row of } (P^{-1}G) = (i\text{th row of } P^{-1})G \tag{23b}$$

If all the rows $\bar{G}_1, \ldots, \bar{G}_n$ are nonzero, then each state variable q_i is driven by u,

$$\dot{q}_i = \lambda_i q_i + \bar{G}_i u \tag{23c}$$

and since the λ_i are distinct, an input u can be found to drive q_i to any desired value [Exercise 17]. However, if some row $\bar{G}_i = 0$, then

$$\dot{q}_i = \lambda_i q_i + 0 \begin{bmatrix} u_1 \\ \vdots \\ u_m \end{bmatrix} \tag{23d}$$

and the input u is disconnected from the state variable q_i and thus cannot control it. We have thus just proved the next theorem.

THEOREM 13

If (F_N, G_N, H_N) is the Jordan canonical representation of (F, G, H) and F has n distinct eigenvalues, then (F, G) is controllable if and only if none of the rows of G_N are zero.   □

The reader should draw the block diagram of (F_N, G_N, H_N) which makes Theorem 13 obvious.

The importance of the hypothesis about distinct eigenvalues in Theorem 13 is illustrated by the next example, which is a generalization of Example 4–3–7.

Example 7

Let $F = \begin{bmatrix} \alpha & 1 \\ 0 & \beta \end{bmatrix}$ and $G = \begin{bmatrix} a \\ b \end{bmatrix}$ as in Figure 7–23.

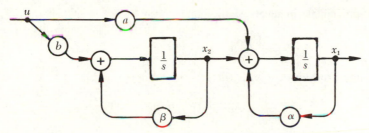

Figure 7–23 A system with eigenvalues α and β.

Since

$$\det[G, FG] = \det\begin{bmatrix} a & a\alpha + b \\ b & b\beta \end{bmatrix}$$

$$= b[a(\beta - \alpha) - b]$$

(F, G) is controllable except when

$$b = 0 \qquad \text{or} \qquad a(\beta - \alpha) - b = 0$$

The eigenvalues of F are α and β, and, if they are equal, F is already in the Jordan form. Clearly, a (which is the first row of G_N) can be zero without affecting controllability.

If $\alpha \neq \beta$, Theorem 13 is applicable but F is not yet in Jordan form. The eigenvectors corresponding to α and β, respectively, are $P_1 = \begin{bmatrix} 1 \\ 0 \end{bmatrix}$ and

$P_2 = \begin{bmatrix} 1 \\ \beta - \alpha \end{bmatrix}$, whence $P = \begin{bmatrix} 1 & 1 \\ 0 & \beta - \alpha \end{bmatrix}$ and $P^{-1} = \dfrac{1}{\beta - \alpha}\begin{bmatrix} \beta - \alpha & -1 \\ 0 & 1 \end{bmatrix}$

yield the Jordan representation

$$F_N = \begin{bmatrix} \alpha & 0 \\ 0 & \beta \end{bmatrix} \text{ and } G_N = \frac{1}{\beta - \alpha}\begin{bmatrix} a(\beta - \alpha) - b \\ b \end{bmatrix}.$$

The vanishing of the rows of G_N yields precisely the same conditions for controllability that we found from the rank of $[G, FG]$.   ◊

Looking again at equations (23a) and (23b), we see that if the matrix G is a column, $G = g$, corresponding to a single-input system, the condition $\bar{G}_i = (i\text{th row of } P^{-1})g = 0$ means that g is orthogonal to the ith row of P^{-1}. But from $P^{-1}P = I$ we know that

$$(i\text{th row of } P^{-1})(j\text{th column of } P) = \delta_{ij} = \begin{cases} 1 & i = j \\ 0 & i \neq j \end{cases}$$

and expanding g in terms of the eigenvector basis

$$g = a_1 P_1 + \cdots + a_i P_i + \cdots + a_n P_n$$

the scalar product of g with the ith row of P^{-1} yields

$$\bar{G}_i = (i\text{th row of } P^{-1})(a_1 P_1 + \cdots + a_n P_n) = a_i$$

the coordinate of g with respect to the ith eigenvector P_i of F. Thus, we have the following result:

COROLLARY 14

If F has n distinct eigenvalues, then the scalar input system (F, g) is controllable if and only if g is not spanned by any proper subset of the eigenvectors of F.   □

Example 8

For $F = \begin{bmatrix} 0 & 6 \\ 1 & -1 \end{bmatrix}$ and $g = \begin{bmatrix} b \\ 1 \end{bmatrix}$, since the eigenvectors of F are $P_1 = \begin{bmatrix} 3 \\ 1 \end{bmatrix}$ and $P_2 = \begin{bmatrix} 2 \\ -1 \end{bmatrix}$, in order for g to be spanned by a proper subset of $\{P_1, P_2\}$, either $\begin{bmatrix} b \\ 1 \end{bmatrix} = c_1 \begin{bmatrix} 3 \\ 1 \end{bmatrix}$ or $\begin{bmatrix} b \\ 1 \end{bmatrix} = c_2 \begin{bmatrix} 2 \\ -1 \end{bmatrix}$, which requires that either $b = 3$ or else $b = -2$. This agrees with the conclusion of Example 4.   ◊

If the eigenvalues of F are not all distinct, things get more complicated and criteria must be stated in terms of conditions on the various sub-blocks of the Jordan canonical form.† At this point the reader should review Exam-

---

† See, for example Chen, *Introduction to Linear System Theory*, pp. 190–196.

ples 4–3–6 and 4–3–7 for the discrete-time case, replace the z^{-1} blocks by $\frac{1}{s}$ blocks, and interpret them as continuous-time examples to develop a feeling for what these criteria mean.

Example 9

$$\text{Let } F = \begin{bmatrix} 1 & 0 & 0 \\ 1 & 2 & 0 \\ 1 & 1 & 2 \end{bmatrix} \quad \text{and} \quad G = \begin{bmatrix} a \\ b \\ c \end{bmatrix}$$

In Example 6–6–2 we saw that this F had the Jordan representation

$$F_N = \begin{bmatrix} 1 & 0 & 0 \\ \hline 0 & 2 & 1 \\ 0 & 0 & 2 \end{bmatrix}$$

under the transformation $x = Pq$ where

$$P = \begin{bmatrix} 1 & 0 & 0 \\ -1 & 0 & 1 \\ 1 & 1 & 0 \end{bmatrix}$$

Under this state transformation we get

$$G_N = P^{-1}G = \begin{bmatrix} 1 & 0 & 0 \\ 0 & 0 & 1 \\ 1 & 1 & 0 \end{bmatrix}\begin{bmatrix} a \\ b \\ c \end{bmatrix} = \begin{bmatrix} a \\ c \\ a+b \end{bmatrix}$$

and the block diagram of (F_N, G_N) is shown in Figure 7–24.

From the controllability matrix of F_N, G_N,

$$[G_N, F_N G_N, F_N^2 G_N] = \begin{bmatrix} a & a & a \\ c & a+b+2c & 4(a+b+c) \\ a+b & 2(a+b) & 4(a+b) \end{bmatrix}$$

subtracting the first column from the second and the third yields the equivalent matrix

$$\begin{bmatrix} a & 0 & 0 \\ c & a+b+c & 4(a+b+c)-c \\ a+b & a+b & 3(a+b) \end{bmatrix}$$

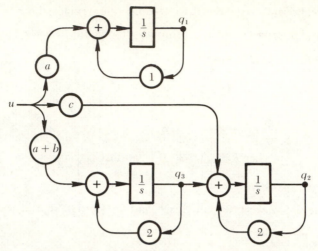

Figure 7–24   Jordan Block Diagram of (F_N, G_N).

and subtracting four times the second column from the third,

$$\begin{bmatrix} a & 0 & 0 \\ 0 & a+b+c & -c \\ a+b & a+b & -(a+b) \end{bmatrix}$$

Finally, adding the fourth column to the first and to the second we get

$$\begin{bmatrix} a & 0 & 0 \\ 0 & a+b & -c \\ 0 & 0 & -(a+b) \end{bmatrix}$$

with determinant $\Delta = -a(a+b)^2$. Thus, (F, G) is controllable as long as $a(a+b) \neq 0$. It is perhaps surprising to see that the gain c has no effect on controllability. Even if $c = 0$, which looks suspicious in Figure 7–24, we are still able to control q_2 as long as $(a+b) \neq 0$, so that the input drives the beginning of the chain of blocks corresponding to $\lambda = 2$. Of course, this chain is just the system we studied in Example 7 for $\alpha = \beta = 2$, so we shouldn't have been too surprised.   ◊

If a constant linear system is controllable, we can use the input to suppress undesirable modes. First, as in Section 6–3, we determine an "initial" state x_λ such that the undesirable mode $e^{\lambda t}$ does not appear in the unforced state response; then we find an input $u_{[0,t_1)}$ to drive the actual initial event $(0, x_0)$ to (t_1, x_λ). Setting $u(t) = 0$ for $t \geq t_1$ lets the system run naturally starting from x_λ with no $e^{\lambda t}$ present.

Example 10

To suppress the unstable mode e^{+2t} in the system

$$F = \begin{bmatrix} 0 & 6 \\ 1 & -1 \end{bmatrix}, \; G = \begin{bmatrix} 0 \\ 1 \end{bmatrix}$$

where the eigenvectors

$$v_1 = \begin{bmatrix} 3 \\ 1 \end{bmatrix} \text{ and } v_2 = \begin{bmatrix} 2 \\ -1 \end{bmatrix}$$

correspond to the eigenvalues $\lambda_1 = 2$, $\lambda_2 = -3$, we can pick $x_\lambda = v_2$. Then, starting from x_λ at time t_1, the natural response

$$x(t) = e^{Ft} x_\lambda \quad t \geq t_1$$

will remain along the eigenvector v_2. Since (F, G) is controllable, the input

$$u(t) = -G^* e^{-F^* t} W^{-1}(0, t_1)[x^0 - e^{-Ft_1} x_\lambda] \tag{24}$$

[Exercise 5] will drive $(0, x^0)$ to (t_1, x_λ).   ◊

If a system is not controllable, it is useful to single out those states which are controllable and to decompose the state space into a controllable and an uncontrollable part. We have already developed all the machinery we need to do this.

Back in the proof of Theorem 10 we showed that $x_2 \in \mathfrak{N}(W(0, t))$ if and only if $x_2^* F^k G = 0$ for all $k \geq 0$, and this last condition was equivalent to $x_2 \in (\mathfrak{R}[G, FG, \ldots, F^{n-1}G])^\perp$; i.e., x_2 is orthogonal to all the columns of the matrix $[G, FG, \ldots, F^{n-1}G]$. But since dim $X = n$, $\mathfrak{N}(W(0, t)) = [\mathfrak{R}(W(0, t))]^\perp$ so $x_2 \in [\mathfrak{R}(W(0, t))]^\perp$ if and only if $x_2 \in (\mathfrak{R}[G, FG, \ldots, F^{n-1}G])^\perp$. Thus, $[\mathfrak{R}(W(0, t))]^\perp = (\mathfrak{R}[G, FG, \ldots, F^{n-1}G])^\perp$ and $\mathfrak{R}(W(0, t)) = \mathfrak{R}[G, FG, \ldots, F^{n-1}G]$, since X is finite-dimensional. We consolidate this discussion into a theorem.

THEOREM 15

If $Q = [G, FG, \ldots, F^{n-1}G]$, then the set of all controllable states of the system (F, G) is just $\mathfrak{R}(W(0, t)) = \mathfrak{R}(Q)$ and is an invariant subspace of F. Moreover, the state space can be decomposed into the direct sum $X = \mathfrak{R}(Q) \oplus \mathfrak{N}(Q^*)$.

Proof

Since $Q: X \to X$, $\Re(Q) = \Re(W(0, t))$ is a subspace, so

$$X = \Re(Q) \oplus [\Re(Q)]^{\perp}$$

But $[\Re(Q)]^{\perp} = \mathfrak{N}(Q^*)$, where Q^* is the adjoint of Q.

To see that $\Re(Q)$ is mapped into itself by F, let $y \in F(\Re(Q))$. Then $y = Fx$ for some $x \in \Re(Q)$. But $x \in \Re(Q)$ means that x is spanned by the columns of $G, FG, \ldots, F^{n-1}G$, so Fx is spanned by the columns of FG, $F^2G, \ldots, F^nG$. By the Cayley-Hamilton theorem, F^n can be written as a linear combination of $I, F, \ldots, F^{n-1}$, so F^nG is spanned by the columns of $G, FG, \ldots, F^{n-1}G$; hence, Fx is also. Thus, $y = Fx \in \Re(Q)$ and $F(\Re(Q)) \subset \Re(Q)$. □

Why do we care whether the set $\Re(Q)$ is an invariant subspace of F or not? The nice thing about splitting a vector space into a direct sum, with an invariant subspace of F as one of the parts, is that we can choose a basis for the space with respect to which the representation of F splits into meaningful subblocks. To see this, let $v_1, \ldots, v_r$ be a basis for the subspace $\Re(Q)$ and let $w_{r+1}, \ldots, w_n$ be a basis of $\mathfrak{N}(Q^*)$, so that $\mathfrak{B} = \{v_1, \ldots, v_r, w_{r+1}, \ldots, w_n\}$ is a basis for $X = \Re(Q) \oplus \mathfrak{N}(Q^*)$. Since $\Re(Q)$ is invariant under F, we have $Fv_i \in \Re(Q)$ for each $i = 1, \ldots, r$, and thus $Fv_i = \alpha_{1i}v_1 + \cdots + \alpha_{ri}v_r + 0w_{r+1} + \cdots + 0w_n$ for appropriate scalars $\alpha_{1i}, \ldots, \alpha_{ri}$. But this shows that the coordinates of $F(v_i)$ with respect to the basis $\mathfrak{B}$ form the n-tuple

$$[F(v_i)]_{\mathfrak{B}} = \begin{bmatrix} \alpha_{1i} \\ \alpha_{2i} \\ \vdots \\ \alpha_{ri} \\ 0 \\ \vdots \\ 0 \end{bmatrix} \begin{array}{l} \left.\rule{0pt}{2.2em}\right\} r \\ \left.\rule{0pt}{2.2em}\right\} n-r \end{array}$$

whose last $n - r$ entries are zero. If we compute $^{\mathfrak{B}}F^{\mathfrak{B}}$, the matrix of F with respect to basis $\mathfrak{B}$, its first r columns are just $[F(v_1)]_{\mathfrak{B}}, \ldots, [F(v_r)]_{\mathfrak{B}}$ so that

$$^{\mathfrak{B}}F^{\mathfrak{B}} = \begin{array}{l} r \left\{ \\ \\ \\ n-r \left\{ \\ \\ \end{array} \begin{bmatrix} \alpha_{11} & \alpha_{12} & & \alpha_{1r} & \vdots & \\ \alpha_{21} & \alpha_{22} & & \alpha_{2r} & \vdots & \bar{F}_{12} \\ \vdots & \vdots & & \vdots & \vdots & \\ \alpha_{r1} & \alpha_{r2} & \cdots & \alpha_{rr} & \vdots & \\ \hline 0 & 0 & & 0 & \vdots & \\ \vdots & & & \vdots & \vdots & \bar{F}_{22} \\ 0 & 0 & \cdots & 0 & \vdots & \end{bmatrix} \\ \underbrace{\hspace{4.5cm}}_{r} \underbrace{\hspace{2cm}}_{n-r} $$

where $\bar{F}_{12}$ and $\bar{F}_{22}$ are the $r \times (n - r)$ and $(n - r) \times (n - r)$ subblocks we get by partitioning the remaining columns of coordinates $[F(w_{r+1})]_{\mathfrak{B}}, \ldots,$ $[F(w_n)]_{\mathfrak{B}}$. Now $G: \mathbf{C}^m \to \mathbf{C}^n$; what does it look like when represented in the $\mathfrak{B}$ basis? Since for $j = 1, \ldots, m$, the jth column of G is in $R(Q)$, its coordinates with respect to $\mathfrak{B}$ will also have the last $n - r$ entries zero, so

$$\mathfrak{B}G^E = \begin{matrix} r \\ n - r \end{matrix} \begin{Bmatrix} \\ \{ \end{Bmatrix} \underbrace{\begin{bmatrix} \bar{G}_1 \\ \hline 0 \end{bmatrix}}_{m}$$

where $\bar{G}_1$ is an $r \times m$ subblock and E denotes the standard basis for $\mathbf{C}^m$.

If we let

$$P = \begin{bmatrix} v_1 & \cdots & v_r & w_{r+1} & \cdots & w_n \end{bmatrix}$$

then P is invertible and $\mathfrak{B}F^{\mathfrak{B}} = P^{-1}FP$ and $\mathfrak{B}G^E = P^{-1}G$; so we have proved the next theorem.

THEOREM 16

If $\dim \mathcal{R}(Q) = r < n$, then there exists an invertible transformation P such that (F, G, H) is algebraically equivalent to $(\bar{F}, \bar{G}, \bar{H})$ of the form

$$\bar{F} = \begin{matrix} r \\ n - r \end{matrix} \begin{Bmatrix} \\ \{ \end{Bmatrix} \underbrace{\begin{bmatrix} \bar{F}_{11} & \bar{F}_{12} \\ \hline 0 & \bar{F}_{22} \end{bmatrix}}_{r \qquad n - r}$$

$$\bar{G} = \underbrace{\begin{bmatrix} \bar{G}_1 \\ \hline 0 \end{bmatrix}}_{m} \begin{matrix} \} & r \\ \} & n - r \end{matrix} \qquad \bar{H} = [\underbrace{\bar{H}_1}_{r} \mid \underbrace{\bar{H}_2}_{n - r}] \qquad \square$$

From this last theorem it follows easily [Exercise 20] that if $\dim[G, FG, \ldots, F^{n-1}G] = r < n$, then the transfer function $\mathcal{S}(s) = H(sI - F)^{-1}G$ is equal to $\bar{H}_1(sI - \bar{F}_{11})^{-1}\bar{G}_1$, the transfer function of the r-dimensional system $(\bar{F}_{11}, \bar{G}_1, \bar{H}_1)$, so that (F, G, H) is zero-state equivalent to a system of lower dimension. This proves that *lack of controllability manifests itself by cancellations in the transfer function*, as claimed in Theorem 9, and that the realization (F, G, H) of $\mathcal{S}(s)$ is not a minimal realization. [See Exercise 21.]

If we draw a block diagram of the $(\bar{F}, \bar{G}, \bar{H})$ of Theorem 16, letting

$$\bar{x} = \begin{bmatrix} \bar{x}_1 \\ \hline \bar{x}_2 \end{bmatrix} \begin{matrix} \} & r \\ \} & n - r \end{matrix}$$

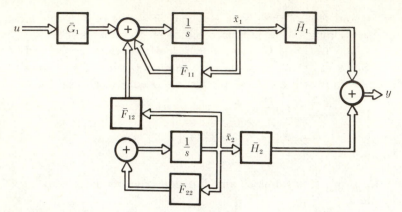

Figure 7–25 Separation of the state space into controllable and uncontrollable parts.

where $\bar{x}_1$ corresponds to the controllable states and $\bar{x}_2$ to the uncontrollable states in the system $\dot{\bar{x}} = \bar{F}\bar{x} + \bar{G}u$ and $y = \bar{H}\bar{x}$, Figure 7–25 results.

The components of $\bar{x}_2$ contribute to the output y and influence the controllable components $\bar{x}_1$, but they are themselves unaffected by the input u.

OBSERVABILITY

Turning now to observability, we recall from Section 4–2 that the states x^1 and x^2 are distinguishable at time t_0 if there is some later time τ and an input $u_{[t_0, \tau)}$ such that

$$\mathcal{S}(\tau, t_0, x^1, u) \neq \mathcal{S}(\tau, t_0, x^2, u)$$

i.e., different outputs result at time $\tau > t_0$ when the system is driven by an appropriate test input u.

We say that x^1 and x^2 are **distinguishable on the interval** $[t_0, t_1]$ if there exists a time τ with $t_0 < \tau \leq t_1$, and an input u such that $\mathcal{S}(\tau, t_0, x^1, u) \neq \mathcal{S}(\tau, t_0, x^2, u)$. A system is called **observable at** t_0 if all its states are distinguishable at t_0, and **observable on** $[t_0, t_1]$ if all states are distinguishable on the interval $[t_0, t_1]$.

For a linear continuous-time system $(F(\cdot), G(\cdot), H(\cdot))$ started in state x^0 at time t_0, the complete response function $\mathcal{S}$ is given by

$$y(t) = \mathcal{S}(t, t_0, x^0, u) = H(t)\Phi(t, t_0)x^0 + \int_{t_0}^{t} g(t, \xi)u(\xi)\, d\xi \qquad (25)$$

where $g(t, \xi) = H(t)\Phi(t, \xi)G(\xi)$, the impulse response of the system, and the state $x(t)$ is given by

$$x(t) = \Phi(t, t_0)x^0 + \int_{t_0}^{t} \Phi(t, \xi)G(\xi)u(\xi) \, d\xi$$

If F, G, and u are known, this last equation shows that $x(t)$ will be known for all $t > t_0$ if and only if we can find its initial value x^0 at time t_0.

From the complete response, equation (25), if we know the output y as well as the input u, then we can find the initial state x^0 precisely whenever we can solve the matrix equation

$$H(\cdot)\Phi(\cdot, t_0)x^0 = \hat{y}(\cdot) \tag{26a}$$

where

$$\hat{y}(t) \overset{\Delta}{=} y(t) - \int_{t_0}^{t} g(t, \xi)u(\xi) \, d\xi \tag{26b}$$

is a known vector determined by F, G, H, u, and y.

Note that

$$\hat{y}(t) \overset{\Delta}{=} \mathcal{S}(t, t_0, x^0, 0) \tag{27}$$

is the response to zero input. Solving equation (26a) is like solving for the initial state of the homogeneous system $(F, 0, H)$, since the read-in matrix G and the input u play no substantive roles.

Example 11

If $F = \begin{bmatrix} 2 & 0 \\ 0 & -3 \end{bmatrix}$ and $H = [a \quad b]$, the zero-input response is

$$\hat{y}(t) = \mathcal{S}(t, t_0, x^0, 0) = H(t)\Phi(t, t_0)x^0 = [a \quad b] \begin{bmatrix} e^{2(t-t_0)} & 0 \\ 0 & e^{-3(t-t_0)} \end{bmatrix} \begin{bmatrix} x_1^0 \\ x_2^0 \end{bmatrix}$$

$$= [ae^{2(t-t_0)} \quad be^{-3(t-t_0)}] \begin{bmatrix} x_1^0 \\ x_2^0 \end{bmatrix}$$

$$\hat{y}(t) = ae^{2(t-t_0)}x_1^0 + be^{-3(t-t_0)}x_2^0 \tag{28}$$

How does one go about solving this last equation for $x^0 = [x_1^0, x_2^0]^T$? One way is to sample the response at two different times, t_1 and t_2. Then we get two equations in the two unknowns, x_1^0 and x_2^0:

$$\hat{y}(t_1) = ae^{2(t_1-t_0)}x_1^0 + be^{-3(t_1-t_0)}x_2^0$$

$$\hat{y}(t_2) = ae^{2(t_2-t_0)}x_1^0 + be^{-3(t_2-t_0)}x_2^0$$

or,

$$\begin{bmatrix} \hat{y}(t_1) \\ \hat{y}(t_2) \end{bmatrix} = \begin{bmatrix} ae^{2(t_1-t_0)} & be^{-3(t_1-t_0)} \\ ae^{2(t_2-t_0)} & be^{-3(t_2-t_0)} \end{bmatrix} \begin{bmatrix} x_1^0 \\ x_2^0 \end{bmatrix}$$

which can be solved uniquely, provided

$$\det \begin{bmatrix} ae^{2(t_1-t_0)} & be^{-3(t_1-t_0)} \\ ae^{2(t_2-t_0)} & be^{-3(t_2-t_0)} \end{bmatrix} = ab[e^{2t_1-3t_2+t_0} - e^{2t_2-3t_1+t_0}] \neq 0$$

Since $t_1 \neq t_2$, the term in brackets is nonzero so we can find the initial state x^0 as long as neither a nor b is zero. A quick sketch of the block diagram of this constant system, already in Jordan canonical form, shows that when either of the gains a or b is zero, one of the modes is disconnected from the output and thus cannot be observed. See Exercise 23 for another approach to solving equation (28) for x^0. ◊

Let us now relate the procedure of solving for the initial state to the property of distinguishability. When we try to distinguish between states x^1 and x^2 in a linear system, the zero-state response term drops out of the difference

$$\mathcal{S}(t, t_0, x^1, u) - \mathcal{S}(t, t_0, x^2, u) = H(t)\Phi(t, t_0)(x^1 - x^2) \tag{29}$$

Thus, linearity reduces this problem to distinguishing x^1 from x^2 under zero input:

$$\mathcal{S}(t, t_0, x^1, 0) - \mathcal{S}(t, t_0, x^2, 0) = H(t)\Phi(t, t_0)(x^1 - x^2) \tag{30}$$

or, equivalently, to distinguishing the state $(x^1 - x^2)$ from 0, the zero state, under zero input:

Calling $\hat{x} = x^1 - x^2$, we see from

$$\mathcal{S}(t, t_0, x^1, u) - \mathcal{S}(t, t_0, x^2, u) = H(t)\Phi(t, t_0)\hat{x}$$

that $\mathcal{S}(\tau, t_0, x^1, u) \neq \mathcal{S}(\tau, t_0, x^2, u)$ if and only if $H(\tau)\Phi(\tau, t_0)\hat{x} \neq 0$. That is, x^1 and x^2 are distinguishable at time t_0 if and only if $\hat{x}$ is not in the nullspace of the matrix $H(\tau)\Phi(\tau, t_0)$ for any time $\tau > t_0$. If x^1 and x^2 can be any distinct states, $x_1 \neq x_2$, then $\hat{x}$ can be any nonzero vector; so this last condition is equivalent to the columns of $H(\cdot)\Phi(\cdot, t_0)$ being linearly independent functions of time on the interval $[t_0, \infty)$. This is precisely the condition under which we may solve the operator equation

$$\hat{y}(\cdot) = H(\cdot)\Phi(\cdot, t_0)x^0$$

uniquely, for the initial state x^0, namely when

$$\mathfrak{N}[H(t)\Phi(t, t_0)] = \{0\} \text{ for some } t > t_0$$

For ease of reference, we summarize this discussion as a theorem.

THEOREM 17

For a continuous-time, linear system $(F(\cdot), G(\cdot), H(\cdot))$ the following conditions are equivalent:

(i)   No states are indistinguishable at time t_0.
(ii) The system is observable at time t_0.
(iii) The initial state x^0 is uniquely recoverable from the zero-input response $\hat{y}$ as the solution of

$$\hat{y}(\cdot) = \mathcal{S}(\cdot, t_0, x^0, 0) = H(\cdot)\Phi(\cdot, t_0)x^0$$

(iv) The columns of the matrix $H(\cdot)\Phi(\cdot, t_0)$, viewed as functions of time on $[t_0, \infty)$, are linearly independent. $\square$

  If we use what we know about adjoints and duality, we can use our study of controllability to generate corresponding theorems and insights about observability. For example, since the columns of the $q \times n$ matrix $H(t)\Phi(t, t_0)$ are just complex conjugates of the rows of the $n \times q$ adjoint matrix

$$[H(t)\Phi(t, t_0)]^* = \Phi^*(t, t_0)H^*(t)$$

we can use Corollary 6 with

$$A(\xi) = \Phi^*(\xi, t_0)H^*(\xi)$$

to get a condition similar to Theorem 7:

THEOREM 18

The system $(F(\cdot), G(\cdot), H(\cdot))$ is observable on $[t_0, t_1]$ and the columns of $H(\cdot)\Phi(\cdot, t_0)$ are linearly independent on $[t_0, t_1]$ if and only if the constant $n \times n$ matrix

$$M(t_0, t_1) \overset{\Delta}{=} \int_{t_0}^{t_1} \Phi^*(\xi, t_0)H^*(\xi)H(\xi)\Phi(\xi, t_0)\, d\xi$$

is non-singular. $\square$

Moreover, analogously to Corollary 5 we have

COROLLARY 19

If $(F(\cdot), G(\cdot), H(\cdot))$ is observable on $[t_0, t_1]$, then the initial state $x^0 = x(t_0)$ is given by

$$x^0 = M^{-1}(t_0, t_1) \int_{t_0}^{t_1} \Phi^*(\xi, t_0) H^*(\xi) \widehat{y}(\xi) \, d\xi$$

where $\widehat{y}(\cdot)$ is the zero input response and $M(t_0, t_1)$ is defined as in Theorem 18.

Proof

To solve the equation

$$\widehat{y}(\xi) = H(\xi) \Phi(\xi, t_0) x^0$$

premultiply by $\Phi^*(\xi, t_0) H^*(\xi)$ and integrate both sides from t_0 to t_1 to get

$$\int_{t_0}^{t_1} \Phi^*(\xi, t_0) H^*(\xi) \widehat{y}(\xi) \, d\xi = M(t_0, t_1) x^0$$

from which the assertion follows.   □

Back in Chapter 5, in Theorem 5 of Section 5–3, we saw that the matrix $\Phi^*(t_0, t)$ is the state transition matrix of the homogeneous *adjoint system* whose state $q(t)$ obeys

$$\frac{dq(t)}{dt} = -F^*(t)q(t)$$

This motivates us to modify our definition in Section 4–5 of the dual of a system $(F(\cdot), G(\cdot), H(\cdot))$ and replace F^* by $-F^*$ to get a similar definition for a continuous-time system.

DEFINITION 1

The **dual** or **adjoint system** of the continuous-time system $(F(\cdot), G(\cdot), H(\cdot))$ is the linear system $(-F^*(\cdot), H^*(\cdot), G^*(\cdot))$ whose state $q(t)$, input $w(t)$, and output $v(t)$ are given by

$$\begin{cases} \dot{q}(t) = -F^*(t)q(t) + H^*(t)w(t) \\ v(t) = G^*(t)q(t) \end{cases} \qquad \bigcirc$$

If we use the subscript D on all symbols pertaining to the dual system and denote the system $(F(\cdot), G(\cdot), H(\cdot))$ by Σ, we may write $\Sigma_D = (F_D(\cdot), G_D(\cdot), H_D(\cdot))$ where

$$F_D(\cdot) = -F^*(\cdot)$$
$$G_D(\cdot) = H^*(\cdot)$$
$$H_D(\cdot) = G(\cdot)$$

and observe that $(\Sigma_D)_D = \Sigma$; the dual of the dual of a system is still the system, even with the minus sign in F_D. Let us now formally relate the observability of Σ to the controllability of Σ_D.

THEOREM 20 (Duality)

A continuous-time linear system Σ is observable on $[t_0, t_1]$ if and only if its dual system Σ_D is controllable on $[t_0, t_1]$.

Proof

The system Σ is observable on $[t_0, t_1]$ if and only if the columns of $H(\cdot)\Phi(\cdot, t_0)$, and hence the rows of $\Phi^*(\cdot, t_0)H^*(\cdot)$, are linearly independent functions on $[t_0, t_1]$. By Theorem 7, the dual Σ_D is controllable on $[t_0, t_1]$ if and only if the rows of $\Phi_D(t_0, \cdot)G_D(\cdot)$ are linearly independent on $[t_0, t_1]$. But the state transition matrix Φ_D of the dual system is given by

$$\Phi_D(t, t_0) = \Phi^*(t_0, t)$$

and the read-in matrix of the dual is given by $G_D(t) = H^*(t)$, so

$$\Phi_D(t_0, t)G_D(t) = \Phi^*(t, t_0)H^*(t)$$

for all t. Thus observability of Σ is equivalent to controllability of Σ_D. $\square$

Since $(\Sigma_D)_D = \Sigma$, we also have the converse:

COROLLARY 21

A linear system is controllable if and only if its dual is observable. $\square$

Turning now to constant systems, we can immediately deduce the following result.

THEOREM 22

The time-invariant linear system (F, G, H) is observable if and only if the $nq \times n$ matrix

$$\mathcal{O} = \begin{bmatrix} H \\ \hline HF \\ \hline HF^2 \\ \hline \vdots \\ \hline HF^{n-1} \end{bmatrix}$$

(the observability matrix of (F, G, H)) has rank n.

Proof

The system (F, G, H) is observable if and only if $(-F^*, H^*, G^*)$ is controllable, and this occurs if and only if the matrix $Q_D = [G_D \,\vdots\, F_D G_D \,\vdots\, \ldots \,\vdots\, F_D^{n-1} G_D]$ of the dual system has rank n. Since

$$Q_D = [H^* \,\vdots\, -F^* H^* \,\vdots\, \ldots \,\vdots\, (-F^*)^{n-1} H^*]$$

$$= \begin{bmatrix} H \\ \hline -HF \\ \hline HF^2 \\ \hline \vdots \\ \hline (-1)^{n-1} HF^{n-1} \end{bmatrix}^*$$

and this last matrix has the same rank as the observability matrix $\mathcal{O}$ (Why?), the theorem follows. $\square$

The matrix $\mathcal{O}$ reveals the observability of the representation (F, G, H), whether it represents a continuous-time linear system or a discrete-time linear system.

The following example, devised by Kalman, provides a physical setting in which to illustrate both observability and controllability.

Example 12

The electrical network of Figure 7–26 is known as a "constant-resistance" network.

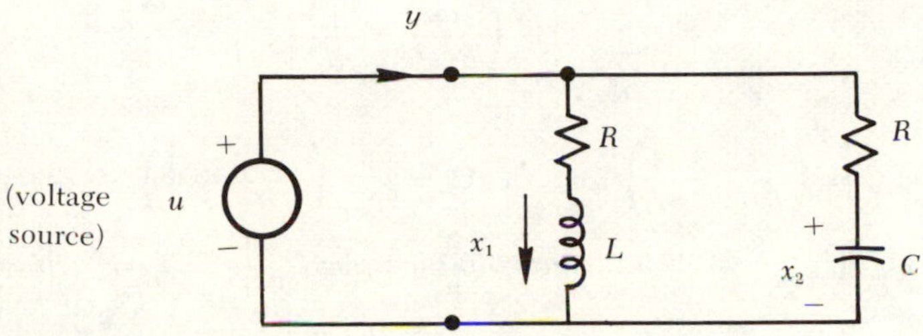

Figure 7–26   A constant-resistance network.

Choosing x_1, the current through the inductor, and x_2, the voltage across the capacitor, as state variables and the output y as the total current through the input voltage source, we can write the circuit equations

$$L\frac{dx_1}{dt} + Rx_1 = u$$

$$C\frac{dx_2}{dt} = \frac{u - x_2}{R}$$

$$y = x_1 + C\frac{dx_2}{dt} = x_1 - \frac{1}{R}x_2 + \frac{1}{R}u$$

In matrix form we get

$$\begin{bmatrix} \dot{x}_1 \\ \dot{x}_2 \end{bmatrix} = \begin{bmatrix} -\dfrac{R}{L} & 0 \\ 0 & -\dfrac{1}{RC} \end{bmatrix}\begin{bmatrix} x_1 \\ x_2 \end{bmatrix} + \begin{bmatrix} \dfrac{1}{L} \\ \dfrac{1}{RC} \end{bmatrix}u$$

$$y = \begin{bmatrix} 1 & -\dfrac{1}{R} \end{bmatrix}\begin{bmatrix} x_1 \\ x_2 \end{bmatrix} + \frac{1}{R}u$$

The observability matrix $\mathcal{O} = \begin{bmatrix} H \\ \hline HF \end{bmatrix}$ is then

$$\mathcal{O} = \begin{bmatrix} 1 & -\dfrac{1}{R} \\ -\dfrac{R}{L} & \dfrac{1}{R^2C} \end{bmatrix}$$

and the controllability matrix $\mathcal{C} = [FG \vdots G]$ is

$$\mathcal{C} = \begin{bmatrix} -\dfrac{R}{L^2} & \dfrac{1}{L} \\[2ex] -\left(\dfrac{1}{RC}\right)^2 & \dfrac{1}{RC} \end{bmatrix}$$

From $\det \mathcal{O} = \left(\dfrac{1}{R^2C} - \dfrac{1}{L}\right)$ and $\det \mathcal{C} = \dfrac{1}{LC}\left(\dfrac{1}{R^2C} - \dfrac{1}{L}\right)$ we see that the system is both observable and controllable unless $\dfrac{1}{R^2C} = \dfrac{1}{L}$ or, equivalently, $\dfrac{1}{RC} = \dfrac{R}{L}$ (i.e., the natural frequencies and hence the "time constants" of the parallel branches are equal). The reader should enjoy playing further with this system and similar physical systems in Exercises 26 and 27.   ◊

If a state x^0 is not observable, then it cannot be distinguished from the zero state at $t_0 = 0$, and it is in the nullspace of $H(t)\Phi(t, 0)$ for some $t > 0$. Since with F and H constant

$$H(t)\Phi(t, 0) = He^{Ft}$$

we have x^0 in the nullspace of He^{Ft} and hence

$$HF^k x^0 = 0 \text{ for all } k \geq 0$$

so

$$0 = \begin{bmatrix} H \\ \hline HF \\ \vdots \\ \hline HF^{n-1} \end{bmatrix} x^0 = \mathcal{O}x^0$$

Thus we have, analogously to Theorem 15, the next theorem.

THEOREM 23

The set of unobservable states of (F, G, H) is the nullspace $\mathfrak{N}(\mathcal{O})$ of the observability matrix $\mathcal{O}$, and is an invariant subspace of X under F.   □

In a fashion similar to Theorems 15 and 16, we may decompose the state space into a direct sum of observable and unobservable parts, and show that the following result is true.

THEOREM 24

If rank $\mathcal{O} = r < n$, then (F, G, H) is algebraically equivalent to $(\widetilde{F}, \widetilde{G}, \widetilde{H})$ where

$$\widetilde{F} = \;^r\!\left\{\underbrace{\left[\begin{array}{c|c} \widetilde{F}_{11} & 0 \\ \hline \widetilde{F}_{21} & \widetilde{F}_{22} \end{array}\right]}_{r \qquad n-r}\right., \quad \widetilde{G} = \left[\begin{array}{c} \widetilde{G}_1 \\ \hline \widetilde{G}_2 \end{array}\right]\!\right\}^r$$

$$\widetilde{H} = \underbrace{[\widetilde{H}_1 \;\vdots\; 0]}_{r}$$

$\square$

From this it follows that if rank $\mathcal{O} = r < n$, then (F, G, H) is zero-state equivalent to the r-dimensional system $(\widetilde{F}_{11}, \widetilde{G}_1, \widetilde{H}_1)$, so that *unobservability also manifests itself by cancellations in the transfer function.* [Be sure to work Exercises 28 and 30.]

It may also be shown† that every system admits a canonical decomposi-

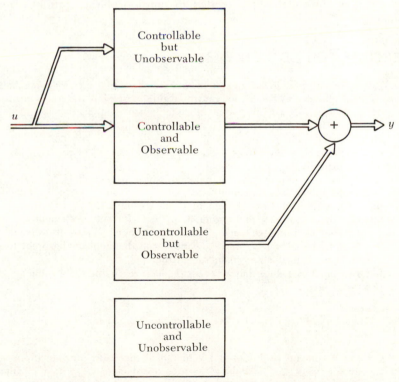

Figure 7–27   Canonical decomposition of the state space.

---

†See Zadeh and Desoer, *Linear System Theory* (McGraw Hill, 1963), Section 11.5.

tion into four subsystems corresponding to the states which are both controllable and observable, either controllable or observable, and neither controllable nor observable, as illustrated symbolically in Figure 7–27.

The subsystem which is both controllable and observable is zero-state equivalent to (F, G, H) and provides a minimal realization of the transfer function $H(sI - F)^{-1}G$.

In closing this lengthy section, we point out that there appear in the literature many different uses of the terms controllability and observability. Our usage of these terms is usually called "complete state controllability or observability" and by some authors "pointwise state controllability" to distinguish them from "modal controllability," "output controllability," "uniform controllability," and other more exotic notions.

Controllability and observability arise in a fundamental way in the attempt to stabilize unstable modes of linear systems, in filtering theory,† and in optimal control theory. We shall pause for a brief introduction to stability theory in the next section and then, in Section 7–5, use these concepts to control and stabilize multiple input-output systems. In Section 7–6 we shall see how controllability is related to solution of the optimal control problem.

EXERCISES FOR SECTION 7–3

1.   Prove, for any time-invariant system, linear or not, that if $(\tau, \tilde{x})$ is reachable, then so is $(\tau', \tilde{x})$ for any τ'. Repeat for controllability. [*Hint:* Use the state transition map $\phi(t_1, t_0, x^0, u_{[t_0, t_1)}) = x(t_1)$.]

2.   Let $a(t) = 1(t)$, the unit step, in Example 1 and find all reachable events $(\tau, \tilde{x})$ for $\tau > 0$.

✓ **3.**   **(a)**   Prove that any two of the vectors $\{x(t_1), x(t_0), \hat{x}\}$ of equation (3) can be specified arbitrarily.

   (b)   Use this to prove that a continuous-time system $(F(\cdot), G(\cdot), H(\cdot))$ is reachable if and only if it is controllable.

   (c)   Use (a) and (b) to write a proof of Theorem 1.

4.   Change the hypotheses of Lemma 3 to "Let $A(\cdot)$ be any smooth matrix function of time assigning to each time t a bounded linear transformation $A(t): U \to X$ of Hilbert spaces U and X," and check the steps of the proof to see that the lemma holds in this more general setting.

5.   Use Lemma 3 to deduce that an event $(t_0, \tilde{x})$ is reachable if, for some $t_1 \leq t_0$, $\tilde{x}$ is in the range of

$$\widehat{W}(t_1, t_0) = \int_{t_1}^{t_0} \Phi(t_0, s) G(s) G^*(s) \Phi^*(t_0, s) \, ds$$

(Note order of arguments in Φ and $\widehat{W}$.) When W is invertible, find an input $u(\cdot)$ which transfers $(t_1, 0)$ to $(t_0, \tilde{x})$. Find a $u(\cdot)$ which transfers (t_0, x^0) to (t_1, x^1). [*Answer:* $u(t) = -G^*(t) \Phi^*(t_0, t) W^{-1}(t_0, t_1)[x^0 - \Phi(t_0, t_1)x^1]$]

6.   Let $A(t) = A$, a constant $n \times m$ matrix in Lemma 3, and let $\tau = 0$ and $t = 1$

---

† See the landmark paper by Kalman and Bucy, "New Results in Linear Prediction and Filtering Theory," *J. Basic Engineering* (Trans. ASME, Ser. D) 83D: 95–100, 1961.

to get a statement about invertibility of $W = AA^*$ in terms of A. Compare with Exercises (27) and (28) of Section 4–4 and Exercise 12 of Section 4–5. What can you say about positive-definiteness of W?

✔ **7.**   This exercise illustrates the problem of trying to control a system to follow a specified path or trajectory between two states.

$$\text{Let } F = \begin{bmatrix} -1 & -2 \\ 1 & -1 \end{bmatrix} \text{ and } G = \begin{bmatrix} 1 \\ 0 \end{bmatrix} \text{ and let } x(0) = \begin{bmatrix} 3 \\ 0 \end{bmatrix}.$$

(a)   Find an input $u(t)$ for $t \geq 0$ such that the state vector $x(t)$ moves from $\begin{bmatrix} 3 \\ 0 \end{bmatrix}$ to the origin $\begin{bmatrix} 0 \\ 0 \end{bmatrix}$ along the spiral path described by

$$\begin{cases} x_1(t) = 3e^{-t} \cos t \\ x_2(t) = 3e^{-t} \sin t \end{cases}$$

(b)   Sketch the path. At what time t_1 does the state pass through $x(t_1) = \begin{bmatrix} 0 \\ 3e^{-\pi/2} \end{bmatrix}$? You just found an input to move $\left(0, \begin{bmatrix} 3 \\ 0 \end{bmatrix}\right)$ to $\left(t_1, \begin{bmatrix} 0 \\ 3e^{-\pi/2} \end{bmatrix}\right)$ along a given path.

(c)   Find $\Phi(t, \tau)$ and compute $\Phi(t_0, \varsigma)G(\varsigma)$.

(d)   Use equations (4), (5), (6) to get a state $\hat{x}$ and an integral equation whose solution gives an (perhaps different) input which moves $\left(0, \begin{bmatrix} 3 \\ 0 \end{bmatrix}\right)$ to $\left(t_1, \begin{bmatrix} 0 \\ 3e^{-\pi/2} \end{bmatrix}\right)$.

Plug your input from part (b) into this integral equation. Does it satisfy it? Discuss.

(e)   Solve the integral equation (6) for a u in any way that you can and compare with the u of part (b).

(f)   Plug the $u(\cdot)$ of part (e) into the solution of

$$\begin{cases} \dot{x} = Fx + Gu \\ x(0) = \begin{bmatrix} 3 \\ 0 \end{bmatrix} \end{cases}$$

and solve for the trajectory which the system follows with this input in moving from $\begin{bmatrix} 3 \\ 0 \end{bmatrix}$ to $\begin{bmatrix} 0 \\ 3e^{-\pi/2} \end{bmatrix}$.

(g)   Go back and study the role that G plays in the controllability of this system. Suppose that we let $G = \begin{bmatrix} a \\ b \end{bmatrix}$ for constants a and b. Can you still do each step (a) through (f)? Just for fun, compute the controllability matrix $\mathcal{C}(F, G) = [G \vdots FG]$. For what values of a and b is it singular? What are the eigenvalues of F?

✔ **8. (a)**   Use Corollary 6 to prove that the two functions given by e^t and te^t are linearly independent on any interval $[a, b]$.

(b)   Sketch the two functions given by $r_1(t) = t$ and $r_2(t) = |t|$ on $[-1, 1]$. Are they independent on $[0, 1]$? On $[-1, 0]$? Prove they are independent on $[-1, 1]$. What did you just discover about controllability on *subintervals*?

(c)   Let $A(t) = \begin{bmatrix} 1 & 0 \\ t & 0 \end{bmatrix}$. Show that $\det A(t) = 0$, its columns are dependent, but that its rows are linearly independent functions. What's going on here? Isn't column rank supposed to equal row rank and determinant rank? Give the column space and row space of $A(\cdot)$ and construct a basis for each.

✔ **9.**   Here we review two other methods of testing a set of $n (1 \times m)$ complex-valued functions $\{r_1(\cdot), \ldots, r_n(\cdot)\}$ as in equations (18a) for linear independence.

I. **(a)** Set $c_1 r_1(\cdot) + c_2 r_2(\cdot) + \cdots + c_n r_n(\cdot) = 0(\cdot)$, where $0(\cdot)$ is the $1 \times m$ zero function. Show that this can be written as

$$CA(\cdot) = 0(\cdot)$$

where $A(\cdot)$ is the matrix whose ith row is $r_i(\cdot)$ as in equation (18b), and $C = [c_1 \ldots c_n]$.

(b) Multiply both sides of $c_1 r_1(t) + c_2 r_2(t) + \cdots + c_n r_n(t) = 0$ on the right by $r_1^*(t)$, then by $r_2^*(t), \ldots$ and finally by $r_n^*(t)$ to get a system of n homogeneous equations. Using the standard inner product (Example 4–4–2) for complex-valued vector functions

$$\langle r_i(\cdot) \,|\, r_j(\cdot) \rangle \triangleq \int_{t_0}^{t_1} r_i^*(\xi) r_j(\xi) \, d\xi$$

integrate each of your n equations from t_0 to t_1 to get a set of n equations with constant coefficients to solve for the "unknowns" $c_1, \ldots, c_n$. [*Answer:* jth equation is $c_1 \overline{\langle r_1 | r_j \rangle} + c_2 \overline{\langle r_2 | r_j \rangle} + \cdots + c_n \overline{\langle r_n | r_j \rangle} = 0$]

(c) Write the system of part (b) as

$$M \begin{bmatrix} c_1 \\ \vdots \\ c_n \end{bmatrix} = \begin{bmatrix} 0 \\ \vdots \\ 0 \end{bmatrix}$$

where $M_{ij} \triangleq \overline{\langle r_j(\cdot) \,|\, r_i(\cdot) \rangle}$ (note order), and M is called the **gram matrix (grammian).**

State the condition on M which decides independence of $r_1(\cdot), \ldots, r_n(\cdot)$ on $[t_0, t_1]$. Compare this matrix M with the W of Corollary 6. Note that this test only requires that the $r_i(\cdot)$ be integrable.

(d) See what happens for the special case $r_i(t) = $ constant.

II. **(a)** Suppose that the functions $r_1(\cdot), \ldots, r_n(\cdot)$ have continuous $(n-1)$st derivatives on $[t_0, t_1]$ and prove the *sufficient* condition:

If for some $\tau \in [t_0, t_1]$ the constant $n \times nm$ matrix $[A(\tau) \,|\, \dot{A}(\tau) \,|\, \ldots \,|\, A^{(n-1)}(\tau)]$ has rank n, then $\{r_1(\cdot), \ldots, r_n(\cdot)\}$ are linearly independent on $[t_0, t_1]$. [*Hint:* $[c_1 \ldots c_n] A(\cdot) = 0$ on $[t_0, t_1] \Rightarrow [c_1 \ldots c_n] \dfrac{d^{(k)}}{dt^{(k)}} A(t) = 0$ for all $t \in [t_0, t_1]$.]

(b) Compare this condition with our proof in Section 5–3 that if a fundamental matrix had rank n at any time, it had rank n for all time.

(c) Apply this condition for $A(\cdot) = \Phi(t_0, \cdot) G(\cdot)$ using what you know about differentiating Φ with respect to its second variable. Specialize to constant systems (F, G) where $\Phi(t, \tau) = e^{F(t-\tau)}$.

10. For the system (F, G) of Exercise 7, compute the matrix $W(t_0, t_1)$ of equation (16).

(a) What is the rank of $W(t_0, t_1)$ where $t_0 = 0$ and $t_1 = \dfrac{\pi}{2}$ as in Exercise 7?

(b) Give the rows of $\Phi(t_0, \cdot) G(\cdot)$ and test for linear independence.

(c) Compute $(W(t_0, t_1))^{-1}$ and find the vector $z = W^{-1}(t_0, t_1) \hat{x}$ as in equation (9), with $\hat{x}$ as found in Exercise 7(d) for equation (6).

(d) From the vector z, compute the input $u(t) = (\Phi(t_0, t) G(t))^* z$, as in equation (11).

(e) Check that this input actually does drive $\left(0, \begin{bmatrix} 3 \\ 0 \end{bmatrix}\right)$ to $\left(\dfrac{\pi}{2}, \begin{bmatrix} 0 \\ 3e^{-\pi/2} \end{bmatrix}\right)$.

Compare with the inputs and corresponding trajectories of Exercise 7.

11.   **(a)**   For the $F(\cdot)$ and $G(\cdot)$ of Example 2, show that with $b(\cdot) \neq 0(\cdot)$, the rows of $\Phi(t_0, \cdot)G(\cdot)$ are linearly dependent on $[t_0, t_1]$ if and only if for some number k we have $a(\zeta) = b(\zeta)[k + \tfrac{1}{5}e^{-5t_0}(1 - e^{-5(\zeta-t_0)})]$ for $t_0 \leq \zeta \leq t_1$.

   (b)   Let $b(t) = 1$, constant, and find an $a(\cdot)$ which makes $(F(\cdot), G(\cdot))$ uncontrollable. Interpret this condition by studying $\dot{x} = Fx + Gu$. Can you find a state that you can't control to the origin?

   (c)   Suppose that $b(t) = 1$ and $a(\cdot)$ is not one of the "bad" functions of part (b), so $(F(\cdot), G(\cdot))$ is controllable. Compute $W(t_0, t_1)$ from equation (16) and construct a formula from (17) for an input $u(\cdot)$ which will transfer any $(t_0, \bar{x})$ to $(t_1, 0)$.

   (d)   Notice in (c) that you had two independent rows with just one independent column. What happened to row rank = column rank?

   (e)   Analyze the controllability of $F(t) = \begin{bmatrix} 1 & e^{-5t} \\ 0 & 1 \end{bmatrix}$ and $G(t) = \begin{bmatrix} a(t) \\ b(t) \end{bmatrix}$. $\left[Hint: \Phi(t, \tau) = e^{(t-\tau)}\begin{bmatrix} 1 & \tfrac{1}{5}(e^{-5\tau} - e^{-5t}) \\ 0 & 1 \end{bmatrix}. \right]$

12.   **(a)**   Let F be a constant matrix, let J be its Jordan canonical form matrix, and let P denote the transformation matrix whose columns are the Jordan basis vectors. Show that the rows of $\Phi(t_0, \cdot)G(\cdot)$ are independent if and only if the rows of $e^{-J(\cdot)}P^{-1}G(\cdot)$ are independent.

   (b)   Illustrate this for the (F, G) of Example 3, where $P = \begin{bmatrix} 3 & 2 \\ 1 & -1 \end{bmatrix}$.

   (c)   Find all values of parameters (a, b) which cause the row functions $r_1(\cdot)$ and $r_2(\cdot)$ of Example 3 to be dependent.

   (d)   Find the one set of parameters (a, b) which makes both brackets vanish simultaneously in Example 3. Prove directly that $r_1(\cdot)$ and $r_2(\cdot)$ are linearly independent for this (a, b). [*Hint:* Set $c_1 r_1(\cdot) + c_2 r_2(\cdot) = 0$ for these values and show that $c_1 = c_2 = 0$.]

✔ **13.**   This exercise gives practice with controllability conditions for constant systems.

   (a)   Show that $G^*(\zeta)\Phi^*(0, \zeta)x_2 \equiv 0$ for all $\zeta \geq 0$ iff $x_2^*\Phi(0, \zeta)G(\zeta) \equiv 0$ for all $\zeta \geq 0$.

   (b)   Let F and G be constant $n \times n$ and $n \times m$ matrices and show that

$$x_2^* e^{-F\zeta}G \equiv 0, \text{ for all } \zeta \geq 0$$

if and only if

$$x_2^* e^{F\zeta}G \equiv 0, \text{ for all } \zeta \geq 0$$

   (c)   By the Cayley-Hamilton theorem, any function of a square matrix A can be written as a polynomial involving only $I, A^1, A^2, \ldots, A^{n-1}$. Write

$$e^{-F\zeta} = \alpha_0(\zeta)I + \alpha_1(\zeta)F + \cdots + \alpha_{n-1}(\zeta)F^{n-1}$$

where the $\alpha_i(\zeta)$ are appropriate scalar functions, and substitute this expression in

$$-x_0 = \int_0^{t_1} e^{-F\zeta}Gu(\zeta)\, d\zeta$$

Let g_j denote the jth column of G, let $u(\zeta) = [u_1(\zeta), \ldots, u_m(\zeta)]^T$, and define

$$r_{ij} = \int_0^{t_1} \alpha_i(\zeta)u_j(\zeta)\, d\zeta$$

Show that

$$-x_0 = \sum_{j=1}^{m} \sum_{i=0}^{n-1} r_{ij} F^i g_j$$

is a linear combination of the columns of $[G, FG, \ldots, F^{n-1}G]$.

(d)   Use (c) to argue that controllability implies

$$\text{rank}[G, FG, \ldots, F^{n-1}G] = n$$

for a different proof of the "if" part of Theorem 10.

(e)   Interpret r_{ij} as the projection of input component $u_j(\cdot)$ on the known function $\alpha_i(\cdot)$, and suppose that you know that

$$\text{rank}[G, FG, \ldots, F^{n-1}G] = n$$

Given an arbitrary vector x_0, see what you can say about finding an input $[u_1(\cdot), \ldots, u_n(\cdot)]^T$ to transfer $(0, x_0)$ to (t_1, x_1).

(f)   Suppose that you only consider constant inputs as in Figure 7–28, where each component $u_j(t)$ has a constant value c_j on the interval $[0, t_1)$. Now try part (e).

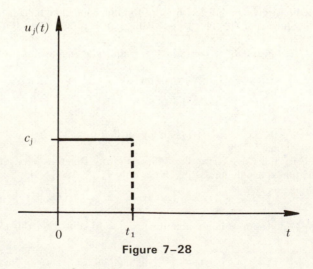

Figure 7–28

✔ **14.**   **(a)**   For the F and G of Example 3, compute $\mathfrak{L}_X(s) = (sI - F)^{-1}G$, the "state transfer function."

$$\left\{ \; Answer: \; (sI - F)^{-1}G = \begin{bmatrix} \dfrac{b(s+1)+6}{(s+3)(s-2)} & \dfrac{s+1+6a}{(s+3)(s-2)} \\[2ex] \dfrac{s+b}{(s+3)(s-2)} & \dfrac{as+1}{(s+3)(s-2)} \end{bmatrix} \right.$$

(b)   What happens in $(sI - F)^{-1}G$ for the "bad" values of (a, b) where the system is not controllable?

(c)   Let $L: V \to W$ be any linear map which is both one to one and onto (a *bijection*). Prove that both L and $L^{-1}: W \to V$ map linearly independent vectors into linearly independent vectors. The Laplace transform is a bijection. Explain how this is used in Corollary 8.

(d)   Prove by direct calculation that any (F, G) is controllable iff there is no cancellation in $(sI - F)^{-1}G$. [If you can't do it in general, do it for the special case where F is diagonal with distinct eigenvalues.]

(e)   Prove that a cancellation in $(sI - F)^{-1}G$ produces a cancellation in the transfer function $\mathscr{S}(s) = H(sI - F)^{-1}G$; hence, uncontrollability implies that (F, G, H) is not a *minimal* realization of $\mathscr{S}(s)$.

15.   **(a)**   Show that $\dfrac{d}{dt} e^{F(t_0 - t)} = e^{Ft_0} \dfrac{d}{dt} e^{-Ft} = e^{Ft_0}(-F)e^{-Ft}$ and   that

$\dfrac{d^j}{dt^j} e^{F(t_0 - t)} = e^{Ft_0}(-1)^j F^j e^{-Ft}$.

(b)   For the F and G of Example 3, let $b = a = 0$ and find an input

$$u(t) = \begin{bmatrix} c_{01} \\ c_{02} \end{bmatrix} \delta(t) + \begin{bmatrix} c_{11} \\ c_{12} \end{bmatrix} \dot{\delta}(t)$$

which drives $x(0^-) = \begin{bmatrix} 5 \\ 3 \end{bmatrix}$ to $x(0^+) = \begin{bmatrix} 0 \\ 0 \end{bmatrix}$. $\left[Answer:$ One possible solution is $u(t) = \begin{bmatrix} 1 \\ 5 \end{bmatrix} \delta(t) + \begin{bmatrix} 0 \\ 2 \end{bmatrix} \dot{\delta}(t). \right]$ What input drives $x(4^-) = \begin{bmatrix} 5 \\ 3 \end{bmatrix}$ to $x(4^+) = \begin{bmatrix} 0 \\ 0 \end{bmatrix}$?

(c)   Again with $a = b = 0$, find an input $u(\cdot)$ which instantaneously "resets" the initial condition of the system (F, G, H) from $x(0^-) = \begin{bmatrix} 5 \\ 3 \end{bmatrix}$ to $x(0^+) = \begin{bmatrix} 5 \\ 0 \end{bmatrix}$. Repeat for $x(0^+) = \begin{bmatrix} 0 \\ 3 \end{bmatrix}$. Repeat for $x(0^+) = \begin{bmatrix} -3 \\ 5 \end{bmatrix}$.

(d)   Take the Laplace transform of the $u(t)$ of part (b). Transform

$$-\bar{x} = \int_0^t e^{F(0 - \zeta)} G(\zeta) u(\zeta) \, d\zeta$$

and solve for c_0 and c_1 when $\bar{x} = \begin{bmatrix} 5 \\ 3 \end{bmatrix}$.

16.   In Example 2-5-9 we studied a mechanical arm (Fig. 2-41) activated by two motor torques, u_1 and u_2, and with state variables chosen as $x_1 = \theta_1$, $x_2 = \dot{\theta}_1$, $x_3 = \theta_2$, $x_4 = \dot{\theta}_2$. We linearized the state equations about two possible reference points and got equations (2-5-3) and (2-5-4), both of the type $\dot{x} = Fx + Gu$. For example, about the point $\left[0, 0, \dfrac{\pi}{2}, 0\right]^T$, corresponding to small $\dot\theta_1$, $\dot\theta_2$, and $\left(\theta_2 - \dfrac{\pi}{2}\right)$, we got

$$F = \begin{bmatrix} 0 & 0 & 1 & 0 \\ 0 & 0 & 0 & 1 \\ 0 & 0 & 0 & 0 \\ 0 & 0 & 0 & 0 \end{bmatrix} \quad \text{and} \quad G = \left(\dfrac{3}{4m\ell^2}\right) \begin{bmatrix} 0 & 0 \\ 0 & 0 \\ 1 & -1 \\ -1 & 5 \end{bmatrix}$$

(a)   See whether the arm is controllable for small excursions about this point.
(b)   Suppose that one of the torque motors fails. Can we control the arm with just one torque? [*Hint:* If $u_2 = 0$, we only have one column in G.]
(c)   What quantities would you monitor for the arm? Give an appropriate read-out matrix H.
(d)   What is the unforced state response of the linearized system? Compute $(sI - F)^{-1}$ by block-partitioning (Section 7–2), and find the poles. Find a feedback matrix K to shift the poles to -2, -3, -4, -5.
17.   Prove that when $\lambda_1, \ldots, \lambda_n$ are distinct, if all the rows $\bar G_1, \ldots, \bar G_n$ of $G_N = P^{-1}G$ are nonzero, then you can find an input $u_{[0, t_1]}$ to drive any $(0, \bar x_0)$ to $(t_1, 0)$. [*Hint:* You can do it with only a one-dimensional input

$$u(t) = \begin{bmatrix} u_1(t) \\ 0 \\ \vdots \\ 0 \end{bmatrix}$$

Try

$$u_1(t) = \sum_{k=1}^{n} c_k e^{-\lambda_k^*(t_1 - t)}$$

and solve for $c_1, \ldots, c_n$ so that

$$-\bar x_{0i} = \int_0^{t_1} e^{-\lambda_i \zeta} \bar G_i u(\zeta)\, d\zeta. \,]$$

Where did you need the distinctness of $\lambda_1, \ldots, \lambda_n$?
18.   **(a)**   For $F = \begin{bmatrix} 0 & 6 \\ 1 & -1 \end{bmatrix}$ and $G = \begin{bmatrix} b & 1 \\ 1 & a \end{bmatrix}$, find the equivalent Jordan realization (F_N, G_N) and draw the block diagram. How do the "bad" values of a and b which lose controllability show up?
(b)   Let

$$F = \begin{bmatrix} 1 & 0 & 0 \\ \hline 0 & 2 & 0 \\ 0 & 0 & 2 \end{bmatrix} \quad \text{and} \quad G = \begin{bmatrix} a \\ b \\ c \end{bmatrix}$$

and see which values of (a, b, c) lose controllability. Draw the block diagram and compare with Example 9.
(c)   Let $F = \begin{bmatrix} \alpha & 0 \\ 0 & \beta \end{bmatrix}$ and $G = \begin{bmatrix} a_1 & a_2 \\ b_1 & b_2 \end{bmatrix}$. Draw the block diagram and study controllability. Compare with Example 4–3–6. What happens if $\alpha = \beta$?

✔ **19.**   Let (F, G, H) be controllable so that it is algebraically equivalent to the controllable canonical form (F_c, G_c, H_c) of Section 7-1. Let P denote the equivalence transformation.

   (a)   Show that

$$P \begin{bmatrix} 1 \\ s \\ \vdots \\ s^{n-1} \end{bmatrix} = \begin{bmatrix} r_1(s) \\ r_2(s) \\ \vdots \\ r_n(s) \end{bmatrix}$$

where $r_1(s), \dots, r_n(s)$ are the rows of $\text{adj}(sI - F)G$.

   (b)   If $(sI - F)^{-1}G$ has a cancellation, then there exists s_0 such that $r_1(s_0) = r_2(s_0) = \cdots = r_n(s_0) = 0$. Show that this contradicts the invertibility of P.

   (c)   Check directly to see that there can be no cancellations in $(sI - F_c)^{-1}G_c$, so that (F_c, G_c, H_c) is controllable.

✔ **20.**   For the $r \times r$ sub-matrix $\bar{F}_{11}$ and the $r \times m$ sub-matrix $\bar{G}_1$ of the matrices $\bar{F}$ and $\bar{G}$ of Theorem 16:

   (a)   Show that $[\bar{G}_1, \bar{F}_{11}\bar{G}_1, \dots, \bar{F}_{11}^{n-1}\bar{G}_1]$ constitute the top r rows of the matrix

$$\bar{Q} = [\bar{G}, \bar{F}\bar{G}, \dots, \bar{F}^{n-1}\bar{G}] = P^{-1}Q$$

and that the bottom $n - r$ rows of $\bar{Q}$ are zero.

   (b)   Show that this means that the r-dimensional system $(\bar{F}_{11}, \bar{G}_{11})$ is controllable.

   (c)   Use the above results to show that if (F, G, H) is not controllable and $\dim R(Q) = r < n$, then $\mathfrak{H}(s) = H(sI - F)^{-1}G = \bar{H}_1(sI - \bar{F}_{11})^{-1}\bar{G}_1$ so that (F, G, H) has an r-dimensional realization $(\bar{F}_{11}, \bar{G}_1, \bar{H}_1)$, where $\bar{H} = [\bar{H}_1 \vdots \bar{H}_2] = HP$.

21.   **(a)**   Since (F, G) and $(\bar{F}, \bar{G})$ of Theorem 16 are algebraically equivalent under P, show their state transfer functions $\mathfrak{H}_X(s)$ and $\bar{\mathfrak{H}}_{\bar{X}}(s)$ are related by

$$\mathfrak{H}_X(s) = P\bar{\mathfrak{H}}_{\bar{X}}(s) = P(sI - \bar{F})^{-1}\bar{G} = \textcircled{P} \begin{bmatrix} sI - \bar{F}_{11} & \vdots & sI - \bar{F}_{12} \\ \cdots & \cdots & \cdots \\ 0 & \vdots & sI - \bar{F}_{22} \end{bmatrix}^{-1} \begin{bmatrix} \bar{G}_1 \\ 0 \end{bmatrix}$$

$$= P \begin{bmatrix} (sI - \bar{F}_{11})^{-1} & \vdots & -(sI - \bar{F}_{11})^{-1}(sI - \bar{F}_{12})(sI - \bar{F}_{22})^{-1} \\ \cdots & \vdots & \cdots \\ 0 & \vdots & (sI - \bar{F}_{22})^{-1} \end{bmatrix} \begin{bmatrix} \bar{G}_1 \\ 0 \end{bmatrix}$$

$$= P \underbrace{\begin{bmatrix} (sI - \bar{F}_{11})^{-1}\bar{G}_1 \\ \cdots \\ 0 \end{bmatrix}}_{m} \begin{matrix} \} \ r \\ \\ \} \ n - r \end{matrix}$$

   (b)   By the invertibility of P, $\mathfrak{H}_X(s)$ and $\bar{\mathfrak{H}}_{\bar{X}}(s)$ have the same number of linearly independent rows. Use Exercise 20 to show that the sub-matrices $\bar{F}_{11}$ and $\bar{G}_1$ are a controllable pair; i.e.,

$$\text{rank } [\bar{G}_1, \bar{F}_{11}\bar{G}_1, \dots, \bar{F}_{11}^{n-1}\bar{G}_1] = r$$

so that all r of the rows of $(sI - \bar{F}_{11})^{-1}\bar{G}_1$ are linearly independent. Thus we can make the following statement:

Corollary 16a: If $\dim \mathfrak{R}(Q) = r < n$, then (F, G) is not controllable and exactly r rows of $(sI - F)^{-1}G$ are linearly independent.

EE   **22.**   Consider the simple electrical circuit shown in Figure 7-29.

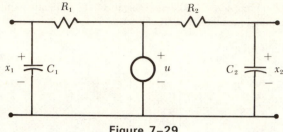

Figure 7–29

(a)   For this system, using capacitor voltages x_1 and x_2 as states with input voltage source u, determine a condition on time constants R_1C_1 and R_2C_2 such that the system is completely state controllable.

(b)   Let $R_1 = 10,000$ ohms
$$R_2 = 20,000 \text{ ohms}$$
$$C_1 = 0.001 \text{ microfarads}$$
$$C_2 = 0.001 \text{ microfarads}$$

and

$$x_1(0) = +60 \text{ microvolts}$$
$$x_2(0) = -30 \text{ microvolts}$$

Determine a piecewise constant control signal u as shown in Figure 7–30 which will

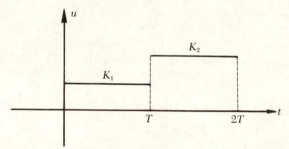

Figure 7–30

bring both capacitor voltages, $x_1(t)$ and $x_2(t)$, to zero for $t > 2T$ seconds, where $T = 10^{-5}$ seconds.

✔ **23. (a)**   For the (F, H) of Example 11, differentiate equation (28) with respect to t and evaluate $\hat{y}(t)$ and $\dfrac{d\hat{y}(t)}{dt}$ at $t = t_0$ to get two equations in the two unknowns, x_1^0 and x_2^0. Find conditions on gains a and b under which the initial state x^0 is uniquely observed.

(b)   Generalize Example 11 by letting $F = \begin{bmatrix} \alpha & 0 \\ 0 & \beta \end{bmatrix}$ and find conditions on the constants α, β, a, b for recovering x^0 from the zero-input response.

(c)   Generalize Example 11 by letting $H = [a(t) \quad b(t)]$, with time-varying gains, and see about solving for x^0. What happens if $a(t) = e^{-2t}$ and $b(t) = e^{+3t}$? Can you distinguish between the initial state $x_1^0 = e^{2t_0}$, $x_2^0 = e^{-3t_0}$, and the zero state?

(d)   For an $(n \times n)$ matrix F and a $(1 \times n)$ read-out matrix H, at how many time instants should you sample the zero-input response to solve for x^0? Generalize to an H that is $(q \times n)$.

   (e)   How many differentiations at $t = t_0$ do you need to get enough equations to solve for x^0 when H is $(1 \times n)$? Give a formula for x^0 in terms of $\mathcal{O}(F, H)$, the observability matrix.

   (f)   Use part (e) to derive an expression for x^0 for a single-output system in the observable canonical form (F_o, G_o, H_o) of Exercise 7–1–9.

✔ **24.**   Let $\Sigma_D = (F_D, G_D, H_D)$ be the dual of $\Sigma = (F, G, H)$ and let $g_D(t, \tau)$ be the impulse response of Σ_D.

   (a)   Show that

$$g_D(t, \tau) = [H(\tau)\Phi(\tau, t)G(t)]^* 1(t - \tau) \qquad \text{(note order)}$$

and that

$$g(t, \tau) = [g_D(\tau, t)]^* 1(t - \tau)$$

   (b)   Use this relationship to see how we may study the effect at a fixed time t of applying an impulse at different times $\tau < t$ by simulating the adjoint system and "running it backwards in time," as control theorists say.

25.   Let F have n distinct eigenvalues and let (F_N, G_N, H_N) be the Jordan canonical form representation of (F, G, H). Prove that (F, H) is observable if and only if none of the columns of H_N are zero.

EE 26.   In the constant resistance circuit of Example 12, let $R = 1$ and let $L = C$.

   (a)   Determine the set of controllable states, as in Theorem 15.

   (b)   As in Theorem 16, determine a lower dimensional system which is zero-state equivalent to our system.

   (c)   Draw a block diagram for our system corresponding to the separation of the state space into controllable and uncontrollable parts, as in Figure 7–25.

   (d)   Give the transfer function for the system for general R, L, and C and then let $R = L = C = 1$ in the result. Do you see why, when $R^2 C = L$, we call it a constant resistance circuit?

   (e)   Use duality to find the observable states and to separate the state space into observable and unobservable parts.

   (f)   Sketch a mechanical system using a mass, a spring, and two viscous damping friction components which obeys the same type of differential equations as does this electrical circuit. Interpret loss of controllability and observability for the system.

EE 27. (a)   Consider the electrical bridge network of Exercise 2–5–8, in which for the indicated choice of states and element values you found the representation (F, G, H), where

$$F = \begin{bmatrix} -\dfrac{1}{2} & 0 & \dfrac{1}{2} \\[6pt] 0 & -\dfrac{1}{2} & \dfrac{1}{2} \\[6pt] \dfrac{1}{2} & 1 & -\dfrac{3}{2} \end{bmatrix}, \ G = \begin{bmatrix} 1 \\[6pt] -\dfrac{1}{2} \\[6pt] 0 \end{bmatrix}, \ H = [1 \ \ -1 \ \ 0]$$

Check for controllability and observability. Can you give a physical interpretation of lack of controllability? Observability? [See *IEEE Transactions on Education*, Feb. 1971, pp. 37–38 for an illuminating electrical interpretation of these phenomena by Robinson and Robinson.]

(b)   Check the observability of the mechanical arm studied in Exercise 16. You will need to select a readout matrix H for the system first. Decide what quantities you would like to observe in this system and experiment with several different choices (hence H's). Try reading out each state variable alone (e.g., $H = \begin{bmatrix} 1 & 0 & 0 & 0 \end{bmatrix}$ reads out $y = x_1$), and see whether or not the system is observable with these H's.

(c)   Using the indicated state variables, write state equations and analyze the electrical circuit of Figure 7–31 for controllability and observability.

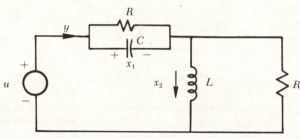

Figure 7–31

For values of L and C such that the system is neither controllable nor observable, show that the system is zero-state equivalent to a pure resistor.

✔ **28.**   Consider the system (F, G, H) with

$$ F = \begin{bmatrix} 0 & 1 \\ 6 & -1 \end{bmatrix}, \; G = \begin{bmatrix} b \\ 1 \end{bmatrix}, \; H = \begin{bmatrix} a & 1 \end{bmatrix} $$

(a)   Draw the wiring diagram; label it clearly. (Notice on the diagram that if you had set $a = 0$, an uneducated person might think that you had "disconnected" state x_1 from the output y and hence that, for this value of a, x_1 is unobservable. Let's clear this up.)

(b)   Compute the controllability and observability matrices $\mathcal{C}$ and $\mathcal{O}$ and determine exactly which values of b and a affect controllability and observability. [Notice that $a = 0$ is *not* one of them.]

(c)   Give the transfer function of (F, G, H) (in terms of b and a). Look at the transfer function when you let b and a take on each of the "bad" values you found in part (b). How does loss of controllability show up in the transfer function? Repeat for loss of observability.

(d)   Transform (F, G, H) to $(\bar{F}, \bar{G}, \bar{H})$, where $\bar{F}$ is diagonal, by a similarity transformation $x = P\bar{x}$.

(e)   Draw the wiring diagram for the $(\bar{F}, \bar{G}, \bar{H})$ of part (d). Notice how "bad" values of a and b are clearly revealed on this "Jordan canonical form" realization $(\bar{F}, \bar{G}, \bar{H})$.

29.   In Exercise 12 of Section 2–5 you wrote the dynamic equations of a stick-balancing cart, illustrated in Figure 2–48. Interpret balancing the stick in terms of controllability. Describe what you need to measure (observe) in order to control the cart. Suppose that the stick is initially greatly out of balance, say with $\theta(0) = 45°$ and $\dot{\theta}(0) = 0$; can you control it back to $\theta = 0$ using linear state feedback?

30.   Use Theorems 16 and 24 to decide the controllability and observability of the composite feedback system of Figures 7–19 and 7–20, which we designed in Section 7–2.

(b)   Write state dynamic equations for the observer of Figure 7-17 and determine whether it is controllable and/or observable.

✓ **31.**   In Exercise 11 of Section 3-4 you linearized the equations of a satellite and, using some simplifying normalizations, got equations of the form

$$\dot{x} = Fx + Gu, \quad y = Hx$$

where

$$F = \begin{bmatrix} 0 & 1 & 0 & 0 \\ 3\omega^2 & 0 & 0 & 2\omega \\ 0 & 0 & 0 & 1 \\ 0 & -2\omega & 0 & 0 \end{bmatrix} \quad G = \begin{bmatrix} 0 & 0 \\ 1 & 0 \\ 0 & 0 \\ 0 & 1 \end{bmatrix}$$

$$H = \begin{bmatrix} 1 & 0 & 0 & 0 \\ 0 & 0 & 1 & 0 \end{bmatrix}$$

Here $u = \begin{bmatrix} u_1 \\ u_2 \end{bmatrix}$, u_1 and u_2 were radial and tangential thrust, and x_1 and x_3 were radial and angular excursions from the reference (circular) orbit, respectively.

(a)   Suppose that you observe the satellite using a radar which displays outputs y_1 and y_2. What does H tell us about what we're trying to measure with the radar?

(b)   Suppose that the radial thruster fails to turn on. Can we control the system using only tangential thrust?

(c)   If the tangential thruster fails, can we control the system with radial thrust alone?

(d)   Suppose that we try to measure only y_1, so that H looks like $[1 \quad 0 \quad 0 \quad 0]$. Is the system observable with this H?

(e)   If we can measure only y_2, so that H becomes $[0 \quad 0 \quad 1 \quad 0]$, is the system observable?

(f)   Discuss the significance of your results from parts (a), (b), (c), (d) and (e).

[Note: This interesting example is due to Brockett, and is discussed in his *Finite Dimensional Linear Systems* (John Wiley & Sons, 1970).]

7-4   An Introduction to Stability Theory

Consider the input-output descriptions

$$y(t_1) = HF^{(t_1-t_0)}x(t_0) + \sum_{j=1}^{t_1-t_0} HF^{(t_1-t_0-j)}Gu(t_0 + j - 1)$$

and

$$y(t_1) = He^{F(t_1-t_0)}x(t_0) + \int_{t_0}^{t_1} He^{F(t_1-t)}Gu(t) \, dt$$

of the discrete-time and continuous-time constant linear systems (F, G, H). If we are meaningfully to relate the input to the output, we should hope that as the time interval between t_0 and t_1 gets larger and larger, so should the contribution of $x(t_0)$ to $y(t_1)$ get smaller and smaller. In other words, we should like small changes in initial state *eventually* to have negligible effects upon the behavior of the system. It could be catastrophic if an unnoticed perturbation in the initial state of the system should "blow up" to overwhelm later input-output behavior. We say that a system is **stable** when it is safe from such a "blow up." There are many different ways of making this informal notion of stability precise—in this section we present just a few of them, and show how they may be related to properties of the eigenvalues in the case of the linear systems noted above.

DEFINITION 1

For any (linear or non-linear) system with state transition function ϕ, we say that x^e is an **equilibrium state** if it does not change under 0 input, i.e.,

$$x^e \text{ is an equilibrium state} \Leftrightarrow x^e = \phi(t_1, t_0, x^e, 0) \text{ for all } t_1 \geq t_0 \qquad \bigcirc$$

For a discrete-time system where

$$x(t_0 + 1) = \phi(t_0 + 1, t_0, x(t_0), 0)$$

if x^e is an equilibrium state we have

$$\phi(t_0 + 1, t_0, x^e, 0) = x^e \text{ for all } t_0$$

For a continuous-time system with

$$x(t) = \phi(t_1, t_0, x(t_0), 0)$$

and

$$\frac{dx(t)}{dt} = f(t, x(t), 0)$$

if $x(t_0) = x^e$ is an equilibrium state, then

$$\frac{dx(t)}{dt} = \frac{d}{dt} \phi(t, t_0, x^e, 0) = \frac{d}{dt} x^e = 0$$

and

$$f(t, x^e, 0) = 0$$

In Section 3-4, when we discussed linearization, we often linearized continuous-time systems in the neighborhoods of equilibrium points because they were physically significant and, mathematically, yielded simpler expansions in the differential approximations.

Example 1

The simple pendulum of Example 3–2–2 is shown in Figure 7–32 for convenience.

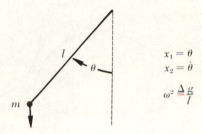

Figure 7–32   The simple pendulum. mg

$x_1 = \theta$
$x_2 = \dot\theta$

$\omega^2 \triangleq \dfrac{g}{l}$

The free motion is given by $\dot{x}(t) = f(t, x(t), 0) = \begin{bmatrix} x_2(t) \\ -\omega^2 \sin(x_1(t)) \end{bmatrix}$.

Since

$$f\left(t, \begin{bmatrix} 0 \\ 0 \end{bmatrix}, 0\right) = \begin{bmatrix} 0 \\ 0 \end{bmatrix}$$

and

$$f\left(t, \begin{bmatrix} \pi \\ 0 \end{bmatrix}, 0\right) = \begin{bmatrix} 0 \\ 0 \end{bmatrix}$$

both $\begin{bmatrix} 0 \\ 0 \end{bmatrix}$ and $\begin{bmatrix} \pi \\ 0 \end{bmatrix}$ are equilibrium points of this non-linear system.   ◊

In the linear case, we have

$$x(t_0 + 1) = \phi(t_0 + 1, t_0, x^e, 0) = F(t_0)x^e = x^e$$

for discrete time, while for continuous time we have

$$\frac{dx(t)}{dt} = f(t, x(t_0), 0) = F(t)x(t_0)$$

whence $F(t)x^e = 0$.

   Thus, for the constant, discrete-time system (F, G, H), x^e is an equilibrium state if and only if $x^e = Fx^e$; that is, if and only if $x^e \in \mathfrak{N}(I - F)$, whereas in the continuous-time case x^e is an equilibrium state if and only if $x^e \in \mathfrak{N}(F)$. Since $0 \in \mathfrak{N}(I - F)$ and $0 \in \mathfrak{N}(F)$, the zero state is always an equilibrium state of a linear system, whether discrete-time or continuous-time, and much of our theory will concentrate on this case.

Let us now motivate three precise concepts related to stability by considering a ball at rest on three different surfaces, as shown in Figure 7–33.

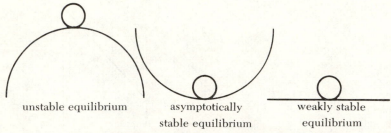

unstable equilibrium · asymptotically stable equilibrium · weakly stable equilibrium

Figure 7–33 Three different types of equilibrium.

In each case, the ball is in equilibrium. But if it is poised on a "hilltop," the slightest displacement will send it hurtling away; if it is started in a "valley," a slight displacement will eventually be overcome and the ball will return to its original position; and a ball resting on a flat surface will neither increase nor decrease a slight initial displacement. These three situations then exemplify the three types of stability whose definitions† we now provide:

DEFINITION 2

Let x^e be an equilibrium point of a system with state transition function ϕ, and, for any initial state x^0 at time t_0, let $\phi_{x^0}(t)$ abbreviate the free motion $\phi(t, t_0, x^0, 0)$. Then we say that:

(i) x^e is an **unstable** equilibrium if for any $\delta > 0$ there exists a state x^0 such that

$$\|x^0 - x^e\| < \delta \text{ yet } \|\phi_{x^0}(t) - x^e\| \to k, \text{ where } \delta < k \leq \infty, \text{ as } t \to \infty$$

(ii) x^e is an **asymptotically stable** equilibrium if there exists some $\delta > 0$ such that

$$\|x^0 - x^e\| < \delta \text{ implies } \|\phi_{x^0}(t) - x^e\| \to 0 \text{ as } t \to \infty$$

(iii) x^e is a **weakly stable** equilibrium if for any $\varepsilon > 0$, there exists some $\delta > 0$

---

†These definitions follow Liapunov and are often said to be "in the sense of Liapunov" to distinguish them from other definitions of stability; see Cesari, *Asymptotic Behavior and Stability Problems in Ordinary Differential Equations*, Third Edition (Springer-Verlag, N.Y., 1971).

such that

$$\|x^0 - x^e\| < \delta \text{ implies } \|\phi_{x^0}(t) - x^e\| < \varepsilon \text{ for all } t \geq t_0$$

We say that a linear system is stable in some sense if its zero state is stable in that sense. ○

The reader should try to state and visualize these definitions geometrically. For example, if a linear system is weakly stable, then we can keep the state trajectory forever within a given ball of radius ε by restricting the initial state x^0 to lie within a ball of radius δ centered at the origin.

Example 2

For the simple non-linear pendulum of Example 1, $x^e = \begin{bmatrix} 0 \\ 0 \end{bmatrix}$ is a weakly stable equilibrium and $x^e = \begin{bmatrix} \pi \\ 0 \end{bmatrix}$ is an unstable equilibrium. There are no asymptotically stable equilibria [see Exercises 1 to 3].

The "small-angle," linearized simple pendulum

$$\begin{bmatrix} \dot{x}_1 \\ \dot{x}_2 \end{bmatrix} = \begin{bmatrix} 0 & 1 \\ -\omega^2 & 0 \end{bmatrix} \begin{bmatrix} x_1 \\ x_2 \end{bmatrix}$$

is weakly stable. ◊

EIGENVALUE CRITERIA FOR STABILITY

Let us return to consideration of the effect of perturbations of the zero state of the system $\dot{x}(t) = Fx(t) + Gu(t)$, whose zero input response is given by

$$x(t) = e^{Ft}x(0)$$

Without loss of generality we may assume that (F, G) has already been transformed to Jordan canonical form, so that

$$x(t) = e^{Jt}x(0)$$

Now, if in fact we set $x(0) = 0$, it then follows that $x(t) = 0$ for all t. However, this is only in the most idealized of all worlds. In any real situation we cannot set the state exactly to zero, but can only set it close to zero. We thus just ask whether the slightly perturbed value of $x(0) \cong 0$ will amplify with time, thus swamping out the system, or will dampen. In fact, we saw in Section 6–6 that the state $x(t)$ is a linear combination of the modes and that, for each state variable x_i, we have for some $k \geq 1$ that

$$x_i(t) = \sum_{j=0}^{k-1} \frac{t^j}{j!} e^{\lambda_i t} x_{j,i}(0)$$

We see that a necessary and sufficient condition that the initial state displacement be dampened out is simply that

$$e^{\lambda_i t} \to 0 \text{ as } t \to \infty \text{ for each } i$$

and this condition will hold if and only if

$$\Re e(\lambda_i) < 0 \text{ for each } i$$

From this discussion and a corresponding analysis of the discrete-time case [Exercise 3], we have:

THEOREM 1

A time-invariant linear *continuous-time* system is asymptotically stable—in the sense that "$x(t) \to 0$ as $t \to \infty$, for any $x(0)$, and for zero input for $t \geq 0$"—if and only if the real parts of the eigenvalues of the matrix F are all negative. A time-invariant linear *discrete-time* system is asymptotically stable if and only if all the eigenvalues of F have magnitudes less than one. □

Putting the stability condition of Theorem 1 another way, it simply says that after using the system for a while, the effect of the initial state will be negligible. In other words, we may so drive the system that it really responds to the inputs we are applying, rather than to the state in which we found it; if the system is stable, we do not have to use our inputs to fight an exponential growth in the state, but can instead devote them to actually obtaining some desired behavior. Thus, in designing a real system, it is extremely important to ensure that it is stable—in other words, to ensure that the real parts of the eigenvalues of its F matrix are all negative. In Section 7–5, we shall see how to use negative† feedback to stabilize a large class of linear systems.

When we take the Laplace transform of the impulse response of a linear system, the eigenvalues of the F matrix reappear as the poles of the system

---

† Traditionally, feedback is called negative when we compare a system's performance with a desired performance and use the discrepancy (error) to drive the system in the "opposite direction," i.e., to reduce the error. Of course, in some applications such as amplification, rather than stabilization or control, positive feedback is useful, and we try to drive a system to some saturation level as quickly as possible by turning up the positive parts of the eigenvalues.

function. Thus, the criterion of Theorem 1 for stability is precisely the classical stability condition, which requires that the poles of the system all lie in the negative half of the complex s-plane; in other words, that they all have negative real parts. Electrical engineering and control theory books are full of ingenious special techniques, such as the *Routh-Hurwitz* criterion and the *Liénard-Chipart* criterion, for determining whether or not the eigenvalues of F have negative real parts without actually having to solve the characteristic polynomial for its roots.[†]

Historically, much of the language and terminology of stability theory grew out of the study of non-linear oscillations and differential equations, and the literature of these fields is rich in examples and qualitative, geometric visualizations. For example, in Figure 7–34 we sketch the trajectories of a mythical two-dimensional zero-input system. It can be shown[‡] that the various features:

> stable focus (F)
> unstable focus (F')
> stable node (N)
> unstable node (N')
> saddle point (S)
> vortex or center (V)
> limit cycle (LC)

of which all but the stable node are illustrated in the diagram, represent all possible classes of equilibrium points for unforced systems $\dot{x} = f(x)$, so long as f is "sufficiently well-behaved."

Thinking of this diagram as a contour map of a region on which the ball of our earlier heuristic discussion is rolling:

A **node** may be thought of as a peak (unstable) or hollow (stable).

A **focus** is also a peak or hollow, but the surface is nearly level, so the path is spiral, each turn closely approximating a contour. Thus, foci are reached after an infinite time (if stable) or were left at time $-\infty$ (if unstable).

Vortices or **centers** are related to a state of rest, or are surrounded by a continuum of closed paths. Thus, motions around a vortex are neither stable nor unstable—but when the system is subject to noise, this "neutral" stability looks more like instability, there being an "aimless drifting."

A **limit cycle** is always surrounded by an annular region where all paths tend toward the cycle, or unwind themselves from it. If the cycle is stable, the motion of the system tends to become ultimately periodic for all initial

---

[†] See, for example, Chen, *Introduction to Linear System Theory* (Holt, Rinehart and Winston, 1970), Section 8–3.

[‡] See for example, Birkhoff and Rota, *Ordinary Differential Equations*, Second Edition (Blaisdell, 1969), Chapter 5; or Sansone and Conti, *Nonlinear Differential Equations* (Pergamon Press, 1964).

Figure 7–34 Some possible types of equilibrium states

conditions interior to the annulus, and this limit is independent of the initial conditions.

A **saddle point** is the equivalent of a pass in the hills, and is the only kind of equilibrium point which can be crossed by a trajectory.

Each closed trajectory contains in its interior at least one equilibrium point.

As $t \to +\infty$, every motion must satisfy one, and only one, of the following conditions:

(i) the motion leads to a stable node, where it stops.
(ii) it leads toward a stable focus.
(iii) it leads toward (or is already on) a limit cycle.
(iv) it goes to infinity (i.e., it proceeds to its "destruction," since there must be bounds on the variables of a physical system).

Let us now give an explicit example of limit cycles:

Example 3

Consider the non-linear system

$$\dot{x} = y + x(1 - x^2 - y^2)$$
$$\dot{y} = -x + y(1 - x^2 - y^2)$$

which becomes, in polar coordinates,

$$\dot{\rho} = \rho(1 - \rho^2)$$
$$\dot{\theta} = -1$$

and has the general solution

$$\rho = (1 + ke^{-2t})^{-1/2}$$
$$\theta = \theta_0 - t$$

where k and θ_0 are determined by the initial conditions.

If $k = 0$, the solution is the circle $\rho = 1$.

If $k > 0$, the solutions represent spirals which move away from the origin $(t = -\infty)$ and approach the circle $\rho = 1$ as t increases.

If $k < 0$, spirals approach $\rho = 1$ from outside. We thus have a stable limit cycle.

These three cases are illustrated in Figure 7–35. If we changed $\dot{\theta} = -1$

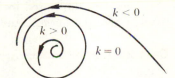

Figure 7–35   A stable limit cycle.

to $\dot{\theta} = 1$, we would have to reverse the arrows, and obtain an unstable limit cycle.   ◊

TWO-DIMENSIONAL CONSTANT LINEAR SYSTEMS

Having developed this general qualitative background, let us see how these various types of equilibria can arise at the zero state of a two-dimensional linear system

$$\begin{bmatrix} \dot{x}_1 \\ \dot{x}_2 \end{bmatrix} = \begin{bmatrix} a & b \\ c & d \end{bmatrix} \begin{bmatrix} x_1 \\ x_2 \end{bmatrix} \tag{1}$$

where a, b, c, and d are real constants. Let the eigenvalues be λ_1 and λ_2 and, for simplicity, let the determinant $(ad - bc) \neq 0$ to keep λ_1 and λ_2 from being zero. We shall study the behavior of our system near the origin for different types of values of λ_1 and λ_2.

Case 1. . Unequal real roots: $\lambda_2 < \lambda_1$.

Let v_1 and v_2 be unit eigenvectors of $F = \begin{bmatrix} a & b \\ c & d \end{bmatrix}$ associated with λ_1 and λ_2 respectively. The general real solution for equation (1) is then given by any choice of real constants c_1 and c_2 in the formula

$$x(t) = c_1 e^{\lambda_1 t} v_1 + c_2 e^{\lambda_2 t} v_2$$

Since $\lambda_1 > \lambda_2$, we see that as $t \to \infty$, the unit vector tangent to $x(t)$ tends to $\pm v_1$ if $c_1 \neq 0$; while as $t \to -\infty$, it tends to $\pm v_2$ if $c_2 \neq 0$. This case arose in Example 6-2-5 and was illustrated in Figure 6-5.

Depending upon the signs of the roots we get one of the following equilibrium points:

Case 1a.   A stable node ($\lambda_2 < \lambda_1 < 0$)

Case 1b.   An unstable node ($0 < \lambda_2 < \lambda_1$)

Case 1c.   A saddle point ($\lambda_2 < 0 < \lambda_1$)

These are illustrated in Figures 7-36(a), 7-36(b) and 7-36(c), respectively, where the lines L_1 and L_2 contain the eigenvectors v_1 and v_2.

Case 2.   Complex roots: $\lambda_1 = \alpha + i\beta$, $\lambda_2 = \alpha - i\beta$, conjugates, since F is real.

Choosing v_1 and v_2 as complex conjugates, the real solutions are given by

$$x(t) = c_1 e^{(\alpha+i\beta)t} v_1 + \overline{c_1} e^{(\alpha-i\beta)t} \overline{v_1} = 2\mathfrak{Re}\{c_1 e^{(\alpha+i\beta)t} v_1\}$$

where c_1 is complex and the bar denotes the conjugate. Depending upon the sign of α, the real part of the roots, we have one of the following equilibrium points:

Case 2a.   A vortex or center (purely imaginary roots, $\alpha = 0$)

Case 2b.   A stable focus (negative real parts, $\alpha < 0$)

Case 2c.   An unstable focus (positive real parts, $\alpha > 0$)

These are shown in Figures 7-36(d), 7-36(e), and 7-36(f), respectively, where the lines A and B contain the real and imaginary parts, respectively, of the complex eigenvector v_1.

Case 3.   Equal roots: $\lambda_1 = \lambda_2 = \lambda$ (purely real).

If there are two real, linearly independent eigenvectors v_1 and v_2 corre-

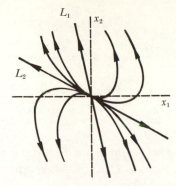

Figure 36a Stable node

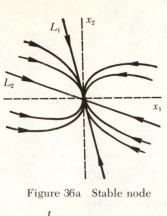

Figure 36b Unstable node

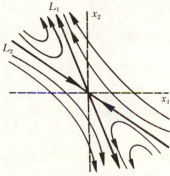

Figure 36c Saddle Point

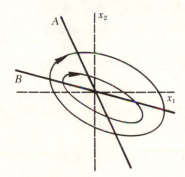

Figure 36d Vortex Point

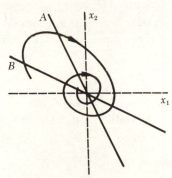

Figure 36e Stable Focus

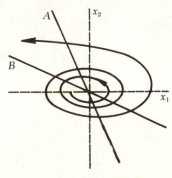

Figure 36f Unstable Focus

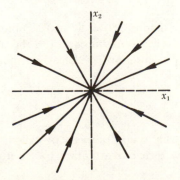

Figure 36g Stable Star Point

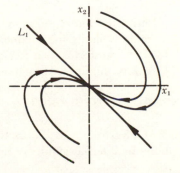

Figure 36h Stable Improper Node

Figure 7–36 The types of equilibria possible in a real two-dimensional constant linear system.

sponding to λ, the solutions are given by

$$x(t) = (c_1 v_1 + c_2 v_2)e^{\lambda t}$$

with c_1 and c_2 arbitrary real constants. This yields the next equilibrium point:

Case 3a. A star point (improper node), illustrated by Figure 7–36(g) for the stable case ($\lambda < 0$), where all trajectories are on straight lines passing through the origin.

If there is only one independent eigenvector v_1 corresponding to λ, the state solution is

$$x(t) = (c_1 + c_2 t)e^{\lambda t}v_1 + c_2 e^{\lambda t}v_2$$

where v_2 is any vector independent of v_1. This yields the last equilibrium point:

Case 3b. An improper node, as shown in Figure 7–36(h) for the stable case $\lambda < 0$, where line L_1 is along v_1.

EXPONENTIAL STABILITY

As we mentioned in our definitions of stability in the sense of Liapunov, there are several other notions of stability to describe the ability of a system to compensate for a slight displacement of its state from equilibrium. We will just give the definition of one of the most useful here, and refer the reader to Exercises 9, 10, and 11 for further exposure.

DEFINITION 3

A linear system is said to be **exponentially stable** if there exist positive constants α and M such that

$$\|\Phi(t, t_0)\| \leq M e^{-\alpha(t-t_0)}$$

for all t and t_0 in the half plane $t > t_0$.   ○

This definition requires that the free motion $x(t) = \Phi(t, t_0)x(t_0)$ die out more rapidly than some exponential, or, as it is sometimes said in the study of Laplace transforms, that $x(t)$ be of "exponential order." Clearly, exponential stability implies asymptotic stability.

INPUT-OUTPUT STABILITY

So far we have concerned ourselves with what might properly be called state stability or internal stability. Let us now turn our attention to the input-output behavior of a system, and define a notion of stability that is useful for characterizing such behavior.

We shall call a system BIBO stable if all bounded inputs produce bounded outputs in the zero-state response. More formally, we have:

DEFINITION 4

A system Σ with response function

$$y(t) = \mathcal{S}(t, t_0, x^0, u)$$

is **BIBO stable** if, for each t_0 and every positive number M_u, there exists a positive number M_y such that

$$\|\mathcal{S}(t, t_0, 0, u)\| \leq M_y$$

for all inputs $u(\cdot)$ such that

$$\|u(t)\| \leq M_u \qquad\qquad \bigcirc$$

What does BIBO stability look like for a linear system? Since the zero-state response of a continuous-time linear system is given by the superposition integral

$$y(t) = \mathcal{S}(t, t_0, 0, u) = \int_{t_0}^{t} g(t, \zeta) u(\zeta)\, d\zeta$$

and the norm of an integral is less than or equal to the integral of the norm,

$$\|\mathcal{S}(t, t_0, 0, u)\| \leq \int_{t_0}^{t} \|g(t, \zeta)\|\, \|u(\zeta)\|\, d\zeta$$

If we bound the input $u(\cdot)$ by M_u, i.e.,

$$\|u(\zeta)\| \leq M_u$$

then

$$\|\mathcal{S}(t, t_0, 0, u)\| \leq M_u \int_{t_0}^{t} \|g(t, \zeta)\|\, d\zeta$$

Clearly, if the area under $\|g(t, \cdot)\|$ is finite, that is, if there exists a number M_g such that

$$\int_{-\infty}^{t} \|g(t, \zeta)\|\, d\zeta \leq M_g$$

for all t, then for each t_0 we will have $\|\mathcal{S}(t, t_0, 0, u)\| \leq M_u \cdot M_g$ as long as $\|u(t)\| \leq M_u$. Thus, we may find the desired number M_y by taking

$$M_y \overset{\Delta}{=} M_u \cdot M_g$$

to see that the existence of a number M_g bounding the term $\int \|g(t, \zeta)\| \, d\zeta$ is *sufficient* to guarantee BIBO stability.

It is somewhat more complicated to show that the existence of M_g is also a necessary condition for BIBO stability:

THEOREM 2

A linear system with impulse response $g(t, r)$ is BIBO stable if and only if there exists a number M_g such that

$$\int_{-\infty}^{t} \|g(t, \zeta)\| \, d\zeta \leq M_g \quad \text{for all } t$$

Proof

We have already proved the "if" part. For the "only if" part, suppose (for hopeful contradiction) that there is no such number M_g. Then for every number m there is a time t_1 such that

$$\int_{-\infty}^{t_1} \|g(t_1, \zeta)\| \, d\zeta > m$$

Now, if $g(\cdot, \cdot)$ were just a scalar function, we could use this last inequality to construct a bounded input which would yield an unbounded output. We could do this by defining

$$u(t) = \text{sgn}[g(t_1, t)]$$

where sgn is the signum function, so that $u(\zeta)$ has the same sign as $g(t_1, \zeta)$ for all ζ.

But for this $u(\cdot)$, the zero-state output at time t_1 is given by

$$y(t_1) = \int_{t_0}^{t_1} g(t_1, \zeta) u(\zeta) \, d\zeta$$

which is just

$$y(t_1) = \int_{t_0}^{t_1} |g(t_1, \zeta)| \, d\zeta$$

Letting $t_0 \to -\infty$, we can get $y(t_1) > m$, an output greater than the arbitrarily chosen number m, and this would imply that the system is not BIBO stable, the desired contradiction.

If $g(\cdot, \cdot)$ is not a scalar function but a matrix function, things get messier but we can still construct from $g(t_1, \cdot)$ a bounded input $u(\cdot)$ which yields an unbounded output at t_1. [See Exercise 8.]   □

Since the change of variable $\tau = t - \zeta$ in $g(t, \zeta)$ yields

$$\int_{-\infty}^{t} \|g(t, \zeta)\| \, d\zeta = \int_{0}^{\infty} \|g(t, t - \tau)\| \, d\tau$$

and since for time-invariant systems $g(t, t - \tau) = g(\tau)$, we have the classical criterion:

COROLLARY 3

A time-invariant linear system with impulse response $g(\cdot)$ is BIBO stable if and only if there exists a number M_g such that

$$\int_{0}^{\infty} \|g(\tau)\| \, d\tau \leq M_g \qquad\qquad □$$

Example 4

Let a be a non-negative real number and let h be a positive real number. The one-dimensional scalar system described by

$$\dot{x} = -ax$$
$$\dot{y} = hx$$

has impulse response

$$g(t) = he^{-at}\mathbf{1}(t)$$

Since $g(\cdot)$ is non-negative, $\|g(t)\| = |g(t)| = g(t)$. Thus,

$$\int_{0}^{\infty} \|g(\tau)\| \, d\tau = h \int_{0}^{\infty} e^{-a\tau} \, d\tau$$

$$= \begin{cases} h\dfrac{1}{-a} & \text{if } a > 0 \\[2mm] \infty & \text{if } a = 0 \end{cases}$$

By Corollary 3, this system is BIBO stable if and only if $a > 0$, for it is only in this case that a bound M_g exists.   ◊

The reader may have noticed in this last example that $\lambda = -a$ was the only eigenvalue of the familiar scalar system with transfer function

$$\mathcal{S}(s) = h\,\frac{1}{s + a}$$

Thus, BIBO stability was equivalent to the pole λ being strictly in the left half of the s-plane.

THEOREM 4

A constant, continuous-time, finite-dimensional linear system (F, G, H) is BIBO stable if and only if the system function

$$\mathcal{S}(s) = H(sI - F)^{-1}G$$

has all of its poles in the open left half of the s-plane.

Proof

For such systems, each entry of the transfer function $\mathcal{S}(s)$ is a proper rational function of s, and the impulse response is $g(t) = He^{Ft}G$. Then by Corollary 3, (F, G, H) is BIBO stable precisely when $\int_0^\infty \|He^{F\tau}G\|\, d\tau < \infty$. If we denote the distinct poles of $\mathcal{S}(s)$ by $\lambda_1, \ldots, \lambda_k$, then each entry of $He^{F\tau}G$ is a linear combination of terms of the form $t^{i-1}e^{\lambda_i t}$, which are absolutely integrable if and only if each λ_i has a negative real part.   □

A system can be BIBO stable without being stable:

Example 5

Consider the system $\begin{cases} \dot{x} = Fx + Gu \\ y = Hx \end{cases}$

with $F = \begin{bmatrix} 0 & 6 \\ 1 & -1 \end{bmatrix}$, $G = \begin{bmatrix} -2 \\ 1 \end{bmatrix}$, and $H = [0 \quad 1]$.

The matrix F has eigenvalues $\lambda_1 = +2$ and $\lambda_2 = -3$, so by Theorem 1 it is not asymptotically stable. But, as we saw in Example 4 of Section 7-3,

$$\mathfrak{H}(s) = H(sI - F)^{-1}G$$

$$= \begin{bmatrix} 0 & 1 \end{bmatrix} \begin{bmatrix} \dfrac{s+1}{(s-2)(s+3)} & \dfrac{6}{(s-2)(s+3)} \\[2ex] \dfrac{1}{(s-2)(s+3)} & \dfrac{s}{(s-2)(s+3)} \end{bmatrix} \begin{bmatrix} -2 \\ 1 \end{bmatrix}$$

$$= \begin{bmatrix} 0 & 1 \end{bmatrix} \begin{bmatrix} \dfrac{-2}{s+3} \\[2ex] \dfrac{1}{s+3} \end{bmatrix} = \dfrac{1}{s+3}$$

so the system is BIBO stable because the eigenvalue $\lambda_1 = +2$ with positive real part is cancelled and does not show up in the zero-state input-output behavior.   ☐

Asymptotic stability of (F, G, H) is determined by the eigenvalues of F, while BIBO stability depends upon the poles of $\mathfrak{H}(s)$. All poles of $\mathfrak{H}(s)$ are eigenvalues of F, so asymptotic stability implies BIBO stability. As we have just been reminded, not all eigenvalues are poles, because cancellation may occur. In Section 7–3, in the discussion following Theorems 16 and 24, we saw that cancellations in $\mathfrak{H}(s) = H(sI - F)^{-1}G$ imply a lack of controllability or of observability. Thus, if each eigenvalue is also a pole of $\mathfrak{H}(s)$, then (F, G, H) is a minimal realization of $\mathfrak{H}(s) = H(sI - F)^{-1}G$ and (F, G, H) is necessarily both controllable and observable. We have proved the following result:

THEOREM 5

If a system (F, G, H) is asymptotically stable, it is BIBO stable. If (F, G, H) is BIBO stable and both controllable and observable, then it is also asymptotically stable.   ☐

It is possible to use the ideas of controllability and observability to relate the state stability and input-output stability of time-varying systems in a manner analogous to Theorem 5, but this requires the notions of *uniform controllability and observability* and exponential stability. The interested reader should see the paper "Controllability, Observability and Stability of Linear Systems" by Silverman and Anderson.†

So far in this section our study of stability theory has been reasonably quantitative and rigorous. In the next two subsections we present an unrigor-

---

†SIAM *Journal on Control*, Vol. 6, No. 1, 1968.

ous, heuristic overview of two extremely important advanced topics, the *Liapunov direct method* and *structural stability and limit cycles*.

We motivate and present, without proof, several famous theorems and illustrate their usefulness with some classical examples. Although it is sketchy and incomplete, we hope that our brief introduction to these concepts will at least provide a feeling for their content and application.

LIAPUNOV'S DIRECT METHOD

Returning to our intuitive example of a ball rolling downhill, illustrated in Figure 7–33, we see that the bottom of a valley is a stable place for its motion. More precisely, we see that the motion is stable if we can find a function V—corresponding, in our picturesque language, to the potential energy or height of the ball—such that in the natural dynamics of the system we have that

$$\frac{d}{dt}\, V(x(t)) \leq 0$$

Again, in our picturesque language, this says that the height of the ball can only get lower (actually, this is inaccurate in this particular example, unless the ball is highly "damped"). If we also have a point which is "at the bottom of the hill" as shown in Figure 7–37, then x_0 is a point of stability for our

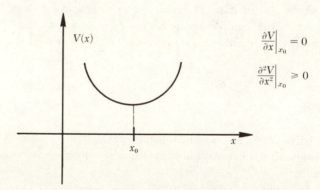

Figure 7–37   A Liapunov function $V(x)$.

dynamical system. Of course, when given a non-linear dynamical system, the tough thing is to find an actual **Liapunov function** (as such a function is called) for the system or else to prove that none exists, should the latter be the case. The use of the existence of such a function is called **Liapunov's Direct Method**† for determining stability, and it works for any non-linear system whenever we are lucky enough to find a V having the right properties.

---

†See LaSalle and Lefschetz, *Stability by Liapunov's Direct Method* (Academic Press, New York, 1961).

THEOREM 6 (LIAPUNOV STABILITY THEOREM)

Let $\dot{x} = f(x)$ have equilibrium point 0, i.e., $f(0) = 0$. Let S be a subset of X containing 0, on which is defined a function $V: S \to \mathbf{R}$ having continuous partial derivatives, such that $V(0) = 0$ and $V(x) > 0$ for $x \neq 0$. Then:

(i) If $\dfrac{d}{dt} V(x(t)) \leq 0$ for any solution $x(t)$ of $\dot{x} = f(x)$, then the origin is a weakly stable equilibrium point.

(ii) If $\dfrac{d}{dt} V(x(t)) < 0$ for $x(t)$ close to 0, then the origin is an asymptotically stable equilibrium point.

(iii) If $\dfrac{d}{dt} V(x(t)) > 0$ for $x(t)$ close to 0, then the origin is an unstable equilibrium point.   □

Notice that when it applies, this theorem lets us determine the stability of the origin without having to solve the differential equation.

Example 6

Consider the one-dimensional constant system $\dot{x} = -x$. Here $f(x) = -x$, so $x_0 = 0$ is an equilibrium point. If we take $S = X = \mathbf{R}$, then the function $V: S \to \mathbf{R}$ defined by $V(x) = x^2$ is a Liapunov function, since $V(0) = 0$ and $V(x) > 0$ for $x \neq 0$. Differentiating $V(x(t))$ with respect to t, we get

$$
\begin{aligned}
\frac{d}{dt} V(x(t)) &= \frac{\partial V(x)}{\partial x} \cdot \frac{dx}{dt} \\
&= 2x(t)[-x(t)] \\
&= -2x^2(t)
\end{aligned}
$$

which is negative for all $x \neq 0$. Thus, condition (ii) of Theorem 6 applies and $x_0 = 0$ is an asymptotically stable equilibrium point. [See Exercise 9.]
◊

In general, when X is n-dimensional, the differentiation of $V(x(t))$ with respect to t will require use of the chain rule

$$
\frac{d}{dt} V(x(t)) = \operatorname{grad} V \cdot \frac{dx}{dt}
$$

where grad $V \overset{\Delta}{=} \left[\dfrac{\partial V}{\partial x_1}, \dfrac{\partial V}{\partial x_2}, \ldots, \dfrac{\partial V}{\partial x_n} \right]^T$, the **gradient** of V, and this is why we want V to have nice partial derivatives on S. It is informative to tie this in with our discussion of mechanical systems in Section 2–5. Consider a mechanical system with Hamiltonian $H(x, p)$, where x is the vector of generalized coordinates and p is the vector of generalized momenta, and let us assume that H can be written in the form

$$H(x, p) = \frac{1}{2} \|p\|^2 + V(x)$$

where $V(x)$ is the potential energy. Then, if V assumes an isolated minimum value h at $x = 0$, we have that

$$H(x, p) - h > 0 \text{ for } (x, p) \text{ near } (0, 0)$$

If the system is conservative, so that $\dfrac{d}{dt} H(x(t), p(t)) = 0$ along trajectories, we see that $H(x, p) - h$ serves as a Liapunov function, and we see that case (i) of Theorem 6 holds. Liapunov functions may thus be thought of as generalized energies. Anyway, the above argument, and the corresponding one when V is a maximum at 0, yield:

THEOREM 7 (LAGRANGE'S THEOREM)

A mechanical system with a critical point at the origin is weakly stable if and only if the potential has a minimum there.   □

THEOREM 8 (LIAPUNOV'S THEOREM)

The origin of a conservative mechanical system is unstable if the potential has an isolated maximum there.   □

There is a corresponding Liapunov theory for time-varying differential systems $\dot{x}(t) = f(t, x(t))$ which we will not go into. The interested reader should see the book *Finite Dimensional Linear Systems* by Brockett [1970], or recent issues of the journals IEEE *Transactions on Automatic Control* and SIAM *Journal on Control*, for references† on this rapidly developing subject.

---

†See, for example, LaSalle, "An Invariance Principle in the Theory of Stability," *Differential Equations and Dynamical Systems*, Hale and LaSalle, editors (Academic Press, N.Y., 1967), and the following survey papers: Kalman and Bertram, "Control system design via the 'second method' of Liapunov," *J. Basic Engineering*, Trans. ASME, Vol. 82, pp. 371–400, 1960; Antosiewicz, "A Survey of Liapunov's Second Method," *Ann. Math. Stud.*, Vol. 41; Lefschetz, *Contributions to the Theory of Nonlinear Oscillations*, Vol. 4 (Princeton University Press, 1958).

LIMIT CYCLES AND STRUCTURAL STABILITY

Usually, the equations with which we "describe" a system yield at best close approximations to the real behavior of the system. Thus, if the mathematical model is to be useful, it must be the case that a slight change in the equations of motion yields only a slight change in the dynamic behavior of the solutions. We say that a system is **structurally stable** if it has this desirable property.

In particular, the linearization of a system about an equilibrium point is an approximation, so we can only expect it to yield information about the original system if the original system is structurally stable.

The whole point of stability studies is that they allow a *qualitative* study of system behavior—for instance, even when linearization gives us poor numerical prediction of actual position, it is still worthwhile if we can use it to tell whether or not a trajectory is stable.

Before giving some general theorems which establish when a system behaves as its linearization predicts near an equilibrium point, let us consider the following cautionary example.

Example 7

Consider the non-linear, scalar system

$$\dot{x} = x^2$$

which has a solution

$$x(t) = \frac{a}{1 - at}$$

This system has an unstable equilibrium at the origin, while its linearization

$$\dot{x} = 0$$

has a stable equilibrium at the origin (Exercise 10).   ◊

By contrast, we now consider the notion of a "slightly non-linear" system which differs from its linearization by a small amount and may be regarded as a perturbation of a linear problem.

THEOREM 9

Consider the non-linear dynamical system

$$\dot{x} = Fx + \psi(x) \tag{NL}$$

where

(i) 0 is an isolated zero of ψ

(ii) $\psi(x) = o(x)$ for small† x

(iii) all eigenvalues of F have nonzero real parts

Then the stability properties of (NL) at the origin are the same as those of its linearization $\dot{x} = Fx$.

Outline of Proof

Change coordinates to place F in Jordan form, and then check that $\|x\|^2$ is a Liapunov function for (NL) [see Exercise 11]. $\qquad\qquad\square$

Example 8 (Prey-Predator)

Consider the equations introduced by Volterra to describe the interaction of two competing populations:

$$\begin{cases} \dot{x} = x - xy \\ \dot{y} = -y + xy \end{cases} \tag{2}$$

These have equilibrium points at $[0 \quad 0]^T$ and $[1 \quad 1]^T$. Near the origin, we have the linearization $\begin{cases} \dot{x} = x \\ \dot{y} = -y \end{cases}$ with eigenvalues 1 and -1, and so the origin is a saddle point. Near $[1 \quad 1]^T$ we get $\begin{cases} \dot{x} = -y \\ \dot{y} = x \end{cases}$ and find $\lambda = \pm i$, so this is a vortex point for the *linearized* system—but Theorem 9 does not help us here, since the λ's have real part 0. So let us analyze $[1 \quad 1]^T$ more carefully by studying $\mathcal{W}(x, y) = -[x + y + \ln x + \ln y]$. Changing variables in equations (2) to $u = x - 1$ and $v = y - 1$, we get

$$\dot{u} = -v - uv$$

$$\dot{v} = u + uv$$

Then $\mathcal{W}(u, v) = -[(u + 1) + (v + 1) + \ln(u + 1) + \ln(v + 1)]$, $\dfrac{d\mathcal{W}}{dt}$ is

constant on trajectories, and $\mathcal{W}$ is negative definite, so $[1 \quad 1]^T$ is stable, by Liapunov. In fact, we have the picture for the Volterra trajectories shown in Figure 7–38, and it can be proven that the center at $[1 \quad 1]^T$ is not structurally stable. $\qquad\qquad\diamond$

---

†This means that $\psi(x)$ tends to zero faster than x tends to zero. We say that "ψ is of order little oh of x" and write $\psi(x) = o(x)$, as in Section 3–4.

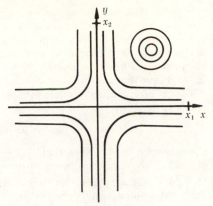

Figure 7–38 Trajectories of the prey-predator problem.

As another classical example, consider the following problem:

Example 9

Consider Van der Pol's relaxation oscillator with the equations

$$\begin{cases} \dot{x} = p \\ \dot{p} = -ax + \varepsilon p \end{cases}$$

with eigenvalues $\lambda = (1/2)(\varepsilon \pm \sqrt{\varepsilon^2 - 4a})$. For $\varepsilon^2 \geq 4a$, we have two positive, real roots (unless $a = 0$) and thus an **unstable node.** For $\varepsilon^2 < 4a$, we have two conjugate complex roots with strictly positive real part, and then 0 is an **unstable spiral point.**

For $\varepsilon = 0$, we get the simple harmonic oscillator, and the origin is a center. Of course, this linearization misses the crucial global fact that Van der Pol's equation has limit cycles. However, it is intuitively clear that the existence of an unstable spiral point is a *necessary* condition for the existence of a single stable limit cycle.   ◊

Now that we have finished our brief survey of stability theory, we next apply our results on controllability, observability, and stability to study the control and stabilization of multiple-input, multiple-output, constant systems.

EXERCISES FOR SECTION 7–4

✔ 1. (a)   For the simple pendulum of Example 1, check that $x^e = \begin{bmatrix} 0 \\ 0 \end{bmatrix}$ is a weakly stable equilibrium and that $x^e = \begin{bmatrix} \pi \\ 0 \end{bmatrix}$ is an unstable equilibrium. Are there any asymptotically stable equilibrium points?

(b) Introduce a viscous friction term in the physical equations of the pendulum and write the state equations. Can you see that with this frictional damping, the zero state will be an asymptotically stable equilibrium?

2. Prove that a continuous-time linear system is weakly stable if and only if all its eigenvalues have non-positive real parts.

✔ **3.** Prove that a *discrete-time* system (F, G, H) is asymptotically stable if and only if all of F's eigenvalues lie inside the unit circle (have absolute value <1), and weakly stable if no eigenvalues lie outside it.

4. Find conditions on the real entries a, b, c, and d of $F = \begin{bmatrix} a & b \\ c & d \end{bmatrix}$ which yield eigenvalues λ_1 and λ_2 such that

    (a) $\lambda_2 < \lambda_1$ (real, unequal roots)

    (b) $\lambda_1 = \overline{\lambda_2}$ (complex conjugates)

✔ **5.** Make up a formal definition of bounded-input–bounded-state (BIBS) stability for a system. What does this definition reduce to for a linear continuous-time system $(F(\cdot), G(\cdot), H(\cdot))$? Specialize it further to constant, linear systems (F, G, H). Does BIBS stability imply BIBO stability?

    [*Extra credit:* How would you define asymptotically regular, bounded-input, bounded-state stability?]

6. Check the following systems for stability, exponential stability, and BIBO stability:

    (a) $\dot{x}(t) = u(t)$, $y(t) = x(t)$

    (b) $\dot{x}(t) = atx(t)$, $y(t) = x(t)$, for $a = +2$ and $a = -2$.

    (c) $\dot{x}(t) = \dfrac{-1}{t+3} x(t) + u(t)$, $y(t) = x(t)$

7. Let $F(\cdot)$ be bounded on $(-\infty, \infty)$ and prove that if $\dot{x}(t) = F(t)x(t)$ is exponentially stable, then there exist constants M_1 and M_2 such that

$$\int_{t_0}^{t} \|\Phi(\zeta, t_0)\|^2 \, d\zeta \leq M_1$$

and

$$\int_{t_0}^{t} \|\Phi(\zeta, t_0)\| \, d\zeta \leq M_2$$

for all $t \geq t_0$. [*Note:* For a proof that the converse is also true, see Brockett, *Finite Dimensional Linear Systems*, pp. 190–193.]

8. As in Theorem 2, use the matrix norm $\|[a_{ij}]\| = \max\limits_{ij} |a_{ij}|$ to show that when $g(\cdot, \cdot)$ is a matrix, $\int_{-\infty}^{t_1} \|g(t_1, \zeta)\| \, d\zeta > m$ implies that $\int_{-\infty}^{t_1} |g_{r,s}(t_1, \zeta)| \, d\zeta > m$ for some entry $i = r$, $j = s$. Use this result to construct a bounded input

$$u(t) = [0, 0, \ldots, \ \mathrm{sgn}[g_{r,s}(t_1, t)], \ldots 0]^T$$
$$\uparrow$$
$$\text{the } s \text{ place}$$

which excites an unbounded output.

9. **(a)** Show that $V(x) = 5x^2$ is also a Liapunov function for the scalar system $\dot{x} = -x$ of Example 6. How about $V(x) = \ln x$?

    (b) Consider $\dot{x} = ax$ and study asymptotic stability of the origin when a is positive, negative, and zero.

    (c) For a physical interpretation of Example 6, let $x(t)$ be the voltage on

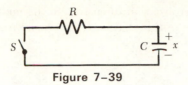

Figure 7-39

the capacitor of the circuit shown in Figure 7-39, with $R = C = 1$, and the switch S is closed at $t = 0$. To what does the Liapunov function $V(x) = x^2$ correspond?

10. Verify in detail that the zero state of $\dot{x} = x^2$ is an unstable equilibrium but that 0 is a stable equilibrium of the linearized system $\dot{x} = 0$, as claimed in Example 7.

11. Let M be a hermitian, positive definite, $n \times n$ matrix. Define the function V by

$$V(x) = \langle x | Mx \rangle = x^*Mx$$

(a) Show that this V is a Liapunov function for the linear system $\dot{x} = Fx$ if M satisfies the matrix equation

$$F^*M + MF = -W$$

where W is *any* positive definite hermitian matrix. [This V is called a **quadratic Liapunov function**, since in linear algebra $\langle x | Mx \rangle$ is called a quadratic form.]

(b) Show that the procedure of part (a) is equivalent to showing that the eigenvalues of F have negative real parts.

✔ **12.** In communication theory we often speak about the **"ideal low-pass filter"** whose frequency response is flat over the band $[-W, W]$ and zero elsewhere, as shown in Figure 7-40. Compute the impulse response $g(t)$ of this filter and show that it is *not* BIBO stable.

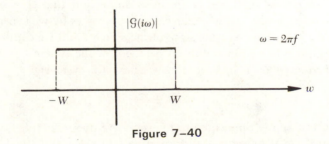

$\omega = 2\pi f$

Figure 7-40

7-5 The Control of Multiple-Input, Multiple-Output Linear Systems

In Sections 7-1 and 7-2 we discussed the notions of pole shifting and state estimation for a system (F, G, H) having a single input and a single output.

In this section we generalize that discussion to the multiple-input, multiple-output case. In the rest of this section, G and H are understood to be $n \times m$ and $q \times n$ matrices, respectively.

STABILIZATION BY USE OF FEEDBACK

We saw in Section 7–4 that the Jordan canonical form of the transition matrix of an arbitrary time-invariant *continuous-time* linear system shows us when the system is asymptotically stable, in that perturbations of the zero state tend to die out under zero input as long as the eigenvalues all have negative real parts. We also saw that if the system was both controllable and observable, then this condition on the transition matrix was equivalent to demanding that all of the poles of the transfer function matrix lie in the left half-plane. Thus, a question of great interest to a control theorist is how he may move the eigenvalues or poles of the system so that they all lie in the left half of the complex plane.

We now wish to stress that—assuming that we have the whole state vector available for our feedback loop—we can stabilize our system by using feedback whenever the system is controllable. In other words, the condition of controllability (which tells us that we may achieve, at least asymptotically, a desired state trajectory by suitable application of control inputs) also tells us that we can damp out any perturbations in the initial state simply by applying an appropriately weighted feedback of the state vector. The feedback scheme that we shall use does, as we have indicated, make use of the whole state vector for feedback. In general, of course, we do not have all of the state variables available, but just the output variables, which provide us with information about only a select few of the components of the state vector. However, we shall see in the next subsection that if a system is observable, then, just as in the scalar case, as time goes by we can make an increasingly good estimate of the state of the system; and we can feed back our state estimate to obtain stabilized behavior of the system.

As a first step, let us imagine that we have the system

$$\dot{x} = Fx + Gu$$

with the output at any time equal to the state. We then take the output and linearly transform it by some transformation

$$K : X \to U$$

and add the resultant contribution Kx to v, the input to the overall feedback system. The result is shown schematically in Figure 7–41.

From the expansion

$$\dot{x} = Fx + G(Kx + v) = (F + GK)x + Gv$$

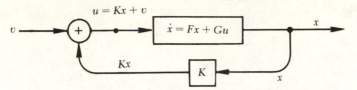

Figure 7–41   A multiple-input state-feedback system.

we see that using feedback with "gain" K, we replace the matrix F by the new matrix $F + GK$. Our task is to show that we may so choose K that $F + GK$ has any desired set of eigenvalues precisely when (F, G) is controllable. Note that if we only make the weaker demand—that K may be chosen so that $F + GK$ is stabilized—then the condition that the pair (F, G) be controllable may not be necessary. For instance, if F itself were a stable matrix, we could then take K equal to zero, regardless of the value of G. In the above discussion, we called K a gain because it generalizes the normal concept of gain. In the more general multi-component case, we may think of the component k_{ij} of K as being the gain when we look at how a signal on the jth input line to the K-box of Figure 7–41 determines the value on the ith output line.

Let us note that for a real matrix, the eigenvalues are the roots of the real polynomial $\det(\lambda I - F) = 0$, and hence either are real or occur in complex conjugate pairs. Thus, if we want to start with real matrices F and G and use real feedback matrices K, we can only obtain eigenvalues which either are real or occur in conjugate pairs. We now wish to generalize Theorem 7–1–5 to prove that if the pair (F, G) is controllable, then we may find a feedback matrix K which obtains an otherwise arbitrary set of eigenvalues. Our treatment evolved from that given by Wonham [1967]. (See also Luenberger [1964, 1966].)

THEOREM 1

Let F be $n \times n$ and G be $n \times m$. The pair (F, G) in $\dot{x} = Fx + Gu$ is controllable if and only if for every choice of a set $\mathcal{E}$ of n scalars such that $\lambda \in \mathcal{E}$ implies its complex conjugate $\bar{\lambda} \in \mathcal{E}$, there is a real matrix K such that $F + GK$ has $\mathcal{E}$ for its set of eigenvalues.

Proof

Let us first prove the sufficiency of the condition—i.e., we show that if to every $\mathcal{E}$ there corresponds a suitable K, then (F, G) must be controllable.

Since there are only n solutions to $\det(\lambda I - F) = 0$, we may certainly choose a set $\mathcal{E}$ of n distinct real numbers, $\lambda_1, \ldots, \lambda_n$, none of which is an eigenvalue of F. Let K then be the real $m \times n$ matrix for which $F + GK$ has eigenvalues $\lambda_1, \ldots, \lambda_n$, so that our state space has a basis $v_1, \ldots, v_n$ for which

$$(F + GK - \lambda_i I)v_i = 0 \qquad 1 \le i \le n$$

and so

$$v_i = -(F - \lambda_i I)^{-1} GK v_i$$

Now, Kv_i is just an m-tuple of numbers, so this last equation shows that v_i can be expressed as a linear combination of columns of the matrices $F^k G$, where $k \ge 0$.† Since the v_i form a basis, the columns of $G, FG, \ldots, F^{n-1}G$ must also span the whole space; therefore, (F, G) is controllable, as was to be proved.

Having proved sufficiency, let us now turn to necessity—in other words, let us show that whenever the pair (F, G) is controllable, we can then choose a feedback matrix to get the desired eigenvalues for the new system. We have already proved this result in Theorem 7-1-5 for a controllable system with a scalar input. In particular, consider such a system which has been put into the controllable canonical form that we presented in Section 7-1:

$$F_c = \begin{bmatrix} 0 & 1 & 0 & \cdots & 0 \\ 0 & 0 & 1 & \cdots & 0 \\ \vdots & \vdots & \vdots & & \vdots \\ 0 & 0 & 0 & \cdots & 1 \\ -a_n & -a_{n-1} & -a_{n-2} & \cdots & -a_1 \end{bmatrix}, \quad g_c = \begin{bmatrix} 0 \\ 0 \\ \vdots \\ 0 \\ 1 \end{bmatrix}$$

The feedback matrix must convert a whole state vector into a scalar input, and thus will have the form of a row vector:

$$k^* = [k_1 \ldots k_n]$$

The local dynamic matrix for a system, when this is used as feedback, will then be

$$F_c + g_c k^* = F_c + \begin{bmatrix} 0 \\ 0 \\ \vdots \\ 0 \\ 1 \end{bmatrix} [k_1 \ldots k_n] = F_c + \begin{bmatrix} 0 & \cdots & 0 \\ 0 & \cdots & 0 \\ \vdots & & \vdots \\ 0 & & 0 \\ k_1 & \cdots & k_n \end{bmatrix}$$

$$= \begin{bmatrix} 0 & 1 & \cdots & 0 \\ 0 & 0 & \cdots & 0 \\ \vdots & \vdots & & \vdots \\ 0 & 0 & & 1 \\ k_1 - a_n & k_2 - a_{n-1} & \cdots & k_n - a_1 \end{bmatrix}$$

---

† Recall Exercise 5(b) and (c) of Section 7-1.

and this just says that we can get the controllable canonical form for a matrix with *any* desired set of eigenvalues by a suitable choice of the feedback matrix k^*.

Our task is now to generalize this result, so that it holds in the case of a system with multidimensional input. Our technique will be first to show that we can use feedback to reduce this system to a new system which is controllable by a scalar input—and then to apply our discussion of the scalar input case to deduce the entire result.

In the scalar case, the input matrix is a single column g, and we saw that the pair (F, g) was controllable iff we could sweep out the whole space simply by hitting g with successive powers of the local transition matrix F; more formally, we saw that the vectors $g, Fg, \ldots, F^{n-1}g$ formed a basis for the state space iff the pair (F, g) was controllable. Let us contrast this with the multidimensional case. Here, our condition for controllability states that the columns of $G, FG, \ldots, F^{n-1}G$ span the entire state space. In particular, let G have columns $g_1, \ldots, g_m$, so that our controllability condition tells us that the set of mn columns $g_1, Fg_1, \ldots, F^{n-1}g_1; g_2, Fg_2, \ldots, F^{n-1}g_2; \ldots; g_m, Fg_m, \ldots, F^{n-1}g_m$ between them span the whole state space—obviously redundantly if $m > 1$—but does not guarantee that there is any single vector here, say $g_k = g$, such that the vectors $g, Fg, \ldots, F^{n-1}g$ by themselves span the whole space. In fact, the best statement that we can make is: (F, G) is controllable iff we can find a basis for the state space of the form

$$g_{i_1}, Fg_{i_1}, \ldots, F^{\alpha_1-1}g_{i_1}; g_{i_2}, Fg_{i_2}, \ldots, F^{\alpha_2-1}g_{i_2}; \ldots; g_{i_p}, Fg_{i_p}, \ldots, F^{\alpha_p-1}g_{i_p}$$

One way of choosing the columns $g_{i_1}, g_{i_2}, \ldots, g_{i_p}$ and the integers $\alpha_1 \geq 1, \ldots, \alpha_p \geq 1$ is as follows: Let g_{i_1} be the first nonzero column of G. Hit it with F. If the result is in the space spanned by g_{i_1}, then $\alpha_1 = 1$ and we let g_{i_2} be the first column of G that is not in the space spanned by g_{i_1}. More generally, having reached the term $F^j g_{i_r}$ in the above sequence of vectors, we next form $F^{j+1}g_{i_r}$. If this vector does not belong to the space spanned by the vectors already written down in the collection, then we add it as a new element of the collection. If, however, $F^{j+1}g_{i_r}$ does lie in the space spanned by the previously noted vectors, then we search for the first column of G which does not lie in that space. If we find such a column, we label it $g_{i_{r+1}}$ and add it as the next column. If, however, no such column exists, and in fact we have revealed that $g_{i_r} = g_{i_p}$, then we see that we must already have a basis for the entire space.

Let us form the $m \times n$ matrix

$$Q = [g_{i_1}, Fg_{i_1}, \ldots, F^{\alpha_1-1}g_{i_1}; g_{i_2}, Fg_{i_2}, \ldots, F^{\alpha_2-1}g_{i_2}; \ldots; g_{i_p}, Fg_{i_p}, \ldots, F^{\alpha_p-1}g_{i_p}]$$

which is non-singular, since its columns form a basis for the state space; and let us denote its columns by $q_1, q_2, q_3, \ldots, q_n$. We shall now find a feedback matrix K_1 such that the resultant local transition matrix $\widehat{F} = F + GK_1$ traces out the whole state space when applied to q_1. More particularly, if we use span$\{\ldots\}$ to denote the space spanned by the columns within the braces, then what we wish to do is to find our matrix K_1 so that

$$\text{span}\{q_1, q_2, \ldots, q_l\} = \text{span}\{q_1, \widehat{F}q_1, \ldots, \widehat{F}^{l-1}q_1\} \tag{1}$$

Let us introduce the abbreviation $t_j = \sum_{r=1}^{j} \alpha_r$. Then, referring to the definition of Q, we see that

$$q_{l+1} = Fq_l \qquad \text{if } l \text{ is not a } t_j$$

while

$$q_{l+1} = g_{i_{j+1}} \qquad \text{if } l = t_j$$

We can certainly satisfy equation (1), then, if we choose K_1 to satisfy

$$
\begin{aligned}
(F + GK_1)q_l &= Fq_l & &\text{if } l \text{ is not a } t_j \\
(F + GK_1)q_l &= Fq_l + g_{i_{j+1}} & &\text{if } l = t_j
\end{aligned}
\tag{2}
$$

since Fq_{t_j} belongs to span$\{q_1, q_2, \ldots, q_{t_j}\}$ by definition of the t_j. Now we can satisfy equations (2) by setting

$$
\begin{aligned}
&[GK_1q_1, \ldots, GK_1q_{t_1}, GK_1q_{t_1+1}, \ldots, GK_1q_{t_2}, \ldots, GK_1q_{t_p}] \\
&= \quad[\;\;0\;\;, \ldots, \;\;g_{i_2}, \qquad 0 \qquad, \ldots, \;\;g_{i_3}\;\;, \ldots, \;\;g_{i_1}\;\;] \\
&= G\;[\;\;0\;\;, \ldots, \;\;e_{i_2}, \qquad 0 \qquad, \ldots, \;\;e_{i_3}\;\;, \ldots, \;\;e_{i_1}\;\;]
\end{aligned}
$$

where e_i is the $n \times 1$ column vector which is all zeros save for a one in the ith position. But this equation is satisfied by taking $K_1Q = [K_1q_1, \ldots, K_1q_{t_p}]$ to be the matrix S which is all zeros save for an entry 1 in the (t_j, t_{j+1}) positions. We may use the non-singularity of Q to deduce that $K_1 = SQ^{-1}$ does the task.

Rephrasing what we have done, we have chosen a feedback matrix K_1 so that the local transition matrix $\widehat{F} = F + GK_1$ of the system with feedback acts upon the columns $q_1, q_2, q_3, \ldots, q_n$ of Q just like the original local transition matrix F, save that when $l = t_j$, $\widehat{F}^l q_{i_1}$ "pops out" a new column $g_{i_{j+1}}$.

Our scalar system (F, g_{i_1}), with an extra feedback loop given by k^*, has the transition matrix

$$\widehat{F} + g_{i_1}k^* = (F + GK_1) + g_{i_1}k^* = F + G(K_1 + e_{i_1}k^*) = F + GK$$

where K is obtained from K_1 by adding k^* to the i_1th row. Since (F, g_{i_1}) is

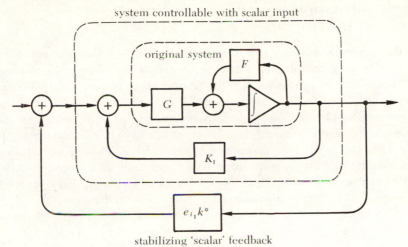

Figure 7-42 Stabilization of a multi-input system.

controllable, we know—by our previous discussion of the scalar input case—that a feedback row k^* can be chosen to yield arbitrary acceptable eigenvalues. We thus conclude the truth of our theorem on pole assignment for arbitrary controllable systems, whether or not they have scalar input. ☐

The configuration we have set up to obtain stability by Theorem 1 is shown in Figure 7-42. The inner feedback loop (through the K_1 matrix) renders the system **cyclic** (i.e., controllable by a scalar input). The outer feedback loop is then chosen by the designer to yield the required assignment of the eigenvalues (poles).

Even though our discussion of stabilization by state feedback has been carried out in terms of *continuous-time* systems, the reader should recognize that Theorem 1 is really just a statement about how the eigenvalues of the matrix $F + GK$ depend upon the matrices F, G, and K. Thus, all of the mathematics is immediately applicable for the stabilization of *discrete-time* systems as well.

We stress again that if we are given just $\mathfrak{H}(s)$ or $\mathfrak{H}(z)$, the transfer function matrix of a continuous or discrete-time system, then we have only information about that part of the system which is both controllable and observable. Thus, we can certainly use feedback to stabilize that part of the system—but we must remember that there may be other parts which the external feedback loop cannot stabilize. If those other parts of the system are stable, then well and good. If they are not stable, however, we are going to be in trouble; although they do not affect the system when it is within its domain of linearity, unrestricted increases in variables of the other part of the system can shift our system out of the domain in which our model is applicable.

STATE ESTIMATION FOR THE MULTIPLE-INPUT, MULTIPLE-OUTPUT CASE

If the reader looks back to Section 7–2, he will see that the entire first subsection on *Simulation and State Estimation* remains valid for the multiple-input, multiple-output case. All of the figures and equations of the subsection were constructed for general $n \times m$ and $q \times n$ matrices G and H. The only apparent specialization for single outputs came just prior to Theorem 7–2–1, when we invoked Theorem 7–1–5. However, we have just finished generalizing Theorem 7–1–5 into Theorem 1 of this section. Thus, if we substitute the incantation "invoking Theorem 7–5–1" for "invoking Theorem 7–1–5" in that subsection, we can use those techniques to design observers which asymptotically estimate the states of multivariable systems.

For convenience, we summarize and rephrase Theorem 7–2–1 here (see Figure 7–43):

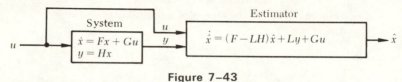

Figure 7–43

THEOREM 2

There exists an asymptotic state estimator for the system

$$\dot{x}(t) = Fx(t) + Gu(t)$$

$$y(t) = Hx(t)$$

if there exists a matrix L^* for which $F^* - H^*L^*$ is stable. Such an L^* certainly exists if the system is observable.   □

Let us recall that for continuous-time stability we demand eigenvalues with negative real parts, while for discrete-time systems they must be within the unit circle. Thus, the fact that F is stable if and only if F^* is stable follows from the easily verified fact that the eigenvalues of F^* are the complex conjugates of the eigenvalues of F:

$$\det(\lambda I - F) = 0 \quad \text{iff} \quad \det(\overline{\lambda I - F}) = 0 \quad \text{iff} \quad \det(\overline{\lambda} I - F^*) = 0$$

Note that if F is real, then its eigenvalues occur in conjugate pairs, so the set $\mathcal{E}$ of eigenvalues is the same for F and for F^*.

A REDUCED-ORDER OBSERVER

As our final development of the observer idea, we consider the fact that the estimator of Figure 7–43 is an n-dimensional dynamic system whose state

$\hat{x}$ approximates the state x of (F, G, H) using observations of y. Suppose that (F, G, H) were so built that its output y was the first component, x_1, of its state vector. There is clearly no point in then approximating x_1 when we have a direct measurement of this state component. Thus, we need only develop a reduced-order observer to approximate the rest of the state, $x_2, \ldots, x_n$.

Since the job of our estimator is only to estimate the state-variable part, we shall ignore the input and simply consider the state estimation problem for this system described by the input-free equations

$$\dot{x} = Fx$$

$$y = Hx$$

We assume, without loss of generality (Exercise 3), that (F, H) is observable and that the output y, which is $q \times 1$, is identical with the first q states. In other words, we so choose coordinates that we may partition the state into two components, $x_a \in \mathbf{R}^q$ and $x_b \in \mathbf{R}^{n-q}$, so that $x_a = y$ is the known output ($H = [I_q \mid 0]$), while we have a standard observer problem with respect to x_b. The equations, in expanded form, are thus

$$-\begin{bmatrix} \dot{x}_a \\ \dot{x}_b \end{bmatrix} - = -\begin{bmatrix} F_{aa} & F_{ab} \\ F_{ba} & F_{bb} \end{bmatrix} -\begin{bmatrix} x_a \\ x_b \end{bmatrix}- \tag{3}$$

while the reduced equations of motion for the system to be estimated are

$$\dot{x}_b = F_{bb}x_b + F_{ba}x_a \tag{4}$$

The x_a term here, being known, is like the input term Gu in the treatment of Theorem 2.

Now we see from equation (3) that $\dot{x}_a = F_{aa}x_a + F_{ab}x_b$. Thus, we may compute $F_{ab}x_b$ as $\dot{x}_a - F_{aa}x_a$. As we shall see below, in discussing equation (8), the apparent need for a differentiator can be avoided. Thus, we may equip the system (4) with the "output equation"

$$y_b = F_{ab}x_b$$

where the "output" y_b is just $\dot{x}_a - F_{aa}x_a = \dot{y} - F_{aa}y$.

If we now use the scheme of Figure 7-43—with x replaced by x_b, $\hat{x}$ by $\hat{x}_b$, F by F_{bb}, Gu by $F_{ba}x_a$, and H by F_{ab}—we obtain an observer for x_b described by the equation

$$\dot{\hat{x}}_b = (F_{bb} - LF_{ab})\hat{x}_b + L(F_{ab}x_b) + F_{ba}x_a \tag{5}$$

The error in the approximation is

$$\tilde{x}_b = x_b - \hat{x}_b$$

and the error obeys the dynamic equation

$$\dot{\tilde{x}}_b = F_{bb}x_b + F_{ba}x_a - [F_{bb}\hat{x}_b + F_{ba}x_a + L(F_{ab}x_b - F_{ab}\hat{x}_b)]$$
$$= (F_{bb} - LF_{ab})\tilde{x}_b \tag{6}$$

Now, the roots of this $(n - q)$-dimensional error system can be made arbitrary by a suitable choice of L if (F_{bb}, F_{ab}) form an observable pair. We must therefore prove that they do, given our original assumption that (F, H) is observable. Let us first form the observability matrix $\mathcal{O}$ for the original system:

$$\mathcal{O} = \begin{bmatrix} H \\ HF \\ HF^2 \\ \vdots \\ HF^{n-1} \end{bmatrix}$$

We need to calculate a few terms:

$$H = [I_q \mid 0]$$

$$HF = [I_q \mid 0] \begin{bmatrix} F_{aa} & F_{ab} \\ F_{ba} & F_{bb} \end{bmatrix} = [F_{aa} \mid F_{ab}]$$

$$HF^2 = [F_{aa} \mid F_{ab}] \begin{bmatrix} F_{aa} & F_{ab} \\ F_{ba} & F_{bb} \end{bmatrix} = [F_{aa}^2 + F_{ab}F_{ba} \mid F_{aa}F_{ab} + F_{ab}F_{bb}]$$

We thus get the matrix

$$\mathcal{O} = \begin{bmatrix} I_q & 0 \\ F_{aa} & F_{ab} \\ F_{aa}^2 + F_{ab}F_{ba} & F_{aa}F_{ab} + F_{ab}F_{bb} \\ \vdots & \vdots \end{bmatrix}$$

If we multiply the block row $[F_{aa} \mid F_{ab}]$ on the left by $-F_{aa}$ and add it to the block row below, the right-hand member below becomes $F_{ab}F_{bb}$. In such a fashion, we can (Exercise 1) reduce $\mathcal{O}$ to the form

$$\mathcal{O}(F, H) = \begin{bmatrix} I_q & 0 \\ ?? & \mathcal{O}(F_{aa}, F_{ab}) \end{bmatrix} \tag{7}$$

Now, the rank of $\mathcal{O}$ is not changed if we add multiples of some rows to others. Since $\mathcal{O}(F, H)$ has full rank n by assumption, we have immediately that

$\Theta(F_{aa}, F_{ba})$ has full rank $(n - q)$. Thus, the reduced order observer error equation (6) can be given arbitrary roots by suitable choice of L.

The final step in the realization of the reduced order observer is to display the equations of motion of the observer (5) in suitable form. In equation (5) the term $F_{ab}x_b$ is used, whereas the term actually available is $\dot{x}_a - F_{aa}x_a$ (where $x_a = y$, the original output). Thus, the real equations of motion to be implemented are

$$\dot{\hat{x}}_b = F_{bb}\hat{x}_b + F_{ba}x_a - LF_{ab}\hat{x}_b + L[\dot{x}_a - F_{aa}x_a]$$

In terms of the system output, y, we have (on combination of terms):

$$\dot{\hat{x}}_b = (F_{bb} - LF_{ab})\hat{x}_b + (F_{ba} - LF_{aa})y + L\dot{y} \tag{8}$$

We wish to implement this system without taking the derivative to obtain $\dot{y}$. This is easy to do. We take a (q-dimensional) integrator with $\dot{\hat{x}}_b - L\dot{y}$ as its input; we form this from $\hat{x}_b$ and y according to equation (8). At the output we add Ly to $\hat{x}_b - Ly$ to obtain $\hat{x}_b$, the required state approximation. The block diagram is shown in Figure 7–44.

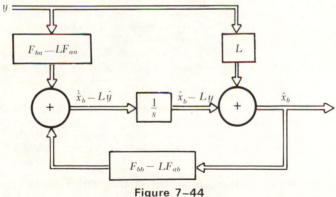

Figure 7–44

When the reduced order observer is used, we obtain a controller which is a dynamical system of order $n - q$, where q is the dimension of the plant output y. Otherwise, all techniques and comments relating to the nth order observer apply virtually unchanged. Details are left to the reader. For more details on the design of reduced order observers, including some numerical examples, the reader should see Gopinath's very readable article, "On the Control of Linear Multiple Input-Output Systems" [1971].

We have thus seen the crucial role played by controllability and observability, not only in the narrow context in which we approach them when discussing the regulator problem, but also because they ensure our ability to use feedback to move the poles of the system around—and this has applications both in the explicit use of feedback to stabilize a controllable system, and in its indirect use, as in the last theorem, to form an asymptotic state estimator for an observable system.

EXERCISES FOR SECTION 7–5

1.   Verify equation (7) by induction on n.

✔ **2.**   Rework Exercise 9 of Section 7–2, using the fact that $y = x_1$ for the realization of Figure 7–2 so that you don't need to estimate x_1. Now design the compensator by employing a reduced order (first order) state estimator instead of the second order estimator you used earlier.

3.   Under what conditions may we *not* assume that a q-dimensional output of a system is identical with the first q dimensions of the state vector in some suitable coordinate system?

✔ **4.**   In Exercise 31 of Section 7–3 you studied a multiple input-output system (F, G, H) representing the motion of a satellite linearized about a desired circular orbit.

     (a)   Find the natural frequencies (eigenvalues) of the system (F, G, H). If no controlling input is applied, will the satellite stay in the circular orbit? Suppose that some unexpected disturbance displaces the satellite slightly from the desired circular orbital condition. Will it settle back down to the circular condition?

     (b)   Suppose that the parameter ω in F has the normalized value of 1. Find the feedback matrices K_1 and k^* of Figure 7–42 which stabilize the satellite system (F, G, H), by relocating eigenvalues so that there are double poles at $(-1 + i1)$ and $(-1 - i1)$, respectively.

     (c)   How would you implement this stabilizing control scheme for the satellite? Notice that a linear state feedback control signal such as $u = Kx$ requires the continuous expenditure of energy. Do you think real satellites are continuously burning fuel to remain in circular orbits?

     (d)   Consider "bang-bang" and pulsed input control philosophies. Could you observe the satellite, see how much it drifted at the end of a "day," and compute an impulsive type control to drive it back to the appropriate "circular" state?

7–6   An Introduction to Optimal Control Theory

Once we have learned how to control a given system, it is only natural to want to control it in some "best" or optimal manner. In this section, we shall formalize this intuitive idea by mathematically formulating the problem of optimal control, and we will illustrate some of the concepts through a number of concrete examples. By showing how the Hamiltonian and Lagrangian formulations of mechanics, which we discussed briefly in Section 2–5, may be viewed as the solutions of optimization problems, we try to motivate two of the most powerful analytical approaches to optimal control: the Lagrange Multiplier Theorem and the Pontryagin Minimum Principle. Because these two approaches are so technical and involve the development of some additional mathematics, we have relegated their developments and proofs to an appendix, A-3, so as not to detract from the brief overview of this section.

How can we tell whether or not a given control strategy is "best" or optimal? Usually we can associate with each control problem some number, called a **figure of merit** or **performance index,** whose value $J(u)$ is a measure of how well a particular control signal $u(\cdot)$ causes the system to perform with respect to specified criteria. For example, on an electric stove, driven by a heating current $u(\cdot)$, are you better off heating your coffee pot at 350°F for 15 minutes or by heating it at 500°F for some shorter time to bring it to a boil? Here, by "better off," we might mean "Will your electricity bill be lower?" if you heat it one way as opposed to the other. Alternatively, if you are in a hurry to catch a plane, by "better off" we might mean "Will the water reach a boil sooner?" In the first instance, our performance index associates with each control signal, u, the cost $J_1(u)$ of the electrical energy in u; in the second case each input signal u costs us a certain amount of time $J_2(u)$.

Of course, not every performance index will be "cost"; sometimes the figure of merit will be a "payoff" measuring the benefit $J(u)$ associated with the control u. *An optimal control signal u, then, will be one which either minimizes some cost function or maximizes some payoff function.* Clearly, the maximum of J is the minimum of $-J$, so that no loss of generality results in our studying only minimization problems formulated in terms of cost functions.

Suppose that, in our homely example of heating coffee water, using the elapsed time taken to bring the water to boil as the cost function, a well-meaning neighbor who studied freshman physics makes a simple calorimetry calculation and informs us that if we heat the coffee pot at 27,306°F, it will boil in just one second. A quick glance at the control dial on our electric stove shows us that its maximum setting is only 575°F, and hence that our neighbor's solution is not an acceptable one since we can't come anywhere near that temperature with our stove. We send him back to his calculations with a more complete statement of the problem, this time including the **control constraint** that his solution must be bounded and should not exceed 575°F.

All physical signals are bounded in amplitude and in energy, so any particular problem may include constraints such as

$$\|u(t)\| \leq 100, \qquad \|u(t)\|^2 \leq 25, \qquad \text{or} \qquad \|\dot{u}(t)\| \leq 16$$

We shall include all such control constraints in our specification of the set Ω of **admissible input functions.** Similarly, we shall let Θ denote the set of all **admissible state trajectories,** and include in its specification any constraints we must impose on the state variables. For example, if x_1, the first component of a state vector x, represents the height of an airplane above the ground, we are not very interested in any "optimal" control strategies which cause x_1 to become negative. The interested reader will find a more detailed treatment of optimal control under specific state constraints outlined in Appendix A-3.

With these preliminaries out of the way, let us now mathematically formulate the optimal control problem. In the discussion which follows, we shall make extensive use of the notion of derivative which we developed for Banach spaces in Section 3–4, for to ensure that a control is optimal, we have to see how changes in the whole control function affect the overall performance—and the natural tool for the study of such changes is the derivative taken with respect to control *functions,* and these lie in an infinite-dimensional function space.

THE OPTIMAL CONTROL PROBLEM

In the typical optimal control problem, we shall study a (possibly non-linear) control system, often called "the plant," whose state dynamics are governed by an equation of the form

$$\dot{x}(t) = f(x(t), u(t)), \quad x(t_0) \text{ fixed, } t \in [t_0, t_1] \tag{1}$$

where we assume that $f: \mathbf{R}^n \times \mathbf{R}^m \to \mathbf{R}^n$ is sufficiently smooth that for each u in our space Ω of admissible control functions $[t_0, t_1] \to \mathbf{R}^m$ there exists a unique $x \in \Theta = C_n[t_0, t_1] \overset{\Delta}{=} \{$the space of continuous functions $[t_0, t_1] \to \mathbf{R}^n\}$ such that x and u together satisfy equation (1). Where it is convenient, we shall denote the state function so determined by u in equation (1) as $x(u)$.

For example, we have devoted a great deal of study in Chapters 5, 6, and 7 to the linear case in which

$$f(x(t), u(t)) = F(t)x(t) + G(t)u(t)$$

for suitable matrix functions $F(t)$ and $G(t)$ defined for all $t \in [t_0, t_1]$.

To determine our criterion of optimality by minimizing a performance index, it is convenient to introduce a **cost rate** or **cost functional** $\ell: \mathbf{R}^n \times \mathbf{R}^m \to \mathbf{R}$ and then to associate with each (state function, input function) pair the total cost functional

$$J(u) \overset{\Delta}{=} g[x(u), u] \tag{2a}$$

where

$$g[x, u] \overset{\Delta}{=} \int_{t_0}^{t_1} \ell(x(t), u(t)) \, dt \tag{2b}$$

For example, we might measure the energy consumed in applying input $u(t)$ over the time interval $[t, t + dt]$ by the quantity $\|u(t)\|^2 \, dt$. In this case, setting $\ell(x(t), u(t)) = \|u(t)\|^2$ would yield

$$g[x, u] = \int_{t_0}^{t_1} \|u(t)\|^2 \, dt \tag{3}$$

which is the total energy consumed by applying control u from time t_0 to time t_1 [Exercise 1].

If, in another problem, we wish to use $(t_1 - t_0)$, the elapsed time in moving the state to the origin, as the performance index to be minimized so that

$$J(u) = g[x, u] = \int_{t_0}^{t_1} \ell(x(t), u(t))\, dt = t_1 - t_0 \tag{4a}$$

this is easily accomplished by using the cost rate

$$\ell(x(t), u(t)) = \begin{cases} 1 \text{ if } x(t) \neq 0 \\ 0 \text{ if } x(t) = 0 \end{cases} \tag{4b}$$

In still another situation, in which we desire the state to be as close to zero as possible, we might take

$$g[x, u] = \int_{t_0}^{t_1} (\|x(t)\|^2 + \|u(t)\|^2)\, dt \tag{5}$$

In any case, our optimal control problem is then to choose u so that the associated cost

$$J(u) = g[x(u), u] = \int_{t_0}^{t_1} \ell(x(u)(t), u(t))\, dt$$

is as small as possible. Usually, we will ask that this optimal u satisfy some further condition. In Appendix A-3, we study the case in which $x(u)$ must satisfy some **terminal constraint**

$$K(x(t_1)) = c$$

where $K: \mathbf{R}^n \to \mathbf{R}^p$ and c is a fixed vector in $\mathbf{R}^p$. For example, if $p = n$ and K is the identity function, we capture the case in which $x(t_1)$ is required to take on a specified target value c. Also in Appendix A-3, we study the case in which each value $u(t)$ of u, for t in $[t_0, t_1]$, must be constrained to lie in some specified subset U_0 of U.

Example 1

Consider the problem of transferring our familiar $\dfrac{1}{s^2}$ plant, the "cart"†

$$\dot{x}_1 = x_2; \quad \dot{x}_2 = u$$

from state $[\tilde{x}_1, 0]^T$ at the time 0 to state $[0, 0]^T$ at time t_1.

---

† Specifically, the cart of unit mass on frictionless wheels acted upon by a force $u(t)$ and located at position $x_1(t)$. See Section 7–1 for a survey of the appearance of this example throughout the book.

Let the cost rate function be the "minimum time" rate

$$\ell(\widehat{x}, \widehat{u}) = \begin{cases} 1 \text{ if } \widehat{x} \neq 0 \\ 0 \text{ if } \widehat{x} = 0 \end{cases}$$

(where $\widehat{x} = [\widehat{x}_1, \widehat{x}_2]^T$). Note that if u_1 first transfers $[\widetilde{x}_1, 0]^T$ to $[0, 0]^T$ at some time $t_2 < t_1$, then the control u_2 with

$$u_2(t) \triangleq \begin{cases} u_1(t) \text{ for } t_1 \leq t \leq t_2 \\ 0 \qquad \text{for } t_2 < t \leq t_1 \end{cases}$$

will also first transfer $[\widetilde{x}_1, 0]^T$ to $[0, 0]^T$ at time t_2, and will have [Exercise 2]

$$J(u_2) = t_2 \leq J(u_1)$$

Thus, if we let Ω be some bounded set of controls sufficiently rich to include at least one control which transfers $[\widetilde{x}_1, 0]^T$ to $[0, 0]^T$, then the control u_0 which minimizes $J(u)$ for $u \in \Omega$ will in fact transfer the cart to its zero state by time t_1, and will do so in minimum time. $\Diamond$

As another example, if we use the cost rate $\|x(t)\|^2 + \|u(t)\|^2$ of equation (5), and require that $u(t)$ be in some bounded set, such as $\|u(t)\| \leq 1$, then a control u which minimizes $J(u)$ will keep (5) small by "bringing $x(t)$ as close to 0 as possible without undue expenditure of energy." If, on the other hand, we remove constraints on the size of $u(t)$ and simply demand a minimum energy control among those constrained to yield $x(t_1) = 0$, we know from the theory of Section 7–3 that we cannot succeed unless state $x(t_0)$ is controllable. Surprisingly, the theory of Section 7–3 actually yields the minimum energy control, as we now show:

Example 2

As a simple example whose tools are ready at hand, we consider the case in which $\ell(x(t), u(t))$ is simply the "input energy" $\|u(t)\|^2$.

In Section 7–3, we associated with the linear system

$$\dot{x}(t) = F(t)x(t) + G(t)u(t)$$

whose state transition matrix is $\Phi(t_0, t)$, the linear transformation

$$W(t_0, t_1) = \int_{t_0}^{t_1} \Phi(t_0, \tau)G(\tau)G^*(\tau)\Phi^*(t_0, \tau) \, d\tau$$

and proved (Corollary 7-3-5) that if for some $t_1 \geq t_0$ the matrix $W(t_0, t_1)$ is invertible, then $(F(\cdot), G(\cdot))$ is controllable, and $(t_0, \bar{x})$ is transferred to $(t_1, 0)$ by the input function u_0 for which

$$u_0(t) = -G^*(t)\Phi^*(t_0, t)W^{-1}(t_0, t_1)\bar{x} \quad \text{for } t_0 \leq t \leq t_1$$

We now prove that u_0 is the *minimum energy control* which transfers $(t_0, \bar{x})$ to $(t_1, 0)$. To see this, suppose that u is any other input function which also transfers $(t_0, \bar{x})$ to $(t_1, 0)$. Then

$$-\Phi(t_1, t_0)\bar{x} = \int_{t_0}^{t_1} \Phi(t_1, t)G(t)u(t)\, dt = \int_{t_0}^{t_1} \Phi(t_1, t)G(t)u_0(t)\, dt$$

Hence, multiplying the second and third expressions by $\Phi(t_0, t_1)$ and subtracting, we obtain

$$\int_{t_0}^{t_1} \Phi(t_0, t)G(t)[u(t) - u_0(t)]\, dt = 0$$

Thus, taking the inner product of this last expression with any vector yields zero. In particular,

$$0 = \langle 0 \mid W^{-1}(t_0, t_1)\bar{x} \rangle = \left\langle \int_{t_0}^{t_1} \Phi(t_0, t)G(t)[u(t) - u_0(t)]\, dt \;\middle|\; W^{-1}(t_0, t_1)\bar{x} \right\rangle$$

By the linearity and continuity of the inner product operation, we can pull the integration outside the inner product to get

$$0 = \int_{t_0}^{t_1} \langle \Phi(t_0, t)G(t)[u(t) - u_0(t)] \mid W^{-1}(t_0, t_1)\bar{x} \rangle \, dt$$

and, using the properties of the adjoint to flip $\Phi(t_0, t)G(t)$ across the inner product, we end up with

$$0 = \int_{t_0}^{t_1} \langle u(t) - u_0(t) \mid G^*(t)\Phi^*(t_0, t)W^{-1}(t_0, t_1)\bar{x} \rangle \, dt$$

But the expression on the right of the inner product is just $-u_0(t)$, so we have shown that

$$0 = \int_{t_0}^{t_1} \langle u(t) - u_0(t) \mid u_0(t) \rangle \, dt \tag{6}$$

Now

$$\int_{t_0}^{t_1} \|u(t)\|^2 \, dt = \int_{t_0}^{t_1} \|u(t) - u_0(t) + u_0(t)\|^2 \, dt$$

$$= \int_{t_0}^{t_1} \langle u(t) - u_0(t) + u_0(t) \,|\, u(t) - u_0(t) + u_0(t) \rangle \, dt$$

$$= \int_{t_0}^{t_1} \|u(t) - u_0(t)\|^2 \, dt + 2 \int_{t_0}^{t_1} \langle u(t) - u_0(t) \,|\, u_0(t) \rangle \, dt$$

$$+ \int_{t_0}^{t_1} \|u_0(t)\|^2 \, dt$$

By equation (6), the second term on the right vanishes and

$$\int_{t_0}^{t_1} \|u(t)\|^2 \, dt = \int_{t_0}^{t_1} \|u(t) - u_0(t)\|^2 \, dt + \int_{t_0}^{t_1} \|u_0(t)\|^2 \, dt$$

and since

$$\int_{t_0}^{t_1} \|u(t) - u_0(t)\|^2 \, dt \geq 0$$

we have

$$\int_{t_0}^{t_1} \|u(t)\|^2 \, dt \geq \int_{t_0}^{t_1} \|u_0(t)\|^2 \, dt$$

Thus, the energy in u is greater than or equal to that in u_0, so u_0 is indeed the minimal energy control, as claimed.   ◊

The last example is a special case of the class of linear control problems with quadratic cost functionals which can be solved by the general methods that we outline in Appendix A-3.

To pave the way for our theoretical development of Appendix A-3, let us now re-express our differential equation (1) in terms of a functional, in much the same way in which, in Section 5–2, we converted the problem of solving the differential equation $\dot{x}(t) = \bar{f}(x(t))$ into the problem of finding a fixed point of the functional S defined by

$$x \mapsto x(t_0) + \int_{t_0}^{(\cdot)} \bar{f}(x(\tau)) \, d\tau$$

If equation (1) is satisfied, then we may integrate each term from t_0 to any t in $[t_0, t_1]$ to obtain

$$x(t) - x(t_0) = \int_{t_0}^{t} f(x(\tau), u(\tau))\, d\tau$$

Thus, if we introduce the mapping $A : \Theta \times \Omega \to \Theta$ defined by the rule

$$(x, u) \mapsto \left[x(\cdot) - x(t_0) - \int_{t_0}^{(\cdot)} f(x(\tau), u(\tau))\, d\tau\right] \tag{7}$$

where the $(\cdot)$ reminds us of the temporal dependence of the function $A[x, u] \in \Theta$, then we see that equation (1) may be rephrased as the condition

$$A[x, u] = 0 \text{ (the zero mapping)} \tag{8}$$

and that for each $u \in \Omega$ there is a unique $x = x(u) \in \Theta$ for which equation (8) holds.

Notice that the mapping defined by equation (7) associates with each pair of functions (x, u) another function which we have denoted by $A[x, u]$. Since $A[x, u]$ is a function in the space Θ, its value at any time t is given by

$$A[x, u](t) = x(t) - x(t_0) - \int_{t_0}^{t} f(x(\tau), u(\tau))\, d\tau \tag{9}$$

Now suppose that we hold the function u fixed and consider the effect on $A[x, u]$ of letting x vary. Denoting this by $A[\cdot, u]$, we have defined a map $A[\cdot, u] : \Theta \to \Theta$ and we may now pose questions like "How does $A[x + h, u]$ compare with $A[x, u]$, where h is some incremental function in Θ?" Similarly, we can define the mapping $A[x, \cdot] : \Omega \to \Theta$ and investigate the relationship of $A[x, u + v]$ to $A[x, u]$, where v is an incremental function in Ω. This leads us quite naturally to consider approximating the map $A[\cdot, u]$ by its derivative map at the point (x, u) (which we will denote by $A_x[x, u]$) and approximating $A[x, \cdot]$ by its derivative map at (x, u), denoted $A_u[x, u]$, as we discussed in Section 3-4.

Using our experience with Banach space derivatives in Section 3-4, we know that the derivatives $A_x[x, u]$ and $A_u[x, u]$, if they exist, are themselves linear transformations $\Theta \to \Theta$ and $\Omega \to \Theta$, respectively, which satisfy the equations

$$A[x + h, u] = A[x, u] + A_x[x, u]h + o(h) \quad \text{for } h \text{ in } \Theta$$

and

$$A[x, u + v] = A[x, u] + A_u[x, u]v + o(v) \quad \text{for } v \text{ in } \Omega$$

Again let us comment upon the notation: the derivative $A_x[x, u]$ is a map

$$A_x[x, u]: \Theta \to \Theta$$

and if we plug into it an increment $h \in \Theta$ we get $A_x[x, u](h)$, another function in Θ. To indicate the value of this last function at any time t we might write

$$A_x[x, u](h)(t)$$

but by now the parentheses would be driving us crazy. Instead of using all the parentheses, we have for clarity left a pair out and written $A_x[x, u]h$ instead of $A_x[x, u](h)$ to denote the action of $A_x[x, u]$ on an incremental function h. Then, if we need to see what value this resulting function has at time t, we can write $A_x[x, u]h(t)$, hopefully with no confusion arising.

Under what conditions on the function f will our mappings $A[\cdot, u]$ and $A[x, \cdot]$, defined by equation (9), actually have the derivatives $A_x[x, u]$ and $A_u[x, u]$ (which look like *partial* derivatives of $A[x, u]$) existing? It is easy to see† that if $f(\widehat{x}, \widehat{u})$ has a partial derivative denoted by $f_x(\widehat{x}, \widehat{u})$ with respect to its first variable $\widehat{x}$, then we have that $A_x[x, u]$ does exist, and

$$A_x[x, u]h = h(\cdot) - \int_{t_0}^{(\cdot)} f_u(x(\tau), u(\tau))v(\tau)\, d\tau \tag{10b}$$

while, by the same token, for $A_u[x, u]$ to exist, it is sufficient that $f_u(\widehat{x}, \widehat{u})$, the partial derivative of f with respect to its second variable $\widehat{u}$, exist, and we then have

$$A_u[x, u]v = -\int_{t_0}^{(\cdot)} f_u(x(\tau), u(\tau))v(\tau)\, d\tau \tag{10b}$$

Similarly, if we were to differentiate in equation (2b),

$$g_x[x, u]h = \int_{t_0}^{t_1} \ell_x(x(t), u(t))h(t)\, dt \tag{11a}$$

and

$$g_u[x, u]v = \int_{t_0}^{t_1} \ell_u(x(t), u(t))v(t)\, dt \tag{11b}$$

We shall make use of these derivatives, and develop some of their properties, in Appendix A-3.

One of the most important approaches to optimal control problems is that of Bellman's **dynamic programming,** which is based on the following obvious, but highly useful, observation.

---

†If it isn't easy to see, work Exercise 3 before going on. In fact, work it anyway.

PRINCIPLE OF OPTIMALITY

If u_0 is optimal on $[t_0, t_1]$, then for any intermediate time t_2 (that is, $t_0 \leq t_2 \leq t_1$), u_0 must be optimal on both $[t_0, t_2]$ and $[t_2, t_1]$ in the specific sense that u_0 minimizes both

$$\int_{t_0}^{t_2} \mathcal{l}(x(u)(t), u(t)) \, dt \text{ for } u \text{ with } x(u)(t_0) = x(u_0)(t_0)$$

$$\text{and } x(u)(t_2) = x(u_0)(t_2)$$

and
$$\int_{t_2}^{t_1} \mathcal{l}(x(u)(t), u(t)) \, dt \text{ for } u \text{ with } x(u)(t_2) = x(u_0)(t_2)$$

$$\text{and } x(u)(t_1) = x(u_0)(t_1)$$

Proof

Exercise 4.   □

The mathematical difficulty and computational complexity of optimal control problems are so prohibitive that Bellman's dynamic programming and Pontryagin's minimum principle, which we present in Appendix A–3, mark the beginnings of real, practical progress in this field.

Bellman's principle of optimality is best utilized with a computer in a "discretized" state space, where a numerical search for an optimal trajectory is ingeniously reduced from an N-stage decision process to a sequence of N single-stage decision processes (where N is the number of possible states). The interested reader should see Chapter 11 of Elgerd [1967] for a simple discussion and illustrative example of the dynamic programming approach, and he should also read the book by Bellman [1957].

The approaches to optimal control that we outline in detail in Appendix A–3 are outgrowths of the approach taken to *classical mechanics* by Joseph-Louis Lagrange in the Eighteenth Century and by William Rowan Hamilton in the Nineteenth. Let us now motivate these approaches. We recall from Section 2–5 the Lagrangian and Hamiltonian formulations:

A system with generalized coordinates (e.g., of position) $q_1, \ldots, q_n$ is said to be **conservative** if there exists a function $V = V(q)$, called the **potential energy** of the system, such that the force F, acting to change q_j, has as its jth component $-\dfrac{\partial V}{\partial q_j}$. Putting this in general derivative terms, we have

$$F = -V_q$$

where V_q denotes the derivative of V with respect to q.

The **kinetic energy** T of the system is expressed as a function of q and the generalized velocities $\dot{q}$ in the Lagrangian formulation. Lagrange introduced the function (now called the **Lagrangian** in his honor)

$$L(q, \dot{q}) = T(q, \dot{q}) - V(q)$$

and observed that the equations of motion could be written in the form of n scalar equations which we may summarize in vector form as

$$\frac{d}{dt}(L_{\dot{q}}) - L_q = 0 \tag{12}$$

where $L_{\dot{q}} = \dfrac{\partial L}{\partial \dot{q}}$. These equations are second order, since the $\dot{q}$'s occurring in $L_{\dot{q}}$ will have their derivatives taken with respect to t in forming $\dfrac{d}{dt}(L_{\dot{q}})$, so that the resulting equation (12) will relate q, $\dot{q}$, and $\ddot{q}$ as we saw in Examples 8 and 9 of Section 2–5.

Hamilton defined the general momenta p_j by setting

$$p_j = \frac{\partial L}{\partial \dot{q}_j} \tag{13}$$

which equals $\dfrac{\partial T}{\partial \dot{q}_j}$, since V has no explicit dependence on $\dot{q}$. Each (q_j, p_j) is called a pair of **conjugate** variables or **adjoint** variables.

Lagrange's equations then take the form

$$L_q = \frac{d}{dt}(p)$$

which is quite close to the version of Newton's $F = ma$, which asserts that the force F equals the time derivative of the momentum mv, where v is velocity, and $a = \dot{v}$ is the acceleration of a particle of mass m.

Now suppose, as is implicit in the version presented above, that L does not depend *explicitly* on t. Then the time derivative of L along a trajectory $(q(t), \dot{q}(t))$ is given by

$$\frac{dL}{dt} = \dot{q}^T L_q + \ddot{q}^T L_{\dot{q}}$$

where $\dot{q}^T$ is the transpose of $\dot{q}$. Then

$$\frac{dL}{dt} = \dot{q}^T \frac{d}{dt}(L_{\dot{q}}) + \ddot{q}^T L_{\dot{q}} \text{ (by Lagrange's equations (12))}$$

$$= \frac{d}{dt}(\dot{q}^T L_{\dot{q}})$$

Thus

$$\frac{d}{dt}\,[\dot{q}^{T}L_{\dot{q}} - L] = 0$$

so that the **Hamiltonian function** (recalling from (13) that $p = L_{\dot{q}}$)

$$H = \dot{q}^{T}p - L \quad \text{where } \dot{q} = \dot{q}(p, q) \tag{14}$$

is a constant of the motion (i.e., $\dot{H} = 0$ along a trajectory) if t does not appear explicitly in L. In most cases of physical interest, H can be identified with the total energy of the system, and so this is an expression of the conservation of energy for conservative systems. Let us now see how we may use H to reformulate Lagrange's equations:

$$dH = \sum_{j} \frac{\partial H}{\partial q_{j}}\,dq_{j} + \sum_{j} \frac{\partial H}{\partial p_{j}}\,dp_{j}$$

and also

$$dH = \sum_{j} \dot{q}_{j}\,dp_{j} + \sum_{j} p_{j}\,d\dot{q}_{j} - dL$$

However,

$$dL = \sum_{j} \frac{\partial L}{\partial q_{j}}\,dq_{j} + \sum_{j} \frac{\partial L}{\partial \dot{q}_{j}}\,d\dot{q}_{j} = \sum_{j} \frac{\partial L}{\partial q_{j}}\,dq_{j} + \sum_{j} p_{j}\,d\dot{q}_{j} \quad \text{by (13)}$$

$$= \sum_{j} \dot{p}_{j}\,dq_{j} + \sum_{j} p_{j}\,d\dot{q}_{j} \quad \text{by (12)}$$

Hence

$$dH = -\sum_{j} \dot{p}_{j}\,dq_{j} + \sum_{j} \dot{q}_{j}\,dp_{j}$$

We thus obtain *Hamilton's canonical form of the equations of motion:*

$$\left.\begin{array}{l} \dot{q} = H_{p} \\ -\dot{p} = H_{q} \end{array}\right\} \tag{15}$$

To bring this into the context of optimization, we note the following theorem, which provides an extremely elegant and philosophically appealing statement of the laws of motion:

HAMILTON'S PRINCIPLE

If a pair of trajectories W_1, W_2 which map $[t_0, t_1]$ into $\mathbf{R}^n$ yield an *extreme value* of

$$\int_{t_0}^{t_1} L(W_3(t), W_4(t)) \, dt \tag{16}$$

among all trajectories for which $W_1(t_0) = W_3(t_0)$, $W_1(t_1) = W_3(t_1)$, $W_2(t_0) = W_4(t_0)$, and $W_2(t_1) = W_4(t_1)$, then they form a solution $W_1 = q$, $W_2 = \dot{q}$ for Lagrange's equations of motion.   □

The proof is a standard exercise in the calculus of variations (see, e.g., Leech [1958]) and need not detain us here. What is of interest here is the general strategy for optimization problems that it suggests. The theory we present in Appendix A–3 may be regarded as motivated by the following identifications:

The Lagrangian L is taken as the negative $-\ell(x(t), u(t))$ of the cost function. [This proves to be more convenient in engineering applications and does not harm our motivation, since minimizing ℓ is the same as maximizing $-\ell$, and Hamilton's principle talks only of extreme values, and does not distinguish minima and maxima.]

The Hamiltonian H is then the function

$$\lambda^T \dot{x} - L = \lambda^T f(x, u) + \ell(x, u) \tag{17}$$

where we introduce a generalized momentum term or adjoint vector λ in addition to x and u.

The point to notice is that x and u are treated on the same footing, since we constrain x to be $x(u)$. The problem is to find the equations which describe the u which maximizes $\int_{t_0}^{t_1} L \, dt$, and we do this by introducing adjoint conjugate variables λ. We then expect that for a u to maximize $\int_{t_0}^{t_1} L \, dt$, it must determine an x and λ which satisfy Hamilton's canonical equations

$$\dot{x} = H_\lambda \text{ and } -\dot{\lambda} = H_x$$

for the Hamiltonian H of equation (17). In other words, not only do we have from $\dot{x} = H_\lambda$ that

$$\dot{x} = f(x, u)$$

as expected, but we must also have that u permits the existence of a function

λ which satisfies $-\dot{\lambda} = H_x$, namely the **adjoint control equation**

$$-\dot{\lambda}(t) = f_x^T(x(t), u(t))\lambda(t) + \ell_x^T(x(t), u(t)) \tag{18}$$

as well as extra conditions depending on the extra demands we make—such as the terminal constraints that we study in the second section of Appendix A–3, or the bounds we place on the control signals in the third section of that appendix.

Before turning to a rigorous analysis of these conditions, let us pay a farewell visit to dynamic programming, by seeing how the principle of optimality implies, under rather strong assumptions, some properties of the Hamiltonian.

Suppose that u_0 yields the minimal cost $J(x)$ for the transition from state $x(t)$ at time t to some fixed state x^1 at time t_1:

$$J(x) = \int_t^{t_1} \ell(x(u_0)(\tau), u_0(\tau)) \, d\tau$$

with

$$x(u_0)(t) = x(t), \quad x(u_0)(t_1) = x^1$$

Using Leibnitz's rule to differentiate under the integral sign, we have

$$\frac{d}{dt} J(x) = -\ell(x(t), u_0(t))$$

But, by the chain rule, we can also write

$$\frac{d}{dt} J(x) = J_x(x)\dot{x}(t)$$

Equating the two expansions, we have the equation

$$\ell(x(t), u_0(t)) + J_x(x)\dot{x}(t) = 0$$

If we set $\lambda^T = J_x(x)$ and substitute for $\dot{x}$, we obtain the familiar expression of equation (17) in the equation

$$\ell(x(t), u_0(t)) + \lambda^T f(x(t), u_0(t)) = 0 \tag{19}$$

Now, suppose that we replaced u_0 by a non-optimal control $\widehat{u}$ over the short time interval $[t, t + dt]$. Then, by the principle of optimality,

$$\ell(x, \widehat{u}) \, dt + J(x + dx) \geq J(x) \tag{20}$$

where

$$x + dx = x + f(x, \widehat{u}) \, dt$$

Thus,

$$J(x + dx) - J(x) = J_x(x)\, dx$$
$$= \lambda^T f(x, \widehat{u})\, dt \tag{21}$$

Combining equations (20) and (21), we have

$$\ell(x, \widehat{u}) - \lambda^T f(x, \widehat{u}) \geq 0 \tag{22}$$

Note that equation (19) says that

$$H(x, u_0(t), \lambda(t)) = 0 \tag{23}$$

for the optimal u_0, while equation (22) says that for any other $\widehat{u}$, we have

$$H(x, \widehat{u}, \lambda(t)) \geq 0 \tag{24}$$

We shall meet this relationship between equations (23) and (24) in more general and more rigorous setting when we prove the Pontryagin Minimum Principle in Theorem 1 of the third section of Appendix A-3.

This concludes our brief motivational discussion of optimal control. With this background the mathematically inclined reader should be able to follow the developments and proofs of the famous Lagrange Multiplier Theorem and the Pontryagin Minimum Principle which we outline in Appendix A-3. For those readers interested in gaining further experience at an introductory, low mathematical level, the excellent texts by Takahashi, Rabins, and Auslander [1970], especially Sections 1-1 and Chapter 14, and by Elgerd [1967], especially Section 7-2 and Chapters 10 and 11, are recommended.

Our proofs of the Lagrange Multiplier Theorem and the Pontryagin Minimum Principle in Appendix A-3 are adapted from the proof offered by Luenberger [1969, Chapter 9], following Luisternik and Sobolev [1961] and Rozonoer [1959], respectively.

Our discussion of classical mechanics is fairly standard (see, e.g., Leech [1958]); surprisingly, this historical motivation is strangely lacking in most accounts of optimal control.

EXERCISES FOR SECTION 7-6

✔ 1.   This exercise shows why, in equation (3), we call the quantity $\displaystyle\int_{t_0}^{t_1} \|u(t)\|^2\, dt$ the energy in the input signal u over the interval $[t_0, t_1]$.

   (a)   Suppose that the signal u is a voltage waveform, and assume that it activates a resistive load as shown in Figure 7-45(a).

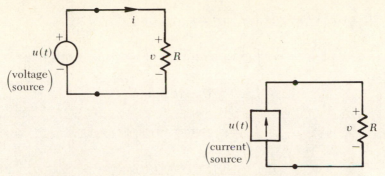

Figure 7-45(a)   A voltage signal, u. **(b)** A current signal, u.

Find the energy supplied by u to the load R over the interval $[t_0, t_1]$. [*Hint:* Let $\omega(t)$ be energy and let $p(t)$ be power at time t. Use $p(t) = \dfrac{d\omega(t)}{dt}$, $p(t) = i(t)v(t)$, and $v(t) = Ri(t)$.]

(b)   Suppose that $u(t)$ is a current waveform as in Figure 7-45(b). Show that the energy supplied by u over $[t_0, t_1]$ is equal to $R \displaystyle\int_{t_0}^{t_1} u^2(t)\, dt$ and hence is directly proportional to $\displaystyle\int_{t_0}^{t_1} \|u(t)\|^2\, dt$.

2.   For the $\dfrac{1}{s^2}$ plant with the cost function of Example 1, carry out all the steps to show that

$$J(u_2) = t_2$$

and

$$J(u_2) \le J(u_1)$$

Why might this cost rate ℓ be called the "a miss is as good as a mile" cost functional?

✔ **3.**   This exercise provides practice in computing the derivative map of functionals of the type

$$J[x_1, x_2] = \int_a^b f(x_1(t), x_2(t))\, dt$$

(a)   Expand the integrand $f(x_1, x_2)$ in a two-dimensional Taylor series

$$f(x_1 + h_1, x_2 + h_2) = f(x_1, x_2) + \frac{\partial f}{\partial x_1}(x_1, x_2)h_1 + \frac{\partial f}{\partial x_2}(x_1, x_2)h_2$$

$$+ \{\text{terms in } h_1^2, h_1 h_2, h_2^2 \text{ and higher}\}$$

to get

$$J[x_1 + h_1, x_2 + h_2] = \int_a^b f(x_1(t) + h_1(t), x_2(t) + h_2(t))\, dt$$

$$= \int_a^b f(x_1(t), x_2(t))\, dt + \int_a^b \frac{\partial f}{\partial x_1}(x_1(t), x_2(t)) h_1(t)\, dt$$

$$+ \int_a^b \frac{\partial f}{\partial x_2}(x_1(t), x_2(t)) h_2(t)\, dt$$

$$+ \int_a^b \{\text{higher order terms of } h_1, h_2\}$$

$$= J[x_1, x_2] + \int_a^b f_{x_1}(x_1(t), x_2(t)) h_1(t)\, dt$$

$$+ \int_a^b f_{x_2}(x_1(t), x_2(t)) h_2(t)\, dt$$

$$+ \{\text{higher order terms in } h_1, h_2\}$$

where f_{x_1} and f_{x_2} denote the partial derivatives of f with respect to its first and second variables respectively.

(b)   From the definition of the derivative map of J with respect to x_1,

$$J[x_1 + h_1, x_2] = J[x_1, x_2] + J_{x_1}[x_1, x_2](h_1) + \{\text{higher order terms in } h_1\}$$

where $J_{x_1}[x_1, x_2]$ is the derivative map of J at the point (x_1, x_2) and $J_{x_1}[x_1, x_2](h_1)$ is the result of plugging the increment h_1 into $J_{x_1}[x_1, x_2]$, use part (a) to show that

$$J_{x_1}[x_1, x_2](h_1) = \int_a^b f_{x_1}(x_1(t), x_2(t)) h_1(t)\, dt$$

(c)   Use part (b) on the map A defined by the rule in equation (7) to get

$$(A_x[x, u]h)(t) = h(t) - \int_{t_0}^t f_x(x(\tau), u(\tau)) h(\tau)\, d\tau$$

as in equation (10a).

(d)   Derive an expression for the derivative of J with respect to x_2, denoted by $J_{x_2}[x_1, x_2](h_2)$, and use it to derive the expression

$$(A_u[x, u]v)(t) = -\int_{t_0}^t f_u(x(\tau), u(\tau)) v(\tau)\, d\tau$$

which is equation (10b).

(e)   Derive the expressions of equations (11a) and (11b).

4.   Prove the principle of optimality.

REFERENCES

1. R. E. BELLMAN, *Dynamic Programming*, Princeton University Press (1957).
2. O. I. ELGERD, *Control Systems Theory*, New York, McGraw-Hill (1967).

3. B. GOPINATH, "On the Control of Linear Multiple Input-Output Systems," *Bell Syst. Tech. J.*, 50 (1971), pp. 1063–1081.

4. R. E. KALMAN, "Contributions to the Theory of Optimal Control," *Boletin de la Sociedad Matematica Mexicana* (1960), pp. 102–119.

5. J. W. LEECH, *Classical Mechanics,* London, Methuen (1958).

6. D. G. LUENBERGER, "Observers of Low Dynamic Order," *IEEE Trans. Military Elec.*, 3, No. 2 (April 1964), pp. 74–80.

7. D. G. LUENBERGER, "Observers for Multivariable Systems," *IEEE Trans. Automatic Control,* AC-11, No. 2 (April 1966), pp. 190–197.

8. L. LUISTERNIK and V. SOBOLEV, *Elements of Functional Analysis,* New York, Ungar (1961).

9. W. T. REID, "A Matrix Differential Equation of the Ricatti Type," *Amer. J. Math.*, 68 (1946), pp. 237–246.

10. L. I. ROZONOER, "L. S. Pontryagin's Maximum Principle in Optimal System Theory," *Automation and Remote Control,* 20 (1959), Part 1, pp. 1288–1302; Part 2, pp. 1405–1421; Part 3, pp. 1517–1532.

11. Y. TAKAHASHI, M. J. RABINS, and D. M. AUSLANDER, *Control and Dynamic Systems,* Reading, Mass., Addison-Wesley (1970).

12. W. M. WONHAM, "On Pole Assignment in Multi-Input Controllable Linear Systems," *IEEE Trans. Automatic Control,* AC-12, No. 6 (December 1967), pp. 660–665.

13. K. YOSIDA, *Functional Analysis* (2nd edition), Berlin, Springer-Verlag (1968).

CHAPTER 8

Algebraic Approaches to System Realization†

In Chapter 6, we saw that the zero-state response of a linear system could be captured in the transfer function $\mathfrak{Z}(s) = H(sI - F)^{-1}G$. We then developed a number of methods for finding *realizations* of a given transfer function; i.e., for going from a given $\mathfrak{Z}(s)$ to three matrices, F, G, and H, such that $\mathfrak{Z}(s) = H(sI - F)^{-1}G$ does indeed hold. In doing so, however, we ignored the question of how we might find $\mathfrak{Z}(s)$ in the first place, and how, given partial data, we may find the F, G, and H which best represent those data. Moreover, we gave no general algebraic setting for our results. In this chapter, we remove all these deficiencies.

In Section 8–1, we discover that the problem of realization for continuous-time constant linear systems is mathematically equivalent to the realization problem for discrete-time constant linear systems. It turns out, then, that we can gain a general perspective by studying the realization problem for *arbitrary* (linear or nonlinear) discrete-time constant systems, the solution of which proves to be surprisingly easy.

In Section 8–2, we take the direct step of extracting the theory for linear

---

†Sections 8–2 and 8–3 were heavily influenced by Arbib's collaboration with R. E. Kalman and H. P. Zeiger. Sections 8–4 through 8–7 emerged from joint work of Arbib and E. G. Manes. L. S. Bobrow helped with the exposition.

624

systems from the general theory of Section 8–1, and then provide a procedure which enables us to recursively update our estimates of F, G, and H as more and more data about $\mathfrak{H}(s)$ become available.

Then, in Section 8–3, we present Kalman's module-theoretic approach to linear systems, placing the ad hoc methods of Chapter 6 in pleasing algebraic perspective by appealing to 19th century matrix theory in the sleek garments of 20th century module theory. We then carry this process of algebraicization even further, in the remainder of the chapter, when we place the discussion of Sections 8–1 and 8–2 into an elegant framework provided by the even more recent mathematical theory of categories: the general input-process approach to system theory discovered by Arbib and Manes in 1972. Our strategy will take us through four more sections as follows:

In Section 8–4 we suggest a plausible hypothesis based on the theory of system realization presented in Sections 8–1 and 8–2, using a formulation that suggests that category theory will provide an appropriate language for our further studies. We discover that the hypothesis is false, but that we still need the vocabulary of category theory to discuss abstract properties of structured objects and of maps which preserve that structure. On this basis we introduce, in Section 8–5, the input-process approach to systems, and then specialize this to get the class of decomposable machines. Finally, we present a general realization theory in Section 8–6, and an alternative form of this theory in Section 8–7.

8–1   Realization of Discrete-Time Systems

We start by considering the relation between the input-output behavior of continuous- and discrete-time constant linear systems. The system

$$\dot{x}(t) = Fx(t) + Gu(t)$$
$$y(t) = Hx(t)$$

has transfer function

$$\mathfrak{H}(s) = H(sI - F)^{-1}G$$

$$= \sum_{l=0}^{\infty} HF^{l}Gs^{-(l+1)}$$

Thus, to know $\mathfrak{H}(s)$ is to know the coefficient A_l of $s^{-(l+1)}$ for $0 \leq l < \infty$, and the realization problem for continuous-time constant linear systems is equivalent to the problem of solving the infinite set of equations

$$A_l = HF^{l}G \text{ for } 0 \leq l < \infty \qquad \textbf{(1)}$$

for the unknown matrices H, F, and G. Further, as we have already discussed in Chapter 6, we wish to do this in such a way that the dimension of F is as small as possible.

The system

$$x(t + 1) = Fx(t) + Gu(t)$$
$$y(t) = Hx(t)$$

has zero-state response

$$\mathcal{S}_0(u_k \ldots u_1 u_0) = \sum_{i=0}^{k} HF^i Gu_i \tag{2}$$

[We have changed the order of the subscripts on the u's from that of Chapter 4. The reason will be apparent in Section 8–2.] Thus, to know $\mathcal{S}_0$ is to know the matrix A_i which operates on each u_i, and the realization problem for *discrete-time* constant linear systems is also equivalent to solving the set (1) of equations for minimal F, G and H.

Again, we make the crucial point: *Very often the mathematical form of a problem associated with a continuous-time system is identical with the mathematical form of a problem associated with the discrete-time system with the same matrices. Thus, we may make use of the powerful intuition available in the discrete-time case to solve the mathematical problem, confident in the knowledge that the result will be applicable in the continuous-time case.* Thus, to solve our original problem of finding F, G, and H such that $H(sI - F)^{-1}G = \mathcal{S}(s)$, we may turn to the study of discrete-time systems.

To get some feel for the problem, let us immediately consider the realization problem for a system of rather simple structure.

Example 1

Consider the system of Figure 8–1. It is a string of n delay elements. The input to the first is the system input u, the input to the jth ($1 < j \leq n$) is the output of the $(j - 1)$th, and the system output y is a weighted sum

$$\sum_{j=1}^{n} a_j x_j$$

of the outputs of the n delay elements. Our task is to find the coefficients $a_1, \ldots, a_n$ via input-output measurements. The solution is simple. Giving the system n zero inputs will allow it to settle down to its zero state. Then if

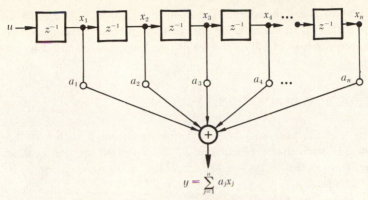

$$y = \sum_{j=1}^{n} a_j x_j$$

Figure 8-1   A simple shift register.

we apply an input of 1 followed by subsequent zeros, the 1 will pass down the line of delay elements reading out the successive a_j's—and so, after n further zero inputs have been applied following the unit input, we shall know exactly what the system parameters are. [This is a very simple example of how a system can be specified in terms of its impulse response (Exercise 1).]

$\Diamond$

It is clear, of course, that this procedure only worked because we knew the structure of the system—in particular, we knew that all the information about the system was contained in a sequence of n outputs. But what if the system had been encased in the proverbial black box, so that even though some spy had told us the type of structure of the system, we did not know the number n of delay elements involved? Suppose, then, that we took our system, guaranteed to be in the zero state, applied a unit input, and watched the output as we applied a sequence of zero inputs, only to observe the sequence

$$0 \quad a_1 \quad a_2 \quad \ldots \quad a_m \quad 0 \quad \ldots \quad 0$$

Could we then deduce that our system was of dimension m? No, for if we had seen k 0's following a_m, this could also be consistent with our system being of dimension greater than $m + k$, but having $a_j = 0$ for $m < j \leq m + k$.

The moral is then clear: *To be able to identify a system, we have not only to know what type of system it is, but also to have an upper bound on how large it is,* so that we may estimate the finite set of parameters which must characterize it. Thus, in Section 8-2, when we present algorithms for the identification of linear systems, we shall find that an upper bound on the dimension of the system plays a crucial role. However, we shall also see that, even when we do not know how large the system is, our methods will still yield useful approximations to the system equations.

Before proceeding any further, the reader should ensure that he is familiar with the discrete-time theory of Sections 4–2, 4–3, and 4–5. To provide a general framework for our realization theory, we must revive our general terminology of Chapter 4 to phrase the *general discrete-time realization problem.*

DISCRETE-TIME REALIZATION PROBLEM

Given a response function, $f: U^* \to Y$, find a system Σ with state x_0 which *realizes f*, i.e., such that the behavior $\mathcal{S}_{x_0}$ of Σ started in state x_0 is precisely *f.* Moreover, do this in such a way that the realization (Σ, x_0) is, in some sense, minimal.

At this stage, a word of apology is in order: The symbol f has become so standard for the response function in the literature on which this chapter is based that it would not be helpful to reserve f any longer for the local transition function, as we have done in the book so far. Instead, we will follow the automata theory literature in using δ for local transition functions and β for output functions. We shall not need impulses or Kronecker deltas in this section, so no confusion should arise from our new use of δ. We shall then use $\delta^*(x, w)$, rather than $\phi(x, w)$, for the state to which a system is sent from state x by iterated application of δ directed by the sequence w of inputs. Thus, $\mathcal{S}_x(w) = \beta(\delta^*(x, w))$.

As our criterion for minimality, we shall demand that the realization Σ be *reachable* and *observable.* Later on, we shall see that this does correspond to our minimal dimension criterion in the case of linear systems. If we denote by $x \cdot w$ the state $\delta^*(x, w)$ of Σ reached by applying input sequence w to Σ started in state x, then our problem may now be formally stated as follows, where we henceforth use f, δ, and β in their new roles:

REALIZATION PROBLEM

Given $f: U^* \to Y$, find a discrete-time system

$$\Sigma_f = (U, X_f, Y, \delta_f, \beta_f)$$

with a state x_0 in X_f such that if we denote by $(\mathcal{S}_f)_x$ the response function of Σ_f started in state x in X_f, we have:

(i) (Σ_f, x_0) is a realization of f, i.e., $(\mathcal{S}_f)_{x_0} = f$.
(ii) Σ_f is reachable: each state in X_f is of the form $x_0 \cdot w$ for some w in U^*.
(iii) Σ_f is observable: the map $x \mapsto (\mathcal{S}_f)_x$ is an injection.

Of course, we do not yet know that, for an arbitrary f, such a Σ_f exists. Our strategy, then, will be to discover what Σ_f must look like *when* it exists, and then show that the resultant description is one that we can in fact follow to build a Σ_f for every f.

Conditions (ii) and (iii) imply that Σ_f (if it exists) has one and only one distinct state for each distinct response function of the form $(\mathcal{S}_f)_{x_0 \cdot w}$. In other words, all states are of the form $x_0 \cdot w$, while

$$x_0 \cdot w_1 = x_0 \cdot w_2 \Leftrightarrow (\mathcal{S}_f)_{x_0 \cdot w_1} = (\mathcal{S}_f)_{x_0 \cdot w_2} \tag{3}$$

We can analyze this condition from Figure 8–2, which shows that the response of Σ_f started in state $x_0 \cdot w_1$ to an input sequence w is just the response

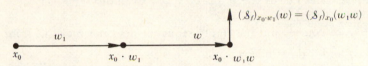

Figure 8–2   The response of a state reachable from x_0.

of Σ_f started in state x_0 to input sequence $w_1 w$. Recalling from (i) that $(\mathcal{S}_f)_{x_0}$ is precisely f, we see that $(\mathcal{S}_f)_{x_0 \cdot w_1}(w) = f(w_1 w)$, and so equation (3) takes the form

$$x_0 \cdot w_1 = x_0 \cdot w_2 \Leftrightarrow f(w_1 w) = f(w_2 w) \text{ for all } w \text{ in } U^*$$

If we define the function $L_{w_j}: U^* \rightarrow U^*: w \mapsto w_j w$ to be that which places the given input sequence w_j to the *Left* of any other sequence w, our last equivalence then takes the simple form

$$x_0 \cdot w_1 = x_0 \cdot w_2 \Leftrightarrow f L_{w_1} = f L_{w_2} \tag{4}$$

since two functions with the same domain and codomain (we call B the **codomain** of a function $g : A \rightarrow B$) are equal if and only if they agree on each argument in the domain.

Example 2

Consider the **parity function** $f : \{0, 1\}^* \rightarrow \{0, 1\}$ with

$$f(w) = \begin{cases} 1 \text{ if } w \text{ contains an odd number of 1's} \\ 0 \text{ if } w \text{ contains an even number of 1's} \end{cases}$$

We can see that f can be realized by the state x_0 of the two-state system of Figure 8–3.

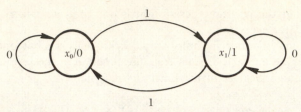

Figure 8-3   A minimal realization of the parity function.

This system is observable, since the output is different for different states. Thus $fL_{w_1} = fL_{w_2}$ will tell us that w_1 and w_2 lead us from the zero state to the *same* state. Let us check this on a few examples.

$fL_{\Lambda} = f$, of course, and the empty sequence does not move us from x_0.

$fL_0 = f$ and 0 does indeed leave us in x_0.

$fL_1 = 1 - f$ and 1 does indeed take us from x_0.

This example reminds us that infinitely many strings may "label" the same state.

If we consider a *non-observable* realization Σ' of f in Figure 8-4, we see that $fL_0 = fL_{00}$ even though $x_0 \cdot 0 = x_1 \neq x_0 = x_0 \cdot 00$. In this case, $f = \mathcal{S}_{x_0} = \mathcal{S}_{x_1}$ even though $x_0 \neq x_1$.   ◊

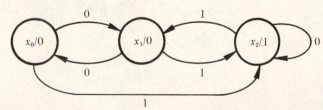

Figure 8-4   A non-observable realization of the parity function.

If we define the equivalence relation (Exercise 2) E_f on U^* by

$$w_1 E_f w_2 \Leftrightarrow fL_{w_1} = fL_{w_2} \tag{5}$$

we deduce from equation (4) that *if Σ_f exists,* its states are in one-to-one correspondence with the equivalence classes of E_f. Expressing this formally, we have

$$X_f \cong U^*/E_f : x_0 \cdot w \mapsto [w]_f \tag{6}$$

where we have written $[w]_f$ for the equivalence class $\{w' \in U^* \mid w' E_f w\}$ of w with respect to E_f. (We call E_f the **Nerode equivalence** of f.)

[We must reiterate that there are many different labels for each element in the equivalence class—but an equivalence class only counts once. To make this more clear, let us consider the set $\mathbf{N}$ of all integers and the equivalence relationship $n_1 \equiv n_2$ which relates two numbers as equivalent iff their difference is an even number. Here, there are only two equivalence classes—namely, the equivalence class of the *even* numbers and the equivalence class of the *odd* numbers. Although we may write $\mathbf{N} / \equiv \; = \{[n] \,|\, n \text{ in } \mathbf{N}\}$, we realize that although we have presented on the right-hand side infinitely many descriptions of the elements of $\mathbf{N} / \equiv$, there are only two elements in that set—it is just that each element has infinitely many descriptions, *odd* by all the odd numbers, and *even* by all the even numbers.]

Finally, let us note that if Σ_f (should it exist) is in state $x_0 \cdot w$ corresponding to $[w]_f$ and receives input u, it will go to state $x_0 \cdot wu$ corresponding to $[wu]_f$. Moreover, as Figure 8–5 makes clear, the output from the state corresponding to $[w]_f$ must be $f(w)$, and we note that x_0 corresponds to $[\Lambda]_f$, where Λ is the empty sequence in U^*.

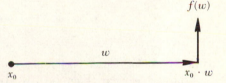

Figure 8–5 Output for a state reachable from a state x_0 with response function f.

So far, so good. We have said that, *if* there is a reachable observable realization of f, then its states can in fact be uniquely labeled by the elements of U^*/E_f. The question now is, "Is there such a system?" We shall show that there is, by giving, for *any* f, a mathematical description of a system which actually has U^*/E_f for its state space. Let us stress that we are giving this realization procedure for an *arbitrary* discrete-time constant system, without assuming that the system is linear. It will only be after we have completed this general analysis that we shall see what advantages we can get, in terms of computational procedures, when we turn our attention to the case of finite-dimensional linear systems.

THEOREM 1 (REALIZATION THEOREM FOR DISCRETE-TIME SYSTEMS)

Given any $f: U^* \to Y$, define the equivalence relation E_f on U^* by

$$w_1 E_f w_2 \Leftrightarrow f L_{w_1} = f L_{w_2}$$

and let $X_f = U^*/E_f$ be the set of equivalence classes of E_f. If we define the functions $\delta_f : X_f \times U \to X_f$ and $\beta_f : X_f \to Y$ by $\delta_f([w]_f, u) = [wu]_f$ and $\beta_f([w]_f) = f(w)$, and define the system Σ_f by

$$\Sigma_f = (U, X_f, Y, \delta_f, \beta_f)$$

then $(\Sigma_f, [\Lambda]_f)$ is a minimal (i.e., reachable and observable) realization of f.

Proof

We must first check that δ_f and β_f are well-defined functions; e.g., for β_f we must check that $\beta_f([w])$—abbreviating $[w]_f$ to $[w]$ henceforth—depends only on the equivalence class $[w]$ and not on the particular w that we chose within it:

$$
\begin{aligned}
w_1 E_f w_2 &\Rightarrow f L_{w_1} = f L_{w_2} \\
&\Rightarrow f L_{w_1}(\Lambda) = f L_{w_2}(\Lambda) \\
&\Rightarrow f(w_1) = f(w_2) \\
&\Rightarrow \beta_f([w_1]) = \beta_f([w_2])
\end{aligned}
$$

as desired. The reader may make the same check on δ_f (Exercise 4(a)).
    Since it is clear (Exercise 4(b)) that

$$\delta_f^*([w], w') = [ww'] \quad \text{for all } w, w' \text{ in } U^*$$

we have that $[w] = \delta_f^*([\Lambda], w)$ and so is reachable from $[\Lambda]$. Hence, Σ_f is reachable. Moreover,

$$
\begin{aligned}
(\mathcal{S}_f)_{[w]}(w') &= \beta_f(\delta_f^*([w], w')) \\
&= \beta_f([ww']) = f(ww') = f L_w(w')
\end{aligned}
\tag{7}
$$

Hence, $(\mathcal{S}_f)_{[\Lambda]}(w') = f(w')$, whence $(\Sigma_f, [\Lambda])$ does indeed realize f. Finally, by (7),

$$(\mathcal{S}_f)_{[w_1]} = (\mathcal{S}_f)_{[w_2]} \Leftrightarrow f L_{w_1} = f L_{w_2} \Leftrightarrow [w_1] = [w_2]$$

and thus $\mathcal{S}_f$ is also observable, and hence is an observable, reachable realization of f, as desired. $\square$

Example 3

We check the result by applying it to the parity function f of Example 2:
    Clearly, $w_1 w$ will have an odd number of 1's if the parity of w_1 differs

from that of w. We thus deduce that

$$f(w_1 w) = f(w_1) \oplus f(w)$$

where $\oplus$ denotes addition modulo 2. Hence

$$w_1 E_f w_2 \Leftrightarrow f(w_1) \oplus f(w) = f(w_2) \oplus f(w) \text{ for all } w \text{ in } U^*$$
$$\Leftrightarrow f(w_1) = f(w_2)$$

Thus, there are two elements in U^*/E_f:

$$x_0 = \{w \lfloor f(w) = 0\}$$
$$x_1 = \{w \lfloor f(w) = 1\}$$

In particular, we have $x_0 = [\Lambda]$ and $x_1 = [1]$ so that we have the table

$$x_0 \cdot 0 = [\Lambda 0] = [0] = x_0 \text{ since } f(0) = 0$$
$$x_0 \cdot 1 = [\Lambda 1] = [1] = x_1 \text{ since } f(1) = 1$$
$$x_1 \cdot 0 = [10] = x_1 \qquad \text{ since } f(10) = 1$$
$$x_1 \cdot 1 = [11] = x_0 \qquad \text{ since } f(11) = 0$$

while the outputs are given by

$$\beta(x_0) = f(\Lambda) = 0$$
$$\beta(x_1) = f(1) = 1$$

We thus recapture the state-transition graph of Figure 8–3.   ◊

Note that Σ_f is reachable because we only consider states which can be labeled by the input sequence that takes us to the given state from the initial state; and that the system is observable because, whenever w_1 and w_2 lead us to states of a realization of f indistinguishable by the experiments specified by function f, we decree that they must lead us to the same state of Σ_f. Note that if we take any system and carry out experiments upon it starting from some initial state, then the only observations that we can obtain will bear upon states reachable from the specified state. If, by interfering with the innards of the machine, we were to obtain that which cannot be obtained by applying input sequences, namely to obtain a nonreachable part of the state space, then the new data obtained in that region of the state space would have to be used in a completely separate realization procedure. Once we have the data from the two realization procedures, we would then have to decide to what extent the two state spaces overlap (Figure 8–6).

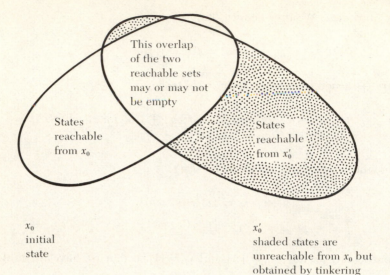

x_0
initial
state

x_0'
shaded states are
unreachable from x_0 but
obtained by tinkering

Figure 8–6

EXERCISES FOR SECTION 8–1

1.   Compute F, G, and H for the discrete-time system of Figure 8–1. (What is the state space?) Write down the formula for

$$A_k = HF^k G$$

Does the sequence $(A_0, A_1, \ldots, A_N)$ for some fixed N:

   (a)   determine the dimension n of F?

   (b)   determine F uniquely if n is known? (Be careful to write down explicitly all your assumptions in answering this part.)

2.   Prove that the relation E_f on U^*

$$w_1 E_f w_2 \Leftrightarrow fL_{w_1} = fL_{w_2}$$

is an equivalence relation on U^*; i.e., that it is reflexive ($wE_f w$), symmetric ($w_1 E_f w_2 \Rightarrow w_2 E_f w_1$), and transitive ($w_1 E_f w_2$ and $w_2 E_f w_3 \Rightarrow w_1 E_f w_3$).

3.   Describe the equivalence classes of E_f when f is $\mathcal{S}_{x_0}$ for the Σ and x_0 diagrammed in Figure 8–7.

4.   **(a)**   For any f, verify that the δ_f of Theorem 1 is well defined. In other words, prove for all w_1, w_2 in U^* and all u in U that

$$w_1 E_f w_2 \Rightarrow w_1 u E_f w_2 u$$

   (b)   Use the inductive definition

$$\delta_f^*([w], \Lambda) = [w]$$
$$\delta_f^*([w], w'u) = \delta_f(\delta_f^*([w], w'), u)$$

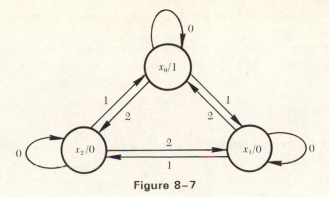

Figure 8–7

to verify, by induction on the length of w', that

$$\delta_f^*([w], w') = [ww'] \text{ for all } w, w' \text{ in } U^*$$

5.   Let x_0 be *any* state of *any* system Σ for which $\mathcal{S}_{x_0} = f$. Show that $\mathcal{S}_{x_0 w} = fL_w$ and thus is independent of the realization (Σ, x_0) of f. Conclude that the states $x_0 \cdot w_1$ and $x_0 \cdot w_2$ are indistinguishable iff $w_1 E_f w_2$.

8–2   Realization of Linear Systems

We have stressed that the realization procedure of Section 8–1 works without any use of linearity. But now the question arises: How do we use this procedure? The method by which we obtained the state space X_f was to take all possible input strings, of arbitrarily great length, and partition them into their equivalence classes under the relationship E_f—a partitioning which presumably requires us to make infinitely many tests on a pair of input sequences to tell whether or not they are indistinguishable! Clearly, for practical purposes, we must replace this infinite description of the state space by some finite algorithm for obtaining it. The purpose of this section is to exhibit finite algorithms for the linear case. Of course, as Example 8–1–1 shows, the finiteness will be of the kind "if you are satisfied with a system of dimension at most N, then D_N is the finite set of data that it will suffice to process (where D_N increases with N)."

In this section, we shall work with modules over a fixed unitary (not necessarily commutative) ring R. For the reader uninterested in the module-theoretic approach of Section 8–3, it will suffice to ignore the generalization—he may think of R as being a field such as **R** or **C**, and of an R-module as a vector space over the field R. The only point of interest to such a reader will be that nowhere through the proof of Corollary 3 will we use the fact that elements r of R have multiplicative inverses r^{-1}. However, the more adventurous reader will find this section a useful opportunity to sharpen his feel for modules, and should at least read Exercises 1 through 3 before proceeding further.

To prepare the way for the abstract theory of Section 8–3, we shall now adopt the following definition of a discrete-time constant linear system.

DEFINITION 1

A discrete-time constant linear system is a system

$$\Sigma = (U, X, Y, \delta, \beta)$$

where U, X, and Y are R-modules (for some fixed unitary ring R) and there exist linear maps $F: X \to X$, $G: U \to X$, and $H: X \to Y$ such that

$$\delta(x, u) = Fx + Gu$$

and

$$\beta(x) = Hx$$

for all x in X and u in U. We may also denote this system as $\Sigma = (X, F, U, G, Y, H)$. ○

Note that our definition does *not* require X to be finite-dimensional.

Just as in equation (1) of Section 8–1, the zero-state response $\mathcal{S}_0$ of Σ is given by the function

$$f: U^* \to Y \quad : \quad u_k \ldots u_1 u_0 \mapsto \sum_{\ell=0}^{k} HF^\ell Gu_\ell \tag{1}$$

Thus, the sequence of linear transformations $HF^\ell G$ for $\ell \geq 0$ determines f completely. Conversely, given f we may determine this sequence: If the input space U is m-dimensional with basis $e_1, \ldots, e_m$, and G has matrix $[g_1, \ldots, g_m] = [Ge_1, \ldots, Ge_m]$ with respect to this basis, then we may measure the matrix $HF^\ell G$ by applying each input sequence $e_j 0^\ell$ to Σ started in its zero state to determine its columns $f(e_j 0^\ell)$ for $1 \leq j \leq m$. Then

$$A_\ell = HF^\ell G = [HF^\ell g_1, \ldots, HF^\ell g_m] = [f(e_1 0^\ell), \ldots, f(e_m 0^\ell)] \tag{2}$$

Thus, the realization problem breaks down into two parts, one empirical and one mathematical. The empirical problem is to measure "sufficiently many" A_ℓ, guided by formula (2). The mathematical problem, for which we shall use the Realization Theorem for Discrete-Time Systems (Theorem 8–1–1) as our key, is to go from the A_ℓ to a set of minimal F, G, and H

such that $A_\ell = HF^\ell G$ for each ℓ—and, along the way, to specify what we mean by "sufficiently many."

Before we treat this discrete-time realization problem for linear systems—confident in the knowledge (Section 8–1) that our results will also be applicable to the continuous-time problem—let us say one more word about the empirical problem. No matter how carefully we measure a system, we can only measure finitely many parameters A_ℓ, and we can only measure these with limited accuracy. Thus, although the mathematical procedure talks of solving equations exactly, in reality we must always solve them approximately. There is a well-developed theory of statistical filtering and control which tells us how to use estimates of the type of noise which is corrupting our measurements to change our identification strategy. For the moment, it will be enough to note that we can do a moderately good job of identification by a cut and try method—for instance, one of the techniques we shall use below is adding a vector to a basis if it is not linearly dependent upon the vectors already in that basis. Due to inaccuracies of measurement, it will be almost certain that no vector is exactly linearly dependent upon preceding vectors—and so we must decide that, for instance, if a vector can be approximated to within 1% by a linear combination of vectors, then it is to be treated as linearly dependent upon those vectors. From now on we shall ignore this "noisy data" problem. Here we wish to provide a firm conceptual understanding of why certain linear realization algorithms work—the reader will find the journals stuffed with computational strategies for dealing with the static which abounds in reality.

Before proceeding further, we should see how to turn U^* into an R-module, for this will allow us to treat the f of equation (1) as a linear function, rather than as just a map, ensuring that the Nerode construction of Section 8–1 will yield linear systems from linear f's. In fact, our approach is more general. Given two R-modules, U and Y, and a sequence of linear transformations $(A_0, A_1, A_2, \ldots, A_\ell, \ldots)$ from U to Y, whether or not explicitly presented in the form (2), we may form the associated function

$$f: U^* \to Y \quad : \quad u_k \ldots u_0 \mapsto \sum_{\ell=0}^{k} A_\ell u_\ell \tag{3}$$

Let us now see that U^* can be changed slightly to yield an R-module $U^\S$ such that f becomes an R-*linear* map from $U^\S$ to Y:

We first observe from equation (3) that $f(w)$ is not changed if we preload w with any sequence of 0's. We are thus led to replace U^* by $U^\S$, the set of all left-infinite sequences of u's, such that only finitely many are non-zero. Using $\mathbf{N}$ to denote the set of non-negative integers, we thus set

$$U^\S = \{w: \mathbf{N} \to U \mid w(\ell) \neq 0 \text{ for only finitely many } \ell\} \tag{4}$$

and associate each function w with the left-infinite sequence $\ldots w(\ell) \ldots$ $w(1)w(0)$. We write supp(w) for the **support** of w, $\{\ell \in \mathbf{N} \mid w(\ell) \neq 0\}$. We may then modify (3) to obtain

$$f: U^\S \to Y \; : \; w \mapsto \sum_{\text{supp}(w)} A_\ell w(\ell) \tag{5}$$

with the definition of $U^\S$ assuring us that the sum in (5) has finitely many terms, and is thus well-defined. We turn $U^\S$ into an R-module simply by defining operations componentwise:

$$(w + w')(\ell) = w(\ell) + w'(\ell) \tag{6a}$$

$$(rw)(\ell) = r(w(\ell)) \tag{6b}$$

where we define the left-hand side operations in $U^\S$ by the already available right-hand side operations in U (Exercise 7).

Then, as we shall now show, f is indeed a linear map. Writing S for supp$(w) \cup$ supp(w') we have

$$f(rw + r'w') = \sum_{\text{supp}(rw+r'w')} A_\ell(rw + r'w')(\ell)$$

$$= \sum_S A_\ell(rw + r'w')(\ell) \qquad \text{since } (rw + r'w')(\ell) = 0$$
$$\text{for } \ell \in S \backslash \text{supp}(rw+r'w')$$

$$= \sum_S A_\ell(rw(\ell) + r'w'(\ell)) \qquad \text{by (6a) and (6b)}$$

$$= \sum_S [rA_\ell w(\ell) + r'A_\ell w'(\ell)] \qquad \text{since } A_\ell \text{ is linear}$$

$$= r\sum_S A_\ell w(\ell) + r'\sum_S A_\ell w'(\ell) \qquad \text{by Exercise 8}$$

$$= r\sum_{\text{supp}(w)} A_\ell w(\ell) + r'\sum_{\text{supp}(w')} A_\ell w'(\ell) \qquad \text{since we are only}$$
$$\text{deleting terms}$$
$$\text{with value 0}$$

$$= rf(w) + r'f(w') \qquad \text{as desired}$$

Thus, giving a sequence $(A_0, A_1, \ldots, A_k, \ldots)$ of linear transformations is equivalent to giving a linear transformation f from $U^\S$ to Y. Note that a given f need not have *any* finite-dimensional realization.

Let us, then, be given a linear response function

$$f: U^\S \to Y: w \mapsto \sum_{\text{supp}(w)} A_\ell w(\ell)$$

and apply to the corresponding $U^* \to Y: u_k \ldots u_1 u_0 \mapsto \sum_{\ell=0}^{k} A_\ell u_\ell$ the method summarized in the Realization Theorem for Discrete-time Systems (Theorem 8–1–1):

$$w_1 E_f w_2 \Leftrightarrow f(w_1 w) = f(w_2 w) \qquad \text{for all } w \text{ in } U^*$$
$$\Leftrightarrow f(w_1 0^{|w|}) = f(w_2 0^{|w|}) \text{ (see Exercise 5) for all } w \text{ in } U^*$$
$$\Leftrightarrow f(w_1 0^n) = f(w_2 0^n) \qquad \text{for all } n \in \mathbf{N}$$

In other words, if we define the map

$$\tilde{f}: U^\S \to Y^{\mathbf{N}}: w \mapsto (n \mapsto f(w0^n))$$

which assigns to each $w \in U^\S$ the infinite sequence

$$(f(w), f(w0), \ldots, f(w0^n), \ldots)$$

in $Y^{\mathbf{N}}$, we have the important result that

$$w_1 E_f w_2 \Leftrightarrow \tilde{f}(w_1) = \tilde{f}(w_2)$$

Now, it is obvious that the linearity of f implies the linearity of $\tilde{f}$, and so (Exercise 3) we have that

$$X_f \simeq U^\S / \ker \tilde{f}$$

is itself an R-module. Note that Λ in U^* corresponds to $0 = (\ldots, 0, \ldots, 0, 0)$ in $U^\S$, so that Σ_f will have state 0 corresponding to response function f.

It remains to analyze the form of the next-state (local transition) function δ_f and the output function β_f. But for any linear system with $\delta(x, u) = Fx + Gu$ and $\beta(x) = Hx$, we have

$$\delta(x, u) = (Fx + G0) + (F0 + Gu)$$
$$= \delta(x, 0) + \delta(0, u)$$

Hence, in the case of our Σ_f with state-space $X_f = U^\S / \ker \tilde{f}$ and linear maps F_f, G_f, and H_f, we have

$$\delta_f([w], u) = \delta_f([w], 0) + \delta_f([\Lambda], u)$$
$$= [w0] + [u]$$

where u in U becomes $(\ldots, 0, \ldots, 0, u)$ in $U^\S$, while, for w in $U^\S$, $w0$ is just

$(\ldots, w(2), w(1), w(0), 0)$. As always (even for nonlinear f), we have $\beta_f([w]) = f(w)$. Hence, we have the general result (irrespective of whether or not f has a finite-dimensional realization):

THEOREM 1 (REALIZATION THEOREM FOR LINEAR SYSTEMS)

Let $f: U^\S \to Y$ be an R-linear map which induces

$$\widetilde{f}: U^\S \to Y^{\mathbf{N}}: w \mapsto (f(w), f(w0), \ldots, f(w0^n), \ldots)$$

Then the pair $(\Sigma_f, 0)$ with

$$\Sigma_f = (X_f, F_f, U, G_f, Y, H_f)$$

where $X_f = U^\S/\ker \widetilde{f}$ with equivalence classes $[w]$, while

$$F_f([w]) = [w0]$$
$$G_f(u) = [u]$$
$$H_f([w]) = f(w)$$

is a minimal realization of f. $\qquad\qquad\qquad\square$

Note the beautiful fact that the *general* method of Section 8–1 yields a *linear* minimal (reachable and controllable) realization of a *linear f*. We shall analyze this "coincidence" more carefully in Section 8–4. The reader should think about it now.

Note that the appeal to Theorem 8–1–1 says that Σ_f is minimal in the sense that it is reachable and observable. It is, in fact, easy to deduce that Σ_f is minimal in the sense of having *least dimension*, where by "least dimension" we mean (for the justification, see Exercise 17) that if X is the state space of any other realization Σ of f, then there is a linear map from X_0, the R-module of states of Σ reachable from 0 [Exercise 16] *onto* X_f. Since any element of X_0 is $\delta^*(0, w)$ (where δ^* extends the next-state function $\delta: X \times U \to X$ of Σ) for some w in U^*, we define the desired map by

$$h: X_0 \to \Sigma_f: \delta^*(0, w) \mapsto [w]$$

To see that it is well-defined, we simply note that

$$\delta^*(0, w_1) = \delta^*(0, w_2) \Rightarrow \delta^*(\delta^*(0, w_1), w) = \qquad \delta^*(\delta^*(0, w_2), w)$$
$$\text{for all } w \text{ in } U^*$$
$$\Rightarrow \delta^*(0, w_1 w) = \delta^*(0, w_2 w) \text{ for all } w \text{ in } U^*$$

$$\Rightarrow \mathcal{S}_0(w_1 w) = \mathcal{S}_0(w_2 w) \quad \text{for all } w \text{ in } U^*, \text{ where}$$
$$\mathcal{S}_0 \text{ is the 0-response of } \Sigma$$
$$\Rightarrow f(w_1 w) = f(w_2 w) \quad \text{for all } w \text{ in } U^*, \text{ since } \Sigma$$
$$\text{is a realization of } f$$
$$\Rightarrow [w_1] = [w_2] \quad \text{as was to be shown}$$

It is obvious that h is onto, and it is straightforward to prove that when f is linear, h is also linear [Exercise 18].

The reader may wish to extend the above argument to conclude that all minimal realizations are isomorphic in the sense that there is a one-to-one onto response-preserving linear map from the state space of one to another.

While beautiful, Theorem 1 does not yield an *algorithm* for computing Σ_f. Following the discussion which introduced Section 8–1, we shall now impose the assumption that we have some bound N on the dimension of Σ_f. After we have used this to obtain an algorithm (Corollary 2), we shall study a partial realization algorithm (Corollary 3) which enables us to make successive approximations to Σ_f if no such bound N is known.

First we recall from Sections 4–3 and 4–5 that, if a linear system has dimension at most N, then

(a) every state reachable from 0 can be reached in at most N steps, and thus can be reached (Exercise 4) in exactly N steps; and
(b) every pair of distinguishable states may be distinguished by testing with sequences (probably depending on the pair of states) of length at most $N - 1$.

Thus, given a guarantee that Σ_f has dimension at most N:

$$w_1 E_f w_2 \Leftrightarrow f(w_1 w) = f(w_2 w) \quad \text{for all } w \text{ with } |w| \leq N - 1$$
$$\Leftrightarrow f(w_1 0^n) = f(w_2 0^n) \text{ for all } n \leq N - 1$$

Hence, two input sequences are equivalent under E_f iff $\langle w_1 \rangle = \langle w_2 \rangle$ where, for any w in U^*, we define the **full-state vector** $\langle w \rangle$ to be

$$\langle w \rangle \triangleq \begin{bmatrix} f(w) \\ f(w0) \\ \vdots \\ f(w0^{N-1}) \end{bmatrix}$$

Now it is clear (recalling comment (a) above) that we can reach every state of Σ_f with a w which is a linear combination of the $e_i 0^k$ where $[e_1, \ldots, e_m]$ is a basis for U, and $0 \leq k < N$, and hence it follows that every $\langle w \rangle$ is a linear combination of such $\langle e_i 0^k \rangle$ (Exercise 6).

Fix k. We then see (recalling equation (2)) that

$$[\langle e_1 0^k \rangle, \ldots, \langle e_m 0^k \rangle] = \begin{bmatrix} f(e_1 0^k) & \cdots & f(e_m 0^k) \\ f(e_1 0^{k+1}) & \cdots & f(e_m 0^{k+1}) \\ \vdots & & \vdots \\ f(e_1 0^{k+N-1}) & \cdots & f(e_m 0^{k+N-1}) \end{bmatrix}$$

$$= \begin{bmatrix} A_k \\ A_{k+1} \\ \vdots \\ A_{k+N-1} \end{bmatrix}$$

when we partition our matrix by rows instead of columns.

Combining the last two observations we conclude that, to determine a basis for the state space of Σ_f if it has dimension at most N, we need only use A_{k+j} for $0 \leq k < N$ and $0 \leq j < N$ in finding a basis for the columns of the following block-partitioned matrix:

$$\mathcal{H}_N = \begin{bmatrix} A_0 & A_1 & \cdots & A_{N-1} \\ A_1 & A_2 & \cdots & A_N \\ \vdots & \vdots & \cdots & \vdots \\ A_{N-1} & A_N & \cdots & A_{2N-2} \end{bmatrix}$$

called the (size N) **Hankel matrix** of the given response function f.

Thus, the problem of finding X_f is reduced to the algorithmic level, and Theorem 1 reduces to the following useful form:

COROLLARY 2 (REALIZATION ALGORITHM FOR LINEAR SYSTEMS)

Let $f: U^\S \to Y$ be an R-linear map $w \mapsto \sum_{\text{supp}(w)} A_\ell w(\ell)$, for which U has basis $e_1, \ldots, e_m$, and for which it is guaranteed that there exists a realization of dimension at most N. Then a minimal realization Σ_f may be constructed as follows:

Select a basis for the space spanned by the array of vectors

$$\mathcal{H}_N = [\langle e_1 \rangle, \ldots, \langle e_m \rangle, \langle e_1 0 \rangle, \ldots, \langle e_m 0 \rangle, \ldots, \langle e_1 0^{N-1} \rangle, \ldots, \langle e_m 0^{N-1} \rangle]$$

Say that this basis consists of the vectors $\langle w_1 \rangle, \ldots, \langle w_{n'} \rangle$ with $n' \leq N$, and let $[w]$ be the column vector whose entries are the coefficients of $\langle w \rangle$ with respect to this basis. Then, with respect to this basis, Σ_f is defined by the matrices

$$F_N = [[w_1 0], \ldots, [w_{n'} 0]]$$
$$G_N = [[e_1], \ldots, [e_m]]$$
$$H_N = [f(w_1), \ldots, f(w_{n'})]$$

Proof

Recall from Theorem 1 that

$$F_f([w]) = [w0]$$
$$G_f(u) = [u]$$
$$H_f([w]) = f(w)$$

Our task is now simply to specify the matrices F_N, G_N, and H_N which represent the linear transformations F_f, G_f, and H_f respectively. Recall that the matrix of a linear transformation $h : A \to B$ is $[h(a_1), \ldots, h(a_k)]$, whose jth column is the image $h(a_j)$ of the jth vector of the basis $(a_1, \ldots, a_k)$ for A with respect to which the matrix is being computed. We thus immediately deduce that

$$F_N = [F_f([w_1]), \ldots, F_f([w_{n'}])] = [[w_1 0], \ldots, [w_{n'} 0]]$$
$$G_N = [G_f(e_1), \ldots, G_f(e_m)] = [[e_1], \ldots, [e_m]]$$
$$H_N = [H_f([w_1]), \ldots, H_f([w_{n'}]) = [f(w_1), \ldots, f(w_{n'})]$$

Note that we use the assumption that f has a realization of dimension at most N to guarantee that the full state vectors $\langle w_1 0 \rangle, \ldots, \langle w_{n'} 0 \rangle$ are linearly dependent on $\langle w_1 \rangle, \ldots, \langle w_{n'} \rangle$, so that the expressions $[w_1 0], \ldots, [w_{n'} 0]$ are well-defined with respect to the basis. Clearly, to ensure that they are well-defined, it suffices to know that rank $\mathcal{H}_N$ = rank $\mathcal{H}_{N+1}$.   □

We reiterate that, although we have used a study of discrete-time systems to motivate our approach, the algorithm solves the purely mathematical problem of finding an observable pair (F, H) and a reachable pair (F, G) which solve the $2N - 1$ equations $A_l = HF^l G$, $0 \le l < 2N - 1$. Thus, although we use discrete-time theory to prove the validity of the algorithm, the resultant algorithm is available to solve equations of the form $A_l = HF^l G$ wherever they may arise—and, in particular, when they arise in the realization of a continuous-time process, where the parameters were obtained from a power series expansion of the transfer function.

We cannot emphasize too greatly the power of the mathematical approach which lets us use whatever intuition we have to obtain a solution to a problem—but then lets us carefully "purify" the statement of the problem into a mathematical form which we may apply to many cases quite different from that in which our intuition was available. For example, we saw that, when F is an $(n \times n)$ matrix, the condition that (F, H) be observable is that $\text{rank}[H^*, F^*H^*, \ldots, (F^*)^{n-1}H^*] = n$, whereas the condition that (F, G) be reachable is expressed by the equality $\text{rank}[G, FG, \ldots, F^{n-1}G] = n$. Thus, whereas from an intuitive point of view we may feel rather at a loss to use knowledge about reachability to answer questions about observability, we know by comparing the above two mathematical expressions that if we have to answer questions about observability of the pair (F, H), we may answer them simply by "pretending" that the question was one about *reachability* of the system $x(t + 1) = F^*x(t) + H^*u(t)$. Of course, no pretense is really involved. We simply state theorems in a clear and crisp mathematical form so that we may apply them whenever they are needed, irrespective of their ancestry.

Example 1

Suppose that we wish to compute the canonical realization of an input-output function

$$f: \mathbf{R}^\S \to \mathbf{R}$$

(i.e., both input and output are real scalars) when we are guaranteed that a three-dimensional realization exists. Identify e_1 and the state $[1]$ to which the system is sent from the zero state by applying a single unit input. Our crucial·matrix is then 3×3 and takes the form

$$\mathfrak{K}_3 = \begin{bmatrix} f(1) & f(10) & f(100) \\ f(10) & f(100) & f(1000) \\ f(100) & f(1000) & f(10000) \end{bmatrix} = [\langle 1 \rangle \langle 10 \rangle \langle 100 \rangle]$$

Let us see how our approach works with the specific values

$$\mathfrak{K}_3 = \begin{bmatrix} 1 & -1 & -7 \\ -1 & -7 & -9 \\ -7 & -9 & 17 \end{bmatrix}$$

We seek a basis for the column space of $\mathfrak{K}_3$—in fact, it is $\langle 1 \rangle$ and $\langle 10 \rangle$, since the third column is linearly dependent, with $\langle 100 \rangle = -5\langle 1 \rangle + 2\langle 10 \rangle$. (Thus, we get by with a two-dimensional realization, and do not need the full three-dimensional state space for which we were prepared.) Recalling that $[w]$ is the vector $\begin{bmatrix} x_1 \\ x_2 \end{bmatrix}$ such that $\langle w \rangle = x_1\langle 1 \rangle + x_2\langle 10 \rangle$, we immediately read from our general formulas that

$$G_3 = [[e_1]] = \begin{bmatrix} 1 \\ 0 \end{bmatrix}$$

$$F_3 = [[10], [100]] = \begin{bmatrix} 0 & -5 \\ 1 & 2 \end{bmatrix}$$

$$H_3 = [f(1), f(10)] = [1 \quad -1]$$

and the reader can easily verify that these do yield the specified values of f.   ◊

Example 2

We find a minimal realization for the impulse response sequence equal to the Fibonacci numbers:

$$\{h_i\} = \{1, 1, 2, 3, 5, 8, 13, \ldots\}$$

We know that the Fibonacci sequence is defined by

$$h_t = h_{t-1} + h_{t-2}$$

This suggests that we take as our state-variables the two most recent values of the sequence, and start at time 0 by plugging a 1 into the present-output component of the state:

$$x(t + 1) = \begin{bmatrix} 0 & 1 \\ 1 & 1 \end{bmatrix} x(t) + \begin{bmatrix} 0 \\ 1 \end{bmatrix} u(t)$$

$$y(t) = [0 \quad 1]x(t)$$

and this is indeed reachable and observable. [Moral: Don't plug in general algorithms when special knowledge allows a short cut.]   ◊

Example 3

We find a realization of the transfer function matrix

$$\mathfrak{Z}(s) = \begin{bmatrix} \dfrac{1}{s} & \dfrac{3s + 8}{s^2 + 4s} \\[2mm] \dfrac{1}{s} & \dfrac{2s + 4}{s^2 + 4s} \end{bmatrix}$$

$$He^{Ft}G = \mathcal{L}^{-1} \begin{bmatrix} \dfrac{1}{s} & \dfrac{2}{s} + \dfrac{1}{s+4} \\[2mm] \dfrac{1}{s} & \dfrac{1}{s} + \dfrac{1}{s+4} \end{bmatrix} = \begin{bmatrix} 1 & 2 + e^{-4t} \\ 1 & 1 + e^{-4t} \end{bmatrix}$$

Equating like powers of t in a power series expansion of $He^{Ft}G = \Sigma A_n \dfrac{t^n}{n!}$, we have $A_0 = HG = \begin{bmatrix} 1 & 3 \\ 1 & 2 \end{bmatrix}$, $A_{n+1} = \begin{bmatrix} 0 & (-4)^n \\ 0 & (-4)^n \end{bmatrix}$. Guessing that a three-dimensional realization will work, we form

$$\mathcal{H}_3 = \begin{bmatrix} 1 & 3 & 0 & -4 & 0 & 16 \\ 1 & 2 & 0 & -4 & 0 & 16 \\ \hline 0 & -4 & 0 & 16 & 0 & -64 \\ 0 & -4 & 0 & 16 & 0 & -64 \\ \hline 0 & 16 & 0 & -64 & 0 & 256 \\ 0 & 16 & 0 & -64 & 0 & 256 \end{bmatrix}$$
$$\langle e_1 \rangle \quad \langle e_2 \rangle \quad \langle e_1 0 \rangle \quad \langle e_2 0 \rangle \quad \langle e_1 00 \rangle \quad \langle e_2 00 \rangle$$

It is clear that a larger matrix will not add to the three vectors $\langle e_1 \rangle$, $\langle e_2 \rangle$, and $\langle e_2 0 \rangle$ we pick for basis. Note that $\langle e_2 00 \rangle = -4 \langle e_2 0 \rangle$.

Then
$$F_3 = [[e_1 0], [e_2 0], [e_2 00]] = \begin{bmatrix} 0 & 0 & 0 \\ 0 & 0 & 0 \\ 0 & 1 & -4 \end{bmatrix}$$

$$G_3 = [[e_1], [e_2]] = \begin{bmatrix} 1 & 0 \\ 0 & 1 \\ 0 & 0 \end{bmatrix}$$

$$H_3 = [f(e_1), f(e_2), f(e_2 0)] = \begin{bmatrix} 1 & 3 & -4 \\ 1 & 2 & -4 \end{bmatrix} \qquad \Diamond$$

Two observations about the Hankel matrix are in order: The first is that since

$$\mathcal{H}_N = \begin{bmatrix} HG & HFG & \cdots & HF^{N-1}G \\ HFG & HF^2G & \cdots & HF^NG \\ \vdots & \vdots & & \vdots \\ HF^{N-1}G & HF^NG & \cdots & HF^{2N-2}G \end{bmatrix}$$

it admits decomposition as the product

$$\mathcal{H}_N = \begin{bmatrix} H \\ HF \\ \vdots \\ HF^{N-1} \end{bmatrix} [G \quad FG \quad \dots \quad F^{N-1}G]$$

of the observability and reachability matrices.

Secondly, note that $\mathcal{H}_N$ does not (of course) change when we change basis—for if F changes to $F' = T^{-1}FT$, then H becomes $H' = HT$ and G yields $G' = T^{-1}G$, so that $A_\ell = HF^\ell G = H'F'^\ell G' = A'_\ell$ for each ℓ.

Any algorithm for finding a basis for

$$\mathcal{H}_N = \begin{bmatrix} A_0 & A_1 & \dots & A_{N-1} \\ A_1 & A_2 & \dots & A_N \\ \hdotsfor{4} \\ A_{N-1} & A_N & \dots & A_{2N-2} \end{bmatrix}$$

yields an algorithm for finding a "canonical (i.e., reachable and observable) realization" of f. For example, Corollary 2 immediately yields the validity of B. L. Ho's algorithm [Exercise 12].

To close this section, then, we give a basis selection algorithm which allows recursive computation of F, G, and H—an algorithm such that, if we have already computed F_N, G_N, and H_N, then relatively little extra work is required to compute F_{N+1}, G_{N+1} and H_{N+1}.

Before giving the algorithm, though, we require an important generalization of Corollary 2, in which we remove the condition that "it is guaranteed that there exists a realization of dimension at most N."

COROLLARY 3 (PARTIAL REALIZATION ALGORITHM FOR LINEAR SYSTEMS)

Let $A_0, A_1, \dots, A_{2N-3}, A_{2N-2}$ be a sequence of R-linear transformations from U to Y, where U has basis $(e_1, \dots, e_m)$. For $1 \leq j \leq N + 1$ we define the Hankel matrix

$$\mathcal{H}_j = \begin{bmatrix} A_0 & A_1 & \dots & A_{j-1} \\ A_1 & A_2 & \dots & A_j \\ \vdots & \vdots & & \vdots \\ A_{j-1} & A_j & \dots & A_{2j-2} \end{bmatrix} = [\langle e_1 \rangle_j, \dots, \langle e_m 0^{j-1} \rangle_j]$$

If $\mathcal{H}_N$ and $\mathcal{H}_{N+1}$ have the same rank, select a basis $\langle w_1 \rangle, \dots, \langle w_{n'} \rangle$ for the column space of $\mathcal{H}_N$ and let $[w]$ be the column vector whose entries are the coefficients of $\langle w \rangle$ with respect to this basis. Then if we define

$$G_N = [[e_1], \ldots, [e_m]]$$
$$F_N = [[w_1 0], \ldots, [w_{n'} 0]]$$
$$H_N = [f(w_1), \ldots, f(w_{n'})]$$

we have that (F_N, G_N, H_N) are matrices of minimal dimension such that

$$A_\ell = H_N F_N^\ell G_N \text{ for } 0 \leq \ell \leq 2N - 2$$

Proof

The condition that $\mathcal{H}_N$ and $\mathcal{H}_{N+1}$ have the same rank simply ensures that F_N is well-defined (compare the proof of Corollary 2), i.e., that each $\langle w_j 0 \rangle$ depends linearly upon the basis $\langle w_1 \rangle, \ldots, \langle w_{n'} \rangle$. Now, if we define the sequence $\bar{A}_\ell$ by

$$\bar{A}_\ell = H_N F_N^\ell G_N \text{ for } all \; \ell$$

it follows from Corollary 1 (and Exercise 13) that we obtain a minimal realization of this sequence; and it is clear that $\bar{A}_\ell = A_\ell$ for $0 \leq \ell \leq 2N - 2$.
□

We now return to an algorithm for forming partial realizations. We first form $\widehat{\mathcal{H}}_N$ from $\mathcal{H}_N$ by the following recursive procedure that ensures that $\widehat{\mathcal{H}}_j$ is a top left-hand block of $\widehat{\mathcal{H}}_k$ wherever $j < k$, and that the rows of $\widehat{\mathcal{H}}_N$ span the row space of (and are a subset of the rows of) $\mathcal{H}_N$:

Basis Step

To form $\widehat{\mathcal{H}}_1$, take in order the rows of $\mathcal{H}_1$ which are not linearly dependent upon preceding rows.

Induction Step

To form $\widehat{\mathcal{H}}_{N+1}$ from $\widehat{\mathcal{H}}_N$, let $\widehat{\mathcal{H}}_N$ have exactly $k(N)$ rows, with row t of $\widehat{\mathcal{H}}_N$ being row j_t^N of $\mathcal{H}_N$ for $1 \leq t \leq k(N)$. [Note that the sequence $(j_1^N, j_2^N, \ldots, j_{k(N)}^N)$ will *not* be monotone, in general.] Then let row t of $\widehat{\mathcal{H}}_{N+1}$ be row j_t^N of $\mathcal{H}_{N+1}$ so that $j_t^{N+1} = j_t^N$ for $1 \leq t \leq k(N)$. Having chosen row j_t^{N+1} for $t \geq k(N)$, let j_{t+1}^{N+1} index the first row of $\mathcal{H}_{N+1}$ that is not linearly dependent on rows j_1^{N+1} through j_t^{N+1} of $\mathfrak{IC}_{N+1}$ and take this row of $\mathfrak{IC}_{N+1}$ as the $(t+1)$th row of $\widehat{\mathfrak{IC}}_{N+1}$. Proceed in this way until a t is reached for which no such row exists. Then $k(N+1)$ is this t, and $\widehat{\mathfrak{IC}}_{N+1}$ has the $k(N+1)$ rows so specified.

It is clear that our realization procedure works just as well applied to $\widehat{\mathfrak{IC}}_N$ as to $\mathfrak{IC}_N$ (Exercise 13). We now present a method for finding a basis for the column space of $\widehat{\mathfrak{IC}}_N$ which is of the form

$$\begin{bmatrix} 1 \\ p_{21} \\ p_{31} \\ \vdots \\ p_{k(N),1} \end{bmatrix} \begin{bmatrix} 0 \\ 1 \\ p_{32} \\ \vdots \\ p_{k(N),2} \end{bmatrix} \begin{bmatrix} 0 \\ 0 \\ 1 \\ \vdots \\ p_{k(N),3} \end{bmatrix} \cdots \begin{bmatrix} 0 \\ 0 \\ 0 \\ \vdots \\ 1 \end{bmatrix}$$

and has the pleasant property that if we form the basis for the column space of $\widehat{\mathfrak{IC}}_M$ for any $M > N$, the first $k(N)$ components of the first N basis vectors are given by the above display. In fact, if we exhibit some factorization, say

$$\widehat{\mathfrak{IC}}_N = \begin{bmatrix} 1 & 0 & & 0 \\ p_{21} & 1 & & 0 \\ \vdots & \vdots & \cdots & \\ p_{k(N),1} & p_{k(N),2} & & 1 \end{bmatrix} \begin{bmatrix} q_{11} & q_{12} & \cdots & q_{1,Nm} \\ \cdots\cdots\cdots\cdots\cdots\cdots\cdots \\ q_{k(N),1} & q_{k(N),2} & \cdots & q_{k(N),Nm} \end{bmatrix}$$

$$= P_N Q_N$$

we shall arrange our algorithm so that neither the p's nor the q's already computed change as N increases—just as the h_{ij} element of $\mathfrak{IC}_N$ equals the h_{ij} element of $\mathfrak{IC}_M$ for any $M \geq N$. *The algorithm does use multiplicative inverses and so requires that R be a field.* It is a straightforward variation on the use of Gaussian elimination in decomposing a square matrix into triangular factors (see, e.g., Gantmakher 1959, vol. I, p. 33).

The procedure now follows. We present it for some fixed value of N, but ask the reader to convince himself that, if we have already computed the p's and q's for $\widehat{\mathfrak{IC}}_N$, we only need to compute p_{ij} for $k(N) < i \leq k(N+1)$ and $1 \leq j < i$, and q_{ij} for $Nm < j \leq (N+1)m$ when $1 \leq i \leq k(N)$, and for $1 \leq j \leq (N+1)m$ when $k(N) < i \leq k(N+1)$. In each step we compute a single row for each of P_N and Q_N. We assume rank $\widehat{\mathfrak{IC}}_N > 0$.

Basis Step

The first row of P_N is already given as $[1 \quad 0 \quad \cdots \quad 0]$, and so we take $q_{1j} = h_{1j}$ for $1 \leq j \leq Nm$.

Induction Step

Suppose that $k > 1$, and that we have already computed the first $(k - 1)$ rows of P_N and Q_N. For each $i < k$, let $s(i)$ be the least j such that $q_{i,s(i)} \neq 0$. Then set $q_{k,s(i)} = 0$ for $1 \leq i < k$. Setting $p_{kk} = 1$ and $p_{kl} = 0$ for $l > k$, we may then compute $p_{k1}, \ldots, p_{k,k-1}$ in turn from the equations:

$$h_{k,s(1)} = p_{k1}q_{1,s(1)}$$
$$h_{k,s(2)} = p_{k1}q_{1,s(2)} + p_{k2}q_{2,s(2)}$$
$$\vdots$$
$$h_{k,s(k-1)} = p_{k1}q_{1,s(k-1)} + \cdots + p_{k,k-1}q_{k-1,s(k-1)}$$

where we make use of the fact that we have forced $q_{l,s(i)} = 0$ for all $l > i$. Once we have formed the p_{kj}, we can then compute $q_{k1}, \ldots, q_{k,Nm}$ from the equations

$$h_{kl} = \sum_{j=1}^{k(N)} p_{kj}q_{jl} = \sum_{j=1}^{k-1} p_{kj}q_{jl} + q_{kl}$$

whence

$$q_{kl} = h_{kl} - \sum_{j=1}^{k-1} p_{kj}q_{jl}$$

It only remains to verify that these equations are consistent with our decree that $q_{k,s(i)} = 0$ for $1 \leq i < k$. But we so defined p_{k1} through p_{ki} that

$$h_{k,s(i)} = p_{k1}q_{1,s(i)} + \cdots + p_{ki}q_{t,s(i)}$$

and so the choice of $q_{k,s(i)}$ as 0 for $k > i$ does indeed satisfy the defining equation for the q_{kl}'s.

This application of this form of matrix decomposition to linear system realization is due to Risannen. Note that the Gaussian method can be applied to any matrix with linearly independent rows—we made no use of the fact that $\mathcal{H}$ was a Hankel matrix.

Example 4

We apply Risannen's decomposition method to the $\mathcal{H}$ of the form

$$\begin{bmatrix} 1 & 3 & 7 & 9 \\ 3 & 7 & 9 & 1 \\ 7 & 9 & 1 & 2 \end{bmatrix}$$

Step 1. Row 1 of P is [1   0   0] and Row 1 of Q is [1   3   7   9]. Thus, $s(1) = 1$.

Step 2.

$$h_{21} = p_{21}q_{11} \text{ implies } p_{21} = 3$$

$$q_{21} = q_{1,s(1)} = 0$$

$$q_{2j} = h_{2j} - p_{21}q_{1j} \text{ for } j > 1$$

Thus $q_{22} = 7 - 3 \cdot 3 = -2$; $q_{23} = -12$; $q_{24} = -26$, and we have as first two rows $\begin{bmatrix} 1 & 0 & 0 \\ 3 & 1 & 0 \end{bmatrix}$ for P, and $\begin{bmatrix} 1 & 3 & 7 & 9 \\ 0 & -2 & -12 & -26 \end{bmatrix}$ for Q. Thus, $s(2) = 2$.

Step 3.

$$h_{31} = p_{31}q_{11} \text{ implies } p_{31} = 7$$

$$h_{32} = p_{31}q_{12} + p_{32}q_{22} \text{ implies } p_{32} = (9 - 7 \cdot 3)/-2 = 6$$

$$q_{31} = q_{32} = 0$$

$$q_{3j} = h_{3j} - (p_{31}q_{1j} + p_{32}q_{2j}) \text{ for } j > 2$$

Thus $q_{33} = 1 - (7 \cdot 7 - 6 \cdot 12) = 24$; $q_{34} = 2 - (7 \cdot 9 - 6 \cdot 26) = 95$. This yields

$$P = \begin{bmatrix} 1 & 0 & 0 \\ 3 & 1 & 0 \\ 7 & 6 & 1 \end{bmatrix} \text{ and } Q = \begin{bmatrix} 1 & 3 & 7 & 9 \\ 0 & -2 & -12 & -26 \\ 0 & 0 & 24 & 95 \end{bmatrix} \qquad \Diamond$$

EXERCISES FOR SECTION 8-2

1.   Let R be a set equipped with two binary operations $(r_1, r_2) \mapsto r_1 + r_2$ and $(r_1, r_2) \mapsto r_1 \cdot r_2$. We say that the triple $(R, +, \cdot)$ is a **ring** iff it satisfies three conditions:

(i) R is an **abelian group** with respect to $+$. Thus, there exists an element 0 in R, and to each r in R there corresponds a $-r$ in R such that for all r_1, r_2, r_3 in R:

   (α) $r_1 + r_2 = r_2 + r_1$
   (β) $r_1 + (r_2 + r_3) = (r_1 + r_2) + r_3$
   (γ) $r_1 + 0 = r_1$
   (δ) $r_1 + (-r_1) = 0$

(ii) R is a **semigroup** with respect to $\cdot$; i.e., $r_1 \cdot (r_2 \cdot r_3) = (r_1 \cdot r_2) \cdot r_3$ for all r_1, r_2, r_3 in R.

(iii)   · distributes over $+$; i.e.,

$$r_1 \cdot (r_2 + r_3) = r_1 \cdot r_2 + r_1 \cdot r_3 \text{ and } (r_2 + r_3) \cdot r_1 = r_2 \cdot r_1 + r_3 \cdot r_1$$

holds for all r_1, r_2, r_3 in R.

We say that R is **unitary** if there is a 1 in R such that $1 \cdot r = r \cdot 1 = r$ for all r in R, and is **commutative** if $r_1 \cdot r_2 = r_2 \cdot r_1$ for all r_1, r_2 in R.

Then check the following:

(a)   Every field is a unitary commutative ring. In fact, $(K, +, \cdot)$ is a **field** iff $(K, +, \cdot)$ is a unitary commutative ring which satisfies the extra condition that · admits inverses on $K\backslash\{0\}$: to each $r \neq 0$ in K there corresponds an r^{-1} in K such that $r \cdot r^{-1} = 1$.

(b)   Verify that the sets **R, Q,** and **C** of real, rational, and complex numbers, respectively, are fields, and thus rings.

(c)   Verify that the integers **Z** form a unitary commutative ring.

(d)   Let 2**Z** be the set of all *even* integers, with the usual definitions of · and $+$. Verify that 2**Z** is a ring. Is it unitary? Is it commutative?

2.   Let R be a unitary ring, and let A be a set which is equipped with a binary operation $A \times A \to A : (a_1, a_2) \mapsto a_1 + a_2$, together with, for each r in R, a map $A \to A : a \mapsto r \cdot a$. Then we call A an R-**module** (short for **left R-module**) iff it satisfies the following conditions:

(i) A is an abelian group with respect to $+$.

(ii) $1 \cdot a = a$

(iii) $(r_1 r_2)a = r_1(r_2 a)$

(iv) $(r_1 + r_2)a = r_1 a + r_2 a$

(v) $r(a_1 + a_2) = ra_1 + ra_2$

for all a, a_1, a_2 in A and all r, r_1, r_2 in R. The operation $a \mapsto ra$ is called scalar multiplication by r, and we call elements of R **scalars.**

Then check the following:

(a)   If R is a field, then an R-module is precisely a vector space over R.

(b)   Prove that $0 \cdot a = 0$ for all a in A (where the 0 on the left is in R, and the 0 on the right is in A). [*Hint:* Consider $(0 + 0) \cdot a$]

(c)   Let A be a **Z**-module. Prove, by induction on n, using (iv) repeatedly, that

(α) $n \cdot a = a + a + \cdots + a$ (n terms) if $n > 0$

(β) $n \cdot a = (-a) + (-a) + \cdots + (-a)$ ($-n$ terms) if $n < 0$

Verify that $n \cdot a$, if it is defined by (α), (β) and $0 \cdot a = 0$ (compare part (b)) must satisfy (ii), (iii), (iv), and (v). Conclude that a **Z**-module is just an abelian group.

3.   Write down the definitions and basic properties of a basis, of linear transformations, and of matrices, verifying when they hold for any R-module and not just for vector spaces over some field F. In particular, if $h : A \to B$ is a linear transformation of the R-module A into the R-module B, we may define

$$h(A) \triangleq \{b \in B \mid b = h(a) \text{ for some } a \text{ in } A\}$$

$$\ker h \triangleq \{a \in A \mid h(a) = 0\}$$

$$A/\ker h \triangleq \{[a] \mid [a] \text{ is the equivalence class of } a \text{ in } A \text{ under}$$
$$\text{the relation } a_1 \equiv_h a_2 \Leftrightarrow a_1 - a_2 \in \ker h\}$$

(a)   Show that the definitions $[a_1] + [a_2] \triangleq [a_1 + a_2]$ and $r[a] \triangleq [ra]$ make $A/\ker h$ into an R-module which is unitary (commutative) if A is unitary (respectively commutative). [*Note:* First check that $[a] = [a'] \Rightarrow [ra] = [ra']$, and so forth.]

(b)   Prove that $h(A) \simeq A/\ker h$ by verifying that the map $h(A) \to$

$A/\ker h : h(a) \mapsto [a]$ is an **isomorphism** (one-to-one onto linear transformation) of R-modules. [*Note:* You must first check (it is easy!) that the formula really yields a well-defined map; that is, that $h(a_1) = h(a_2) \Rightarrow [a_1] = [a_2]$.]

4.   Prove that an x is of the form $\displaystyle\sum_{k=0}^{n} F^k Gu_k$ for some $n \le N$ iff x is of the form

$$\sum_{k=0}^{N} F^k Gu'_k$$

5.   Let $f(u_n \ldots u_0) = \displaystyle\sum_{k=0}^{n} A_k u_k$ for all $u_n \ldots u_0$ in U^*. Pick any m with $0 \le m \le n$

and observe that

$$\sum_{k=0}^{n} A_k u_k = \sum_{k=m+1}^{n} A_k u_k + \sum_{k=0}^{m} A_k u_k$$

$$= \sum_{k=0}^{n} A_k u'_k + \sum_{k=0}^{m} A_k u_k$$

where $u'_k = u_k$ if $m < k \le n$, while $u_k = 0$ if $0 \le k \le m$. Deduce that:
(a) $f(w_1 w) = f(w_1 0^{|w|}) + f(w)$ for all w_1, w in U^*, where $w_1 0^{|w|}$ is the sequence w_1 followed by an all-zero sequence of the same length as w.
(b) $f(w_1 w) = f(w_2 w) \Leftrightarrow f(w_1 0^{|w|}) = f(w_2 0^{|w|})$.

6.   Let $w = u_{N-1} \ldots u_0$ be any U-sequence of length N. Let $u_k = \displaystyle\sum_{i=1}^{m} \alpha_{ki} e_i$ be the

representation of u_k with respect to the basis $\{e_i\}$ for U. Then use the full-state vector formula

$$\langle w \rangle = \begin{bmatrix} f(w) \\ f(w0) \\ \vdots \\ f(w0^N) \end{bmatrix}$$

to deduce that

$$\langle w \rangle = \sum_{k=0}^{N-1} \sum_{i=1}^{m} \alpha_{ki} \langle e_i 0^k \rangle$$

7.   Let U be any R-module. Prove that the $(U^{\S}, +, u \mapsto ru)$ defined by equations (4) and (6) is also an R-module.
8.   Let A be an R-module, let r and r' be scalars, and let $a_1, \ldots, a_n, a'_1, \ldots, a'_n$ be elements of A. Use the module axioms to verify, by induction on n, that

$$\sum_{k=1}^{n} (ra_k + r'a'_k) = r \sum_{k=1}^{n} a_k + r' \sum_{k=1}^{n} a'_k$$

9. Find a two-dimensional realization for the $\mathbf{Z}$-linear map

$$f:\mathbf{Z}^\S \to \mathbf{Z}:w \mapsto \sum_{\mathrm{supp}(w)} (-k)w(k)$$

[*Hint:* Let H be of the form $[1 \quad 0]$.]

10. Find a realization for the transfer function of Example 4 by "visual inspection," drawing a signal graph with three integrators. Why do we need three integrators when s^2 is the highest power of s in the denominator of this $\mathfrak{B}(s)$? Verify that the realization is reachable and observable.

11. For the F, G and H of Example 3, compute the reachability and observability matrices and verify that their product is indeed $\mathfrak{K}$.

12. The following linear realization algorithm is due to B. L. Ho: Given a linear map $f:(\mathbf{R}^m)^\S \to \mathbf{R}^p$ for which a realization exists of dimension $\leq N$, proceed in three stages:

(a) Form the two $Np \times Nm$ matrices

$$\mathfrak{K}_N = \begin{bmatrix} A_0 & A_1 & \cdots & A_{N-1} \\ A_1 & A_2 & \cdots & A_N \\ \cdots\cdots\cdots\cdots\cdots\cdots\cdots\cdots \\ A_{N-1} & A_N & \cdots & A_{2N-2} \end{bmatrix}$$

$$\sigma\mathfrak{K}_N = \begin{bmatrix} A_1 & A_2 & \cdots & A_N \\ A_2 & A_3 & \cdots & A_{N+1} \\ \cdots\cdots\cdots\cdots\cdots\cdots\cdots\cdots \\ A_N & A_{N+1} & \cdots & A_{2N-1} \end{bmatrix}$$

(b) Find a non-singular $pN \times pN$ matrix P and a non-singular $mN \times mN$ matrix M over $\mathbf{R}$ such that

$$P\mathfrak{K}_N M = \begin{bmatrix} I_n^n & 0_{mN-n}^n \\ 0_n^{pN-n} & 0_{mN-n}^{pN-n} \end{bmatrix} = E_n^{pN} E_{mN}^n$$

where

$$E_n^m = \begin{cases} [I_m^m \quad 0_{n-m}^m] & \text{if } m < n \\ I_m^m & \text{if } m = n \\ \begin{bmatrix} I_n^n \\ 0_n^{m-n} \end{bmatrix} & \text{if } m > n \end{cases}$$

Here, I_n^n is the $n \times n$ unit matrix, and 0_n^m is the $m \times n$ zero matrix.

(c) Then

$$F = E_{pN}^n P[\sigma\mathfrak{K}_N]ME_n^{mN}$$
$$G = E_{pN}^n P\mathfrak{K}_N E_m^{mN}$$
$$H = E_{pN}^p \mathfrak{K}_N ME_n^{mN}$$

yields a canonical realization of f.

Deduce the validity of this algorithm from Corollary 2, showing that it simply involves explicating the choice of basis $\langle w_1 \rangle, \ldots, \langle w_{n'} \rangle$ in the matrices P and M.

13. Show that the validity of Corollaries 2 and 3 is not affected if we replace

$\mathcal{H}_N$ by any matrix $\widehat{\mathcal{H}}$ whose rows (respectively columns) are linear combinations of rows (columns) of $\mathcal{H}$, and which span the row space (column space) of $\mathcal{H}_N$.

14.   Apply Gaussian elimination to decompose

(a)

$$\mathcal{H}_2 = \begin{bmatrix} 1 & 0 & 2 & 1 \\ 0 & 1 & 3 & 2 \\ 2 & 1 & 2 & 4 \\ 3 & 2 & 4 & 0 \end{bmatrix}$$

(b)

$$\mathcal{H} = \begin{bmatrix} 1 & 2 & 3 & 4 \\ 2 & 3 & 1 & 0 \\ 3 & 5 & 0 & 4 \\ 3 & 5 & 4 & 4 \end{bmatrix}$$

15.   Assume that $\mathcal{H}_N$ has the same rank as $\mathcal{H}_{N+1}$. Then express a minimal partial realization (F_N, G_N, H_N) in terms of the $\widehat{\mathcal{H}}_N$, P_N, and Q_N of Risannen's decomposition, and $\sigma\widehat{\mathcal{H}}_N$, which is obtained from $\widehat{\mathcal{H}}_{N+1}$ by including only those rows which were used in forming $\widehat{\mathcal{H}}_N$ from $\mathcal{H}_N$. [*Hint:* Assume $Y = K^p$, and note that the first p rows of $\mathcal{H}_N$ are linear combinations of the first r rows of $\widehat{\mathcal{H}}_N$ for some $r \leq p$.]

16.   Let (X, F, U, G, Y, H) be a linear system. Verify that the set

$$X_0 = \left\{ \delta^*(0, u_k \ldots u_0) = \sum_{\ell=0}^{k} F^\ell G u_\ell \;\middle|\; u_k \ldots u_0 \in U^* \right\}$$

of states reachable from the zero state is a submodule of X; i.e., that $x_1, x_2 \in X_0$ and $r_1, r_2 \in R \Rightarrow r_1 x_1 + r_2 x_2 \in X_0$, so that X_0 is itself an R-module (where addition and scalar multiplication take place as in X).

17.   Let A, B, and C be finite-dimensional real vector spaces, with $A \simeq \mathbf{R}^a, B \simeq \mathbf{R}^b$, and $C \simeq \mathbf{R}^c$.

    (a)   Verify that if A is a subspace of B, then $a \leq b$.

    (b)   Verify that if $h: A \to C$ is a linear map of A *onto* C, then $c \leq a$.

18.   Verify that the map $h: X_0 \to \Sigma_f : \delta^*(0, w) \mapsto [w]$ of the discussion following Theorem 1 is linear and onto.

8–3   Modules and System Structure

We have seen that the "heart" of a linear system is the R-module X which represents its state-space, and the linear transformation F which represents the change of state induced by input 0. To complete the description, we add $G: U \to X$ which reads inputs into the state space, and $H: X \to Y$ to read outputs from the state space. Such a triple (F, G, H) yields the **zero-state extended response map**

$$\bar{f}: U^\S \to Y^\mathbf{N} \quad : \quad w \mapsto (f(w), f(w0), f(w0^2), \ldots) \tag{1}$$

where $f(\ldots u_k \ldots u_1 u_0) = \displaystyle\sum_k HF^k G u_k$.

Conversely, given any linear map $\bar{f}: U^{\S} \to Y^{\mathbf{N}}$ of the form (1) for some linear $f: U^{\S} \to Y$, we have seen in Theorem 8-2-1 how to recapture the "heart" of the minimal (reachable and observable) system which realizes it:

The canonical module for $\bar{f}$ is $X_f = U^{\S}/\ker \bar{f}$

The 0 input action is given by $F_f: X_f \to X_f: [w] \mapsto [w0]$, while $G_f: U \to X_f: u \mapsto [u]$ and $H_f: X_f \to Y: [w] \mapsto f(w)$.

While the study of linear systems (F, G, H) is a fairly recent development, algebraists have long been interested in providing structure theorems for linear transformations $F: X \to X$. In Chapter 6, we developed many structure theorems for real and complex matrices on a rather ad hoc basis, with the emphasis more on plausible manipulations of signal flow graphs than on rigorous algebraic theory. Our aim in this section, then, is to see how structure theorems for linear F have been formulated by algebraists as extremely powerful results on R-modules in general, and then see what these theorems tell us about the structure of our linear systems (F, G, H).

Given a unitary ring R, we may regard R itself as an R-module, where $r + r'$ and rr' are just as in R. [Check Exercises 1 and 2 of Section 8-2 to see that the definitions go through.] Hence we can define $R^{\S}$ to be

$$\{w: \mathbf{N} \to R \,|\, \mathrm{supp}(w) \text{ is finite}\}$$

and have it be the carrier of an R-module. Given an element

$$(\ldots u_k \ldots u_1 u_0)$$

of $R^{\S}$ we may represent it as a *formal* polynomial

$$\cdots + u_k z^k + \cdots + u_1 z + u_0 \tag{2}$$

where we only write down the terms with non-zero coefficients. We say that the polynomial is *formal* because we have not said that z represents a number which is to be substituted into the expression (2). Rather, z just "sits there" as an uninterpreted symbol or *indeterminate*. In our study of linear systems, we shall be most interested in the case in which z can be interpreted as a linear transformation.

We adopt the notation $R[z]$ for $R^{\S}$ viewed as a collection of polynomials (2). The importance of the form (2) is that it immediately suggests how to *multiply* elements of $R[z]$. The familiar formula

$$(a_2 z^2 + a_1 z + a_0)(b_2 z^2 + b_1 z + b_0)$$
$$= (a_2 b_2)z^4 + (a_2 b_1 + a_1 b_2)z^3 + (a_0 b_2 + a_1 b_1 + a_2 b_0)z^2$$
$$+ (a_0 b_1 + a_1 b_0)z + (a_0 b_0)$$

immediately extends to the general formula

$$\left(\sum_j a_j z^j \right) \left(\sum_l b_l z^l \right) = \sum_k \left(\sum_{j+l=k} a_j b_l \right) z^k \tag{3a}$$

$R[z]$ is thus equipped with addition

$$\left(\sum_j a_j z^j \right) + \left(\sum_j b_j z^j \right) = \sum_j (a_j + b_j) z^j \tag{3b}$$

as well as the multiplication (3a), which we call convolution. It is easy to check that $R[z]$ is a ring with respect to these operations (Exercise 3).

We remind the reader that an $R[z]$-module can be thought of as a vector space whose scalars are polynomials from $R[z]$—save that we must do without the assumption that non-zero scalars have multiplicative inverses. The reader of this section should feel comfortable with Exercises 1 to 3 of Section 8–2 before proceeding further. In any case, it is now natural to ask "What does an $R[z]$-module X look like?" Since each element of R appears in $R[z]$ as a constant polynomial, and since addition and multiplication in R are just addition and multiplication of constant polynomials by (3), we see that X must certainly be an R-module, with rx defined for every r in R and x in X. Again, since $z = 1 \cdot z$ is an element of $R[z]$, we must have zx defined for each x in X. By the module axioms (Exercise 8–2–2) we must have

$$\begin{aligned} z(r_1 x_1 + r_2 x_2) &= z(r_1 x_1) + z(r_2 x_2) \\ &= (r_1 z) x_1 + (r_2 z) x_2 \\ &= r_1 (z x_1) + r_2 (z x_2) \end{aligned}$$

on twice noting that the polynomial $r_j z$ is the product of the polynomial $z = 1 \cdot z$ and the constant polynomial r_j. Thus *the map* $x \mapsto zx$ *is a linear transformation of* X *considered as an* R*-module.* Let us call this R-linear map F. Then the R-module structure on X and the map F determine the $R[z]$-module structure on X, since a simple induction on the degree of the polynomial verifies that

$$\left(\sum_j a_j z^j \right) x = \sum_j a_j F^j x \tag{4}$$

where $\sum_j a_j z^j$ is in $R[z]$, x is in X, and the sum on the right-hand side is over elements in X. We thus have the following result:

THEOREM 1

Let R be a unitary ring. Then there is a bijection (one-to-one and onto map)

$$(R[z] \times X \to X) \mapsto (R \times X \to X, x \mapsto zx)$$

from the collection of all $R[z]$-modules to the collection of all R-modules equipped with an R-linear transformation. $\qquad\square$

From here through Theorem 6, we follow Kalman's $R[z]$ approach[†] to linear systems. Given the R-module X and the zero-input state-transition map F, we make X into an $R[z]$ module by defining zx to be Fx for all x in X. We then fix a basis $[e_1, \ldots, e_m]$ for U, which we thus take to be R^m.

Then if we represent each u_j in U by its vector $\begin{bmatrix} u_{j1} \\ u_{j2} \\ \vdots \\ u_{jm} \end{bmatrix}$, we have the bijection

$$w = (\ldots u_n \ldots u_1 u_0) \leftrightarrow \begin{pmatrix} \cdots + u_{n1}z^n + \cdots + u_{11}z + u_{01} \\ \cdots + u_{n2}z^n + \cdots + u_{12}z + u_{02} \\ \vdots \\ \cdots + u_{nm}z^n + \cdots + u_{1m}z + u_{0m} \end{pmatrix} = \begin{pmatrix} \pi_1^w(z) \\ \pi_2^w(z) \\ \vdots \\ \pi_m^w(z) \end{pmatrix}$$

between $U^{\S}$ and $R[z]^m$. If we now let the input map G have matrix $G = [Ge_1, \ldots, Ge_m] = [g_1, \ldots, g_m]$ with respect to this basis, we then have the formula

$$\sum_j F^j G u_j = \sum_j F^j \left(\sum_{k=1}^m g_k u_{jk} \right)$$

$$= \sum_{k=1}^m \left(\sum_j u_{jk} F^j \right) g_k$$

$$= \sum_{k=1}^m \pi_k^w(z) g_k$$

recalling that the action of F on g_k in X simply yields zg_k.

The output map is an R-**linear** map (i.e., a linear map of R-modules) $H: X \to Y$. However, we must *not* expect it to be $R[z]$-**linear** (i.e., a linear map of $R[z]$-modules), since Y is only an R-module, so that $R[z]$-linearity

---

[†] For further details, see Chapter 10 of R. E. Kalman, P. L. Falb, and M. A. Arbib, *Topics in Mathematical System Theory*, New York, McGraw-Hill [1969].

does not make sense here. [Note that if Y *were* an $R[z]$-module with $z \sim F'$, an R-linear $A: X \to Y$ would have to satisfy the extra condition $AF = F'A$ to be $R[z]$-linear.] We write $f(w)$ for $\sum_{k=1}^{m} H\pi_k^w(z)g_k$.

Thus, a linear system may be completely described by the display

$$(X, \{g_1, \ldots, g_m\}, H) \tag{5}$$

where X is an $R[z]$-module, the g's are elements of X, and $H: X \to Y$ is an R-linear map.

To fit our realization story into this picture, we note that just as $\mathbf{R}^m$ is a real vector space, so $R[z]^m$ is an $R[z]$-module under the obvious operation

$$w'(z) \cdot \begin{bmatrix} \pi_1^w(z) \\ \vdots \\ \pi_m^w(z) \end{bmatrix} = \begin{bmatrix} w'(z) \cdot \pi_1^w(z) \\ \vdots \\ w'(z) \cdot \pi_m^w(z) \end{bmatrix} \tag{6}$$

We should also note that $Y^\mathbf{N}$, which we already know to be an R-module under componentwise operations, is an $R[z]$-module if, just as in Section 4–6, we define z to be the **left shift**:

$$z: (y_0, y_1, y_2, \ldots, y_n, \ldots) \mapsto (y_1, y_2, y_3, \ldots, y_{n+1}, \ldots) \tag{7}$$

Note also that expression (2) implies that $z^n w$ is another way of writing $w0^n$.

The crucial observation, then, is that the map f of (1) becomes

$$\tilde{f}: K[z]^m \to Y^\mathbf{N} : w \mapsto (f(w), f(zw), f(z^2w), \ldots, f(z^nw), \ldots) \tag{8}$$

which is not only R-linear, but also $R[z]$-linear, for it is clear that

$$\tilde{f}(zw) = (f(zw), f(z^2w), f(z^3w), \ldots, f(z^{n+1}w), \ldots) = z\tilde{f}(w)$$

where the last equality follows by (7), which defines the action of z on $Y^\mathbf{N}$. Thus the image

$$X_f \cong K[z]^m / \ker \tilde{f}$$

is not only an R-module but also an $R[z]$-module (on applying Exercise 8–2–3 to $R[z]$-modules). We recapture F_f, then, simply as the action of z on X_f

$$F_f[w] = [zw] = z[w] \tag{9}$$

as before. We capture the columns of G_f as $[e_j]$, for $1 \le j \le m$, where $e_j = (0, \ldots, 1 \text{ in } j\text{th place}, \ldots, 0)^T$. Finally, H_f is the R-linear map which reads the first component $f(w)$ from $\tilde{f}(w)$.

We shall discover in Section 8–6 that this algebraically insightful recasting of the realization process is available in far more general settings than that of R-modules. However, before setting off on the path of greater generality in the next section, we devote the remainder of this section to a consideration of how the algebraist's structure theory for modules yields structure theorems for linear systems.

From now on, we shall assume that R is a *commutative* ring.

We first recall some basic definitions and three useful results from modern algebra (for proofs and a more leisurely exposition see Birkhoff and MacLane; or Bobrow and Arbib, Chapter 8).

We say that a subset I of R is an **ideal** if for each r_1, r_2 in I and r in R, we have that $r_1 + r_2$ and rr_1 are both in I. In other words, I is an R-module under the usual operations of R. We say that the ideal is **principal** if there exists some distinguished element $\hat{r}$ of I such that $I = R\hat{r} = \{\hat{rr} \,|\, r \in R\}$. We say that R is a **principal ideal domain** if every ideal of R is principal. The first useful result we state without proof is the following:

THEOREM 2

If R is a field, then $R[z]$ is a principal ideal domain.   □

We say that an R-module X is **finitely generated** if there exists a finite set $\{x_1, \ldots, x_n\}$ of elements of X such that every element of X can be expressed in the form $r_1 x_1 + \cdots + r_n x_n$ for some choice of elements $r_1, \ldots, r_n$ from R. The R-**dimension** of X is then the smallest such n.

We say that the R-module X is **cyclic** if it is generated by one element: $X = Rx$ for some x in X.

Given any element ψ of the R-module X, we may define a congruence (Exercise 1) on X by setting

$$x \equiv x' \bmod \psi \Leftrightarrow x - x' = r\psi \text{ for some } r \in R \tag{10}$$

and the set $X/\equiv$ [which we denote by $X/(\psi)$] inherits the R-module structure of X.

Note, in particular, that if X is cyclic, with $X = Rx$, then we may define a linear map

$$k : R \to X : r \mapsto rx$$

of R onto X. The kernel of k, $\ker k = \{r \in R \,|\, rx = 0\}$, is an ideal (Exercise 2). Thus, if R is a principal ideal domain, there exists ψ in R such that ker $k = (\psi)$, the ideal ψR generated by ψ, and it follows that (Exercise 3 of Section 8–2)

$$X = k(R) \cong R/\ker k \cong R/(\psi)$$

Conversely, it is clear that $R/(\psi) \cong R[1]$ is cyclic for any ψ in R.

   With this background, we can present a second useful result without proof:

THEOREM 3

Let R be a principal ideal domain, and let X be finitely generated as an R-module. Then X is isomorphic to a direct sum of cyclic R-modules

$$X \cong R/(\psi_1) \oplus \cdots \oplus R/(\psi_q) \oplus R \oplus \cdots \oplus R$$

where the number of terms in the direct sum does not exceed the number of generators of X, and where $\psi_{i+1} | \psi_i$ (i.e., $\psi_i = \psi_{i+1}\psi$ for some $\psi \in R$) for each i from 1 through $q - 1$. Moreover, the ψ_i are essentially unique in that if

$$X \cong R/(\psi_1') \oplus \cdots \oplus R/(\psi_{q'}') \oplus R \oplus \cdots \oplus R$$

with $\psi_i' | \psi_{i+1}'$ for $i = 1, \ldots, q'$, then $q = q'$; the number of copies of R is equal in both expressions, and each ψ_i differs from ψ_i' only by an invertible factor: $\psi_i = \psi_i'\psi$ where ψ has an inverse ψ^{-1} with $\psi\psi^{-1} = \psi^{-1}\psi = 1$. $\square$

   Let us see what becomes of this result when R is $K[z]$ for some field K (and so is certainly a principal ideal domain, by Theorem 2).

THEOREM 4

Let K be a field. Then every finitely generated $K[z]$-module X is the direct sum of cyclic modules

$$X \cong \sum_{j=1}^{m} K[z]/\psi_j$$

where each polynomial ψ_k divides ψ_j for $j \geq k$. Noting that X, and each $K[z]/\psi_i$, is also a K-module, and noting that $K[z]$ is infinite-dimensional as a K-module, we see that if X is finite-dimensional over K, then no ψ_k can be 0 (for then $K[z]/(\psi_k) = K[z]$) and we then have that the K-dimension of X is $\sum\limits_{j=1}^{m} \deg \psi_i$. □

To see what this means in concrete terms that relate back to Chapter 6, we must study in more detail the action of z on the $K[z]$-module $K[z]/\psi_i$:

Let the $K[z]$-module X have finite dimension n *as a K-module*. [Its dimension is thus *at most* n as a $K[z]$-module.] Pick any $x \neq 0$ in X, and consider the sequence

$$x, zx, \ldots, z^n x$$

Since dim $X = n$ as a K-module, there exists a set of coefficients $a_0, a_1, \ldots, a_n$, not all equal to 0, such that

$$a_0 x + a_1 zx + \cdots + a_n z^n x = 0$$

But this just says that $\psi^x(z)x = 0$, where $\psi^x(z)$ is the polynomial $(a_0 + a_1 z + \cdots + a_n z^n)$ in $K[z]$.

Let $x_1, \ldots, x_n$ be a basis for X as a K-module. Then it is clear that $\psi(z) = \psi^{x_1}(z) \ldots \psi^{x_n}(z)$ has the property that

$$\psi(z)x = 0 \text{ for all } x = \sum_{j=1}^{n} k_j x_j \text{ in } X$$

Hence the set

$$A_X = \{ \pi \in K[z] \,|\, \pi(z)x = 0 \text{ for all } x \text{ in } X \}$$

does not equal (0). Moreover, A_X is clearly an ideal of $K[z]$, which we call the **annihilating ideal** for X. Moreover, since $K[z]$ is a principal ideal domain, we deduce that

$$A_X = K[z]\psi \text{ for some suitable } \psi \in K[z]$$

It is clear from our general discussion that the following result holds:

LEMMA 5

If X is cyclic as a $K[z]$-module, with annihilating ideal $A_X = K[z]\psi$, then $X \cong K[z]/(\psi)$. Thus, the dimension of X as a K-module equals the degree of ψ. □

Note that a $K[z]$-cyclic module of K-dimension n may be thought of as the state-space of a one-input linear system, whose states are in bijective correspondence with the input sequences of length n which reach them from the 0 state.

Given any linear map $F: K^n \to K^n$, we define the $K[z]$-module X_F to be that with carrier (underlying set) K^n and z-action $x = Fx$. We say that F is **cyclic** if X_F is cyclic. Now, the **minimal polynomial** of F is, by definition, the polynomial of least degree for which $\psi_F(F) = 0$, and so it is clear that for any F we must have

$$A_{X_F} = K[z]\psi_F$$

Thus, it follows that X_F is cyclic iff $X_F \cong K[z]/\psi_F$, and so has K-dimension equal to $\deg \psi_F$. We may now prove the following:

LEMMA 6

F is cyclic iff the characteristic polynomial of F equals the minimal polynomial of F.

Proof

The characteristic polynomial of F is, of course,

$$\chi_F(z) = \det(zI - F)$$

and by the Cayley-Hamilton Theorem we have that $\chi_F(F) = 0$. Hence $\chi_F \in A_{X_F}$, and so ψ_F divides χ_F. But it is clear that K-dim $X_F = \deg \chi_F$. Thus, F is cyclic iff $\chi_F = \psi_F$.   □

Let then $F: K^n \to K^n$ be cyclic. Pick e_1 in K^n, and then set

$$e_j = F^{j-1}e_1 = z^{j-1}e_1 \text{ for } 1 < j \leq n$$

Since F is cyclic, the e_j form a basis. We then have

$$Fe_j = e_{j+1} \text{ for } 1 \leq j < n$$

while

$$Fe_n = z^n e_1$$
$$= [z^n - \chi_F(z)]e_1 \text{ since } \chi_F \in A_{X_F}$$
$$= [-a_1^{n-1} - \cdots - a_n]e_1$$

where $\chi_F(z) = z^n + a_1 z^{n-1} + \cdots + a_n$. Thus, with respect to this basis, F is in *companion form:*

$$\begin{bmatrix} 0 & 0 & \ldots & 0 & -a_n \\ 1 & 0 & \ldots & 0 & -a_{n-1} \\ 0 & 1 & \ldots & 0 & -a_{n-2} \\ \vdots & \vdots & \ldots & & \vdots \\ 0 & 0 & \ldots & 1 & -a_1 \end{bmatrix}$$

Thus, Theorem 3 simply says that every matrix F is similar to a matrix

$$\begin{bmatrix} A_1 & 0 & \ldots & 0 \\ 0 & A_2 & \ldots & 0 \\ & & \ldots & \\ 0 & 0 & \ldots & A_m \end{bmatrix}$$

where each matrix A_j is in companion form

$$A_j = \begin{bmatrix} 0 & 0 & \ldots & -a_{jn_j} \\ 1 & 0 & \ldots & -a_{j,n_j-1} \\ & & \ldots & \\ 0 & 0 & \ldots & -a_{j1} \end{bmatrix}$$

where $\psi_j(z) = z^{n_j} + a_{j1} z^{n_j-1} + \cdots + a_{jn_j}$.

We now return to general module theory, to cite one more useful theorem. For any R, we say that an R-module X is a **torsion** module if there exists a nontrivial annihilation element ψ for X—i.e., $\psi \neq 0$, while $\psi x = 0$ for all x in X. We have already seen that every $K[z]$-module which is finite-dimensional as a K-module must be a torsion module.

Given any unitary ring R, we call $r \in R$ a **unit** if it has an inverse, i.e., if there exists r^{-1} such that $rr^{-1} = r^{-1}r = 1$. Thus, every non-zero element of a field is a unit, while (and this motivates the name) 1 and -1 are the only units in **Z**. Then we say that a non-unit r' in R is a **prime** if, whenever we have $r' = r_1 r_2$ with r_1 and r_2 both in R, it must be the case that r_1 or r_2 is a unit. Again, this reduces to the normal usage in case $R = \mathbf{Z}$. Then, just as in **Z**, we have that every element of a principal ideal domain has a factorization as a finite product of powers of primes. We may now cite one more result:

THEOREM 7

Let X be a torsion module over a principal ideal domain R. Then if X has annihilating polynomial

$$\psi = \rho_1^{n_1} \ldots \rho_s^{n_s}$$

where each ρ_j is a prime, then

$$X \cong X_{\rho_1} \oplus \cdots \oplus X_{\rho_s}$$

where X_{ρ_j} contains all elements of X which are annihilated by some power of ρ_j.   □

In $K[z]$, a prime is called an **irreducible polynomial.** It is clear that the units of $K[z]$ are just the non-zero elements of K. Using Theorem 7 to further decompose each term $K[z]/(\psi_j)$—which has annihilating polynomial ψ_j—in Theorem 4, we immediately deduce the following:

THEOREM 8

Let K be a field, and let X be a $K[z]$-module of finite dimension over K, so that

$$X \cong \sum_{j=1}^{m} K[z]/\psi_j$$

where each ψ_k divides ψ_j for $j \geq k$, and no ψ_j is zero. Then M is the direct sum of submodules X_{θ_l}, where the θ_l are the irreducible factors of the ψ_i, and each X_{θ_l} consists of those elements of X which are annihilated by some power of θ_l.   □

When $K = \mathbf{C}$, the complex field, the only irreducible polynomials are (modulo a constant multiple) the linear functions $z - \lambda$. Since X is finite dimensional, it is certainly true that $X_{z-\lambda}$ is finite dimensional. Let us consider, then, a basis of the form

$$v_1, (z - \lambda)v_1, \ldots, (z - \lambda)^{m_1-1}v_1, \ldots, v_k, (z - \lambda)v_k, \ldots, (z - \lambda)^{m_k-1}v_k \quad \textbf{(11)}$$

where each v_j is chosen to be linearly independent of the preceding elements, while $(z - \lambda)^{m_j}v_j = 0$. Such an m_j exists, since we know that v_j is in $X_{z-\lambda}$. Then, as we shall spell out below, with respect to this basis, z has the matrix

$$\begin{bmatrix} J_{m_1}(\lambda) & \cdots & 0 \\ \vdots & \ddots & \vdots \\ 0 & \cdots & J_{m_k}(\lambda) \end{bmatrix}$$

so that Theorem 6 just tells us that every matrix may be placed in Jordan canonical form.

Let us conclude this section, then, by giving a somewhat more leisurely algebraic analysis of the Jordan canonical form. We shall start with a careful discussion of a $k \times k$ Jordan block

$$J_k(\lambda) = \begin{bmatrix} \lambda & 1 & 0 & \ldots & 0 & 0 \\ 0 & \lambda & 1 & \ldots & 0 & 0 \\ 0 & 0 & \lambda & \ldots & 0 & 0 \\ \cdots & \cdots & \cdots & \cdots & \cdots & \cdots \\ 0 & 0 & 0 & \ldots & \lambda & 1 \\ 0 & 0 & 0 & \ldots & 0 & \lambda \end{bmatrix}$$

Let $e_1, e_2, \ldots, e_k$ be the basis with respect to which the above matrix is obtained. The only eigenvector in this basis is e_1—and every eigenvector of $A = J_k(\lambda)$ is in fact a multiple of e_1. More explicitly, we may note that the action of A upon the basis vectors is given by the following equations:

$$Ae_1 = \lambda e_1$$
$$Ae_2 = e_1 + \lambda e_2$$
$$\vdots$$
$$Ae_k = e_{k-1} + \lambda e_k$$

and rearranging these equalities in terms of the operator $A - \lambda I$, which plays so important a role in our study of eigenvectors, we have

$$[A - \lambda I]e_1 = 0$$
$$[A - \lambda I]e_2 = e_1 \quad (\Rightarrow [A - \lambda I]^2 e_2 = 0)$$
$$\vdots$$
$$[A - \lambda I]e_k = e_{k-1}$$

Noting that all the e_j's are nonzero, forming part of a basis, we can rewrite these last equations in the following very elegant and compact form:

$$[A - \lambda I]^r e_j = \begin{cases} e_{j-r} & \text{if } r < j \\ 0 & \text{if } r \geq j \end{cases}$$

This expression suggests the following generalization of our notion of an eigenvector:

DEFINITION 1

A vector x is a **supereigenvector** of the linear transformation A of **index** j

(an integer ≥ 1) with respect to the eigenvalue λ if

$$(A - \lambda I)^r x = 0 \text{ iff } r \geq j \qquad \text{(12)}$$

Note immediately that if the expression (12) is to be true for x, then λ must certainly be an eigenvalue of A, since (12) implies that $(A - \lambda I)^{r-1}x$ is a nonzero solution of the equation $(A - \lambda I)[(A - \lambda I)^{r-1}x] = 0$—and this equation just tells us that $(A - \lambda I)^{r-1}x$ is an eigenvector of A. But if it is an eigenvector of A, then λ must be an eigenvalue. Note also that an eigenvector of A is simply a supereigenvector of index 1.

Thus we see that even though we cannot find a basis of eigenvectors of our linear transformation $J_k(\lambda)$, we can find a basis of supereigenvectors. In fact, we have done more than that, because the complete basis can be obtained from the single supereigenvector e_k of highest index among the k basis vectors by repeated application of the operator $J_k(\lambda) - \lambda I$, for in fact we have

$$e_j = [J_k(\lambda) - \lambda I]^{k-j}e_k \quad \text{for} \quad 1 \leq j \leq k$$

(where the 0th power yields the identity operator, for $j = k$). We formalize this observation in the following definition:

DEFINITION 2

Let x be a supereigenvector of A of index j with respect to the eigenvalue λ. Then the **span** of x is the set of j vectors.

$$x, (A - \lambda I)x, (A - \lambda I)^2 x, \ldots, (A - \lambda I)^{j-1}x$$

[Note that $(A - \lambda I)^i x$ is a supereigenvector of index $j - i$.]

We may now say that if a linear transformation can be put in the form of the $k \times k$ Jordan block $J_k(\lambda)$, then it has a supereigenvector of index k whose span provides a basis for the whole space.

Before proving the converse, we must check that the j vectors in the span of a supereigenvector of index j are linearly independent. Suppose, then, that we have x of index j and that we have j scalars $c_i (0 \leq i < j)$ such that

$$\sum_{i=0}^{j-1} c_i (A - \lambda I)^i x = 0$$

Multiplying this by $(A - \lambda I)^{j-1}$ we obtain

$$\sum_{i=0}^{j-1} c_i (A - \lambda I)^{j-1+i} x = 0$$

But $(A - \lambda I)^r x = 0$ iff $r \geq j$, and so we obtain

$$c_0 (A - \lambda I)^{j-1} x = 0$$

to deduce that $c_0 = 0$. The reader may check by induction, applying $(A - \lambda I)^r$ for decreasing values of r, that all the coefficients c_i are zero.

We have now proved most of the following theorem, which includes our observations upon the Jordan block, and also the converse of these observations:

THEOREM 9

A linear transformation may be placed in the form of a Jordan block $J_k(\lambda)$ iff it has characteristic polynomial $(z - \lambda)^k$ and a supereigenvector x of index k. The span of x yields the basis with respect to which our transformation has $J_k(\lambda)$ for matrix.

Proof

All that remains is to check that if indeed a linear transformation does have a supereigenvector x of index k, then the basis which we get from the span of x does place our transformation in the desired form. But in fact this is immediate—if we call our linear transformation A, we are told that a basis for our space is

$$e_1, e_2, \ldots, e_k$$

where $e_k = x$ and $e_j = (A - \lambda I)^{k-j} x$ for $1 \leq j < k$. But then the matrix of A with respect to this basis has for its jth column the column vector which represents $A e_j$ with respect to this basis. But $(A - \lambda I) e_j = e_{j-1}$ and so $A e_j = e_{j-1} + \lambda e_j$ for $1 < j \leq k$, whereas $(A - \lambda I) e_1 = (A - \lambda I)^k x = 0$ and so $A e_1 = \lambda e_1$. $\square$

We now state a generalization of the above theorem which can be proved by an easy application of the techniques we have already developed—its proof is thus left to the reader.

THEOREM 10

An n-dimensional linear transformation A may be placed in **Jordan canonical form**

$$
J = \begin{bmatrix}
J_{k_1}(\lambda_1) & 0 & \cdots & 0 \\
0 & J_{k_2}(\lambda_2) & \cdots & 0 \\
& \cdots & \ddots & \\
0 & 0 & \cdots & J_{k_i}(\lambda_i)
\end{bmatrix}
$$

(a block diagonal matrix, with Jordan blocks on the diagonal, zero blocks off the diagonal) if and only if A has i supereigenvectors $x_1, \ldots, x_i$ (with x_j having index k_j and eigenvalue λ_j) such that $n = \sum_{j=1}^{i} k_j$ and the n vectors in the span of the i supereigenvectors form a basis. It is with respect to this basis in the order

$$(A - \lambda_1 I)^{k_1 - 1} x_1, \ldots, (A - \lambda_1 I) x_1, x_1,$$
$$(A - \lambda_2 I)^{k_2 - 1} x_2, \ldots, x_2, \ldots, (A - \lambda_i I) x_i, x_i$$

that A has the Jordan form matrix J. □

  Finally, we show that given *any* linear transformation A, it can indeed be put into Jordan canonical form—and to do this we need simply note that Theorem 8, and the corollary equation (11), has shown us that we can go from any linear transformation to a basis consisting entirely of its super-eigenvectors. Let us spell this out in algorithmic form: We compute all the eigenvalues of the linear transformation, and then find as many linearly independent eigenvectors as we can. If these are not sufficient in number to give us a basis, then we look for supereigenvectors of index 2—but we do not look for supereigenvectors of an eigenvalue if the number of linearly independent eigenvectors already obtained equals the multiplicity of that eigenvalue. We continue in this way generating more and more supereigen-vectors of higher and higher index. Theorem 8 guarantees that this process will eventually terminate with a basis of n supereigenvectors. We may for-malize the above procedure in an algorithm, and the following theorem states that algorithm and guarantees that it will work.

THEOREM 11 (Algorithm for Jordan Canonical Form)

Any n-dimensional linear transformation A can be put into Jordan canonical form

$$J = \begin{bmatrix} J_{k_1}(\lambda_1) & 0 & \cdots & 0 \\ 0 & J_{k_2}(\lambda_2) & \cdots & 0 \\ \vdots & \vdots & & \vdots \\ 0 & 0 & \cdots & J_{k_i}(\lambda_i) \end{bmatrix}$$

by the following procedure:

(1) Determine the eigenvalues of A—say they are μ_1 with multiplicity n_1, μ_2 with multiplicity $n_2, \ldots, \mu_\ell$ with multiplicity n_ℓ.

(2) For each eigenvalue μ_j, determine a maximal set of linearly independent eigenvectors. Place them in the set SE_j of supereigenvectors of A with respect to the eigenvalue μ_j.

(3) (i) Set $r = 2$. Go to (ii).

   (ii) If the union of the sets SE_i contains n vectors, go to (4). If not, go to (iii).

   (iii) (a) Set $i = 1$. Go to (b).

   (b) Does $|SE_i| = n_i$? Yes: Go to (d). No: Go to (c).

   (c) Find a maximal set of linearly independent supereigenvectors of index r with respect to eigenvalue λ_i. Add these to SE_i. Go to (d).

   (d) Replace i by $i + 1$. If this new value exceeds ℓ, replace r by $r + 1$ and go to (ii). If not, go to (b).

(4) Select from each SE_i the supereigenvectors which are *maximal* in the sense that they are linearly independent of the vectors in the span of all the other supereigenvectors in SE_i.

(5) Express A with respect to the basis obtained by taking all the vectors in the spans of the maximal supereigenvectors produced in (4).

Let $\lambda_1, \ldots, \lambda_m$ be the distinct eigenvalues of A. Let

$$SE_{\lambda_1} \oplus SE_{\lambda_2} \oplus \cdots \oplus SE_{\lambda_m}$$

be the *sum* of the supereigenspaces SE_{λ_i} (i.e., the smallest subspace containing all the SE_{λ_i}'s). The sum is direct—i.e., each x in the sum can be uniquely written in the form $x_1 + x_2 + \cdots + x_m$ with each x_i in SE_{λ_i}; and, in fact, the sum equals the whole space X:

$$SE_{\lambda_1} \oplus SE_{\lambda_2} \oplus \cdots \oplus SE_{\lambda_m} = X$$

Proof

We have already seen that this follows from Theorem 6. However, since we stated that theorem without proof, a direct proof of the existence of an eigenvector basis may be of some interest:

First, to see that the sum is direct, suppose that the 0 vector has the representation

$$0 = u_1 + u_2 + \cdots + u_m$$

where for each i we have $(A - \lambda_i I)^{r_i} u_i = 0$ for some suitable integer $r_i \geq 1$. If at least one u_i, say u_1, is nonzero, we can in fact choose r_1 so that $(A - \lambda_1 I)^{r_1} u_1 = 0$ but $(A - \lambda_1 I)^{r_1 - 1} u_1 \neq 0$. Thus, replacing each u_i by $(A - \lambda_1 I)^{r_1 - 1} u_1 = v_1$, we have

$$0 = v_1 + v_2 + \cdots + v_m$$

with $v_1 \neq 0$. If we now form the product $p(\lambda) = (\lambda - \lambda_2)^{r_2} \ldots (\lambda - \lambda_m)^{r_m}$, we see that

$$0 = p(A) \cdot 0 = p(A) \cdot \Sigma v_1 = p(A) v_1$$

But since $A v_1 = \lambda v_1$, we see that $p(A) \cdot v_1 = (\lambda_1 - \lambda_2)^{r_2} \ldots (\lambda_1 - \lambda_m)^{r_m} v_1$ which is non-zero. Thus v_1 must be zero—after all—and we see that the sum is direct.

We now check that the sum of the SE_{λ_i} equals X:

Let E_{λ_i} comprise all the eigenvectors of a linear transformation A on X corresponding to an eigenvalue λ_i. Then E_{λ_i} and SE_{λ_i} are clearly subspaces of X:

$$(A - \lambda I)x = 0 \text{ and } (A - \lambda I)x' \Rightarrow (A - \lambda I)(kx + k'x') = 0$$

for all scalars k and k', while $(A - \lambda I)^r x = 0$ and $(A - \lambda I)^{r'} x'$, with $r \geq r'$, say, implies that

$$(A - \lambda I)^r (kx + k'x') = 0$$

for all scalars k and k'. We thus have $E_{\lambda_i} \subset SE_{\lambda_i} \subset X$.

We may factorize the characteristic polynomial of A as

$$\chi_A(\lambda) = (\lambda - \lambda_1)^{\alpha_1}(\lambda - \lambda_2)^{\alpha_2} \ldots (\lambda - \lambda_m)^{\alpha_m}$$

From this we obtain the polynomial $q_i(\lambda)$ by deleting the $(\lambda - \lambda_i)$ terms:

$$q_i(\lambda) = (\lambda - \lambda_1)^{\alpha_1} \ldots (\lambda - \lambda_{i-1})^{\alpha_{i-1}}(\lambda - \lambda_{i+1})^{\alpha_{i+1}} \ldots (\lambda - \lambda_m)^{\alpha_m}$$

Then the g.c.d. of $q_1, \ldots, q_m$ equals 1, and so there exist polynomials $t_1, \ldots, t_m$ such that

$$\Sigma t_i(\lambda) q_i(\lambda) = 1$$

and so

$$\Sigma t_i(A) q_i(A) = I$$

Hence any x can be written as

$$x = \sum_{i=1}^{m} t_i(A)q_i(A)x$$

and we simply verify that $u_i = t_i(A)q_i(A)x$ belongs to SE_{λ_i}. But this is immediate from the Cayley-Hamilton Theorem:

$$(A - \lambda_i I)^{\alpha_i}u_i = t_i(A)[(A - \lambda_i I)^{\alpha_i}q_i(A)]x$$
$$= t_i(A) \cdot \chi_A(A) \cdot x$$
$$= 0$$

Hence any x lies in $\text{SE}_{\lambda_1} \oplus \cdots \oplus \text{SE}_{\lambda_m}$.   □

EXERCISES FOR SECTION 8–3

1.   Given an R-module X, we say that an equivalence relation $\equiv$ on X is a **congruence** if for all r_1, r_2 in R, and all x_1, x_1', x_2, x_2' in X, we have

$$x_1 \equiv x_1' \text{ and } x_2 \equiv x_2' \Rightarrow r_1 x_1 + r_2 x_2 \equiv r_1 x_1' + r_2 x_2'$$

(a)   Verify that the $\equiv$ of equation (10) is a congruence.
(b)   Let $\equiv$ be any congruence on X. Verify that $X/\equiv$ becomes an R-module if we set $r[x] \triangleq [rx]$ and $[x_1] + [x_2] \triangleq [x_1 + x_2]$.
2.   Let $k: X_1 \rightarrow X_2$ be any linear map of R-modules. Verify that the **kernel** of k,

$$\ker k \triangleq \{x \in X_1 | f(x) = 0\}$$

is an ideal of X_1.
3.   Verify that $R[z]$ is a ring under the operations (2) and (3).

8–4   The Need for a More General Approach

In Section 8–1 we proved an important theorem that told us how to realize *any* (linear or non-linear) response function f which specifies how a discrete-time system, started in a fixed state, will respond to a sequence w in U^* of inputs with an output y in Y:

THEOREM 1

Given any $f: U^* \rightarrow Y$, define the equivalence of relation E_f on U^* by

$$w_1 E_f w_2 \Leftrightarrow fL_{w_1} = fL_{w_2}$$

and let $X_f = U^*/E_f$ be the set of equivalence classes of E_f. Then the machine

$$\Sigma_f = (U, X_f, Y, \delta_f, \beta_f)$$

with

$$\delta_f : X_f \times U \to X_f \quad : \quad ([w], u) \mapsto [wu]$$

and

$$\beta_f : X_f \to Y \quad : \quad [w] \mapsto f(w)$$

is reachable and observable, and realizes $f : (\mathcal{S}_f)_{[\Lambda]} = f$.   □

In Section 8–2, we explored the case where U and Y had the linear structure of R-modules. We then studied f's which were *linear* response functions in the sense that they could be factored in the form

$$U^* \to U^\S \to Y$$

where $U^* \to U^\S$ simply identifies $(u_k \ldots u_1 u_0)$ with the infinite sequence w for which

$$w(\ell) = \begin{cases} u_\ell & \text{if } 0 \le \ell \le k \\ 0 & \text{if } k \le \ell \end{cases}$$

by "preloading with zeros," and where $U^\S \to Y$ is an R-linear map from the R-module $U^\S$ [where $(rw + r'w')(\ell) = rw(\ell) + r'w'(\ell)$] to the R-module Y. We found that Theorem 1, when applied to such a linear f, yielded a Σ_f which was itself a linear system:

THEOREM 2

If U and Y are R-modules and $f : U^\S \to Y$ is linear, then the

$$\Sigma_f = (U, X_f, Y, \delta_f, \beta_f)$$

of f given by Theorem 1 is such that X_f can be given the structure of an R-module, and that δ_f and β_f are then R-linear maps; in fact,

$$\delta_f([w], u) = F_f[w] + G_f u$$
$$\beta_f([w]) = H_f[w]$$

where the maps $F_f : X_f \to X_f : [w] \mapsto [w0]$, $G_f : U \to X_f : u \mapsto [u]$, and $H_f : X_f \to Y : [w] \mapsto f(w)$ are all R-linear.   □

Before proceeding further, the reader should carefully review the proofs of Theorems 8–1–1 and 8–2–1 and fully understand that they do yield the slightly modified restatements of the theorems given above. Despite its abstraction, all that follows is essentially simple—but this simplicity may elude the reader who does not have a good feel for the Nerode approach to system realization.

If we take R to be $\mathbf{Z}$ [Exercise 8–2–2], then R-modules reduce to *abelian groups* [Exercise 1], and we refer to R-linear maps as (abelian group) *homomorphisms;* i.e., they are simply maps for which $h(a + b) = h(a) + h(b)$ for all a and b in the domain. This suggests a definition:

DEFINITION 1

An **abelian group machine** is a system

$$\Sigma = (U, X, Y, \delta, \beta)$$

for which U, X and Y are abelian groups, and there exist homomorphisms $F: X \to X$, $G: U \to X$, and $H: X \to Y$ for which

$$\delta(x, u) = Fx + Gu$$
$$\beta(x) = Hx$$

for all x in X and u in U.   ○

Again, U^* may be mapped onto the abelian group $U^\S$, in which addition is defined componentwise, by sending $(u_k, \ldots, u_0)$ to $(\ldots, 0, \ldots, 0, u_k, \ldots, u_0)$, and the 0-state response of Σ then factors as

$$U^* \to U^\S \to Y$$

where $U^\S \to Y: (\ldots u_k \ldots u_0) \mapsto \sum_{k=0}^{\infty} HF^k Gu_k$ is well-defined, since only finitely many u_k's are non-zero, and is clearly a homomorphism. Conversely, Theorem 2 tells us that any f which is additive in the sense of having such a factorization will have, via Theorem 1, a Σ_f which is an abelian group machine.

Our success with linear systems and with abelian group machines suggests a broad generalization:

We let $\mathcal{K}$ be any collection of sets with structure, called $\mathcal{K}$-objects (such as R-modules, or abelian groups), together with a collection of structure-preserving maps (such as R-linear maps and homomorphisms, respectively, in the above examples) called $\mathcal{K}$-morphisms. We then make the following definition:

DEFINITION 2

By a $\mathcal{K}$-**machine** we mean a system

$$\Sigma = (U, X, Y, \delta, \beta)$$

for which U, X, and Y, are $\mathcal{K}$-objects, with X having a distinguished binary operation $(x_1, x_2) \mapsto x_1 \cdot x_2$, and for which there exist $\mathcal{K}$-morphisms $F: X \to X$, $G: U \to X$, and $H: X \to Y$ which yield

$$\delta(x, u) = Fx \cdot Gu$$
$$\beta(x) = Hx$$

for all x in X and u in U. $\bigcirc$

This immediately suggests how we might generalize Theorem 2. However, it will turn out that the "obvious" generalization is false!

FALSE CONJECTURE

Let U and Y be $\mathcal{K}$-objects. Then there is a $\mathcal{K}$-object $\widehat{U}$ depending only on U, and a standard map $U^* \to \widehat{U}$ of U^* onto $\widehat{U}$ such that whenever $f: U^* \to Y$ has a factorization of the form

$$U^* \to \widehat{U} \to Y$$

with $\widehat{U} \to Y$ a $\mathcal{K}$-morphism, then the

$$\Sigma_f = (U, X_f, Y, \delta_f, \beta_f)$$

of f given by Theorem 1 is such that X_f can be given the structure of a $\mathcal{K}$-object, and that δ_f and β_f can be written as

$$\delta_f([w], u) = F_f[w] \cdot G_f u$$
$$\beta_f([w]) = H_f[w]$$

where the maps F_f, G_f, and H_f are all $\mathcal{K}$-morphisms. $\bigcirc$

AN INTRODUCTION TO CATEGORY THEORY

We shall see below that we can form a counter-example to this generalization by simply taking $\mathcal{K}$-objects to be groups. Before giving the examples, however, we give the general vocabulary of category theory which abstracts the crucial properties that a collection of $\mathcal{K}$-objects and $\mathcal{K}$-morphisms must satisfy in the general theory, which will lead us to a correct general realization theory in the next two sections.

DEFINITION 3

A **category** $\mathcal{K}$ consists of the following three things:

I. A class of objects.

II. For each ordered pair (A, B) of objects, there is a (possibly empty) set $\mathcal{K}(A, B)$, an element of which is referred to as a $\mathcal{K}$-**morphism** from A to B. [Although a $\mathcal{K}$-morphism is not necessarily a map, it is an abstraction of the concept of a map; so we write $f: A \to B$ or $A \xrightarrow{f} B$ to indicate that $f \in \mathcal{K}(A, B)$, i.e., that f is a $\mathcal{K}$-morphism.]

III. For each ordered triple (A, B, C) of objects, there is a map (in the ordinary set-theoretic sense)

$$\text{comp: } \mathcal{K}(A, B) \times \mathcal{K}(B, C) \to \mathcal{K}(A, C) : (A \xrightarrow{f} B, B \xrightarrow{g} C) \mapsto A \xrightarrow{gf} C$$

called **composition of $\mathcal{K}$-morphisms,** where:

(a) for all $\mathcal{K}$-morphisms $A \xrightarrow{f} B$, $B \xrightarrow{g} C$, and $C \xrightarrow{h} D$, we have the associative law $h(gf) = (hg)f$

(b) for each object A, there is a $\mathcal{K}$-morphism $id_A \in \mathcal{K}(A, A)$, called the identity morphism of A, such that for all $A \xrightarrow{f} B$, we have

$$A \xrightarrow{id_A} A \xrightarrow{f} B = A \xrightarrow{f} B = A \xrightarrow{f} B \xrightarrow{id_B} B$$

i.e.,

$$\text{comp} (A \xrightarrow{id_A} A, A \xrightarrow{f} B) = A \xrightarrow{f} B = \text{comp} (A \xrightarrow{f} B, B \xrightarrow{id_B} B)$$

or $f(id_A) = f = (id_B)f$.   ○

Example 1

The class of all sets is a category and is denoted by $\mathcal{S}$. The objects of the class are sets, and the $\mathcal{S}$-morphisms constituting $\mathcal{S}(A, B)$ are all the maps from A to B. Composition of $\mathcal{S}$-morphisms is given by the composition operation for maps, while for id_A we select the identity map $A \to A : a \mapsto a$. It is precisely for this reason (that the morphism id_A is an abstraction of the identity map) that we use the symbol id_A in the definition of a category as we do.   ◊

Example 2

The class of all monoids is **Mon,** the category of monoids, when **Mon** (A, B) is the collection of all monoid homomorphisms (Exercise 3) from A to B. Since composition of monoid homomorphisms yields a monoid homomorphism, and the identity map on a monoid is certainly a monoid homomorphism, we do indeed have another category. Similarly, groups with the same notion of homomorphism form a category **Gp.**   ◊

Example 3

R-Mod is the category whose objects are R-modules and whose **R-Mod**-morphisms are linear maps. This conclusion is based on the fact that when f and g are linear maps, then gf is; and the identity map is certainly linear.   ◊

Examples 1, 2, and 3 all conform to the intuition that a category is a collection of structured sets together with maps which preserve that structure. However, we shall now give an example (which may be omitted at a first reading) which shows that the concept of a category is much broader than this intuition.

Example 4

Let $\mathfrak{K}$ have for objects all of the integers; i.e., each object is a single element of the set $\mathbf{Z}$. For each m and n in $\mathbf{Z}$, we define the set $\mathfrak{K}(m, n)$ by

$$\mathfrak{K}(m, n) = \begin{cases} \{1_{mn}\} & \text{if } m \leq n \\ \varnothing & \text{if } m > n \end{cases}$$

Thus, there is at most one $\mathfrak{K}$-morphism for each ordered pair (m, n) of objects, and it is by no means a map. If $m \leq n$, the name of the $\mathfrak{K}$-morphism is 1_{mn}; while if $m > n$, there is no $\mathfrak{K}$-morphism for the pair (m, n).

In defining composition for $\mathfrak{K}$-morphisms for this $\mathfrak{K}$ as

$$\mathfrak{K}(m, n) \times \mathfrak{K}(n, p) \to \mathfrak{K}(m, p)$$

we need only spell out what happens when both $\mathfrak{K}(m, n)$ and $\mathfrak{K}(n, p)$ are nonempty. Thus, there is only one possible way to define composition, that being

$$1_{np} \cdot 1_{mn} = 1_{mp}$$

whenever $n \leq p$ and $m \leq n$, so that $m \leq p$ as required for 1_{mp} to exist. It is then clear that

$$(1_{pr}) \cdot (1_{np} \cdot 1_{mn}) = 1_{pr} \cdot 1_{mp} = 1_{mr}$$

and

$$(1_{pr} \cdot 1_{np}) \cdot (1_{mn}) = 1_{nr} \cdot 1_{mn} = 1_{mr}$$

so that associativity holds. Finally, by taking id_m to be 1_{mm}, we have that

$$1_{mn} \cdot 1_{mm} = 1_{mn} = 1_{nn} \cdot 1_{mn}$$

or

$$m \xrightarrow{1_{mm}} m \xrightarrow{1_{mn}} n = m \xrightarrow{1_{mn}} n = m \xrightarrow{1_{mn}} n \xrightarrow{1_{nn}} n$$

Hence, $\mathcal{K}$ is a category.   ◊

In much of system theory, we are concerned with systems that are unique up to "relabelling" of states. To make such a concept available in our general study of systems in a category, we must first abstract the notion of an isomorphism in any category.

We know (Exercise 6) that for two sets X and Y, the following two conditions are equivalent:

(i) There exists a map $f: X \to Y$ which is bijective.
(ii) There exist two maps $f: X \to Y$ and $g: Y \to X$ such that $gf = id_X$ and $fg = id_Y$; i.e., gf and fg are the identity maps on X and Y, respectively.

Furthermore, a necessary condition for two sets to be isomorphic is that there exists a bijection between them. Unfortunately, however, condition (i) above does not extend to an arbitrary category, since $\mathcal{K}$-morphisms are not necessarily maps. Thus, we take an approach suggested instead by condition (ii).

DEFINITION 4

In a category $\mathcal{K}$, we say that the $\mathcal{K}$-morphism $f: A \to B$ is an **isomorphism** whenever there exists a $\mathcal{K}$-morphism $g: B \to A$ such that the diagram

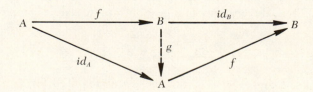

commutes. If there exists an isomorphism $f: A \rightarrow B$, then we say that A and B are **isomorphic** and write $A \cong B$.   ○

Reiterating, to say that a diagram *commutes* is to say that we get the same overall composition regardless of which route we take from a given place to a given destination. Thus, for the diagram of Definition 4, commutativity simply says that

$$gf = id_A \quad \text{and} \quad fg = id_B$$

(as well as $f(id_A) = (id_B)f$; however, for a category this is trivially true by definition).

Since $id_A: A \rightarrow A$ is an isomorphism (Exercise 8), it is now easy to check (Exercise 9) that $\cong$ is an equivalence relation on the objects of a category. We may think, heuristically, that $A \cong B$ if "A and B have the same 'abstract structure.'"

COPRODUCTS

We now want to see how an important construction in the category $\S$ of sets can be defined in the general language of category theory:

Example 5

Let us be given any family $\{K_\alpha \mid \alpha \in I\}$ of sets, one for each $\alpha \in I$. Then their disjoint union is obtained by relabelling elements so the sets have no overlap, and then taking the union:

$$\{a, b\} \cup \{c, b\} = \{a, b, c\}, \text{ordinary union}$$

while

$$\{a, b\} \amalg \{c, b\} = \{a, b_1, b_2, c\}, \text{disjoint union}$$

where we indicate that b_1 is the b from the first set, while b_2 is the b from the second set. More generally, we may define

$$C = \coprod_{\alpha \in I} K_\alpha = \{(x, \alpha) \mid x \in K_\alpha, \alpha \in I\}$$

where the map $K_\alpha \xrightarrow{in_\alpha} C$ injects each $x \in K_\alpha$ into the disjoint union by relabelling it as (x, α) so that it will not be confused with x in any other K_β, $\beta \neq \alpha$. The collection of maps $\{K_\alpha \xrightarrow{in_\alpha} C \mid \alpha \in I\}$ has an interesting property, namely that given any family f_α of maps from the K_α into an arbitrary fixed set X, there exists a unique map $C \xrightarrow{f} X$ such that each f_α factors as $f \cdot in_\alpha$—we see that for each α the diagram

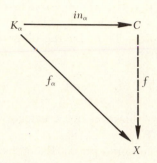

commutes; clearly, that map must be given by

$$f(x, \alpha) = f_\alpha(x) \qquad\qquad \Diamond$$

We seize upon the diagram as the basis for the general definition of a *coproduct*. In fact, the whole strategy of category theory is to show that many results can be obtained in great generality if we "chase arrows around diagrams" rather than by using element-by-element computations:

DEFINITION 5

Let $\{K_\alpha \mid \alpha \in I\}$ be a family of objects in the category $\mathcal{K}$. A **coproduct** of $\{K_\alpha\}$ is an object C together with a family of $\mathcal{K}$-morphisms $\{K_\alpha \xrightarrow{in_\alpha} C\}$ with the property that for any family $\{K_\alpha \xrightarrow{f_\alpha} X\}$ of $\mathcal{K}$-morphisms there exists exactly one f for which the following diagram commutes:

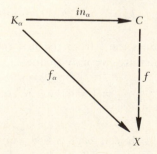

We write $C = \coprod_{\alpha \in I} K_\alpha$. $\qquad\qquad \bigcirc$

Thus, coproducts exist for *any* family in S, namely as disjoint unions. However, note that we wrote above

$$\{b, c\} \amalg \{a, b\} = \{c, b_1, b_2, a\}$$

while our general formula yields

$$\{b, c\} \amalg \{a, b\} = \{(a, 2), (b, 1), (b, 2), (c, 1)\}$$

This suggests that, at least in S, the disjoint union is unique up to relabelling of the elements. In fact, a corresponding result holds in any category.

We shall now prove briefly that the coproduct C is unique up to isomorphism. A more expository proof of this style is given for Theorem 1 below.

Suppose, then, that $K_\alpha \xrightarrow{f_\alpha} X$ was also a coproduct. We could then define $g: X \longrightarrow C$ by the diagram

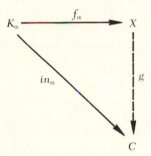

We infer from the diagrams obtained by splicing together the last two diagrams in two different ways

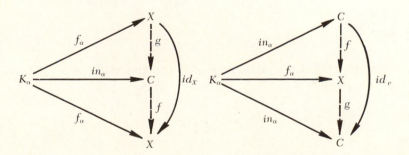

and from the uniqueness condition in the definition of coproduct that

$$fg = id_X \qquad \text{and} \qquad gf = id_C$$

so that f and g are isomorphisms, and $f_\alpha = f \cdot in_\alpha$ while $in_\alpha = g \cdot f_\alpha$.

In fact, most categories of interest to us have coproducts:

Example 6

In the category **R-Mod** of modules over the ring R, the coproduct of a family $\{K_\alpha | \alpha \in I\}$ of R-modules is the direct sum

$$C = \coprod_{\alpha \in I} K_\alpha = \bigoplus_{\alpha \in I} K_\alpha = \{(x_\alpha) | x_\alpha \neq 0 \text{ for only finitely many } \alpha\}$$

with $K_\beta \xrightarrow{in_\beta} C : x_\beta \mapsto (\delta_{\alpha\beta} x_\beta)$. This is because, given any family of R-linear maps $K_\alpha \xrightarrow{f_\alpha} X$, we may define $C \xrightarrow{f} X$ uniquely by $(x_\alpha) \mapsto \sum_{\substack{\alpha \in I \\ x_\alpha \neq 0}} f_\alpha(x_\alpha)$ to

ensure the commutativity of

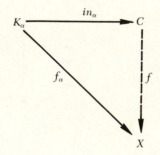

$\diamond$

Example 7

In the category **Gp** of groups, the coproduct of a family $\{K_\alpha | \alpha \in I\}$ of groups is the "free product"

$$C = \coprod_{\alpha \in I} K_\alpha$$

defined by the following rather elaborate construction:

1. The elements of C consist of all finite sequences (including the empty one, Λ) of elements of the form

$$(\kappa, \alpha) \text{ for which } \alpha \in I \text{ and } \kappa \in K_\alpha$$

subject to the conditions:

(i) No (κ, α) in the string has $\kappa =$ identity e_α of K_α
(ii) No string $(\kappa_1, \alpha_1)(\kappa_2, \alpha_2) \ldots (\kappa_n, \alpha_n)$ has $\alpha_j = \alpha_{j+1}$ for any j, $1 \leq j < n$.

2. Multiplication in C is obtained by concatenation of sequences, followed by application of the following operations:

(iii) Replace consecutive elements of the form $(\kappa, \alpha)(\kappa', \alpha)$ (for the same α)
    by the single element $(\kappa\kappa', \alpha)$, using multiplication in K_α.
(iv) Delete elements of the form (e_α, α)
until obtaining a sequence (possibly empty) which satisfies (i) and (ii).

It is clear that Λ is the identity, and that $(\kappa_1, \alpha_1) \ldots (\kappa_n, \alpha_n)$ has inverse $(\kappa_n^{-1}, \alpha_n) \ldots (\kappa_1^{-1}, \alpha_1)$. One simply checks associativity to confirm that $\amalg K_\alpha$ is indeed a group.

Then, given any family of homomorphisms $K_\alpha \xrightarrow{f_\alpha} X$, we may define $C \xrightarrow{f} X$ uniquely by

$$f[(\kappa_1, \alpha_1) \ldots (\kappa_n, \alpha_n)] = f_{\alpha_1}(\kappa_1) \ldots f_{\alpha_n}(\kappa_n)$$

to ensure that

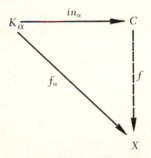

commutes.   ◇

With these three examples before him, the reader should have little trouble in following our derivation of the following elementary properties of coproducts in *any* category:

If we consider $\{K_\alpha | \alpha \in I\}$ for I the empty set $\varnothing$, the condition for C to be their coproduct is simply that for *any* X in $\mathcal{K}$, there is a *unique* map $C \xrightarrow{f} X$. Such a C is called **initial**. For $\mathcal{K} = \mathcal{S}$, we have $C = \varnothing$; for $\mathcal{K} =$ **R-mod**, it is the one-element module $\{0\}$; and for $\mathcal{K} =$ **Gp**, it is the one-element group $\{e\}$.

Where no ambiguity can result, we may write $\varnothing$ for the initial object in any category ($\varnothing$ is thus unique up to isomorphism *if it exists*). We shall often write $K + L$ for the coproduct of two objects K and L. It can easily be checked that

1. $K + \varnothing \cong K$
2. $K + L \cong L + K$
3. $(K + L) + M \cong K + (L + M) \cong K + L + M$.

GROUP MACHINES

With these concepts, we are now in a position not only to show the falsity of our conjecture of p. 675, but also to analyze wherein the conjecture fails:

Specializing Definition 2 to $\mathcal{K} = \mathbf{Gp}$, a **group machine** is one for which U, X, and Y are groups, and

$$\delta(x, u) = Fx \cdot Gu; \quad \beta(x) = Hx$$

for suitable homomorphisms $F : X \to X$, $G : U \to X$, and $H : X \to Y$.

Mimicking the construction of $U^\S$ for R-modules, we may form a group $\widehat{U}$ by setting

$$\widehat{U} = \{w : \mathbf{Z} \to U \,|\, \{\ell \,|\, w(\ell) \neq e\} \text{ is finite}\}$$

and defining multiplication by $(w_1 \cdot w_2)(\ell) = w_1(\ell) \cdot w_2(\ell)$, and then defining

$$U^* \to \widehat{U} : (u_k, \ldots, u_0) \mapsto (\ldots, e, \ldots, e, u_k, \ldots, u_0)$$

It is clear that every identity-state response

$$(u_k, \ldots, u_0) \mapsto HF^k Gu_k \cdot \cdots \cdot HGu_0$$

factors through $U^* \to \widehat{U}$. However, it turns out that even in the simplest case the corresponding factor $\widehat{U} \to Y$ is not a homomorphism:

Example 8

Let $X = U = Y$ be any nonabelian group. Let F, G, and H be the identity maps. The identity-state response function of the resultant group machine is then given by

$$f(u_k \ldots u_0) = u_k \ldots u_0$$

However, this does *not* induce a homomorphism $\widehat{f} : \widehat{U} \to Y$, for, even under componentwise multiplication,

$$w = (\ldots, e, \ldots, e, u_1, u_0) = w_1 w_2 = w_2 w_1$$

where

$$w_1 = (\ldots, e, \ldots, e, u_0) \text{ and } w_2 = (\ldots, e, \ldots, e, u_1, e)$$

But if f induced a homomorphism, we would have

$$f(w) = f(w_1)f(w_2) = u_0 u_1$$

and

$$f(w) = f(w_2)f(w_1) = u_1 u_0$$

but this could only hold for arbitrary u_0 and u_1 in U if U were abelian, which it is not.   ◊

Thus, the obvious candidate for $\widehat{U}$ fails. However, the situation is even worse, for the following example shows that if we restrict the state space of a group machine to contain only states reachable from the identity, then the resulting space may only be a *subset*, and not a subgroup, of the original group. In other words, irrespective of the choice of $\widehat{U}$, we cannot expect the Σ_f of Theorem 1 to be a group if we apply the Nerode approach directly to the f of a group machine.

Example 9

Let $\mathbf{D}_4$ be the so-called **dihedral group** whose set of elements is

$$\{e, y, x, xy, x^2, x^2 y, x^3, x^3 y\}$$

where e is the identity, and where multiplication is computed by repeated application of the cancellation rules

$$x^4 = y^2 = e \text{ and } xyx = y$$

For example, we have that

$$xy \cdot xy = (xyx)y = y^2 = e$$

and

$$y \cdot x = e \cdot y \cdot x = x^4 \cdot yx = x^3 \cdot xyx = x^3 y$$

Now consider the group machine with

$$U = Y = \mathbf{Z}_2 \text{ (i.e., } \{0, 1\} \text{ under addition modulo 2)}$$
$$X = \mathbf{D}_4$$

while F, G, and H are given by

$$Fx = e, Fy = xy \text{ (so that } Fe = e, Fxy = e \cdot xy = xy, \text{ etc.)}$$
$$G(0) = e, G(1) = y$$
$$H(x) = 0, H(y) = 1$$

Then the only states reachable from e are (Exercise 12) those of

$$\mathfrak{R} = \{e, y, x, xy\}$$

which is not a subgroup of X, since $yx = x^3 y \notin \mathfrak{R}$.   ◊

In our false conjecture, we have U^* mapping onto $\widehat{U}$, so that $\widehat{U}$ was in some sense *smaller* than U^*. But Example 9 suggests that, for group machines, U^* itself is too small. The question immediately arises: Is there some way to mimic the R-module construction of $U^\S$ from U which, in the category **Gp,** yields a structure which is *larger* than U^*? The answer is yes—the reader should ponder Example 6 to find the answer (Exercise 13). If inspiration fails, the answer, and a correct realization theory based on that answer, will be found in Section 8-6.

FUNCTORS AND ADJOINTS

To close this section, we must present some more general concepts from category theory. We first introduce a functor as a structure-preserving map from one category to another.

DEFINITION 6

A **functor** from category $\mathfrak{A}_1$ to category $\mathfrak{A}_2$ is a function H defined for objects

$$H: \mathfrak{A}_1 \to \mathfrak{A}_2 : A \mapsto AH$$

and for morphisms

$$H: \mathfrak{A}_1(A, B) \to \mathfrak{A}_2(AH, BH) : (A \xrightarrow{f} B) \mapsto (AH \xrightarrow{fH} BH)$$

such that
$$(fg)H = (fH)(gH) \text{ for all composable } f \text{ and } g \text{ in } \mathfrak{A}_1$$
and
$$id_A H = id_{AH} \text{ for all } A \text{ in } \mathfrak{A}_1$$

We denote a functor simply by either $H: \mathfrak{A}_1 \to \mathfrak{A}_2$ or $\mathfrak{A}_1 \xrightarrow{H} \mathfrak{A}_2$.   ○

Example 10

Suppose that $\mathcal{Q}_1 = \mathcal{Q}_2 = \mathcal{Q}$ and that the function H is given by

$$H:\mathcal{Q} \to \mathcal{Q}:A \mapsto AH = A$$

and

$$H:\mathcal{Q}(A, B) \to \mathcal{Q}(A, B):(A \xrightarrow{f} B) \mapsto (A \xrightarrow{fH = f} B)$$

Since

$$(fg)H = fg = (fH)(gH) \text{ for all composable } f \text{ and } g \text{ in } \mathcal{Q}$$

and

$$id_A H = id_A = id_{AH} \text{ for all } A \text{ in } \mathcal{Q}$$

then this $H:\mathcal{Q} \to \mathcal{Q}$ is a functor, called the **identity functor,** which is denoted by $id_{\mathcal{Q}}$. ◊

Example 11

From Exercise 8–2–2, we know that for a ring R, an R-module is an abelian group $(A, +)$ together with a function $\lambda:R \times A \to A:(r, x) \mapsto rx$ subject to suitable axioms. Thus, let us denote an R-module by the triple $(A, +, \lambda)$.

Consider the function H defined by

$$H:\textbf{R-Mod} \to \textbf{Mon}:(A, +, \lambda) \mapsto (A, +, \lambda)H = (A, +)$$

and

$$H:\textbf{R-Mod}\,[(A_1, +, \lambda_1), (A_2, +, \lambda_2)] \to \textbf{Mon}\,[(A_1, +), (A_2, +)]:$$
$$[(A_1, +, \lambda_1) \xrightarrow{f} (A_2, +, \lambda_2)] \mapsto [(A_1, +) \xrightarrow{fH = f} (A_2, +)]$$

Since

$$(fg)H = fg = (fH)(gH) \text{ for all composable } f \text{ and } g \text{ in } \textbf{R-Mod}$$

and

$$id_{(A, +, \lambda)}H = id_{(A, +)} = id_{(A, +, \lambda)H} \text{ for all } (A, +, \lambda) \text{ in } \textbf{R-Mod}$$

then $H:\textbf{R-Mod} \to \textbf{Mon}$ is a functor. ◊

Example 12

Consider $H:\textbf{Mon} \to \mathcal{S}$ defined by

$$H:(S, \cdot) \mapsto (S, \cdot)H = S$$

and

$$H:[(S_1, \cdot) \xrightarrow{f} (S_2, \cdot)] \mapsto (S_1 \xrightarrow{fH = f} S_2)$$

Then, similarly to Example 11, we have that H is a functor. We call such a functor a **forgetful functor** since it "forgets" the monoid operation and only remembers that a monoid homomorphism is a mapping of sets. ◊

Example 13

Consider $\mathcal{S}$, the category of sets, and let us arbitrarily select some object in this category, say the set U. Let us define a functor which crosses a set with U, denoted by $- \times U$, as follows:

$$- \times U : \mathcal{S} \to \mathcal{S} : X \mapsto X \times U$$

and

$$- \times U : \mathcal{S}(X, X') \to \mathcal{S}(X \times U, X' \times U) : (X \xrightarrow{f} X') \mapsto$$
$$(X \times U \xrightarrow{f \times id_U} X' \times U)$$

i.e.,

$$
\begin{array}{ccc}
X & \xrightarrow{\quad - \times U \quad} & X \times U \\
\downarrow{f} & & \downarrow{f \times id_U} \\
X' & & X' \times U
\end{array}
$$

where

$$f \times id_U : (\widehat{x}, \widehat{u}) \mapsto (f(\widehat{x}), \widehat{u})$$

Since

$$fg \times id_U : (\widehat{x}, \widehat{u}) \mapsto (fg(\widehat{x}), \widehat{u})$$

and

$$(f \times id_U)(g \times id_U) : (\widehat{x}, \widehat{u}) \mapsto (f[g(\widehat{x})], \widehat{u}) = (fg(\widehat{x}), \widehat{u})$$

then

$$fg \times id_U = (f \times id_U)(g \times id_U) \text{ for all } f \text{ and } g \text{ composable in } \mathcal{S}$$

Also,

$$id_X \times id_U : (\widehat{x}, \widehat{u}) \mapsto (\widehat{x}, \widehat{u})$$

and

$$id_{X \times U} : (\widehat{x}, \widehat{u}) \mapsto (\widehat{x}, \widehat{u})$$

so

$$id_X \times id_U = id_{X \times U}$$

Hence, $\mathcal{S} \xrightarrow{- \times U} \mathcal{S}$ is indeed a functor. ◊

We have already defined what we mean when we say that two objects of a category are isomorphic. We now define what it means to say that two categories are isomorphic.

DEFINITION 7

We say that a functor $H:\mathcal{K}_1 \to \mathcal{K}_2$ is an **isomorphism** if either of the following two equivalent conditions holds:

(i) $H:A \mapsto AH$ is a bijection on objects and $H:f \mapsto fH$ is a bijection on morphisms.

(ii) There exists a functor $H^{-1}:\mathcal{K}_2 \to \mathcal{K}_1$ such that the diagram

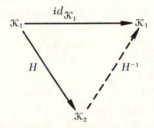

commutes. If there exists an isomorphism $H:\mathcal{K}_1 \to \mathcal{K}_2$, then we say that $\mathcal{K}_1$ and $\mathcal{K}_2$ are **isomorphic**.   ○

The final ingredient that we need for our general theory of automata is the concept of "freeness." We start with monoids and then generalize.

DEFINITION 8

Given a set U, then the monoid U^* is called the **free monoid** on the basis U of generators.   ○

We may characterize the notion of a free monoid as follows:

THEOREM 3

A monoid S is isomorphic to U^* if and only if there exists a map $\eta:U \to S$ such that for any monoid S' and any map $f:U \to S'$ there exists a unique monoid homomorphism $\psi:S \to S'$ such that

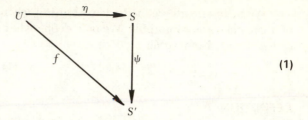

(1)

commutes; i.e., $\psi\eta(u_j) = f(u_j)$ for all $u_j \in U$.

Proof

(i) Suppose that $S \cong U^*$, with $\alpha : U^* \to S$ an isomorphism. Define η to be the restriction of α to U so that $\eta(\hat{u}) = \alpha(\hat{u})$ for each x in U. Suppose that we are given $f : U \to S'$. Then by (1) we must have

$$\psi\alpha(\hat{u}) = f(\hat{u}) \tag{2}$$

Now, let $\alpha(u_1 \ldots u_n)$ be any element of S. Then, if we define

$$\psi(\alpha(u_1 \ldots u_n)) \stackrel{\Delta}{=} f(u_1) \ldots f(u_n)$$

it is clear that ψ is a homomorphism, since

$$\psi(\alpha(u_1 \ldots u_n) \cdot \alpha(u'_1 \ldots u'_m))$$
$$= \psi(\alpha(u_1 \ldots u_n u'_1 \ldots u'_m)) \text{ since } \alpha \text{ is a homomorphism}$$
$$= f(u_1) \ldots f(u_n) f(u'_1) \ldots f(u'_m) \text{ by definition}$$
$$= \psi(\alpha(u_1 \ldots u_n))\psi(\alpha(u'_1 \ldots u'_m))$$

Moreover, this ψ is clearly the only homomorphism $S \to S'$ which satisfies (2).

(ii) Conversely, suppose that S is such that every $U \xrightarrow{f} S'$ has a unique homomorphic extension as defined by (1). In particular, taking $f = i_U : U \to U^* : \hat{u} \mapsto \hat{u}$, we have a unique homomorphism ψ such that

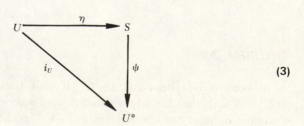

(3)

commutes. But, by part (i), U^* also has the property which assures us that there exists a unique homomorphism ψ' such that the following diagram commutes:

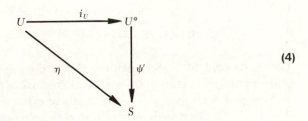

(4)

Splicing (3) and (4) together in two ways, we obtain the two commutative diagrams

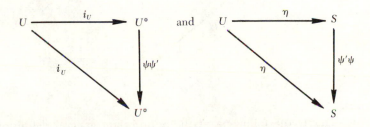

But we clearly also have the commutative diagrams

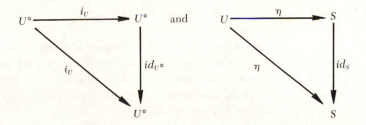

By the uniqueness property of the homomorphic extensions, we see that

$$\psi\psi' = id_{U^*}$$

since both are homomorphic extensions of i_U, while

$$\psi'\psi = id_S$$

since both are homomorphic extensions of η. We thus deduce that $\psi : S \rightarrow U^*$ is an isomorphism, so that $S \cong U^*$, as was to be proved. $\square$

This theorem suggests the following formulation for the concept of freeness:

Suppose that we denote the forgetful functor by $\mathcal{U}: \mathbf{Mon} \to \mathbf{S}$; i.e.,

$$\mathcal{U}: (S, \cdot) \mapsto (S, \cdot)\mathcal{U} = S$$

$$\mathcal{U}: [(S_1, \cdot) \xrightarrow{f} (S_2, \cdot)] \mapsto (S_1 \xrightarrow{f\mathcal{U} = f} S_2)$$

Then a monoid $M = (S, \cdot)$, with $U \subset S$ and equipped with any "inclusion of the generators" map $\eta: U \to M\mathcal{U}$, is **free** over U if for each $M' = (S', \cdot)$ in **Mon** and each map $f: U \to M'\mathcal{U}$, there exists a unique homomorphism $\psi: M \to M'$ in **Mon** such that $(\psi\mathcal{U})\eta = f$; i.e., such that the diagram

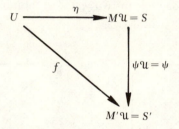

commutes.

This suggests the following as the appropriate general definition (we provide heuristics afterward):

DEFINITION 9

If we are given categories $\mathcal{A}$ and $\mathcal{B}$ with A an object of $\mathcal{A}$, B an object of $\mathcal{B}$, a functor $H: \mathcal{A} \to \mathcal{B}$, and $\eta: B \to AH$ a $\mathcal{B}$-morphism, then we say that the pair (A, η) is **free** over B with respect to H if for any object A' in $\mathcal{A}$ and any $\mathcal{B}$-morphism $f: B \to A'H$, there exists a unique $\mathcal{A}$-morphism $\psi: A \to A'$ such that $(\psi H)\eta = f$; i.e., such that the diagram

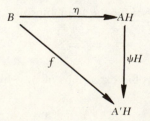

commutes. We refer to ψ as the **unique $\mathcal{A}$-morphic extension** of f.

Given a functor $H: \mathcal{A} \to \mathcal{B}$, if for all B in $\mathcal{B}$ there exists a pair (A, η)

[with A in $\mathcal{Q}$ and $\eta:B \to AH$ in $\mathcal{B}$] which is free over B with respect to H, then we say that H has a **left adjoint**. $\bigcirc$

Heuristic Meanings

Think of $\mathcal{Q}$ as a category of $\mathcal{B}$-objects with additional structure, and of H as the forgetful functor that throws away this structure. If B is a "basis" object and A is in $\mathcal{Q}$, we generalize "B is a subset of A," which really says "B is a subset of AH," by specifying a map $\eta:B \to AH$. We then think of A as a "free object with η as inclusion of generators," although η need not be "one-to-one" in the general setting. The definition then asserts that every map f on B extends uniquely to an $\mathcal{Q}$-morphism ψ.

We close the section by showing that B determines the A which is free over B up to isomorphism (Theorem 4) and then by showing that if H has a left adjoint, there exists a functor which provides the free objects (Theorem 5). However, we shall not need to use these facts, and some readers may wish to move straight on to the next section.

THEOREM 4

If (A, η) and (A', η') are both free over B with respect to H, then $A \cong A'$ in the strong sense that there exists an isomorphism $\phi:A \to A'$ such that $(\phi H)\eta = \eta'$.

Proof

We state the proof concisely, leaving it to the reader to fill in the details, using part (ii) of the proof of Theorem 3 as a model:

By freeness, there exist unique morphisms $A \xrightarrow{\varphi} A'$ and $A' \xrightarrow{\psi} A$ such that we have the commutative diagram

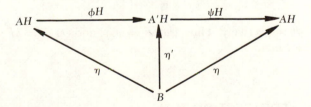

We may thus deduce, using the uniqueness assertion of Definition 9, that $\psi\phi = id_A$. Similarly, $\phi\psi = id_{A'}$. Thus, ϕ is an isomorphism, and we do indeed have that $(\phi H)\eta = \eta'$. $\square$

THEOREM 5

Let $\mathcal{C} \xrightarrow{H} \mathcal{B}$ have a left adjoint. Define an object mapping $\mathcal{B} \to \mathcal{C} : B \mapsto BF$ by letting (BF, η) be our choice (unique up to isomorphism) of a free pair over B with respect to H. Given a $\mathcal{B}$-morphism $B \xrightarrow{f} B'$, we define fF to be the unique morphism which renders commutative the diagram

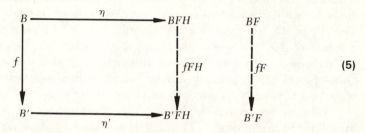

$$(5)$$

Then the pair $(B \mapsto BF, f \mapsto fF)$ defines a functor.

Proof

If $f = id_B : B \to B$, then we may take $fF = id_{BF}$ to render (5) commutative, for since H is a functor $(id_{BF})H = id_{BFH}$, the left-hand diagram reduces to the truism $\eta \cdot id_B = id_{BFH} \cdot \eta$, which just says that $\eta = \eta$.

If $f = B \xrightarrow{g} B'' \xrightarrow{h} B'$, we may take $fF = BF \xrightarrow{gF} B''F \xrightarrow{hF} B'F$ to render (5) commutative, for since H is a functor $(hF \cdot gF)H = hFH \cdot gFH$, so that the left-hand diagram reduces to

$$\eta' \cdot (hg) = hFH \cdot gFH\eta$$

which follows immediately from the defining properties

$$\eta' \cdot h = hFH \cdot \eta''$$

and

$$\eta'' \cdot g = gFH \cdot \eta$$

for hF and gF, respectively. Thus, $(id_B)F = id_{BF}$, and $(hg)F = hF \cdot gF$, so that F is indeed a functor. $\qquad\square$

EXERCISES FOR SECTION 8–4

1. We say that a set G equipped with a binary map $G \times G \to G : (g_1, g_2) \mapsto g_1 \cdot g_2$ is a **group** if

(α) $(g_1 \cdot g_2) \cdot g_3 = g_1 \cdot (g_2 \cdot g_3)$ for all g_1, g_2, g_3 in G.

(β)   There exists e in G such that

$$e \cdot g = g = g \cdot e \text{ for all } g \text{ in } G.$$

(γ)   There is a map $G \to G : g \mapsto g^{-1}$ such that $g \cdot g^{-1} = g^{-1} \cdot g = e$ for all g in G.

(a)   Verify that an **abelian group** is a group for which $g_1 \cdot g_2 = g_2 \cdot g_1$ for all g_1, g_2 in G. [*Hint:* Write $g_1 + g_2$ for $g_1 \cdot g_2$; 0 for e; and $-g$ for g^{-1}.]

(b)   We call a map $h : G \to H$ from one group to another a **homomorphism** if $h(g_1 \cdot g_2) = h(g_1) \cdot h(g_2)$ for all g_1, g_2 in G. Verify that if $h : G \to H$ and $k : H \to K$ are homomorphisms, then so too is $kh : G \to K : g \mapsto k(h(g))$. Note that the identity map $id_G : G \to G : g \mapsto g$ is always a homomorphism.

2.   Given a category $\mathcal{K}$ and any object A, show that id_A is unique.

3.   Recall that a map $f : S \to S'$ from one monoid to another is a **monoid homomorphism** if

$$f(s_1 \cdot s_2) = f(s_1) \cdot f(s_2) \text{ for all } s_1, s_2 \in S$$

and

$$f(1) = 1'$$

where 1 and $1'$ are the identities for S and S', respectively. Verify that if $f : S \to S'$ and $g : S' \to S''$ are monoid homomorphisms, then their composition $gf : S \to S'' : s \mapsto g(f(s))$ is also a monoid homomorphism. Also verify that the identity map $id_S : S \to S$ is always a monoid homomorphism.

4.   Describe **Gp,** the category of groups.

5.   Let $(P, \leq)$ be any partially ordered set. Let $\mathcal{K}$ have the elements of P for its class of objects, and define

$$\mathcal{K}(a, b) = \begin{cases} \{1_{ab}\} & \text{if } a \leq b \\ \varnothing & \text{if not} \end{cases}$$

Turn $\mathcal{K}$ into a category and verify your construction.

6.   Given the sets X and Y, suppose that the map $f : X \to Y$ is a bijection. Prove that $f^{-1} : Y \to X$ is such that $ff^{-1} = id_Y$ and $f^{-1}f = id_X$. Conversely, prove that if $f : X \to Y$ and $g : Y \to X$ are maps for which $fg = id_Y$ and $gf = id_X$, then f and g are bijections and $f^{-1} = g$.

7.   Using Definition 2, prove that a group homomorphism is an isomorphism if and only if the homomorphism is one-to-one and onto. State and verify the corresponding result for **R-Mod,** and for the category of Exercise 5.

8.   In a category $\mathcal{K}$, show that $id_A : A \to A$ is an isomorphism, and hence, $A \cong A$.

9.   Show that $\cong$ is an equivalence relation (i.e., A is related to B iff A is isomorphic to B) on the objects of a category $\mathcal{K}$.

10.   Describe a functor $H : \textbf{R-Mod} \to \textbf{Gp}$. (See Exercise 4.)

11.   Write out a more detailed proof of Theorem 4, using part (ii) of the proof of Theorem 1 as a model.

12.   Verify, for the machine of Example 9, that $\mathcal{R} = \{e, y, xy, x\}$.

13.   Apply the method of Example 6 to define $U^{\S}$ for R-modules by an appropriate coproduct construction. Use Example 7 to suggest an appropriate definition of $U^{\S}$ for groups.

8–5   Systems in a Category

Let us start by presenting the notion of a dynamorphism in a way which extends our discussion of equivalent systems in Section 5–4:

Suppose that we are given two systems, $\Sigma = (X, U, Y, \delta, \beta)$ and $\Sigma' = (X', U, Y', \delta', \beta')$, having the same input set U. We refer to the maps $\delta : X \times U \to X$ and $\delta' : X' \times U \to X'$ as the **dynamics** of their respective systems. We say that a map $h : X \to X'$ is a **dynamorphism** if the following diagram commutes:

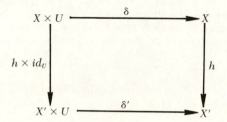

i.e., if $h[\delta(\hat{x}, \hat{u})] = \delta'(h(\hat{x}), \hat{u})$ for all $\hat{x} \in X$ and $\hat{u} \in U$.

Now recall (Example 8–4–14) that by the definition of the functor $-\times U : \mathbb{S} \to \mathbb{S}$, the map $h : Q \to Q'$ is sent by this functor to the map $h \times U : X \times U \to X' \times U$, where $h \times U$ is defined to be $h \times id_U$; i.e., $h \times U : (\hat{x}, \hat{u}) \mapsto (h(\hat{x}), \hat{u})$ for all $\hat{x} \in X$ and $\hat{u} \in U$. Thus, we may express the commutative diagram above in the equivalent form:

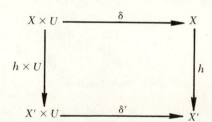

i.e., $\delta'(h \times U) = h\delta$.

Consider all the systems that have U as their input set. We form the class whose objects are the dynamics of these systems. To each pair of dynamics $(\delta : X \times U \to X, \delta' : X' \times U \to X')$ we associate the set of all dynamorphisms from X to X'. We claim that this results in a category, which will be denoted by **Dyn**(U). To verify this contention, we proceed as follows: If $h : X \to X'$ and $h' : X' \to X''$ are dynamorphisms, let us define (morphism) composition by using the usual definition of composition of maps. We must, then, first check that $h'h : X \to X''$ is indeed a dynamorphism.

Let us take the commutative diagrams corresponding to the dynamorphisms $h : X \to X'$ and $h' : X' \to X''$ and combine them to form the following commutative diagram:

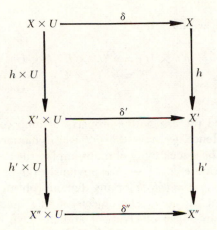

Since $h: X \to X'$ and $h': X' \to X''$ are maps, we can form the composite map $h'h: X \to X''$ and then the map $h'h \times U: X \times U \to X'' \times U$. Representing these maps in the previous diagram, we have:

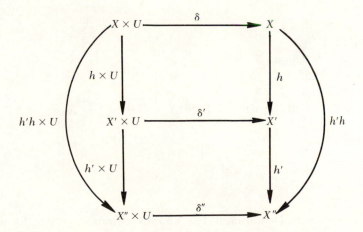

We must demonstrate that this diagram commutes [i.e., $\delta''(h'h \times U) = h'h\delta$] in order to show that $h'h$ is a dynamorphism. Since $- \times U: \mathcal{S} \to \mathcal{S}$ is a functor, we have that

$$h'h \times U = (h' \times U)(h \times U) \tag{1}$$

Since h and h' are dynamorphisms, we have

$$\delta'(h \times U) = h\delta \tag{2}$$

and

$$\delta''(h' \times U) = h'\delta' \tag{3}$$

respectively. Hence, we have that

$$
\begin{aligned}
\delta''(h'h \times U) &= \delta''(h' \times U)(h \times U) &&\text{by (1)}\\
&= h'\delta'(h \times U) &&\text{by (3)}\\
&= h'h\delta &&\text{by (2)}
\end{aligned}
$$

where we have used the associative property of map composition without explicit mention. Hence, we have that $h'h$ is a dynamorphism.

Clearly, now, the associativity of maps implies the associativity of dynamorphisms. Furthermore, $id_X : X \to X$ is obviously a dynamorphism with the property that $h(id_X) = h = (id_{X'})h$ for any dynamorphism $h : X \to X'$. Hence, **Dyn**(U) is indeed a category. The objects of this category are dynamics $[\delta : X \times U \to X$ is an object of **Dyn**$(U)]$ and the **Dyn** (U)-morphisms are dynamorphisms $[h : X \to X' \in \mathbf{Dyn}(U)(\delta : X \times U \to X, \delta' : X' \times U \to X')]$.

The crucial result for our general theory of systems in a category is that, among the class of all systems with input set U, there always exist "free" dynamics in the sense of Definition 8–4–9. We state this in technical language as follows:

THEOREM 1

The forgetful functor

$$\mathfrak{U} : \mathbf{Dyn}(U) \to \mathcal{S} : [\delta : X \times U \to X] \mapsto X$$

has a left adjoint.

Proof

In accordance with Definition 8–4–9, we must demonstrate that for any set X in $\mathcal{S}$, there exist a dynamics (call it δ_X) in **Dyn**(X) and a map ($\mathcal{S}$-morphism) $\eta_X : X \to \delta_X \mathfrak{U}$ such that (δ_X, η_X) is free over X with respect to $\mathfrak{U}$. We shall show that this can be done by defining the dynamics δ_X on the state set $\delta_X \mathfrak{U} = X \times U^*$ as follows:

$$\delta_X : (X \times U^*) \times U \to X \times U^* : ((\hat{x}, w), \hat{u}) \mapsto (\hat{x}, w\hat{u})$$

Thus, the corresponding system "remembers" (via registers, say) an element of X and a string from U^*, and reacts to each input symbol from U simply by adjoining it to the end of the U^*-register. We select the inclusion of generators map $\eta_X : X \to \delta_X \mathfrak{U} = X \times U^*$ to be $\eta_X : \widehat{x} \mapsto (\widehat{x}, \Lambda)$. This simply inserts $\widehat{x}$ into the X-register and does not place anything in the U^*-register.

To verify that (δ_X, η_X) is free over X with respect to $\mathfrak{U}$, we must check (see Definition 8-4-9) that for any dynamics $\delta' : X' \times U \to X'$ in $\mathbf{Dyn}(U)$ and any map $f : X \to X'$, there exists a unique dynamorphism $\psi : X \times U^* \to X'$ such that the diagram

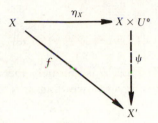

commutes. However, since ψ is a dynamorphism it must also be true (by the definition of dynamorphism) that the diagram

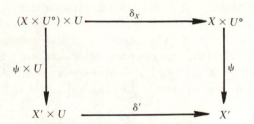

commutes. Thus, by the former diagram we require

$$\psi(\eta_X(\widehat{x})) = f(\widehat{x})$$

or

$$\psi(\widehat{x}, \Lambda) = f(\widehat{x}) \tag{4}$$

for all $\widehat{x} \in X$, while by the latter diagram we want

$$\psi \delta_X((\widehat{x}, w), \widehat{u}) = \delta'(\psi \times U)((\widehat{x}, w), \widehat{u})$$

or

$$\psi(\widehat{x}, w\widehat{u}) = \delta'(\psi(\widehat{x}, w), \widehat{u}) \tag{5}$$

for all $\widehat{x} \in X$, $w \in U^*$, and $\widehat{u} \in U$. Setting $w = \Lambda$ in equation (5), we get

$$\psi(\widehat{x}, \Lambda\widehat{u}) = \delta'(\psi(\widehat{x}, \Lambda), \widehat{u})$$

and by (4)

$$\psi(\widehat{x}, \widehat{u}) = \delta'(f(\widehat{x}), \widehat{u}) \tag{6}$$

Next, setting $w = u_1$ in (5) yields

$$\psi(\widehat{x}, u_1\widehat{u}) = \delta'(\psi(\widehat{x}, u_1), \widehat{u})$$
$$= \delta'(\delta'(f(\widehat{x}), u_1), \widehat{u}) \quad \text{by (6)}$$
$$= (\delta')^*(f(\widehat{x}), u_1\widehat{u})$$

One may then verify (Exercise 1) by induction on the length of w that we must have

$$\psi(\widehat{x}, w) = (\delta')^*(f(\widehat{x}), w)$$

for all $\widehat{x} \in X$ and $w \in U^*$. Thus, ψ is uniquely specified. Since this ψ results in the commutivity of the previous two diagrams, the pair (δ_X, η) is indeed free over X with respect to $\mathfrak{U}$. $\qquad\qquad\square$

Let us now generalize the notions given above by replacing the functor $-\times U: \mathfrak{S} \to \mathfrak{S}$ by a functor $P: \mathfrak{K} \to \mathfrak{K}$. Firstly, if X is an object of the category $\mathfrak{K}$, we define a P-**dynamics** to be any $\mathfrak{K}$-morphism $\delta: XP \to X$. (For the case $\mathfrak{K} = \mathfrak{S}$, by setting $P = -\times U: \mathfrak{S} \to \mathfrak{S}$, the X-dynamics $\delta: XP \to X$ recaptures the dynamics $\delta: X \times U \to X$.) Because of this more general definition, we will be able to consider dynamics for systems which "live" in other categories $\mathfrak{K}$ in addition to $\mathfrak{S}$. (The astute reader may well suspect that linear systems "live" in **R-Mod,** the category of R-modules.) Secondly, given P-dynamics $\delta: XP \to X$ and $\delta': X'P \to X'$, we say that the $\mathfrak{K}$-morphism $h: X \to X'$ is a P-**dynamorphism** if the diagram

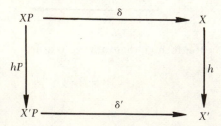

commutes. Note that if we specifically let $\mathfrak{K} = \mathfrak{S}$ and $P = -\times U$, then this definition of P-dynamorphism reverts to the original definition of dynamorphism.

Now, the class of all P-dynamics along with the associated P-dynamorphisms forms a category and is denoted by **Dyn**(P). The reader may formally establish this fact by using the procedure given for verifying that **Dyn**(U) is a category as a guide (Exercise 2).

For an arbitrary functor $P: \mathfrak{K} \to \mathfrak{K}$, we cannot in general expect the forgetful functor

$$\mathfrak{U}: \mathbf{Dyn}(P) \to \mathfrak{K}: [\delta: XP \to X] \mapsto X$$

to have a left adjoint. However, the central idea of the Arbib-Manes theory of systems is that we can extend many results of automata theory to any functor P which has the property that was established for the functor $-\times U$ in Theorem 1; namely, that P has a left adjoint. We call any such P an *input process*.

In summary, we will make the validity of Theorem 1 the axiom for our general definition, after making the crucial change in viewpoint of thinking of the input for a dynamics in **Dyn**(U) not as the set U but as the functor $-\times U:\mathbb{S}\to\mathbb{S}$. We shall reserve the symbol P for such an entire "process." More formally, we have the full definition:

DEFINITION 1

A **process** in an arbitrary category $\mathcal{K}$ is a functor $P:\mathcal{K}\to\mathcal{K}$. The **category of P-dynamics,** denoted **Dyn**(P), has as objects P-**dynamics** and has as morphisms P-**dynamorphisms.** The P-dynamics are the $\mathcal{K}$-morphisms $\delta:XP\to X$, where X is in $\mathcal{K}$. The P-dynamorphisms from given P-dynamics $\delta:XP\to X$ to $\delta':X'P\to X'$, are the $\mathcal{K}$-morphisms $h:X\to X'$ for which the diagram

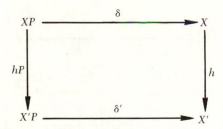

commutes. Composition and identities are defined as in $\mathcal{K}$ so that **Dyn**(P) is a category equipped with the obvious forgetful functor

$$\mathcal{U}:\mathbf{Dyn}(P)\to\mathcal{K}:[\delta:XP\to X]\mapsto X$$

which extracts the "state-object" of a P-dynamics. We say that P is an **input process** iff $\mathcal{U}$ has a left adjoint. In this case, the free dynamics over X will be denoted $\delta_Q:(XP^@)P\to XP^@$ and the "inclusion of generators" will be denoted $\eta_X:X\to XP^@$. ○

In other words, $XP^@$ is an object in $\mathcal{K}$ (which corresponds to the set $X\times U^*$ when $P=-\times U$), while $\delta_X:(XP^@)P\to XP^@$ is a $\mathcal{K}$-morphism (which corresponds to $\delta_X:(X\times U^*)\times U\to X\times U^*:((\widehat{x},w),\widehat{u})\mapsto(\widehat{x},w\widehat{u})$ when $P=-\times U$), and inclusion of generators $\eta_X:X\to XP^@$ is a $\mathcal{K}$-morphism (which corresponds to $\eta_X:X\to X\times U^*$ when $P=-\times U$). Thus, the condition that (δ_X, η_X) is free over X is simply that for any P-dynamics $\delta':X'P \to X'$ in $\mathbf{Dyn}(P)$ and any $\mathcal{K}$-morphism $f:X \to X'$, there exists a unique P-dynamorphism $\psi:XP^@ \to X'$ such that the diagram

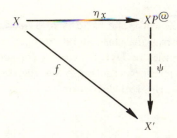

commutes, where the condition that ψ is a P-dynamorphism is simply that the diagram

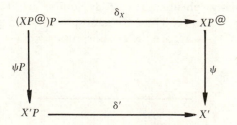

commutes. We refer to ψ as the **unique dynamorphic extension** of f.

Now let us remind ourselves of what enters the definition of a system with input *set* U so that we may formally define what we mean by a system with input *process* P. We start with the dynamics $\delta:X \times U \to X$, which generalizes to a P-dynamics $\delta:XP \to X$. We specify an output *map* $\beta:X \to Y$, and we generalize this to a $\mathcal{K}$-*morphism* $\beta:X \to Y$, so that the output *set* Y in $\mathcal{S}$ is generalized to an output *object* Y in $\mathcal{K}$. We specify an initial state $\widehat{x}_0$ in X in much of system theory, but there do exist applications (e.g., tree automata [Bobrow and Arbib, 1973, Chapter 4]) in which we specify a whole *set* of initial states, and this specification may be given by a map $\tau:I \to X$ which assigns a state $\tau(i)$ to each i in some set of labels for initial states. We generalize this to call for a $\mathcal{K}$-morphism $\tau:I \to X$. Then, in ordinary system theory, we can take $I = \{0\}$ and refer to $\tau(0) = x_0$ as *the* initial state. Putting all this together, we obtain our official definition:

DEFINITION 2

A **system in the category** $\mathcal{K}$ is a septuple

$$\Sigma = (P, X, \delta, I, \tau, Y, \beta)$$

where:

P is an input process

$\delta : XP \to X$ is a P-dynamics in $\mathbf{Dyn}(P)$, and the object X of $\mathcal{K}$ is called the **state object**

I is an object of $\mathcal{K}$ called the **initial state object**

$\tau : I \to X$ is a $\mathcal{K}$-morphism called the **initial state morphism**

Y is an object of $\mathcal{K}$ called the **output object**

$\beta : X \to Y$ is a $\mathcal{K}$-morphism called the **output morphism**   ○

Why free systems? This is because the realization theory of Section 8–1 noted that it was always trivial to provide a *free* realization of a response function $f : U^* \to Y$, namely the system

$$F_f = (U^*, U, Y, \mathrm{conc}, f)$$

and we see that the dynamics

$$\mathrm{conc} \colon U^* \times U \to U^* \colon (w, \widehat{u}) \mapsto w\widehat{u}$$

is just the free dynamics on a one-input set X, for if $X = \{\widehat{x}_0\}$, then

$$X \times U^* \to U^* \colon (\widehat{x}_0, w) \mapsto w$$

is in fact not only a bijection of states for F_f and δ_X, but a dynamorphism; i.e., the dynamics of F_f "is" the free dynamics on a one-element X. Since the existence of F_f proved to be the starting point for our classical realization theory, it will come as no surprise to the reader that the ability to construct free systems will be a crucial property for any functor P which is to act as input process for systems with an interesting realization theory.

We shall now demonstrate how linear systems (discussed in Section 7–5) fit into this framework.

We begin by selecting as $\mathcal{K}$ the category R-**Mod** of R-modules and linear maps, and consider the P-dynamics when $P = id_{R\text{-}\mathbf{Mod}}$. By definition, the P-dynamics are the R-**Mod**-morphisms (linear maps) of the form

$$\delta : XP \to X$$

which for this P are just linear maps

$$F : X \to X : \widehat{x} \mapsto F\widehat{x} \tag{7}$$

To specify the P-dynamorphisms when P is the identity process for R-**Mod**, suppose that $h : X \to X'$ is a P-dynamorphism. Then, by definition, we must have that the diagram

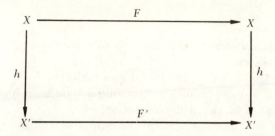

commutes. Thus, h is a P-dynamorphism from (X, F) to (X', F') if and only if

$$hF = F'h \qquad (8)$$

We shall now demonstrate that P is an input process by exhibiting the free dynamics

$$\delta_X : (XP^@)P \to XP^@$$

and the inclusion of generators morphism

$$\eta_X : X \to XP^@$$

for each X. We first recall from Section 8–2 that for any R-module X, we can form the R-module, denoted $X^\S$, that consists of all finite-support, left-infinite sequences of elements of X. In other words,

$$X^\S = \{w : \mathbf{N} \to X \mid w(n) \neq 0 \text{ for only finitely many } n\}$$

where we may write w in the form $(\ldots, w_n, \ldots, w_2, w_1, w_0)$ for $w_n = w(n)$, and the operations for $X^\S$ are defined componentwise by

$$(w + w')(n) = w(n) + w'(n)$$
$$(rw)(n) = rw(n)$$

where the left-hand-side operations in $X^\S$ are defined in terms of the known right-hand-side operations in the R-module X. In a similar manner, we can form the R-module $U^\S$. Additionally, let us define the map

$$z : X^\S \to X^\S : (\ldots, x_n, \ldots, x_1, x_0) \mapsto (\ldots, x_{n-1}, \ldots, x_0, 0)$$

which shifts sequences to the left and adjoins a 0; and the map

$$m : X \to X^\S : \widehat{x} \mapsto (\ldots, 0, \ldots, 0, \widehat{x})$$

which inserts an element into the zero position of a left-infinite sequence. Thus, z is the **left shift** (or successor), and m is the zero-place **injection**.

To show that P is an input process, we now claim that $XP^@ = X^\S$ is the state space for a free dynamics $F_X:(XP^@)P \to XP^@$ given by the linear map

$$z:X^\S \to X^\S$$

and inclusion of generators $\eta_X:X \to XP^@$ is given by the linear map

$$m:X \to X^\S$$

To substantiate our claim above, we must check that (δ_X, η_X) is indeed free over X with respect to the forgetful functor

$$\mathfrak{U}:\mathbf{Dyn}(P) \to R\text{-}\mathbf{Mod}:(X, F) \mapsto X$$

According to Definition 8-4-9 and the definition of an X-dynamorphism, we have to show that for each linear map $h:X \to X'$ (i.e., $\mathcal{K}$-morphism for $\mathcal{K} = R\text{-}\mathbf{Mod}$) and each P-dynamics (X', F'), there exists a unique P-dynamorphism

$$\psi:X^\S \to X'$$

such that each of the two diagrams

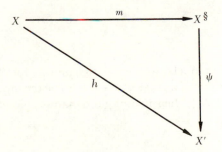

and

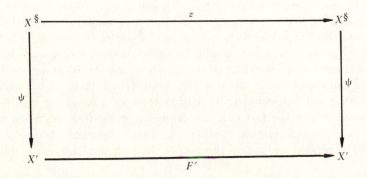

commutes. To say that the former diagram commutes is to say that

$$h(\widehat{x}) = \psi(m\widehat{x}) \tag{9}$$

The second diagram yields

$$F'\psi(w) = \psi(zw) \tag{10}$$

We now use linearity to see that (9) and (10) specify ψ completely:

$$\psi(\ldots, x_n, \ldots, x_1, x_0) = \psi\left(\sum_n (\ldots, x_n, 0, \ldots, 0)\right)$$

$$= \psi\left(\sum_n z^n(\ldots, 0, \ldots, x_n)\right)$$

$$= \sum_n \psi(z^n m x_n) \qquad \text{by linearity of } \psi$$

$$= \sum_n (F')^n \psi(m x_n) \qquad \text{by repeated application of (10)}$$

$$= \sum_n (F')^n h(x_n) \qquad \text{by (9)}$$

Thus, ψ is defined uniquely, so the identity process for R-**Mod** is indeed an input process. Note that if we take $h : X \to X'$ to be the input map $G' : U \to X'$ of a system (F', G', H'), then its unique dynamorphic extension is just the state-response map

$$(\ldots, u_n, \ldots, u_1, u_0) \mapsto \sum_n (F')^n G' u_n$$

The reader, while relieved that such constructs of linear system theory have proved to live in the category R-**Mod**, may nonetheless feel that since linear systems are so clearly a special case of ordinary systems, it is hardly necessary to go to so much trouble to unify the two theories. Perhaps the most dramatic demonstration of the power of the Arbib-Manes method is given a full exposition in Bobrow and Arbib [1973, Chapter 9], where it is shown that our general definition of systems in a category includes tree automata, despite the fact that the definition so far has only been seen to apply to sequential systems. However, the most important observation here is that the general realization theorem of the next section has the following important property: Recall that a *group system* $\Sigma = (X, U, Y, \delta, \beta)$ is one for which X, U, and Y are groups, and for which there exist homomorphisms $F:X \to X$, $G:U \to X$ and $H:X \to Y$ such that

$$\delta(\hat{x}, \hat{u}) = F(\hat{x}) \cdot G(\hat{u}) \text{ and } \beta(\hat{x}) = H(\hat{x}) \text{ for all } \hat{x} \text{ in } X \, \hat{u} \text{ in } U$$

Then the ordinary theory of Section 8–1 applied to the response $\mathcal{S}_e$ of the identity state in the state-group G of a group system does *not* in general yield a group system as the minimal system (Example 8–4–9 [p. 685] shows why), whereas the categorical theory of the next section *does* always yield a group system. This is in striking contrast to the situation found in Section 8–2, in which we saw that the ordinary theory when applied to a linear response function $f:U^\S \to Y$ did always yield a linear system as the minimal system.

Let us now put the definitions of linear and group systems into a form which suggests the general notion of a decomposable system within our framework of systems in a category:

A **linear system** is an R-linear map $F:X \to X$ together with R-linear maps $G:U \to X$ and $H:X \to Y$.

A **group system** is a homomorphism $F:X \to X$ together with homomorphisms $G:U \to X$ and $H:X \to Y$.

That is, they are both systems of the identity process in their category. In the next lemma and theorem, we verify that under very general conditions, the identity process has (as we have just verified in the special case of R-modules) the "free system" property which makes it an input process—and thus allows us to embed linear systems ($\mathcal{K} = R\text{-}\mathbf{Mod}$) and group systems ($\mathcal{K} = \mathbf{Gp}$) into our general setting of systems in a category.

The crucial key to our generalization is to note (recall Definition 8–4–5 and Example 8–4–6) that $X^\S$ in $R\text{-}\mathbf{Mod}$ is the coproduct of denumerably many copies of X:

LEMMA 2

Suppose that $\mathcal{K}$ is such that the coproduct of denumerably many copies of I (i.e., each of $I_0, I_1, I_2, \ldots, I_n, \ldots$ is a copy of I) exists; we refer to the coproduct as the **countable copower** of I, and write $I \xrightarrow{in_n} I^\S$ for the map of the nth copy of I into this copower. Then, if we define $I \xrightarrow{m} I^\S$ to be $I \xrightarrow{in_0} I^\S$, and define $I^\S \xrightarrow{z} I^\S$ to be the map defined uniquely† by

---

†The definition (8–4–5) of the coproduct tells us that the set of equations $f \cdot in_n = f_n : I \to I^\S$ has a unique solution for each sequence $(f_0, f_1, f_2, \ldots)$. Here we have chosen $f_n = in_{n+1}$ for each $n \in \mathbf{N}$.

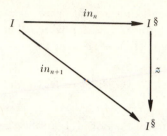

then for each pair $I \xrightarrow{G} X$ and $X \xrightarrow{F} X$ of $\mathcal{K}$-morphisms, there exists a unique morphism $I^{\S} \to X$ such that

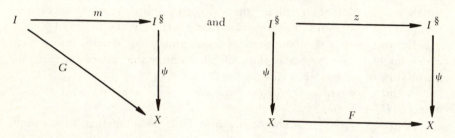

commute.

Proof

(The reader may first wish to study Example 1.) We have to check that, given $I \xrightarrow{G} X \xrightarrow{F} X$, there exists exactly one ψ such that

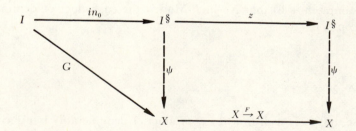

commute. But from the left-hand triangle we read

$$\psi \cdot in_0 = G \tag{11}$$

while from the right-hand square, and the fact that $z \cdot in_n = in_{n+1}$ by definition, we read

$$F \cdot \psi \cdot in_n = \psi \cdot z \cdot in_n = \psi \cdot in_{n+1} \text{ for each } n \in \mathbf{N} \tag{12}$$

This defines $\psi \cdot in_n$ for each $n \in \mathbf{N}$

$$\psi \cdot in_n = F^n G \tag{13}$$

and so defines ψ uniquely by applying the coproduct property to

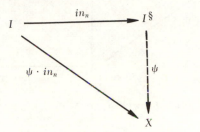

Let us check that this does indeed yield our previous result when $\mathcal{K} = R\text{-}\mathbf{Mod}$.

Example 1

Let U be an R-module. Then the countable copower $U^\S$ is indeed the R-module of all U-sequences

$$(\ldots 0, \ldots, 0, u_k, u_{k-1}, \ldots, u_1, u_0)$$

of finite support, with

$$in_n : U \to U^\S : \widehat{u} \mapsto (\ldots, 0, \ldots, 0, \widehat{u}, 0, \ldots, 0)$$

sending u to the sequence with $u_n = \widehat{u}$ while $u_j = 0$ for $j \neq n$. In particular, $U \xrightarrow{m} U^\S$ sends $\widehat{u}$ to $(\ldots, 0, \ldots, 0, \widehat{u})$. The defining equation

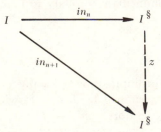

says that $z(\ldots, \widehat{u}$ in nth place$, \ldots) = (\ldots, \widehat{u}$ in $(n+1)$st place$, \ldots)$ and so z is just the *shift* operator (or *successor* operator) which sends $(\ldots, u_k, \ldots, u_1, u_0)$ to $(\ldots, u_{k-1}, \ldots, u_0, 0)$.

Suppose, then, that we are given R-linear maps $I \xrightarrow{G} X$ and $X \xrightarrow{F} X$, and

seek an R-linear ψ such that

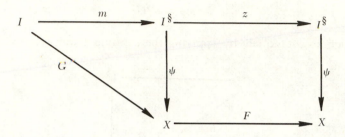

commutes. Were such a ψ to exist, we would have (by linearity)

$$\psi(\ldots, u_n, \ldots, u_0) = \sum_{n \in \mathbf{N}} \psi \cdot in_n(u_n)$$

By (1),

$$\psi \cdot in_0(u_0) = G(u_0)$$

By (2),

$$\psi \cdot in_{n+1}(u_{n+1}) = F\psi in_n(u_{n+1})$$
$$= F^{n+1}G(u_0) \qquad \text{by induction}$$

so $\psi(\ldots, u_n, \ldots, u_0) = \sum_n F^n Gu_n$. This looks like the updated state-contributions of a string of inputs. $\diamond$

The above process of building ψ from G and F is called **simple recursion.** Thus, we say that $\mathcal{K}$ is a **simple recursive category**† if the countable copower $(I \xrightarrow{in_n} I^\S : n \in \mathbf{N})$ exists for all $I \in \mathcal{K}$, and if we have in $\mathcal{K}$ all binary coproducts $K + L$ and an initial object $\varnothing$. We then have the theorem promised above, as an immediate corollary to Lemma 2:

THEOREM 3

The identity functor is an input process in any simple recursive category. The free dynamics is determined by $IP^@ = I^\S$ with dynamics $z : I^\S \to I^\S$, while the inclusion of generators is $m : I \to I^\S$. $\square$

With this, we can now make the definition which will be central to the rest of this chapter:

---

†Actually, this definition is more restrictive than that used in the literature, but it will do for now.

DEFINITION 3

A **decomposable system** in the simple recursive category $\mathcal{K}$ is a sextuple

$$(X, F, U, G, Y, H)$$

where

$$F: X \to X \text{ is an } id_{\mathcal{K}}\text{-dynamics}$$
$$G: U \to X \text{ is the initial state}$$
$$H: X \to Y \text{ is the output map}$$

We denote the system by (F, G, H) for short. ○

We now define the response, or the behavior, of any system in a category, and then define the realization problem to be that of finding a system with a specified behavior. In Section 8–6, we shall work within this framework and use the intuition obtained in Section 8–3 to build a realization theory for decomposable systems.

We begin by defining a number of concepts for the general system

$$\Sigma = (P, X, \delta, I, \tau, Y, \beta)$$

in a category $\mathcal{K}$ of Definition 2, after motivating them by the usual concepts given in Chapter 4.

First, we note that the construction of a free dynamics $\delta_X: (XP^@)P \to XP^@$, with the corresponding inclusion of generators $\eta_X: X \to XP^@$, is available for any object X in $\mathcal{K}$. In particular, then, it is available for the initial state object I. Let us start by examining $IP^@$ for $P = - \times U$.

Example 2

If $P = - \times U$ in $\mathcal{S}$, from the proof of Theorem 1 we know that $XP^@ = X \times U^*$. If we want the classical case, we take I to be a one-element set; say $I = \{0\}$. Then we have that

$$IP^@ = \{0\} \times U^* \cong U^*$$

where the obvious bijection is $(0, w) \leftrightarrow w$. $IP^@$ is, then, just the set of all input sequences and is referred to as the "object of inputs."

Given the system $\Sigma = (- \times U, X, \delta, I, \tau, Y, \beta)$, where $I = \{0\}$ and $\tau(0) = x_0 \in X$; let us study r, the unique dynamorphic extension of τ; i.e., the unique r for which each of the two diagrams

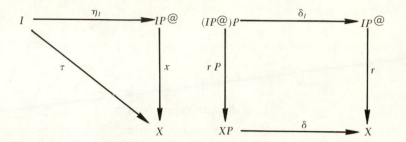

commute. In the present case, by referring to the proof of Theorem 1, we see that these two diagrams take the form

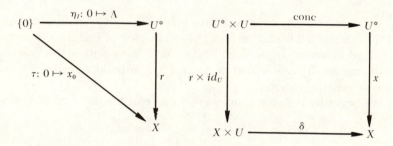

Thus, we have that

$$r\eta_I(0) = \tau(0)$$
$$r(\Lambda) = x_0$$

and

$$r(\text{conc})(w, \widehat{u}) = \delta(r \times id_U)(w, \widehat{u})$$

so that

$$r(w\widehat{u}) = \delta(r(w), \widehat{u})$$

for all $w \in U^*$ and $\widehat{u} \in U$. Hence, just as in the proof of Theorem 1, we conclude that

$$r: U^* \to X: w \mapsto \delta^*(x_0, w)$$

We thus call r the **reachability map,** since it sends the input string w to the state which w causes Σ to reach from the initial state x_0.

Since $r: U^* \to X$ and $\beta: X \to Y$, we can define $f: U^* \to Y$ by $f = \beta r$; and

$$f(w) = \beta r(w) = \beta[\delta^*(x_0, w)] = \mathcal{S}_{x_0}(w)$$

which is the response to w of Σ when started in state x_0. We call $f = \mathcal{S}_{x_0}$ the **response,** or **behavior,** of Σ. Note that in Section 8–1 we spoke of $\mathcal{S}_{\widehat{x}}$ as the response of Σ *when started in state $\widehat{x}$.* If we do not specify $\widehat{x}$, it is to be understood that x_0 is intended, since the specification of Σ that we now use actually includes the initial state. Further, in Section 8–1 we called any function f of the form $f: U^* \to Y$ a behavior and said that an automaton Σ with initial state x_0 **realizes** a behavior $f: U^* \to Y$ if and only if $f = \mathscr{S}_{x_0}$. ◊

We saw in Section 8–1 that any behavior $U^* \to Y$ has a (not necessarily finite-state) realization, just as we saw in Section 8–2 that any linear behavior $U^\$ \to Y$ has a (not necessarily finite-dimensional) realization as the behavior of a linear system.

In the next section, we shall prove a general realization theorem which covers group systems and linear systems as special cases. For the general theorem which yields a realization theory for sequential machines and tree automata as well, see Arbib and Manes [1973a].

We now formalize the notions observed in Example 2.

DEFINITION 4

Suppose that $\Sigma = (P, X, \delta, I, \tau, Y, \beta)$ is a system in a category $\mathcal{K}$ with $\delta_I : (IP^@)P \to IP^@$ being the free dynamics over I, and with inclusion of generators $\eta_I : I \to IP^@$. We then call $IP^@$ the **object of inputs** for Σ. The **reachability morphism** $r: IP^@ \to X$ of Σ is the unique dynamorphic extension of the initial state morphism $\tau: I \to X$, i.e., the unique $\mathcal{K}$-morphism r for which both of the diagrams

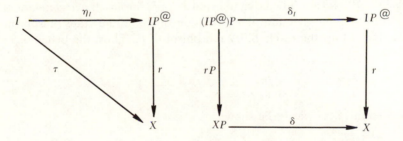

commute. The **response,** or **behavior,** of Σ is the composition $\beta r: IP^@ \to Y$; i.e.,

$$IP^@ \xrightarrow{\ \ r\ \ } X \xrightarrow{\ \beta\ } Y$$

For fixed I, P, and Y, a **response morphism** is any $\mathcal{K}$-morphism $f: IP^@ \to Y$. Finally, we say that a machine Σ **realizes** f, or is a **realization** of f, if and only if $f = \beta r$ for Σ. ◯

Before we turn, in Section 8–6, to the *minimal* realization of behaviors $f: IP^@ \to Y$ for identity input processes P, let us just note that every behavior has at least one realization, no matter what the input process P may be.

PROPOSITION 4

Let P be an input process and $f: IP^@ \to Y$ a behavior. Let $\delta_I : (IP^@)P \to IP^@$ be the free dynamics over I, with inclusion of generators $\eta_I : I \to IP^@$. Then the system

$$F_f = (P, IP^@, \delta_I, I, \eta_I, Y, f)$$

is a realization of f, called the **free realization** of f.

Proof

Since we have the diagram

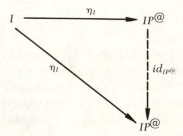

and since $id_{IP^@}$ is certainly a dynamorphism of δ_I into itself, we conclude that $id_{IP^@}$ is the unique dynamorphic extension of the initial state morphism η_I of F_f, and is thus the reachability morphism of F_f. Thus, the behavior of F_f is

$$IP^@ \xrightarrow{\ id_{IP^@}\ } IP^@ \xrightarrow{\ f\ } Y$$

which equals f, as was to be shown. $\square$

EXERCISES FOR SECTION 8–5

1.   Verify that the **Dyn**(P) of Definition 2 is indeed a category.

2.   Let **Mon** have as objects the *monoids*—sets S equipped with an associative $S \times S \to S : (s_1, s_2) \mapsto s_1 \cdot s_2$, and an identity $e \in S$ such that $e \cdot s = s = s \cdot e$ for all $s \in S$—and let it have as morphisms the monoid homomorphisms, i.e., maps $h: S_1 \to S_2$ such that $h(s_1 \cdot s_2) = h(s_1) \cdot h(s_2)$, while $h(e_1) = e_2$, where e_1 and e_2 are the identities in S_1 and S_2 respectively. Verify that **Mon** is a category when we define composition and identity morphisms as in $\mathcal{S}$; and that U^* can be made into a monoid. What is the multiplication and identity for U^*?

8–6   Realization of Decomposable Systems

We start by considering the response, or system behavior, of any decomposable system:

Example 1

For the decomposable system

$$\Sigma = (X, F, U, G, Y, H)$$

of Definition 8–5–3, the input map is $G: U \to X$. By equation (17) on page 709 we see that the reachability map $U^\S \xrightarrow{r} X$ satisfies $r \cdot in_j = F^j G$ for all $j \in \mathbf{N}$.

The system behavior f equals Hr, and so is the unique $U^\S \xrightarrow{f} Y$ satisfying

$$f \cdot in_j = HF^j G \text{ for all } j \in \mathbf{N}$$

a formula very familiar from linear system theory. $\Diamond$

In our earlier studies, we have said that a system Σ is *reachable* if and only if $r: U^* \to X$ is *onto*, i.e., iff every state of X is representable in the form $\delta^*(x_0, w)$ for some w in U^*. To extend our concept of reachability, we must find an appropriate "arrow-theoretic" form for the concept of an onto map. This will prove to be the notion of a *coequalizer*, which we now motivate:

Let $A \xrightarrow{f} B$ be an onto map in $\S$. Define the subset E of $A \times A$ to consist of all pairs with equal image—it is the equivalence relation of f:

$$E = \{(a_1, a_2) \mid f(a_1) = f(a_2)\}$$

Then we may define two projections

$$E \xrightarrow{\alpha} A : (a_1, a_2) \mapsto a_1 \quad \text{and} \quad E \xrightarrow{\gamma} A : (a_1, a_2) \mapsto a_2$$

and we note that the definition of E ensures that

$$f\alpha(a_1, a_2) = f\gamma(a_1, a_2) \quad \text{for all} \quad (a_1, a_2) \in E$$

Now suppose that we are given any other map $A \xrightarrow{g} C$ which shares with f the property that $g\alpha = g\gamma$. This says that $f(a_1) = f(a_2)$ implies $g(a_1) = g(a_2)$. Since f is onto, we may then define a map $h: B \to C$ which sends $f(a)$ to $g(a)$, since the choice of a for which $b = f(a)$ will not affect the value of $h(b)$.

What we have proved, then, is that every onto map in $\S$ is a coequalizer, as we next define it:

DEFINITION 1

A $\mathcal{K}$-morphism $A \xrightarrow{f} B$ is said to be a **coequalizer** if there exists a pair $E \xrightarrow{\alpha} A$ of morphisms with the property that, for any $A \xrightarrow{g} C$ for which $g\alpha = g\gamma$, there exists a unique $B \to C$ such that $hf = g$:

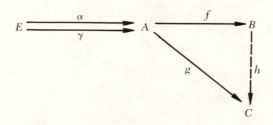

We call f the **coequalizer** of α and γ, and write $f = \text{coeq}(\alpha, \gamma)$.   ○

DEFINITION 2

We say that an arbitrary system

$$M = (P, X, \delta, I, \tau, Y, \beta)$$

in a category $\mathcal{K}$ is **coequalizer-reachable** iff the reachability map $IP^{@} \xrightarrow{r} X$ is a coequalizer; i.e., iff there exists an object E and a pair of maps $E \xrightarrow{\to} IP^{@}$ such that $r\alpha = r\gamma$, and such that for every map r' for which $r'\alpha = r'\gamma$, there is a unique map ϕ such that the following diagram commutes:

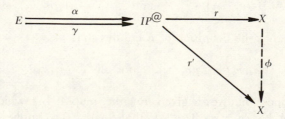

Given a morphism $h:A \to B$, we may extend it by "component-by-component action" to a morphism $\hat{h}:A^{\S} \to B^{\S}$ by specifying that

$$\hat{h} \cdot in_j = in_j \cdot h : A \to B^\S \text{ for each } j \in N$$

Note that if h is a coequalizer, then so too is $\hat{h}$. Suppose that $h = \text{coeq}(a, b)$, where $C \underset{b}{\overset{a}{\rightrightarrows}} A$. We verify that $\hat{h} = \text{coeq}(\hat{a}, \hat{b})$. It is certainly clear that $\hat{h}\hat{a} = (\widehat{ha}) = (\widehat{hb}) = \hat{h}\hat{b}$. Now suppose that $k\hat{a} = k\hat{b}$:

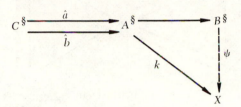

Then $(k \cdot in_j) \cdot a = k \cdot \hat{a} \cdot in_j = k \cdot \hat{b} \cdot in_j = (k \cdot in_j)b$. Thus, since $h = \text{coeq}(a, b)$, there is a unique ψ_j such that

$$k \cdot in_j = \psi_j h$$

But if we now define $\psi : B^\S \to X$ by $\psi \cdot in_j = \psi_j$, we have $(\psi\hat{h}) \cdot in_j = \psi\, in_j\, h = \psi_j h = k \cdot in_j$ for each $j \in N$, and thus $\psi\hat{h} = k$, so that ψ is clearly the unique map with this property.

LEMMA 1

In **Gp** and R-**Mod,** the coequalizers are precisely the onto maps, so that linear systems and group systems (X, F, U, G, Y, H) are coequalizer-reachable iff they are reachable in the conventional sense that $r(U^\S) = X$.

Proof

It is straightforward to check that each onto map is a coequalizer (Exercise 3). We now prove that every coequalizer is onto. (The proof is so constructed that it may be read both as a proof for the R-**Mod** case, and as a proof for the **Gp** case.)

We have already seen that an onto map is a coequalizer. Conversely, suppose that $f : A \to B$ is the coequalizer of $\alpha, \gamma : E \to A$. Define

$$E' = \{(x, y) \mid x, y \text{ in } A \text{ such that } f(x) = f(y)\}$$

and let p and q be the projections $(x, y) \mapsto x$ and $(x, y) \mapsto y$, respectively.

Let $\theta : A \rightarrow A/E' : a \mapsto [a]$ be the projection which sends each a in A to its equivalence class $[a] = \{a' \mid (a, a') \in E'\}$.

Splicing together the coequalizer diagram for f and the congruence diagram for θ, we obtain the diagram

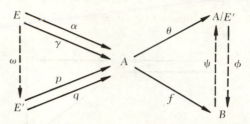

whose dashed arrows we now begin filling in.

We set $\omega : E \rightarrow E' : e \mapsto (\alpha(e), \gamma(e))$, which is well-defined since $f\alpha = f\gamma$.

Then $\theta\alpha = \theta p\omega = \theta q\omega = \theta\gamma$, and so, since $f = \mathrm{coeq}(\alpha, \gamma)$, there is a unique ψ with $\psi\theta = f$.

We set $\phi : A/E' \rightarrow B : [a] \rightarrow f(a)$, which is clearly well-defined and satisfies $f = \phi\theta$.

But we then have $\phi\psi f = f$, which implies that $\phi\psi = id_B$ since f is a coequalizer (and thus an epimorphism). But id_B is onto, and so ϕ must be onto. We know that ψ is onto by definition, and so $f = \phi\psi$ is also onto, as was to be proved. $\square$

Recalling the discussion immediately following the proof of Theorem 8-2-1, we see that the following problem does indeed reduce to the usual problem for linear machines:

REALIZATION PROBLEM FOR DECOMPOSABLE SYSTEMS

Given a response morphism $f : U^\S \rightarrow Y$, find a decomposable system $E_f = (X_f, F, U, G, Y, H)$ which is the minimal coequalizer-reachable realization of f in the sense that:

(i) E_f realizes f; i.e., $f = H \cdot r_f$.
(ii) The reachability map $U^\S \xrightarrow{r_f} X$ of E_f is a coequalizer.
(iii) E_f is minimal in that for all $M = (X, F', U, G', Y, H')$ satisfying (i) and (ii) there exists a unique dynamorphism ψ for which we have the commutativity of

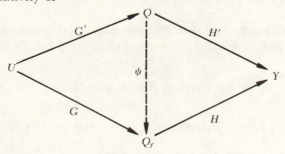

To solve this realization problem, we shall provide a general treatment of three stages into which the presentation of the theory of Section 8-3 may be broken:

Stage 1: We build $U^\S$ with associated maps

$$in_j: U \to U^\S : \hat{u} \mapsto (\ldots, 0, \hat{u} \text{ in } j\text{th place}, 0, \ldots, 0)$$

and $Y^{\mathbf{N}}$ with associated maps

$$\pi_k: Y_\S \to Y: (y_0, y_1, \ldots, y_k, \ldots) \mapsto y_k$$

in such a way that the reachability map

$$r: U^\S \to X$$

is uniquely defined by the equation

$$r \cdot in_j = F^j G$$

while the observability map

$$\sigma: X \to Y_\S$$

is uniquely defined by the equation

$$\pi_n \cdot \sigma = HF^n$$

We then define the total system response by

$$\tilde{f} = \sigma \cdot r: U^\S \to Y_\S$$

and the system response itself by

$$f = \pi_0 \tilde{f}: U^\S \to Y$$

Stage 2: Conversely, given $\tilde{f}: U^\S \to Y^{\mathbf{N}}$, we take the state space of its minimal realization to be

$$X_f = \tilde{f}(U^\S)$$

If we denote $\tilde{f}(U^\S)$ by $\text{Im}(\tilde{f})$, we are making use of a factorization

$$U^\S \overset{\tilde{f}}{\to} Y_\S = U^\S \overset{p}{\to} \text{Im}(\tilde{f}) \overset{i}{\to} Y_\S$$

where the map p is *onto* (all states of the minimal realization are *reachable*) and the map i is *one-to-one* (all states of the minimal realization are *observable*).

Stage 3: We then define F_f so that $F_f \cdot \tilde{f} = \tilde{fz}$; define G_f so that $G_f = p \cdot in_0$; and define H_f to be simply $\pi_0 \cdot i$. We then check that (F_f, G_f, H_f) really is a minimal realization of $\tilde{f}$.

We now proceed to the three stages of our generalized analysis:

Stage 1: Reachability and Observability Maps

We have already set up the reachability map $r: U^\S \to X$ by taking $U^\S$ to be the countable copower of U. It turns out that we can obtain $Y_\S$ by reversing all the arrows in the discussion of countable copowers (cf. Lemma 8-5-2 of p. 707):

DEFINITION 3

Let K be an object in the category $\mathcal{K}$. A countable copower of K is an object C together with a family of $\mathcal{K}$-morphisms $\{C \xrightarrow{\pi_n} K \,|\, n \in \mathbf{N}\}$ with the property that for any family of $\mathcal{K}$-morphisms $\{X \xrightarrow{f_n} K \,|\, n \in N\}$ there exists exactly one $f: X \to C$ for which the following diagram commutes.

C is unique up to isomorphism (Exercise 4) and will be denoted by $K_\S$. ○

Then the reversed-arrow proof of Lemma 8-5-2 immediately yields:

LEMMA 2

Suppose that $\mathcal{K}$ is such that the countable copower $Y_\S$ of Y exists. If we define $Y_\S \xrightarrow{z} Y_\S$ to be defined uniquely by

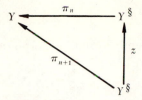

then for each pair $X \xrightarrow{H} Y$ and $X \xrightarrow{F} X$ of $\mathcal{K}$-morphisms, there exists a unique morphism $X \xrightarrow{\sigma} Y_{\S}$ such that

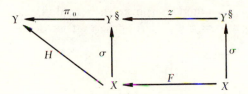

commutes.   □

If we have a decomposable machine (F, G, H) for which the countable copower of Y exists, then we call the $\sigma : X \to Y_{\S}$ defined by Lemma 2 the **observability map.**

In $\S$, **Gp,** and R-**Mod,** the countable copower exists for every object Y, and is just the denumerable cartesian coproduct

$$\{(y_0, y_1, \ldots, y_k, \ldots) \,|\, y_k \in Y \text{ for each } k \in \mathbf{N}\}$$

endowed with the usual componentwise operations. It is also immediate that in each case we have

$$\pi_k \cdot \sigma(x) = HF^k x \text{ for each } k \in \mathbf{N}.$$

We have already chosen coequalizers as our categorical generalization of the onto maps of $\S$. We now offer *monomorphisms* as the generalization of one-to-one maps:

DEFINITION 4

The map $f : A \to B$ is a **monomorphism** in the category $\mathcal{K}$ if two morphisms $g, h : C \to A$ satisfy $f \cdot g = f \cdot h$ iff $g = h$.

LEMMA 3

In the categories $\mathcal{S}$, **Gp**, and *R*-**Mod**, a morphism is one-to-one iff it is a monomorphism.

Proof

We handle $\mathcal{S}$, leaving **Gp** and *R*-**Mod** to the reader (Exercise 5). Suppose that $f \cdot g = f \cdot h$. Then for every $c \in C$, the fact that $f \cdot g(c) = f \cdot h(c)$ would imply $g(c) = h(c)$ were f one-to-one, so that f would be a monomorphism.

Conversely, if f is a monomorphism there could be no pair a and a' such that $a \neq a'$ while $f(a) = f(a')$, for suppose that we define

$$g = id_A : A \to A$$

while
$$h(\widehat{a}) = \begin{cases} \widehat{a} \text{ if } \widehat{a} \neq a' \\ a \text{ if } a = a' \end{cases}$$

Then we would have $f \cdot g = f \cdot h$ while $g \neq h$, contradicting the fact that f is a monomorphism. Thus f must be one-to-one. $\qquad\square$

We thus say that a decomposable system (F, G, H) is **observable** iff its observability map $\sigma : X \to Y_\mathcal{S}$ is a monomorphism.

Combining our studies of reachability and observability we may, as soon as the countable copower $U^\mathcal{S}$ and the countable power $Y_\mathcal{S}$ are defined, define the **total system response** of the decomposable system (F, G, H) to be

$$\widetilde{f} = \sigma \cdot r : U^\mathcal{S} \to Y_\mathcal{S}$$

Note, also, that r is a dynamorphism $(U^\mathcal{S}, z) \to (X, F)$ [since $(rz) \cdot in_j = F^{j+1}G = F(F^j G) = F(r \cdot in_j) = (Fr) \cdot in_j$] and that σ is a dynamorphism $(X, F) \to (Y_\mathcal{S}, z)$ [since $\pi_k \cdot (z\sigma) = HF^{k+1} = (\pi_k \cdot \sigma)F = \pi_k(\sigma F)$]. Thus, $\widetilde{f} = \sigma \cdot r$ is a dynamorphism.

Given $\widetilde{f}$, we form the **system response**

$$f = \pi_0 \cdot \widetilde{f} : U^\mathcal{S} \to Y$$

Conversely, given *any* $\mathcal{K}$-morphism $f : U^\mathcal{S} \to Y$, we may define a corresponding $\widetilde{f} : U^\mathcal{S} \to Y_\mathcal{S}$ by

$$\pi_k \widetilde{f} = f z^k \text{ for } k \in \mathbf{N} \text{ (with } z^0 = id_{Y_\mathcal{S}})$$

Then $\widetilde{f}$ is immediately a dynamorphism $(U^\mathcal{S}, z) \to (Y_\mathcal{S}, z)$ since

$$\pi_k(\widetilde{f}z) = (\pi_k \widetilde{f})z = (fz^k)z = fz^{k+1} = \pi_{k+1}\widetilde{f} = \pi_k(z\widetilde{f})$$

which is just the dynamorphism condition.

Stage 2: Defining the Image of a Morphism

We start by providing conditions on a category $\mathcal{K}$ which allow us to associate with each morphism $f : A \to B$ its "image" $\text{Im}(f)$. In the case of the category $\mathcal{S}$ of sets, forming $\text{Im}(f)$ gives rise to a factorization

$$f = A \to \text{Im}(f) \xrightarrow{i} B$$

where $p(a) = f(a)$ for all $a \in A$ and i is the inclusion map. Clearly, p is surjective (onto) and i is injective (one-to-one). It is easy to check that if $f = i'p'$ is another surjective-injective factorization of f, then there exists a unique bijection $\psi : \text{Im}(f) \to C$ such that $\psi \cdot p = p'$ and $i' \cdot \psi = i$;

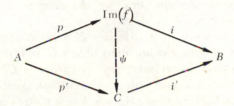

that is, surjective-injective factorizations are unique up to isomorphism in $\mathcal{S}$.

We say f has a **coequalizer-mono factorization** (in any category $\mathcal{K}$) if it factors as $f = ip$ with p a coequalizer and i mono.

LEMMA 4

Coequalizer-mono factorizations are unique up to isomorphism.

Proof

Let $f = ip = i'p'$ be two coequalizer-mono factorizations of f:

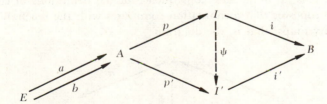

There exist a and b with $p = \text{coeq}(a, b)$. Since i' is mono, $p'a = p'b$ which induces a unique $\psi : I \to I'$ with $\psi \cdot p = p'$. Since p is an epimorphism,

$i' \cdot \psi = i$. Then $\psi^{-1} : I' \to I$ is induced symmetrically. That ψ and ψ^{-1} are mutually inverse is easy to show, because i and i' are mono. $\square$

As a result of the above lemma, we may write *the* coequalizer-mono factorization of $f : A \to B$ as

$$f = A \xrightarrow{p} \text{Im}(f) \xrightarrow{i} B$$

Stage 3: Defining the Minimal Realization

We may now define the minimal realization Σ_f of a response morphism $f : U^\S \to Y$ and prove that it exists uniquely up to isomorphism. When $\mathcal{K}$ has coequalizer-mono factorizations, we characterize Σ_f as a reachable, observable realization of f.

We continue to work in a category $\mathcal{K}$ with countable powers and copowers and coequalizers. Fix an input object U and an output object Y. A **response morphism** is an arbitrary morphism $f : U^\S \to Y$. A system $\Sigma = (F, G, H)$ **realizes** f if and only if its response coincides with f. Σ is **reachable** if its reachability map $r : U^\S \to X$ is a coequalizer in $\mathcal{K}$.

Consider the category $\mathfrak{M}$ whose objects are decomposable systems $\Sigma = (X, F, G, H)$ with dynamics (X, F), input map $G : U \to X$, and output map $H : X \to Y$ (we henceforth hold U and Y fixed, though X may vary); and whose morphisms are **simulations** $\psi : \Sigma \to \Sigma'$ (we say Σ **simulates** Σ'), meaning dynamorphisms $\psi : X \to X'$ (so that $\psi F = F' \psi$) which commute with the input and output; that is, the diagram

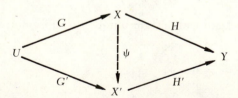

commutes. It is an immediate consequence of the definitions of countable power and copower that a simulation commutes with the reachability and observability maps; that is, the diagram

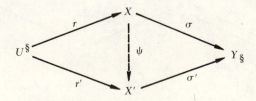

commutes. In particular, then, the existence of a simulation guarantees that the two machines have the same (total) response.

A system Σ is a **minimal realization** of f if Σ is a reachable realization of f and if Σ is simulated by every other reachable realization of f; that is, given any other reachable realization, Σ', of f, then there exists a (unique) simulation $\psi:\Sigma' \to \Sigma$. In other words, σ is a terminal object in the category whose objects are reachable realizations of f and whose morphisms are simulations (composition and identities being at the level of $\mathcal{K}$). Thus Σ *is unique up to isomorphism*. This recaptures the definition in linear system theory discussed after Theorem 8–2–1. From now on, we speak of *the* minimal realization of f and denote it by $\Sigma_f = (X_f, F_f, G_f, H_f)$. Of course, we must first prove that Σ_f exists.

Before presenting the minimal realization of a given $f:U^\S \to Y$, we present two useful lemmas: the first is trivial, the second crucial.

LEMMA 5

Let the $id_{\mathcal{K}}$-dynamics (X_1, F_1), (X_2, F_2), and (X_3, F_3) be such that there exist dynamorphisms $f:(X_1, F_1) \to (X_2, F_2)$ and $g \cdot f:(X_1, F_1) \to (X_3, F_3)$

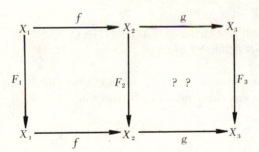

If f is an epimorphism in $\mathcal{K}$, then g is a dynamorphism $(X_2, F_2) \to (X_3, F_3)$.

Proof

Since gf is a dynamorphism,

$$F_3 gf = gfF_1$$
$$= gF_2 f \quad \text{since } f \text{ is a dynamorphism}$$

Thus $F_3 g = gF_2$, since f is epi, and so g is a dynamorphism. $\square$

LEMMA 6

If $h:(X, F) \to (X', F')$ is a dynamorphism, and h has a coequalizer-mono factorization

$$X \xrightarrow{p} \mathrm{Im}(f) \xrightarrow{i} X'$$

then there exists a unique dynamical structure F'' on $\mathrm{Im}(f)$ making $p:(X, F) \to (\mathrm{Im}(f), F'')$ and $i:(\mathrm{Im}(f), F'') \to (X', F')$ dynamorphisms.

Proof

We see that $ipF = hF = F'h = F'ip$, since h is a dynamorphism. If $p = \mathrm{coeq}(\alpha, \gamma)$, we have $p\alpha = p\gamma$ so that $ipF\alpha = ipF\gamma$, whence $pF\alpha = pF\gamma$ because i is mono. Hence, there is a unique F'' such that $pF = F''p$; i.e., a unique $F'':\mathrm{Im}(f) \to \mathrm{Im}(f)$ such that $p:(Q, F) \to (\mathrm{Im}(f), F'')$ is a dynamorphism. By the previous lemma, the fact that p is an epimorphism then ensures that $i:(\mathrm{Im}(f), F'') \to (Q', F')$ is also an epimorphism. $\square$

We have now accomplished most of the hard work for our main theorem:

THEOREM 7 (MINIMAL REALIZATION THEOREM FOR DECOMPOSABLE SYSTEMS)

Given a dynamorphism $\tilde{f}: U^\S \to Y_\S$, form the unique dynamics F_f such that p and i in the coequalizer-mono factorization

$$\tilde{f} = i \cdot p : U^\S \xrightarrow{p} X_f \xrightarrow{i} Y_\S$$

are dynamorphisms. Define

$$G_f = p \cdot in_0 : U \to X_f$$

and

$$H_f = \pi_0 \cdot i : X_f \to Y$$

Then $\Sigma_f = (X_f, F_f, G_f, H_f)$ is a minimal realization of $\tilde{f}$.

Proof

First note that Σ_f has reachability map p and observability map i. Since $\tilde{f} = i \cdot p$, Σ_f does indeed realize f. Since p is a coequalizer, Σ_f is reachable; since i is mono, Σ_f is observable:

Let, now, $\Sigma = (X, F, G, H)$ be another reachable realization of $\tilde{f}$, with reachability map r and observability map σ. Since p and r are both coequalizers, we may set

$$p = \text{coeq}(a, b)$$
$$r = \text{coeq}(\alpha, \gamma)$$

as shown in the diagram

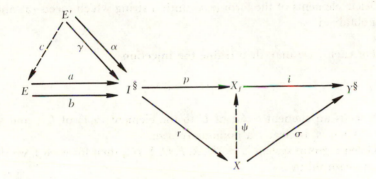

Since $r\alpha = r\gamma$ and $ip = \sigma r$, we have

$$i p \alpha = \sigma r \alpha = \sigma r \gamma = i p \gamma$$

Since i is mono, we deduce that

$$p\alpha = p\gamma$$

Hence, since $r = \text{coeq}(\alpha, \gamma)$, there exists a unique morphism $\psi : X \to X_f$ such that $\psi r = p$. By Lemma 6, ψ is a dynamorphism. Thus, Σ_f is a reachable realization of f which is simulated by every other realization of f and so is minimal.   □

This not only yields the usual realization theory for linear systems, but also yields the more recent realization theory for group systems (Arbib [1972]):

Given a **group** U, the elements of $U^\S$ may be taken to be strings of the form

$$(u_{i_1}, i_1)(u_{i_2}, i_2) \ldots (u_{i_n}, i_n) \text{ with each } u \in \mathbf{U}, i \in \mathbf{N}$$

(we use Λ to denote the empty string) subject to the restrictions:

(a) No u_{i_j} is the identity.
(b) For each j, $1 \leq j < n$, we have that $i_j \neq i_{j+1}$.

Multiplication is simply concatenation, save that if the string

$$(v_{k_1}, k_1)(v_{k_2}, k_2) \ldots (v_{k_m}, k_m)$$

so formed does not satisfy conditions (a) and (b) we apply the following operations:

(c) Replace consecutive elements of the form $(v, n)(v', n)$ for some n by the single element (vv', n).
(d) Delete elements of the form (e, n) until a string which meets (a) and (b) is obtained.

For each j we may then define the **injection**

$$in_j: U \to U^\S$$

which sends an element $u \neq e$ of U to the element (u, j) of $U^\S$, and sends e to Λ. Then in_j is clearly a homomorphism.

Given a group system $\Sigma = (U, X, F, G, Y, H)$, then for each j, we define the homomorphism

$$r_j: U \to X: U \mapsto F^j Gu$$

The reason that coproducts were invented is that this yields, as we have seen, a unique homomorphism

$$r: U^\S \to X$$

which is the **reachability map** of Σ.

Since $U^\S$ is a group and r is a homomorphism, it follows that $r(U^\S)$ is a subgroup of X. Note, however, that since U^*, considered as sequences of the form $(u_{i_1}, i_1)(u_{i_2}, i_2) \ldots (u_{i_n}, i_n)$ for which $i_1 > i_2 > \cdots > i_n$, is not a subgroup of $U^\S$, it follows that there is no guarantee that $r(U^*)$ is a subgroup of X, as indeed we saw in Example 8-4-9.

Example 2

Consider the group system with $U = Y = \mathbf{Z}_2$, $X = \mathbf{D}_4$, $G(0) = e$, $G(1) = u$, $F(x) = e$, and $F(y) = xy$. Then $r(U^\S) = \mathbf{D}_4$, since, for example,

$$x^3 y = x \cdot x \cdot xy = r((1, 1)(1, 0)) \, r((1, 1)(1, 0)) \, r((1, 1))$$
$$= r((1, 1)(1, 0)(1, 1)(1, 0)(1, 1)) \qquad \qquad \Diamond$$

Next we define the **identity-state response function** of the system to be

$$f = Hr : U^\S \to Y$$

which is the unique homomorphism for which $f \cdot in_j = HF^j G$.

Example 3

Consider the example in which $U = X = Y$ is a nonabelian group, and F, G, and H are identity maps. Then

$$(u_1, 1)(u_2, 0) \text{ and } (u_2, 0)(u_1, 1)$$

are different elements of $U^\S$, and we have

$$f((u_1, 1)(u_2, 0)) = f(u_1, 1)f(u_2, 0) = HFGu_1 \cdot HGu_2 = H(FGu_1 \cdot Gu_2)$$
$$f((u_2, 0)(u_1, 1)) = f(u_2, 0)f(u_1, 1) = HGu_2 \cdot HFGu_1 = H(Gu_2 \cdot FGu_1) \quad \Diamond$$

Now, for abelian groups we reduced the Nerode equivalence to the simultaneous satisfaction of the equivalences

$$f^\S(w_1 0^n) = f^\S(w_2 0^n)$$

for each $n \in \mathbf{N}$. We now set up the corresponding sequence of equivalences for the group case.

For each n, we define the successor homomorphism by

$$z \cdot in_n : U \to U^\S : u \mapsto (u, n + 1)$$

which yields

$$z : U^\S \to U^\S : (u_{i_1}, i_1) \ldots (u_{i_n}, i_n) \mapsto (u_{i_1}, i_1 + 1) \ldots (u_{i_n}, i_n + 1)$$

Given any homomorphism $f : U^\S \to Y$, we then define the congruence E_f on $U^\S$ by

$$w_1 E_f w_2 \Leftrightarrow fz^n(w_1) = fz^n(w_2) \text{ for all } n \in \mathbf{N}$$

Let X_f be the factor group $U^\S/E_f$, and let $\eta_f : U^\S \to U^\S/E_f$ be the canonical epimorphism. Then we may define three homomorphisms as follows:

$$F_f : X_f \to X_f : [w] \mapsto [zw]$$
$$G_f : U \to X_f : u \mapsto [in_0(u)]$$
$$H_f : X_f \to Y : [w] \mapsto f(w)$$

and we have that

$$\Sigma_f = (X_f, F_f, U, G_f, Y, H_f)$$

is indeed the minimal realization of f which is a group system.

It is of some interest to define a system which simulates the response of a group system to all of $U^\S$:

Given U, we define the set $\tilde{U}$ to be $U \cup \{\rho\}$, where ρ is a new symbol, indicating a reset. We then define a map $e: U^\S \to \tilde{U}^*$ inductively by taking $e(\Lambda) = \Lambda$ and $e(u, n) = u$, and then setting

$$e[w \cdot (u, n) \cdot (u', n')] = \begin{cases} e[w \cdot (u, n)] \cdot u' & \text{if } n > n' \\ e[w \cdot (u, n)] \cdot \rho \cdot u' & \text{if } n < n' \end{cases}$$

Then, given the group system $\Sigma = (X, F, U, G, Y, H)$, we define its **cumulator** $\tilde{\Sigma}$ to be the system

$$\tilde{\Sigma} = (X \times X, \tilde{U}, Y, \tilde{\delta}, \tilde{\beta})$$

for which

$$\tilde{\delta}((x_1, x_2), u) = \begin{cases} (x_1 x_2, 1) & \text{if } u = \rho \\ (x_1, Fx_2 \cdot Gu) & \text{if } u \neq \rho \end{cases}$$

$$\tilde{\beta}(x_1, x_2) = H(x_1 \cdot x_2)$$

If $\tilde{f}: \tilde{U}^* \to Y$ is the $(1, 1)$-state response of $\tilde{\Sigma}$ while f is the identity-state response of $\tilde{\Sigma}$, it is then straightforward to verify that the following diagram commutes:

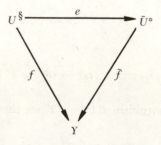

$\diamond$

We close by noting that the realization theory for decomposable machines is but a very special case of the general realization theory for machines in a category due to Arbib and Manes [1973a]. A careful exposition is given by Bobrow and Arbib [1974, Chapter 9], but we chose to focus on decomposable machines here because they offer the framework most appropriate for general algebraic studies in that area of system theory focused upon linear systems. We see, to our surprise, that much of linear system theory depends not on the linearity of vector space theory but on the far more general algebraic properties of powers, copowers, and so forth.

EXERCISES FOR SECTION 8–6

1.   In a category $\mathcal{K}$, we say that a morphism $f:K \to L$ is an **epimorphism** if whenever $g:L \to M$ and $h:L \to M$ satisfy $gf = hf$, it must follow that $g = h$. Prove that every coequalizer is an epimorphism. *Hint:* Consider the diagram

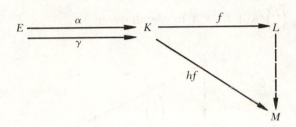

2.   Let $\varnothing$ be an initial object in the category $\mathcal{K}$. Prove that $K + \varnothing = K$ for each $\mathcal{K}$-object $\mathcal{K}$. Prove that $\varnothing^{\S} = \varnothing$.

3.   Prior to Definition 1, we showed that every onto map in $\S$ is a coequalizer.

   (a)   Check that if the original f was an R-linear map in R-**Mod**, then E would be a submodule of $A \times A$, that α, γ, and h would be linear, and so forth, to deduce that every onto map in R-**Mod** is a coequalizer.

   (b)   Verify that every onto map (epimorphism) in **Gp** is a coequalizer.

4.   Emulate the discussion following Definition 8–4–6 to verify that the countable copower of $\mathcal{K}$ is unique up to isomorphism.

5.   Prove Lemma 3 for **Gp** and R-**Mod**.

8–7   AN ALTERNATIVE FORMULATION OF DECOMPOSABLE SYSTEMS

To conclude, it will prove insightful to give a uniform format for linear and group machines which yields a second general notion of a decomposable machine within our framework of machines in a category. We first note some further properties of the coproduct:

If we consider $\{K_\alpha \mid \alpha \in I\}$ for I the empty set $\varnothing$, the condition for C to be their coproduct is simply that for *any* X in $\mathcal{K}$, there is a unique map $C \xrightarrow{\sim} X$. Such a C is called **initial**. For $\mathcal{K} = \S$, we have $C = \varnothing$; for $\mathcal{K} = R$-**Mod**, it is the one-element module $\{0\}$; and for $\mathcal{K} = $ **Gp**, it is the one-element group $\{e\}$.

Where no ambiguity can result, we may write $\varnothing$ for the initial object in any category ($\varnothing$ is thus unique up to isomorphism *if it exists*). We shall often write $K + L$ for the coproduct of two objects K and L. It can easily be checked that the following properties hold:

(1) $K + \varnothing \cong K$
(2) $K + L \cong L + K$
(3) $(K + L) + M \cong K + (L + M) \cong K + L + M$

We now show how to combine maps using the coproduct construction:

DEFINITION 1

(i) Given $K \xrightarrow{f} X$ and $L \xrightarrow{g} X$, and some specific choice of $K + L$, we define the map $\binom{f}{g} : K + L \to X$ by the diagram

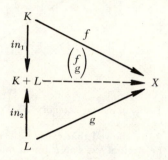

(ii) Given $K \xrightarrow{f} K'$ and $L \xrightarrow{g} L'$, we define

$$K + L \xrightarrow{f + g} K' + L'$$

by $f + g = \binom{f\, in'_1}{g\, in'_2}$; i.e., we have

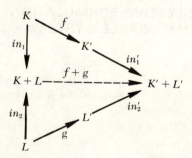

Example 1

For R-modules,

$$\binom{f}{g}(k, \ell) = f(k) + g(\ell) \text{ in } X$$

$$(f + g)(k, \ell) = (f(k), g(\ell)) \text{ in } K \oplus L$$

For groups,

$$\binom{f}{g}(k_1\ell_1k_2\ell_2\ldots) = f(k_1)g(\ell_1)f(k_2)g(\ell_2)\ldots \quad \text{in } X$$

$$(f + g)(k_1\ell_1k_2\ell_2\ldots) = f(k_1)g(\ell_1)f(k_2)g(\ell_2)\ldots \quad \text{in } K + L \qquad \Diamond$$

As an interesting aside, we note that if $U^\S$ and $U + U^\S$ exist, then $\binom{in_0}{z}:U + U^\S \to U^\S$ is an isomorphism, which may be interpreted by saying that U is an "infinite" object (think of $\S$, and U a non-empty set).

With these definitions, we may succinctly state our alternative general format:

A **linear machine** is an R-linear map $\binom{F}{G}:X + U \to X$ together with an R-linear map $H:X \to Y$.

A **group machine** is a homomorphism $\binom{F}{G}:X + U \to X$ together with an R-linear map $H:X \to Y$.

We suddenly understand a source of great confusion in general system theory. For R-modules, the product $X \times U$ and the coproduct $X + U$ are isomorphic, and thus we could always treat linear systems as if the next-state map were from $X \times U$ to X. However, it is clear that $X \times U$ and $X + U$ are *not* isomorphic for groups (*unless they are abelian*—can you see why?), and it was the attempt to provide a realization theory for group machines that led Arbib and Manes to see that coproducts provided the appropriate algebraic setting.

With this motivation, we are led to consider the following process in *any* category $\mathcal{K}$: Pick an object $U \in \mathcal{K}$, and let P be the process $P = - + U:\mathcal{K} \to \mathcal{K}:X \mapsto X + U$. We assume that each $X + U$ exists; but then, the action on morphisms is given by

$$
\begin{array}{ccc}
X & \xrightarrow{\;in_1\;} & X + U \\
\Big\downarrow{\scriptstyle f} & & \Big\downarrow{\scriptstyle f + id_U} \\
X' & \xrightarrow[\;in'_1\;]{} & X' + U
\end{array}
$$

It is clear that P is indeed a functor. It is in fact an input process (i.e., it admits free machines):

THEOREM 1

Let X be any object of a simple recursive category $\mathcal{K}$ with finite products. Then the functor $P = - + U$ is an input process on $\mathcal{K}$. Moreover, the free dynamics on X is determined as follows:

$$XP^@ = X^\S + U^\S$$

Denoting the zero and successor maps, respectively, by

$$X \xrightarrow{in_0} X^\S \quad \text{and} \quad X^\S \xrightarrow{z} X^\S \quad \text{for } X$$

$$U \xrightarrow{in_0} U^\S \quad \text{and} \quad U^\S \xrightarrow{z} U^\S \quad \text{for } U$$

we have

$$(XP^@)P \xrightarrow{\delta_x} XP^@ \quad \text{is} \quad X^\S + U^\S + U \xrightarrow{z + \binom{z}{in_0}} X^\S + U^\S$$

$$X \xrightarrow{\eta_x} XP^@ \quad \text{is} \quad X \xrightarrow{in_0} X^\S \xrightarrow{in_1} X^\S + U^\S$$

Proof

Given any $X' + U \xrightarrow{\binom{F}{G}} X'$ and $X \xrightarrow{f} X'$, we must check that there exists

a unique $X^\S + U^\S \xrightarrow{\binom{\alpha}{\beta}} X'$ such that the following diagrams commute:

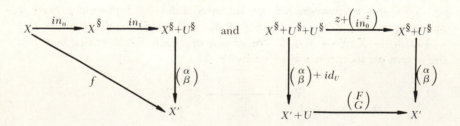

Reading off the "top" of the two diagrams spliced together, and off the "middle" and "bottom" of the right-hand diagram, we obtain

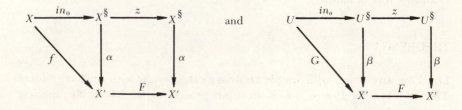

and these are clearly simple recursions; thus, they define α and β uniquely.

$\square$

Having assured ourselves that $- + U$ is an input process, we can then make our alternative definition:

DEFINITION 2

A **decomposable machine** (Mark II) in the simple recursive category $\mathcal{K}$ with finite coproducts is a 6-tuple

$$(X, F, U, G, Y, H)$$

where $X + U \xrightarrow{\binom{F}{G}} X$ is a $P = - + U$ dynamics, $X \xrightarrow{H} Y$ is the output map, and it is understood that we take the initial state to be the unique map $\varnothing \xrightarrow{\tau} X$. We denote the machine by (F, G, H) for short. $\bigcirc$

This indeed fits in with our convention of taking 0 as the initial state of a linear machine and e as the initial state of a group machine.

Now it is easy to check that $\varnothing^\S \cong \varnothing$ so that

$$\varnothing P@ = \varnothing^\S + U^\S \cong \varnothing + U^\S \cong U^\S$$

Thus, the reachability map of (F, G, H) is simply the map

$$r : U^\S \rightarrow X$$

which is the "β" for $f = \tau$ in the last proof, and this satisfies

$$rin_n = F^n G \text{ for all } n \in \mathbf{N}$$

The system behavior f equals Hr, and is the unique $f : U^\S \rightarrow Y$ satisfying

$$fin_n = HF^n G \text{ for all } n \in \mathbf{N}$$

This formula is very familiar from linear system theory (which is a special case).

To conclude, then, let us check that our two notions of a decomposable machine mesh. Given a fixed input process P, initial state object U, and output object Y, we may define a category $\mathfrak{M}_{P,U,Y}$ whose objects are the machines

$$(P, X, \delta, U, \tau, Y, \beta)$$

for arbitrary X, δ, τ, and β, and whose morphisms $M' \overset{\psi}{\to} M$ are the X-dyna-morphisms $\psi : X' \to X$ which satisfy

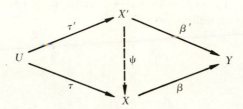

Noting that the identity process in a category $\mathcal{K}$ with finite products equals $-+\varnothing$, we have our equivalence result in the following form:

THEOREM 2

Let $\mathcal{K}$ be a simple recursive category with finite coproducts, and let U and Y be fixed objects of $\mathcal{K}$. Set

$$\mathfrak{M}_1 = \mathfrak{M}_{-+\varnothing, U, Y} \quad \text{and} \quad \mathfrak{M}_2 = \mathfrak{M}_{-+U, \varnothing, Y}$$

Then the maps

$$(-+\varnothing, X, F, U, G, Y, H) \mapsto \left(-+U, X, \binom{F}{G}, \varnothing, \tau, Y, H\right)$$

$$\psi : X' \to X \mapsto \psi : X' \to X$$

define a behavior and reachability preserving isomorphism Φ from $\mathfrak{M}_1$ to $\mathfrak{M}_2$.

Proof

The crucial point is that

$$U(-+\varnothing)^{@} = U^{\S} + \varnothing^{\S} \cong U^{\S} \cong \varnothing^{\S} + U^{\S} = \varnothing(-+U)^{@}$$

The rest follows by routine calculation. $\qquad\qquad\square$

In particular, it follows that a machine M is a minimal coequalizer-reachable realization of $f : U^{\S} \to Y$ in $\mathfrak{M}_1$ iff $\Phi(M)$ is a minimal coequalizer-reachable realization of f in $\mathfrak{M}_2$.

REFERENCES AND FURTHER READING

The approach of Sections 8–1 and 8–2 develops that of Arbib and Zeiger [1969]. The standard account of Kalman's approach to linear systems is given in Chapter 10 of Kalman, Falb and Arbib [1969]. For more on group machines, see Brockett and Willsky [1962] and Arbib [1973]. Arbib and Manes [1973a,b,c] develop their input-process approach to systems—other applications of category theory to system theory are given by Goguen [1972] and Bainbridge [1973]. For more on category theory, see MacLane [1972].

M. A. Arbib [1973] Coproducts and Group Machines, *J. Comp. Syst. Sci.* **7**, 278–287.

M. A. Arbib and E. G. Manes [1973a] Machines in a Category, *SIAM Review,* in press.

M. A. Arbib and E. G. Manes [1973b] Adjoint Machines, State-Behavior Machines, and Duality, Technical Report 73B-1, Computer and Information Science, University of Massachusetts, Amherst.

M. A. Arbib and E. G. Manes [1973c] Foundations of System Theory, I, *Automatica,* in press.

M. A. Arbib and H. P. Zeiger [1969] On the Relevance of Abstract Algebra to Control Theory, *Automatica* **5**, 589–606.

E. S. Bainbridge [1973] "A Unified Minimal Realization Theory, with Duality, for Machines in a Hyperdoctrine" (Ph.D. Dissertation, Univ. of Michigan).

G. Birkhoff and S. MacLane [1967] *Algebra,* New York: The Macmillan Company.

L. S. Bobrow and M. A. Arbib [1974] *Discrete Mathematics: Applied Algebra for Computer and Information Science,* Philadelphia: W. B. Saunders.

R. W. Brockett and A. S. Willsky [1972] Finite-State Homomorphic Sequential Machines, *IEEE Trans. Aut. Control,* **AC-17,** 483–490. (See also the related note by Arbib in the same issue, pp. 554–555.)

F. R. Gantmacher [1959] *The Theory of Matrices* (Translated from the Russian by K. A. Hirsch), New York: Chelsea.

J. A. Goguen [1972] Minimal Realization of Machines in Closed Categories, *Bulletin of the American Math. Soc.,* **78**, 777–783.

R. E. Kalman, P. L. Falb and M. A. Arbib [1969] *Topics in Mathematical System Theory,* McGraw-Hill. [See especially Chapter 10.]

S. MacLane [1972] *Categories for the Working Mathematician,* Springer-Verlag.

J. Rissanen [1971] Recursive Identification of Linear Systems, *SIAM J. Control* **9**, no. 3.

APPENDIX A-1

The Laplace Transform

The one-sided Laplace Transform* was defined in Section 6-3 as

$$\mathcal{L}\{f(t)\} = F(s) \triangleq \int_{0^-}^{\infty} f(t)e^{-st}\, dt$$

where this definition differs from that given in many older texts in that the lower limit of integration is taken as $t = 0^-$ rather than as $t = 0^+$ to permit $f(t)$ to contain impulses or their derivatives at the origin.

In the following table we summarize some of the more useful properties and transform pairs:

	TIME FUNCTION	LAPLACE TRANSFORM
1.	$f(t)$	$F(s)$
2.	$af(t) + bg(t)$	$aF(s) + bG(s)$
3.	(unit-impulse at $t = 0$) $\delta(t)$	1
4.	(unit-impulse at $t = a$) $\delta(t - a),\ a \geq 0$	e^{-as}
5.	(unit-step at $t = a$) $1(t - a),\ a \geq 0$	$\dfrac{1}{s} e^{-as}$
6.	$f(t - a)1(t - a)$ (shifting), $a \geq 0$	$e^{-as}F(s)$
7.	$\displaystyle\int_0^t f(t - \tau)g(\tau)\, d\tau$ (convolution) [set $f(\xi)$ to 0 for $\xi < 0$]	$F(s)G(s)$
8.	$\dfrac{d}{dt} f(t)$	$sF(s) - f(0^-)$
9.	$\dfrac{d^n}{dt^n} f(t)$	$s^n F(s) - \displaystyle\sum_{j=1}^{n} s^{n-j} f^{(j-1)}(0^-)$
10.	$\displaystyle\int_{0^-}^t f(\tau)\, d\tau$	$\dfrac{1}{s} F(s)$
11.	$t^n f(t)$	$(-1)^n \dfrac{d^n}{ds^n} F(s)$
12.	t (unit-ramp)	$\dfrac{1}{s^2}$
13.	t^2	$\dfrac{2}{s^3}$

*Zadeh and Desoer, *Linear Systems Theory* (McGraw-Hill, 1963), Appendix B.

	TIME FUNCTION	LAPLACE TRANSFORM
14.	$\dfrac{t^n}{n!}$	$\dfrac{1}{s^{n+1}}$
15.	$\dfrac{d^n}{dt^n}\,\delta(t)$	s^n
16.	e^{-at}	$\dfrac{1}{s+a}$
17.	$\dfrac{1}{a-b}\,(e^{-at}-e^{-bt})$	$\dfrac{1}{(s+a)(s+b)}$
18.	$\sin at$	$\dfrac{a}{s^2+a^2}$
19.	$\cos at$	$\dfrac{s}{s^2+a^2}$
20.	$e^{-at}\cos bt$	$\dfrac{s+a}{(s+a)^2+b^2}$
21.	$e^{-at}\sin bt$	$\dfrac{b}{(s+a)^2+b^2}$
22.	$e^{-at}f(t)$	$F(s+a)$

23. Initial Value Theorem: $f(0^+) = \lim\limits_{t\to 0^+} f(t) = \lim\limits_{s\to\infty} sF(s)$, provided $f(t)$ has at worst a step discontinuity at $t = 0$.

24. Final Value Theorem: $\lim\limits_{t\to\infty} f(t) = \lim\limits_{s\to 0} sF(s)$, provided $F(s)$ has no poles in the right half of the s plane.

APPENDIX A–2

The $\mathbb{Z}$-transform

The one-sided $\mathbb{Z}$-transform* was defined in Section 4–6 as

$$\mathbb{Z}\{f(k)\} = F(z) \triangleq \sum_{k=0}^{\infty} f(k)z^{-k}$$

In the following table we summarize some of the more useful properties and transform pairs:

	DISCRETE-TIME FUNCTION	$\mathbb{Z}$-TRANSFORM
1.	$f(k)$	$F(z)$
2.	$af(k) + bg(k)$	$aF(z) + bG(z)$
3.	$f(k - m)$ for $m \geq 0$	$z^{-m}F(z)$
4.	$f(k + m)$ for $m \geq 0$	$z^m F(z) - \sum_{j=0}^{m-1} f(j)z^{m-j}$
5.	$\sum_{j=0}^{k} f(k - j)g(j)$	$F(z)G(z)$
6.	$k^m f(k)$	$-z\left(\dfrac{d}{dz}\right)^m F(z)$
7.	$a^k f(k)$	$F(a^{-1}z)$
8.	$\sum_{j=0}^{k} f(j)$	$\dfrac{z}{z - 1}F(z)$
9.	(unit-pulse at $k = 0$) $\delta(k)$	1
10.	(unit pulse at $k = m$) $\delta(k - m)$	z^{-m}
11.	(discrete unit-step) $1(k)$	$\dfrac{z}{z - 1}$
12.	(discrete unit-ramp) k	$\dfrac{z}{(z - 1)^2}$
13.	k^2	$\dfrac{z(z + 1)}{(z - 1)^2}$
14.	a^k	$\dfrac{z}{z - a}$

---

*Cadzow and Martens, *Discrete-Time and Computer Control Systems* (Prentice-Hall, Inc., 1970).

DISCRETE-TIME FUNCTION	Z-TRANSFORM
15. e^{ak}	$\dfrac{z}{z - e^a}$
16. ka^k	$\dfrac{za}{(z - a)^2}$
17. $\sin \omega k$	$\dfrac{z \sin \omega}{z^2 - 2z \cos \omega + 1}$
18. $\cos \omega k$	$\dfrac{z(z - \cos \omega)}{z^2 - 2z \cos \omega + 1}$
19. $e^{-ak} \sin \omega k$	$\dfrac{ze^{-a} \sin \omega}{z^2 - 2ze^{-a} \cos \omega + e^{-2a}}$

20. Initial Value Theorem: $f(0) = \lim\limits_{z \to \infty} F(z)$

21. Final Value Theorem: $f(\infty) = \lim\limits_{k \to \infty} f(k) = \lim\limits_{z \to 1} (z - 1)F(z)$

APPENDIX A-3

An Outline of Optimal Control Theory

In Section 7-6 we presented an introductory overview of optimal control theory, and contented ourselves there with formulating the problem and setting up a convenient notation. In this appendix we outline the mathematical development and introduce the three major tools of optimal control theory: the linear regulator, bang-bang control, and the minimum (maximum) principle. In Section A-3-1 we develop—as a generalization of the theory of Section 4-4—the notion of conjugate spaces and adjoint transformations for Banach spaces instead of just inner product spaces. These provide the appropriate setting for the optimization theory that we present in the final sections. In Section A-3-2 we generalize the Language Multiplier approach to minimization and maximization, hopefully familiar from freshman calculus, which provides optimal control for the situation in which the target state at the end of a prescribed time interval must satisfy certain linear equalities. Then, in Section A-3-3, we consider the Pontryagin Minimum Principle, which provides sufficient conditions for optimal control when we must place constraints upon the possible inputs which may be applied as control signals. Our treatment derives much from that of Luenberger [1969], to which the reader is referred for a wealth of further material on optimality.

A-3-1   Conjugate Spaces and Adjoint Transformations

To develop our main theorems in Sections A-3-2 and A-3-3, we will need to make use of the idea of an adjoint transformation not, as in Section 4-4,

743

in the setting of Hilbert spaces, but in the more general setting of Banach spaces. In a Hilbert space X we had available an inner product $\langle \hat{x} | x \rangle$, and we defined the adjoint A^* of a linear transformation A by requiring that $\langle A\hat{x} | x \rangle = \langle \hat{x} | A^*x \rangle$ hold for all x and $\hat{x}$. We also have that every bounded linear functional $X \to \mathbf{R}$ on X is of the form $x \mapsto \langle \hat{x} | x \rangle$ for some $\hat{x} \in X$. In a Banach space X, there may be no inner product available, but the above observations suggest that we can introduce a space X^* of the bounded linear functionals $X \to \mathbf{R}$ and then define the adjoint $A^*: X^* \to X^*$ of a linear transformation $A: X \to X$ by requiring that $(A^*f)(x) = f(Ax)$ for all $f \in X^*$ and $x \in X$.

This can in fact be done. Recall that a linear function $f: X \to \mathbf{R}$ is *bounded* if there exists a constant M such that $|f(x)| \leq M\|x\|$ for all $x \in X$. The smallest possible such M is then called the *norm* of f, and denoted by $\|f\|$. Thus, we may write $|f(x)| \leq \|f\| \, \|x\|$ for all $x \in X$. The space X^* of all bounded linear functionals can be seen to be a Banach space with respect to this norm as follows:

We make X^* into a vector space by the rules

$$(f_1 + f_2)(x) = f_1(x) + f_2(x)$$

$$(rf)(x) = rf(x)$$

and we then have $\|(rf)(x)\| = \|rf(x)\| = |r| \, \|f(x)\|$, from which it is clear that $\|rf\| = |r| \, \|f\|$ and $rf \in X^*$, while $\|f_1(x) + f_2(x)\| \leq \|f_1(x)\| + \|f_2(x)\| \leq (\|f_1\| + \|f_2\|)\|x\|$ so that $(f_1 + f_2) \in X^*$, and $\|f_1 + f_2\| \leq \|f_1\| + \|f_2\|$. Finally, if $\|f\| = 0$, then $\|fx\| \leq 0$ for all $x \in X$, so that $f(x) = 0$ for all $x \in X$. It only remains to check that X^* is complete. Suppose then that $(f_1, f_2, f_3, \ldots, f_n, \ldots)$ is a Cauchy sequence in X^*, so that for each $\varepsilon > 0$ there exists N such that $\|f_m - f_n\| < \varepsilon$ for all $m, n > N$. This implies that for each x we have $\|f_m(x) - f_n(x)\| < \varepsilon\|x\|$ for $m, n > N$. Thus, the sequence $(f_1(x), f_2(x), f_3(x), \ldots, f_n(x), \ldots)$ has a limit in $\mathbf{R}$, which we shall call $f(x)$. It is then straightforward (Exercise 1) to verify that the map $f: X \to \mathbf{R}$ defined by $x \mapsto f(x)$ does indeed belong to X^* with $\|f\| = \lim_{n \to} \|f_n\|$.

Having verified that X^* is also a Banach space, let us now verify that, for each linear transformation $A: X \to X$, there is a unique linear transformation $A^*: X^* \to X^*$ defined by the equation

$$(A^*f)(x) = f(Ax)$$

for all $f \in X^*$ and x in X. But this equation defines (A^*f) at every x in X and so specifies it uniquely. It only remains to check that A^* is linear:

$$(A^*(r_1 f_1 + r_2 f_2))(x) = (r_1 f_1 + r_2 f_2)(Ax) \text{ by definition of } A^*$$
$$= r_1 f_1(Ax) + r_2 f_2(Ax) \text{ by definition of the}$$
$$\text{linear operations in } X^*$$

$$= r_1(A^*f_1)(x) + r_2(A^*f_2)(x)$$
$$= (r_1A^*f_1 + r_2A^*f_2)(x)$$

so

$$A^*(r_1f_1 + r_2f_2) = r_1A^*f_1 + r_2A^*f_2, \text{ as was to be shown.}$$

The following result generalizes our observations on the adjoint of a linear transformation $A : \mathbf{R}^n \to \mathbf{R}^n$:

THEOREM 1

Let $A : X \to X$ be a linear transformation of Banach spaces, for which $\mathcal{R}(A)$ is closed. Then

$$\mathcal{N}(A)^\perp = \mathcal{R}(A^*)$$

where for any $Y \subset X$, $Y^\perp = \{ f \in X^* \mid f(x) = 0 \text{ for all } x \in Y \}$.

Proof

$$f \in \mathcal{N}(A)^\perp \Leftrightarrow f(x) = 0 \text{ for all } x \in \mathcal{N}(A)$$
$$\Leftrightarrow [Ax = 0 \Rightarrow f(x) = 0]$$
$$f \in \mathcal{R}(A^*) \Leftrightarrow f = A^*g \text{ for some } g \in X^*$$
$$\Leftrightarrow f(x) = g(Ax) \text{ for all } x \in X$$

Thus $f \in \mathcal{R}(A^*) \Rightarrow [Ax = 0 \Rightarrow f(x) = 0] \Rightarrow f \in \mathcal{N}(A)^\perp.$

The proof that $\mathcal{N}(A)^\perp \subset \mathcal{R}(A^*)$ requires far more analytic machinery, and the reader interested in the details is referred to Yosida [1968, Section VII.5] or Luenberger [1969, Section 6.6]. ☐

We now turn to a conjugate space which is most important in our theory of optimal control. Let $C_n[t_0, t_1]$ be the space of all continuous maps from the finite interval $[t_0, t_1]$ to $\mathbf{R}^n$. We present a representation, due to Riesz, of all the linear functionals on $C_n[t_0, t_1]$.

We say that a function $\alpha : [t_0, t_1] \to \mathbf{R}$ is of **bounded variation** if there is a bound M such that for any set of points $t_0 = \tau_1 < \tau_2 < \tau_3 < \cdots < \tau_n = t_1$, the sum

$$\sum_{j=1}^{n-1} |\alpha(\tau_{j+1}) - \alpha(\tau_j)| < M$$

and we call the least such M the **total variation** of α.

Given any continuous function $f : [t_0, t_1] \to \mathbf{R}$, and an α of bounded variation, we may form the sum

$$\sum_{j=1}^{n-1} f(\tau_j)(\alpha(\tau_{j+1}) - \alpha(\tau_j)) \tag{1}$$

It can be shown that as we choose subdivisions $\{\tau_i\}$ with $|\tau_{j+1} - \tau_j| \to 0$, this sum (1) approaches a unique limit, which we denote by

$$\int_{t_0}^{t_1} f(t)\, d\alpha(t) \tag{2}$$

and call the **Stieltjes integral** of f with respect to α. (Note that we recapture the ordinary integral if $\alpha(t) = t$ for each $t \in [t_0, t_1]$.) It is clear that the map

$$f \mapsto \int_{t_0}^{t_1} f(t)\, d\alpha(t) \tag{3a}$$

is a bounded linear functional, with $\left| \int_{t_0}^{t_1} f(t)\, d\alpha(t) \right| \leq \| f \|$(total variation of α). Then we owe to Riesz (see, e.g., Riesz and Nagy [1955, Chapter III]) the observation that every bounded linear functional on $C[t_0, t_1]$ is of this form. It is then immediate that every bounded linear functional on $C_n[t_0, t_1]$ is of the form

$$\alpha^* f = \int_{t_0}^{t_1} d\alpha^T(t) f(t) = \sum_{j=1}^{n} \int_{t_0}^{t_1} d\alpha_j(t) f_j(t) \tag{3b}$$

where $\alpha(t) = \begin{bmatrix} \alpha_1(t) \\ \vdots \\ \alpha_n(t) \end{bmatrix} \in BV^n[t_0, t_1]$, the space of all n-tuples of functions of bounded linear variation from $[t_0, t_1]$ to $\mathbf{R}$.

To give some practice in the use of this space, and to provide a lemma of great importance for Sections A–3–2 and A–3–3, we now show how the derivative $A_x[x, u]$ introduced in Section 7–6 can be used to give a functional specification of the adjoint control equation (18) of Section 7–6 (which section the reader should review for notation, etc., before proceeding further):

$$-\dot{\lambda}(t) = f_x^T(x(t), u(t))\lambda(t) + \ell_x^T(x(t), u(t)) \tag{4}$$

It will help to recall formulas (10) and (11) of Section 7–6, and to note that for the terminal constraint $\widehat{K}[x] = K(x(t_1)) - c$ we have $\widehat{K}_x[x]h = K_x(x(t_1))h(t_1)$.

LEMMA 2

If $\lambda \in BV^n[t_0, t_1]$ satisfies $\lambda(t_1) = 0$, and the equation

$$\lambda^T A_x[x, u] + g_x[x, u] + \mu^T \widehat{K}_x[x] = 0 \tag{5}$$

holds, then λ is a differentiable function on (t_0, t_1), with a jump of magnitude $-\mu^T K_x(x(t_1))$ at t_1, satisfying the differential equation

$$-\dot{\lambda}(t) = f_x^T(x(t), u(t))\lambda(t) + \ell_x^T(x(t), u(t)) \tag{4}$$

on (t_0, t_1).

Proof

We recall that

$$A_x[x, u]h = h(\cdot) - \int_{t_0}^{(\cdot)} f_x(x(\tau), u(\tau))h(\tau) \, d\tau \tag{6a}$$

$$g_x[x, u]h = \int_{t_0}^{t_1} \ell_x(x(t), u(t))h(t) \, dt \tag{6b}$$

$$\widehat{K}_x[x]h = K_x(x(t_1))h(t_1) \tag{6c}$$

If we let both sides of the given operator equation (5) operate on a typical function h, we get

$$\lambda^T A_x[x, u](h) + g_x[x, u](h) + \mu^T \widehat{K}_x[x](h) = 0 \tag{7}$$

where the last two terms on the left side are easily evaluated by equations (6b) and (6c). To evaluate the first term on the left of (7), we use the Riesz representation (3b) of λ^T:

$$\lambda^T A_x[x, u](h) = \int_{t_0}^{t_1} d\lambda^T(t)(A_x[x, u]h)(t) \tag{8}$$

But $(A_x[x, u]h)(t)$ can be evaluated from (6a) as

$$(A_x[x, u]h)(t) = h(t) - \int_{t_0}^{t} f_x(x(\tau), u(\tau))h(\tau) \, d\tau$$

Thus, our equation (7) can be written

$$\int_{t_0}^{t_1} d\lambda^T(t) \left[h(t) - \int_{t_0}^{t} f_x(x(\tau), u(\tau))h(\tau) \, d\tau \right]$$

$$+ \int_{t_0}^{t_1} \ell_x(x(t), u(t))h(t) \, dt + \mu^T K_x(x(t_1))h(t_1) = 0 \quad \text{(9)}$$

and (9) can be integrated by parts to yield

$$\int_{t_0}^{t_1} \ell_x(x(t), u(t))h(t) \, dt + \int_{t_0}^{t_1} d\lambda^T(t)h(t)$$

$$+ \int_{t_0}^{t_1} \lambda^T(t) f_x(x(t), u(t))h(t) \, dt + \mu^T K_x(x(t_1))h(t_1) = 0 \quad \text{(10)}$$

Now, this equality is to hold for all h. Suppose, then, by way of contradiction, that λ has a jump in $[t_0, t_1)$ of a at the point t_2. Then the second term would contribute $ah(t_2)$ to the sum, and this could be made very large by making $h(t_2)$ very large. But the other terms are not sensitive to a change in $h(t_2)$, provided that the integral of the distortion is kept very small; and since the overall sum is zero, we conclude that $a = 0$. Similarly, to avoid the fourth term making an undue difference, we conclude that λ does have a jump at t_1, of magnitude $-\mu^T K_x(x(t_1))$.

Since (10) holds for all $h \in C_n[t_0, t_1]$, it certainly holds for all continuously differentiable h which vanish at t_1. Thus, we may integrate the second term of (10) by parts to determine that

$$\int_{t_0}^{t_1} \{ \ell_x(x(t), u(t))h(t) - \lambda^T(t)\dot{h}(t) + \lambda^T(t) f_x(x(t), u(t))h(t) \} \, dt = 0$$

for all such h. Then, by Lemma 3 of this section (see below), $\lambda^T(t)$ is differentiable on (t_0, t_1) and satisfies the adjoint control equation

$$\dot{\lambda}^T(t) = -f_x^T(x(t), u(t))\lambda(t) - \ell_x^T(x(t), u(t)) \qquad \square$$

We now note two elementary but useful lemmas. Let $C^1[t_0, t_1]$ be the space of all functions $[t_0, t_1] \to \mathbf{R}$ with continuous derivatives.

LEMMA 3

If $\alpha(t)$ and $\beta(t)$ are continuous in $[t_0, t_1]$ and

$$\int_{t_0}^{t_1} [\alpha(t)h(t) + \beta(t)\dot{h}(t)] \, dt = 0 \qquad \text{(11)}$$

for every $h \in C^1[t_0, t_1]$ with $h(t_0) = h(t_1) = 0$, then β is differentiable and $\dot{\beta}(t) \equiv \alpha(t)$ in $[t_0, t_1]$.

Proof

Defining $A(t) = \int_{t_0}^{t} \alpha(\tau) \, d\tau$ we have

$$\int_{t_0}^{t_1} \alpha(t) h(t) \, dt = - \int_{t_0}^{t_1} A(t) \dot{h}(t) \, dt$$

by integration by parts. Thus, condition (11) may be restated as

$$\int_{t_0}^{t_1} [-A(t) + \beta(t)] \dot{h}(t) \, dt = 0 \qquad \textbf{(12)}$$

Setting $a(t) = -A(t) + \beta(t)$, define $c = \dfrac{1}{t_1 - t_0} \int_{t_0}^{t_1} a(t) \, dt$, and consider the choice for $h(t)$ of $\int_{t_0}^{t} [a(\tau) - c] \, d\tau$, which certainly lies in $C^1[t_0, t_1]$ and satisfies $h(t_0) = h(t_1) = 0$. Then

$$\int_{t_0}^{t_1} [a(t) - c]^2 \, dt = \int_{t_0}^{t_1} [a(t) - c] \dot{h}(t) \, dt \text{ by choice of } h$$

$$= \int_{t_0}^{t_1} a(t) \dot{h}(t) \, dt + c[h(t_1) - h(t_0)]$$

$$= 0 \text{ by (12) and the fact that } h(t_1) = h(t_0) = 0$$

Thus, $a(t) \equiv c$ and so $\beta(t) \equiv A(t) + c$. Hence, $\dot{\beta}(t) \equiv \alpha(t)$ as was to be shown.

□

LEMMA 4

If $\alpha(t)$ is continuous on $[t_0, t_1]$ and $\int_{t_0}^{t_1} h(t) \alpha(t) \, dt = 0$ for every $h \in C^1[t_0, t_1]$ with $h(t_0) = h(t_1) = 0$, then $\alpha(t) \equiv 0$ on $[t_0, t_1]$.

Proof

Assume by way of contradiction that $\alpha(t_2) \neq 0$, say $\alpha(t_2) > 0$. Then $\alpha(t)$ is positive on some interval $[t_0', t_1'] \subset [t_0, t_1]$. Let $h(t)$ be any function which looks like Figure A–3–1.

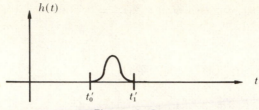

Figure A–3–1

Then $\int_{t_0}^{t_1} h(t)\alpha(t)\,dt > 0$, contradicting our claim on α.   □

EXERCISES FOR SECTION A–3–1

1.   Verify that the limit $f\colon X \to \mathbf{R}$ of a Cauchy sequence $(f_1, f_2, f_3, \ldots, f_n, \ldots)$ of elements of X^* does itself belong to X^*.

2.   Verify that the map (3a) is a bounded linear functional $C_n[t_0, t_1] \to \mathbf{R}$ when α is a function of bounded variation.

3.   Go through the steps of the proof of Lemma 2 and see where we used the hypothesis that $\lambda(t_1) = 0$. (*Hint:* Do the integration by parts.)

A–3–2   Optimal Control with Terminal Constraints

In this section, we wish to find a u which minimizes

$$J(u) = \int_{t_0}^{t_1} \mathcal{l}(x(t), u(t))\,dt$$

when x satisfies the equivalent conditions

$$\dot{x}(t) = f(x(t), u(t)), \; x(t_0) \text{ fixed}$$
$$A[x, u] = \mathbf{0}$$

and also satisfies a terminal constraint of the form

$$\widehat{K}[x] = 0$$

where $\widehat{K}[x] = K(x(t_1)) - c$, with $K\colon \mathbf{R}^n \to \mathbf{R}^P$ differentiable and c fixed in $\mathbf{R}^P$. Clearly we have, as noted in the previous section, that

$$\widehat{K}_x[x]h = K_x(x(t_1))h(t_1) \tag{1}$$

To provide necessary conditions for u to be minimal, we first introduce the concept of a regular point, and then provide very general conditions under which a regular point is an extremum among those satisfying certain constraints.

DEFINITION 1

Let $T:B \to Y$ be continuously differentiable, where B and Y are Banach spaces, and let $T_b[b_0]$ denote the derivative map of T at the point $b = b_0$. We say that b_0 in B is a **regular point** of T if $T_b[b_0]$ maps B *onto Y*.   ○

In the one-dimensional case $T:\mathbf{R} \to \mathbf{R}$, we recall that

$$T_b[b_0]:\mathbf{R} \to \mathbf{R} \text{ by the formula } h \mapsto T'(b_0)h$$

where $T'(b_0)$ is the derivative of T at b_0 in the classical sense. Hence, in this case, b_0 is a regular point if and only if $T'(b_0) \neq 0$. Thus, in Figure A–3–2, b_0 is regular but b_1 and b_2 are not.

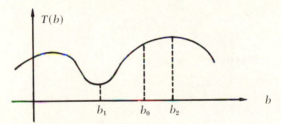

Figure A–3–2

It can then be shown (Luenberger [1969, pp. 240–242]) that if b_0 is regular with $T(b_0) = y_0$, there exist a $\delta > 0$ and a constant K such that for each y in Y with $\|y - y_0\| < \delta$ the equation $T(b) = y$ has a solution satisfying $\|b - b_0\| \leq K\|y - y_0\|$. In Figure A–3–2, this is certainly true for b_0 on taking $0 < \delta < T(b_2) - T(b_0)$, and $K = \sup_{|b-b_0|<(b_2-b_0)} (|b - b_0|/|T(b) - T(b_0)|)$.

The reader may recall (but it does not matter if he does not) the Lagrange Multiplier theorem from his study of ordinary calculus: If we want to find the minimum of a function $f(x_1, x_2)$ among those points for which $H(x_1, x_2) = 0$, we find an x and a constant λ (called a **Lagrange multiplier**), such that $f(x_1, x_2) + \lambda H(x_1, x_2)$ has zero partial derivatives.

Example 1

To find the smallest value of x_2 subject to the constraint that $x_1^2 + x_2^2 = 4$ shown in Figure A–3–3, it is no good to take the derivative of x_2 and equate

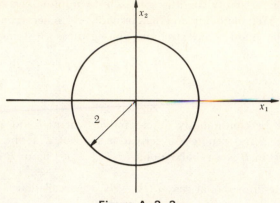

Figure A–3–3

it to 0—this yields the contradiction $1 = 0$, since x_2 has no *un*constrained minimum.

But if we take the partial derivatives of

$$x_2 + \lambda(x_1^2 + x_2^2 - 4)$$

and set them equal to zero, we obtain

$$2x_1\lambda = 0$$

and

$$1 + 2x_2\lambda = 0$$

The first equation says that either x_1 or λ is 0. The second equation says that λ cannot be zero. Thus, $x_1 = 0$, and then the constraint $x_1^2 + x_2^2 = 4$ implies that the extreme value of x_2 is ±2.   ◊

We may now state the very general version of this result, in which the Lagrange multiplier λ is replaced by a linear functional z_0^*:

THEOREM 1 (THE LAGRANGE MULTIPLIER THEOREM)

Let B and Z be Banach spaces, and let $f: B \to \mathbf{R}$ and $H: B \to Z$ have continuous derivatives. Then if f has a local extremum under the constraint $H(b) = 0$ at the regular point b_0 of H, then there exists an element $z_0^* \in Z^*$ such that the functional

$$L(b) = f(b) + z_0^* H(b)$$

is stationary at b_0, i.e., such that

$$f_b[b_0] + z_0^* H_b[b_0] = 0$$

Proof

We start by showing that $f_b[b_0] \in \mathfrak{N}(H_b[b_0])^\perp$.

Consider the map $T: B \to \mathbf{R} \times Z$ by $b \mapsto (f(b), H(b))$. If there were an h such that $H_b[b_0]h = 0$ but $f_b[b_0]h \neq 0$, then $T_b[b_0]: B \to \mathbf{R} \times Z$ would be onto $\mathbf{R} \times Z$, since we would then have

$$T_b[b_0](h_1 + rh) = (f_b[b_0]h_1 + r f_b[b_0]h, \quad H_b[b_0](h_1)) = (r_1, z)$$

and, by regularity of b_0, we may choose h_1 to get any desired value of z, and then may manipulate r to get the desired r_1 without affecting z.

But if b_0 is a regular point of T, it follows by Luenberger's theorem that for any $\varepsilon > 0$, no matter how small, there is a $b \in B$ and a $\delta > 0$ such that $T(b) = (f(b_0) - \delta, 0)$, contradicting the local minimality of b_0.

Since $H_b[b_0]$ is continuous, $\mathfrak{N}(H_b[b_0])^\perp = \mathfrak{R}[H_b[b_0]^*]$, and so $f_b[b_0] = -z_0^* H_b[b_0]$ for some $(-z_0^*) \in Z^*$.   □

Let us now try to relate this theorem to the optimal control problem with terminal constraints which introduced the section:

The function $f: B \to \mathbf{R}$ whose minimum we wish to study becomes

$$g: \Theta \times \Omega \to \mathbf{R}$$

so that $\Theta \times \Omega$ is our current B. The constraint $H(b) = 0$ now takes the form of a pair $A[x, u] = 0$ and $\widehat{K}[x] = 0$, so that $H: B \to Z$ becomes

$$(A, \widehat{K}): \Theta \times \Omega \to \Theta \times \mathbf{R}^p$$

so that $\Theta \times \mathbf{R}^p$ is our current Z.

The Lagrange multiplier theorem, then, will give us a *necessary* condition (we discuss necessity vs. sufficiency after Theorem 2 below) for (x, u) to yield a constrained minimum of g, provided that we can guarantee that (x, u) is a regular point of $(A, \widehat{K})$. Let us, then, establish conditions under which

$$\Theta \times \Omega \to \Theta \times R^p \text{ by the rule } (h, v) \mapsto (A_x[x, u]h + A_u[x, u]v, \widehat{K}_x[x]h) \quad \textbf{(2)}$$

is onto.

From (1), we have that

$$\widehat{K}_x[x]h = K_x(x(t_1))h(t_1)$$

Thus, we require that rank $K_x(x(t_1)) = p$ for (2) to map onto its $\mathbf{R}^p$ component.

Condition L_1: rank $K_x(x(t_1)) = p$.

We now wish to establish a plausible condition under which, for any $\bar{x} \in \Theta$, we may find $(h, v) \in \Theta \times \Omega$ such that

$$A_x[x, u]h + A_u[x, u]v = \bar{x} \tag{3}$$

and such that, moreover, h has its value specified at t_1 so that $\hat{K}_x[x]h = K_x(\bar{x}(t_1))h(t_1)$ takes on the desired value in $\mathbf{R}^p$. Expanding (3) via the formulas (10) of Section 7-6 (assuming that the derivatives f_x and f_u exist), we obtain

$$h(t) - \int_{t_0}^{t} f_x(x(\tau), u(\tau))h(\tau) \, d\tau - \int_{t_0}^{t} f_u(x(\tau), u(\tau))v(\tau) \, d\tau = \bar{x}(t) \tag{4}$$

for all $t \in [t_0, t_1]$. Setting $v = 0$, we can, by the theory of Section 5-2, find a function $\bar{h}$ such that

$$\bar{h}(t) - \int_{t_0}^{t_1} f_x(x(\tau), u(\tau))\bar{h}(\tau) \, d\tau = \bar{x}(t) \tag{5}$$

so long as f_x is continuous. If we then subtract (5) from (4), and set $w(t) = h(t) - \bar{h}(t)$, we obtain

$$w(t) - \int_{t_0}^{t} f_x(x(\tau), u(\tau))w(\tau) \, d\tau - \int_{t_0}^{t} f_u(x(\tau), u(\tau))v(\tau) \, d\tau = 0 \tag{6}$$

Consider now our original system

$$\dot{x}(t) = f(x(t), u(t))$$

If we make a slight perturbation $u + \delta u$ about u, we obtain a perturbation $x + \delta x$ about x which satisfies

$$\begin{aligned}
\frac{d}{dt}(x + \delta x)(t) &= f((x + \delta x)(t), (u + \delta u)(t)) \\
&= f(x(t), u(t)) + f_x(x(t), u(t))\, \delta x(t) + f_u(x(t), u(t))\, \delta u(t) \\
&\quad + o(\delta x(t), \delta u(t))
\end{aligned}$$

Thus, the behavior of δx is *approximately* described by the linearized equation

$$(\dot{\delta x})(t) = f_x(x(t), u(t))\, \delta x(t) + f_u(x(t), u(t))\, \delta u(t) \tag{7}$$

and we observe that (6) is an integrated form of (7). Thus, we stipulate the second condition.

Condition L_2: The linearized system (7) for perturbations about (x, u) is *controllable.*

By controllability, then, we may find a $v \in \Omega$ such that the $w \in \Theta$ which it determines via (6) takes on any desired value at time t_1. Requiring, then, that $w(t_1) = h(t_1) - \overline{h}(t_1)$, where $h(t_1)$ was specified at the outset and $\overline{h}(t_1)$ was determined by our solution of (5), we are done: (x, u) is a regular point of $(A, \widehat{K})$ if it satisfies Conditions L_1 and L_2. With this, we have established the "fine point" which allows us to state and prove our theorem:

THEOREM 2 (NECESSARY CONDITIONS FOR OPTIMAL CONTROL WITH TERMINAL CONSTRAINTS)

Let $f: \mathbf{R}^n \times \mathbf{R}^m \to \mathbf{R}^n$ and $\ell: \mathbf{R}^n \times \mathbf{R}^m \to \mathbf{R}$ possess continuous partial derivatives. Let (x_0, u_0) satisfy the regularity conditions L_1 and L_2, and minimize

$$g[x, u] = \int_{t_0}^{t_1} \ell(x(t), u(t)) \, dt$$

subject to the constraints

$$\dot{x}(t) = f(x(t), u(t)), \text{ for } t \in [t_0, t_1], \quad x(t_0) \text{ fixed}$$

and

$$K(x(t_1)) = c$$

Then there exists a continuous function $\lambda: [t_0, t_1] \to \mathbf{R}^n$ and a vector $\mu \in \mathbf{R}^p$ such that for all $t \in [t_0, t_1]$ we have

$$-\dot{\lambda}(t) = f_x^T(x_0(t), u_0(t))\lambda(t) + \ell_x^T(x_0(t), u_0(t)) \tag{8}$$
$$\lambda(t_1) = \mu^T K_x(x(t_1)) \tag{9}$$

and

$$\lambda^T(t) f_u(x_0(t), u_0(t)) + \ell_u(x_0(t), u_0(t)) = 0 \tag{10}$$

Proof

By the Lagrange multiplier theorem, there exists $z_0^* \in Z^* = (\Theta \times \mathbf{R}^p)^*$ such that

$$f_b[b_0] + z_0^* H_b[b_0] = 0 \tag{11}$$

where b_0 is now (x_0, u_0), f is g, and H is $(A, \widehat{K})$.

Now, any $z_0^* \in (\Theta \times \mathbf{R}^p)^*$ is, by Section A–3–1, of the form (λ, μ), with $\lambda \in BV^n[t_0, t_1]$ and $\mu \in \mathbf{R}^p$. Thus, we may unpack (11) to obtain (on equating the operators on h and u to zero separately)

$$g_x[x, u] + \lambda^* A_x[x, u] + \mu^T \widehat{K}_x[x] = 0 \tag{12}$$

and

$$g_u[x, u] + \lambda^* A_u[x, u] = 0 \tag{13}$$

By Lemma 2 of Section A–3–1, we have that λ satisfies (8), and has a jump of $-\mu^T K_x(x_0(t_1))$ at t_1. Recalling the definition of λ^* from Section A–3–1 we may rewrite (13) as

$$\int_{t_0}^{t_1} \ell_u(x_0(t), u_0(t)) v(t)\, dt - \int_{t_0}^{t_1} d\lambda^T(t) \int_{t_0}^{t} f_u(x_0(\tau), u_0(\tau)) v(\tau)\, d\tau = 0 \tag{14}$$

for all $v \in \Omega$. Taking $v(t_1) = 0$, we may integrate (14) by parts to obtain

$$\int_{t_0}^{t_1} [\ell_u(x_0(t), u_0(t)) + \lambda^T(t) f_u(x_0(t), u_0(t))] v(t)\, dt = 0$$

which clearly implies (10) by Lemma 4 of Section A–3–1.

Finally, to say that there exists a λ with $\lambda(t_1) = 0$ and a jump of $-\mu^T K_x(x_0(t_1))$ at t_1 which satisfies (8) and (10) is to say that there exists a continuous λ with $\lambda(t_1) = \mu^T K_x(x_0(t_1))$ which satisfies (8) and (10), and so we are done. ☐

We now study the most important practical application of Theorem 2, namely that in which the cost function $\ell(\widehat{x}, \widehat{u})$ has the *quadratic form*

$$\ell(\widehat{x}, \widehat{u}) = \frac{1}{2} [\widehat{x}^T Q \widehat{x} + \widehat{u}^T R \widehat{u}]$$

where Q is an $n \times n$ symmetric positive semidefinite matrix and R is an $m \times m$ symmetric positive definite matrix; and in which the system is linear,

$$f(\widehat{x}, \widehat{u}) = F\widehat{x} + G\widehat{u}$$

for some $n \times n$ matrix F and $n \times m$ matrix G.

To say that Q is **positive semidefinite** is to say that $\widehat{x}^T Q \widehat{x} \geq 0$, so that some states need not be costly; while to say that R is **positive definite** is to say that $\hat{u}^T R \hat{u} > 0$ unless $\hat{u} = 0$, so that no non-zero control is cost-free. Forming derivatives, we obtain (Exercise 3)

$$f_x = F; \qquad f_u = G \tag{15}$$

$$\ell_x = \hat{x}^T Q; \quad \ell_u = \hat{u}^T R \tag{16}$$

Theorem 2 then tells us that for u to be optimal there must exist a continuous function $\lambda : [t_0, t_1] \to \mathbf{R}^n$ such that

$$-\dot{\lambda}(t) = F^T \lambda(t) + Qx(t) \quad \text{(recall that } Q^T = Q\text{)} \tag{17}$$

$$\lambda(t_1) = 0 \tag{18}$$

and

$$\lambda^T(t)G + u^T(t)R = 0 \tag{19}$$

Since R is positive definite, it must be invertible; and since R is symmetric, R^{-1} is also symmetric (Exercise 4). Thus, (19) yields

$$u(t) = -R^{-1} G^T \lambda(t) \tag{20}$$

The pair (17) and (18) as they stand are of little practical value—they ask us to solve (17) starting at time t_1 and working backward in time, whenever we need λ to compute the control u via (20) starting at time t_0 and working forward in time. Fortunately, there is a way around this, namely by introducing the matrix differential equation known as the **Riccati equation:**

$$\dot{P}(t) = -P(t)F - F^T P(t) + P(t)GR^{-1}G^T P(t) - Q, \text{ with } P(t_1) = 0 \quad \textbf{(21)}$$

It can be shown (see, for example, Reid [1946]) that this equation has a unique symmetric positive semidefinite solution on $[t_0, t_1]$. Note that $P(t)$ does not depend on $x(t)$, and so can be precomputed for $[t_0, t_1]$. We now see that the λ defined by

$$\lambda(t) = P(t)x(t)$$

satisfies equation (17). Certainly $\lambda(t_1) = 0$, since $P(t_1) = 0$. If we take $u(t) = -R^{-1}G^T P(t)x(t)$ by (20), then we obtain

$$\dot{x}(t) = Fx(t) - GR^{-1}G^T P(t)x(t)$$

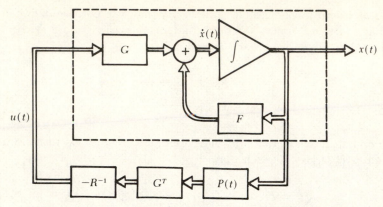

Figure A–3–4 Optimal quadratic control of a linear system.

so that

$$\dot{\lambda}(t) = \dot{P}(t)x(t) + P(t)\dot{x}(t)$$
$$= -[P(t)F + F^TP(t) - P(t)GR^{-1}G^TP(t) + Q]x(t)$$
$$\quad + P(t)[Fx(t) - GR^{-1}G^TP(t)x(t)]$$
$$= F^TP(t)x(t) - Qx(t)$$
$$= F^T\lambda(t) - Qx(t)$$

so that this choice of λ does indeed solve our problem. Pleasingly, the control

$$u(t) = -R^{-1}G^TP(t)x(t)$$

given by this scheme is a linear feedback law, as shown in Figure A–3–4.

Is the caption of Figure A–3–4 justified? The reader should give his own answer before reading on. Recalling the words of caution which preceded equation (2), we see that the answer is "Not yet." This is because the conditions for optimality given in Theorem 2 are necessary but *not* sufficient. What this means is that *if there exists an optimal control* for our present problem, then Figure A–3–4 supplies it; but *it does not guarantee that an optimal control exists.*

This point is beautifully made by L. C. Young[†] on page 22 of his idiosyncratic, delightful, and important book (which particularly addresses the question of what to do when there is no optimal control that is piecewise continuous, but there do exist optimal "chattering" controls, i.e., controls which switch *infinitely often*):

Here is Perron's Paradox: "Let N be the largest positive integer. Then for $N \neq 1$ we have $N^2 > N$ contrary to the definition of N as largest. Therefore $N = 1$."

Of course, we know that 1 is not the largest integer. What the paradox

---

[†]L. C. Young, *Lectures on the Calculus of Variations and Optimal Control Theory* (W. B. Saunders Co., 1969).

yields is the truth that "If there is a largest integer, it is *necessary* that that integer be 1." But there is no largest integer. Similarly, the engineer will (as indicated parenthetically above) encounter situations in which Ω contains no control which is optimal for his problem, and then there is no reason to expect the solution of the adjoint control equation to yield anything of particular value.

Fortunately, however, we can, in the present case, verify the optimality of the control presented to us by the necessary conditions, simply by emulating Example 2 of Section 7–6. However, we shall not give the details here.

THEOREM 3

The control for the linear system

$$\dot{x}(t) = Fx(t) + Gu(t), \; x(t_0) \text{ fixed}$$

which is optimal over the time period $[t_0, t_1]$ with respect to the cost functional

$$g[x, w] = \frac{1}{2} \int_{t_0}^{t_1} [x(t)^T Qx(t) + u(t)^T Ru(t)] \, dt$$

is given by the feedback control law

$$u(t) = -R^{-1}G^T P(t)x(t) \qquad \square$$

If F and G are constant, as in the treatment we have given, it is then possible to solve the problem of infinite time control, on taking $t_0 = 0$ and $t_1 = \infty$. Recalling the Riccati equation

$$\dot{P}(t) = -P(t)F - F^T P(t) + P(t)GR^{-1}G^T P(t) - Q, \text{ with } P(t_1) = 0 \quad \textbf{(21)}$$

we let $t_1 \to \infty$, and it can be shown (Kalman [1960]) that the solution tends to a unique matrix $\mathcal{P}$, independent of t, which satisfies the equilibrium equation

$$\mathcal{P}F + F^T\mathcal{P} - \mathcal{P}GR^{-1}G^T\mathcal{P} + Q = 0 \qquad \textbf{(22)}$$

Thus, if t_1 is to be very large, our optimal control reduces to a linear feedback control *with constant coefficients*.

Example 2

Consider the two-dimensional linear system with scalar input described by

$$F = \begin{bmatrix} 0 & 1 \\ 0 & -1 \end{bmatrix} \quad G = \begin{bmatrix} 0 \\ 1 \end{bmatrix}$$

$$Q = \begin{bmatrix} 1 & 0 \\ 0 & 0 \end{bmatrix} \quad R = [k] \quad \text{for some } k > 0$$

The value of k is left open to be used in a study of the relationship between the cost of control and the shape of the resulting transient response. Let the symmetric matrix $\mathcal{P}$ be written as

$$\begin{bmatrix} p & q \\ q & r \end{bmatrix}$$

Substituting into (22) we obtain

$$\begin{bmatrix} 0 & p-q \\ 0 & q-r \end{bmatrix} + \begin{bmatrix} 0 & 0 \\ p-q & q-r \end{bmatrix} - \begin{bmatrix} q^2/k & qr/k \\ qr/k & r^2/k \end{bmatrix} + \begin{bmatrix} 1 & 0 \\ 0 & 0 \end{bmatrix} = 0$$

Reading off each component of the sum of these matrices gives the equations

$$1 - \frac{q^2}{k} = 0$$

$$p - q - \frac{qr}{k} = 0$$

$$2(q - r) - \frac{r^2}{k} = 0$$

The positive definite solution to these equations is given by the particular values

$$q = \sqrt{k}$$

$$r = \left(\sqrt{1 + \frac{2}{\sqrt{k}}} - 1 \right) k$$

$$p = \sqrt{k} \left(\sqrt{1 + \frac{2}{\sqrt{k}}} \right)$$

The control law for this problem is then given by

$$u(t) = Cx(t)$$

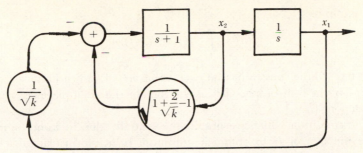

Figure A–3–5   Closed-loop system with optimal control for the problem of Example 2.

where

$$C = -R^{-1}G^T \mathcal{P}$$

$$= -\frac{1}{k} [0 \;\; 1] \begin{bmatrix} p & q \\ q & r \end{bmatrix}$$

$$= -\frac{1}{k} [q \;\; r]$$

$$= \left[-\frac{1}{\sqrt{k}}, -\left(\sqrt{1 + \frac{2}{\sqrt{k}}} - 1 \right) \right]$$

as shown in Figure A–3–5. Note that this control law becomes unbounded as the cost on control, k, is reduced toward zero.

EXERCISES FOR SECTION A–3–2

1.   Apply the "classical" form (i.e., using a single multiplier λ) of the Lagrange multiplier theorem to minimize the surface area $A = 2(x_1x_2 + x_2x_3 + x_3x_1)$ of a rectangular parallelopiped subject to the constraint that the volume $x_1x_2x_3$ be V.
2.   The proof of Theorem 1 covers only the case in which the constrained extremum of f is a minimum. Rewrite the proof to cover the case of a local maximum.
3.   Verify formulas (15) and (16).
4.   Consider the equation $R\hat{x} = 0$ to deduce that if R is positive definite, then R is invertible. Transpose $RR^{-1} = I = R^{-1}R$ to verify that if R is also symmetric, then R^{-1}, too, is symmetric.

A–3–3   The Pontryagin Minimum Principle

We now consider the problem of finding a control u for the system

$$\dot{x}(t) = f(x(t), u(t)), \; x(t_0) \text{ fixed and } t \in [t_0, t_1]$$

which minimizes

$$J(u) = \int_{t_0}^{t_1} \ell(x(u)(t), u(t))\, dt$$

subject only to the constraint that each value $u(t)$ of u must lie in some fixed subset U_0 of the control space $\mathbf{R}^m$. Ω will then be the set of piecewise continuous functions $[t_0, t_1] \to U_0$.

The condition which we obtain is essentially due to Pontryagin—we change the sign of H to obtain a minimum principle instead of a maximum principle. Let $g: \Theta \times \Omega \to \mathbf{R}$ be the functional defined by

$$g(x, u) = \int_{t_0}^{t_1} \ell(x(t), u(t))\, dt$$

Condition A: Let $\ell : \mathbf{R}^n \times \mathbf{R}^m \to \mathbf{R}$ have a continuous derivative ℓ_x with respect to the $\mathbf{R}^n$ variable.

Condition B: Let $f : \mathbf{R}^n \times \mathbf{R}^m \to \mathbf{R}^n$ have a continuous derivative f_x and satisfy the uniform Lipschitz condition

$$\| f(\hat{x}, \hat{u}) - f(\overline{x}, \overline{u}) \| \le M[\|\hat{x} - \overline{x}\| + \|\hat{u} - \overline{u}\|]$$

In particular, this ensures that

$$\dot{x}(t) = f(x(t), u(t)), \quad x(t_0) \text{ fixed}$$

has a unique solution $x(u)$, say, on $[t_0, t_1]$ for each $u \in \Omega$.

THEOREM 1

Given J, ℓ, and f satisfying Conditions A and B, define the Hamiltonian $H : \mathbf{R}^n \times \mathbf{R}^m \times \mathbf{R}^n \to \mathbf{R}$ by

$$H(\hat{x}, \hat{u}, \hat{\lambda}) = \hat{\lambda}^T f(\hat{x}, \hat{u}) + \ell(\hat{x}, \hat{u})$$

Constraining the $\hat{u}$ for each t to be $u(t)$ for each $t \in [t_0, t_1]$ for some fixed $u \in \Omega$, we may then obtain the Hamiltonian equations

$$\dot{x}(t) = H_\lambda(x(t), u(t), \lambda(t)) = f(x(t), u(t)) \tag{1}$$

$$-\dot{\lambda}(t) = H_x(x(t), u(t), \lambda(t)) \tag{2}$$
$$= f_x^T(x(t), u(t))\lambda(t) + \ell_x^T(x(t), u(t))$$

Let $x(u) \in \Theta$ be the solution of (1) with fixed $x(t_0)$ and let $\lambda \in \Theta$ then be the solution of (2) with $\lambda(t_1) = 0$. Then in order for $u_0 \in \Omega$ to minimize

$$J(u) = g(x(u), u) = \int_{t_0}^{t_1} \mathscr{l}(x(u)(t), u(t)) \, dt$$

it is *necessary* that for all $t \in [t_0, t_1]$ and all $\widehat{u} \in U_0$ we have

$$H(x_0(t), u(t), \lambda(t)) \leq H(x_0(t), \widehat{u}, \lambda(t)) \tag{3}$$

Proof

Defining $A[x, u]$ and $g[x, u]$ as usual, we may note that the continuity of f_x and $\mathscr{l}_x$ assures us that $A_x[x, u]$ and $g_x[x, u]$ are continuous functions of $(x, u) \in \Theta \times \Omega$. Let $x + \delta x = x(u + \delta u)$ where $x = x(u)$. Then, since $x(u)(t) = x(t_0) + \int_{t_0}^t f(x(u)(\tau), u(\tau)) \, d\tau$, we have

$$\|\delta x(t)\| \leq \int_{t_0}^t M[\|\delta x(\tau)\| + \|\delta u(\tau)\|] \, d\tau$$

Hence $\|\delta x(t)\| \leq M e^{M(t_1 - t_0)} \int_{t_0}^{t_1} \|\delta u(\tau)\| \, d\tau$ so that $\|\delta x\| \leq K\|\delta u\|$, where $K = M e^{M(t_1 - t_0)}(t_1 - t_0)$.

In other words, solutions of $A[x, u] = 0$ satisfy a Lipschitz condition of the form

$$\|x(u) - x(v)\| \leq K\|u - v\| \tag{4}$$

Note, too, that our differential equation (2) for λ is equivalent (by applying Lemma 2 of Section A-3-1 to the case $\widehat{K}_x[x] = 0$) to

$$\lambda^T A_x[x, u_0] + g_x[x, u_0] = 0 \tag{5}$$

We define $L[x, u, \lambda] = \lambda^T A[x, u] + g[x, u]$. Let us now show that if λ satisfies (5), then for any $u \in \Omega$

$$J(u_0) - J(u) = L[x(u_0), u_0, \lambda] - L[x(u_0), u, \lambda] + o(\|u_0 - u\|) \tag{6}$$

In fact, we shall verify this using only the continuity of A_x and g_x, and the Lipschitz condition (4):

$$J(u_0) - J(u) = g[x(u_0), u_0] - g[x(u), u]$$
$$= (g[x(u_0), u_0] - g[x(u_0), u]) + (g[x(u_0), u] - g[x(u), u])$$

Now

$$g[x(u_0), u] - g[x(u), u] = g_x[x(u_0), u](x(u_0) - x(u)) + o(\|x(u_0) - x(u)\|)$$

But by the Lipschitz condition, $o(\|x(u_0) - x(u)\|) = o(\|u_0 - u\|)$, while by the continuity of g_x, $g_x[x(u_0), u] = g_x[x(u_0), u_0] + o(\|u_0 - u\|)$. Therefore, $g_x[x(u_0), u](x(u_0) - x(u)) = g_x[x(u_0), u_0](x(u_0) - x(u)) + o(\|u_0 - u\|)$. Putting this all together,

$$J(u_0) - J(u) = (g[x(u_0), u_0] - g[x(u_0), u]) \\ + g_x[x(u_0), u_0](x(u_0) - x(u)) + o(\|u_0 - u\|)$$

Replacing g by A, and noting that $A[x(u_0), u_0] = 0 = A[x(u), u]$, we obtain

$$0 = (A[x(u_0), u_0] - A[x(u_0), u]) + A_x[x(u_0), u_0](x(u_0) - x(u)) + o(\|u_0 - u\|)$$

Hence; if λ is such that $\lambda^T A_x[x, u_0] + g_x[x, u_0] = 0$ we have, since $L[x, u, \lambda] = \lambda^T A[x, u] + g[x, u]$, that

$$\begin{aligned} J(u_0) - J(u) &= (g[x(u_0), u] - g[x(u_0), u]) \\ &\quad + g_x[x(u_0), u](x(u_0) - x(u)) + o(\|u_0 - u\|) \\ &= (g[x(u_0), u_0] - g[x(u_0), u]) \\ &\quad - \lambda^T A_x[x(u_0), u_0](x(u_0) - x(u)) + o(\|u_0 - u\|) \\ &= (g[x(u_0), u_0] - g[x(u_0), u]) \\ &\quad + \lambda^T (A[x(u_0), u_0] - A[x(u_0), u]) + o(\|u_0 - u\|) \\ &= L[x(u_0), u_0, \lambda] - L[x(u_0), u, \lambda] + o(\|u_0 - u\|) \end{aligned}$$

To conclude our proof, then, it suffices to show that (6) implies (3). Note that

$$\begin{aligned} L[x, u, \lambda] &= \lambda^T A[x, u] + g(x, u) \\ &= \int_{t_0}^{t_1} d\lambda^T(t) \left[x(t) - x(t_0) - \int_{t_0}^{t} f(x(\tau), u(\tau)) \, d\tau \right] + \int_{t_0}^{t_1} \ell(x(t), u(t)) \, dt \\ &= \int_{t_0}^{t} d\lambda^T(t)[x(t) - x(t_0)] + \int_{t_0}^{t_1} \lambda^T(t) f(x(t), u(t)) \, dt + \int_{t_0}^{t_1} \ell(x(t), u(t)) \, dt \\ &= \int_{t_0}^{t} d\lambda^T(t)[x(t) - x(t_0)] + \int_{t_0}^{t_1} H(x(t), u(t), \lambda(t)) \, dt \end{aligned}$$

Thus,

$$J(u_0) - J(u) = \int_{t_0}^{t_1} [H(x_0(t), u_0(t), \lambda(t)) \\ - H(x_0(t), u(t), \lambda(t))] \, dt + o(\|u - u_0\|) \quad \textbf{(7)}$$

Now, suppose by way of contradiction, that there exists a $\bar{u} \in \Omega$ and $\bar{t} \in [t_0, t_1]$ such that

$$H(x(\bar{t}), u_0(\bar{t}), \lambda(\bar{t})) > H(x(\bar{t}), \bar{u}, \lambda(\bar{t}))$$

Since x, f, λ, and l are continuous, and u is piecewise continuous, we can find $\varepsilon > 0$ and an interval around $\bar{t}$ such that

$$H(x(t), u_0(t), \lambda(t)) - H(x(t), \bar{u}, \lambda(t)) > \varepsilon$$

on the interval. Now if we let $u(t)$ equal $\bar{u}$ on that interval, and $u_0(t)$ outside the interval, we have from (7) that

$$J(u_0) - J(u) > r\varepsilon + o(\|u - u_0\|)$$

where r is the length of the interval. But $\|u - u_0\| = o(r)$. Thus by making r sufficiently small, $J(u_0) - J(u)$ can be made positive; contradicting the optimality of u_0.   □

Example 1

Consider the problem, discussed in detail in Section 7–1, of bringing the $\dfrac{1}{s^2}$ plant, the "cart," described by

$$\dot{x}_1 = x_2; \quad \dot{x}_2 = u$$

from a given initial state $\hat{x}$ to the origin in the shortest possible time, subject to the constraint $|u| \le 1$. Thus,

$$H(\hat{x}, \hat{u}, \hat{\lambda}) = \hat{\lambda}_1 \hat{x}_2 + \hat{\lambda}_2 \hat{u} + 1$$

which yields the adjoint equations

$$\dot{\lambda}_1 = 0 \text{ and } \dot{\lambda}_2 = -\lambda_1$$

Thus, λ_1 is constant, say equal to c_1, and this yields $\lambda_2 = c_2 - c_1 t$ for a suitable constant c_2. Thus, to maximize

$$H(x(t), u(t), \lambda(t)) = c_1 x_2(t) + (c_2 - c_1 t)u(t) + 1$$

subject to the constraint $|u(t)| \le 1$, we must take

$$u(t) = \text{sgn}(c_2 - c_1 t)$$

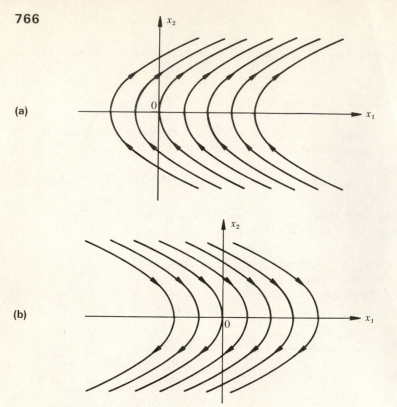

Figure A–3–6 (a) Paths when $u = +1$. **(b)** Paths when $u = -1$.

When $u \equiv 1$ we have the trajectory

$$x_2(t) = t + \widehat{c}_2; \quad x_1(t) = \frac{1}{2} t^2 + \widehat{c}_2 t + \widehat{c}_1$$

so that

$$x_1 = \frac{1}{2} x_2^2 + \widehat{c} \text{ for a suitable constant } \widehat{c}$$

as shown in Figure A–3–6a. Similarly, $u \equiv -1$ yields the $x_1 = -\frac{1}{2} x_2^2 + \widehat{c}$ curves of Figure A–3–6b.

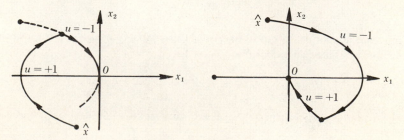

Figure A–3–7(a), (b)

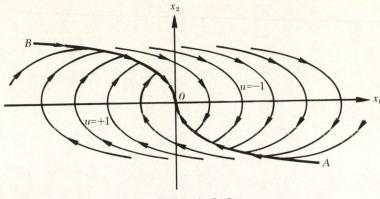

Figure A–3–8

Since the function $c_2 - c_1 t$ changes sign at most once, the optimal trajectory (it *is* optimal, though we do not prove sufficiency here) must consist of at most two arcs of these parabolas, as shown in Figures A–3–7a and A–3–7b.

Thus, the optimal control is given by

$$u(t) = v(x(t))$$

where

$$v(\hat{x}) = \begin{cases} +1 \text{ if } \hat{x} \text{ is below the switching curve } AOB, \text{ or on the arc } AO \\ -1 \text{ if } \hat{x} \text{ is above the switching curve } AOB, \text{ or on the arc } OB \end{cases}$$

as shown in Figure A–3–8.   ◊

REFERENCES

R. E. BELLMAN [1957] *Dynamic Programming,* Princeton University Press.

R. E. KALMAN [1960] "Contributions to the Theory of Optimal Control," Boletin de la Sociedad Matematica Mexicano, 102–119.

J. W. LEECH [1958] *Classical Mechanics,* London: Methuen.

D. G. LUENBERGER 1969 *Optimization by Vector Space Methods,* New York: Wiley.

L. LUISTERNIK and V. SOBOLEV [1961] *Elements of Functional Analysis,* New York: Ungar.

W. T. REID [1946] "A Matrix Differential Equation of the Riccati Type," Am. J. Math., **68,** 237–246.

F. RIESZ and B. SZ.-NAGY 1955 *Functional Analysis* (translated from the 2nd French Edition by Leo F. Boron), New York: Ungar.

L. I. ROZONOER [1959] "L. S. Pontryagin's Maximum Principle in Optimal System Theory," Automation and Remote Control, **20,** Part 1, 1288–1302; Part 2, 1405–1421; Part 3, 1517–1532.

K. YOSIDA [1968] *Functional Analysis* (2nd Edition), Berlin: Springer-Verlag.

List of Symbols

Symbols of basic set theory are summarized on page 11.

Index